Introduction to Modern Biophysics

This textbook provides an introduction to the fundamental and applied aspects of biophysics for advanced undergraduate and graduate students of physics, chemistry, and biology.

The application of physics principles and techniques in exploring biological systems has long been a tradition in scientific research. Biological systems hold naturally inbuilt physical principles and processes which are popularly explored. Systematic discoveries help us understand the structures and functions of individual biomolecules, biomolecular systems, cells, organelles, tissues, and even the physiological systems of animals and plants. Utilizing a physics-based scientific understanding of biological systems to explore disease is at the forefront of applied scientific research.

This textbook covers key breakthroughs in biophysics whilst looking ahead to future horizons and directions of research. It contains models based on both classical and quantum mechanical treatments of biological systems. It explores diseases related to physical alterations in biomolecular structures and organizations alongside drug discovery strategies. It also discusses the cutting-edge applications of nanotechnologies in manipulating nanoprocesses in biological systems.

Key Features:

- Presents an accessible introduction to how physics principles and techniques can be used to understand biological and biochemical systems.
- Addresses natural processes, mutations, and their purposeful manipulation.
- Lays the groundwork for vitally important natural scientific, technological, and medical advances.

Mohammad Ashrafuzzaman, a biophysicist and condensed matter scientist, is passionate about investigating biological and biochemical processes utilizing physics principles and techniques. He is a professor of biophysics at King Saud University's Biochemistry Department in the College of Science, Riyadh, Saudi Arabia; the co-founder of MDT Canada Inc., and the founder of Child Life Development Institute, Edmonton, Canada. He has authored *Biophysics and Nanotechnology of Ion Channels*, *Nanoscale Biophysics of the Cell*, and *Membrane Biophysics*. He has also published about 50 peer-reviewed articles and several patents, edited two books, and has been serving on the editorial boards of Elsevier and Bentham Science journals. Dr. Ashrafuzzaman has held research and academic ranks at Bangladesh University of Engineering & Technology, University of Neuchatel (Switzerland), Helsinki University of Technology (Finland), Weill Medical College of Cornell University (USA), and University of Alberta (Canada). During 2013–2018 he also served as a Visiting Professor at the Departments of Oncology, and Medical Microbiology and Immunology, of the University of Alberta. Dr. Ashrafuzzaman earned his highest academic degree, Doctor of Science (D.Sc.) in condensed matter physics from the University of Neuchatel, Switzerland in 2004.

Introduction to Modern Biophysics

Mohammad Ashrafuzzaman

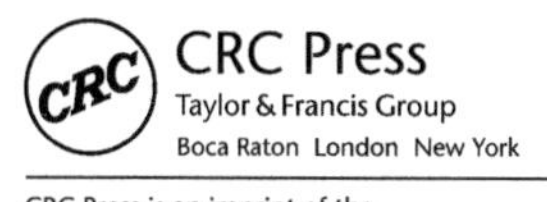

CRC Press is an imprint of the
Taylor & Francis Group, an **informa** business

First edition published 2024
by CRC Press
2385 NW Executive Center Drive, Suite 320, Boca Raton FL 33431

and by CRC Press
4 Park Square, Milton Park, Abingdon, Oxon, OX14 4RN

CRC Press is an imprint of Taylor & Francis Group, LLC

Library of Congress Cataloging-in-Publication Data
Names: Ashrafuzzaman, Mohammad, author.
Title: Introduction to modern biophysics / Mohammad Ashrafuzzaman.
Description: First edition. | Boca Raton, FL : CRC Press, 2024. |
Includes bibliographical references and index. |
Identifiers: LCCN 2023031947 | ISBN 9781032256702 (hbk) |
ISBN 9781032263342 (pbk) | ISBN 9781003287780 (ebk)
Subjects: LCSH: Biophysics–Textbooks.
Classification: LCC QH505 .A84 2024 | DDC 612/.014–dc23/eng/20230831
LC record available at https://lccn.loc.gov/2023031947

ISBN: 9781032256702 (hbk)
ISBN: 9781032263342 (pbk)
ISBN: 9781003287780 (ebk)

DOI: 10.1201/9781003287780

Typeset in Times
by Newgen Publishing UK

Contents

Preface

The history of the philosophical and scientific understanding of the living world traces the eras from ancient to modern times. The human quest to understand biological life has always been at the forefront of research. The concept of life around Aristotle's era (384–322 BC) was mostly based on philosophical thoughts and traditional beliefs (religions). Today we find that life is strongly a biological entity, which contains physical mechanisms following fundamental laws of physics. Not only life but also almost all aspects of living systems need interdisciplinary approaches to reveal their inner meanings.

The modern-era explorations of biological systems using fundamental physics principles and techniques started in the hands of Robert Hooke and Antonie van Leeuwenhoek, who discovered cells and bacteria, respectively. Both discoveries happened during the second half of the seventeenth century with the use of specialized microscopes that they developed or modified to achieve certain resolutions so that they could see the tiny cells and bacteria. The set of these low micrometer (μm) level biological observations initiated the creation of a multitude of avenues leading to the discovery of biological processes that were previously unknown. Looking through lenses was popularized by Galileo Galilei at the beginning of the seventeenth century when he focused his vision toward heaven in search of distant objects in the universe. Both Hooke and Leeuwenhoek did the opposite, pointing toward the tiny matters in living systems. We may consider biophysics to have emerged in the beginning by utilizing the physics of light in living systems. Since then almost all physics laws, including classical and quantum mechanical principles, have been found useful in understanding biology.

Attempts at applying physics principles to living systems can be traced to the early creators of modern science. Galileo could analyze the structure of animal bones using physical principles. Galileo's application of mechanics to biological problems was so inspiring that he is even considered the founder of biomechanics. Newton's handling of light was so powerful and inspiring that his applications revealed the perception of color. Volta and Cavendish studied animal electricity. Volta, in the eighteenth century, observed frog contractions by connecting two points of the same nerve without any contact with the muscle. Galvani found animal electricity to exist in a state of "disequilibrium," and an animal may move in response to internal stimuli or external influences. Lavoisier demonstrated that the process of respiration is an oxidative chemical reaction. Robert Mayer in his physiological studies formulated the first law of thermodynamics.

Physics was successfully applied to physiology in hydrodynamics by Poiseuille, who analyzed blood flow using his 1838-discovered law on the flow of liquids. The airflow in the lungs follows the laws of aerodynamics. In the nineteenth century, Hermann von Helmholtz created the foundations for understanding vision and hearing. Among others especially mentionable are Delbrück, Kendrew, von Bekesy, Crick, Meselson, Hartline, Gamow, Schrödinger, Hodgkin, Huxley, Fröhlich, Davydov, Cooper, and Szent-György who worked tirelessly to push the frontier of the life sciences using physical analysis.

If we are to pick a few breakthroughs generating biological discoveries where new fronts of biophysics were initiated, we need to mention various discoveries, including especially the discovery of cells by Hooke and the understanding of the double helix model of DNA by Watson and Crick. The first X-ray and NMR protein structural discoveries created platforms for understanding vital biomolecular structures and functions which are important in medical discoveries, including addressing mutations and drug discovery. Genomics, proteomics, and bolomics are growing on the backbone of biophysics techniques. The tools utilized in these discoveries have mostly been developed using fundamental physics principles and engineering techniques. The applications of these tools with their ever-enriching sophistications in the ever-growing number of biological fields are beyond imagination. The trend continues at an accelerating pace up until now.

The biological environment is understood now in ways that were not available even a century ago. The state of the aqueous phase, and other associated physicochemical properties thereof which biomolecules encounter in living systems, have a lot to contribute in raising the biostructures' collective physical properties. Water molecules in general and the arrangement of atoms in a water molecule in particular play considerable roles in creating and maintaining macromolecular structures. Repetitive patterns of water bridges are found to help in linking oxygen atoms both within a single peptide chain, between different chains, and even between different triple helices. DNA stability and rigidity are found to be linked to water fixed by AT pairs in the minor groove, enforcing a competition between local enthalpy and entropy associated with nucleotide base pairs AT versus GC. Loss of water drives transcription factor binding to the minor groove. The role of water in raising physical biostructure states is therefore crucial and requires thorough analysis from biophysics perspectives.

An alternative to understanding biomolecules using structural analysis is to simulate them in computers and supercomputers, a trend that is being enriched in terms of enhancing resolution by orders of magnitude every decade. Research in the simulation area has been found possible utilizing minimum supports and means. So instead of nurturing big lab concepts, research has also been conducted popularly at the individual level among both rich and poor nations. In these *in silico* approaches, nowadays machine learning and artificial intelligence algorithms are rapidly being adopted to enhance our performance in dealing with big data as we try to accommodate big biomolecular structures and fluctuations thereof in supercomputers.

Modern biophysics now covers aspects of the economy, environment, social development, etc. Biophysical modeling of wealth distribution among societies for ensuring healthy and justified GDP growth is very popular now. The flow of energy and low emissions for reducing environmental impacts are other areas where biophysical techniques are applied.

As we envision going to other planets and space and trying to search for alternative habitats, we need to work toward creating biology-compatible environments there. We have to create artificial environments that host physiology- and biology-friendly ingredients that are reproducible with the help of recycling protocols. Biophysics techniques are emerging to deal with this.

Biological fields are probably the scientific areas where proportionately the highest number of scientific minds and explorers from various basic science, applied engineering, and social science fields have been engaged. Hardcore physics principles, including those of quantum mechanics, are now finding their applications in exploring biological problems. For example, techniques like X-ray, crystallography, etc. are regularly used in biological system explorations. The concept of many-body interactions, charge particle interactions, stochastic principles, principles of electrostatics and electrodynamics, etc., are found quite applicable to addressing various phenomena in cellular systems. Classical mechanics principles have long been used in exploring mechanisms behind cellular systems' functions. Quantum mechanics principles have recently been found important in exploring cell systems.

The main objectives of biophysics research lie in understanding three aspects of life: what it is, how life works, and why life fails. In doing so we need to cover two main areas: the natural processes of life, and their distortion as biology experiences mutations. Biophysical techniques regularly deal with both by inspecting principally the structural aspects of vital biomolecules that are responsible for normal functioning of life processes and help us address diseases by leading to drug discovery.

Introduction to Modern Biophysics includes 19 chapters elaborating on distinguishable topics. Each chapter focuses on a specific topic that is of modern-day interest. Besides presenting a rigorous background of respective fields, I have added a lot of material describing the ongoing research activities and findings and presented materials that are guiding us towards achieving futuristic goals. A historical perspective of the systematic technological breakthroughs that took place over centuries has also been presented, while analyzing the associated scientific glories that were achieved by applying those technologies in biological and medical sciences. Readers are encouraged to passionately read all chapters chronologically to get a complete picture of this fascinating subject. This book will hopefully be found suitable as a textbook for undergraduate and graduate studies.

Special Note: I have added three appendices to elaborate on thermodynamics of the living system, nanotechnology of biological systems, and biophysics of evolution, respectively. These three appendices were initially planned as independent chapters but due to limited space in the book I decided to webhost them as appendices. Readers are especially encouraged to read these extra chapters hosted on the website www.routledge.com/Introduction-to-Modern-Biophysics/Ashrafuzzaman/p/book/9781032256702.

Acknowledgments

I am thankful to Professor Jack Tuszynski and 2020 Medicine Nobel Laureate Professor Michael Houghton of Alberta University, Professor Olaf Sparre Andersen of Weill Medical College of Cornell University, and Dr. Chih-Yuan Tseng of MDT Canada Inc. for many insightful discussions on various topics that helped develop some of the ideas incorporated in this book. Writing this book would have been impossible without using a great deal of experimental and theoretical data from publications of various authors (all are quoted in references). It's a great pleasure to thank all of them for their contributions to the field. Hundreds of discussions with colleagues, academic friends, and research group members helped me to shape ideas while writing this book during the last three years. Editorial assistance and encouragement provided by the staff members at CRC Press, Taylor & Francis Group, especially Dr. Rebecca Hodges-Davies and Dr. Danny Kielty, are thankfully acknowledged. I am thankful to my parents Mohammad Shahjahan Ali and Marzia Khatun whose encouragements and blessings are always with me. I especially value the emotional support given by my wife Anwara, two sons Imtihan Ahmed and Yakin Ahmed, and my daughter-in-law Emma. Anwara has always accompanied me on trips between my Canadian and Saudi residences and workstations to provide much-needed social support in my academic activities.

Mohammad Ashrafuzzaman

Edmonton, Alberta, Canada and Riyadh, Saudi Arabia

About the Author

Mohammad Ashrafuzzaman, a biophysicist and condensed matter scientist, is passionate about investigating biological and biochemical processes utilizing principles and techniques of physics. His theoretical and experimental works have created a generalized template to address the membrane interactions of cell-targeted agents, including drugs and nanoparticles. He has generalized the screened Coulomb interaction method to apply in biological systems and has developed a set of techniques (US patented) to help design novel drug molecules, especially aptamers, to validate the drugs' biological target binding potency, and to help address drug efficacy against specific diseases.

Before joining the King Saud University's Biochemistry Department in the College of Science, Riyadh, Saudi Arabia, he held academic positions at Bangladesh University of Engineering and Technology, Neuchatel University (Switzerland), Helsinki University of Technology (Finland), Cornell University (United States), and Alberta University (Canada). Between 2013 and 2018 Prof. Ashrafuzzaman worked at the Departments of Oncology and Medical Microbiology & Immunology, University of Alberta (Canada) as a Visiting Professor. He is the co-founder of MDT Canada Inc., and the founder of Child Life Development Institute, Edmonton, Canada. Besides publishing many peer-reviewed articles he is currently serving on the editorial boards of four highly reputed journals. He has also authored *Biophysics and Nanotechnology of Ion Channels*, *Nanoscale Biophysics of the Cell*, and *Membrane Biophysics*.

1 Physical Understanding of Biological Systems

Past, Present, and the Future

Energy and entropy are key players behind the creation, existence, and function of biological systems. Right at the beginning of life's formation (Cavalazzi et al., 2021) these two quantitative physical properties took part in constructing biological systems. The combination of physics, geology, and chemistry suggests a 4.36±0.1 billion years' timeline for the formation of ribonucleic acid (RNA) (Benner et al., 2020). The early evolution of life is hypothesized to have been dominated by RNA (Trevino et al., 2011). Most organisms have deoxyribonucleic acid (DNA) made genes while some viruses use RNA. Both RNA and DNA are known to use physical principles and properties to get constructed, stabilized, and replicated (Chen and Dill, 2000; Vologodskii and Frank-Kamenetskii, 2018; Hajiaghayi et al., 2012). Genome complexity relies on considerable energetics taking place in biological cells (Lynch and Marinov, 2015; Remacle et al., 2020). The double-helical DNA structure was modeled based on physical energetic concepts behind biomolecular structural stability (Watson and Crick, 1953b). Likewise, energy and entropy also play equally important roles in the genetics of life (Hasegawa and Yano, 1974; Saha and Sarkar, 2020; Ben-Naim, 2022). Both life's complexity and evolution are associated with a renewed state of entropy (Ulanowicz and Hannon, 1987; Brooks and Wiley, 1988; Crofts, 2007). Major transitions in evolution, including the origin of life, represent specific types of physical phase transitions, a concept found to provide a phenomenological description of evolution utilizing thermodynamic concepts (Vanchurin et al., 2022). Biomolecule conformations such as protein folding rely on specific thermodynamic order parameters in biological systems (Chong and Ham, 2018). Nucleic acid interactions with ligands have been addressed using thermodynamics concepts (Lane and Jenkins, 2000).

Biological energy conversion in cellular systems is catalyzed by membrane-bound proteins that transduce chemical or light energy into physical energy forms to power endergonic cellular processes. The catalytic processes are known to involve elementary electron-, proton-, charge-, and energy-transfer reactions taking place in molecular machines associated with cell respiration and photosynthesis (Kaila, 2021). Physical energetics are active in biomolecular folding and structural alterations (Baldwin, 2007; Naganathan et al., 2006; Meli et al., 2020).

Impaired physical energetics in mutated biomolecular structures regulate their stability and function in cellular systems, knowledge of which helps us understand the origins of diseases (Cang and Wei, 2017; Beal, 2000). Increasing entropy appears as a fundamental driving cause of neural and cognitive decline with aging (Drachman, 2006). Cancer cells' progression may be predicted via entropy generation (Movahed et al., 2021).

Modern physics techniques began to be applied in understanding biological systems at micrometer scale resolutions in the seventeenth century. We know the very discovery of cells happened through the application of optophysics techniques in compound microscopy by physicist Robert

DOI: 10.1201/9781003287780-1

Hooke in 1665 (Hooke, 1665; Donaldson, 2010). This discovery initiated a revolution in our understanding of biology and medicine.

Physics in the modern era started to participate in modeling our economy (Kondratenko, 2009), environments (Karev, 2019), ecosystems (Hogan et al., 1989), and many other applied fields. Physics and especially biophysics have enormous roles in our understanding of the origin of life, of how life may exist in harsh conditions such as exoplanetary regions, of running our everyday life's essential activities, and in dealing with abnormal conditions of life's processes like those associated with diseases. This chapter is dedicated to creating a platform on which subsequent chapters of the book will be constructed.

Following Hooke's and Leeuwenhoek's seventeenth-century observations of microscopic cells and bacteria, respectively, our thirst for understanding life started to slowly grow, which took us eventually to the molecular level with the famous discovery of DNA's double helical structure, modeled by Watson and Crick (Watson and Crick, 1953a). It's not just life that we are interested to know about in this book. We aim at elaborating our knowledge of biology and the environment by seeing things through the lenses of physics. Biophysics of the past mostly had a history of primitive information and predictive knowledge up until the 1950s. Since then, we have experienced a tremendous amount of progress over the last seven decades. During this latest era, we have not only been able to understand many crucial biochemical and biological processes active in biology with the intervention of physics principles and techniques but have also been developing biophysical means to intervene in abnormalities in biomolecular systems' structures and functions that we find in disease conditions. These matters will be addressed in respective chapters of this book. As this book is titled *Modern Biophysics*, I wish to push my readers to think ahead of time. I shall provide some routes through which biophysical applications are expected to penetrate the future when we shall have to deal with enormous issues related to energy production, utilizing biomaterials, and regulating our economy among versatile societies in diversified ecosystems on our planet and elsewhere. Readers will find that each chapter demonstrates specific issues utilizing corresponding parameters, materials, and information so that readers with diversified expertise and interests will find chapters to interest them and to explore independently. The book is constructed as a genuine interdisciplinary biophysics textbook for academia, industries, and general readerships who envision breaking the borders of their core academic training and practices and going beyond.

1.1 EARLY BREAKTHROUGHS IN PHYSICAL UNDERSTANDING OF BIOLOGICAL SYSTEMS

A few sectors and breakthrough findings that revolutionized the beginning of our understanding of the fundamentals of biology will be addressed here as example cases. Primarily, we shall explore aspects of biology that we could gain knowledge of using physics laws and principles, and associated technologies. Though primitive, th knowledge created the background of modern-day understanding of biology under the umbrella of biophysics.

1.1.1 Micrometer to Nanometer Resolution Observations of the Cell

Following the seventeenth-century microscopic imaging of cells and bacteria, scientists interested in biology started taking advanced and ambitious approaches in exploring biosystems, which eventually led to breakthroughs in both understanding biology and discovering techniques and tools to manipulate biological structures and regulate physical processes active in biological systems. Hooke's discovery of cells (Figure 1.1A presents a sketch of the compound microscope that Hooke used to image cork cells, presented in Figure 1.1B) paved the way for imaging biosystems, while Leeuwenhoek's discovery of bacteria (see details in a recent article, Lane, 2015), initiated the understanding of microorganisms hosted by our environments (Prussin and Marr, 2015) that regulate the health of our those environments (Cavicchioli et al., 2019).

A.

B.

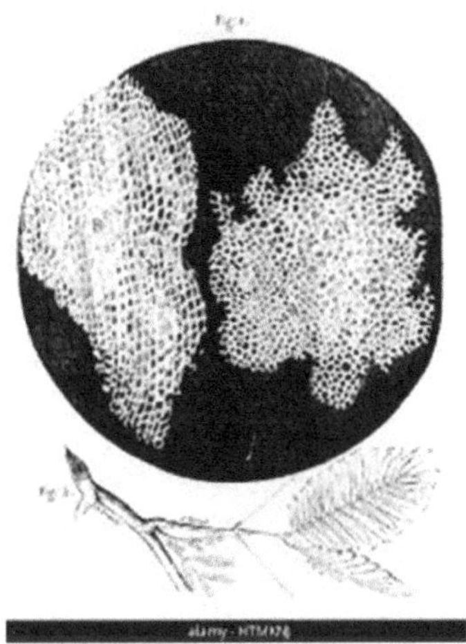

FIGURE 1.1 Around 1665 Robert Hook refined the design of the compound microscope and published a book titled *Micrographia* which illustrated his findings, mostly at low micrometer (μm) resolution, using the instrument. (A) This microscope (https://micro.magnet.fsu.edu/primer/museum/hooke.html) was used by Robert Hooke to obtain the first ever picture of cells (B).

***Source*: Hooke (1665).**

The era between the 1665 discovery of cells and the understanding of the role of the cell in heredity or genetics (Cobb, 2013), based mainly on the 1953 discovery of DNA's double helical structure (Watson and Crick, 1953a), experienced huge progress in understanding the versatile roles of cells (Müller-Wille, 2010). The understanding of cells got a boost following especially the nineteenth-century development of cell theory (Schwann, 1847). There are "three tenets" of cell theory:

> that all plants and animals are made of cells, that cells possess all the attributes of life (assimilation, growth, reproduction) (these parts were based on a conclusion made by Schwann and Matthias Schleiden in 1838, after comparing their observations of plant and animal cells), and that all cells arise from the division of preexisting cells (Rudolf Virchow described it in 1858).
>
> **(Müller-Wille, 2010)**

With the availability of improved microscopes (having suitable lenses, higher resolution and magnification without aberration, and more satisfactory illumination) during the 1830s, Robert Brown could observe the cell nucleus (Gerould, 1922; Wilson, 1944; Nicholson, 2010). Thus, after the seventeenth-century microscopic discovery of cells we had to wait almost two centuries to discover the cellular organelle "nucleus," also using microscopic imaging techniques. It took another century to discover the nuclear DNA structure using another imaging technique, "X-ray imaging," which led to modeling DNA as a double helical structure (Watson and Crick, 1953b). We are now even able to produce direct images of single molecules at ultra-low angstrom (Å) dimension resolution. A high-resolution transmission electron microscope (TEM) was recently operated at 80 keV with 1.5 Å resolution in order to produce direct images of a single DNA molecule (see Figure 1.2). A single suspended DNA molecule is imaged here in contrast to the traditional imaging of a crystal as done in X-ray imaging. The direct image allows for performing a quantitative evaluation of all relevant characteristic lengths present in a biomolecule such as DNA.

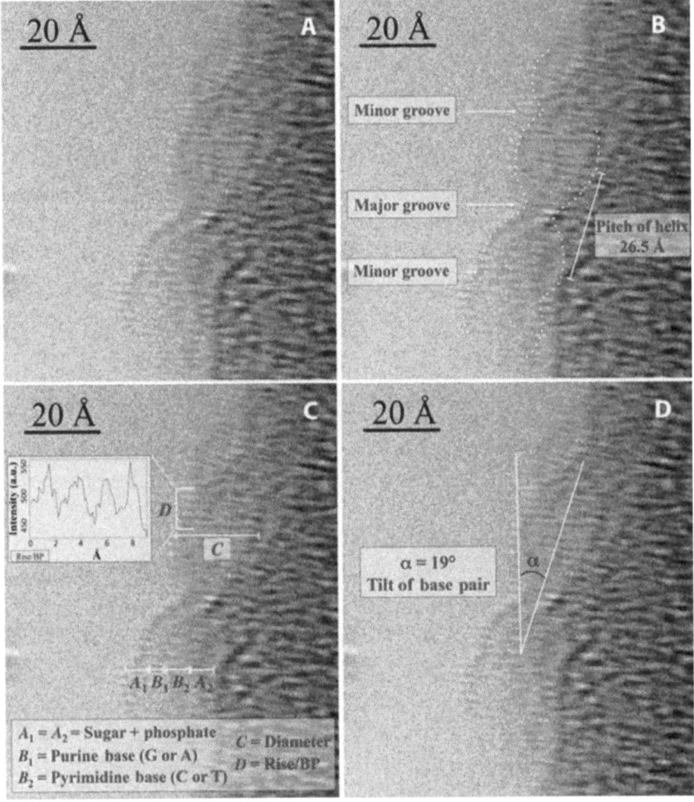

FIGURE 1.2 A-DNA direct image and metrology. (A) High-resolution transmission electron microscope (HRTEM) phase-contrast image of a single A-DNA helix bound to a 100 Å DNA bundle obtained by stacking two images acquired with 50 e/s Å^2 at 80 keV. (B) Dotted line sharpens the DNA location. Major and minor grooves and the helix pitch of 26.5 Å are highlighted. (C) The principal lengths (the backbones, the base pairs (BPs), the diameter, and the rise per base pair) are indicated and reported. The length difference between the purine and pyrimidine bases is also shown: a.u., arbitrary unit. (D) The tilt of the base pairs concerning the helix axis is reported and measures 19°.

***Source*: Marini et al. (2015).**

Versatile applications of principles of electromagnetic radiation or light are to be solely credited for vital biological discoveries, such as cells, cellular organelles like nuclei, and genetic materials like nucleic acids. These discoveries appear as fundamental opto-biophysics breakthroughs revolutionizing our understanding of biology (Krueger et al., 2019; Singh and Siddhanta, 2021).

1.1.2 Microscopic Discovery of Bacteria and Its Modern Biophysical Understanding

The 1677 paper of Leeuwenhoek, a letter on the protozoa, provides the first description of protists and bacteria that live in various environments (Lane, 2015). Following a period of ignorance of the remarkable discovery, it reappeared dramatically on the scientific scene in the twentieth century when research findings confirmed Leeuwenhoek's discovery of bacterial cells, with a resolution of less than 1 μm. The era between the seventeenth-century microscopic imaging-based discovery of bacteria and the twentieth-century discovery of penicillin by Alexander Fleming (Tan and Tatsumura, 2015) to combat bacteria (Gaynes, 2017) experienced real challenges in dealing with various microbes: algae, protozoa, slime moulds, fungi, bacteria, archaea, viruses, etc. (Pitt and Barer, 2012; Monod, 1949; Burmeister et al., 2021). During this era, there were various scientific breakthroughs, including especially the famous "germ theory" providing the understanding that important diseases are caused by infection with microorganisms (Steere-Williams, 2015; Scott et al., 2020). Germ theory created versatile avenues for understanding microorganisms; the knowledge indirectly helped Fleming in his discovery of life-saving penicillin (Fleming, 1929). Fletcher and Florey injected the first human patient, 43-year-old Albert Alexander, with penicillin on 12 February 1941, which initiated our ongoing modern era of antibiotic therapy. Penicillin's biophysical characterization (Caneva et al., 1993) and genetic basis analysis using advanced technological adaptation (Bagcigil et al., 2012; Haveri et al., 2005; van den Berg, 2010) have made it possible to understand the basis of the function or nonfunction of penicillin, including its physical resistances. Figure 1.3 presents a schematic representation of the penicillin biosynthesis pathway that follows the cloning of the biosynthetic genes encoding the biosynthetic enzymes (Carr et al., 1986).

Like Hooke's cork cell discovery, the discovery of bacteria also emerged as an opto-biophysics breakthrough, which created the basis for further life-saving scientific discoveries, like the discovery of penicillin. Penicillium notatum, a mold that Fleming observed through a microscope, looked brush-like, hence its Latin name, Penicillium or "small brush" (Houbraken et al., 2011).

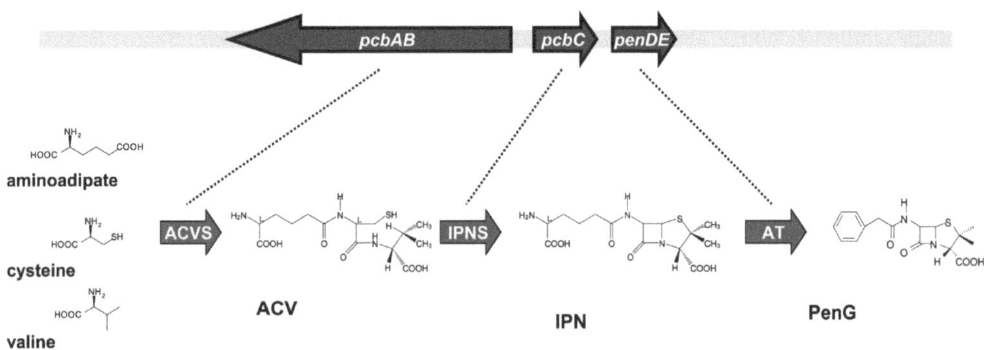

FIGURE 1.3 Schematic representation of the penicillin biosynthetic pathway. Top panel, biosynthetic gene cluster; bottom panel, biosynthetic pathway. ACVS = α-aminoadipoyl-L-cysteinyl-D-valine synthetase; IPNS = Isopenicillin N synthase; AT = acyl-CoA: isopenicillin N acyltransferase.

***Source*: van den Berg (2010).**

The primary step leading to bacterial infection is bacterial adhesion to the surface of host cells, which is mediated by a multitude of molecular-level active physical interactions (Parreira and Martins, 2021). During recent decades, at the interface of life sciences and physics, we have experienced advances in atomic force microscopy (AFM)-based spectroscopy techniques, which have made it possible to measure the forces actively driving bacteria-cell and bacteria-materials interactions on a single molecule/cell basis (single molecule/cell force spectroscopy). Microscopic techniques and other imaging techniques have been made available to assist us in physically addressing the molecular mechanisms that are active behind penicillin- and versatile antibiotics-induced control of bacterial growth. We are also able to develop novel biophysical techniques and protocols which should reveal dynamic processes that infections follow at the molecular level (Leake, 2016).

Following the discovery of penicillin, we, in the modern day, came to know how to deal with infectious diseases, especially those caused by bacteria. We have resources helping us to perform biophysical characterization of the penicillin interaction with bacterial cell membranes (Danelon et al., 2006). Therefore, our use of antibiotics is now made following standard physical protocols to ensure correct technology-driven delivery (Fadaka et al., 2021) in therapy alongside following necessary medical standards.

The use of antibiotics is found to date back to 350–550 CE, as traces of tetracycline, for example, have been found in human skeletal remains from ancient Sudanese Nubia (Bassett et al., 1980; Nelson et al., 2010; Aminov, 2010). Late Roman period skeletons from the Dakhleh Oasis, Egypt, also show evidence of antibiotic exposure (Cook et al., 1989). Discrete fluorochrome labeling was found here, which is consistent with the presence of tetracycline in the diet during that era. In the pre-antibiotic era, antimicrobial use could have been through the remedies used for millennia in traditional/alternative medicine (Cui and Su, 2009). Despite all these historical perspectives on bacterial infections and means to fight them, biophysical approaches to dealing with both understandings of bacteria, especially various aspects of the bacterial cells, and understanding how we could work on discovering antibiotics have flourished greatly during the last couple of decades, though initiatives started about half a century ago (Parreira and Martins, 2021; Jeckel et al., 2019; Kavčič et al., 2021). The future will see an era of excellence when biophysics is expected to have enhanced roles in structural and functional understanding and discovering means to fight bacteria (Thaker and Wright, 2015; Allen and Waclaw, 2016; Mitcheltree et al., 2021).

1.1.3 Physics Revolutionized the Understanding of Viruses

Although bacteria were discovered in the seventeenth century (using the opto-physics imaging technique), the discovery of the first virus, tobacco mosaic virus (TMV), was reported during the late nineteenth century (Lecoq, 2001). Ivanoski reported in 1892 that extracts from infected leaves were still infectious after filtration through a Chamberland filter candle. In 1898, Beijerinck was the first to use the word "virus" for the incitant of the tobacco mosaic. Ivanovski and Beijerinck are both credited with bringing unequal but decisive and complementary contributions to the discovery of viruses. The first physical understanding of viruses perhaps happened through the 1930s work of Wendell Stanley, who produced the TMV in solid crystalline form. Thus the smallest microbes, viruses, appeared as a type of particle (Donchenko et al., 2017; Bhat et al., 2022) and not a liquid contagion. Viruses are of various types, while a complete virus particle is called a virion, which delivers DNA or RNA genome into the host cell so that the genome can be expressed (transcribed and translated) by the host cell (Gelderblom, 1996).

Physical methods are popularly used for both virus diagnosis and analysis of various structural and functional aspects of viruses (Pease III et al., 2011; Richert-Pöggeler et al., 2019). Microscopy has a huge application in the detection of viruses (Goldsmith and Miller, 2009). The first electron microscope (EM) visualization of viruses came after the EM was developed by Ernst Ruska and Max Knoll in 1931. Ruska, Kausche, and Pfankuch then created the first group to visualize TMV with

EM (Kausche et al., 1939). Detailed analysis of various virus structures soon appeared using various physics (mostly imaging) techniques (Crick and Watson, 1956; Klug and Franklin, 1957; Klug et al., 1957; Mattern, 1962). The high-resolution crystal structure of TMV is presented in Figure 1.4, which includes an internal glutamic acid residue (GLU97/106, blue) and an external tyrosine residue (TYR139, red) that can be functionalized using the well-established copper-catalyzed azide-alkyne cycloaddition (CuAAC) strategy, also known as click chemistry (Bruckman et al., 2014; Bruckman and Steinmetz, 2014). The internal TMV surface was modified with alkyne groups, then conjugated with the macrocyclic T_1-MRI contrast agent Gd-DOTA-azide via a click reaction.

Analytical physical methods have been developed to characterize the structure and fundamental properties of virus capsids, which are the protein shells of viruses, enclosing their genetic materials (Kondylis et al., 2019). Being particles, viruses resemble most of the physical properties that particles reveal in an environment (Kiss et al., 2020). These bioactive particles are found to be highly dynamic physical structures (Bruinsma et al., 2021). General application of physics has been making important contributions to structural virology during the last half century or more (Caspar and Klug, 1962; Lauffer, 1981; Baker et al., 1999; Grunewald et al., 2003; Roos et al., 2010; Wan et al., 2017). Biophysical virology is thus seen to be a little less than a century old, while medical virology has a history of just over a century (Burrell et al., 2017).

1.1.4 Early Biophysical Understanding of Photosynthesis

Photosynthesis is considered a physicochemical process (Franck, 1955). An almost five decades old review addressed the then-understood biophysical aspects of photosynthesis by detailing the energy and electron transfer processes associated with this vital process of plant biology (Barber, 1978). Photosynthesis helps plants, algae, and certain bacteria to capture solar energy and efficiently convert it into a storable chemical form. Almost seven decades ago, interest in photosynthesis arose strongly among scientists from many disciplines, including especially physicists. However, the origin of the subject dates back to 1771 when Joseph Priestley reported that green plants were able to purify air that had been "contaminated" by the respiration of animals. Within a few years following Priestley's discovery, John Ingenhousz showed that light was important in oxygen production, but it was not until 1937 that Robert Hill found that this gas was derived from the photolysis of water (Hill, 1937). Physical concepts associated with photosynthesis even date back to 1845, when Julius Robert Mayer realized that photosynthesis was an energy-converting process; the observation and concept helped him to formulate the concept of the first law of thermodynamics. For a detailed understanding of Mayer's works (Mayer, 1842; Mayer, 1845), see Lindsay (1973) and Aquilini et al. (2021), which provide a historical perspective.

The underlying mechanism of photosynthesis involves a physical process causing the removal of electrons and protons from a donor and their transfer to an acceptor. Light appears to cause this electron/proton transfer process. The resulting charge transfer species appear as the energy storage products. A series of charge transfers are seen to take place and the final electron acceptor is carbon dioxide (CO_2). This electron/proton donation gives rise to reduced CO_2, which then acts as the basic building block for the synthesis of the end product carbohydrate (sugar). Photosynthesis theory emphasizes that the oxygen molecules released into the atmosphere originate from water oxidation, not from carbon dioxide, as established in 1941 using ^{18}O-labelled water (Ruben et al., 1941). The whole process from light absorption to energy production and storage happens very fast, within femtosecond (fs) to nanosecond (ns) time range. Light energy-driven transfer of charges certainly reflects vital physical mechanisms, so photosynthesis is also a physical process besides being a chemical one. Figure 1.5 presents a diagram of the photosynthetic apparatus and electron transport (ET) pathways in plants and algae.

Mathematical modeling is a highly valuable tool for understanding and making predictions regarding various aspects of photosynthesis (Stirbet et al., 2020). Existing hypotheses and physical

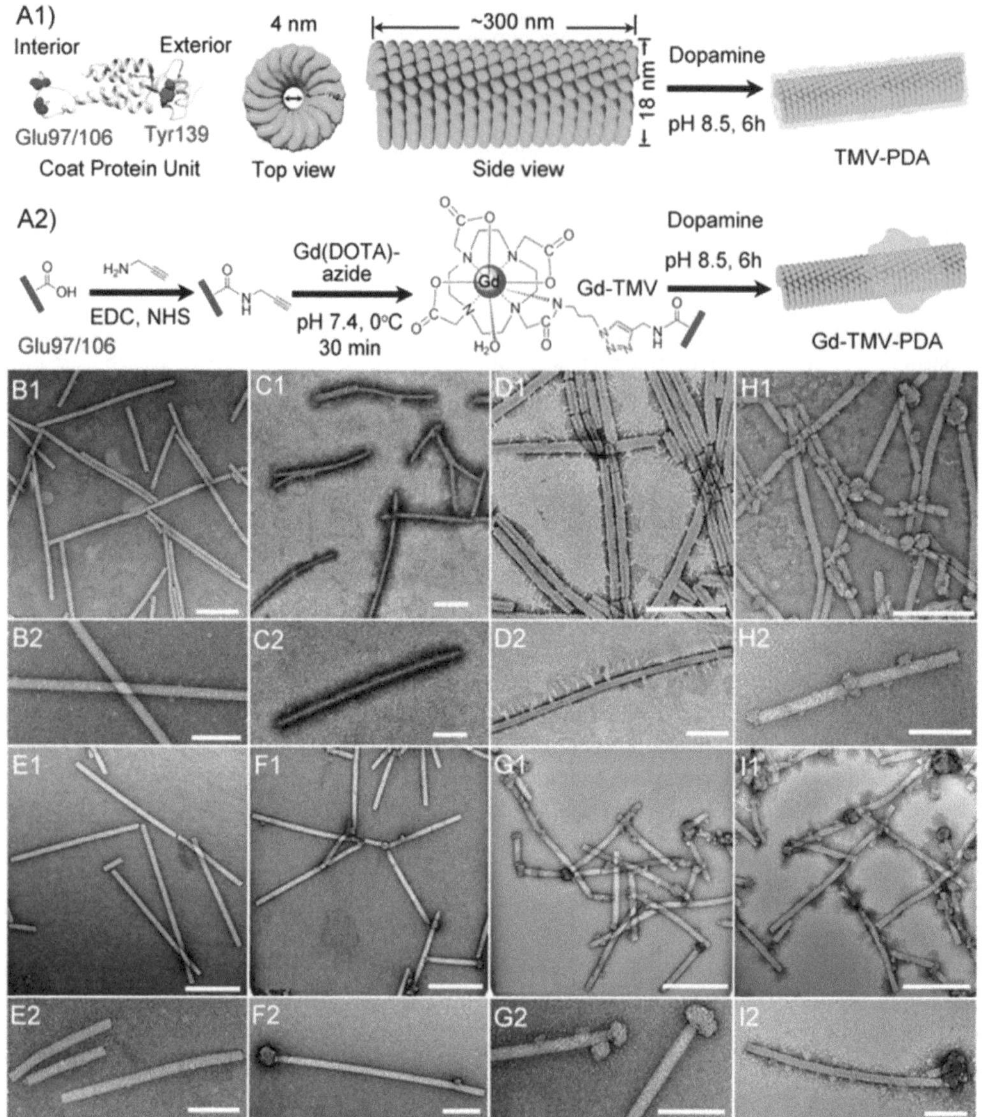

FIGURE 1.4 (**A1**) Structure of the Tobacco mosaic virus (TMV) coat protein, showing the surface-exposed internal glutamic acid (blue) and external tyrosine (red) residues, and the structure of the assembled wild-type capsid coated with polydopamine (TMV-PDA). (**A2**) Strategy for internal loading of Gd-DOTA (Gd-TMV) and partial decoration with PDA (Gd-TMV-PDA). Images were created using UCSF Chimera software, PDB entry 2TMV and ChemDraw v15.0. Transmission electron micrographs (TEM) of (B1–2) native wild-type TMV and TMV-PDA with different reaction mass ratios of TMV:dopamine: (C1–2) 2:2.5 and (D1–2) 2:3.5. TEM of (E1–2) Gd-TMV and Gd-TMV-PDA with different mass ratios of Gd-TMV: dopamine: (F1–2) 2:2.5, (G1–2) 2:3.5, (H1–2) 2:5, and (I1–2) 2:7. Scale bar = 200 nm (B1-I1) and 100 nm (B2-I2). The color contrast in the figure here and all subsequent ones will be clear in the online version of the book.

***Source*: Hu et al. (2019).**

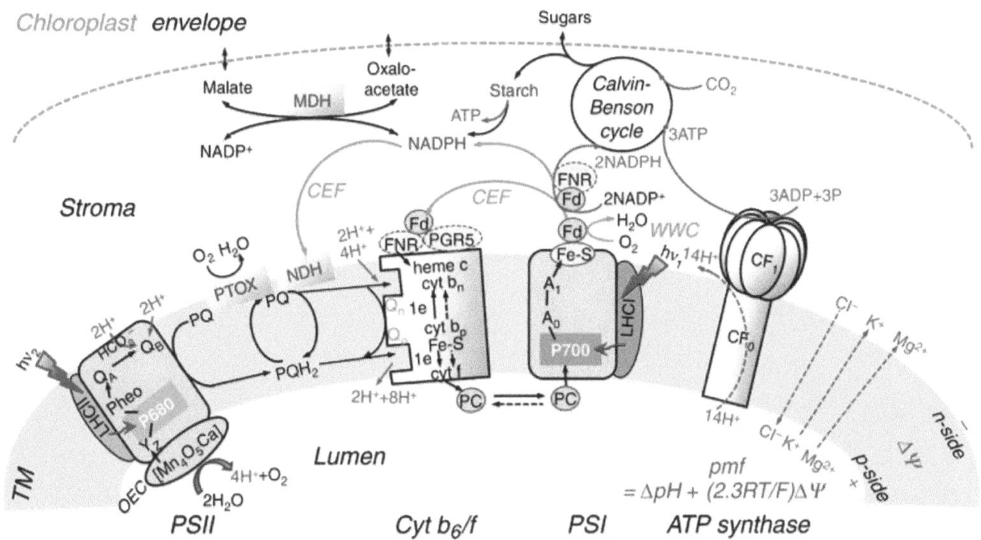

FIGURE 1.5 Diagram of the photosynthetic apparatus and ET pathways in plants and algae. Four major protein complexes in the thylakoid membrane (TM) participate in the production of ATP and nicotinamide adenine dinucleotide phosphate (NADPH) in reduced form, needed for the Calvin–Benson cycle to fix CO_2 to produce sugars: two photosystems (PSII and PSI) connected in series via the cytochrome (Cyt) b_6/f, and the ATP synthase. Light is absorbed simultaneously by pigments in the light harvesting complexes of PSI and PSII (LHCI and LHCII); excitation energy is transferred to reaction centre (RC) P700 (in PSI) and P680 (in PSII), where primary charge separation takes place, initiating a chain of redox reactions. PSII functions as a water/PQ (photo)-oxidoreductase, which has a manganese complex [Mn_4O_5Ca], and a tyrosine-161 (Y_Z), located on D1 protein on the electron donor side, as well as pheophytin (Pheo), plastoquinones Q_A and Q_B, and a non-haem (heme) iron binding a bicarbonate ion (HCO_3^-) on the electron acceptor side. By contrast, PSI is a plastocyanin (PC)/ferredoxin (Fd) (photo)-oxidoreductase; it uses reduced PC as an electron donor, and a particular Chl *a* molecule (A_0), vitamin K_1 (A_1), and three non-haem iron–sulfur centres (shown in the figure as Fe-S) are on the acceptor side of PSI. The Cyt b_6/f complex includes a Cyt f, a Rieske iron–sulfur protein (Fe-S), two cytochromes b (Cyt b_p and Cyt b_n) that participate in the oxidation and reduction of PQH_2 and PQ: PQH_2 is oxidized at the Q_p-site by Cyt b_p, while PQ is reduced at the Q_n-site by Cyt b_n. The Q_p- and Q_n-sides are also called Q_o- and Q_i-sides, respectively. Besides the linear ET flow from water to $NADP^+$, there are several pathways leading to electron donation to alternative electron acceptors: cyclic electron flow (CEF) around PSI mediated by Fd (involving Fd-$NADP^+$-reductase, FNR, and a proton gradient regulator, PGR5), or NADPH (via NADPH dehydrogenase, NDH); water–water cycle (WWC); chlororespiration (through the plastid terminal oxidase, PTOX); and the malate valve (through malate dehydrogenase, MDH). The proton motive force (*pmf*), consisting of the proton concentration difference (ΔpH) and the electric potential (ΔΨ) across TM, is used by ATP synthase to produce ATP from ADP and phosphate (P_i); in the *pmf* formula, *R* is the gas constant, *F* is the Faraday constant, and *T* is the absolute temperature (in K).

***Source*: Stirbet et al. (2020), which was modified from an earlier published version in Alric (2010).**

theories on diverse photosynthetic processes have also been mathematically modeled, which helps validate through simulation(s) of available experimental data, such as chlorophyll a fluorescence induction, measured with fluorometers using continuous (or modulated) exciting light, and absorbance changes at 820 nm (ΔA_{820}) related to redox changes in P700, the reaction center of photosystem I. High-performance computing capability and knowledge of various steps of photosynthesis provide opportunities to adopt mathematical modeling to better understand the dynamics of this process (Lazár nd Schansker, 2009; Jablonsky et al., 2011). It is now strongly believed (Zhu et al., 2010; Simkin et al., 2019) that the photosynthetic mechanism can be regulated by applying genetic engineering to increase crop yields (Rosenthal et al., 2011; McGrath and Long, 2014). Physical and mathematical modeling may be utilized for performing computational predictions on achieving opportunities for specific genetic modifications and devising optimized engineering designs to improve photosynthesis (Zhu et al., 2007). Future goals in artificial regulation of apparently one of the most important natural physicochemical processes, photosynthesis, lie in the rigorous use of methods of physics, mathematics, computational algorithms, and various engineering tools.

1.2 EARLY ESTABLISHMENTS OF BIOPHYSICS AND EMERGENCE OF NEW BIOPHYSICS DISCIPLINES

Over the last two centuries, we have experienced tremendous developments in innovations and upgradation of biophysics techniques and approaches, enabling biophysics research to cover new areas, leading to the emergence of interdisciplinary and multidisciplinary subjects around the core biophysics discipline. I shall pinpoint a few distinguished examples here, although they will be elaborated on in respective chapters. The emergence of these disciplines happened chronologically because most of them are interlinked and dependent on the precedent background. Therefore, I wish to briefly consider a chronological timeline of their developments.

To construct a biophysics timeline demonstrating key historical elements I wish to mention all that is "first" and/or breakthrough-creating aspects. Physics is known to have played key roles in biology since the beginning of the history of life, as we have already learned briefly at the beginning of this chapter. The known history of human intervention in biology using physics started basically in the seventeenth century when Robert Hooke first imaged cells in 1665 using his compound microscope (see Figure 1.1). This big event was immediately followed by another history-creating discovery – the first ever discovery of bacteria or living cells – by Leeuwenhoek in 1676.

Since the discovery of cells and bacteria utilizing optics techniques, biophysics explorations didn't progress reasonably until the middle of the nineteenth century. Physics principles were first thought to be applicable in biological systems in 1847 by four scientists, Emil du Bois-Reymond, Ernst von Brücke, Hermann von Helmholtz, and Carl Ludwig (Cranefield, 1957; Andersen, 2016). These four giants imagined constituting physiology on a chemicophysical foundation, which would give it equal scientific rank with physics (Cranefield, 1957). In 1856, the first-ever biophysics book was published in German, authored by Adolf Fick (1858). In 1892, biophysics got its name and appeared as an independent field. Karl Pearson introduced the term "Bio-Physics" in *The Grammar of Science* (see www.cell.com/biophysj/collections/introduction-to-biophysics, accessed March 9, 2023; Pearson, 1900; Andersen, 2016).

Toward the end of the nineteenth century, the emergence of biomedical imaging initiated a remarkable beginning of the applications of physical technologies to help us understand medical issues that we encounter in almost all aspects of diagnosis. In 1895, the German physicist Wilhelm Röntgen produced the first-ever X-ray image of human finger bones (which brought him the inaugural Nobel Prize in physics in 1901) (Tubiana, 1996). The concept of medical imaging of our body's internal solid structures was initiated vy the discovery of X-rays based on the principle of passing ionizing radiation through the body and having the images projected on a photosensitive plate placed behind it. Doctors and medical professionals are now able to deal with broken and fractured bones and many other structures based on X-ray images taken using modern diagnostic setups.

Crystal structures of biomolecules are X-ray imaged to help discover the biomolecular structures. This structural study helps us understand biomolecular processes besides revealing the structure in a crystal form. The first successful X-ray crystallography structure analysis at single biomolecule level happened through the discovery of DNA (Watson and Crick, 1953a) and then protein structures (Perutz et al., 1960; Kendrew et al., 1958). We know that most of the structural biology research breakthroughs happened by utilizing X-ray crystallography imaging; see Figure 1.6 for a brief history (Shi, 2014).

Besides versatile applications of X-ray techniques in structural biology, another physical technique, nuclear magnetic resonance (NMR) (Bloch et al., 1946), has been found very successful to address biomolecule structures, fluctuations, dynamics, and interactions in solutions (Marion, 2013; Purslow et al., 2020). Discovered as a physical technique in 1946 (Bloch et al., 1946; Purcell et al., 1946), NMR developed in diversified stages to help us explore various aspects of structures (Becker, 1993). These biological explorations were revolutionized a little later. In 1957, the first spectrum of bovine pancreatic ribonuclease was recorded at 40 MHz (Saunders et al., 1957), but the real breakthrough in the application of NMR in biology became apparent with the 1985 discovery of a small globular protein structure (Williamson et al., 1985). The three-dimensional (3D) NMR was introduced first on unlabeled proteins, followed quickly by a new set of triple resonance experiments using 15N and 13C labeled samples (Bax, 1994). High-resolution NMR (Glenn et al., 2018; Mureddu and Vuister, 2019) is now regularly applied in solving structural biology problems.

A century ago biophysics entered into a primarily social science area, economics. "Biophysical economics" was a term coined in the 1920s by mathematician, physical chemist, and statistician Alfred J. Lotka. The biophysical reading of the economic process has also at times been named differently, e.g. "thermoeconomics" or "bioeconomics." The roots can be traced to the eighteenth-century physiocrats (see https://biophyseco.org/biophysical-economics/what-biophysical-economics-is-not, accessedFebruary 4, 2023). Around the same time as Alfred Lotka (1922), Frederick Soddy strongly advanced the concept of biophysical economics with a growing body of literature devoted to the analysis of the role of natural resources in human affairs, and particularly in economic production. Frederick Soddy, a notable author and Nobel laureate in chemistry, applied the laws of thermodynamics (Atkins, 2010) to economic systems and devoted himself significantly to a critique of standard economic theory. Like the physiocrats, Soddy maintained that a comprehensive theory of economic wealth has biophysical laws as first principles because

> life derives the whole of its physical energy or power not from anything self-contained in living matter, and still less from an external deity, but solely from the inanimate world. It is dependent for all the necessities of its physical continuance upon the principles of the steam engine. The principles and ethics of all human conventions must not run counter to those of thermodynamics.

Biophysical economics is popularly characterized by a wide range of analyses from diverse fields that use basic ecological and thermodynamic principles for analyzing the economic process (Cleveland, 1987). Considering the importance of this field, I have written an independent chapter detailing the latest progress.

Biophysics saw the beginning of its new extension in the field of quantum physics when the great physicist Erwin Schrödinger addressed life in his book *What is Life?* (1944). Here Schrödinger discussed applications of quantum mechanics in biology, which we know as quantum biology. He introduced the idea of an "aperiodic crystal" that contained genetic information in its configuration of covalent chemical bonds. He further suggested that mutations are introduced by "quantum leaps" (Margulis and Sagan, 1995).

Before Schrödinger's quantum biology idea, Franck and Teller proposed a quantum coherent mechanism for excitation energy transfer in photosynthesis in 1938 (Franck and Teller, 1938). They considered the diffusion of a Frenkel exciton, which is a coherent superposition of the electronic excitations of the individual photosynthetic pigments (see Figure 1.7). Frenkel excitons are bound states of electron-hole pairs that both reside on the same lattice cell or molecule. The femtosecond (fs) transient absorption spectroscopy application in the 1990s helped detect the long-lived vibrational coherences, lasting on the order of a picosecond (ps) for bacterial and plant light-harvesting

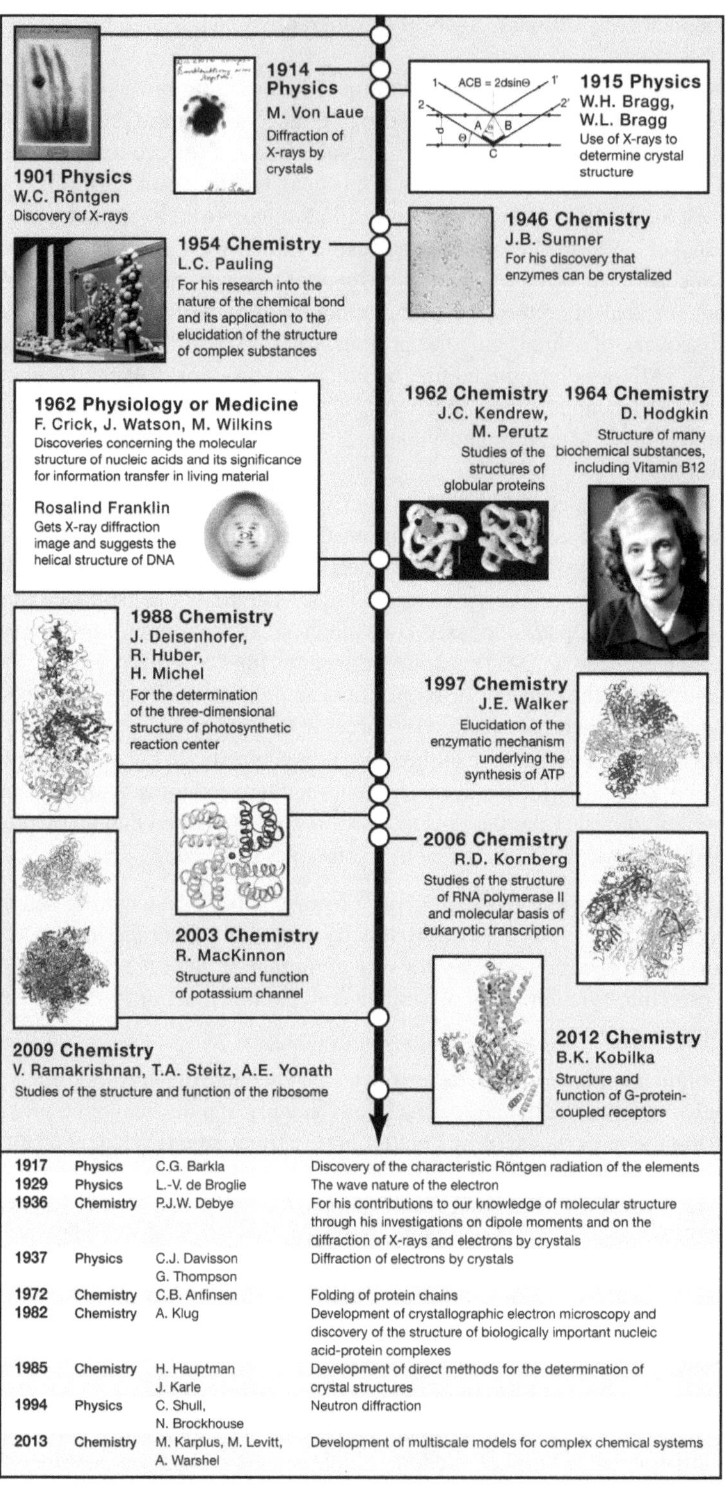

1917	Physics	C.G. Barkla	Discovery of the characteristic Röntgen radiation of the elements
1929	Physics	L.-V. de Broglie	The wave nature of the electron
1936	Chemistry	P.J.W. Debye	For his contributions to our knowledge of molecular structure through his investigations on dipole moments and on the diffraction of X-rays and electrons by crystals
1937	Physics	C.J. Davisson G. Thompson	Diffraction of electrons by crystals
1972	Chemistry	C.B. Anfinsen	Folding of protein chains
1982	Chemistry	A. Klug	Development of crystallographic electron microscopy and discovery of the structure of biologically important nucleic acid-protein complexes
1985	Chemistry	H. Hauptman J. Karle	Development of direct methods for the determination of crystal structures
1994	Physics	C. Shull, N. Brockhouse	Neutron diffraction
2013	Chemistry	M. Karplus, M. Levitt, A. Warshel	Development of multiscale models for complex chemical systems

FIGURE 1.6 The history of X-ray crystallography illustrated through Nobel prizes. X-ray crystallography-based discoveries have been recognized by at least 14 Nobel prizes, which started with Röntgen receiving first ever Nobel prize in physics in 1901.

***Source*: Shi (2014).**

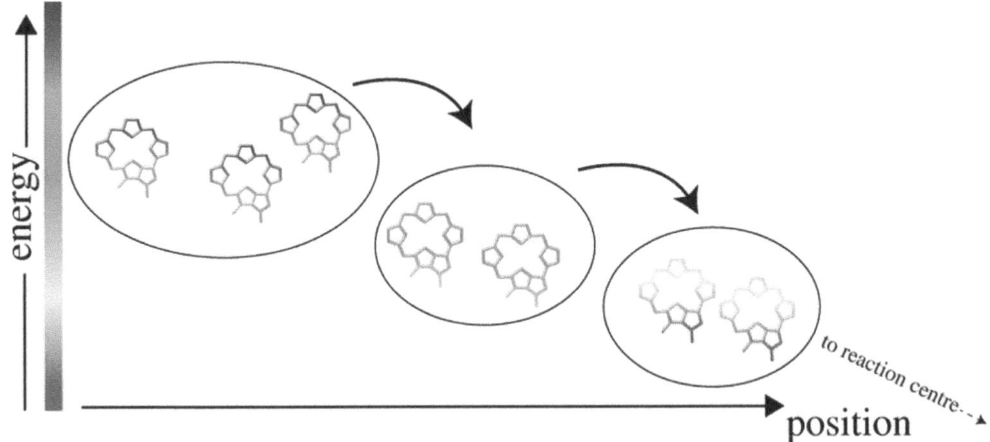

FIGURE 1.7 A schematic illustrating the concept of excitation energy transfer in photosynthetic light-harvesting complexes. The ovals depict pigment clusters in which strong excitonic coupling occurs while the arrows represent incoherent transfer of the energy amongst the clusters. Within each cluster, the energy is delocalized over all pigments and is transferred coherently.

***Source*: Marais et al. (2018).**

complexes (Chachisvilis et al., 1994; Vos et al., 1994; Kumble et al., 1996; Agarval et al., 2000; Novoderezhkin et al., 2000; Novoderezhkin et al., 2004).

In various examples of biological systems and processes the quantum phenomena are evident. The role of quantum phenomena, such as tunneling and quantum coherence, has now been widely accepted in the activities of living cells (Ghasemi and Shafiee, 2019). The role of the ion channel selectivity filter is crucial in channel conduction mechanisms and it is assumed to exhibit quantum coherence (Seifi et al., 2022). Quantum interference is assumed to be a solution to explain the selectivity mechanism in ion channels since interference happens between similar ions through the same size of ion channels (Salari et al., 2017). Just recently, the relationship between hopping rate and maintaining coherence in ion channels was investigated (Seifi et al., 2022). The distillable coherence and the second-order coherence function of the system were studied here. The oscillation of distillable coherence from zero, after the decoherence time, and also the behavior of the coherence function are claimed to show the point that the system is coherent in ion channels with high throughput rates.

When quantum biology was still at the infant level of its theoretical beginning, single biomolecule imaging emerged to asist the most important biophysical breakthrough by helping discover DNA structure in 1953. Although micrometer (μm) resolution biological imaging (biological cell) was already celebrating its third century, X-ray crystallography of DNA structure appeared as the first-ever nanometer (nm) resolution imaging. In this discovery, a novel concept, "biomolecular modeling," appeared to create actual understanding through processing information drawn from Rosalind Franklin's X-ray image of DNA. It took only another five years to produce an image of a whole fetus. First-ever ultrasound images of the fetus and also gynaecological masses were taken in 1958 (Donald et al., 1958). The sonogram technique is now one of the most utilized medical imaging protocols to understand the dynamics and fluctuations of our functional inner soft masses (The, 2012; Kaeley et al., 2020).

Many of those who are less informed still participate in unnecessary debates regarding crediting Rosalind Franklin and/or the Watson-Crick duo with the DNA double helix structure discovery. Although the experimentalist Franklin produced X-ray images, it was the geniuses of the

Watson-Crick duo who could process the information drawn from X-ray images to model the double helical form of two-stranded DNA molecules. The modeling was as an undeniably excellent method combining information with theory and producing a predicted outcome. The modeling trend then continued to develop. Within a decade of DNA modelling, in 1965, we come across another form of research field, "bioinformatics," which relies heavily on computer-based modeling, what we prefer to call *in silico* modeling. Dayhoff's 1965 book, *Atlas of Protein Sequence and Structure*, is considered to be the founding text for bioinformatics (Dayhoff et al., 1965; Dayhoff, 1969a). Here, she reported all 65 known protein sequences, organized by gene families. During the 1960s, Dayhoff initiated use of computers and algorithms to address biomolecules' various aspects *in silico* (Dayhoff, 1964; Dayhoff, 1969b; Dayhoff and Ledley, 1962; Dayhoff et al., 1964), which later evolved as an independent field of research where computer scientists, physicists, mathematicians, chemists, and biologists gather together to exchange and merge ideas. The first-ever Nobel prize in this field was awarded in 2013 in chemistry (see www.nobelprize.org/prizes/chemistry/2013/summary, accessed February 5, 2023). During the 1970s and 1980s, the Karplus, Levitt, and Warshel trio laid the foundation for powerful programs and continued using them to understand and predict chemical processes. They showed successfully that computer models using appropriate algorithms could mirror real life, which became crucial for many of the advancements made in chemistry and biology today (see Callaway, 2020).

Following the first ever modeling of a single DNA biomolecule in the early 1950s, we experienced another great modeling-based understanding of a biological structure, the "fluid mosaic model of membranes," in the early 1970s (Singer and Nicolson, 1972). Singer and Nicolson were the first to successfully realize a dynamical liquid crystal biological structure (Zhong-Can et al., 1999: 1–27). The fluid mosaic model described the gross organization and structure of the proteins and lipids of biological membranes, which work under thermodynamics restrictions. The fluid mosaic structure is analogous to a two-dimensional (2D) oriented solution of integral proteins or lipoproteins in the viscous phospholipid bilayer solvent. The fluid mosaic model led to the discovery of additional aspects of the membrane's structures and transport mechanisms over the last half century; for details, readers may consult another of my books, *Membrane Biophysics* (Ashrafuzzaman and Tuszynski, 2012).

1.3 LATEST BIOPHYSICS FIELDS WITH FUTURISTIC GOALS

The application of versatile biophysics principles and techniques has been in practice for centuries. And there are promising areas where biophysics is seeing enhanced use, which will help develop newer avenues. I shall pinpoint these areas and provide brief information.

1.3.1 Biophysics of Drug Discovery

Biophysics methods are found quite essential in versatile drug discovery research areas (Holdgate et al., 2013; Renaud et al., 2016; Holdgate et al., 2019; Holdgate and Bergsdorf, 2021; Gavriilidou et al., 2022). Over the past four decades, X-ray crystallography, NMR spectroscopy, surface plasmon resonance (SPR) spectroscopy, and isothermal titration calorimetry (ITC) have been found to play key roles in drug discovery platforms (Renaud et al., 2016). Since the early stages of academic- and industry-based drug discovery, mass spectrometry, SPR, NMR, and differential scanning isothermal titration calorimetry have become well-established biophysical techniques that are regularly utilized (Holdgate et al., 2019).

Biophysical measurements have particular impacts in various areas, as follows (Renaud et al., 2016):

(i) enabling drug discovery for relatively challenging targets, such as protein–protein interactions (Arkin et al., 2014; Higueruelo et al., 2013);

(ii) identifying drug binding kinetics as a crucial factor for efficacy and selectivity (Holdgate and Gill, 2011; Copeland, 2016; Copeland et al., 2006); and
(iii) providing the foundation for fragment-based drug discovery (FBDD) (Jhoti et al, 2013; Silvestre et al., 2013; Erlanson and Zartler, 2015).

The above-mentioned technologies have enabled studies of the thermodynamics of drug binding (Klebe, 2015). Biophysical data appear to provide important complements to those on biochemical and cellular activity as well as aggregation, solubility, and cell permeability (Leeson and Springthorpe, 2007). They have traditionally formed the basis of the hit and the lead discovery, prioritization, and optimization.

Taken together, these developments have enabled rational, rigorous problem-solving approaches to initial drug discovery phases. Although the first successful biomolecule structural discovery in the early 1950s (Watson and Crick, 1953b) created the basis for the roles of X-ray, its applied roles started being visible about three decades ago (Renaud et al., 2016; Carvalho et al., 2010). Single-molecule and cellular biophysical measurements are advancing at an unpredictable rapid pace and enabling a high degree of characterization (Leake, 2013), helping drug discovery research achieve huge successes.

Besides utilizing biophysical techniques, consideration of physicochemical properties of the biomolecules appears to help create novel drug design methods as well as designing novel compounds (Ashrafuzzaman, 2021; Sliwoski et al., 2014; Ashrafuzzaman et al., 2020). Biophysical and biochemical analyses of structural parameters and charge profiles have been found, for the first time by us (Ashrafuzzaman, 2021), to provide vital physical ingredients/parameters in designing novel agents as drug candidates. This suggests that inherent physical properties of biomolecules have important roles in designing drugs.

1.3.2 Quantum Biophysical Microscopy

Quantum microscopy helps measure and image the microscopic properties of matter and quantum particles. In this kind of microscopy, quantum principles are utilized, such as the quantum properties of light (Malik and Boyd, 2014). We now see their increased biological applications (Taylor et al., 2014; Genovese, 2016; Casacio et al., 2021; Liu et al., 2022). Quantum paradigm concepts are now finding their utility in highly medically utilized bioimaging such as MRI, PET, etc. (Allen, 2017; Siemens Healthcare, 2023, www.siemens-healthineers.com/molecular-imaging/news/mso-quantum-leap-biograph-vision-quadra, accessed on April 4, 2023).

The development of the quantum-enhanced microscope and its applications in biological microscopy was rigorously demonstrated a decade ago (Taylor, 2015). This microscope can track quantum particles, which makes it a powerful applied tool in science and medicine. Taylor used the microscope to perform quantum-enhanced biological measurements. Sub-diffraction-limited quantum imaging was achieved for the first time with a scanning probe imaging configuration at 10 nanometer (nm) resolution (Taylor et al., 2014). Taylor and colleagues could track nanoparticles with quantum-correlated light as they diffused through an extended region of a living cell in a quantum-enhanced photonic-force microscope. This allowed spatial structure within the cell to be mapped at length scales down to 10 nm. This application of quantum imaging techniques to subdiffraction-limited biological imaging demonstrates that nonclassical light can improve spatial resolution in biological applications. The viscoelastic structure within a living yeast cell was sampled along the trajectory of a thermally driven nanoparticle, which revealed the spatial structure at low nm dimensions. The quantum microscopic imaging results confirmed that quantum-correlated light could enhance spatial resolution at the nanoscale level in biology. Predictions are made here that further improvements might ensure an order of magnitude improvement in resolution over similar classical imaging techniques (Taylor et al., 2014).

Casacio and colleagues recently showed experimentally (Casacio et al., 2021) that quantum correlations would allow a signal-to-noise ratio beyond the photodamage limit of conventional microscopy (Camp and Cicerone, 2015; Fu et al., 2006). They used a coherent Raman microscope that offers subwavelength resolution and incorporates bright quantum-correlated illumination (Casacio et al., 2021). The correlations allowed imaging of the cell-based molecular bonds with a 35 percent improved signal-to-noise ratio compared with conventional microscopy, corresponding to a 14 percent improvement in concentration sensitivity. Although coherent Raman microscopes are known to allow highly selective biomolecular fingerprinting in unlabeled specimens (Cheng and Sunney, 2015; Wei et al., 2017), its photodamage issue is unsustainable. The quantum microscopy of Casacio and colleagues has demonstrated that the photodamage limit could be overcome which would enable order-of-magnitude improvements in the signal-to-noise ratio and the imaging speed (Casacio et al., 2021). Figure 1.8 presents the sketches of the microscopic setup and example spectra produced by Casacio and colleagues. Figure 1.8a presents a custom-designed stimulated Raman microscope that uses excitation lasers at both pump and Stokes frequencies (Cheng et al., 2015). Figure 1.8b shows a typical Raman spectrum, exhibiting several Raman bands.

Figure 1.9 presents quantum-enhanced imaging of biological samples (Casacio et al., 2021). Figure 1.9a shows a typical quantum-enhanced image of a collection of 3-µm polystyrene beads, where a signal-to-noise ratio was enhanced by 23 percent compared with the shot-noise limit. Figure 1.9b shows the equivalent image for a single yeast cell, in this case, recorded at a Raman shift of 2,850 cm^{-1} to target the CH_2 bonds that are most prevalent in lipids. Improved technical arrangements ensured a 35 percent enhancement in the signal-to-noise ratio, which increased the contrast of the image. The visible cell damage was observed at higher pump intensities (Figure 1.9c).

Recently, Liu and colleagues introduced intrinsic spatial differentiation into heralded single-photon microscopy imaging technique, which made the structure of pure-phase objects visible at low photon levels, avoiding any biophysical damage to living cells (Liu et al., 2022). The use of heralded single photons is known to ensure the background counts are virtually eliminated from the

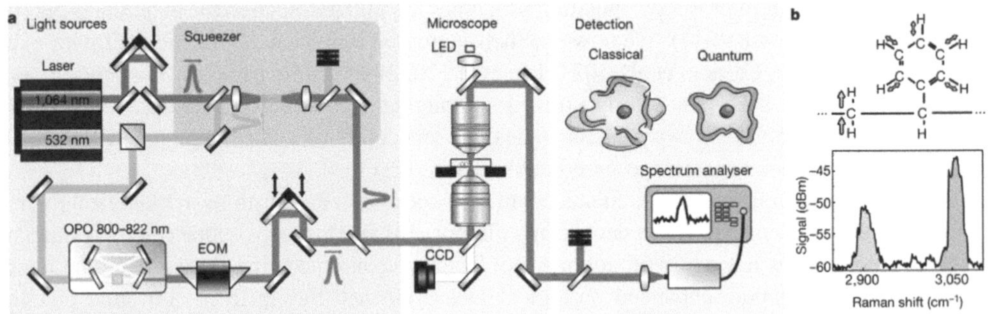

FIGURE 1.8 (a), Setup schematics. Left, preparation of the pump beam (purple) via an optical parametric oscillator (OPO) and 20-MHz modulation from an electro-optic modulator (EOM), and of the Stokes beam (red) that is amplitude-squeezed in a periodically poled $KTiOPO_4$ crystal pumped with 532-nm light. Middle, stimulated Raman scattering is generated in samples at the microscope focus, with raster imaging performed by scanning the sample through the focus. A charge-coupled device (CCD) camera and a light-emitting diode (LED) allow simultaneous bright-field microscopy. Right, after filtering out the pump, the Stokes beam is detected and the signal processed using a spectrum analyzer. For all experiments, 3 mW of detected Stokes power was used. (b) Raman spectra measured from a 3-µm polystyrene bead, showing the CH2 antisymmetric stretch (purple) and CH aromatic stretch (green) resonances. Spectra taken with 100-kHz spectrum analyzer resolution bandwidth (RBW). Color contrast will be visible in the online version of the book.

***Source*: Casacio et al. (2021).**

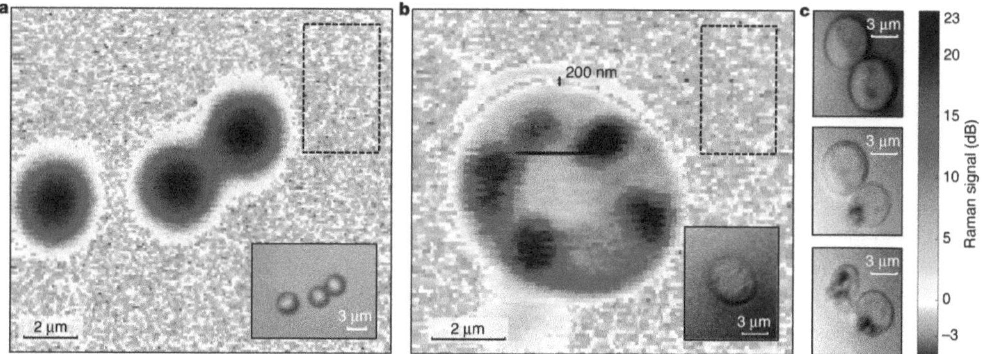

FIGURE 1.9 Quantum-enhanced imaging. (a) Image of 3-µm polystyrene beads at a Raman shift of 3,055 cm^{-1} obtained with 6 mW of pump power at the sample. The background (colored green) has no Raman signal, and is limited by measurement noise, which is 0.9 dB below shot noise, providing a 23 percent increase in the signal-to-noise ratio. (b) Image of a live yeast cell (*Saccharomyces cerevisiae*) in aqueous buffer at 2,850 cm^{-1} Raman shift. Several organelles are clearly visible. The faint outline of what may be the cell membrane or wall is also visible, showing that the microscope has a resolution of around 200 nm. Here the measurement noise is reduced by 1.3 dB below shot noise, corresponding to a 35 percent signal-to-noise ratio improvement. This image was recorded with around 30 mW of pump power at the sample. The pump intensity of 210 W μm^{-2} was beneath that at which visible cell damage was observed. Dashed rectangular boxes in (a), and (b) show the regions used to determine the measurement noise and the insets are bright-field microscopy images. RBW, 1 kHz; video bandwidth, 10 Hz. (c), A sequence of images in which two cells are illuminated with the same pump power as in (b) but focused to roughly a factor of two higher intensity. Visible photodamage is observed after only a few seconds of exposure (middle and bottom images). Color contrast will be visible in the online version of the book.

***Source*: Casacio et al. (2021).**

recorded images (Morris et al., 2015). Using the polarization entanglement mechanism, the switch between dark-field imaging and bright-field imaging was remotely controlled in the heralding arm. This approach is found to enrich both fields of optical analog computing and quantum microscopy, opening avenues to nondestructive imaging of living biological samples.

Development in quantum microscopy is ongoing and during the coming decades we may expect to see considerable progress. This field is undoubtedly going to explode at an unprecedented pace and help us address biological imaging at a higher resolution. It may help us achieve a transition, at least partially, from conventional imaging to quantum imaging, which will reveal many unknown quantum mechanical phenomena in biology.

1.3.3 Biophysical Proteomics

Biophysical proteomics provides systematic information about changes in dynamic structures that are directly linkable to protein functions (Mateus et al., 2021). Single-molecule optical proteomics is capable of probing whole biological systems (Shashkova and Leake, 2018). The proteome refers to the complete set of proteins expressed by a cell, tissue, or organism. Omics classified areas include proteomics, transcriptomics, genomics, metabolomics, lipidomics, and epigenomics, which correspond to global analyses of proteins, RNA, genes, metabolites, lipids, and methylated DNA or modified histone proteins in chromosomes, respectively. Proteomics and associated complementary analysis methods are essential components of the "systems biology" approach seeking to describe biological systems through the integration of diverse types of data to ultimately allow computational simulations of complex biological systems (Patterson and Aebersold, 2003). Almost half a century ago some of the first studies were conducted to encompass proteomics (Suran, 2022). The term "proteomics" was first suggested in 1997 (James, 1997) as an analogy with genomics (the study of genes). Biophysical proteomic techniques focus mainly on protein separation techniques, including two-dimensional (2D) gel electrophoresis (2DE) and liquid chromatography (LC), and protein characterization techniques, such as mass spectrometry. Biophysical proteomics has accomplished the greatest advancements so far using mass spectrometry (Yates et al., 2009; Aslam et al., 2016; Rozanova et al., 2021).

Biophysical proteomic techniques are helping to advance our understanding of diseases and leading us to understand means of drug discovery (Xu et al., 2012; Mateus et al., 2021; Schirle et al., 2012; Dias et al., 2016; Wenzel et al., 2021). Biophysical proteomic techniques utilized in the study of disease include the delineation of altered protein expression, not only at the whole-cell or tissue levels, but also in subcellular structures, protein complexes, and biological fluids. Biophysical proteomics techniques are also applied for the discovery of novel disease biomarkers, exploration of the pathogenesis of diseases, development of new diagnostic methodologies, and identification of new targets for therapeutics. Quantitative proteomic methods are applied in translational pharmacology offering recommendations for selecting and implementing proteomic techniques (El-Khateeb et al., 2019).

1.3.4 Biophysical Genomics and Understanding Evolution

Biophysics techniques are actively involved in addressing various types of predicted and observed evolutions in our timeline of the history of life. Life is expected to have experienced a tremendous amount of development since its earliest forms as microscopic organisms, "microbes," which left signals of their presence in rocks (Dodd et al., 2017). Dodd and colleagues described putative fossilized microorganisms that are believed to be at least 3,770 million and possibly 4,290 million years old in ferruginous sedimentary rocks, interpreted as seafloor-hydrothermal vent-related precipitates, from the Nuvvuagittuq belt in Canada. Biophysical inspection of the evolutionary process in life's fundamental units is capable of revealing crucial information about our development process (Sikosek and Chan, 2014).

Biophysics finds itself applicable in diversified areas concerning evolution (Baltazar-Soares et al., 2018; Schwartz, 2019; Zepp et al., 2021). Biophysics treatment of evolutionary genetics addresses specific binding sites of biomolecules such as proteins, DNA, etc. Genomic analysis predicts that they undergo a stochastic evolutionary process governed by selection, mutations, and genetic drift (Lässig, 2007). Biophysics deals with crucial molecular aspects of evolutionary genomics covering a wide range of subjects investigating the evolution of species' genomes (Ridley et al., 2008), including helping to create models of evolution in bacteria (Lin, 2016), plants (Allaby et al., 2015), and humans (Lopez et al., 2016; Currat et al., 2016). The genetic code similarity (ies) across all types of organisms allows for comparisons of DNA sequences between and within species (Diekmann and Pereira-Leal, 2016; Nunes, 2016). Biophysical understanding of molecular evolution, especially in the area of genetic structures and functions, may resolve some (certainly not all) of the predictions and controversies continuously emerging on evolution; see model demonstrations in Figure 1.10 suggesting especially that hominins originated in Africa from Miocene ape ancestors unlike any living species.

Evolution itself has a history of billions of years during which all living animals and plants have matured genetically (Avise, 2006). Due to this long evolution, humans emerged sharing genes with all living organisms. The percentage of genes that all organisms share records their similarities. The proportion of human genetic variation due to differences between populations is modest, and individuals from different populations can be genetically more similar than individuals from the same population (Witherspoon et al., 2007). Neanderthals that were 99.7 and 98.8 percent genetically

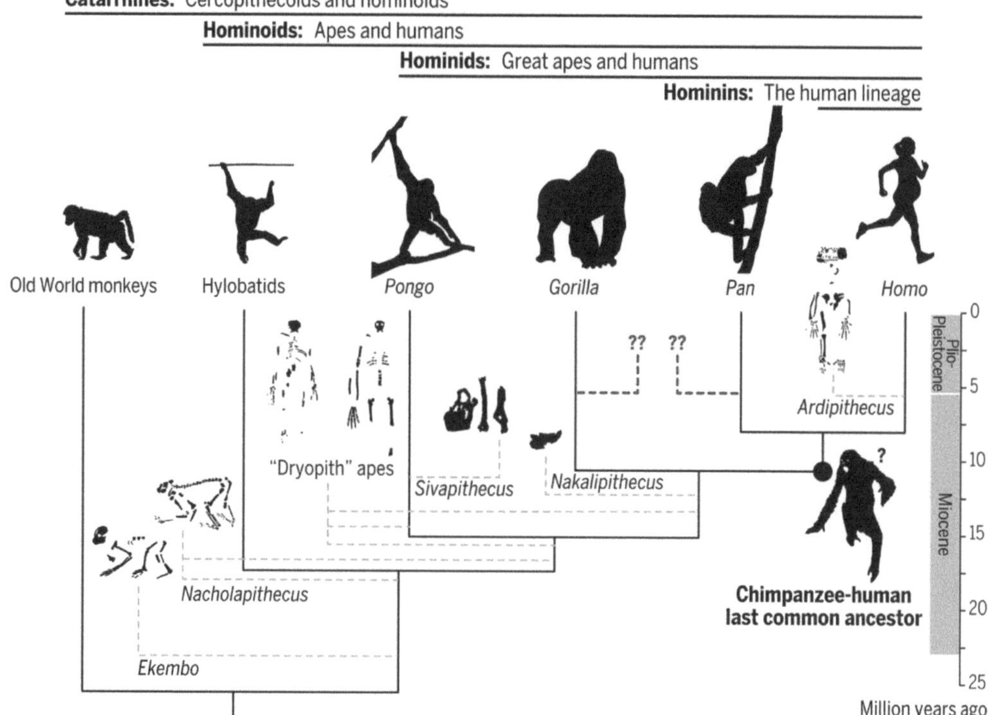

FIGURE 1.10 The evolutionary history of apes and humans is largely incomplete. Whereas the phylogenetic relationships among living species can be retrieved using genetic data, the position of most extinct species remains contentious. Surprisingly, complete enough fossils that can be attributed to the gorilla and chimpanzee lineages remain to be discovered. Assuming different positions of available fossil apes (or ignoring them owing to uncertainty) markedly affects reconstructions of key ancestral nodes, such as that of the chimpanzee–human last common ancestor (LCA).

***Source*: Almécija et al. (2021).**

identical to present-day human and chimpanzee DNA, respectively, lacked essential modern human characteristics (Noonan, 2010). Neanderthals went extinct 30,000 years ago (Green et al., 2010). The divergence of human and chimpanzee genomes was earlier estimated at roughly 1 percent (King and Wilson, 1975), based on the comparison of protein-coding sequences. Later the non-coding (major) part of DNA was considered, and eventually the idea of ~ 99 percent similarity between these two species' genomes was challenged with the emergence of nearly complete initial sequencing of both humans (Lander, 2001) and chimpanzees (Consortium, 2005). Human DNA is 99.9 percent identical while human and chimpanzee DNAs are 98.8 percent identical. The bonobo genome shares almost the same percentage of its DNA with us as the chimpanzee genome does (see www.science.org/content/article/bonobos-join-chimps-closest-human-relatives (accessed April 11, 2023).

Neanderthals were intelligent enough that they could recognize modern-day human tools and lived in established communities (Noonan, 2010). Neanderthals were reported to share more genetic variants with present-day humans in Eurasia than with present-day humans in sub-Saharan Africa. This geography-centered unequal correlation perhaps suggests that gene flow from Neanderthals into the ancestors of non-Africans occurred before the divergence of Eurasian groups from each other (Green et al., 2010). Apart from humans, chimpanzees show the greatest diversity of tool use in the animal kingdom, which is influenced by genetic or environmental variations (McGrew, 2004).

Recently, an international collaborative group analyzed cerebral organoids from humans and chimpanzees using immunohistofluorescence, live imaging, and single-cell transcriptomics (Mora-Bermúdez et al., 2016). The cytoarchitecture, cell type composition, and neurogenic gene expression programs of humans and chimpanzees were reported here as remarkably similar. Notably, however, live imaging of apical progenitor mitosis uncovered a lengthening of prometaphase-metaphase in humans compared to chimpanzees that is specific to proliferating progenitors and not observed in non-neural cells. Consistently, the small set of genes more highly expressed in human apical progenitors points to increased proliferative capacity, and the proportion of neurogenic basal progenitors is lower in humans. These subtle differences in cortical progenitors between humans and chimpanzees are predicted to have consequences for human neocortex evolution.

We see here that among species, variations in genetics play crucial roles in their appearance, performance, and adaptations. Modern biophysics tools (Manhart, 2014; McCammon, 2009; Liberles et al., 2012) are capable of addressing many of them molecularly. Information from species' genome projects may now be provided as inputs into the biophysical understanding of evolution (Durmaz et al., 2015); Schwartz, 2019).

Note: An in-depth analysis on this topic has been made in an independent chapter "Appendix C: Biophysics of Evolution" (see online version of the book).

1.3.5 AI and ML in Biophysics

Artificial intelligence (AI) is a relatively new technique applied in biological research (Richards et al., 2022). Over the last decade, the application of AI in various sectors such as the economy, society, and science has been found to grow at a remarkable pace. AI's role in addressing general biology and specifically neuroscience problems is quite amazing (Richards et al., 2022). Neuroscience has emerged as a source of AI research inspiration (Ullman, 2019). For analyzing neural data and understanding the computation of the brain, AI is utilized (Landhuis, 2020; Richards et al., 2022). Biophysics-inspired AI for addressing biological problems, including issues related to diseases, has started to grow considerably (Hardy et al., 2021). Biophysics problems, like all other science problems, require thorough scientific understanding based on available experimental and computational data, where fundamental AI approaches are found utilizable (Krenn et al., 2022). Machine learning (ML), AI, and data science are now popularly utilized in understanding the spectrum of diseases and discovering drugs (Peña-Guerrero et al., 2021). Google's AlphaFold, a novel ML approach, incorporates physical and biological knowledge about protein structures, leveraging multi-sequence alignments, into the design of the deep learning (DL) algorithm (Jumper et al., 2021). The AlphaFold AI system helps to predict the 3D topology of complex proteins and identify novel protein knots (Brems et al., 2022). DL relies on a class of machine-learnable models constructed using "differentiable programs," which has been found quite applicable in biophysics and functional genomics problems (AlQuraishi and Sorger, 2021). Algorithm-based modeling of biological structures that are otherwise addressed in mainly biophysical techniques may help our knowledge to advance faster. Therefore, an independent subject in this area seems guaranteed to emerge strongly within the next decade.

1.3.6 Computational Biophysics

Modeling unknown or semi-known biomolecular structures is a popular alternative way to predict the biomolecules' structural and functional roles. Knowledge-based versus physics-based methods and hardware versus software advances in propelling biomolecular modeling have been detailed by Schlick and Portillo-Ledesma (2021). The biomolecular modeling field has been flourishing during the last half century through rapid adaptation and tailoring of state-of-the-art technology. Modeling has progressed based on key developments in algorithms, software, and hardware, which is demonstrated chronologically in Figure 1.11.

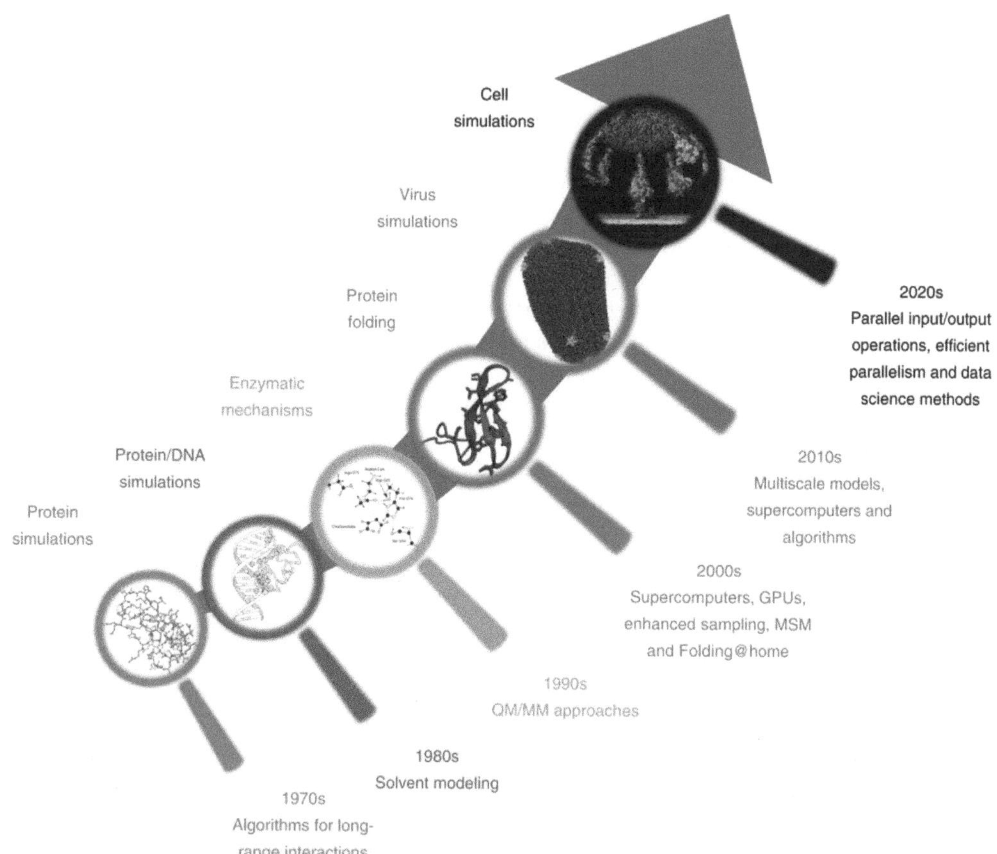

FIGURE 1.11 Chronological advancement phases in the field of modeling during last five decades since the inception of the successful modeling ideas. 1970s: simulations of <1,000-atom systems and few picoseconds in vacuo were possible due to the development of digital computers and algorithms to treat long-range Coulomb interactions. The image depicts the structure of the small protein BPTI that was simulated for 8.8 ps without hydrogen atoms and with four water molecules (McCammon et al., 1977). 1980s: simulations that considered solvent effects became possible, and algorithms such as SHAKE, to constrain covalent bonds involving hydrogen atoms, allowed the study of systems with explicit hydrogens. The image depicts the 125 ps simulation of the 12 bp DNA in complex with the lac repressor protein (de Vlieg et al., 1989) in aqueous solution using the simple three-point charge water model. 1990s: QM/MM methods can perform geometry optimizations, MD and Monte Carlo simulations (Field et al., 1990). The image depicts the acetyl-CoA enolization mechanism by the citrate synthase enzyme studied with AM1/CHARMM. 2000s: GPU-based MD simulations, specialized supercomputers such as Anton, shared resources such as Folding@home, enhanced sampling algorithms, and Markov state models (MSMs) all contributed to advance protein folding (Lane et al., 2013). The image depicts the long 100 μs simulation of the Fip35 folding conducted in Anton (red) compared with the X-ray structure (blue). 2010s: all-atom and coarse-grained MD simulations of viruses performed on supercomputers such as Blue Waters became common (Ode et al., 2012). The image depicts the MD simulation of the all-atom HIV capsid using the MDFF method that uses cryo-electron microscopy data to guide simulations. 2020s: physical whole-cell models are being developed to fully understand how biomolecules behave inside cells and to study interactions between them, for example, within viruses and cells. The image depicts the model of the interaction between the spike protein in the SARS-CoV-2 surface and the ACE2 receptor on a human cell surface being developed in the Amaro Lab.

***Source*: Schlick and Portillo-Ledesma (2021). From left to right, images adapted with permission from: first image, McCammon et al. (1977); second image, de Vlieg et al. (1989), Wiley; third image, Mulholland and Richards (1997), Wiley; fourth image, Shaw et al. (2010), AAAS; fifth image, Zhao et al. (2013), Springer Nature Ltd; sixth image, https://amarolab.ucsd.edu/news.php, Amaro Lab.**

Biophysical modeling of physical states of mutant biomolecules associated with various diseases is in a progressing phase (Wylie and Shakhnovich, 2011; Stein et al., 2019). Mutation-driven thermodynamic instability in protein structures is physically modeled to predict further changes (Li et al., 2020). This kind of modeling is capable of measuring the fundamental thermodynamic quantity Gibbs free energy for the biophysical process of protein folding. The research thus aids in drug discovery (Wang et al., 2022) and combating associated mutation-based diseases. Computational research also helps find general drugs' targets in biological systems (Li et al., 2011). Li and colleagues have developed a computational drug repositioning pipeline for performing large-scale molecular docking of small molecule drugs against protein drug targets, for mapping the drug–target interaction space and finding novel interactions. The computational method here emphasizes removing false positive interaction predictions using criteria from known interaction docking, consensus scoring, and specificity.

Biophysical modeling has been found successful in addressing general evolutionary features (Serohijos and Shakhnovich, 2014). The modeling addresses aspects of viral evolution and response to antibiotic stresses (Chéron et al., 2016). I have dedicated a chapter of this book to various features of biomolecular computations. Therefore, I shall not go into detail here.

1.3.7 Clinical Biophysics

Biophysics techniques are popularly applied in a variety of clinical research (Göçmen, 2021). Biophysical methods are found especially applicable at many points during the early stages of drug discovery (Holdgate et al., 2019; Holdgate and Bergsdorf, 2021). Projects in early clinical research phases often benefit from the combined use of biophysical and biochemical assays, which help derive information that is vital in late clinical research phases. Biophysical methods are especially considered an integral part of any protein-centric drug discovery project (Coyle and Walser, 2020).

Clinical biophysics, which is a branch of applied medical sciences, studies the process of the action and the effects of non-ionizing physical energies that are utilized for therapeutic purposes (Anbar, 1981; Aaron et al., 2006). Clinical biophysics approaches and classifications, including their clinical applications, have already been addressed quite rigorously (Göçmen, 2021). The clinical practice of orthopedics has empirically created a variety of biophysical environments in attempts to optimize skeletal repair. We are beginning to understand the biological effects of physical stimulation and are now poised to replace empiricism with treatment paradigms based on physiologic understandings of dose and biological response (Aaron et al., 2006).

Biophysical techniques have recently emerged as important clinical tools to help us both understand viral interactions with biological cells and discover theranostic (therapeutic and diagnostic) means to deal with the viruses through designing agents for detecting viral pathways and regulating the viral infection potency (Preston et al., 2021; Brisse et al., 2020; Myatt et al., 2022). This is a recent clinical pathway where modern biophysics has been applied with excellence to deal with one of the most serious ongoing socioeconomic issues (Lee and Huang, 2022; Deb et al., 2022).

Modern biophysics recently started dealing with cosmetics of biological life, such as understanding how we get the biological determination of diversified body surface color (Juzeniene et al., 2009; Jablonski, 2021), quality of hair (Bhushan, 2010; Philpott, 2021), quality of sustained body strength and performance beyond youth stages (Warden et al., 2014), maintenance of sex (Birdsell and Wills, 2003; Hörandl, 2009), etc. Physical understanding of these biological processes helps to find biophysical means for scientific interventions (Schwartz, 2019) for obtaining enhanced quality of life.

1.3.8 Exoplanetary Biophysics

The physics of life on planets other than our Earth or in exoplanetary regions will be explored alongside the exploration of life itself there. Probably even the physics of life in environments other than that of our planet Earth is to be researched before exploring life itself if we are to create or simulate an environment welcoming to life that migrates from the Earth.

Our foremost interest in this modern era is to work toward having a second habitable environment on the moon, as we face various environmental parametric disasters in our habitable environment on our primary habitat Earth. Whether life ever existed on the moon is currently answered based on scientific guesses, not scientific facts (see https://edition.cnn.com/2018/07/23/us/moon-life-study/index.html, accessed April 11, 2023), as scientists have found no evidence of life there yet. Similar claims are valid for any exoplanetary region. Despite having various unfavorable issues working against processes of life, creating a long-term self-regulating biological ecosystem may be one of the most important solutions we can now conceive of for a sustained stay on the moon (see https://universemagazine.com/en/what-difficulties-awaiting-humans-on-the-moon-physiological-problems-of-extraterrestrial-life, accessedApril 11, 2023).

So far scientists have categorized exoplanets into the following types: gas giant, Neptunian, super-Earth, and terrestrial. Exoplanets lie beyond our solar system whereas planets are worlds that orbit our Sun, like Mars, Jupiter, and our own Earth. We need to understand the physics of life on Mars ahead of working toward fulfilling our far-distant dream of the creation of a future civilization there. This should follow after our success in creating our first habitat on the surface of the moon, as the moon is our nearest neighbor and resembles our earth more than any other known region in the universe. But the creation of these non-Earth surface habitats requires certain scientific and technological accomplishments (McKay, 2014). Our dream of creating these new habitats may be boosted if we could find evidence of microbial life throughout the solar system (Mann, 2018). On our earth we now not only understand the biophysics of microbes, we have developed artificial means to regulate their biological architecture (Miyata et al., 2012). We may be able to revise our approaches on Earth and recreate technologies to deal with them in the exoplanetary regions, and in our solar system. This new physics may be called "exoplanetary biophysics," which in the future our scientists will have to become heavily involved in. Biophysics of exoplanets is expected to be part of the multidisciplinary approaches our future scientists will be taking seriously to understand a multitude of issues, including searching for life in the exoplanets (Howell, 2020). Human intervention in the exoplanetary environment is a futuristic socio-scientific goal.

1.4 ACCOMPLISHMENTS IN MODERN-ERA BIOPHYSICS RESEARCH

Although not conclusive, we may consider the discovery of the molecular structure of DNA (Watson and Crick, 1953b) as the beginning of a multitude of modern biophysics excellences, which initiated the modern era of biophysics research. From the 1950s to the1970s, application of biophysical techniques ensured other vital structural biology discoveries (Brooks-Bartlett and Garman, 2015). Especially mentionable are the first protein structures (Perutz et al., 1960; Kendrew et al., 1958) and the first enzyme structure (Steinrauf, 1959). We see that during just a decade, starting in 1953 (Watson and Crick, 1953b; Wilkins et al., 1953), we experienced an explosion in single biomolecular structural understanding and that happened using X-ray crystallography technique (Bragg, 1913).

The structural elucidation of nucleic acid-protein complexes and viruses was first addressed using crystallography and electron microscopy in 1975 (Crowther and Klug, 1975). Following the 1972 discovery of physical modeling, with the fluid-mosaic membrane model of lipid bilayer membranes by Singer and Nicolson (1972), another new dimension-making discovery happened in 1984 when the structure of the photosynthetic reaction center, a large membrane-embedded protein complex, was determined (Deisenhofer et al., 1984). In the 1980s, biophysical structural explorations using nuclear magnetic resonance (NMR) was initiated in the hands of Kurt Wüthrich (Wüthrich et al., 1982; Kumar et al., 1980; Havel and Wüthrich, 1984), which created a new dimension in biophysical structural analysis of biomolecules in solution. In the 1990s, we experienced two highly cross-examined biophysical discoveries, the structures of F_1 ATPase (Abrahams et al., 1994) and potassium channels (Doyle et al., 1998). Since the beginning of the current century, biophysical techniques have continued to reveal new glories, such as the determination of the structure of RNA polymerase and elucidation of the mechanism of transcription (Cramer and Bushnell, 2001), elucidating the structure of ribosome (Schluenzen et al., 2000; Wimberly et al., 2000; Ban et al., 2000). In 2007, another important biophysics glory was achieved when Brian Kobilka and colleagues solved the structure of the transmembrane signal carrier, G-protein-coupled receptor (GCPR) (Rasmussen et al., 2007). During the last half century we have experienced a revolution in the use of theoretical biophysics principles in computation to address biomolecular systems or simulate complex biomolecules (McCammon et al., 1977), initiated primarily by Martin Karplus (Macuglia et al., 2021). Based on all these groundbreaking structural discoveries, the scientific community found biophysical tools and means to develop applied medical science research and engaged in addressing diseases and drug discoveries (Waigh, 2007; Rando and Ambrosio, 2018; Bakhtiyarovna, 2021).

Biophysical methods that are applied to soft matter, surface physics, and biomimetic nanotechnology (Wei et al., 2022) have been found implementable in developing alternative therapeutic strategies to combat malaria (Mognetti et al., 2019; Saint-Sardos et al., 2020). Introini and colleagues recently provided an in-depth analysis of the state-of-the-art technologies developed for probing the mechanics of malaria infection, specifically during Plasmodium invasion, maturation, and cytoadhesion, and how these would shape the future directions of malaria research (Introini et al., 2022). Promising bioengineering, soft matter physics, and immunomechanics techniques that could be beneficial for studying malaria, including understanding the malaria lifecycle and apicomplexan infections with complex host–pathogen interactions, are detailed here. MalDA provides a unique forum to accelerate malaria target-based drug discovery. Malaria Drug Accelerator (MalDA), a public–private partnership, formed in 2012, is a consortium of 15 academic and industrial laboratories whose collective expertise and resources include various biophysical approaches to accelerate malaria drug development (Yang et al., 2021). Under the MalDA initiative, identifying novel validated targets, advanced target-based drug discovery, etc. are underway. Hit-to-lead compound optimization on inhibitors of many chemically validated targets, including Pf phenylalanyl-tRNA synthetase (PheRS), Pf acetyl-CoA synthetase (AcAS), and Pf Niemann–Pick type C1-related protein (NCR1), is in research phases.

Biophysical methods have been found quite successful in both understanding the characteristics of bacteria (Ērglis et al., 2007; Jeckel et al., 2019; Viljoen et al., 2022) and combatting them through discovering antibiotics (Rodrigues et al., 2016; Zhang et al., 2018; Kavčič et al., 2021). Fleming's 1928 discovery of penicillin marked the beginning of the antibiotic era (Fleming, 1929), but the biophysical technology-based intervention in antibiotic research progressed strongly following all those groundbreaking cellular and molecular structural discoveries during the 1950s to 1980s. I avoid elaborating on them here as there are various texts already available (Vekshin, 2011).

Cancer is another example which has drawn a huge amount of biophysical exploration in understanding the associated disorders (Suresh, 2007; Dutta et al., 2020; Vasilaki et al., 2021) and malfunctions in cell signaling (Sever and Brugge, 2015), imaging the distorted cell sociology (Enfield et al., 2019), etc. Cancer cells have many therapeutic targets, e.g. P53 gene (Zhu et al., 2020), Bcl 2 proteins (Yip and Reed, 2008), etc. In cancer cells plasma membrane externalization of phosphatidylserine (PS) is compromised, which is a hallmark of apoptosis, and hence appears as a diagnostic agent in understanding the state of cancer. In both the former and latter cases biophysical methods have been rigorously applied in finding therapeutic (Alves et al., 2016; Campbell and Tait, 2018; Kuznetsov et al., 2021) and diagnostic agents (Tseng et al., 2011; Ashrafuzzaman et al., 2013; Ashrafuzzaman, 2021), respectively.

I could elaborate by pinpointing almost all areas in therapeutic and diagnostic drug discovery research where modern biophysics methods are applied. There are many articles and textbooks addressing this topic (Renaud et al., 2016; O'Brien et al., 2017; Coyle et al., 2020; Huddler and Zartler, 2017).

1.5 CONCLUDING REMARKS

Biophysics has a glorious history. In this chapter, I have attempted to address this history in a chronological fashion. Both basic science and applied technological research accomplishments have been achieved since the first ever physics technique was applied in the seventeenth century when both cork cells and bacterial cells were discovered. Until the discovery of DNA's double helical structure, we lacked an understanding of actual biomolecule structures in physiological conditions. Before the understanding of these molecular-level structures, we accomplished a huge amount of biophysics research-based understanding of various aspects of biological systems which have been briefly addressed here. In the modern era, since 1953, we have been achieving research breakthroughs that we may associate with the essence of life itself. Biophysics research is very much focused now on developing novel techniques and finding novel means that are deliverable to our daily life such as understanding the origin of life, evolutionary pathways, genetics, etc., as well as dealing with various types of disorders and diseases, and discovering agents and means to treat those physiological issues. Not only helping us understand the biology of life, biophysics also attempts to develop socioeconomic models that are essential for policymakers, social leaders, governing parties, and others concerned to help them take essential decisions and develop policies for dealing with continued growth, energy production, managing of the environment, etc. Due to fast growth in biophysics research techniques, especially with the use of computer-associated algorithms and artificial intelligence, we are now experiencing the emergence of new areas, fields, and even subjects under the big biophysics umbrella. I have summarized the chronological development of biophysics that took place during the era of human intervention in biology using physics principles in a timeline sketch in Figure 1.12. This figure presents only the fields that have provided

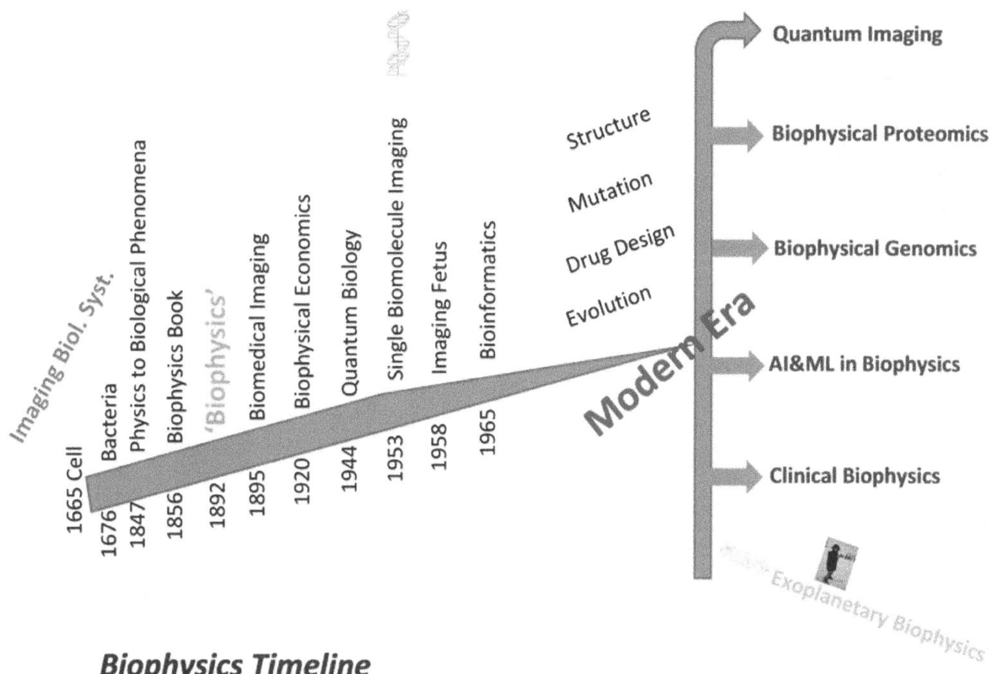

FIGURE 1.12 Biophysics timeline since the breakthrough discovery of biological cells using microscopy imaging. This chronology has been drawn to cover the era of human interventions. It mainly summarizes groundbreaking biophysics events that have happened since the discovery of cells, and projects forward into a future where we hope to find or create environments outside our planet Earth and hopefully someday settle there.

great impacts so far and the fields that are expected to evolve in the future with greater interest and attention. Most of these fields are elaborated in subsequent chapters.

REFERENCES

Aaron, R.K., Ciombor, D.M., Wang, S., & Simon, B. (2006). Clinical biophysics: the promotion of skeletal repair by physical forces. *Ann. N. Y. Acad. Sci. 1068*, 513–531

Abrahams, J.P., Leslie, A.G., Lutter, R., & Walker, J.E. (1994). Structure at 2.8 Å resolution of F1-ATPase from bovine heart mitochondria. *Nature*, *370*, 621–628

Agarval, R., Krueger, B.P., Scholes, G.D., Yang, M., Yom, J., Mets, L., & Fleming, G.R. (2000). Ultrafast energy transfer in LHC-II revealed by three-pulse photon echo peak shift measurements. *J. Phys. Chem. B*, *104*, 2908–2918. doi: 10.1021/jp9915578

Allaby, R.G., Kitchen, J.L., & Fuller, Q.D. (2015). Surprisingly low limits of selection in plant domestication. *Evolutionary Bioinformatics*, *11*, EBO-S33495

Allen, K. (2017). Quantum mechanics, imaging and other technologies: a method that will improve the standards of medical care for millions of Americans. New York: Nova Science

Allen, R., & Waclaw, B. (2016). Antibiotic resistance: a physicist's view. *Physical Biology*, *13*(4), 045001

Almécija, S., Hammond, A.S., Thompson, N.E., Pugh, K.D., Moyà-Solà, S., & Alba, D.M. (2021). Fossil apes and human evolution. *Science*, *372*(6542), eabb4363

AlQuraishi, M., & Sorger, P.K. (2021). Differentiable biology: using deep learning for biophysics-based and data-driven modeling of molecular mechanisms. *Nat. Methods*, *18*, 1169–1180. https://doi.org/10.1038/s41592-021-01283-4

Alric, J. (2010). Cyclic electron flow around photosystem I in unicellular green algae. *Photosynthesis Research 106*, 47–56

Alves, A.C., Ribeiro, D., Nunes, C., & Reis, S. (2016). Biophysics in cancer: the relevance of drug-membrane interaction studies. *Biochimica et Biophysica Acta (BBA) – Biomembranes*, *1858*(9), 2231–2244

Aminov, R.I. (2010). A brief history of the antibiotic era: lessons learned and challenges for the future. *Front. Microbiol.*, *1*, 134. doi: 10.3389/fmicb.2010.00134; PMID: 21687759; PMCID: PMC3109405

Anbar, M. (1981). Clinical biophysics: a new concept in undergraduate medical education. *J. Medical Education*, *56*, 443–444

Andersen, O.S. (2016). Introduction to biophysics week: what is biophysics? *Biophys J.*, *110*(5), E01–3. doi: 10.1016/j.bpj.2016.02.012. Erratum in: *Biophys J.* (2017 May 9), 112(9), 2019. PMID: 26958896; PMCID: PMC4788750

Aquilini, E., Cosentino, U., Pasqualetti, N., et al. (2021). Julius Robert Mayer and the principle of energy conservation. *ChemTexts 7*, 22. https://doi.org/10.1007/s40828-021-00147-w

Arkin, M.R., Tang, Y., & Wells, J.A. (2014). Small-molecule inhibitors of protein–protein interactions: progressing toward the reality. *Chem. Biol. 21*, 1102–1114

Ashrafuzzaman, M. (2021). Energy-based method for drug design. US Patent, US10916330B1, https://patents.google.com/patent/US10916330B1/en

Ashrafuzzaman, M., Tseng, C.Y., Kapty, J., Mercer, J.R., & Tuszynski, J.A. (2013). A computationally designed DNA aptamer template with specific binding to phosphatidylserine. *Nucleic Acid Therapeutics*, *23*(6), 418–426

Ashrafuzzaman, M., Tseng, C.Y., & Tuszynski, J.A. (2020). Charge-based interactions of antimicrobial peptides and general drugs with lipid bilayers. *Journal of Molecular Graphics and Modelling*, *95*, 107502

Ashrafuzzaman, M., & Tuszynski, J.A. (2012). *Membrane biophysics*. Berlin and Heidelberg: Springer. https://doi.org/10.1007/978-3-642-16105-6

Aslam, B., Basit, M., Nisar, M.A., Khurshid, M., & Rasool, M.H. (2016). Proteomics: technologies and their applications. *Journal of Chromatographic Science*, *55*, 182–196

Atkins, P. (2010). *The laws of thermodynamics: a very short introduction*. Oxford: Oxford University Press.

Avise, J.C. (2006). *Evolutionary pathways in nature: a phylogenetic approach*. Cambridge: Cambridge University Press

Bagcigil, A.F., Taponen, S., Koort, J., Bengtsson, B., Myllyniemi, A.L., & Pyörälä, S. (2012). Genetic basis of penicillin resistance of S. aureus isolated in bovine mastitis. *Acta Vet. Scand.*, *54*(1), 69. doi: 10.1186/1751-0147-54-69; PMID: 23176642; PMCID: PMC3575348

Baker, T.S., Olson, N.H., & Fuller, S.D. (1999). Adding the third dimension to virus life cycles: three-dimensional reconstruction of icosahedral viruses from cryo-electron micrographs. *Microbiol. Mol. Biol. Rev.*, *63*, 862

Bakhtiyarovna, N.F. (2021). Organization and methodology laboratory works on biophysics for dental direction. *Annals of the Romanian Society for Cell Biology*, 597–607

Baldwin, R.L. (2007). Energetics of protein folding. *Journal of Molecular Biology*, *371*(2), 283–301

Baltazar-Soares, M., Hinrichsen, H.H., & Eizaguirre, C. (2018). Integrating population genomics and biophysical models towards evolutionary-based fisheries management. *ICES Journal of Marine Science*, *75*(4), 1245–1257

Ban, N., Nissen, P., Hansen, J., Moore, P.B., & Steitz, T.A. (2000). The complete atomic structure of the large ribosomal subunit at 2.4 Å resolution. *Science*, *289*(5481), 905–920

Barber, J. (1978). Biophysics of photosynthesis. *Reports on Progress in Physics*, *41*(8), 1157

Bassett, E.J., Keith, M.S., Armelagos, G.J., Martin, D.L., & Villanueva, A.R. (1980). Tetracycline-labeled human bone from ancient Sudanese Nubia (AD 350). *Science*, *209*(4464), 1532–1534

Bax, A. (1994). Multidimensional nuclear magnetic resonance methods for protein studies. *Curr. Opin. Struc. Biol.*, 4, 738–744

Beal, M.F. (2000). Energetics in the pathogenesis of neurodegenerative diseases. *Trends in Neurosciences*, *23*(7), 298–304

Becker, E.D. (1993). A brief history of nuclear magnetic resonance. *Analytical Chemistry*, *65*(6), 295A–302A

Ben-Naim, A. (2022). Information, entropy, life, and the universe. *Entropy*, *24*(11), 1636

Benner, S.A., Bell, E.A., Biondi, E., Brasser, R., Carell, T., Kim, H.J., ... & Trail, D. (2020). When did life likely emerge on Earth in an RNA-first process? *ChemSystemsChem*, *2*(2), e1900035

Bhat, T., Cao, A., & Yin, J. (2022). Virus-like particles: measures and biological functions. *Viruses*, *14*(2), 383. doi: 10.3390/v14020383; PMID: 35215979; PMCID: PMC8877645

Bhushan, B. (2010). *Biophysics of human hair: structural, nanomechanical, and nanotribological studies*. Berlin and Heidelberg: Springer

Birdsell, J.A., & Wills, C. (2003). The evolutionary origin and maintenance of sexual recombination: a review of contemporary models. *Evolutionary biology*, 33, 27–138

Bloch F., Hansen, W.W., & Packard M. (1946). The nuclear induction experiment. *Phys. Rev.*, *70*, 474–485

Bragg, W.L. (1913). The structure of some crystals as indicated by their diffraction of X-rays. *Proc. Roy. Soc. London*, 89, 248–277

Brems, M.A., Runkel, R., Yeates, T.O., & Virnau, P. (2022). AlphaFold predicts the most complex protein knot and composite protein knots. *Protein Science*, *31*(8), e4380

Brisse, M., Vrba, S.M., Kirk, N., Liang, Y., & Ly, H. (2020). Emerging concepts and technologies in vaccine development. *Frontiers in Immunology*, *11*, 583077

Brooks, D.R., & Wiley, E.O. (1988). *Evolution as entropy*. Chicago: University of Chicago Press

Brooks-Bartlett, J.C., & Garman, E.F. (2015). The nobel science: one hundred years of crystallography. *Interdisciplinary Science Reviews*, *40*(3), 244–264

Bruckman, M.A., Jiang, K., Simpson, E.J., Randolph, L.N., Luyt, L.G., Yu, X., & Steinmetz, N.F. (2014). Dual-modal magnetic resonance and fluorescence imaging of atherosclerotic plaques in vivo using VCAM-1 targeted tobacco mosaic virus. *Nano Letters*, *14*(3), 1551–1558

Bruckman, M.A., & Steinmetz, N.F. (2014). Chemical modification of the inner and outer surfaces of Tobacco Mosaic Virus (TMV). In *Virus Hybrids as Nanomaterials: Methods and Protocols*, ed. B. Lin & B. Ratna, pp. 173–185. Totowa, NJ: Humana Press. doi: 10.1007/978-1-62703-751-8_13

Bruinsma, R.F., Wuite, G.J.L. & Roos, W.H. (2021). Physics of viral dynamics. *Nat. Rev. Phys.*, *3*, 76–91. https://doi.org/10.1038/s42254-020-00267-1

Burmeister, A.R., Hansen, E., Cunningham, J.J., Rego, E.H., Turner, P.E., Weitz, J.S., & Hochberg, M.E. (2021). Fighting microbial pathogens by integrating host ecosystem interactions and evolution. *Bioessays*, *43*(3), 2000272

Burrell, C.J., Howard. C.R., & Murphy, F.A. (2017). History and impact of virology. *Fenner and White's Medical Virology*, *2017*, 3–14. doi: 10.1016/B978-0-12-375156-0.00001-1. PMCID: PMC7150216

Callaway, E. (2020). "It will change everything": DeepMind's AI makes gigantic leap in solving protein structures. *Nature*, *588*(7837), 203–205

Camp, C.H., Jr, & Cicerone, M.T. (2015). Chemically sensitive bioimaging with coherent Raman scattering. *Nat. Photon.*, *9*, 295–305

Campbell, K.J., & Tait, S.W. (2018). Targeting BCL-2 regulated apoptosis in cancer. *Open Biology*, *8*(5), 180002

Caneva, E., Di Gennaro, P., Farina, F., Orlandi, M., Rindone, B., & Falagiani, P. (1993). Synthesis and characterization of a penicillin-poly (L-lysine) which is recognized by human IgE anti-penicillin antibodies. *Bioconjugate Chemistry*, *4*(5), 309–313

Cang, Z., & Wei, G.W. (2017). Analysis and prediction of protein folding energy changes upon mutation by element specific persistent homology. *Bioinformatics*, *33*(22), 3549–3557

Carr, L.G., Skatrud, P.L., Scheetz II, M.E., Queener, S.W., & Ingolia, T.D. (1986). Cloning and expression of the isopenicillin N synthetase gene from Penicillium chrysogenum. *Gene*, *48*(2–3), 257–266

Carvalho, A.L., Trincão, J., & Romão, M.J. (2010). X-Ray crystallography in drug discovery. In *Ligand-Macromolecular Interactions in Drug Discovery*, ed. A. Roque, Methods in Molecular Biology 572, pp. 31–56. Totowa, NJ: Humana Press. https://doi.org/10.1007/978-1-60761-244-5_3

Casacio, C.A., Madsen, L.S., Terrasson, A., Waleed, M., Barnscheidt, K., Hage, B., ... & Bowen, W.P. (2021). Quantum-enhanced nonlinear microscopy. *Nature*, *594*(7862), 201–206

Caspar, D.L.D., & Klug, A. (1962). Physical principles in construction of regular viruses. *Cold Spring Harb. Symp. Quant. Biol.*, *27*, 1–24

Cavalazzi, B., Lemelle, L., Simionovici, A., Cady, S.L., Russell, M.J., Bailo, E., ... & Hofmann, A. (2021). Cellular remains in a~ 3.42-billion-year-old subseafloor hydrothermal environment. *Science Advances*, *7*(29), eabf3963

Cavicchioli, R., Ripple, W.J., Timmis, K.N., Azam, F., Bakken, L.R., Baylis, M., ... & Webster, N.S. (2019). Scientists' warning to humanity: microorganisms and climate change. *Nature Reviews Microbiology*, *17*(9), 569–586

Chachisvilis, M., Pullerits, T., Jones, M.R., Hunter, C.N., & Sundström, V. (1994). Vibrational dynamics in the light-harvesting complexes of the photosynthetic bacterium Rhodobacter sphaeroides. *Chem. Phys. Lett.*, *224*, 345–354. doi: 10.1016/0009-2614(94)00560-5

Chen, S.J., & Dill, K.A. (2000). RNA folding energy landscapes. *Proceedings of the National Academy of Sciences*, *97*(2), 646–651

Cheng, J.-X., & Sunney, X. (2015). Vibrational spectroscopic imaging of living systems: an emerging platform for biology and medicine. *Science*, *350*, aaa8870

Chéron, N., Serohijos, A.W., Choi, J.M., & Shakhnovich, E.I. (2016). Evolutionary dynamics of viral escape under antibodies stress: a biophysical model. *Protein Sci.*, *25*(7), 1332–40. doi: 10.1002/pro.2915; PMID: 26939576; PMCID: PMC4918420

Chong, S.H., & Ham, S. (2018). Examining a thermodynamic order parameter of protein folding. *Scientific Reports*, *8*(1), 7148.

Cleveland, C.J. (1987). Biophysical economics: historical perspective and current research trends. *Ecological Modelling*, *38*(1–2), 47–73

Cobb, M. (2013). 1953: when genes became "information". *Cell*, *153*(3), 503–506

Consortium, C.S.a.A. (2005). Initial sequence of the chimpanzee genome and comparison with the human genome. *Nature*, *437*(7055), 69–87

Cook, M., Molto, E.L., & Anderson, C. (1989). Fluorochrome labelling in Roman period skeletons from Dakhleh Oasis, Egypt. *American Journal of Physical Anthropology*, *80*(2), 137–143

Copeland, R.A. (2016). The drug-target residence time model: a 10-year retrospective. *Nat. Rev. Drug Discov.*, *15*, 87–95

Copeland, R.A., Pompliano, D.L., & Meek, T.D. (2006). Drug-target residence time and its implications for lead optimization. *Nat. Rev. Drug Discov.*, *5*, 730–739

Coyle, J., & Walser, R. (2020). Applied biophysical methods in fragment-based drug discovery. *SLAS DISCOVERY: Advancing the Science of Drug Discovery*, *25*(5), 471–490

Cramer, P., & Bushnell, D.A. (2001). Kornberg RD structural basis of transcription: RNA polymerase II at 2.8 angstrom resolution. *Science*, 292, 1863–1876

Cranefield, P.F. (1957). The organic physics of 1847 and the biophysics of today. *J. Hist. Med. Allied Sci.*, *12*, 407–423

Crick, F.H., & Watson, J.D. (1956). Structure of small viruses. *Nature*, *177*(4506), 473–475

Crofts, A.R. (2007). Life, information, entropy, and time: vehicles for semantic inheritance. *Complexity*, *13*(1), 14–50. doi: 10.1002/cplx.20180; PMID: 18978960; PMCID: PMC2577055

Crowther, R.A., & Klug, A. (1975). Structural analysis of macromolecular assemblies by image reconstruction from electron micrographs. *Ann. Rev. Biochem.*, *44*, 161–182

Cui, L., & Su, X.Z. (2009). Discovery, mechanisms of action and combination therapy of artemisinin. *Expert Review of Anti-infective Therapy*, *7*(8), 999–1013

Currat, M., Gerbault, P., & Sanchez-Mazas, A. (2016). SELECTOR: a program to simulate genetic lineages under selection in a spatially-explicit population framework. *Evolutionary Bioinformatics*, *2015*, 11s2

Danelon, C., Nestorovich, E.M., Winterhalter, M., Ceccarelli, M., & Bezrukov, S.M. (2006). Interaction of zwitterionic penicillins with the OmpF channel facilitates their translocation. *Biophysical Journal*, *90*(5), 1617–1627

Dayhoff, M.O. (ed.). (1969a). *Atlas of protein sequence and structure* (Vol. 4). Silver Spring, MD: National Biomedical Research Foundation

Dayhoff, M.O. (1969b). Computer analysis of protein evolution. *Scientific American*, *221*, 86–95

Dayhoff, M.O. (1964). Computer search for active site configurations. *Journal of the American Chemical Society*, *86*(11), 2295–2297

Dayhoff, M.O, Eck, R.V., Chang, M.A., & Sochard, M.R. (1965). *Atlas of Protein Sequence and Structure*. Silver Spring, MD: National Biomedical Research Foundation

Dayhoff, M.O., and Ledley, R.S. (1962). Comprotein: a computer program to aid primary protein structure determination. *Proceedings of the Fall Joint Computer Conference*. Santa Monica: American Federation of Information Processing Societies

Dayhoff, M.O, Lippincott, E.R., & Eck, R.V. (1964). Thermodynamic equilibria in prebiological atmospheres. *Science*, *146*, 1461–1464

de Vlieg, J., Berendsen, H.J.C., & van Gunsteren, W.F. (1989). An NMR-based molecular dynamics simulation of the interaction of the lac repressor headpiece and its operator in aqueous solution. *Proteins*, *6*, 104–127

Deb, P., Furceri, D., Jimenez, D., et al. (2022). The effects of COVID-19 vaccines on economic activity. *Swiss J. Economics Statistics*, *158*, 3. https://doi.org/10.1186/s41937-021-00082-0

Deisenhofer, J., Epp, O., Miki, K., Huber, R., & Michel, H. (1984). X-ray structure analysis of a membrane protein complex: Electron density map at 3 Å resolution and a model of the chromophores of the photosynthetic reaction center from Rhodopseudomonas viridis. *J. Mol. Biol.*, *180*, 385–398

Dias, M.H., Kitano, E.S., Zelanis, A., & Iwai, L.K. (2016). Proteomics and drug discovery in cancer. *Drug Discovery Today*, *21*(2), 264–277

Diekmann Y, & Pereira-Leal, J.B. (2016). Gene tree affects inference of sites under selection by the branch-site test of positive selection. *Evol. Bioinform. Online*, *18*(11) (Suppl. 2), 11–7. doi: 10.4137/EBO.S30902; PMID: 26819542; PMCID: PMC4718152

Dodd, M.S., Papineau, D., Grenne, T., Slack, J.F., Rittner, M., Pirajno, F., ... & Little, C.T. (2017). Evidence for early life in Earth's oldest hydrothermal vent precipitates. *Nature*, 543(7643), 60–64

Donald, I., MacVicar, J., & Brown, T.G. (1958). Investigation of abdominal masses by pulsed ultrasound. *Lancet*, 1, 1188–1195

Donaldson, I.M. (2010). Robert Hooke's Micrographia of 1665 and 1667. *Journal of the Royal College of Physicians of Edinburgh*, *40*(4), 374–376

Donchenko, E.K., Pechnikova, E.V., Mishyna, M.Y., Manukhova, T.I., Sokolova, O.S., Nikitin, N.A., ... & Karpova, O.V. (2017). Structure and properties of virions and virus-like particles derived from the coat protein of Alternanthera mosaic virus. *PLoS One*, *12*(8), e0183824

Doyle, D.A., Cabral, J.M., Pfuetzner, R.A., Kuo, A., Gulbis, J.M., Cohen, S.L., ... & MacKinnon, R. (1998). The structure of the potassium channel: molecular basis of K+ conduction and selectivity. *Science*, *280*(5360), 69–77.

Drachman, D.A. (2006). Aging of the brain, entropy, and Alzheimer disease. *Neurology*, 67(8), 1340–1352

Durmaz, A.A., Karaca, E., Demkow, U., Toruner, G., Schoumans, J., & Cogulu, O. (2015). Evolution of genetic techniques: past, present, and beyond. *Biomed. Res. Int*, *2015*, 461524. doi: 10.1155/2015/461524; PMID: 25874212; PMCID: PMC4385642

Dutta, D., Palmer, X.L., Ortega-Rodas, J., Balraj, V., Dastider, I.G., & Chandra, S. (2020). Biomechanical and biophysical properties of breast cancer cells under varying glycemic regimens. *Breast Cancer: Basic and Clinical Research*, 14, 1178223420972362

El-Khateeb, E., Vasilogianni, A.M., Alrubia, S., Al-Majdoub, Z.M., Couto, N., Howard, M., ... & Achour, B. (2019). Quantitative mass spectrometry-based proteomics in the era of model-informed drug development: applications in translational pharmacology and recommendations for best practice. *Pharmacology & Therapeutics*, 203, 107397

Enfield, K.S., Martin, S.D., Marshall, E.A., Kung, S.H., Gallagher, P., Milne, K., … & Guillaud, M. (2019). Hyperspectral cell sociology reveals spatial tumor-immune cell interactions associated with lung cancer recurrence. *Journal for Immunotherapy of Cancer*, *7*(1), 1–13

Ērglis, K., Wen, Q., Ose, V., Zeltins, A., Sharipo, A., Janmey, P.A., & Cēbers, A. (2007). Dynamics of magnetotactic bacteria in a rotating magnetic field. *Biophysical Journal*, *93*(4), 1402–1412

Erlanson, D.A., & Zartler, E. (2015). Fragments in the clinic: 2015 edition. *Practical Fragments* (blog), http://practicalfragments.blogspot.fr/2015/01/fragments-in-clinic-2015-edition.html

Fadaka, A.O., Sibuyi, N.R.S., Madiehe, A.M., & Meyer, M. (2021). Nanotechnology-based delivery systems for antimicrobial peptides. *Pharmaceutics*, *13*(11), 1795. doi: 10.3390/pharmaceutics13111795; PMID: 34834210; PMCID: PMC8620809

Fick, A. (1858). Die *medizinische physik*. Braunschweig: Vieweg

Field, M.J., Bash, P.A. & Karplus, M. (1990). A combined quantum mechanical and molecular mechanical potential for molecular dynamics simulations. *J. Comput. Chem.*, *11*, 700–733

Fleming, A. (1929). On the antibacterial action of cultures of a penicillium, with special reference to their use in the isolation of B. influenzae. *British Journal of Experimental Pathology*, *10*(3), 226

Franck, J. (1955). Physical problems of photosynthesis. *Daedalus*, *86*(1), 17–42

Franck, J., & Teller, E. (1938). Migration and photochemical action of excitation energy in crystals. *J. Chem. Phys.*, *6*, 861–872. doi: 10.1063/1.1750182

Fu, Y., Wang, H., Shi, R. & Cheng, J.-X. (2006). Characterization of photodamage in coherent anti-Stokes Raman scattering microscopy. *Opt. Express*, 14, 3942–3951

Gavriilidou, A.F., Sokratous, K., Yen, H.Y., & De Colibus, L. (2022). High-throughput native mass spectrometry screening in drug discovery. *Frontiers in Molecular Biosciences*, *9*, 837901

Gaynes, R. (2017). The discovery of penicillin—new insights after more than 75 years of clinical use. *Emerging Infectious Diseases*, *23*(5), 849.

Gelderblom, H.R. (1996). Structure and classification of viruses. In *Medical Microbiology* (4th edn), ed. S. Baron, Chapter 41. Galveston: University of Texas Medical Branch at Galveston. Available from www.ncbi.nlm.nih.gov/books/NBK8174

Genovese, M. (2016). Real applications of quantum imaging. *Journal of Optics*, *18*(7), 073002

Gerould, J.H. (1922). The dawn of the cell theory. *Scientific Monthly*, *14*(3), 268–277

Ghasemi, F., & Shafiee, A. (2019). A quantum mechanical approach towards the calculation of transition probabilities between DNA codons. *BioSystems*, *184*, 103988

Glenn, D., Bucher, D., Lee, J., et al. (2018). High-resolution magnetic resonance spectroscopy using a solid-state spin sensor. *Nature*, *555*, 351–354. https://doi.org/10.1038/nature25781

Göçmen, S.U. (2021). Clinical applications of biophysics – clinical biophysics. *Biophysical Journal*, *120*(3), 14a

Goldsmith, C.S., & Miller, S.E. (2009). Modern uses of electron microscopy for detection of viruses. *Clin. Microbiol. Rev.*, *22*(4), 552–563. doi: 10.1128/CMR.00027-09; PMID: 19822888; PMCID: PMC2772359

Green, R.E., Krause, J., Briggs, A.W., Maricic, T., Stenzel, U., Kircher, M., … & Pääbo, S. (2010). A draft sequence of the Neandertal genome. *Science*, *328*(5979), 710–722

Grunewald, K., Desai, P., Winkler, D.C., Heymann, J.B., Belnap, D.M., Baumeister, W., & Steven, A.C. (2003). Three-dimensional structure of herpes simplex virus from cryo-electron tomography. *Science*, *302*(5649), 1396–1398

Hajiaghayi, M., Condon, A., & Hoos, H.H. Analysis of energy-based algorithms for RNA secondary structure prediction. *BMC Bioinformatics*, *13*, 22 (2012). https://doi.org/10.1186/1471-2105-13-2

Hardy, N.P., Dalli, J., Mac Aonghusa, P., Neary, P.M., & Cahill, R.A. (2021). Biophysics inspired artificial intelligence for colorectal cancer characterization. *Artificial Intelligence in Gastroenterology*, *2*(3), 77–84. www.wjgnet.com/2644-3236/full/v2/i3/77.htm

Hasegawa, M., & Yano, T.A. (1974). Entropy of the genetic information and evolution. In *Cosmochemical evolution and the origins of life: Proceedings of the Fourth International Conference on the Origin of Life and the First Meeting of the International Society for the Study of the Origin of Life, Barcelona, June 25–28, 1973, Volume II: Contributed papers*, pp. 219–227. Dordrecht: Springer Netherlands

Havel, T.F., & Wüthrich, K. (1984). A distance geometry program for determining the structures of small proteins and other macromolecules from nuclear magnetic resonance measurements of intramolecular 1H-1H proximities in solution, *Bull. Math. Biol.*, *46*, 673–678

Haveri, M., Suominen, S., Rantala, L., Honkanen-Buzalski, T., & Pyörälä, S. (2005). Comparison of phenotypic and genotypic detection of penicillin G resistance of Staphylococcus aureus isolated from bovine intramammary infection. *Vet. Microbiol.*, *106*(1–2), 97–102. doi: 10.1016/j.vetmic.2004.12.015; PMID: 15737478

Higueruelo, A.P., Jubb, H., & Blundell, T.L. (2013). Protein–protein interactions as druggable targets: recent technological advances. *Curr. Opin. Pharmacol.*, *13*, 791–796

Hill, R. (1937). Oxygen evolved by isolated chloroplasts. *Nature*, 139(3525), 881–882

Hogan, J.A., Miller, F.C., & Finstein, M.S. (1989). Physical modeling of the composting ecosystem. *Applied and Environmental Microbiology*, *55*(5), 1082–1092

Holdgate, G.A., & Bergsdorf, C. (2021). Applications of biophysics in early drug discovery. *SLAS DISCOVERY: Advancing the Science of Drug Discovery*, 26(8), 945–946

Holdgate, G., Embrey, K., Milbradt, A., & Davies, G. (2019). Biophysical methods in early drug discovery. *ADMET DMPK*, *7*(4), 222–241. doi: 10.5599/admet.733; PMID: 35359617; PMCID: PMC8963580

Holdgate, G., Geschwindner, S., Breeze, A., Davies, G., Colclough, N., Temesi, D., & Ward, L. (2013). Biophysical methods in drug discovery from small molecule to pharmaceutical. *Methods in Molecular Biology*, *1008*, 327–355

Holdgate, G.A., & Gill, A.L. (2011). Kinetic efficiency: the missing metric for enhancing compound quality? *Drug Discov. Today*, *16*, 910–913

Hooke, R. (1665). *Micrographia:* or some physiological descriptions of minute bodies made by magnifying glasses. *With* observations and inquiries thereupon. London: Royal Society

Hörandl, E. (2009). A combinational theory for maintenance of sex. *Heredity (Edinb)*, *103*(6), 445–457. doi: 10.1038/hdy.2009.85; PMID: 19623209; PMCID: PMC2854797

Houbraken, J., Frisvad, J.C., & Samson, R.A. (2011). Fleming's penicillin producing strain is not Penicillium chrysogenum but P. rubens. *IMA Fungus*, *2*, 87–95. https://doi.org/10.5598/imafungus.2011.02.01.12

Howell, S.B. (2020). The grand challenges of exoplanets. *Frontiers in Astronomy and Space Sciences*, *7*, 10

Hu, H., Yang, Q., Baroni, S., Yang, H., Aime, S., & Steinmetz, N.F. (2019). Polydopamine-decorated tobacco mosaic virus for photoacoustic/magnetic resonance bimodal imaging and photothermal cancer therapy. *Nanoscale*, *11*(19), 9760–9768

Huddler, D., & Zartler, E.R. (eds.). (2017). *Applied biophysics for drug discovery*. Hoboken, NJ: John Wiley & Sons

Introini, V., Govendir, M.A., Rayner, J.C., Cicuta, P., & Bernabeu, M. (2022). Biophysical tools and concepts enable understanding of asexual blood stage malaria. *Frontiers in Cellular and Infection Microbiology*, *12*, 908241

Jablonski, N.G. (2021). The evolution of human skin pigmentation involved the interactions of genetic, environmental, and cultural variables. *Pigment Cell Melanoma Res.*, *34*(4), 707–729. doi: 10.1111/pcmr.12976; PMID: 33825328; PMCID: PMC8359960

Jablonsky, J., Bauwe, H., & Wolkenhauer, O. (2011). Modeling the Calvin–Benson cycle. *BMC Systems Biology*, 5, 185

James, P. (1997). Protein identification in the post-genome era: the rapid rise of proteomics. *Q. Rev. Biophys.*, *30*(4), 279–331. doi: 10.1017/S0033583597003399

Jeckel, H., Matthey, N., & Drescher, K. (2019). Common concepts for bacterial collectives. *Elife*, *8*, e47019.

Jhoti, H., Williams, G., Rees, D.C., & Murray, C.W. (2013). The "rule of three" for fragment-based drug discovery: where are we now? *Nat. Rev. Drug Discov.*, *12*, 644–645

Jumper, J., Evans, R., Pritzel, A., et al. (2021). Highly accurate protein structure prediction with AlphaFold. *Nature*, *596*, 583–589. https://doi.org/10.1038/s41586-021-03819-2

Juzeniene, A., Setlow, R., Porojnicu, A., Steindal, A.H., & Moan, J. (2009). Development of different human skin colors: a review highlighting photobiological and photobiophysical aspects. *J. Photochem. Photobiol. B.*, *96*(2), 93–100. doi: 10.1016/j.jphotobiol.2009.04.009; PMID: 19481954

Kaeley, G.S., Bakewell, C., & Deodhar, A. (2020). The importance of ultrasound in identifying and differentiating patients with early inflammatory arthritis: a narrative review. *Arthritis Res. Ther.*, *22*, 1. https://doi.org/10.1186/s13075-019-2050-4

Kaila, V.R. (2021). Resolving chemical dynamics in biological energy conversion: long-range proton-coupled electron transfer in respiratory complex I. *Accounts of Chemical Research*, *54*(24), 4462–4473

Karev, V.I., Klimov, D. & Pokazeev, K. (eds.). (2019). *Physical and mathematical modeling of earth and environment processes*. Cham: Springer. https://link.springer.com/book/10.1007/978-3-319-77788-7

Kausche, G., Pfankuch, E., and Ruska, H. (1939). Die Sichtbarmachung von pflanzlichem Virus im Übermikroskop. *Naturwissenschaften*, *27*, 292–299

Kavčič, B., Tkačik, G., & Bollenbach, T. (2021). Minimal biophysical model of combined antibiotic action. *PLoS Computational Biology*, *17*(1), e1008529

Kendrew, J.C., Bodo, G., Dintzis, H.M., Parrish, R.G., Wyckoff, H., & Phillips, D.C. (1958). A three-dimensional model of the myoglobin molecule obtained by x-ray analysis. *Nature*, *181*(4610), 662–666

King, M.C., & Wilson, A.C. (1975). Evolution at two levels in humans and chimpanzees. *Science*, *188*(4184), 107–116

Kiss, B., Mudra, D., Török, G., Mártonfalvi, Z., Csík, G., Herényi, L., & Kellermayer, M. (2020). Single-particle virology. *Biophysical Reviews*, *12*(5), 1141–1154

Klebe, G. (2015). Applying thermodynamic profiling in lead finding and optimization. *Nat. Rev. Drug Discov.*, *14*, 95–110

Klug, A., Finch, J.T., & Franklin, R.E. (1957). Structure of turnip yellow mosaic virus. *Nature*, *179*(4561), 683–684

Klug, A., & Franklin, R.E. (1957). The reaggregation of the A-protein of tobacco mosaic virus. *Biochimica et biophysica acta*, *23*, 199–201

Kondratenko, A.V. (2009). Physical modeling of economic systems: classical and quantum economies. SSRN, http://dx.doi.org/10.2139/ssrn.1304630

Kondylis, P., Schlicksup, C.J., Zlotnick, A., & Jacobson, S.C. (2019). Analytical techniques to characterize the structure, properties, and assembly of virus capsids. *Anal. Chem.*, *91*(1), 622–636. doi: 10.1021/acs.analchem.8b04824; PMID: 30383361; PMCID: PMC6472978

Krenn, M., Pollice, R., Guo, S.Y., et al. (2022). On scientific understanding with artificial intelligence. *Nat. Rev. Phys.*, *4*, 761–769. https://doi.org/10.1038/s42254-022-00518-3

Krueger, D., Izquierdo, E., Viswanathan, R., Hartmann, J., Pallares Cartes, C., & De Renzis, S. (2019). Principles and applications of optogenetics in developmental biology. *Development*, *146*(20), dev175067

Kumar, A., Ernst, R.R., & Wüthrich, K. (1980). A two-dimensional nuclear Overhauser enhancement (2D NOE) experiment for the elucidation of complete proton-proton cross-relaxation networks in biological macromolecules. *Biochem. Biophys. Res. Commun.*, *95*, 1–6

Kumble, R., Palese, S., Visschers, R.W., Dutton, P.L., & Hochstrasser, R.M. (1996). Ultrafast dynamics within the B820 subunit from the core (LH-1) antenna complex of Rs. rubrum. *Chem. Phys. Lett.*, *261*, 396–404. doi:10.1016/0009-2614(96)01021-4

Kuznetsov, M., Clairambault, J., & Volpert, V. (2021). Improving cancer treatments via dynamical biophysical models. *Physics of Life Reviews*, *39*, 1–48

Lander, E.S., et al. (2001). Initial sequencing and analysis of the human genome. *Nature*, *409*(6822), 860–921

Landhuis, E. (2020). Probing fine-scale connections in the brain – artificial intelligence and improved microscopy make it feasible to map the nervous system at ever-higher resolution. *Nature*, *586*, 631–633. doi: 10.1038/d41586-020-02947-5

Lane, A.N., & Jenkins, T.C. (2000). Thermodynamics of nucleic acids and their interactions with ligands. *Quarterly Reviews of Biophysics*, *33*(3), 255–306

Lane, N. (2015). The unseen world: reflections on Leeuwenhoek (1677) "Concerning little animals". *Philosophical Transactions of the Royal Society B: Biological Sciences*, *370*(1666), 20140344

Lane, T.J., Shukla, D., Beauchamp, K.A., & Pande, V.S. (2013). To milliseconds and beyond: challenges in the simulation of protein folding. *Curr. Opin. Struct. Biol.*, *23*, 58–65

Lässig, M. (2007). From biophysics to evolutionary genetics: statistical aspects of gene regulation. *BMC Bioinformatics*, *8* (Suppl 6), S7. https://doi.org/10.1186/1471-2105-8-S6-S7

Lauffer, M.A. (1981). Biophysical methods in virus research. In *Comprehensive Virology*, ed. H. Fraenkel-Conrat & R.R. Wagner, pp. 1–82. Boston, MA: Springer. https://doi.org/10.1007/978-1-4615-6693-9_1

Lazár, D., & Schansker, G. (2009). Models of chlorophyll a fluorescence transients. In *Photosynthesis in silico: understanding complexity from molecules to ecosystems*, ed. A. Laisk, L. Nedbal, & Govindjee, Advances in photosynthesis and respiration 29, pp. 85–123. Dordrecht: Springer

Leake, M.C. (2016). The biophysics of infection. In *Biophysics of Infection*, ed. M.C. Leake, pp. 1–3. Cham: Springer

Leake, M.C. (2013). The physics of life: one molecule at a time. *Phil. Trans. R. Soc. B*, *368*, 20120248

Lecoq, H. (2001). Découverte du premier virus, le virus de la mosaïque du tabac: 1892 ou 1898? [Discovery of the first virus, the tobacco mosaic virus: 1892 or 1898?]. *C. R. Acad. Sci. III*, *324*(10), 929–933. doi: 10.1016/s0764-4469(01)01368-3; PMID: 11570281

Liberles, D.A., Teichmann, S.A., Bahar, I., Bastolla, U., Bloom, J., Bornberg-Bauer, E., ... & Whelan, S. (2012). The interface of protein structure, protein biophysics, and molecular evolution. *Protein Science*, *21*(6), 769–785

Lee, J., & Huang, Y. (2022). COVID-19 vaccine hesitancy: the role of socioeconomic factors and spatial effects. *Vaccines (Basel)*, *10*(3), 352. doi: 10.3390/vaccines10030352; PMID: 35334984; PMCID: PMC8950417

Leeson, P.D., & Springthorpe, B. (2007). The influence of drug-like concepts on decision-making in medicinal chemistry. *Nat. Rev. Drug Discov.*, *6*, 881–890

Li, B., Yang, Y.T., Capra, J.A., & Gerstein, M.B. (2020). Predicting changes in protein thermodynamic stability upon point mutation with deep 3D convolutional neural networks. *PLoS Computational Biology*, *16*(11), e1008291

Li, Y.Y., An, J., & Jones, S.J. (2011). A computational approach to finding novel targets for existing drugs. *PLoS Computational Biology*, *7*(9), e1002139

Lin, K. (2016). Phylogeny inference of closely related bacterial genomes: combining the features of both overlapping genes and collinear genomic regions. *Evolutionary Bioinformatics*, *11*(Suppl. 2), 1–9

Lindsay, R.B. (1973). *Julius Robert Mayer: prophet of energy*. Oxford: Pergamon Press

Liu, J., Yang, Q., Chen, S., Xiao, Z., Wen, S., & Luo, H. (2022). Intrinsic optical spatial differentiation enabled quantum dark-field microscopy. *Physical Review Letters*, *128*(19), 193601

Lopez, S., Van Dorp, L., & Hellenthal, G. (2016). Genetic studies supporting the human migration out of Africa and waves of dispersal to other continents. *Evolutionary Bioinformatics*, *11*(Suppl. 2), 57–68

Lynch, M., & Marinov, G.K. (2015). The bioenergetic costs of a gene. *Proceedings of the National Academy of Sciences*, *112*(51), 15690–15695

Macuglia, D., Roux, B., & Ciccotti, G. (2021). The breakthrough of a quantum chemist by classical dynamics: Martin Karplus and the birth of computer simulations of chemical reactions. *EPJ H*, *46*, 12. https://doi.org/10.1140/epjh/s13129-021-00013-w

Malik, M., & Boyd, R.W. (2014). Quantum imaging technologies. *La Rivista del Nuovo Cimento*, *37*, 273–332

Manhart, M. (2014). *Biophysics and stochastic processes in molecular evolution*. PhD dissertation, Rutgers University

Mann, A. (2018). Hunting for microbial life throughout the solar system. *Proceedings of the National Academy of Sciences*, *115*(45), 11348–11350. www.pnas.org/doi/10.1073/pnas.1816535115

Marais, A., Adams, B., Ringsmuth, A.K., Ferretti, M., Gruber, J.M., Hendrikx, R., ... & Van Grondelle, R. (2018). The future of quantum biology. *Journal of the Royal Society Interface*, *15*(148), 20180640

Margulis, L., & Sagan, D. (1995). *What Is Life?* Berkeley: University of California Press

Marini, M., Falqui, A., Moretti, M., Limongi, T., Allione, M., Genovese, A., ... & Di Fabrizio, E. (2015). The structure of DNA by direct imaging. *Science Advances*, *1*(7), e1500734

Marion, D. (2013). An introduction to biological NMR spectroscopy. *Mol. Cell. Proteomics*, *12*(11), 3006–3025. doi: 10.1074/mcp.O113.030239; PMID: 23831612; PMCID: PMC3820920

Mateus, A., Savitski, M.M., & Piazza, I. (2021). The rise of proteome-wide biophysics. *Mol. Syst. Biol*, *17*(7), e10442. doi: 10.15252/msb.202110442; PMID: 34293219; PMCID: PMC8297615

Mattern, C.F. (1962). Electron microscopic observations of tobacco mosaic virus structure. *Virology*, *17*(1), 76–83

Mayer, J.R. (1845). Die organische Bewegung in ihrem Zusammenhang mit dem Stoffwechsel. In *Die Mechanik der Wärme* in *Gesammelte Schriften*, pp. 13–126. Stuttgart: J.G. Cotta'sche Verlagsbuchhandlung

Mayer, J.R. (1842). Bemerkungen über die Kräfte der unbelebten Natur. *Annalen der Chemie und Pharmacie*, *42*, 233–240

McCammon, J.A. (2009). Darwinian biophysics: electrostatics and evolution in the kinetics of molecular binding. *Proceedings of the National Academy of Sciences*, *106*(19), 7683–7684

McCammon, J.A., Gelin, B.R., & Karplus, M. (1977). Dynamics of folded proteins. *Nature*, *267*, 585–590

McGrath, J.M., & Long, S.P. (2014). Can the cyanobacterial carbon-concentrating mechanism increase photosynthesis in crop species? A theoretical analysis. *Plant Physiology*, *164*(4), 2247–2261

McGrew, W.C. (2004). *The cultured chimpanzee: reflections on cultural primatology*. Cambridge: Cambridge University Press

McKay, C.P. (2014). Requirements and limits for life in the context of exoplanets. *Proceedings of the National Academy of Sciences*, *111*(35), 12628–12633. www.pnas.org/doi/10.1073/pnas.1304212111

Meli, M., Morra, G., & Colombo, G. (2020). Simple model of protein energetics to identify ab initio folding transitions from all-atom MD simulations of proteins. *Journal of Chemical Theory and Computation*, *16*(9), 5960–5971

Mitcheltree, M.J., Pisipati, A., Syroegin, E.A., Silvestre, K.J., Klepacki, D., Mason, J.D., … & Myers, A.G. (2021). A synthetic antibiotic class overcoming bacterial multidrug resistance. *Nature*, *599*(7885), 507–512

Miyata, K., Nishiyama, N., & Kataoka, K. (2012). Rational design of smart supramolecular assemblies for gene delivery: chemical challenges in the creation of artificial viruses. *Chemical Society Reviews*, *41*(7), 2562–2574

Mognetti, B.M., Cicuta, P., & Di Michele, L. (2019). Programmable interactions with biomimetic DNA linkers at fluid membranes and interfaces. *Rep. Prog. Phys.*, *82*(11), 116601. doi: 10.1088/1361-6633/ab37ca

Monod, J. (1949). The growth of bacterial cultures. *Annual Review of Microbiology*, *3*(1), 371–394

Mora-Bermúdez, F., Badsha, F., Kanton, S., Camp, J.G., Vernot, B., Köhler, K., Voigt, B., Okita, K., Maricic, T., He, Z., Lachmann, R., Pääbo, S., Treutlein, B., and Huttner, W.B. (2016). Differences and similarities between human and chimpanzee neural progenitors during cerebral cortex development. *Elife*, *5*, e18683. doi: 10.7554/eLife.18683; PMID: 27669147; PMCID: PMC5110243

Morris, P., Aspden, R., Bell, J., et al. (2015). Imaging with a small number of photons. *Nat. Commun.*, *6*, 5913. https://doi.org/10.1038/ncomms6913

Movahed, T.M., Bidgoly, H.J., Manesh, M.H.K., & Mirzaei, H.R. (2021). Predicting cancer cells progression via entropy generation based on AR and ARMA models. *International Communications in Heat and Mass Transfer*, *127*, 105565

Mulholland, A.J., & Richards, W.G. (1997). Acetyl-CoA enolization in citrate synthase: a quantum mechanical/molecular mechanical (QM/MM) study. *Proteins*, *27*, 9–25

Müller-Wille, S. (2010). Cell theory, specificity, and reproduction, 1837–1870. *Studies in History and Philosophy of Science Part C: Studies in History and Philosophy of Biological and Biomedical Sciences*, *41*(3), 225–231

Mureddu, L., & Vuister, G.W. (2019). Simple high-resolution NMR spectroscopy as a tool in molecular biology. *FEBS J.*, *286*(11), 2035–2042. doi: 10.1111/febs.14771; PMID: 30706658; PMCID: PMC6563160

Myatt, D.P., Wharram, L., Graham, C., Liddell, J., Branton, H., Pizzey, C., … & Shattock, R. (2022). Biophysical characterisation of the structure of a SARS-CoV-2 self-amplifying-RNA (saRNA) vaccine. *Biol. Methods Protoc.*, *8*(1), bpad001

Naganathan, A.N., Doshi, U., Fung, A., Sadqi, M., & Muñoz, V. (2006). Dynamics, energetics, and structure in protein folding. *Biochemistry*, *45*(28), 8466–8475

Nelson, M.L., Dinardo, A., Hochberg, J., & Armelagos, G.J. (2010). Brief communication: mass spectroscopic characterization of tetracycline in the skeletal remains of an ancient population from Sudanese Nubia 350–550 CE. *American Journal of Physical Anthropology*, *143*(1), 151–154

Nicholson, D.J. (2010). Biological atomism and cell theory. *Studies in History and Philosophy of Science Part C: Studies in History and Philosophy of Biological and Biomedical Sciences*, *41*(3), 202–211

Noonan, J.P. (2010). Neanderthal genomics and the evolution of modern humans. *Genome Res.*, *20*(5), 547–553. doi: 10.1101/gr.076000.108; PMID: 20439435; PMCID: PMC2860157

Novoderezhkin, V., Monshouwer, R., & van Grondelle, R. (2000). Electronic and vibrational coherence in the core light-harvesting antenna of Rhodopseudomonas v iridis. *Journal of Physical Chemistry B*, *104*(50), 12056–12071

Novoderezhkin, V., Yakovlev, A.G., van Grondelle, R., & Shuvalov, V.A. (2004). Coherent nuclear and electronic dynamics in primary charge separation in photosynthetic reaction centers: a Redfield theory approach. *J. Phys. Chem. B*, *108*, 7445–7457. doi:10.1021/jp0373346

Nunes, J.M. (2016). Using uniformat and gene[rate] to analyze data with ambiguities in population genetics. *Evol. Bioinform. Online*. *11*(Suppl 2), 19–26. doi: 10.4137/EBO.S32415; PMID: 26917942; PMCID: PMC4762493

O'Brien, R., Markova, N., & Holdgate, G.A. (2017). Thermodynamics in drug discovery. In *Applied biophysics for drug discovery*, ed. D. Huddler & E.R. Zartler, 7–28. Hoboken, NJ: John Wiley & Sons

Ode, H., Nakashima, M., Kitamura, S., Sugiura, W., & Sato, H. (2012). Molecular dynamics simulation in virus research. *Front. Microbiol.*, *3*, 258

Parreira, P., & Martins, M.C.L. (2021). The biophysics of bacterial infections: adhesion events in the light of force spectroscopy. *The Cell Surface*, *7*, 100048

Patterson, S.D., & Aebersold, R.H. (2003). Proteomics: the first decade and beyond. *Nature Genetics*, *33*(3), 311–323

Pearson, K. (1900). The Grammar of Science (2nd edn). London: Adam and Charles Black

Pease III, L.F., Tsai, D.H., Brorson, K.A., Guha, S., Zachariah, M.R., & Tarlov, M.J. (2011). Physical characterization of icosahedral virus ultra structure, stability, and integrity using electrospray differential mobility analysis. *Analytical Chemistry*, *83*(5), 1753–1759

Peña-Guerrero, J., Nguewa, P.A., & García-Sosa, A.T. (2021). Machine learning, artificial intelligence, and data science breaking into drug design and neglected diseases. *WIREs Computational Molecular Science*, *11*(5), e1513

Perutz, M.F., Rossmann, M.G., Cullis, A.F., Muirhead, H., Will, G., & North, A.C.T. (1960). Structure of hæmoglobin: a three-dimensional Fourier synthesis at 5.5-Å. resolution, obtained by X-ray analysis. *Nature*, *185*, 416–422

Philpott, M.P. (2021). Pigmentation: watching hair turn grey. *eLife*, *10*, e70584. https://doi.org/10.7554/eLife.70584

Pitt, T.L., & Barer, M.R. (2012). Classification, identification and typing of micro-organisms. *Medical Microbiology*, *2012*, 24–38. doi: 10.1016/B978-0-7020-4089-4.00018-4; PMCID: PMC7171901

Preston, K.B., Wong, T.A.S., To, A., Tashiro, T.E., Lieberman, M.M., Granados, A., ... & Randolph, T.W. (2021). Single-vial filovirus glycoprotein vaccines: biophysical characteristics and immunogenicity after co-lyophilization with adjuvant. *Vaccine*, *39*(39), 5650–5657

Prussin, A.J. & Marr, L.C. (2015). Sources of airborne microorganisms in the built environment. *Microbiome*, *3*, 78. https://doi.org/10.1186/s40168-015-0144-z

Purcell, E.M., Torrey, H.C., & Pound, R.V. (1946). Resonance absorption by nuclear magnetic moments in a solid. *Phys. Rev.*, *69*, 37–38

Purslow, J.A., Khatiwada, B., Bayro, M.J., & Venditti, V. (2020). NMR methods for structural characterization of protein-protein complexes. *Frontiers in Molecular Biosciences*, *7*, 9

Rando, T.A, & Ambrosio, F. (2018). Regenerative rehabilitation: applied biophysics meets stem cell therapeutics. *Cell Stem Cell*, *22*(3), 306–309. doi: 10.1016/j.stem.2018.02.003. Erratum in: *Cell Stem Cell* (2018), *22*(4), 608. PMID: 29499150; PMCID: PMC5931336

Rasmussen, S.G., Choi, H.J., Rosenbaum, D.M., Kobilka, T.S., Thian, F.S., Edwards, P.C., ... & Kobilka, B.K. (2007). Crystal structure of the human β2 adrenergic G-protein-coupled receptor. *Nature*, *450*(7168), 383–387

Remacle, F., Graeber, T.G., & Levine, R.D. (2020). Thermodynamic energetics underlying genomic instability and whole-genome doubling in cancer. *Proceedings of the National Academy of Sciences*, *117*(31), 18880–18890

Renaud, J.P., Chung, C.W., Danielson, U.H., Egner, U., Hennig, M., Hubbard, R.E., & Nar, H. (2016). Biophysics in drug discovery: impact, challenges and opportunities. *Nature Reviews Drug Discovery*, *15*(10), 679–698

Richards, B., Tsao, D., & Zador, A. (2022). The application of artificial intelligence to biology and neuroscience. *Cell*, *185*(15), 2640–2643

Richert-Pöggeler, K.R., Franzke, K., Hipp, K., & Kleespies, R.G. (2019). Electron microscopy methods for virus diagnosis and high resolution analysis of viruses. *Frontiers in Microbiology*, *9*, 3255

Ridley, C.P., Lee, H.Y., & Khosla, C. (2008). Evolution of polyketide synthases in bacteria. *Proceedings of the National Academy of Sciences*, *105*(12), 4595–4600

Rodrigues, J.V., Bershtein, S., Li, A., Lozovsky, E.R., Hartl, D.L., & Shakhnovich, E.I. (2016). Biophysical principles predict fitness landscapes of drug resistance. *Proceedings of the National Academy of Sciences*, *113*(11), E1470–E1478

Roos, W.H., Bruinsma, R., & Wuite, G.J.L. (2010). Physical virology. *Nature Physics*, *6*(10), 733–743

Rosenthal, D.M., Locke, A.M., Khozaei, M., Raines, C.A., Long, S.P., & Ort, D.R. (2011). Over-expressing the C3 photosynthesis cycle enzyme Sedoheptulose-1-7 Bisphosphatase improves photosynthetic carbon gain and yield under fully open air CO2 fumigation (FACE). *BMC Plant Biology*, *11*(1), 1–12

Rozanova, S., Barkovits, K., Nikolov, M., Schmidt, C., Urlaub, H., & Marcus, K. (2021). Quantitative mass spectrometry-based proteomics: an overview. *Methods Mol. Biol.*, *2228*, 85–116. doi: 10.1007/978-1-0716-1024-4_8; PMID: 33950486

Ruben, S., Randall, M., Kamen, M., & Hyde, J. L. (1941). Heavy oxygen (O^{18}) as a tracer in the study of photosynthesis. *Journal of the American Chemical Society*, *63*, 877–879

Saha, P., & Sarkar, B.K. (2020). Entropy based analysis of genetic information. *Journal of Physics: Conference Series*, 1579(1), 012003

Saint-Sardos, A., Sart, S., Lippera, K., Brient-Litzler, E., Michelin, S., Amselem, G., et al. (2020). High-throughput measurements of intra-cellular and secreted cytokine from single spheroids using anchored microfluidic droplets. *Small*, *16*(49), e2002303. doi: 10.1002/smll.202002303

Salari, V., Naeij, H., & Shafiee, A. (2017). Quantum interference and selectivity through biological ion channels. *Sci. Rep.*, *7*, 41625. https://doi.org/10.1038/srep41625

Saunders, M., Wishnia, A., & Kirkwood, J.G. (1957). The nuclear magnetic resonance spectrum of ribonuclease. *J. Am. Chem. Soc.*, *79*, 3289–3290

Schirle, M., Bantscheff, M., & Kuster, B. (2012). Mass spectrometry-based proteomics in preclinical drug discovery. *Chemistry & biology*, *19*(1), 72–84

Schlick, T., & Portillo-Ledesma, S. (2021). Biomolecular modeling thrives in the age of technology. *Nat. Comput. Sci.*, *1*, 321–331. https://doi.org/10.1038/s43588-021-00060-9

Schluenzen, F., Tocilj, A., Zarivach, R., Harms, J., Gluehmann, M., Janell, D., … & Yonath, A. (2000). Structure of functionally activated small ribosomal subunit at 3.3 Å resolution. *Cell*, *102*(5), 615–623

Schrödinger, E. (1944). *What is Life*. Cambridge: Cambridge University Press

Schwann, T.H. (1847). *Microscopical researches into the accordance in the structure and growth of animals and plants*. Рипол Классик, originally published in 1839

Schwartz, D.C. (2019). Biophysics and the genomic sciences. *Biophysical Journal*, *117*(11), 2047–2053

Scott, E.A., Bruning, E., Nims, R.W., Rubino, J.R., & Ijaz, M.K. (2020). A 21st century view of infection control in everyday settings: moving from the germ theory of disease to the microbial theory of health. *American Journal of Infection Control*, *48*(11), 1387–1392

Seifi, M., Soltanmanesh, A., & Shafiee, A. (2022). Quantum coherence on selectivity and transport of ion channels. *Sci. Rep.*, *12*, 9237. https://doi.org/10.1038/s41598-022-13323-w

Serohijos, A.W., & Shakhnovich, E.I. (2014). Merging molecular mechanism and evolution: theory and computation at the interface of biophysics and evolutionary population genetics. Curr. Opin. Struct. Biol., *26*, 84–91. doi: 10.1016/j.sbi.2014.05.005; PMID: 24952216; PMCID: PMC4292934

Sever, R., & Brugge, J.S. (2015). Signal transduction in cancer. *Cold Spring Harb. Perspect. Med.*, *5*(4), a006098. doi: 10.1101/cshperspect.a006098; PMID: 25833940; PMCID: PMC4382731

Shashkova, S., & Leake, M.C. (2018). Systems biophysics: single-molecule optical proteomics in single living cells. *Current Opinion in Systems Biology*, *7*, 26–35

Shaw, D.E., Maragakis, P., Lindorff-Larsen, K., Piana, S., Dror, R.O., Eastwood, M. P., … & Wriggers, W. (2010). Atomic-level characterization of the structural dynamics of proteins. *Science*, *330*(6002), 341–346

Shi, Y. (2014). A glimpse of structural biology through X-ray crystallography. *Cell*, *159*(5), 995–1014

Sikosek, T., & Chan, H.S. (2014). Biophysics of protein evolution and evolutionary protein biophysics. *Journal of the Royal Society Interface*, *11*(100), 20140419

Silvestre, H.L., Blundell, T.L., Abell, C., & Ciulli, A. (2013). Integrated biophysical approach to fragment screening and validation for fragment-based lead discovery. *Proc. Nat. Acad. Sci. USA*, 110, 12984–12989

Simkin, A.J., López-Calcagno, P.E., & Raines, C.A. (2019). Feeding the world: improving photosynthetic efficiency for sustainable crop production. *Journal of Experimental Botany*, *70*(4), 1119–1140

Singer, S.J., & Nicolson, G.L. (1972). The fluid mosaic model of the structure of cell membranes: cell membranes are viewed as two-dimensional solutions of oriented globular proteins and lipids. *Science*, *175*(4023), 720–731. doi: 10.1126/science.175.4023.720; PMID: 4333397

Singh, S.P., & Siddhanta, S. (2021). Optical imaging in biology: basics and applications. In *Modern Techniques of Spectroscopy: Basics, Instrumentation, and Applications*, ed. D.K. Singh, M. Pradhan, & A. Materny, pp. 637–660. Singapore: Springer Nature Singapore

Sliwoski, G., Kothiwale, S., Meiler, J., & Lowe, E.W. (2014). Computational methods in drug discovery. *Pharmacological Reviews*, *66*(1), 334–395

Steere-Williams, J. (2015). The Germ Theory. In *A companion to the history of American science*, ed. G.M. Montgomery & M.A. Largent, pp. 397–407. Chichester: Wiley-Blackwell.

Stein, A., Fowler, D.M., Hartmann-Petersen, R., & Lindorff-Larsen, K. (2019). Biophysical and mechanistic models for disease-causing protein variants. *Trends Biochem. Sci*, *44*(7), 575–588. doi: 10.1016/j.tibs.2019.01.003; PMID: 30712981; PMCID: PMC6579676

Steinrauf, L.K. (1959). Preliminary X-ray data for some new crystalline forms of β-lactoglobulin and hen-egg-white lysozyme. *Acta Crystallographica*, *12*(1), 77–79

Stirbet, A., Lazár, D., Guo, Y., & Govindjee, G. (2020). Photosynthesis: basics, history and modelling. *Ann. Bot.*, *126*(4), 511–537. doi: 10.1093/aob/mcz171; PMID: 31641747; PMCID: PMC7489092

Suran, M. (2022). After the genome – a brief history of proteomics. *JAMA*, *328*(12), 1168–1169

Suresh, S. (2007). Biomechanics and biophysics of cancer cells. *Acta Biomater.*, *3*(4), 413–438. doi: 10.1016/j.actbio.2007.04.002: PMID: 17540628; PMCID: PMC2917191

Tan, S.Y., & Tatsumura, Y. (2015). Alexander Fleming (1881–1955): discoverer of penicillin. *Singapore Med. J.*., *56*(7), 366–367. doi: 10.11622/smedj.2015105; PMID: 26243971; PMCID: PMC4520913

Taylor, M. (2015). *Quantum microscopy of biological systems*. Cham: Springer International Publishing. https://doi.org/10.1007/978-3-319-18938-3

Taylor, M.A., Janousek, J., Daria, V., Knittel, J., Hage, B., Bachor, H.A., & Bowen, W.P. (2014). Subdiffraction-limited quantum imaging within a living cell. *Physical Review X*, *4*(1), 011017

Thaker, M.N., & Wright, G.D. (2015). Opportunities for synthetic biology in antibiotics: expanding glycopeptide chemical diversity. *ACS Synthetic Biology*, *4*(3), 195–206

The, J. (2012). Ultrasound of soft tissue masses of the hand. *J. Ultrason.*, *12*(51), 381–401. doi: 10.15557/JoU.2012.0028; PMID: 26673615; PMCID: PMC4603238

Trevino, S.G., Zhang, N., Elenko, M.P., Lupták, A., & Szostak, J.W. (2011). Evolution of functional nucleic acids in the presence of nonheritable backbone heterogeneity. *Proceedings of the National Academy of Sciences*, *108*(33), 13492–13497

Tseng, C.Y., Ashrafuzzaman, M., Mane, J.Y., Kapty, J., Mercer, J.R., & Tuszynski, J.A. (2011). Entropic fragment-based approach to aptamer design. *Chemical Biology & Drug Design*, *78*(1), 1–13

Tubiana, M. (1996). Wilhelm Conrad Röntgen et la découverte des rayons X [Wilhelm Conrad Röntgen and the discovery of X-rays]. *Bull. Acad. Natl. Med*, *180*(1), 97–108. PMID: 8696882

Ulanowicz, R.E., & Hannon, B.M. (1987). Life and the production of entropy. *Proceedings of the Royal Society of London: Series B: Biological Sciences*, *232*(1267), 181–192

Ullman, S. (2019). Using neuroscience to develop artificial intelligence. *Science*, *363*, 692–693. doi: 10.1126/science.aau6595

van den Berg, M.A. (2010). Functional characterization of penicillin production strains. *Fungal Biology Reviews*, *24*(1–2), 73–78

Viljoen, A., Dufrêne, Y.F., & Nigou, J. (2022). Mycobacterial adhesion: from hydrophobic to receptor-ligand interactions. *Microorganisms*, *10*(2), 454

Vologodskii, A., & Frank-Kamenetskii, M.D. (2018). DNA melting and energetics of the double helix. *Physics of Life Reviews*, *25*, 1–21

Vos, M., Jones, M.R., Hunter, C.N., Breton, J., Lambry, J.-Ch., & Martin, J.-L. (1994). Coherent nuclear dynamics at room temperature in bacterial reaction centers. *Proc. Natl. Acad. Sci. USA*, *91*(26), 12701–12705. doi:10.1073/pnas.91.26.12701

Vanchurin, V., Wolf, Y.I., Koonin, E.V., & Katsnelson, M.I. (2022). Thermodynamics of evolution and the origin of life. *Proceedings of the National Academy of Sciences*, *119*(6), e2120042119

Vasilaki, D., Bakopoulou, A., Tsouknidas, A. et al. (2021). Biophysical interactions between components of the tumor microenvironment promote metastasis. *Biophys. Rev.*, *13*, 339–357. https://doi.org/10.1007/s12551-021-00811-y

Vekshin, N. (2011). *Biophysics of DNA-antibiotic complexes*. New York: Nova Biomedical Books

Waigh, T.A. (2007). *Applied biophysics: a molecular approach for physical scientists*. Chichester: John Wiley & Sons

Wan, W., Kolesnikova, L., Clarke, M., Koehler, A., Noda, T., Becker, S., & Briggs, J.A. (2017). Structure and assembly of the Ebola virus nucleocapsid. *Nature*, *551*(7680), 394–397

Wang, G., Bai, Y., Cui, J., Zong, Z., Gao, Y., & Zheng, Z. (2022). Computer-aided drug design boosts RAS inhibitor discovery. *Molecules*, *27*(17), 5710

Warden, S.J., Mantila Roosa, S.M., Kersh, M.E., Hurd, A.L., Fleisig, G.S., Pandy, M.G., & Fuchs, R.K. (2014). Physical activity when young provides lifelong benefits to cortical bone size and strength in men. *Proceedings of the National Academy of Sciences*, *111*(14), 5337–5342

Watson, J.D., & Crick, F.H. (1953a). Genetical implications of the structure of deoxyribonucleic acid. *Nature*, *171*, 964–967

Watson, J.D., & Crick, F.H. (1953b). Molecular structure of nucleic acids: a structure for deoxyribose nucleic acid. *Nature*, *171*(4356), 737–738.

Wei, L., Chen, Z., Shi, L., Long, R., Anzalone, A.V., Zhang, L., … & Min, W. (2017). Super-multiplex vibrational imaging. *Nature*, *544*(7651), 465–470

Wei, W., Cheng, W., Dai, W., Lu, F., Cheng, Y., Jiang, T., et al. (2022). A nanodrug coated with membrane from brain microvascular endothelial cells protects against experimental cerebral malaria. *Nano Lett.*, *22*(1), 211–219. doi: 10.1021/acs.nanolett.1c03514

Wenzel, C., Drozdzik, M., & Oswald, S. (2021). Mass spectrometry-based targeted proteomics method for the quantification of clinically relevant drug metabolizing enzymes in human specimens. *Journal of Chromatography B*, *1180*, 122891

Wilkins, M.F.H., Stokes, A.R., & Wilson, H.R. (1953). Molecular structure of deoxypentose nucleic acids. *Nature*, *171*, 738–740

Williamson, M.P., Havel, T.F., & Wüthrich, K. (1985). Solution conformation of proteinase inhibitor IIA from bull seminal plasma by 1H nuclear magnetic resonance and distance geometry. *J. Mol. Biol.*, *182*, 295–315

Wilson, J.W. (1944). Cellular tissue and the dawn of the cell theory. *Isis*, *35*(2), 168–173

Wimberly, B.T., Brodersen, D.E., Clemons, Jr, W.M., Morgan-Warren, R.J., Carter, A.P., Vonrhein, C., … & Ramakrishnan, V. (2000). Structure of the 30S ribosomal subunit. *Nature*, *407*(6802), 327–339.

Witherspoon, D.J., Wooding, S., Rogers, A.R., Marchani, E.E., Watkins, W.S., Batzer, M.A, & Jorde, L.B. (2007). Genetic similarities within and between human populations. *Genetics*, *176*(1), 351–359. doi: 10.1534/genetics.106.067355; PMID: 17339205; PMCID: PMC189302

Wüthrich, K., Wider, G., Wagner, G., & Braun, W. (1982). Sequential resonance assignments as a basis for determination of spatial protein structures by high resolution proton nuclear magnetic resonance. *Journal of Molecular Biology*, *155*(3), 311–319.

Wylie, C.S., & Shakhnovich, E.I. (2011). A biophysical protein folding model accounts for most mutational fitness effects in viruses. *Proceedings of the National Academy of Sciences*, *108*(24), 9916–9921

Xu, Q., Cui, Z., Venkatraman, G., & Gomes, A.V. (2012). The use of biophysical proteomic techniques in advancing our understanding of diseases. *Biophys. Rev.*, *4*(2), 125–135. doi: 10.1007/s12551-012-0070-2; PMID: 28510094; PMCID: PMC5418381

Yang, T., Ottilie, S., Istvan, E.S., Godinez-Macias, K.P., Lukens, A.K., Baragaña, B., … & Winzeler, E.A. (2021). MalDA, accelerating malaria drug discovery. *Trends in Parasitology*, *37*(6), 493–507

Yates, J.R., Ruse, C.I., & Nakorchevsky, A. (2009). Proteomics by mass spectrometry: approaches, advances, and applications. *Annual Rev. Biomed. Eng.*, *11*, 49–79. doi: 10.1146/annurev-bioeng-061008-124934

Yip, K.W., & Reed, J.C. (2008). Bcl-2 family proteins and cancer. *Oncogene*, *27*(50), 6398–6406. doi: 10.1038/onc.2008.307; PMID: 18955968

Zepp, J.A., Morley, M.P., Loebel, C., Kremp, M.M., Chaudhry, F.N., Basil, M.C., … & Morrisey, E.E. (2021). Genomic, epigenomic, and biophysical cues controlling the emergence of the lung alveolus. *Science*, *371*(6534), eabc3172

Zhang, Q., Liu, H., Liu, X., Jiang, D., Zhang, B., Tian, H., … & Rao, Z. (2018). Discovery of the first macrolide antibiotic binding protein in Mycobacterium tuberculosis: a new antibiotic resistance drug target. *Protein & Cell*, *9*(11), 971–975

Zhao, G., Perilla, J.R., Yufenyuy, E.L., Meng, X., Chen, B., Ning, J., … & Zhang, P. (2013). Mature HIV-1 capsid structure by cryo-electron microscopy and all-atom molecular dynamics. *Nature*, *497*(7451), 643–646.

Zhong-Can, O.Y., Ji-Xing, L., & Yu-Zhang, X. (1999). *Geometric methods in the elastic theory of membranes in liquid crystal phases*. Advanced Series on Theoretical Physical Science 2. Singapore: World Scientific Publishing Co.

Zhu, G., Pan, C., Bei, J.X., Li, B., Liang, C., Xu, Y., & Fu, X. (2020). Mutant p53 in cancer progression and targeted therapies. *Frontiers in Oncology*, *10*, 595187

Zhu, X.G., de Sturler, E., & Long, S.P. (2007). Optimizing the distribution of resources between enzymes of carbon metabolism can dramatically increase photosynthetic rate: a numerical simulation using an evolutionary algorithm. *Plant Physiology*, *145*, 513–526

Zhu, X.G., Long, S.P., & Ort, D.R. (2010). Improving photosynthetic efficiency for greater yield. *Annual Review of Plant Biology*, *61*, 235–261

2 Biophysics of Life

Biological systems are ever-active, dynamic, and fluctuating in nature. They are mainly compositions of physical and chemical systems contained in thermodynamic baths. Biological systems exist in either living conditions or in a state of nonliving status. The main difference between their living and nonliving status is known as "life." When life exists, biological systems show measurable or describable psychological and behavioral statuses. Living systems are therefore considered to appear with mental health that helps to process and understand information and experiences. The other vital one is emotional health which involves the capability of expressing emotions and managing situations. Hierarchy in the living system is mainly determined by these mental and emotional health statuses. The lifeless biological systems may continue showing vital material properties, although their altered physicochemical states lose almost all of the gross dynamic properties and fluctuations thereof. Life is therefore not just a concept, but a measurable entity. Laws of physics play vital roles in leading biological systems to coordinate expressions, regulations, and distributions of biochemicals within and between various compartments that are interlinked structurally and/or functionally. Life thus owes a lot to physics laws in becoming defined. Energy, entropy, and time are vital physics parameters involved in determining biological life. The flow of energy through plants, animals, and bacteria is known to drive the primary processes of life, such as metabolism, movement, and ion transport. Living systems ensure the supply of energy for their various compartments by continuously consuming foods (animals) or through photosynthesis (plants). Unlike the whole universe (viewed as an isolated system) having increasing entropy, all living things are found to maintain a highly ordered, low entropy structure. The energy of the living system is partly wasted, partly given off as heat, partly utilized to continue cellular processes, etc. For maintaining order in living systems, a constant input of energy is required. A steady flow of energy within the system is necessary. In order to sustain itself, a living system increases the overall entropy while the entropy may decrease locally within an organism. That means organisms are assumed to live by continuously decreasing their local entropy and increasing the entropy of the surroundings. Aging is predicted to be a consequence of thermodynamics and entropy. The second law of thermodynamics suggests that all species progress through changes in their living systems, and they experience aging as their ability to resist entropy declines. It is well established that most of the processes of life originate at cellular levels. We see a biological cell as an open system maintaining continuous communication with its surroundings, and capable of avoiding excessive local entropic changes, thus the cell's life is sustained for a course of time until other signaling pathways altogether lead it to apoptosis. Intervening in cellular-level disorders (due to natural reasons, specific diseases, or mutations in genes) with drugs may slow the process of entropic changes. Therefore, medication is essential to deal with diseases causing stress in life.

DOI: 10.1201/9781003287780-2

2.1 BIOLOGICAL PERSPECTIVES ASSOCIATED WITH LIFE

Understanding biological systems requires addressing systems at their cellular level. Cells are the building blocks of biological systems. The biological explanation of life's processes are mainly found in cells' versatile structures and functions. But that doesn't exclude the possibility of finding any vital role that might be played outside cells concerning life's processes. I am going to pinpoint a few of them.

2.1.1 The Cell Hosts Life's Processes

Since the discovery of the cell by Robert Hooke in 1665 (Hooke, 1665) the cell has been considered the fundamental unit of life. Most of the processes of life are contained in various sections or compartments inside cells. The process of cell division and cell reproduction is unique to living organisms. This process follows deterministic time scales and statistics (Kozawa et al., 2016). But the initiation happens at molecular levels inside cells following sets of molecular mechanisms involving various biomolecules. The life of any organism may be grossly defined by considering several common and key characteristics or functions, namely order, sensitivity and response to the environment, reproduction capability, growth and development, regulation, homeostasis, and energy processing. These gross observational aspects originate at the molecular structural and functional levels, which are available mostly inside cells. Cells are therefore biological systems' vital compartments that need to be explored thoroughly to understand life's various processes. We may certainly not pinpoint any specific aspect or cellular process which is unique to define life. Instead, we find a collective mode of various actions constructing the cause of life. So the failure of life may be due to any or many causes contained in cells. Aging may be found to be linked to alterations of many cellular processes and functions of specific cellular systems. The cell health condition is perhaps therefore directly or indirectly an indication of one's life expectancy. The life span of an organism certainly depends on many medical, social, and economic conditions (Ohlhorst et al., 2013; Yildiz et al., 2021; Li et al., 2018), but the role of the cell health condition (Madeo et al., 2015) is perhaps one of the most crucial determinants of one's life span. The inhibition of the major nutrient and growth-related signaling pathways and the upregulation of anti-aging pathways may cause the induction of autophagy and mediate life span extension (Madeo et al., 2015). On the other hand, adopting a proper diet practice may also help extend life span through mostly metabolic means. If athletes start consuming the prescribed retirement diet after 50 years of age, the longevity of cyclists, weightlifters, rugby players, and golf players may increase by 7, 22, 30, and 8 years, respectively (Yildiz et al., 2021). Lifestyle and controlled consumption of nutrients, besides natural cellular processes, may be associated with various defining aspects of life, including life span.

Cellular health is controlled at various stages and points in the cell, starting in the nucleus through the chromosome structure or organization, the regulation of the transcription, and the nuclear export/import, ranging outward to the protein translation and quality control, the autophagic recycling of organelles, the maintenance of the structure of the cytoskeletal, and finally the maintenance of the extracellular matrix and the extracellular signaling. Figure 2.1 demonstrates major cellular functionalities at an early stage of the birth of a cell and its aged stage (DiLoreto and Murphy, 2015). All regulatory systems receive information from each other and establish a complex state of bioinformatics, resulting in an intricate interplay of regulation controlling cellular aging.

2.1.1.1 Autophagy is Essential for Maintaining Correct Cell Health

All cells are made of nucleic acids, proteins, carbohydrates, and lipids. Proteins, which are chains of amino acids, are now considered to be the building blocks of life. Protein synthesis and degradation are both essential components of cell health. Yoshinori Ohsumi discovered how cells recycle their content, a process known as autophagy, a Greek term for "self-eating." The degradation of the

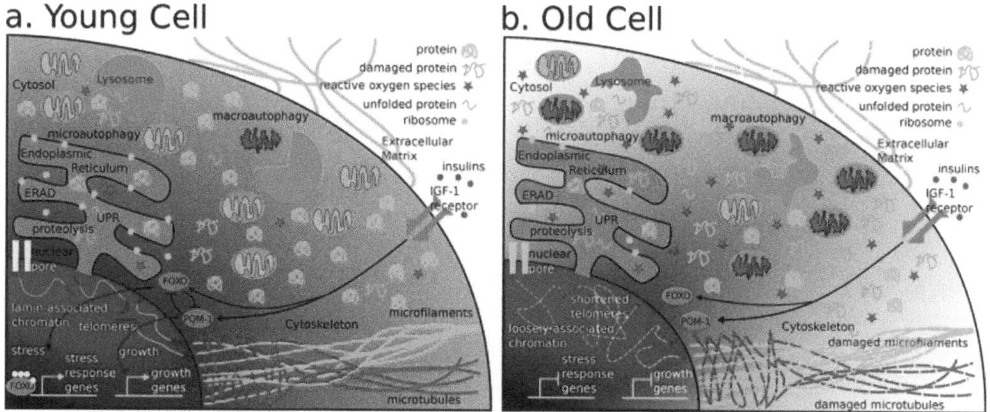

FIGURE 2.1 Major features of cellular aging. With aging, translational defects and entropy progressively increase the amount of cellular damage, and the clearance and quality control mechanisms grow less effective. Left panel (a): in a young cell, most organelles are naturally very healthy, and when proteins are translated and misfolded or acquire damage in the cytosol, they are cleared either by ERAD (in the ER) or autophagy (in the cytosol). When organelles become too damaged, they are degraded to component parts by macroautophagy. Right panel (b): in an older cell, the accumulated damage leads to a less healthy cell. ROS build up from damaged mitochondria and contribute to a greater fraction of the proteome consisting of damaged proteins and protein aggregates. For details, see DiLoreto and Murphy (2015). Life, as it is associated with cell health condition, is therefore linked to various cellular sections' structural and functional conditions. A few of them are briefed here to help understand life better. Color contrast here and in all subsequent figures is distinguishable in the digital version of the book.

***Source*: DiLoreto and Murphy (2015).**

cell via the removal of unnecessary or dysfunctional components through a lysosome-dependent regulated mechanism is considered autophagy (Klionsky, 2008). Autophagy is the main intracellular degradation system by which cytoplasmic materials are delivered to and degraded in the lysosome. Besides the elimination of materials, autophagy importantly serves as a dynamic recycling system helping to produce new building blocks and energy required for cellular renovation and homeostasis. Different types of autophagy and the steps therein have been modeled in Figure 2.2. In mammalian cells, we observe the following three types of autophagy:

1. macroautophagy
2. chaperone-mediated autophagy
3. microautophagy

These three classes of autophagy experience distinguishable structural and functional alterations happening inside cells. Autophagy thus contributes to raising crucial cellular processes that altogether determine vital cell health conditions. Ohsumi summarized this by stating "everyone believed protein synthesis was so important to life, but in my opinion, protein degradation is equally important" (Hornyak, 2017). He further elaborated by saying "if you have a defect in autophagy activity, the cell might have many defects and some could be related to disease."

Induction of autophagy may be triggered, in an organ- or tissue-specific manner, due to external factors such as starvation, hypoxia, high temperatures, and cellular stress (Feng et al., 2014). Besides, the process of autophagy was observed to be initiated in SH-SY5Y cells after treatment with adenosine triphosphate (Lu et al., 2014). Within one hour of the exposure of 6 mmol/L

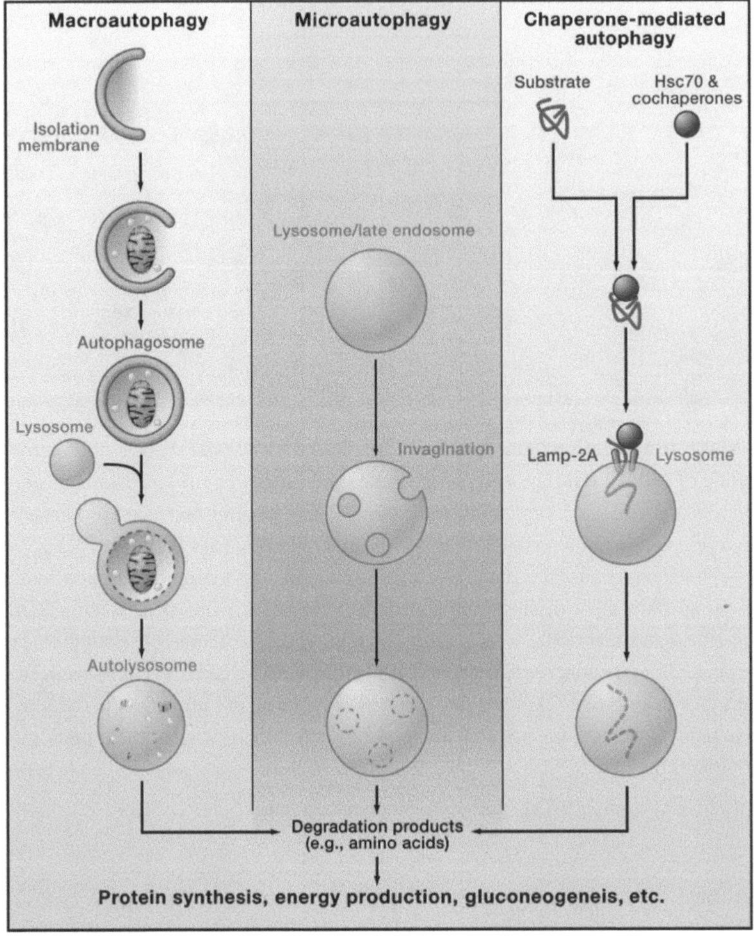

FIGURE 2.2 Different types of autophagy.

***Source*: Mizushima and Komatsu (2011).**

adenosine triphosphate, a strong presence of autophagic vacuoles was observed, which was found to decline with prolonged treatment times, tested up to six hours. After three hours of adenosine triphosphate treatment, the cell viability was significantly reduced to almost 50 percent, suggesting enhanced apoptosis. This result suggests that induction of autophagy may be targeted for therapeutics, resulting in apoptosis.

Autophagy is a cellular process that is partly driven by the membrane and regulated tightly by membrane-associated proteins. Some details on this matter are provided in a model demonstration in Figure 2.3.

Using the model demonstration in Figure 2.3, we see that cells go through a few steps in order to complete an autophagy cycle involving various membrane-associated proteins (Li et al., 2021). These are as follows:

1. autophagy initiation (signals activate autophagy) and nucleation of the phagophore/isolation membrane (IM, another name of the phagophore)
2. phagophore elongation
3. closure to form the autophagosome

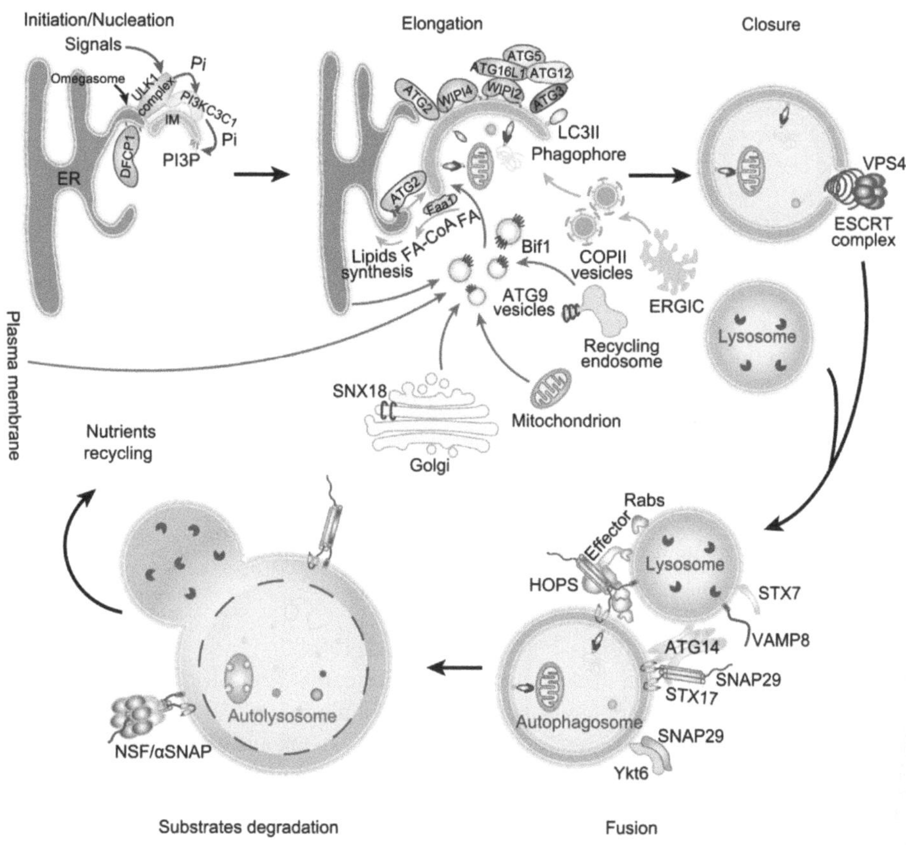

FIGURE 2.3 Autophagy and roles of membrane-associated proteins are highlighted. For details see Li et al. (2021). Color contrast is distinguishable in the digital version of the book.

***Source*: Li et al. (2021).**

4. fusion between the autophagosome and lysosome
5. degradation of substrates in autolysosomes

Figure 2.3 demonstrates a few functional steps that autophagy is seen to follow. Autophagy begins when cells sense the stimulation signals. The omegasome (a PI3P-enriched subdomain of ER where DFCP1 localizes through binding to PI3P) is the platform for the nucleation of the phagophore. This step involves two important complexes, the ULK1 complex, and the PI3KC3C1 complex. The ULK1 complex phosphorylates and activates the PI3KC3C1 complex. The activated PI3KC3C1 complex generates PI3P from PI. Then, PI3P recruits WIPIs, which in turn recruit more autophagy machinery proteins. ATG12~ATG5-ATG16L1 recruited by WIPI2 catalyze ATG3-mediated conjugation of ATG8 family proteins with membrane resident PE, generating products like LC3II, which is the characteristic signature of autophagic membranes and is involved in ATG9 vesicle sequestration of cargo. There are multiple membrane sources of the autophagosome, including endoplasmic reticulum (ER), Golgi, mitochondria, endosome, ERGIC, and plasma membrane. There are a few possible means of lipid transport, including ATG9 vesicle-mediated transport, COPII vesicle-mediated transport, ATG2-mediated lipid transport, etc. The cargos are sequestered while the autophagosomal membrane expands. Then, the sealing of this membrane structure by scission proteins such as ESCRT and other regulators gives rise to a double-membrane structure called the autophagosome. After becoming fully sealed, the autophagosome will recruit tethering proteins and SNARE proteins for fusion. Once fused, the acidic hydrolases in the lysosome degrade the autophagic cargos, salvaged nutrients are released back to the cytoplasm to be used by the cells, and the cis-SNARE complex is disassembled and recycled by NSF/αSNAP complex.

Autophagy and membrane organization/reorganization are directly associated. Cellular membranes facilitate diffusion, transfer, and controlled transport of materials and information that are vital processes of life (Ashrafuzzaman and Tuszynski, 2012a). Besides the presence of versatile chemical compositions, cellular membranes hold various physical properties that are utilized to participate in crucial cellular processes including that of autophagy. Three important membrane remodeling processes in autophagy are provided in Li et al. (2021), which may help in understanding the role of membranes in autophagy:

- lipid transfer for phagophore elongation (Figure 2.4),
- membrane scission for phagophore closure (Figure 2.5),
- autophagosome-lysosome membrane fusion (Figure 2.6).

2.1.1.2 Mitochondrial Function is Associated with Aging

Mitochondria are now standing at the epicenter of aging research. The roles of mitochondria or specifically processes of mitochondria contributing to determining the definition of life aren't totally clear to date. Mitochondrial energy metabolism and aging are so linked (Bratic and Trifunovic, 2010) that it is not surprising to construct hypotheses linking mitochondrial role and life. Mitochondria are considered one of the key regulators of longevity. Mitochondrial energy metabolism and longevity are related by two seemingly opposing theories. The "rate of living hypothesis," proposed by Pearl in 1928, finds a direct link between the metabolic output of an organism and its longevity (Pearl, 1928). The rate of living theory is now rejected as not being a valid overall explanation for why we and most other species age. Although the metabolic rate can affect aging, scientists believe that doesn't mean that it always does so. Caloric restrictions are found to help extend life in many different species, and they does so without reducing the animal's metabolic rate. The "uncoupling to survive" theory recently proposed that energy metabolism and longevity are positively correlated (Brand, 2000). This theory is based on the notion that inefficiency in the generation of mitochondrial ATP may be necessary to reduce reactive oxygen species (ROS) generation in the cell. High proton motive force drives an efficient ATP synthesis and thus comes at the cost of the production of ROS. However, the general role of mitochondria or their ATP production in determining life is still confusing. Different genetic and dietary manipulations, which are known to prolong life span, have not always shown a unique type of result. They are found to both decrease and increase ATP production in cells. The molecular mechanism behind this phenomenal dualism is not known (Bratic and Trifunovic, 2010).

It is, however, now thought that the role of mitochondria goes well beyond their dogmatically held function as the "powerhouse of the cell." The mitochondrial role may be linked to the extension of life span through activities: see the hypothetical outline schematized for humans in Figure 2.7 (for detailed description see Lanza and Nair, 2010). It is likely that mitochondrial decay that occurs with age cannot be counteracted in humans who are not on caloric restriction unless physical activity is voluntarily enhanced.

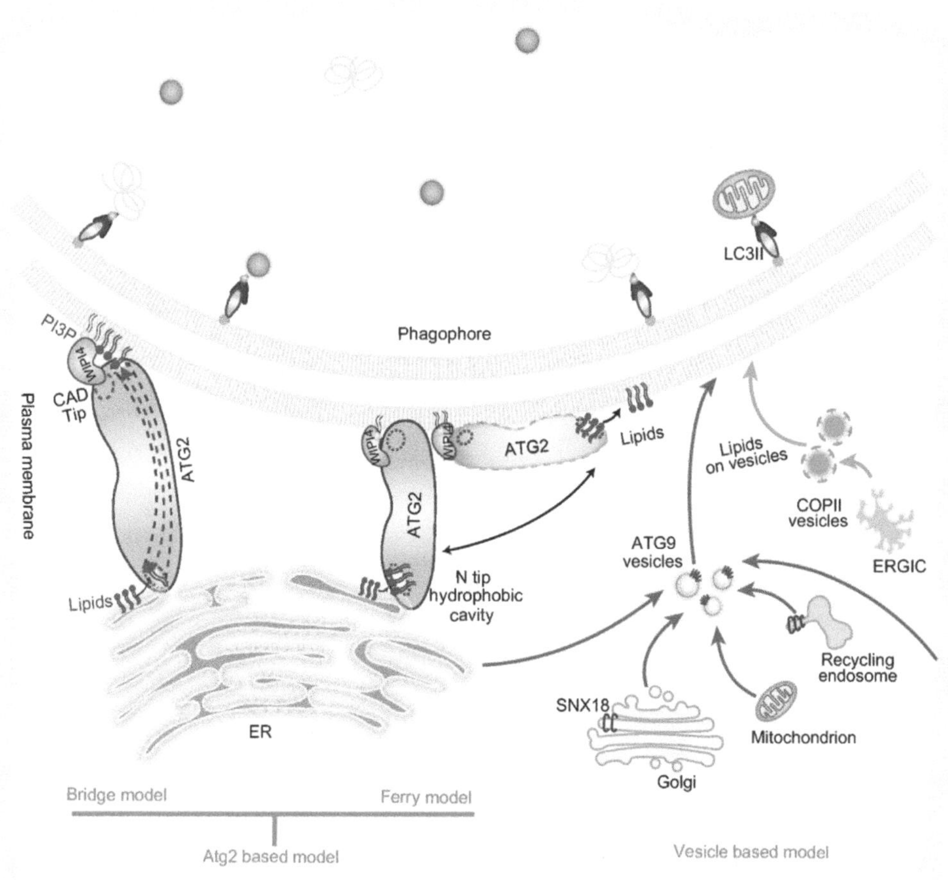

FIGURE 2.4 Hypothetic model for ATG2-mediated lipid transfer during autophagy. ATG2, ATG9-vesicles, and COPII-vesicles help to provide membrane sources for phagophore membrane elongation. ATG2 contains a CAD tip that binds to the PI3P interacting protein WIPI4 and an N-terminal hydrophobic cavity that binds to lipids. Inside the protein, there is an extended cavity or a series of cavities along the length of it. Two models are presented for the mechanism of lipid transfer for ATG2, the bridge-tunnel model (left), and the ferry model (middle). In the bridge model, ATG2 stably tethers the phagophore and ER, where its CAD tip interacts with WIPI4, a PI3P effector on the phagophore, and its N tip hydrophobic cavity binds to the ER. The lipids could transfer in the tunnel inside ATG2 from the ER to the phagophore. In the ferry model, the CAD tip anchors on the phagophore membrane through WIPI4, while the N tip hydrophobic cavity takes lipids from ER, and swings like a ferry boat between the ER and the phagophore to transfer lipids. Simultaneously, ATG9-vesicles and COPII-vesicles can act in a vesicle-mediated membrane fusion to deliver lipids from many cellular organelles to the phagophore for its elongation. For details see Li et al. (2021). Color contrast is distinguishable in the digital version of the book.

***Source*: Li et al. (2021).**

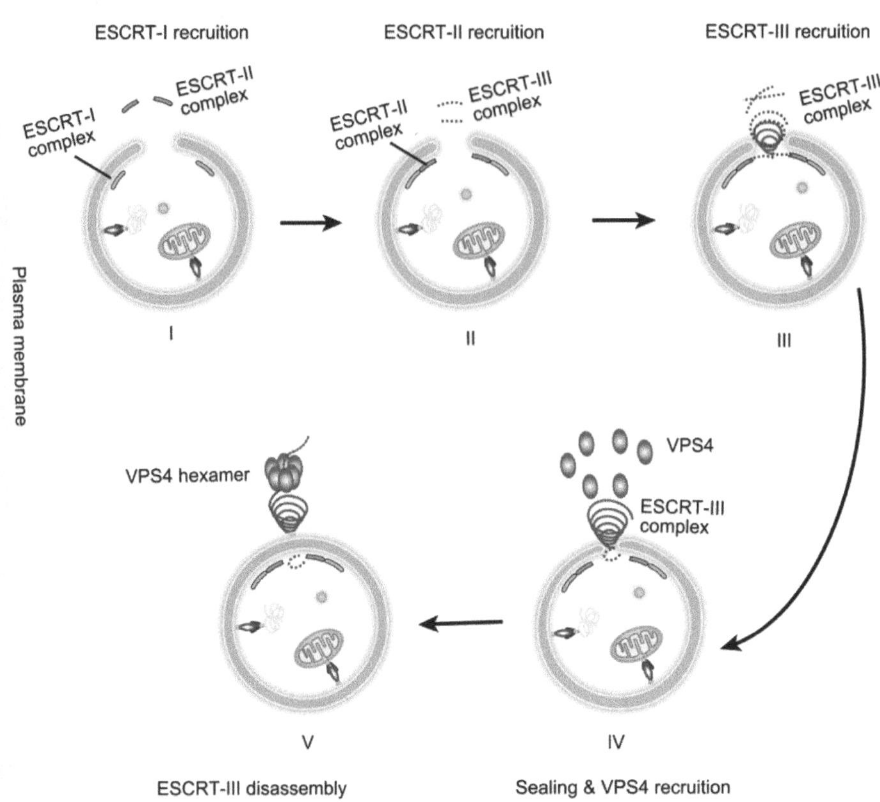

FIGURE 2.5 Hypothetic model for ESCRT-mediated autophagosome pore closure. The phagophore pore closure model was plotted according to the ESCRT-mediated membrane scission, as the pore closure might share the same regulators and mechanism with the classic ESCRT model. The first step of phagophore closure is the recruitment of the ESCRT-I complex and other related machinery to the phagophore pore. ESCRT-I recruits the ESCRT-II and ESCRT-III complex. ESCRT-III will form oligomeric filaments, flat spirals, tubes, and conical funnels to remodel the membrane shape on the neck, and finally, scissor the membrane to form a closed double-membrane autophagosome. After scission, the AAA+ ATPase VPS4 is recruited and forms a hexamer to disassemble and recycle the ESCRT-III complex. For details see Li et al. (2021). Color contrast is distinguishable in the digital version of the book.

***Source*: Li et al. (2021).**

The mitochondrial function is altered with old age and may underlie many age-related changes in physical function, protein synthesis, and muscle mass. A few of the effects of aging on mitochondrial function may be linked to the following molecular mechanisms and/or specific mitochondrial structural components:

1. **Gene expression**. Aging affects the expression of genes encoding mitochondrial proteins, evidenced by decreased messenger RNA (mRNA) transcript levels (Short et al., 2005; Barazzoni et al., 2000), possibly due to reduced gene transcription or mRNA instability with aging.

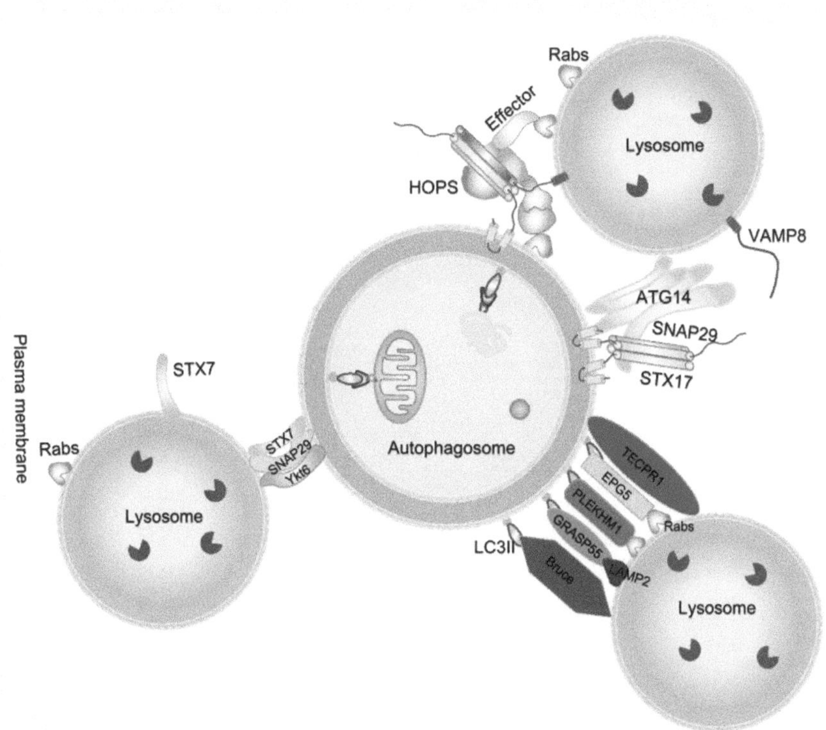

FIGURE 2.6 Model for SNARE-mediated autophagosome-lysosome fusion. Autophagosome-lysosome fusion is the key step of autophagy and is highly regulated by SNARE proteins, tethering factors, Rab GTPase, SM proteins, and other proteins. The fusion SNAREs identified so far for autophagy include STX17-SNAP29-VAMP8 and YKT6/SNAP29/STX7. Sealed autophagosome recruits the SNARE binary complex together with lysosomal SNARE protein to form a four-helix bundle to mediate autophagosome-lysosome fusion. Besides, tethering between autophagosomes and lysosomes can promote this fusion. Proteins involved in tethering include ATG14, Rab GTPase, HOPS, PLEKHM1, UVRAG, EPG5, BRUCE, TECPR1, GRASP55, etc. For details see Li et al. (2021). Color contrast is distinguishable in the digital version of the book.

***Source*: Li et al. (2021).**

2. **Mitochondrial DNA**. Mitochondrial DNA (mtDNA) copy number decreases with age (Short et al., 2005; Barazzoni et al., 2000), which could account for the reduction of mitochondrial gene transcripts and, therefore, the proteins encoded by these genes.
3. **Protein synthesis**. Both availability of the mRNA template and the rate of protein synthesis may affect protein synthesis, which declines with aging (Rooyackers et al., 1996; Jaleel et al., 2008).
4. **Whole body proteolysis**. Proteolysis, a process for a protein to break down partially, into peptides, or completely, into amino acids, decreases with aging (Henderson et al., 2009).
5. **Accumulation of dysfunctional proteins**. Aging-induced decreased protein turnover may lead to the accumulation of oxidatively damaged dysfunctional proteins, such as protein carbonylation (Hepple et al., 2008) and nitrotyrosine-modified proteins (Fugere et al., 2006).
6. **mtDNA damage**. Aging effects on mitochondrial DNA (mtDNA) abundance, gene expression, and other downstream effects may cause alterations and mutations to portions of mtDNA encoding mitochondrial proteins (Linnane et al., 1990; Zhang et al., 1993).

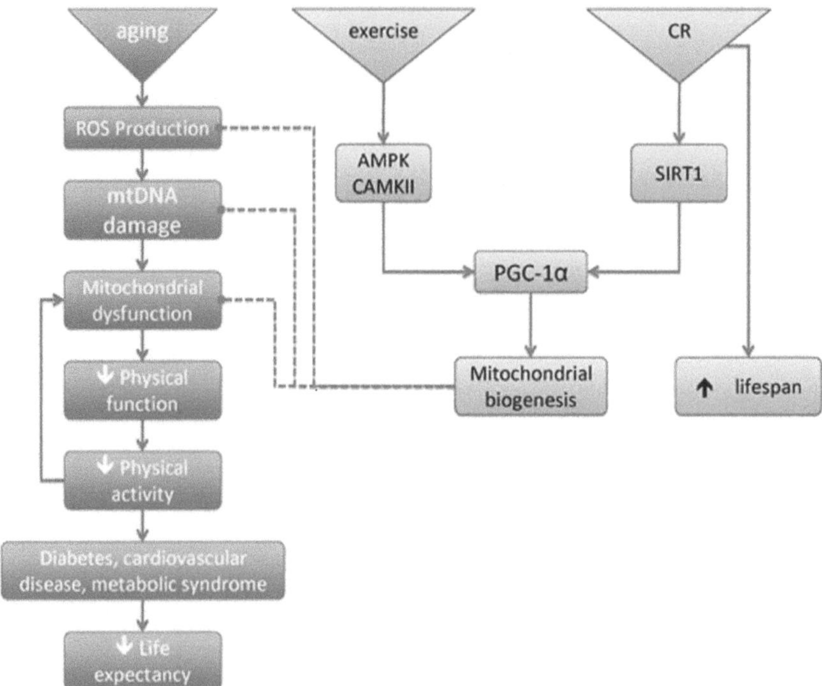

FIGURE 2.7 The free radical theory of aging considers that a senescent phenotype is induced by accumulation of oxidative damage resulting from reactive oxygen species. Exercise and caloric restriction are two interventions that induce mitochondrial biogenesis through peroxisome proliferator-activated receptor-γ coactivator-1α (PGC-1α). Although exercise and caloric restriction (CR) increase average life expectancy by protecting against age-related comorbidities, only CR has been shown to increase maximal life span; an effect that seems to require the activation of sirtuins. Here Sirtuin 1 (SIRT1).

***Source*: Lanza and Nair (2010).**

An obvious question may still be asked whether mitochondrial dysfunction is more of a cause or consequence of aging. We have just listed a few effects. However, an accelerated aging phenotype is evident in mice with increased mtDNA mutations and genetic mitochondrial diseases (Kujoth et al., 2005; Trifunovic et al., 2005).

2.1.1.3 Aging-induced Telomere Length Shortening Leads to Decreased Life Span

Life span among species is over-distributed, ranging between as low as a day and over 400 years. Although no concrete genetic explanations are found that can predict the correlation between nucleic acid structure, length or function, and aging, there is specific information (as research findings) involving telomeres emerging that might help us understand aging. There is no strong correlation between initial telomere length and the life span of any species. But a strong correlation between the telomere shortening rate and the life span of a species is reported (see Figure 2.8) (Whittemore et al., 2019; Vera et al., 2012). It is unlikely that species die when their telomeres are completely eroded since the life spans predicted by complete telomere erosion are longer than the observed life spans for most species. The telomere length, when species die at the age of the maximum life span, has been found to ~50 percent of the original telomere length for that particular species. We should consider that the rate of the annual loss of base pairs (bps/year) is different for different species (see Figure 2.8).

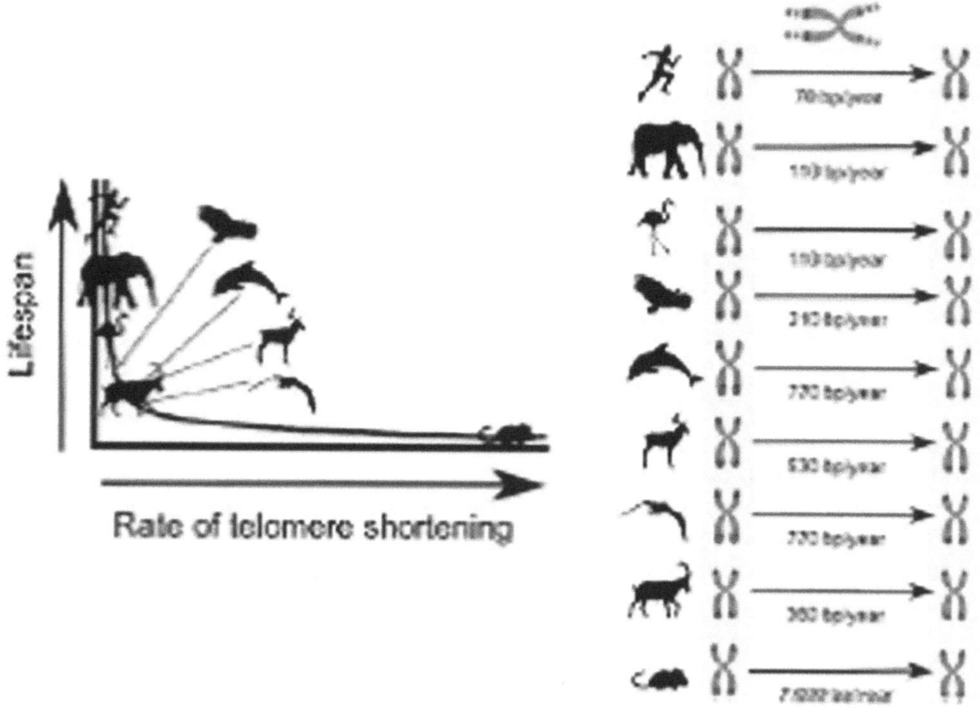

FIGURE 2.8 Graphical illustration which shows that faster telomere shortening rates (right panel) result in shorter species life spans (left panel).

***Source*: Whittemore et al. (2019).**

The following facts have been obtained so far in various research findings (Whittemore et al., 2019):

1. Life span is negatively correlated with telomere shortening rate
2. Life span is negatively correlated with the initial telomere length
3. Life span is positively correlated with body weight
4. Life span is negatively correlated with heart rate

Among all of these variables (1–4), the first one (the telomere shortening rate) appears to have the greatest power to predict life span.

2.1.1.4 An Animal's Lifespan is Written in DNAs

Analysis of DNA may help to predict an animal's life span. Recently, a new method was developed for estimating the life spans of different species by analyzing their DNA, which claimed humans have a "natural" lifespan of around 38 years (Mayne et al., 2019). Here a predictive life span clock has been derived based on cytosine-phosphate-guanosine (CpG) density in a selected set of promoters. Aging is associated with several epigenetic changes involving DNA methylation (DNAm) (Booth and Brunet, 2016). DNAm of CpG sites involves a covalent modification to cytosine to form 5-methylcytosine. This modification to DNA appears to have the potential to regulate gene expression, including of genes critical for longevity, without altering the underlying sequence. Involving CpG while creating a life span clock is therefore an excellent strategy. To estimate life span from CpG density, maximum life span data available from AnAge were used (Tacutu et al., 2018).

The life span clock (Mayne et al., 2019) accurately predicts the maximum life span in vertebrates from the CpG site density within only 42 selected promoters. This clock provides a new method for accurately estimating life span using genome sequences alone and enables the estimation of this challenging parameter for both poorly understood and extinct species. This method looked at how DNA changes as an animal ages. The study found that it varies from species to species and is related to the length of the animal's longevity. Extrapolating from genetic studies of species with known life spans, it found that the extinct woolly mammoth probably lived around 60 years and bowhead whales can expect to live more than 250 years.

2.1.2 The Non-Cellular Processes of Life

Life which exists without requiring any cellular structures is considered non-cellular life. Any organism is traditionally considered to have a cell, although cell structures can vary considerably, in order to hold life. This excludes things like viruses from a list of "living" organisms. However, this classification might be erroneous, and viruses could be considered life forms of "non-cellular life" to distinguish them from cellular life like bacteria, fungi, archaea, Protists, animals, and plants.

Traditionally viruses cannot be considered alive, because they don't hold all of the seven processes of life, namely movement, sensitivity, respiration, nutrition, excretion, reproduction, and growth. But viruses hold genetic information coded in DNA or RNA, an essential characteristic of all cellular life. So the question remains whether viruses can be considered "alive." Despite the lack of ability to move, grow, convert nutrients into energy, or excrete waste products, viruses reproduce (in a host living cell), infect people, and cause illnesses. During replication, viruses recruit cellular ribosomes for translating viral mRNAs. To translate mRNAs, viruses use various strategies, pirated from their cellular hosts, to obtain key factors helping to initiate, elongate, and terminate the translation (Walsh and Mohr, 2011). Regarding causing damage, viruses may neutralize translation machinery in infected cells, thus forcing the host cells to compromise normal cellular functioning.

The discovery of giant mimiviruses, two decades ago (Scola et al., 2003), added further details to our understanding of non-cellular life's additional complexities. *Acanthamoeba polyphaga* mimivirus (APMV) was first observed in amoebae resembling Gram-positive bacteria associated with pneumonia. These giant viruses of amoebae appear to have genetic, proteomic, and structural complexities, thought not to exist among the microscopic viruses, comparable to those of bacteria, archaea, and small eukaryotes. They contain mRNA and more than 100 proteins, and they have gene repertoires that are broader than those of other viruses and, notably, some encode translation components (Colson et al., 2017). These biomaterials and processes resemble some major characteristics of cell-based life processes. Overall, giant viruses of amoebae, including mimiviruses, marseilleviruses, pandoraviruses, pithoviruses, faustoviruses, and molliviruses, challenge the definition and classification of viruses, and are now detected increasingly often in humans. The existence of genes in these giant viruses responsible for translation, metabolism, DNA repair, and protein folding raises timely questions about the history and evolution of viruses. So non-cellular life's complexities, e.g. as found in viruses, are yet to be understood, requiring advanced research in the future.

2.2 STATES OF PHYSICAL STRUCTURES OF CELLULAR SYSTEMS ARE ASSOCIATED WITH LIFE

Life's processes mostly exist in biological cells. The cell is full of structures, molecular machinery, and processes. Many of these biological structures and associated functions are understood from their biochemical structural classifications and the perspectives of their physical states. When it comes to understanding matter utilizing its physical state properties, state stability appears as an important parameter. Cellular structures, such as membranes, proteins, nucleic acids, etc. are dynamic entities. Their structures always fluctuate considering local energetics and global environmental conditions, including especially biological thermodynamics. Physical structural stability is therefore associated with cellular functions which are linked to determining the basic parameters of life (addressed in an earlier section). In the previous section, we learned that various cell-based processes are associated with various sections within cells. In understanding these cellular and subcellular systems regarding their structures and functions that are associated with life, we probably need to consider various associated parameters, such as energy, entropy, stability, etc. Erwin Schrödinger, in "What is Life?," explained the purpose of life as to rely on creating entropy, and therefore concluded living things should not be considered as just "self-reproducing" entities, as all living cells involve more than just replication of DNA (Schrödinger, 1945). Life may be defined as an organized matter that provides genetic information metabolism (Tetz and Tetz, 2019). To understand life correctly, we need to explore the various aspects of associated cellular structures as physical entities. Cell metabolism is popularly characterized by the fundamental concept of energetics (Fernandez-de-Cossio-Diaz and Vazquez, 2018). Information on physical states, energies, entropies, and stabilities of cell-based physical structures helps us understand the biology of the cell, hence life. As an example, we may briefly discuss autophagy from the physical perspective, besides its versatile biological perspectives (Yim and Mizushima, 2020; Parzych and Klionsky, 2014), and try to understand how certain physical properties, e.g. energy, may help us understand crucial aspects of life.

In mammalian cells, macroautophagy relies on de novo formation of cytosolic double-membrane vesicles, autophagosomes, to sequester and transport cargo to the lysosome. Chaperone-mediated autophagy transports individual unfolded proteins directly across the lysosomal membrane. Microautophagy involves the direct uptake of cargo through the invagination of the lysosomal membrane. All three types of autophagy (see Figure 2.2) lead to the degradation of cargo and release of the breakdown products back into the cytosol for reuse by the cell.

All three types of autophagy (Figure 2.2) complete various structural alteration phases in a thermodynamic condition. These structural phase transitions undergo energetic and entropic changes. Figure 2.9 demonstrates how different catabolic pathways, e.g. in macroautophagy, converge in lysosomes, and that while promoting metabolic activity through the breakdown of complex molecules (proteins, lipids, etc.), the release of energy is ensured.

Besides various protein breakdowns, autophagy is especially known to contribute to the mobilization of diverse cellular energy stores (Singh and Cuervo, 2011). We shall perhaps agree without any doubt that muscle fitness can be maintained using proper exercise, besides maintaining an appropriate nutrition supply for the body. Muscle fitness and autophagy are found to be associated through the control of energy metabolism (Sebastián and Zorzano, 2020). The metabolic cues, originating due to nutrient conditions and physiological conditions such as muscle contraction and exercise, appear as modulators of autophagy in the muscle. Functional autophagy is necessary for correct metabolic adaptation, muscle function, and exercise performance. The metabolic pathways controlling autophagy in the plastic tissue "skeletal muscle" undergo major changes in energy demands. The alteration in the autophagy-induced regulation of energy metabolism may affect energy homeostasis in the body, leading to the development of metabolic diseases and aging. From this analysis we may conclude that energetics inside cellular systems is an essential process associated with autophagy, thus linked to defining life.

2.3 ENERGETIC AND ENTROPIC FLUCTUATIONS OF CELLULAR SYSTEMS LINKED TO THE MILIEU OF LIFE

On a molecular scale level, the cell is considered a vast metropolis, inhabited by billions of throbbing machines, interacting with trillions of molecules producing continuous chaos (Brown, 1999). Life is, therefore, projected at and from these multitudes of chaos in general, and energetics and entropic

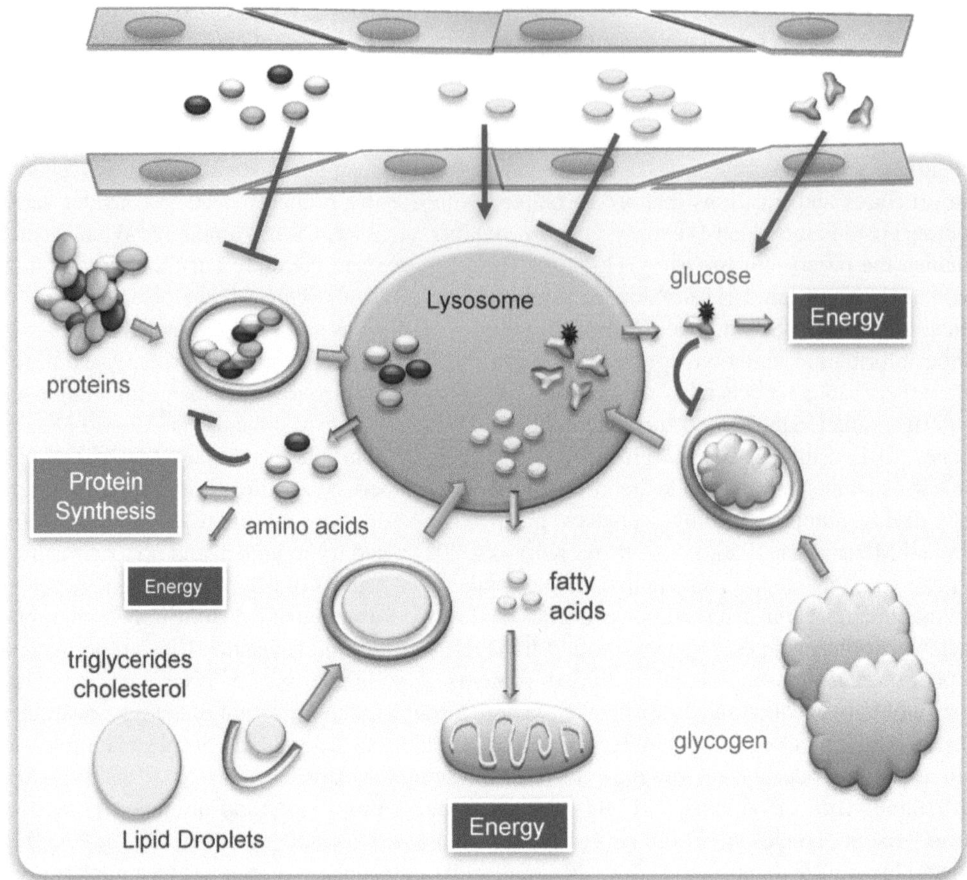

FIGURE 2.9 Different catabolic pathways converge in lysosomes. Macroautophagy contributes to the delivery of proteins, lipid stores, and glycogen for breakdown into lysosomes. The constituent components of these macromolecules exit the lysosome and become available for production of energy. In the case of protein breakdown, the resulting amino acids may have less energetic value and be preferentially utilized for the synthesis of new proteins. Levels of amino acids, free fatty acids, and sugars circulating in blood or in the extracellular media have a direct impact on intracellular macroautophagy. The role of "energy" is especially demonstrated in various phases of the autophagy. For details, see Singh and Cuervo (2011).

***Source*: Singh and Cuervo (2011).**

fluctuation are involved in the associated chaotic systems and underlying processes. Chaos, or exponential sensitivity of the system to even modest perturbations in the components' structures or functions, may appear everywhere in nature, including living systems (Toker et al., 2020). Life's status thus originated partly at the scale of chaos in cellular systems.

The laws of thermodynamics are known not to pose major restrictions to the onset of overall order in biological systems (Pontes, 2016). The construction of living matter is therefore not made at the expense of significant decreases of entropy (Blumenfeld, 1981), thus living matter is found not to present an "anti-entropic" tendency. Entropy is not known to capture the intuitive idea of order we find in living beings. However, the fact is that entropy decreases due to the establishment of ordering of living systems. It is now predicted that the entropy decrease may be associated with physical and chemical processes within the systems in order to fulfill the overall thermodynamic restrictions.

Biological systems are known to self-organize at low thermodynamic costs. In cancer, molecular changes can be quantified in terms of entropy gain or information loss (Frieden and Gatenby, 2011; Davies et al., 2012). The entropic changes can affect embryogenesis at the intracellular reorganization level of the cytoskeleton, and various biochemical interactions there can overcome entropic disorder (Tuszynski and Gordon, 2012).

The energetics of molecular structural alterations, collisions, interactions, etc. are natural cellular processes. Association/dissociation among ion channel building blocks, channel-membrane building blocks, etc. are prime examples of cellular processes where energetics appears as the determinant process of underlying phenomena. Specific measurements of the amount of energy flow can help achieve an understanding of the structural changes (Andersen and Koeppe, 2007; Ashrafuzzaman and Tuszynski, 2012b; Ashrafuzzaman et al., 2014). Binding energies due to creation or breakup of actual bonds (Gohlke et al., 2003; Horowitz and Trievel, 2012; Herschlag and Pinney, 2018) and due to both short- and long-range active interactions (Ashrafuzzaman et al., 2014; Ashrafuzzaman et al., 2020), etc. contribute to energetics taking place inside cells. In this kind of energetics, we discovered that consideration of only the mechanical properties of the participating biological structures isn't sufficient, as considered by Andersen and Koeppe (2007). Instead, in Ashrafuzzaman and Tuszynski (2012a; 2012b), it has been discovered that the charge properties of the participating biomolecules (specifically, the distribution of polarized electric charges in the associated structural complex) contribute primarily while the mechanical properties appear to be secondary contributors to the energetics of association-dissociation of biomolecules. Most of these molecular bonding-based and interaction-based energetic changes happen within a few to less than a hundred kcal/mole free energy flow.

Recently, a study has provided an interesting approach to simulate protein–ligand binding and measure the free energy of binding (Duan et al., 2016). In addressing protein–ligand binding free energy the main challenge lies in the calculation of entropic contribution to protein–ligand binding or interaction systems. A new interaction entropy method has been proposed (Duan et al., 2016) combining rigorous theoretical analysis and efficient computations, helping to deduce reliable numerical parameters to calculate entropic contribution to free energy in protein–ligand binding and other possible biomolecular interaction processes. The normal mode method is usually employed for calculating entropy change in protein–ligand binding, but this method is found to be extremely expensive, whereas the new method, advertised as vastly superior to the standard normal mode approach, calculates the entropic component, which may be considered as "interaction entropy" ($= -T\Delta S$), of the binding free energy directly from molecular dynamics simulation, involving no additional computational costs. Here T is the absolute temperature representing the thermodynamic condition of the biological system and ΔS is the physical entropic change. After performing an extensive study of over a dozen randomly selected protein–ligand binding systems, the interaction entropy, as derived from the free energy of interaction (ΔE_{int}), has been found to be correlated with simulation time (see Figure 2.10). This plot suggests that after achieving equilibrium condition the interaction entropy is found to be quite deterministic. The calculated free energy of biomolecular association using the theoretical approach that considers screened Coulomb interactions (SCIs) among distributed charges in a biomolecule complex (Ashrafuzzaman and Tuszynski, 2012a; 2012b) measures energetics which is identical to ΔE_{int} measured in simulations (Duan et al., 2016). Therefore, it is understandable that the mentioned theoretical approach relying on screened Coulomb interactions may also appear as a powerful method to address the physics of biomolecular interactions. This method helps produce universal parameters that are related to life. For example, the state of entropy of the biomolecular systems and fluctuations thereof due to association/dissociation among biomolecules may be understood indirectly.

In any protein–protein interactions, the recent trend is to determine both binding free energies and entropic contributions (Gohlke et al., 2003) to understand the underlying molecular processes correctly, as these hallmark parameters dictate the transitions between these distinguishable physical

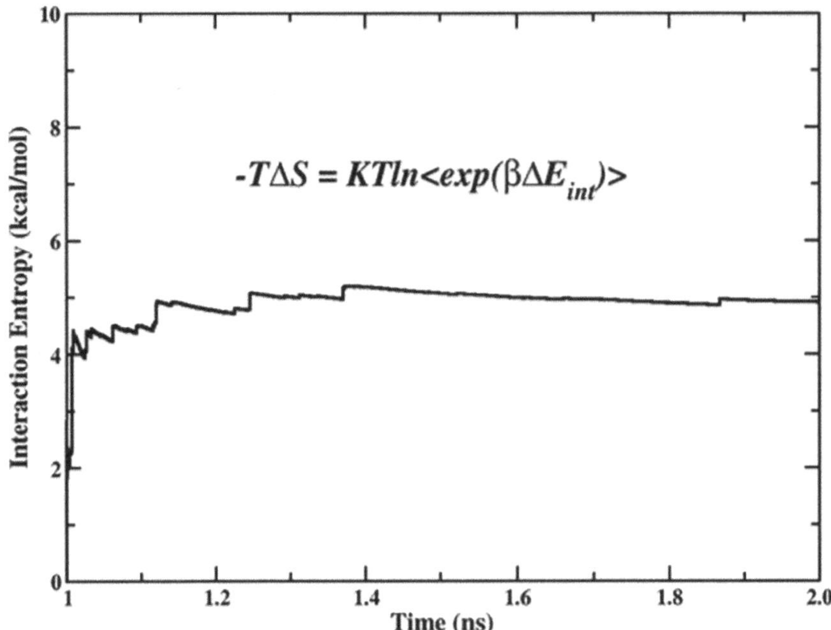

FIGURE 2.10 Interaction entropy progression with time.

Source: Duan et al. (2016).

states, namely the free protein state(s) and their associated one(s). The alterations (mostly toward reducing mode) in the modes of localized protein structural fluctuation may also appear as considerable contributors to the energetics of protein–protein binding. The conformational changes in the participating protein structures upon complex formation may trigger larger structural changes in the component proteins of the complex, hence in the energetics (Gohlke et al., 2003). Besides, hydration effects (Chong and Ham, 2017), viscous effects (Woldeyes et al., 2020), etc. contribute substantially to the energetics of biomolecular interactions.

In the molecular dynamics (MD) simulations addressing the energetics of Ras–Raf and Ras–RalGDS association and complex formation (Figure 2.11), a few bold points have been found which may help us understand the biomolecular complex creation and associated energetics using actual values of a few free energies involved.

5 ns MD simulations were carried out for the unbound proteins Ras, Raf, and RalGDS as well as for the complexes Ras–Raf and Ras–RalGDS (Figure 2.11). The following are major contributors to the binding free energies:

- Gas-phase energies (including Coulombic energy, van der Waals energy as determined by a Lennard–Jones potential, and internal energy)
- Solvation free energies (including a non-polar part and an electrostatic part)
- Entropic terms resulting from translational, rotational, and vibrational contributions

Inspecting the interface regions of the complexes, the Ras part appears with larger conformational changes compared to the effector parts. These conformational changes, even after making the complex, contribute to modulating the vibrational mode of the complex, so contribute also to energetics in the post structural association phase.

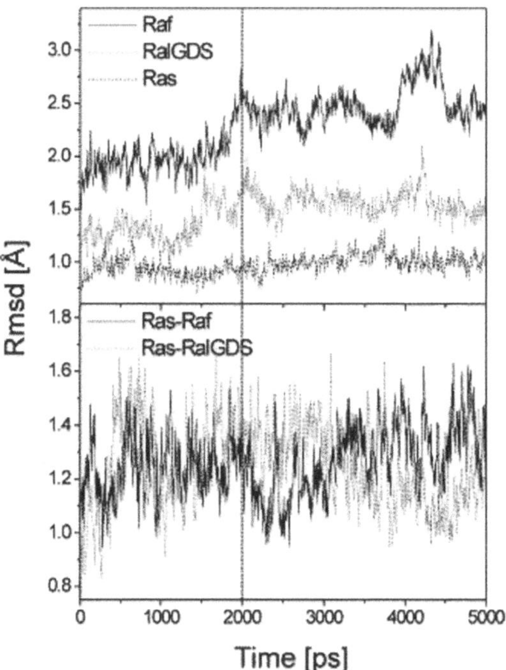

FIGURE 2.11 Time-series of rmsd of backbone atoms from the starting structures for the unbound proteins and the complexes over 5 ns of MD simulations. The equilibration phase is not included. The vertical broken line indicates the time after which snapshots were extracted for binding free energy calculations. For details see Gohlke et al. (2003).

***Source*: Gohlke et al. (2003).**

The binding free energies have been reported as –15.0(±6.3) kcal mol^{-1} for Ras–Raf, and –19.5(±5.9) kcal mol^{-1} for Ras–RalGDS in the MD simulation studies (Gohlke et al., 2003). These values appear in fair agreement with experimentally determined values for them, –9.6 kcal mol^{-1} and –8.4 kcal mol^{-1}, respectively (Rudolph et al., 2001).

Above we explained how cellular energetics (as originated from various biomolecular interactions, conformational changes, etc.) and entropic fluctuations may appear as determinants of life's processes. Behind all these we need to focus on a very important aspect of "cell metabolism," which is characterized by three fundamental energy demands, mainly to sustain cell maintenance, to trigger aerobic fermentation, and to achieve maximum metabolic rate (Fernandez-de-Cossio-Diaz and Vazquez, 2018). As said earlier, billions of molecular processes are linked to cell health conditions. All three processes making energy demand (Fernandez-de-Cossio-Diaz and Vazquez, 2018) are somehow dependent on functional energetics in molecular interaction levels. The supply and maintenance of the cellular energy demand are somehow associated with the normal functioning of parameters associated with biological life. The energy demand is somehow associated with the energy expended by various molecular motors continuously working inside cells. A model demonstration of the function of molecular motors is presented in Figure 2.12. Here the combined role of the motors is to ensure fluidization in the intracellular milieu to ensure necessary strength in cell metabolism.

Cellular metabolism relies on sets of chemical reactions that may also occur following specific energetics of physical interactions (Ashrafuzzaman and Tuszynski, 2012a; 2012b). These metabolic chemical reactions (Muchowska et al., 2020) and physical interactions (Muley and Ranjan, 2013)

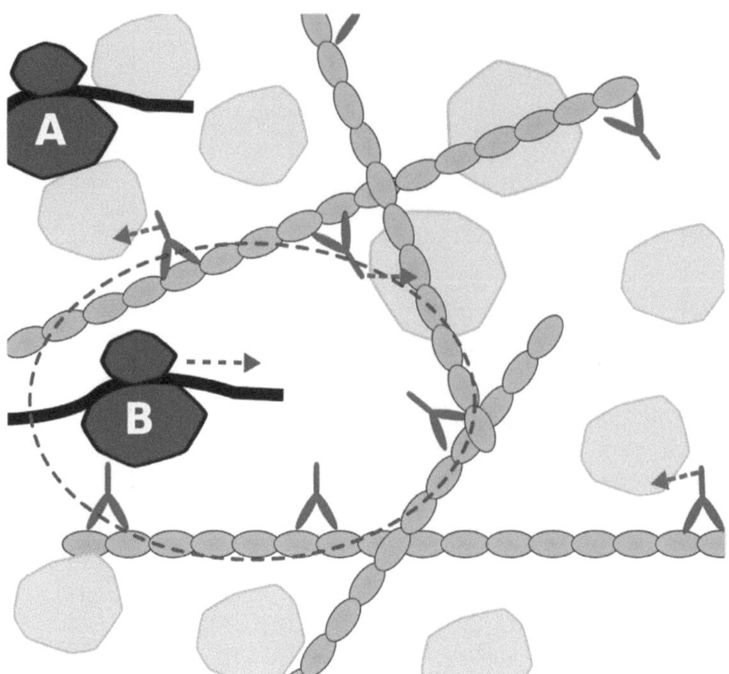

FIGURE 2.12 Molecular motors counteracting molecular crowding. Molecular crowders (gray) impede cellular processes such as protein synthesis. (A) A ribosome (blue) translating a mRNA strand must displace the crowders before it can move to the next codon. (B) Molecular motors (red) increase the propensity of cavities with lower density of molecular crowders where processes like translation can proceed freely. For details, see Fernandez-de-Cossio-Diaz and Vazquez (2018).

***Source*: Fernandez-de-Cossio-Diaz and Vazquez (2018).**

occurring in living organisms work collectively to maintain life. The cell metabolism operates in an intracellular milieu crowded with various macromolecules and organelles (van den Berg et al., 2017), and the crowding needs to maintain a chemical balance through molecular processes and interactions, as modeled in Figure 2.12. Besides maintaining the chemical balance (partly through maintaining sensible energetics of physicochemical interactions), the cell metabolism needs to address an entropic pressure as a result of the molecular crowding.

The molecular machinery associated with life operates on a background consisting of a gel-like substance having distinct properties that often differ from any ideal solutions (Bausch and Kroy, 2006). The molecular crowding mechanism hinders the movement and diffusion of large macromolecules. As a trace biomolecule is confined by the surrounding biomolecules, its movement is impeded, so a physical hole needs to be created to make some space so that the trace biomolecule can move and diffuse (Figure 2.12, molecule A). The creation of such a hole is expected to lower the entropy of the surrounding system by reducing the number of microscopic configurations of crowding molecules and free space. Based on this undeniable hypothesis, the entropic pressure (P_S) associated with the creation of a hole of size equal to m sites can be estimated as (for details see Fernandez-de-Cossio-Diaz and Vazquez, 2018):

$$P_S = \frac{T\Delta S}{V_h} = \frac{k_B T}{V_c} ln \frac{\Phi}{\Phi - \varphi} \tag{2.1}$$

Here V_c is the typical volume of molecular crowders (considered as the hard spheres), $V_h = mV_c$ the hole volume, $\varphi = nV_c/V$ the excluded volume fraction by the molecular crowders, $\Phi = NV_c/V$ the maximum packing density and V the cell volume. Here we have considered a regular lattice of N sites where n hard spheres can be placed. The entropic pressure thus entails a few rules, as follows:

1. The entropic pressure diverges as φ approaches the maximum packing density Φ, indicating that at maximum packing the creation of holes becomes unfeasible.
2. The entropic pressure is inversely proportional to the crowder volume V_c.

The logarithmic dependency of entropic pressure for molecular crowding on associated parameters (Equation 2.1) may not be universally applicable for other cases, such as for molecular motor pressure (P_M), which follows a linear relation with associated parameters, as follows:

$$P_M = \frac{pFd}{12} n_M \tag{2.2}$$

Here F and d are the motor's force and displacement per kick, n_M is the number of molecular motors per cell volume, and p is the motor's persistence, a non-dimensional parameter quantifying the motors' tendency to maintain their direction of motion upon contact with macromolecules. When motors approach and come in contact with macromolecules, they can transmit a large impulse (Figure 2.12, arrows). Introducing this modification to the classical kinetic theory of gases, the motor pressure has been deducted (for details see Fernandez-de-Cossio-Diaz and Vazquez, 2018).

2.4 QUANTUM MECHANICS OF LIFE

So far I have analyzed the existing classical mechanical concepts of life. Life is now predicted to hold processes and phenomena that are explainable using quantum mechanics (Davies, 2005). In a series of lectures before publishing his book (Schrödinger, 1945), Erwin Schrödinger described how he believed that quantum mechanics, or some variant of it, would soon solve the riddle of life (Davies, 2005). This physics-based modern prediction of life, which was then one of the most investigated molecular biology problems, opened up renewed interest among scientists engaged in interdisciplinary research. The best example was the 1953 discovery of DNA structure, the first-ever complete and successful biomolecule structure model, discovered using physical modeling (Watson and Crick, 1953). Life's various processes are now accepted as relying on a multitude of quantum phenomena (Ball, 2011).

Quantum mechanical behavior was first proven to exist in biological systems that are involved in photosynthesis (Thyrhaug et al., 2018). The idea that excitonic (electronic) coherences are of fundamental importance to natural photosynthesis gained popularity when slowly dephasing quantum beats (QBs) were observed in the two-dimensional electronic spectra of the Fenna–Matthews–Olson (FMO) complex at 77 K (Engel et al., 2007; Panitchayangkoon et al., 2010). Thyrhaug and colleagues recently revisited the coherence dynamics of the FMO complex with the use of polarization-controlled two-dimensional (2D) electronic spectroscopy, which was supported by theoretical modeling (Thyrhaug et al., 2018). Here the long-lived QBs were found to be exclusively vibrational in origin, whereas the dephasing of the electronic coherences was completed within 240 fs even at 77 K (see Figures 2.13 and 2.14). It was further reported here that specific vibrational coherences are produced via vibronically coupled excited states. The presence of such states suggests that vibronic coupling is relevant for photosynthetic energy transfer. For detailed analysis, including elaboration on the theoretical expressions, readers may consult the original article of Thyrhaug et al. (2018).

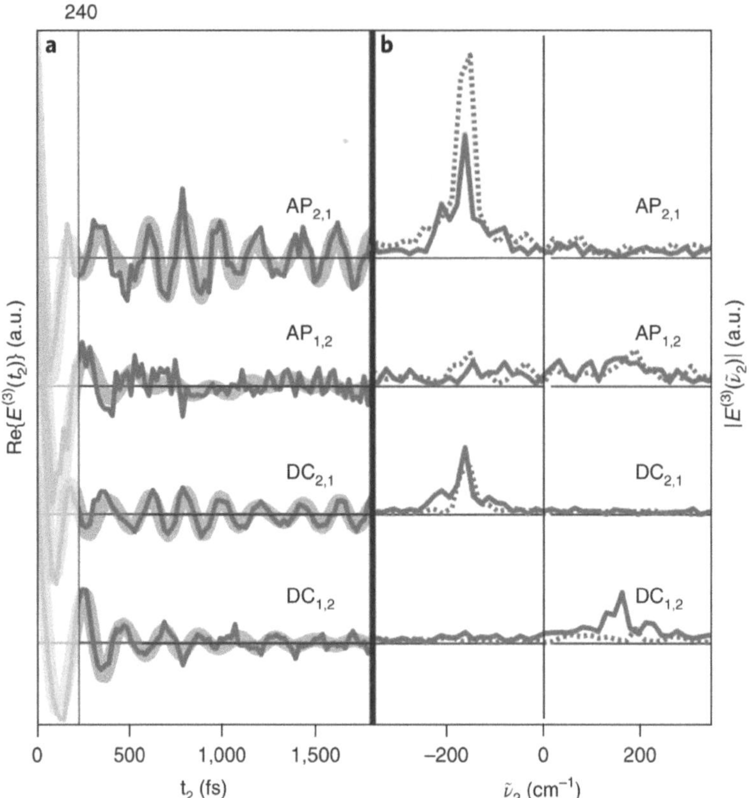

FIGURE 2.13 Selected QBs in FMO. (a) Measured real-part rephasing t_2 traces (thin lines) at the cross-peak locations labelled in Figure 2.14 after subtraction of multiexponential dynamics. Long-lived (picoseconds) QBs are clearly visible in the dynamics. Individual time-domain fits (thick lines), which correspond to the oscillatory terms in the theory (see Thyrhaug et al., 2018), are overlaid onto each trace. The traces are not normalized, but are vertically offset for clarity. (b), Fourier transform amplitudes of the experimental data (t_2 > 240 fs) shown in (a) and the theoretical vibronic exciton model data extracted from the same points (broken lines).

***Source*: Thyrhaug et al. (2018).**

It is well accepted that genetics is determinant of quality of life (Bin-Jumah et al., 2022). Recently, proton tunneling in DNA has been explained (Slocombe et al., 2022). A theoretical analysis of the hydrogen bonds between the Guanine-Cytosine (G-C) nucleotide has been made, which includes an accurate model of the structure of the base pairs, the quantum dynamics of the hydrogen bond proton, and the influence of the decoherent and dissipative cellular environment. Slocombe and colleagues determined that the quantum tunneling contribution to the proton transfer rate is several orders of magnitude larger than the classical over-the-barrier hopping. Due to the significance of the quantum tunneling even at biological temperatures, they reported the canonical and tautomeric forms of G-C inter-convert over timescales far shorter than biological ones and hence thermal equilibrium is rapidly reached.

The evolution of life and the development of disease have been explained by applying quantum mechanics in biological systems (Feng et al., 2022). I have pinpointed a few pieces of evidence of quantum phenomena active in biological systems that we have come across during recent decades. More research in this area will generate a further understanding of biological systems. We certainly

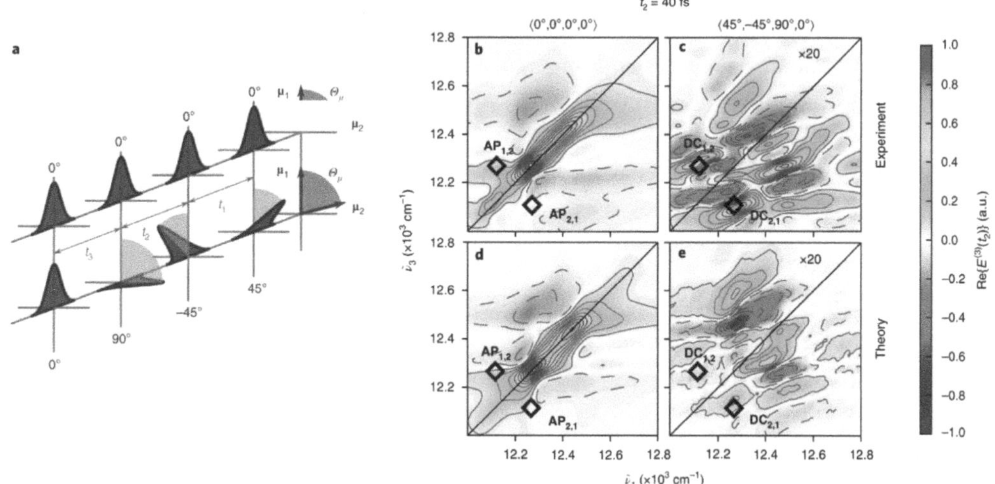

FIGURE 2.14 Polarization-controlled two-dimensional electronic spectroscopy (2DES) of the FMO complex. (a) Schematic representation of the AP pulse sequence (top), which favors interaction pathways that involve parallel transition dipoles (μ1 only), and the DC pulse sequence (bottom), which favours pathways that involve non-parallel transition dipoles (μ1 and μ2). The two pulse sequences were used to provide detection selectivity of the various QBs present in the 2D spectra of the FMO complex. (b–e) Real part of the rephasing 2D spectra at t2 = 40 fs. Experimental (b and c) and theoretical (d and e) spectra that result from AP (b and d) and DC (c and e) pulse sequences are shown. The spectra are normalized to the exciton 2 diagonal peak of the AP spectra and the DC spectra are scaled by a factor of 20 for clarity. Markers AP and DC denote the exciton 1–2 and 2–1 cross-peak positions in the appropriate spectra at which the dynamics presented in Figure 2.12 are taken.

***Source*: Thyrhaug et al. (2018).**

need to wait at least a few decades more before being able to appreciate the possible quantum mechanical understanding of life with confidence. A chapter has been dedicated to explaining the quantum mechanics of biological systems in this book. Readers may go through this chapter for enhanced understanding.

2.5 CONCLUDING REMARKS

Scientific understanding of life provided key components that could be targeted to improve both life expectancy and life span during the last century. We have discussed here various structures and processes in the living body that are associated with defining life. Energetics and entropic fluctuations in the living body are found to be related to normal and abnormal functioning, and so are participants in qualifying life from biophysical perspectives. All these life-associated biological and biophysical components of living systems are medically dealt with these days to ensure a healthy life, especially for mankind. We can now deal with mutations and also induce genetic alterations. As a result, we see an altered scenario in both the life expectancy and life span of humans.

In this chapter, I have addressed various biological, biochemical, and physical systems, phenomena, and processes of the living body that participate in determining and qualifying, and sometimes quantifying, biological life in general. Besides the traditional biological and biochemical understanding, we now believe that life requires to be understood using the obvious physical laws and principles as well as modelings and hypotheses that especially involve force, energy, and entropy. The whole living system or compartments thereof continuously experiences motion due to

TABLE 2.1
Correlation of cellular processes and mechanisms with biological life or generally speaking life expectancy. ↗ and ↘ stand for up (or favoring) and down (or disfavoring) regulation, respectively.

Cellular processes or mechanisms	Biological life (expected impact)	Primary nature of effects: biological (B), biochemical (C), biophysical (P)
Autophagy	↗	B
Apoptosis	↗	B
Aging	↘	B
Fast telomere shortening	↘	C
Slow telomere shortening	↗	C
Controlled energetics	↗	P
Controlled entropic change	↗	P

the collective actions of various types of forces and energetics. Entropic fluctuation is an undeniable reality in these chaotic systems. Therefore, the role of physical interpretation is as important as that of biological and biochemical concepts to understand life. To summarize, I wish to construct a fun table demonstrating the roles of various parameters (explained in this chapter) linked to mainly cellular or other biological systems' functions in defining biological life. Table 2.1 contains only a few parameters, but many more can be added as we progress.

The correlation of biological life with cellular processes and mechanisms, mentioned in Table 2.1, may differ among people experiencing different sociocultural, economic, and medical factors providing deterministic roles in biological life expectancy. Moreover, the correlation patterns in Table 2.1 may also flip in individual cases depending on health conditions. Every correlation may be inspected regarding all of the possible effects, as presented in column 3 (Table 2.1). Although I have proposed here the primary nature of effects, the other ones may also play important or secondary roles to create underlying molecular mechanisms.

Despite having an in-depth understanding of life using scientific studies, data, and evidence, we may also want to see the scenario from archaeological perspectives. A few years ago the BBC published a report summarizing much historical data (www.bbc.com/future/article/20181002-how-long-did-ancient-people-live-life-span-versus-longevity). This report covered many aspects, but to satisfy our curiosity, let us focus on only two parameters, life expectancy, and life span. Life expectancy experienced tremendous progress over the last centuries or millennia. In 2016, Gazzaniga published her archaeological research data on more than 2,000 ancient Roman skeletons, all working-class people who were buried in common graves (Caldarini et al., 2015). The average age of death was 30, many showing the effects of trauma from hard labor, as well as diseases we would associate with later ages, like arthritis. We know the world's average life expectancy has now reached over 70 years. This is higher (by almost 25 percent) in economically rich nations, where 100 years is considered achievable. The current child death rate is way lower than ancient rates. Discrepancies in access to foods and medicines lead to major differences among life expectancies. However, life span might not necessarily increase tremendously for individual groups over centuries or even millennia. Elites' life span in ancient times wasn't considerably different than their peers today. The Roman emperor Tiberius was killed at the age of 77 years, and the Roman empress Livia, wife of Augustus, died in her late 80s. Islam's prophet Muhammad died at 62 years. Britain's Queen Elizabeth I and Queen Victoria died at the age of 70 and 81, respectively. Here we don't see many changes over several thousand years. This indicates that elites have always maintained quality living by eating healthy foods, enjoying a peaceful life, and getting the best possible medical care. As a result, their cellular processes associated with life, as summarized in Table 2.1, may be naturally maintained. Of course, this conclusion requires more research combining various natural science, social scientific, anthropological, and archaeological approaches.

REFERENCES

Andersen, O S., & Koeppe, R.E. (2007). Bilayer thickness and membrane protein function: an energetic perspective. *Annual Rev. Biophys. Biomol. Struct.*, *36*, 107–130

Ashrafuzzaman, M., Tseng, C.Y., & Tuszynski, J.A. (2020). Charge-based interactions of antimicrobial peptides and general drugs with lipid bilayers. *Journal of Molecular Graphics and Modelling*, *95*, 107502

Ashrafuzzaman, M., Tseng, C.Y., & Tuszynski, J.A. (2014). Regulation of channel function due to physical energetic coupling with a lipid bilayer. *Biochemical and Biophysical Research Communications*, *445*(2), 463–468

Ashrafuzzaman, M., & Tuszynski, J.A. (2012a). *Membrane biophysics*. Berlin and Heidelberg: Springer. https://doi.org/10.1007/978-3-642-16105-6]

Ashrafuzzaman, M., & Tuszynski, J. (2012b). Regulation of channel function due to coupling with a lipid bilayer. *Journal of Computational and Theoretical Nanoscience*, *9*(4), 564–570

Ball, P. (2011). Physics of life: the dawn of quantum biology. *Nature*, *474*, 272–274. https://doi.org/10.1038/474272a

Barazzoni, R, Short, K.R., & Nair, K.S. (2000). Effects of aging on mitochondrial DNA copy number and cytochrome c oxidase gene expression in rat skeletal muscle, liver, and heart. *J. Biol. Chem.*, *275*, 3343–3347

Bausch, A.R., & Kroy, K. (2006). A bottom-up approach to cell mechanics. *Nature Physics*, *2*, 231–238

Bin-Jumah, M.N., Nadeem, M.S., Gilani, S.J., Al-Abbasi, F.A., Ullah, I., Alzarea, S. I., … & Kazmi, I. (2022). Genes and longevity of lifespan. *International Journal of Molecular Sciences*, *23*(3), 1499

Blumenfeld, L.A. (1981). *Problems of biological physics*. Berlin: Springer.

Booth, L.N., & Brunet, A. (2016). The aging epigenome. *Molecular Cell*, *62*, 728–744. https://doi.org/10.1016/j.molcel.2016.05.013

Brand, M.D. (2000). Uncoupling to survive? The role of mitochondrial inefficiency in ageing. *Exp. Gerontol.*, *35*, 811–820

Bratic, I., & Trifunovic, A. (2010). Mitochondrial energy metabolism and ageing. *Biochimica et Biophysica Acta (BBA) – Bioenergetics*, *1797*(6–7), 961–967

Brown, G.C. (1999). *The energy of life: the science of what makes our minds and bodies work*. New York: Free Press

Caldarini, C., Catalano, P., Gazzaniga, V., Marinozzi, S., & Zavaroni, F. (2015). The study of ancient bone remains. In A. Piccioli, V. Gazzaniga, and P. Catalano, *Bones*, pp. 3–38. Cham: Springer. https://doi.org/10.1007/978-3-319-19485-1_1

Chong, S.H., & Ham, S. (2017). Dynamics of hydration water plays a key role in determining the binding thermodynamics of protein complexes. *Sci. Rep.*, *7*, 8744. https://doi.org/10.1038/s41598-017-09466-w

Colson, P., La Scola, B., Levasseur, A., et al. (2017). Mimivirus: leading the way in the discovery of giant viruses of amoebae. *Nat. Rev. Microbiol.*, *15*, 243–254

Davies, P. (2005). A quantum recipe for life. *Nature*, *437*, 819. https://doi.org/10.1038/437819a

Davies, P.C., Demetrius, L., & Tuszynski, J.A. (2012). Implications of quantum metabolism and natural selection for the origin of cancer cells and tumor progression. *AIP Adv.*, *2*, 011101

DiLoreto, R., & Murphy, C.T. (2015). The cell biology of aging. *Molecular Biology of the Cell*, *26*(25), 4524–4531

Duan, L., Liu, X., & Zhang, J.Z. (2016). Interaction entropy: a new paradigm for highly efficient and reliable computation of protein–ligand binding free energy. *Journal of the American Chemical Society*, *138*(17), 5722–5728

Engel, G.S., Calhoun, T.R., Read, E.L., Ahn, T.K., Mančal, T., Cheng, Y.C., … & Fleming, G.R. (2007). Evidence for wavelike energy transfer through quantum coherence in photosynthetic systems. *Nature*, *446*(7137), 782–786

Feng, Y., He, D., Yao, Z., & Klionsky, D.J. (2014). The machinery of macroautophagy. *Cell Res.*, *24*, 24–41

Feng, J., Song, B., & Zhang, Y. (2022). Semantic parsing of the life process by quantum biology. *Progress in Biophysics and Molecular Biology*, *175*, 79–89

Fernandez-de-Cossio-Diaz, J., & Vazquez, A. (2018). A physical model of cell metabolism. *Sci. Rep.*, *8*, 8349. https://doi.org/10.1038/s41598-018-26724-7

Frieden, B.R., & Gatenby, R.A. (2011). Information dynamics in living systems: prokaryotes, eukaryotes, and cancer. PLoS ONE, *6*(7), e22085

Fugere, N.A., Ferrington, D.A., & Thompson, L.V. (2006). Protein nitration with aging in the rat semimembranosus and soleus muscles. *Journals of Gerontology Series A: Biological Sciences and Medical Sciences*, *61*(8), 806–812

Gohlke, H., Kiel, C., & Case, D.A. (2003). Insights into protein–protein binding by binding free energy calculation and free energy decomposition for the Ras–Raf and Ras–RalGDS complexes. *Journal of Molecular Biology*, *330*(4), 891–913

Henderson, G.C., Dhatariya, K., Ford, G.C., Klaus, K.A., Basu, R., Rizza, R.A., … & Nair, K.S. (2009). Higher muscle protein synthesis in women than men across the lifespan, and failure of androgen administration to amend age-related decrements. *FASEB Journal*, 23(2), 631

Hepple, R.T., Qin, M., Nakamoto, H., & Goto, S. (2008). Caloric restriction optimizes the proteasome pathway with aging in rat plantaris muscle: implications for sarcopenia. *American Journal of Physiology – Regulatory, Integrative and Comparative Physiology*, *295*(4), R1231–R1237

Herschlag, D., & Pinney, M.M. (2018). Hydrogen bonds: simple after all? *Biochemistry*, *57*(24), 3338–3352

Hooke, R. (1665). *Micrographia: or some physiological descriptions of minute bodies made by magnifying glasses. With observations and inquiries thereupon.* London: Royal Society

Hornyak, T. (2017). Profile: Yoshinori Ohsumi: the rise and rise of a biology superstar. *Nature*, 543, S19. https://doi.org/10.1038/543S19a

Horowitz, S., & Trievel, R.C. (2012). Carbon-oxygen hydrogen bonding in biological structure and function. *Journal of Biological Chemistry*, *287*(50), 41576–41582

Jaleel, A., Short, K.R., Asmann, Y.W., Klaus, K.A., Morse, D.M., Ford, G.C., & Nair, K.S. (2008). In vivo measurement of synthesis rate of individual skeletal muscle mitochondrial proteins. *Am. J. Physiol Endocrinol Metab.*, *295*, E1255–E1268

Klionsky, D.J. (2008). Autophagy revisited: a conversation with Christian de Duve. *Autophagy*, *4*(6), 740–743. doi: 10.4161/auto.6398. PMID 18567941

Kozawa, S., Akanuma, T., Sato, T., et al. (2016). Real-time prediction of cell division timing in developing zebrafish embryo. *Sci. Rep.*, *6*, 32962. https://doi.org/10.1038/srep32962

Kujoth, G.C., Hiona, A., Pugh, T.D., Someya, S., Panzer, K., Wohlgemuth, S.E., … & Prolla, T.A. (2005). Mitochondrial DNA mutations, oxidative stress, and apoptosis in mammalian aging. *Science*, *309*(5733), 481–484

Lanza, I.R., & Nair, K.S. (2010). Mitochondrial function as a determinant of life span. *Pflügers Archiv – Eur. J. Physiol.*, *459*, 277–289. https://doi.org/10.1007/s00424-009-0724-5

Li, L., Tong, M., Fu, Y., et al. (2021). Lipids and membrane-associated proteins in autophagy. *Protein Cell*, *12*, 520–544. https://doi.org/10.1007/s13238-020-00793-9

Li, Y., Pan, A., Wang, D.D., Liu, X., Dhana, K., Franco, O.H., … & Hu, F.B. (2018). Impact of healthy lifestyle factors on life expectancies in the US population. *Circulation*, *138*(4), 345–355

Linnane, A.W., Baumer, A., Maxwell, R.J., Preston, H., Zhang, C.F., & Marzuki, S. (1990). Mitochondrial gene mutation: the ageing process and degenerative diseases. *Biochemistry International*, *22*(6), 1067–1076

Lu, N., Wang, B., Deng, X., Zhao, H., Wang, Y., & Li, D. (2014). Autophagy occurs within an hour of adenosine triphosphate treatment after nerve cell damage: the neuroprotective effects of adenosine triphosphate against apoptosis. *Neural Regen. Res.*, *9*(17), 1599–605. doi: 10.4103/1673-5374.141811; PMID: 25368646; PMCID: PMC4211201

Madeo, F., Zimmermann, A., Maiuri, M.C., & Kroemer, G. (2015). Essential role for autophagy in life span extension. *Journal of clinical Investigation*, *125*(1), 85–93

Mayne, B., Berry, O., Davies, C., et al. (2019). A genomic predictor of lifespan in vertebrates. *Sci. Rep.*, *9*, 17866. https://doi.org/10.1038/s41598-019-54447-w

Mizushima, N., & Komatsu, M. (2011). Autophagy: renovation of cells and tissues. *Cell, 147*(4), 728–741. doi: 10.1016/j.cell.2011.10.026; PMID: 22078875

Muchowska, K.B., Varma, S.J., & Moran, J. (2020). Nonenzymatic metabolic reactions and life's origins. *Chemical Reviews, 120*(15), 7708–7744

Muley, V.Y., & Ranjan, A. (2013). Evaluation of physical and functional protein-protein interaction prediction methods for detecting biological pathways. *PLoS One, 8*(1), e54325

Ohlhorst, S.D., Russell, R., Bier, D., Klurfeld, D.M., Li, Z., Mein, J.R., ... & Konopka, E. (2013). Nutrition research to affect food and a healthy life span. *Journal of Nutrition, 143*(8), 1349–1354

Panitchayangkoon, G., Hayes, D., Fransted, K.A., Caram, J.R., Harel, E., Wen, J., ... & Engel, G.S. (2010). Long-lived quantum coherence in photosynthetic complexes at physiological temperature. *Proceedings of the National Academy of Sciences, 107*(29), 12766–12770

Parzych, K.R., & Klionsky, D.J. (2014). An overview of autophagy: morphology, mechanism, and regulation. *Antioxid. Redox Signal., 20*(3), 460–473. doi: 10.1089/ars.2013.5371; PMID: 23725295; PMCID: PMC3894687

Pearl, R. (1928). *The rate of living, being an account of some experimental studies on the biology of life duration.* New York: Alfred A. Knopf

Pontes, J. (2016). Determinism, chaos, self-organization and entropy. *Anais da Academia Brasileira de Ciências, 88*(2), 1151–1164

Rooyackers, O.E., Adey, D.B., Ades, P.A., & Nair, K.S. (1996). Effect of age on in vivo rates of mitochondrial protein synthesis in human skeletal muscle. *Proc. Natl. Acad. Sci. USA, 93*, 15364–15369

Rudolph, M.G., Linnemann, T., Grünewald, P., Wittinghofer, A., Vetter, I.R., & Herrmann, C. (2001). Thermodynamics of Ras/effector and Cdc42/effector interactions probed by isothermal titration calorimetry. *Journal of Biological Chemistry, 276*(26), 23914–23921

Schrödinger, E. (1945). What is Life? The physical aspect of the living cell. *American Naturalist, 79*, 554–555

Scola, B.L., Audic, S., Robert, C., Jungang, L., de Lamballerie, X., Drancourt, M., ... & Raoult, D. (2003). A giant virus in amoebae. *Science*, 299(5615), 2033

Sebastián, D., & Zorzano, A. (2020). Self-eating for muscle fitness: autophagy in the control of energy metabolism. *Dev. Cell., 54*(2), 268–281. doi: 10.1016/j.devcel.2020.06.030; PMID: 32693059

Short, K.R., Bigelow, M.L., Kahl, J., Singh, R., Coenen-Schimke, J., Raghavakaimal, S., & Nair, K.S. (2005). Decline in skeletal muscle mitochondrial function with aging in humans. *Proceedings of the National Academy of Sciences, 102*(15), 5618–5623

Singh, R., & Cuervo, A.M. (2011). Autophagy in the cellular energetic balance. *Cell Metab., 13*(5), 495–504. doi: 10.1016/j.cmet.2011.04.004; PMID: 21531332; PMCID: PMC3099265

Slocombe, L., Sacchi, M. & Al-Khalili, J. (2022). An open quantum systems approach to proton tunnelling in DNA. *Commun. Phys., 5*, 109. https://doi.org/10.1038/s42005-022-00881-8]

Tacutu, R., Thornton, D., Johnson, E., Budovsky, A., Barardo, D., Craig, T., ... & De Magalhães, J.P. (2018). Human ageing genomic resources: new and updated databases. *Nucleic Acids Research, 46*(D1), D1083–D1090

Tetz, V.V., and Tetz, G.V. (2019). A new biological definition of life. *Biomolecular Concepts, 11*(1), 1–6. https://doi.org/10.1515/bmc-2020-0001

Thyrhaug, E., Tempelaar, R., Alcocer, M.J.P., et al. (2018). Identification and characterization of diverse coherences in the Fenna–Matthews–Olson complex. *Nature Chem., 10*, 780–786. https://doi.org/10.1038/s41557-018-0060-5

Toker, D., Sommer, F.T., & D'Esposito, M. (2020). A simple method for detecting chaos in nature. *Commun. Biol., 3*, 11. https://doi.org/10.1038/s42003-019-0715-9

Trifunovic, A., Hansson, A., Wredenberg, A., Rovio, A.T., Dufour, E., Khvorostov, I., ... & Larsson, N.G. (2005). Somatic mtDNA mutations cause aging phenotypes without affecting reactive oxygen species production. *Proceedings of the National Academy of Sciences, 102*(50), 17993–17998.

Tuszynski, J.A., & Gordon, R. (2012). A mean field Ising model for cortical rotation in amphibian one-cell stage embryos. *Biosystems, 109*, 381–389

van den Berg, J., Boersma, A.J., & Poolman, B. (2017). Microorganisms maintain crowding homeostasis. *Nature Reviews Microbiology, 15*, 309–318

Vera, E., Bernardes de Jesus, B., Foronda, M., Flores, J.M., & Blasco, M.A. (2012). The rate of increase of short telomeres predicts longevity in mammals. *Cell Rep., 2*, 732–737

Walsh, D., & Mohr, I. (2011). Viral subversion of the host protein synthesis machinery. *Nat. Rev. Microbiol.*, *9*, 860–875. https://doi.org/10.1038/nrmicro2655

Watson, J.D., & Crick, F.H. (1953). Molecular structure of nucleic acids: a structure for deoxyribose nucleic acid. *Nature*, *171*(4356), 737–738

Whittemore, K., Vera, E., Martínez-Nevado, E., Sanpera, C., & Blasco, M.A. (2019). Telomere shortening rate predicts species life span. *Proceedings of the National Academy of Sciences*, *116*(30), 15122–15127

Woldeyes, M.A., Qi, W., Razinkov, V.I., Furst, E.M., & Roberts, C.J. (2020). Temperature dependence of protein solution viscosity and protein–protein interactions: insights into the origins of high-viscosity protein solutions. *Molecular Pharmaceutics*, *17*(12), 4473–4482

Yildiz, C., Öngel, M.E., Yilmaz, B., & Özilgen, M. (2021). Diet-dependent entropic assessment of athletes' lifespan. *Journal of Nutritional Science*, *10*, e83

Yim, W.W.Y., & Mizushima, N. (2020). Lysosome biology in autophagy. *Cell Discov.*, *6*(6). https://doi.org/10.1038/s41421-020-0141-7

Zhang, C., Linnane, A.W., & Nagley, P. (1993). Occurrence of a particular base substitution (3243 A to G) in mitochondrial DNA of tissues of ageing humans. *Biochem. Biophys. Res. Commun.*, *195*, 1104–1110

3 Stochasticity in Biological Systems

Biological systems hold both stochastic and deterministic processes. Biological cells are naturally inherently dynamic. Physical interactions, forces, energetics, and entropic fluctuations contribute continuously to the cause of construction and liquidation of structural components in living systems. The dynamics of cells originate in their various compartments. The cellular dynamics are not necessarily smooth, as they result due to forces relying on often indeterministic free energetics of interactions among participating components that construct individual systems inside cells. The inherent ripple effects of the dynamics in cellular or biological systems propagate to create fluctuations or physical noises. Random stochasticity or noise exists in both microbial and eukaryotic cells (Eldar and Elowitz, 2010). Both deterministic and stochastic processes are involved in constructing biological systems, leading to the creation of fully functional living bodies. Stochastic noise in specific cellular systems may regulate their functions, e.g. stochastic gene expression is related to molecular decision-making (Blake and Collins, 2005), stochastic fluctuations in genetic circuit components regulate cellular functions (Eldar and Elowitz, 2010), etc. On the other hand, regulated noise in the epigenetic landscape is found to be associated with the rise of possible developmental disorders and diseases. E.g., the variability of DNA methylation and gene expression is increased in cancer, leading to tumor cell heterogeneity (Pujadas and Feinberg, 2012). In drug discovery, epigenetic stochasticity may therefore be targeted by therapeutic agents. Noise-modulating chemicals appear to provide probes allowing control over diverse cell-fate decisions, thus helping discover drugs to fight cellular diseases (Dar et al., 2014).

3.1 BIOLOGICAL STOCHASTICITY AND ASSOCIATED FEATURES

Living systems resist their decay into thermal equilibrium by expending entropy for maintaining necessary cellular processes and functions (Lestas et al., 2010; Paulsson, 2004). They continue maintaining or increasing local orders by working against the second law of thermodynamics (Skinner and Dunkel, 2021). According to the second law of thermodynamics, during the transfer of energy, there will always be less energy available at the end of the transfer process than at the beginning, and the entropy of the system increases with the transfer of energy. Usually, this entropy accounts for the number of disorders induced in the systems. The thermodynamic consistency is restored as the systems consume free energy associated with various biomolecular or biochemical interactions, thereby increasing the net entropy of the environment. Entropy, a dimensionless quantity, is used for measuring the amount of uncertainty about the state of a system. It can also imply a certain physical quality, where entropy is referred to as explaining the number of missing orders in a system. Wiener could connect Shannon's concept of entropy as a measure of uncertainty in communication with Schrödinger's concept of negative entropy as a source of order in living organisms

DOI: 10.1201/9781003287780-3

(Wiener, 1948). The biological information transmitted in intra-biological systems is equivalent to the amount of entropy of the source of the information – the higher the initial uncertainty resulting in entropy, the higher the amount of resultant information achieved. The Shannon entropy accounts for information whereas the Schrödinger entropy indicates less order in the system. Wiener redefined the notion of entropy, related not to the initial uncertainty (as in Shannon's definition) but to the degree of uncertainty remaining after the message has been received. Higher entropy, associated with higher noise, entails less information (Stonier, 1990). The concepts of entropy, negative entropy, information processing, and entropy increase or decrease often have very different specific meanings in different disciplines and for different scientists. The meanings of these concepts are unified in essence and the differences between them are generated when the same kinds of concept are applied to the study of different directions of the same kind of phenomena: either to the static structuralization degree of the system or to the change of the degree of dynamic structuralization of the system (Wu et al., 2020). In living systems, disorder, entropy, information, noise, stochasticity, etc. are all expressions of natural processes that need to be addressed collectively to understand the systems properly. This chapter does so.

3.1.1 Biological System-Specific Stochasticity

Biological system heterogeneity evolves from the following three major sources; see the diagrammatic modeling in Figure 3.1:

- *genetic*, decided naturally (quite deterministic)
- *environmental*, decided naturally but may vary due to differences in its exposure
- *stochastic* (chance due to inherent fluctuations or noise)

Genetic heterogeneity is due to genetic variations among different species. Therefore, cells and animals in the wild with different genes are expected to be different. However, in the case of convergent evolution, different species might be found to evolve similar phenotypic features independently

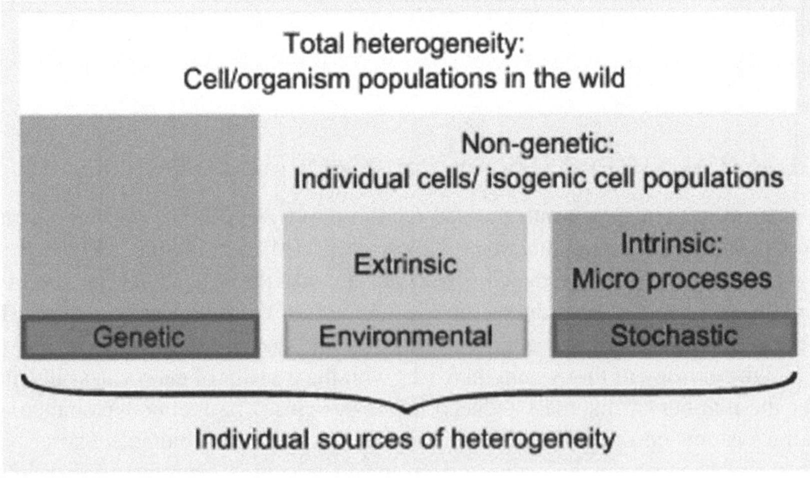

FIGURE 3.1 Sources of heterogeneity. The total observed heterogeneity in a natural system broken down into its three components: genetic, environmental, and stochastic. Different sources of heterogeneity are present at different scales, and the typical scales are given. For details, see Székely Jr and Burrage (2014).

***Source*: Székely Jr and Burrage (2014).**

(Losos, 2011). The phenotypic similarity usually transpires to have a different genetic basis and the relationship between genotype and phenotype appears complex (Manceau et al., 2010). However, considerable investigations show that a surprising number of cases may be found of convergent (phenotypic) evolution appearing with a corresponding convergence in genotype (Stern, 2013; Parker et al., 2013). The non-genetic or phenotypic heterogeneity (Huang, 2009; Avery, 2006) has been found to originate due to various reasons, such as environment-induced or extrinsic causes (Volfson et al., 2006; Johnston et al., 2012), and living systems hosted intrinsic noise-induced causes. The noise-induced stochasticity arises from individual molecular-level random thermal fluctuations (McAdams and Arkin, 1999; Kaern et al., 2005). This intrinsic stochasticity may affect structural integrity in vital biomolecules such as DNA, RNA, protein, etc. This intrinsic noise may also help promote randomness in biomolecular interactions and in biomolecular dynamics in the physiological environment (via the Brownian motion). The effects of noise are propagated into the gene expression level. As a result, even identical genes in an identical intracellular environment may be expressed differently, despite the fact that the genes might not draw any genotypic or environmental effects to contribute to raising the heterogeneity (Kaern et al., 2005; Maheshri and O'Shea, 2007). The extrinsic heterogeneity, in contrast, arises from outside sources affecting all genes inside a cell identically. The distinguishable intracellular gene expression may therefore be predicted to inherit randomness due to only the intrinsic stochasticity or noise. The intrinsic stochasticity-induced processes in the living system have been identified. Individual bacteria from an isogenic population were found to maintain different swimming patterns throughout their entire lives, manifesting the inherent noise-induced intrinsic heterogeneity (Avery, 2006). In this study, Avery created a chart on heterogeneity (Table 3.1) inspecting the microbial cell individuality where enormous stochastic noise-induced phenotyping is found.

The timescales over which any phenotype fluctuates can provide an indication as to the likely source of heterogeneity. Figure 3.2 presents a comparable chart, also from Avery (2006).

3.1.2 Mutations in Biomolecular Structures Regulate Stochasticity

Since the discovery of the model addressing the double-helical DNA structure by Watson and Crick seven decades ago (Watson and Crick, 1953) scientists have been investing all-out efforts in understanding versatile structures and functions of nucleic acids in various physiological conditions (Minchin and Lodge, 2019). We now have quite a clear understanding of the structure of the DNA molecule, the way it is packaged into chromosomes, and how it experiences replication before cell division. We also know how DNA is copied into RNA, what we call transcription, and translated into protein, what we call translation. On the ground of this understanding, recently the goal of scientists has substantially shifted toward understanding DNA-based mutations as this phenomenon is found to be associated with the rise of most cell-originated diseases.

DNA mutations can be simplistically modeled as presented in Figure 3.3. (Slote et al., 2019).

We can simply say that the insertion, deletion, duplication, or inversion of nucleic acid building blocks adenine (A), guanine (G), cytosine (C), and thymine (T) in the DNA structure may lead to mutations. Some of the frequencies of sunlight provide enough energies to sometimes cause mutations (Figure 3.4). Sunlight creates structures called thymine dimers or T-T dimers. This mutation occurs on a single strand of DNA, when two thymines become connected, instead of to adenines on the second strand of DNA. This means that two T bases on the same DNA strand become connected abnormally, instead of correctly attaching to the complementary base A on the opposite strand. The T-T dimers create kinks in the DNA shape (Ackerman and Horton, 2018).

Now I wish to correlate the mechanisms causing the dimer mutations shown in Figure 3.4 with the fundamental DNA structures that are associated with DNA replications. We know telomere is a sequence "AGGGTT" conserved at the replication terminal. The terminal ends of chromosomes contain this highly repeated sequence AGGGTT; for example, in humans the sequence is repeated

TABLE 3.1
Summary of the underlying drivers identified for major heterogeneously expressed microbial phenotypes

Phenotype	Comments	Underlying driver (and organisms investigated)
Variable expression of cell-surface antigens	Different epigenetic mechanisms operate in different systems	Epigenetic (*Plasmodium falciparum*, *Trypansoma brucei*,*Saccharomyces cerevisiae* and *Escherichia coli*)*
Phenotypic switching	White? opaque switching requires heterozygosity or hemizygosity at the mating-type locus. Histone deacetylases modulate switch frequency	Probably epigenetic (*Candida* spp.)
Antibiotic persistence	The proportion of persisters is influenced by the *hipBA* toxin-antitoxin module	Stochastic? (*E. coli*)
Heterogeneous heat resistance	Variable expression of heat-shock-protein-encoding genes	Cell cycle and stochastic? (*S. cerevisiae* and *Salmonella enterica* serovar Typhimurium)
Heterogeneous DNA damage	Several mechanisms suggested	Ageing and ultradian rhythm? (*S. cerevisiae*) Stochastic? (*E. coli*)
Heterogeneous resistance to metals and pro-oxidants	Several deterministic mechanisms demonstrated	Cell cycle, ultradian rhythm and ageing (*S. cerevisiae*)
Bacterial motility and chemotaxis	Variable activity in *E. coli* of the CheR methyltransferase, a key upstream component of the chemotaxis signalling network	Stochastic? (*E. coli* and *Bacillus subtilis*)
Phage λ decision	The decision hinges primarily on the activity of the cII activator protein	Probably stochastic
Individuality in competence and sporulation	Heterogeneity requires bistable expression of the major transcriptional regulators, ComK and Spo0A	Probably stochastic (*B. subtilis*)

* In many cases, where spontaneous epigenetic modification is involved, there remains the underlying question of what determines which cells will switch and when (see text).

? "?" Indicates speculative, which is generally the case where stochasticity is indicated.

Source: Avery (2006). See Avery for details.

in tandem 100 to over 1,000 times. Repeated rounds of DNA replication result in the shortening of these telomeric sequences, that is, the number of repeats will reduce. Telomerase, an RNA-containing enzyme, can add additional copies of the repeat sequence to the 3′ end, replacing those lost during DNA replication (see Figure 3.5).

About three decades ago we discovered a harmonic oscillation pattern in the telomere repetition at the DNA replication end. The frequency of this oscillation falls close to the infrared light frequency (Ashrafuzzaman and Shafee, 2003). The interaction energies between different types of bases of a single strand of DNA molecule were calculated. Using those values of energies, the harmonic behavior of a number of base patterns of DNA was studied. We then investigated, with simple models, the harmonic behavior of the telomere pattern of bases considering the inter-base interaction energies. The telomere pattern shows harmonic frequencies which are closest to the frequencies of the electromagnetic radiation coming from sunlight. Possible resonance of the telomere frequency with such radiation may be responsible for damage to the reproductive ability of the cells and consequent aging and other problems.

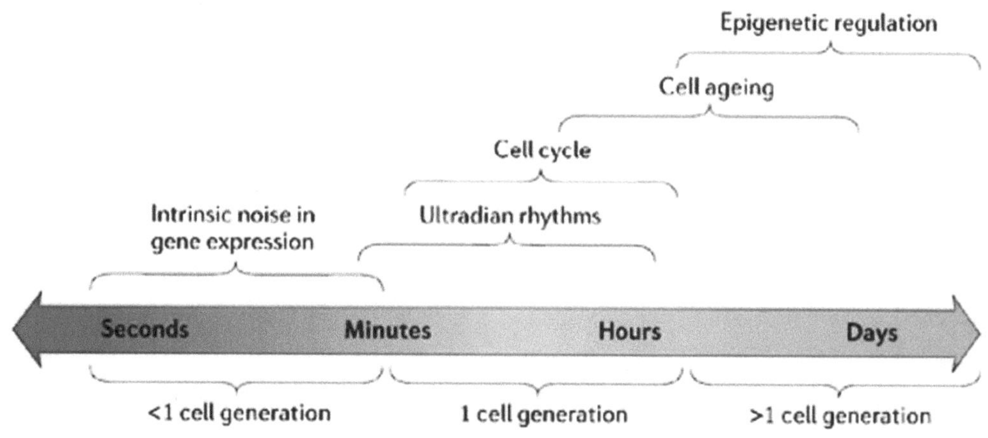

FIGURE 3.2 Temporal information on the stability of heterogeneous phenotypes is generally lacking. However, the timescales of such stability (for example, for how long a particular phenotype is maintained by individual cells) can facilitate prediction of the source of the heterogeneity. The figure is intended to reflect the hierarchical nature of the timescales involved. The timescale values indicated are approximations and are generally organism-dependent and condition-dependent. Note that although epigenetically regulated phenotypes might be sustained over many generations of cells, it could be stochastic events operating at the opposite end of the timescale that, with time, trigger a spontaneous switch to an alternative epigenetic state. See Avery (2006) for details.

***Source*: Avery (2006).**

Now we may correlate this possible resonance between the undeniable environmental (e.g. sunlight) exposure to energies and inherited telomere frequencies as one of the inherited causes behind rising damages over time, leading to mutations in the DNA structures, like those shown in Figure 3.4! Among the possible sources of fluctuations in gene expression, the effects of time-dependent DNA replication were investigated recently (Peterson et al., 2015), which showed its correlation with the noise in mRNA levels. Stochastic simulations of constitutive and regulated gene expression are used to analyze the time-averaged mean and variation in each case. However, the DNA replication fidelity remains quite conserved as a combined effect of multiple processes, as outlined in Figure 3.6. But when various mechanisms turn defective, buffering fails and the rate of mutations increases significantly, e.g. a mutation as a result of base substitutions discovered in Schmidt et al. (2017). We may conclude that stochasticity may not only be an issue that might regulate the time-dependent dynamics associated with DNA replications (see the simulation results in Yousefi and Rowicka, 2019), the stochastic noise may also lead to mutations right in the DNA structures (e.g. maintenance of base sequences, as we see in Figure 3.4).

The noise or variability in mRNA level in a biological cell is found to be due to the transient relaxation of the mRNA from a low- to a high-copy steady state after a gene replication event (Peterson et al., 2015). The location of the gene, the degradation rate of the mRNA, and the cell doubling time are all found to contribute to observed noise. As the chromosome replication is tightly controlled and thus fails to exhibit considerable noise, it is theoretically deduced that the crucial molecular features of the gene replication need to be taken into account while modeling the gene expression (Peterson et al., 2015), and that their regulations due to possible mutations may be predicted to contribute in regulating the stochastic noise in associated biomolecules such as the mRNA.

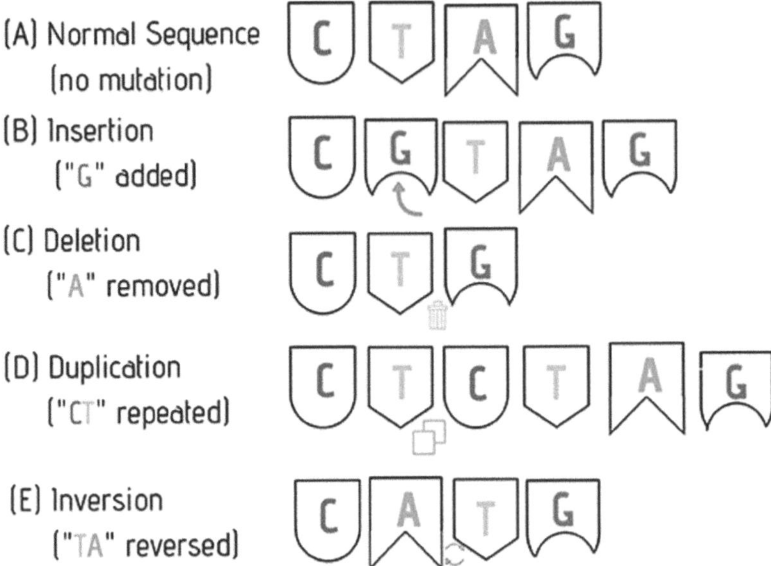

FIGURE 3.3 Examples of common types of mutations. (A) The "normal" DNA sequence without a mutation, (B) an insertion, where a base is added, (C) a deletion, where a base is removed, (D) a duplication, where some bases are repeated, and (E) an inversion, where the order of bases is reversed.

***Source*: Slote et al. (2019).**

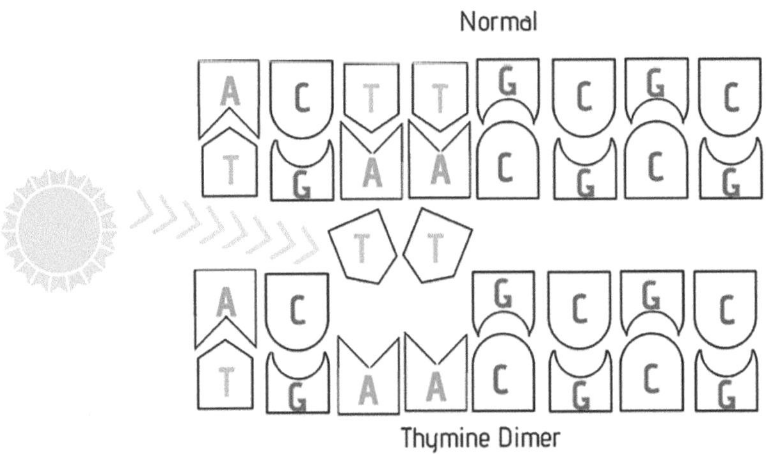

FIGURE 3.4 The ultraviolet (UV) rays in sunlight cause the formation of thymine dimer mutations, which is when two thymine molecules on the same strand of DNA bond together instead of correctly bonding with adenines on the opposite strand.

***Source*: Slote et al. (2019).**

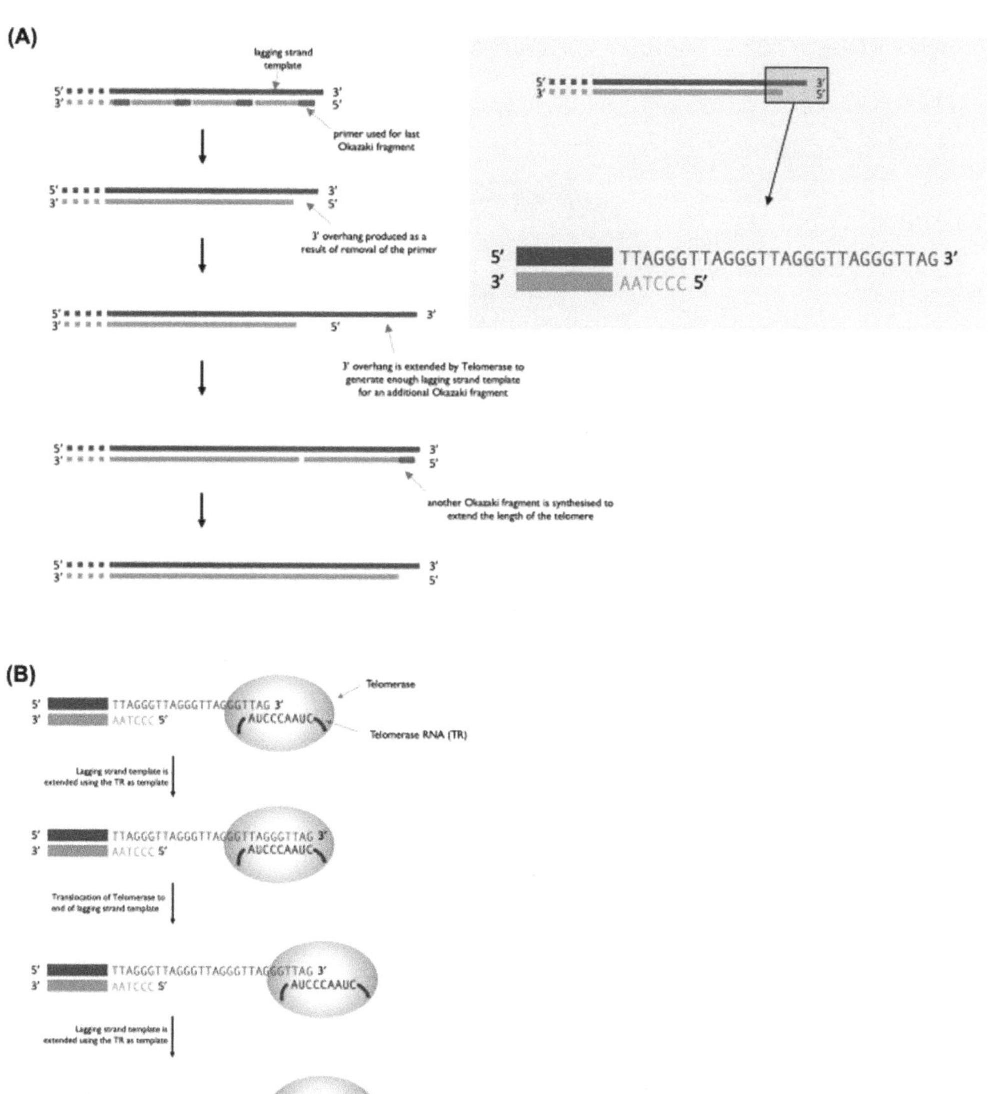

FIGURE 3.5 Telomeres and telomerase. (A) Following DNA replication and removal of the primer for the last Okazaki fragment of the lagging strand, there will be a region at the 3′ end that is not base paired, called a 3′ overhang. (B) Telomerase binds and uses the RNA it contains to act as a template to extend the 3′ overhang. This extends the 3′ end sufficiently for a new RNA primer to bind and the final Okazaki fragment to be made.

***Source*: Minchin and Lodge (2019).**

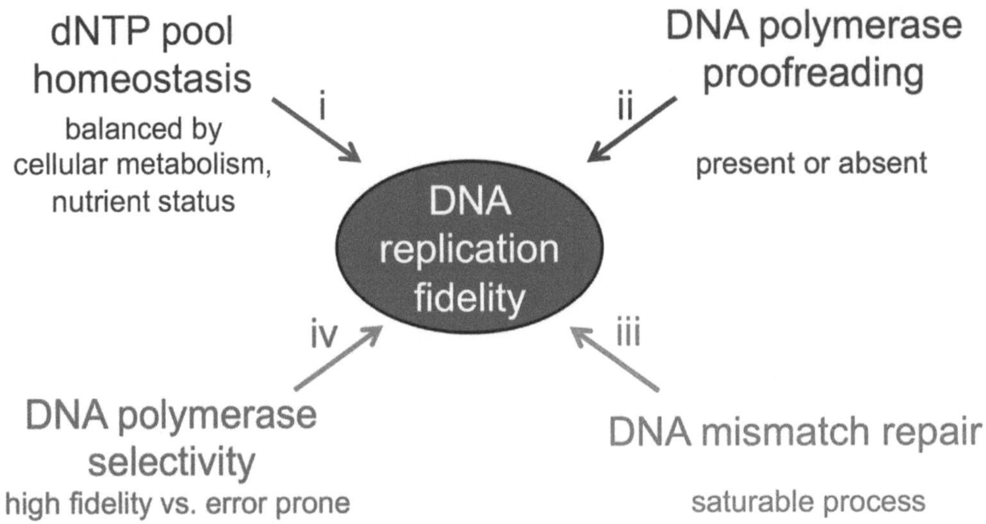

FIGURE 3.6 Removing the buffering capabilities of DNA replication and DNA MMR uncovers new factors that contribute to nucleotide pool homeostasis. Multiple mechanisms contribute to the overall fidelity of DNA replication. For example, each of the four mechanisms outlined in this figure can compensate for defects in the others. (*i*) A homeostasis mechanism responds to cellular metabolism to regulate the biosynthesis of dNTP and dNTP precursors to maintain a balanced concentration of dNTPs. (*ii*) DNA polymerase proofreading activities: During eukaryotic DNA replication, the major replicative polymerases Polε and Polδ both contain proofreading activities that enable them to excise misincorporation errors. However, Polα, which initiates DNA synthesis at origins and at Okazaki fragments, does not contain such a function. (*iii*) DNA MMR: MMR identifies mismatches resulting primarily from DNA replication errors. Nascent DNA is excised by the MMR machinery and is subsequently repaired by DNA synthesis. The MMR protein machinery can become saturated when there is an abundance of replication errors, allowing mismatches to escape repair. (*iv*) DNA polymerase selectivity: Polε and Polδ are high-fidelity replicative polymerases but can differ by ~10-fold in their ability to incorporate the correct base (St Charles et al., 2015). For details, see Manhart and Alani (2017).

***Source*: Manhart and Alani (2017).**

Cells and organisms that are genetically identical may also exhibit remarkable diversity even when they have identical histories of environmental exposure. This observed phenotype may be due to the inherent noise or variability in gene expression.

The variation in gene expression may be originated from four potential sources, as follows (Raser and O'Shea, 2005):

(1) The inherent stochasticity of biochemical processes that are dependent on infrequent molecular events involving small numbers of molecules
(2) The differences in the internal states of a population of cells, either from predictable processes e.g. cell cycle progression, or a random process such as partitioning of mitochondria during cell division
(3) Subtle environmental differences, e.g. morphogen gradients in multicellular development
(4) Ongoing random or directed genetic mutations

Due to the effects of individual or all of the above-mentioned sources of noise, identical twin humans or cloned cats may differ in appearance and behavior (see Figure 3.7). As with other genetically identical animals with multicolored coats, the cloned kitten's color patterning is found to be

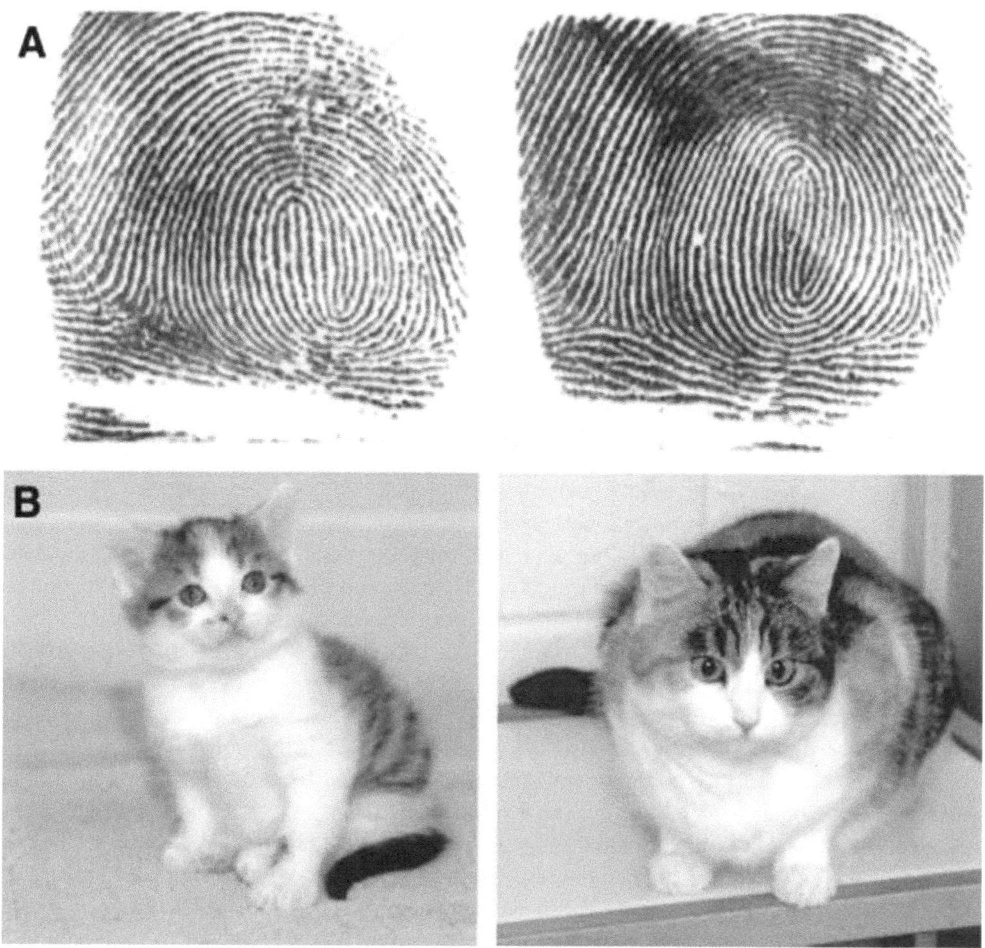

FIGURE 3.7 Examples of possible stochastic influences on phenotype. (A) The fingerprints of identical twins are readily distinguished on close examination. Reprinted from Jain et al. (2002) with permission from Elsevier. (B) Cc, the first cloned cat (left) and Rainbow, Cc's genetic mother (right), display different coat patterns and personalities For details, see Raser and O'Shea (2005).

***Source*: (A) Reprinted from Jain et al. (2002) with permission from Elsevier. (B) Shin et al. (2002), photo credit College of Veterinary Medicine and Biomedical Sciences, Texas A&M University.**

not the same as that of the nuclear donor (Shin et al., 2002). This is because the pattern of pigmentation in multicolored animals is the result of both genetic and developmental factors, including the mutations thereof.

3.2 MODELING AND SIMULATING BIOLOGICAL STOCHASTICITY

DNA replication is one of the key molecular mechanisms controlling our biological origins. DNA structures, DNA replication, and associated functions not only help distinguish animal species and plants at the molecular level but also host origins of mutations leading to versatile forms of genetic disorders and gene-originated diseases (Bellelli and Boulton, 2021). DNA replication is found clearly to be associated with mRNA noise (Peterson et al., 2015). The noise associated with DNA replication, and mRNA processes, is known to lead to phenotype variation at cellular and species levels. I shall evaluate a few cases here.

3.2.1 Stochasticity Associated with DNA Replication

DNA replication-transcription collisions may cause replication fork arrest, premature transcription termination, breaking of the DNA, and recombination intermediates threatening the genome integrity (Sankar et al., 2016; French, 1992; Liu and Alberts, 1995). Modeling the dynamics of DNA replication to understand different dynamical parameters is therefore of great interest in biomolecular simulation science (Yousefi and Rowicka, 2019). Here Yousefi and Rowicka have found using modeling that the stochasticity of the forks' speed plays a crucial role in the DNA replication dynamics. This study has shown that consideration of the stochasticity in forks' speed may help to accurately reconstruct the movement of individual replication forks, measured by DNA combing (Yousefi and Rowicka, 2019). As the completion of the replication on time is a challenge due to stochasticity, an empirically derived modification to replication speed was proposed based on the distance to the approaching fork, which could promote the timely completion of the replication.

DNA replication in eukaryotes is found to start at multiple sites, usually known as replication origins. Despite showing population-wide reproducibility, the replication timing at individual sites is stochastic. The genome is indeed replicated exactly once and the replication is finished in time. DNA replication, due to these characteristics, appears to make computational modeling a useful tool to study the replication mechanisms. The model of Yousefi and Rowicka (2019) is superior due to consideration of the stochasticity in forks' speed, while other earlier models assumed a constant replication forks' speed (considering no variation between forks), contradicting actual molecular mechanisms (see e.g. Barberis et al., 2010). Considerable stochasticity of the replication forks' speeds was observed in *in vitro* biophysical studies of individual forks and in DNA combing and two-dimensional gel analysis in S. cerevisiae (Lewis et al., 2017).

The Repli-Sim of Yousefi and Rowicka (2019) is a probabilistic numerical model for DNA replication, which simulates DNA replication in S. cerevisiae genome-wide assuming stochastic replication forks' speeds. The Repli-Sim is set here to include local parameters specific to each origin inferred from the experimental data and the global parameters that are assigned to origins with the use of a Monte Carlo method, which is then optimized through a genetic algorithm. A model of the DNA replication encoded in Repli-Sim is presented in Figure 3.8. Both data on distances traveled by individual replication forks (DNA combing) and cell population-wide measurements (DNA copy number data) were used to validate the Repli-Sim model.

In Figure 3.8 two forks have been formed which elongate bidirectionally across the genome for forming the DNA tracks (Δx). For each origin i in a cell population, at time t_{exp} (measured from the beginning of DNA replication), the distribution of Δx was derived here based on following two assumptions.

1. Firstly, the firing time of the origin, t_0^i, is derived from a normal distribution with a mean firing time μ_t^i, which is specific to that origin and derived from the experimental data, and with the global standard deviation σ_t:

$$t_0^i \leftarrow N\left(\mu_t^i \middle| \sigma_t^2\right) \tag{3.1}$$

2. Secondly, individual forks are assigned with different speeds, v^i, that are derived from the same probability distribution with a mean speed, μ_v and the standard deviation σ_v:

$$v^i \leftarrow N\left(\mu_v \middle| \sigma_v^2\right) \tag{3.2}$$

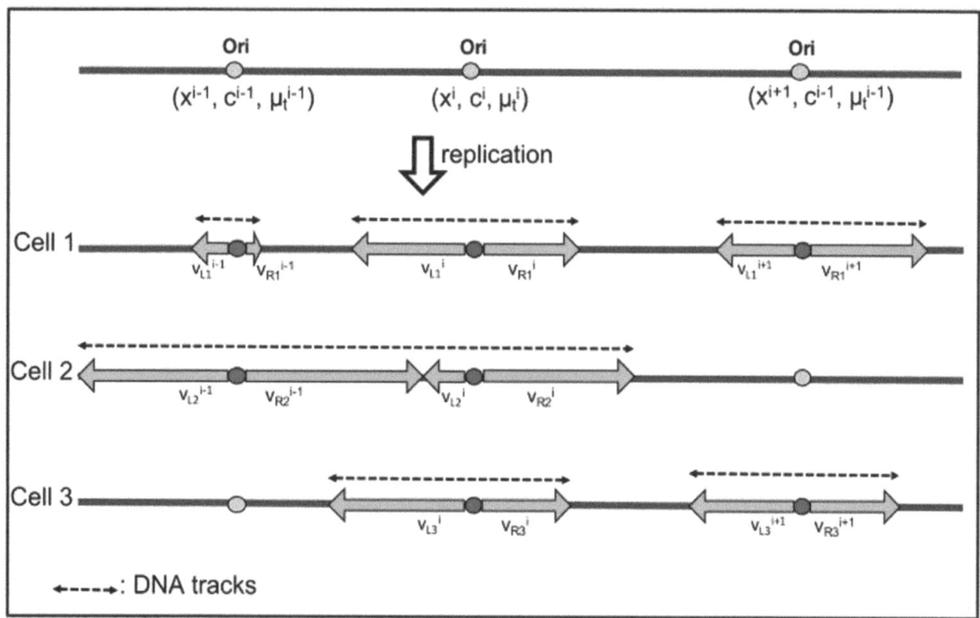

FIGURE 3.8 A schematic of the mechanism of DNA replication encoded in Repli-Sim. Repli-Sim includes local origin parameters (position x^i, competence c^i, and mean firing time μ^i_t) and global parameters (firing time variance σ_t, mean forks' speed μ_v and its variance σ_v). When an origin of replication activates, two forks are formed and elongate bidirectionally until they meet an approaching fork. The speed of an individual fork is constant, but speed varies between forks, even if they emanate from the same origin. The continuous length of the replicated DNA (Δx, DNA tracks) is shown with the dashed lines.

***Source*: Yousefi and Rowicka (2019).**

The DNA replication dynamics draw impacts from the forms' speed stochasticity. The impact of forks' speed variance on the DNA tracks distribution can be illustrated by analyzing a single origin of the DNA replication. Figure 3.9 shows the distribution of DNA tracks (Δx) for two cases, constant ($\sigma_v = 0$) and variable ($\sigma_v \neq 0$) forks' speeds and for a single origin. The differences between the variable and the constant forks' speeds are pronounced later in the S phase, while the average DNA track length remains identical for both of the models.

The stochasticity of the distribution of DNA tracks, $\sigma_{\Delta x}$, can be measured from the following equation (Yousefi and Rowicka, 2019):

$$\sigma^2_{\Delta x} = \mu_v^2 \sigma_t^2 + \mu_{\Delta t}^2 \sigma_v^2 \tag{3.3}$$

The DNA tracks' distribution stochasticity depends on the following parameters:

1. The average forks' speed, μ_v
2. The average firing time, $\mu_{\Delta t}$
3. The degree of randomness of both of forks' speed and firing time, σ_v and σ_t

Equation (3.3) suggests that if we assume a constant forks' speed implying $\sigma_v = 0$, we see $\sigma_{\Delta x}$ to be over-simplistically relying on just the term $\mu_v \sigma_t$, resulting in artificially increased stochasticity of origin firing (σ_t), which manifests itself e.g. by known late origins to fire early in S phase, as if

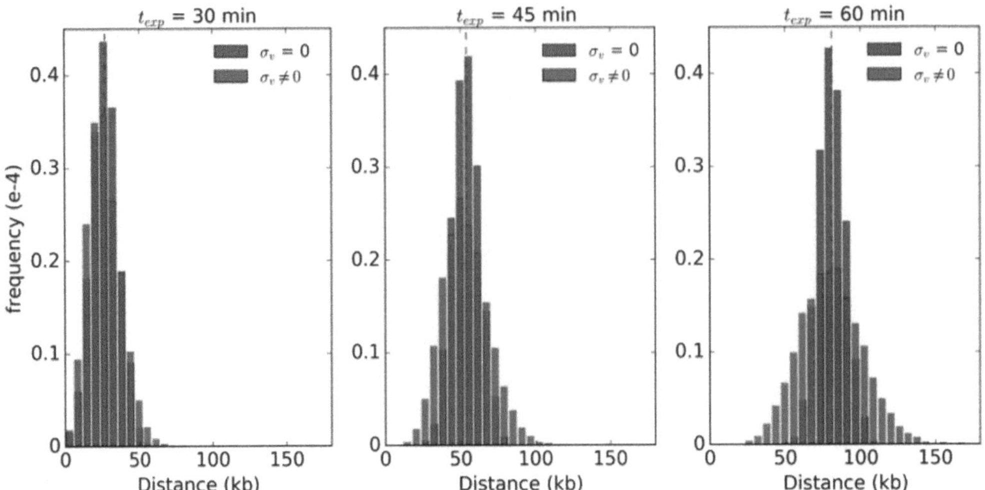

FIGURE 3.9 The impact of stochasticity of replication forks' speeds on the distribution of DNA tracks. The distributions of DNA tracks for constant (σ_v = 0, blue) and variable ($\sigma_v \neq 0$, orange) replication forks' speeds at three different time points within the S phase: 30, 45, and 60 minutes. The differences in DNA track distributions between constant and variable forks' speed models become most pronounced at later times. The dashed red line denotes the mean value. The color contrast in the figure here and all subsequent ones will be clear in the online version of the book.

they were early origins, as described in other models in Hawkins et al. (2013). Therefore, inclusion of parameters designated to address the randomness regarding both forks' speed and firing time in Equation (3.3) appears a smarter choice.

The applicability of the above-explained models on the forks' speeds stochasticity has been cross-examined in hydroxyurea-treated wild type (wt) yeast cells (see Figure 3.10).

In the above model and experimental data demonstrations it has become clear that the stochasticity in the forks' speeds is key to reconstructing the DNA replication dynamics in single cells, as measured by DNA combing. The traditionally used constant forks' speeds models may be found incapable of accurately reconstructing distribution of distances traveled by individual replication forks, as measured by DNA combing. An individual fork speed may be found to depend on the distance to the approaching fork. Such kinds of fork speed modification may promote timely completion of the DNA replication, which is usually a molecular functional challenge due to the nature of stochasticity.

3.2.2 mRNA Noise: Origins and Associated Phenotyping

Now we wish to address the possible effects of the noise in the DNA replication mechanisms on mRNA noise (Peterson et al., 2015) and subsequent effects at the cellular level. The variability in mRNA levels or the mRNA noise across a population is modest and the noise is due to the transient relaxation of the mRNA from a low- to a high-copy steady state after a gene replication event. The gene location, the mRNA degradation rate, and the cell doubling time are found to contribute to observed noise. It is common sense to accept the hypothesis that the DNA replication time may depend on the associated parameters involved in raising noise or stochasticity in the DNA replication process, as explained earlier here. Hence, we may hypothesize that the mRNA noise may be found to be associated with this DNA replication noise too. Therefore, it is essential to account for gene replication when modeling the gene expression or when interpreting any associated *in vitro* experimental or biological results (Peterson et al., 2015).

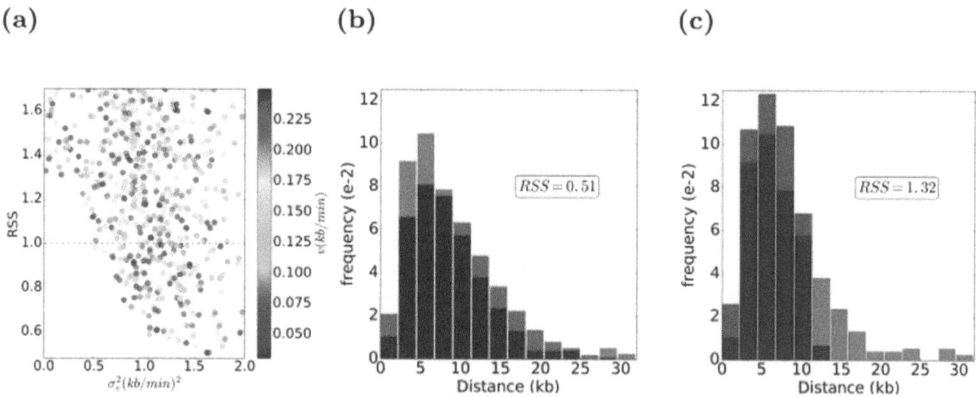

FIGURE 3.10 Results of Repli-Sim for HU-treated cells for both constant and variable fork speed models. (a) Some models (parameter sets) considered. The stochasticity of the replication forks' speeds, σ_v^2, is shown on the horizontal axis and residual sum of squares (RSS) with the experimental data taken from Poli et al. (2012) shown on the vertical axis; the average forks' speed v is color-coded (side bar). Best fitting models (smallest RSS) are characterized by more stochastic forks' speeds. Best fitting constant speed model is ($\sigma_t^2 = 11(min^2)$, $t_{exp} = 52\ (min)$, $v = 0.07\ (kb/min)$, $\sigma_v^2 = 0\ (kb/min)^2$), the best selected variable speed model is ($\sigma_t^2 = 7.7\ (min^2)$, $t_{exp} = 52\ (min)$, $v = 0.12(kb/min)$, $\sigma_v^2 = 0.07(kb/min)^2$, $p_{end} = 2e - 6$). (b, c) The distribution of DNA tracks for both best stochastic (b, orange) and constant (c, blue) forks' speeds models is shown along with the distribution of DNA tracks from experimental data (gray), which shows a better fit for the stochastic forks' speeds model. For details, see Yousefi and Rowicka (2019). The mentioned color contrast may be clearer in the online version of the book.

***Source*: Yousefi and Rowicka (2019).**

The inherited stochasticity in the gene expression is found to cause profound effects on cellular behavior, enabling switching between phenotypes both by individual cells (Friedman et al., 2006; Schultz et al., 2007; Choi et al., 2008) and possibly for entire populations to divergently adapt to multiple environmental niches (Macneil and Walhout, 2011). Both theoretical and experimental works have been dedicated to biophysically and biologically understanding and quantifying various sources of biological stochasticity. As a result, we know a lot about stochastic gene expression eliciting steady-state distributions of proteins and mRNA (Friedman et al., 2006; Raj et al., 2006; Shahrezaei and Swain, 2008; Taniguchi et al., 2010; Peccoud and Ycart, 1995).

Jones and colleagues (2014) were probably the first to show that the duplication of a gene during replication can directly contribute to the observable noise in mRNA copy number. They could successfully partition the experimentally observed mRNA noise into contributions associated with the gene duplication, the variability in RNA polymerase copy numbers, and the possible experimental error. Recently, Peterson and colleagues (2015) performed stochastic simulations, exactly sampling chemical master equations (CME) that explicitly account for chromosome replication, for studying how gene duplication contributes to the variability in the mRNA expression. The simulated results differed consistently and often significantly from the predictions of Jones and colleagues (2014). Peterson et al. showed that after gene duplication, a cell's mRNA count relaxes slowly from a low state associated with the initial gene copy number to a high state associated with the copy number after replication at a rate proportional to the mRNA half-life, a transition that can take up to several minutes and account for a significant portion of the overall cell cycle (see Figure 3.11). The seemingly minor effect may lead to divergence between the predicted and simulated mRNA Fano factors, which is a measure of the "noisiness" of the transcribed mRNA and equal to the variance over the mean, of 20 percent to more than 80 percent, depending on the cell doubling time, the location of the

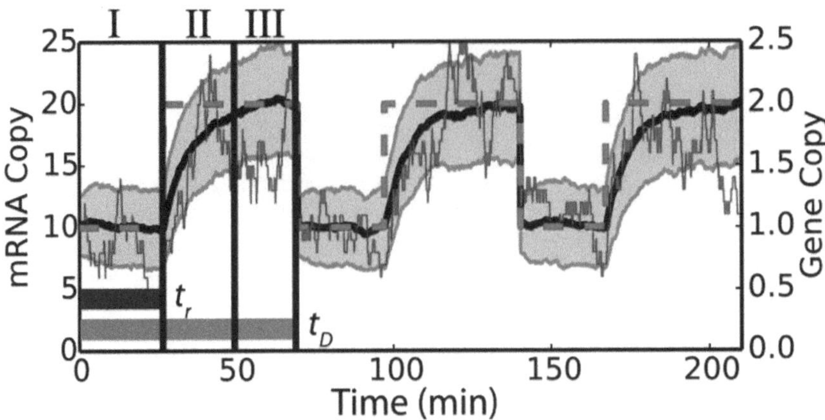

FIGURE 3.11 Simulation schematic composed of 200 simulation replicates shows the progress of the average mRNA count (black line) before and after a gene duplication event (traced by a green dashed line). The area encompassing the average $\pm 1\sigma$ (blue) is shown along with a representative simulation trace (red). Gene duplication is followed by a transient period where the mRNA relaxes from an initially low to a high count at a rate proportional to the degradation rate of the mRNA. Three regions exist and are delineated by vertical lines: a preduplication state (I), wherein the mRNA is in a low copy number steady state, and a relaxation period just after duplication (II), where the mRNA relaxes up to a new equilibrium steady state (III). In these simulations the doubling time (t_D) was taken to be 70 min., the total DNA replication time was taken to be 45 min., the gene was positioned 55 percent of the way from the origin to the terminus ($t_r \sim 27$ min.), the transcription rate k_t was 1.26 molecules/min., and the degradation rate k_d was 0.126/min. For details, see Peterson et al. (2015). Color contrast will be visible in the online version of the book.

***Source*: Peterson et al. (2015).**

gene on the chromosome, and the mRNA degradation rate. The failure to account for the slow relaxation of the messenger distribution to its post-duplication steady state results in the overestimation of the associated noise, and the overestimation may have a profound impact on the interpretation of experimental and theoretical results.

As the gene duplication-associated mRNA noise scales directly proportionally with the (mean) messenger expression level, the misattribution's potential is greatest among the highly expressed genes. For example, in *E. coli*, these include genes involved in key cellular processes, such as the translation including those encoding the ribosomal proteins, the ATP synthesis including the ATP synthase genes, the transcriptional regulation, and the central metabolism including the glycolytic genes *gapA* and *eno* (Taniguchi et al., 2010; Earnest et al., 2015). The potential for noise misattribution is also related to the fraction of the cell cycle after gene duplication, f and the messenger decay rate, k_d. Figure 3.12 shows the relative error between the time-dependent Fano factor expression and that of the time-independent theory, computed as $(F_{TI} - F_{TD})/F_{TI}$, for a cell doubling in 70 min. For highly expressed long-lived transcripts the error can easily be greater than 100 percent, whereas even in moderate cases the error can be in the range of 20–50 percent; most of this divergence is model-specific, mostly originating from deviation of the time-independent (TI) model, as the time-dependent (TD) model agrees well with simulation. For details, see Peterson et al. (2015), which provides especially the comparison of the TD and TI theories applied for explaining the constitutively expressed genes. Interestingly, as the error can change dramatically over a narrow range of values of f (i.e. $0.4 < f < 0.7$; see Figure 3.12), and as f itself is a function of the cell's growth rate, small differences in cell doubling times may lead to a profound effect on mRNA noise interpretation. This study is indicative that the time dependence of gene duplication and mRNA relaxation

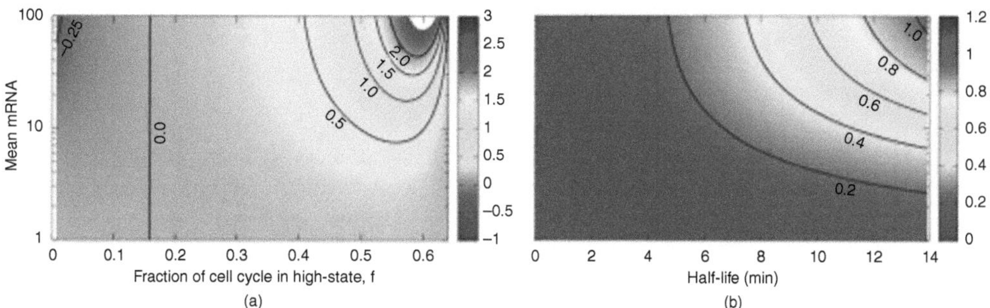

FIGURE 3.12 (*A* and *B*) Deviation of the time-dependent from time-independent theory of the estimated Fano factor $((F_{TI} - F_{TD})/F_{TI})$ when neglecting time dependence of the mRNA relaxation as a function of (*A*), the mean mRNA count and fraction of cell cycle after gene replication, and (*B*) mean mRNA and messenger half-life. Here a slow-growing cell was considered ($t_d \sim 70$ min.). In *A* the mRNA half-life was the average in *E. coli* of 5.5 min. In *B* the fraction of the cell cycle after replication was taken to be 0.7. Scale bars indicate the value of the deviation. Contours are indicated with lines and the value along the contour is denoted. For details, see Peterson et al. (2015). Color contrast will be clear in the digital version of the book.

***Source*: Peterson et al. (2015).**

cannot be ignored during the modeling of the stochastic gene expression or analyzing the experimental data.

3.3 STOCHASTICITY IN CELL FATE DECISIONS AND GENERATION OF CELLULAR DIVERSITY

Cellular functions are mostly determined due to the collected effects of underlying molecular mechanisms. These mechanisms responsible for creating distinguished structures, ensuring stability, causing alterations, and leading transitions among genes, DNAs, RNAs, proteins, etc., are sets of both biochemical and physical processes. These active processes inside cells often work collectively (and occasionally independently) to play set roles to help qualify the biological cells in the performance of specific duties toward creating corresponding living systems. This is simply genetics-based cellular sociology (Chandebois, 1976; Robinson et al., 2007). Cellular sociology is known to regulate the hierarchical spatial patterning and organization of biological cells in specific organisms (Ganesh et al., 2020). Cellular sociology is associated with cellular genetics, a naturally inbuilt biological program focused on creating the cell's fate. Cellular genetics helps us to understand the cellular identity, the underlying processes that regulate cellular functions, the inherited norm of the relationships among them in a biological vicinity, and how the cells may modify to adapt to changed conditions during development, considering healthy state, abnormal disease, and aging conditions and stages. We may construct a flow chart to address the internal links among various leading stages, altogether participating in determining the cellular sociology of any specific species, as modeled in Figure 3.13.

Fluctuations in the IBMs/cells and/or cells/organism connectivity, modeled in Figure 3.13, may lead to raising fluctuating conditions in the cellular decision-making process (Balázsi et al., 2011). We wish to inspect whether these kinds of fluctuations in the process of cellular decision-making may lead to creating gross biological fluctuations (Lehner and Kaneko, 2011). It is generally argued that the natural behavior of biological systems may not necessarily be concerned with their intimate molecular details (Lehner and Kaneko, 2011).

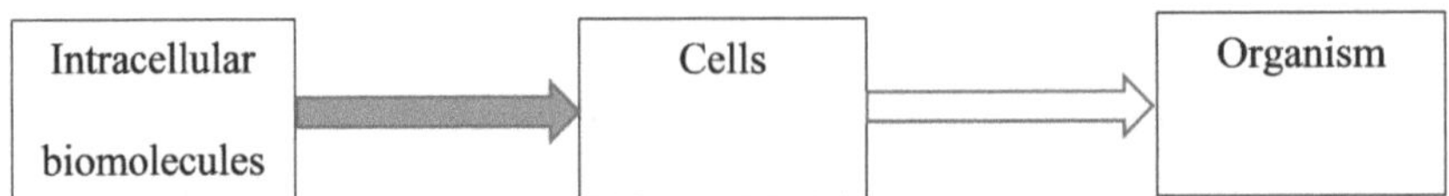

FIGURE 3.13 Modeling the successive steps leading to cellular sociology leading to creating an organism. The filled right-directed arrow connecting intracellular biomolecules (IBMs) and cells represents collective sets of underlying biochemical and physical processes active in various cell-based biomolecules which are determined by inherited genetics. As a result, cells get constructed and function. The empty right-directed arrow connecting cells and an organism represents cellular processes determining the characteristics of the organism. The whole process of connectivity determines the biological sociology behind constructing a deterministic organism that characteristically originates from the biomolecules inside cellular levels.

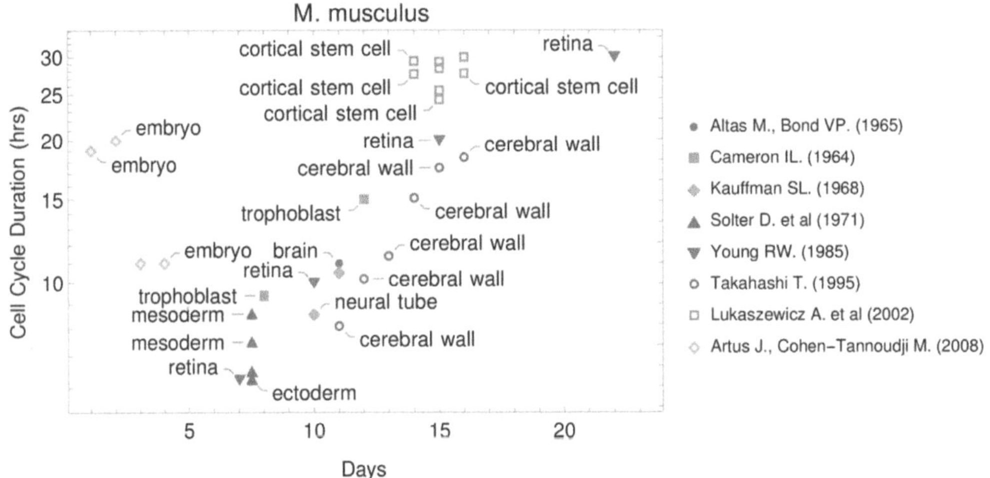

FIGURE 3.14 Cell cycle duration changes during mouse development. The data was curated from several publications (PubMed identifiers: 5859018, 14105210, 5760443, 5542640, 4041905, 7666188, 12151540, 18164540), shown in the legend as authors and (year).

***Source*: Abou Chakra et al. (2021).**

In section 3.2, we have addressed the biological stochasticity originated in the DNA replication fluctuations leading to the rise of mRNA noise and subsequently causing mutations (Bellelli and Boulton, 2021; Peterson et al., 2015; Sankar et al., 2016; French, 1992; Liu and Alberts, 1995). These processes may be associated with the function of the leftmost block in my modeling in Figure 3.13.

The noise in DNA replication, mRNA processes, etc. are all found to lead to causing phenotype variation at cellular and species levels, which connects the function of the leftmost block with that of the middle and rightmost block in Figure 3.13. Some of the aspects, modeled in the central block in Figure 3.13, are addressed in Abou Chakra et al. (2021), which concludes that cell cycle dynamics may be important across multicellular animals to control gene transcript expression and cell fate. Figure 3.14 presents a summary of cell cycle duration in mouse model experiments.

Figure 3.14 shows that the cell cycle duration experiences clear fluctuations at various stages of the development of the mouse. The cell cycle duration may be found to also control the cell diversity (see Figure 3.15).

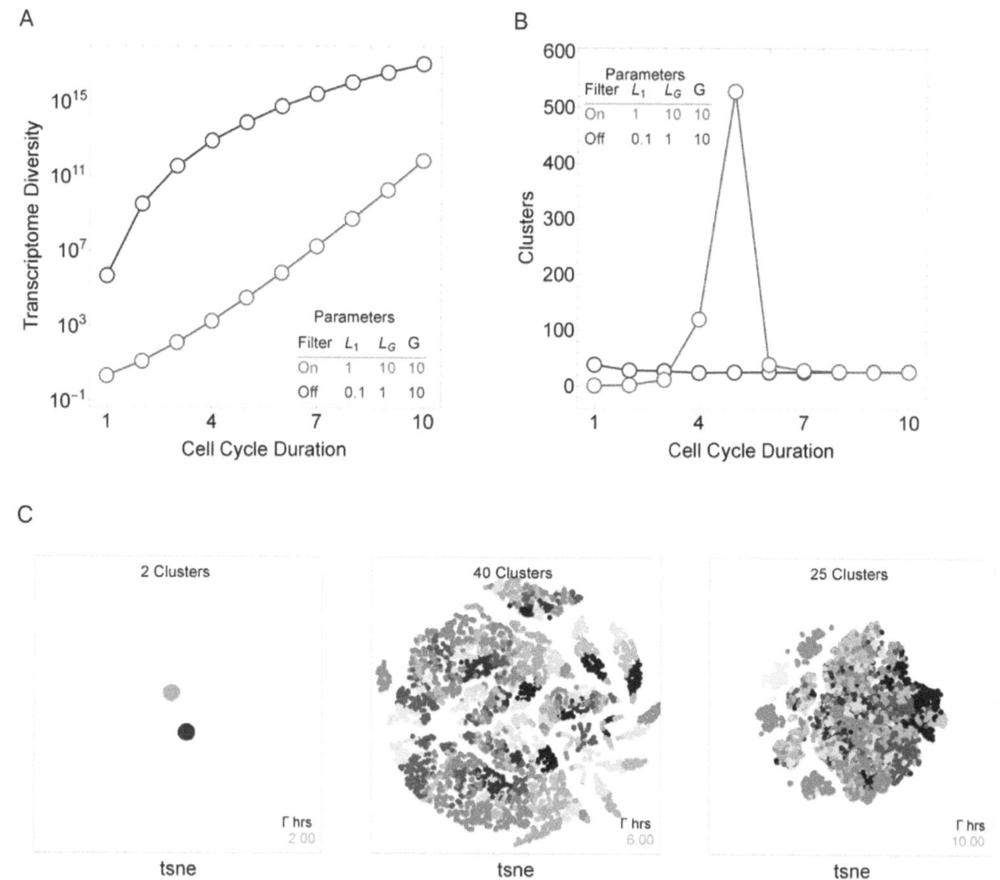

FIGURE 3.15 Cell cycle duration may control the cell diversity. Simulations explore the effects of cell cycle duration, Γ, gene number, G, and gene length distribution. (A) Simulations show that cell diversity (transcriptome diversity) increases as a function of cell cycle duration. Short cell cycle durations can constrain the effects of gene number as long as a transcriptional filter is active (gene length distributions are broad, $L1<\ldots(\Gamma*\lambda)\ldots<LG$). When $LG<(\Gamma*\lambda)$, cell cycle duration does not control cell diversity. Cell cycle duration effects are relative to the gene length distribution in the genome. (B) We use Seurat to cluster the simulated single-cell transcriptomes (10,000 cells) using default parameters and report the number of cell clusters over the simulations. This shows that cell diversity increases with gene number, but the number of clusters identified decreases when all the gene transcripts can be expressed similarly among all cells. (C) Representative examples (10,000 cells) of t-SNE visualizations (RunTSNE using Seurat version 3.1.2) are shown for simulations with cell cycle durations 2, 6, and 10 hr (genome G = 10, $gene^{L_1-L_{10}}$, ploidy n = 1, and transcription rate, λ = 1 kb/hr, RNA polymerase II re-initiation, Ω = 0.25kb). For details, see Abou Chakra et al. (2021).

***Source*: Abou Chakra et al. (2021).**

The cell cycle duration may act as a transcriptional filter that constrains the transcription (Rothe et al., 1992; Shermoen and O'Farrell, 1991). If the cell cycle progresses relatively fast, the transcription of long genes will be found to be interrupted. Typically, the gene transcription rate in cells is between 1.4 and 3.6 kb per minute (Ardehali and Lis, 2009). The 8 min. cell cycle is expected to only allow transcription of the shortest genes ~ 10 kb genomic length, including introns and exons, whereas a 10 hr cell cycle should allow transcription of genes ~ a megabase long genome. Cell cycle-dependent transcriptional filtering has been proposed to be important in cell fate control

FIGURE 3.16 Short and long genes exhibit different pathways in a wide range of species. The shortest 5 percent quantile was selected as a list of short genes (top panels in blue) and genes above the 95 percent

FIGURE 3.16 *(Continued)*

quantile was selected to define a list of long genes (bottom panels in gray). *Saccharomyces cerevisiae* (short < 0.24 kb, long > 3.5 kb), *Ashbya gossypii* (short < 0.36 kb, long > 3.5 kb), *Komagataella pastoris* (short < 0.37 kb, long > 3.3kb), *Yarrowia lipolytica* (short < 0.39 kb, long > 3.5 kb), *Caenorhabditis elegans* (short < 0.47 kb, long > 9.6 kb), *Drosophila melanogaster* (short < 0.56 kb, long > 29 kb), *Danio rerio* (short < 1.3 kb, long > 127 kb), *Takifugu rubripes* (short < 0.72 kb, long > 27 kb), *Xenopus tropicalis* (short < 0.93 kb, long > 83 kb), *Gallus gallus* (short < 0.67 kb, long > 104 kb), *Mus musculus* (short < 1.2 kb, long > 183 kb), and *Sus scrofa* (short < 0.57 kb, long > 197 kb). For each gene group, all corresponding Gene Ontology biological process terms were identified from the Ensembl genome database (100) and resulting term frequencies were visualized as word clouds using Mathematica. For details, see Abou Chakra et al. (2021).

***Source*: Abou Chakra et al. (2021).**

(Bryant and Gardiner, 2016; O'Farrell et al., 2004). We know that multicellular eukaryotic animals mostly start embryogenesis with short cell cycle durations and a limited transcription state (O'Farrell et al., 2004) with typically short zygotic transcripts (Heyn et al., 2014). These cells are found to allocate their majority cycle time to S-phase (synthesis), where the transcription process is inhibited (Newport and Kirschner, 1982), and M-phase (division), with little to no time for transcription in the gap phases. When the cell cycle slows down, the time available for the transcription increases (Edgar et al., 1986), enabling the longer genes to be transcribed (Djabrayan et al., 2019). Short and long genes are found to exhibit different pathways, and this trend is consistent across a wide species range (see Figure 3.16).

The genotypes-affected transcript levels are generally assumed to affect phenotypes, which is also found to work in reverse order, that is, the transcripts associated with the phenotypes are found to be associated with genotypes (Lee et al., 2017). This forward and backward connectivity or dependence hints at spreading biological fluctuations all over if the fluctuations originate at any point (or block otherwise in the model) (Figure 3.13) Therefore, it may be considered more correct if we replaced both right-directional arrows with bidirectional arrows in the model representation in Figure 3.13. All these interconnected genotype and cellular cycle fluctuations, and noises along the lines of associated pathways, are expected to work in a correlated manner leading to the rise of phenotypes in species (Robinson and Webber, 2014; Schwander et al., 2014). It may also be predicted how phenotypic heterogeneity arises from molecular mechanisms in single cells where physiological responses emerge from intrinsic fluctuations of biochemical reactions (Thomas et al., 2018).

3.4 CONCLUDING REMARKS

The role of biological stochasticity is assumed to be central in evolutionary biology (Lenormand et al., 2009). Stochastic processes, e.g. genetic drift, may hinder adaptation (DeLong and Cressler, 2023). DeLong and Cressler showed that stochasticity may alter population dynamics and lead to evolutionary outcomes that are not predicted by classic eco-evolutionary modeling approaches. Their results revealed that considering population processes during transient periods can greatly improve our understanding of the path and pace of evolution. Three types of stochasticity – stochasticity of mutation and variation, individual life histories, and environmental change – were reported by Lenormand and colleagues (2009). They could also explain when stochasticity matters in evolution, distinguishing four broad situations: stochasticity contributes to maladaptation or limits adaptation; it drives evolution on flat fitness landscapes (evolutionary freedom); it might promote jumps from one fitness peak to another (evolutionary revolutions); and it might shape the selection pressures themselves.

The evolutionary dynamics of mutations in stem cells were detailed about a decade ago (Dingli and Pacheco, 2011) taking into consideration the impact of mutations on the reproductive fitness of cells. Stochastic effects may explain clinical observations, including the extinction of acquired clonal stem cell disorders. Drug discovery targeting biological stochasticity is emerging as a popular strategy (Mori and Ben Amar, 2023; Jensen and Lynch, 2020).

REFERENCES

Abou Chakra, M., Isserlin, R., Tran, T.N., & Bader, G.D. (2021). Control of tissue development and cell diversity by cell cycle-dependent transcriptional filtering. *Elife*, *10*, e64951.

Ackerman, S., and Horton, W. (2018). Chapter 2.4–effects of environmental factors on DNA: damage and mutations. *Green Chem.*, *1*, 109–28. doi: 10.1016/B978-0-12-809270-5.00005-4

Ardehali, M.B., & Lis, J.T. (2009). Tracking rates of transcription and splicing in vivo. *Nature Dtructural & Molecular Biology*, *16*(11), 1123–1124

Ashrafuzzaman, M., & Shafee, A. (2003). Investigation by physical methods of the possible role of telomeres in DNA in aging process. *arXiv: Physics: Biological Physics*. https://doi.org/10.48550/arXiv.physics/0301031

Avery, S.V. (2006). Microbial cell individuality and the underlying sources of heterogeneity. *Nat. Rev. Microbiol.*, *4*, 577–587

Balázsi, G., Van Oudenaarden, A., & Collins, J.J. (2011). Cellular decision making and biological noise: from microbes to mammals. *Cell*, *144*(6), 910–925

Barberis, M., Spiesser, T.W., & Klipp, E. (2010). Replication origins and timing of temporal replication in budding yeast: how to solve the conundrum? *Curr. Genomics*, *11*(3), 199–211

Bellelli, R., & Boulton, S.J. (2021). Spotlight on the replisome: aetiology of DNA replication-associated genetic diseases. *Trends in Genetics*, *37*(4), 317–336

Blake, W.J., & Collins, J.J. (2005). And the noise played on: stochastic gene expression and HIV-1 infection. *Cell*, *122*(2), 147–149

Bryant, S.V., & Gardiner, D.M. (2016). The relationship between growth and pattern formation. *Regeneration*, *3*(2), 103–122

Chandebois, R. (1976). Cell sociology: a way of reconsidering the current concepts of morphogenesis. *Acta Biotheoretica*, *25*(2–3), 71–102

Choi, P.J., Cai, L., Frieda, K., & Xie, X.S. (2008). A stochastic single-molecule event triggers phenotype switching of a bacterial cell. *Science*, *322*(5900), 442–446

Dar, R.D., Hosmane, N.N., Arkin, M.R., Siliciano, R.F., & Weinberger, L.S. (2014). Screening for noise in gene expression identifies drug synergies. *Science*, *344*(6190), 1392–1396

DeLong, J.P., & Cressler, C.E. (2023). Stochasticity directs adaptive evolution toward nonequilibrium evolutionary attractors. *Ecology*, *104*(1), e3873

Dingli, D., & Pacheco, J.M. (2011). Stochastic dynamics and the evolution of mutations in stem cells. *BMC Biology*, *9*, 1–7

Djabrayan, N.J.V., Smits, C.M., Krajnc, M., Stern, T., Yamada, S., Lemon, W.C., … & Shvartsman, S.Y. (2019). Metabolic regulation of developmental cell cycles and zygotic transcription. *Current Biology*, *29*(7), 1193–1198

Earnest, T.M., Lai, J., Chen, K., Hallock, M.J., Williamson, J.R., & Luthey-Schulten, Z. (2015). Toward a whole-cell model of ribosome biogenesis: kinetic modeling of SSU assembly. *Biophysical Journal*, *109*(6), 1117–1135

Edgar, B.A., Kiehle, C.P., & Schubiger, G. (1986). Cell cycle control by the nucleo-cytoplasmic ratio in early Drosophila development. *Cell*, *44*(2), 365–372

Eldar, A., & Elowitz, M.B. (2010). Functional roles for noise in genetic circuits. *Nature*, *467*(7312), 167–173

French, S. (1992). Consequences of replication fork movement through transcription units in vivo. *Science*, *258*, 1362–1365

Friedman, N., Cai, L., & Xie, X.S. (2006). Linking stochastic dynamics to population distribution: an analytical framework of gene expression. *Phys. Rev. Lett.*, *97*(16), 168302

Ganesh, S., Utebay, B., Heit, J., & Coskun, A.F. (2020). Cellular sociology regulates the hierarchical spatial patterning and organization of cells in organisms. *Open Biology*, *10*(12), 200300

Hawkins, M., Retkute, R., Müller, C., Saner, N., Tanaka, T., de Moura, A.P.S., and Nieduszynski, C.A. (2013). High-resolution replication profiles define the stochastic nature of genome replication initiation and termination. *Cell Reports*, *5*(4), 1132–1141.

Heyn, P., Kircher, M., Dahl, A., Kelso, J., Tomancak, P., Kalinka, A.T., & Neugebauer, K.M. (2014). The earliest transcribed zygotic genes are short, newly evolved, and different across species. *Cell Reports*, *6*(2), 285–292

Huang, S. (2009). Non-genetic heterogeneity of cells in development: more than just noise. *Development*, *136*, 3853–3862

Jain, A.K., Prabhakar, S., & Pankanti, S. (2002). On the similarity of identical twin fingerprints. *Pattern Recognition*, *35*(11), 2653–2663

Jensen, J.D., & Lynch, M. (2020). Considering mutational meltdown as a potential SARS-CoV-2 treatment strategy. *Heredity*, *124*(5), 619–620

Johnston, I.G., Gaal, B. das Neves, R.P., Enver, T., Iborra, F.J., & Jones. N.S. (2012). Mitochondrial variability as a source of extrinsic cellular noise. *PLoS Comput. Biol.*, *8*, e1002416

Jones, D.L., Brewster, R.C., & Phillips, R. (2014). Promoter architecture dictates cell-to-cell variability in gene expression. *Science*, *346*(6216), 1533–1536

Kaern, M., Elston, T., Blake, W., & Collins. J. (2005). Stochasticity in gene expression: from theories to phenotypes. *Nat. Rev. Genet.*, *6*, 451–464

Lee, S., Wang, H., & Xing, E.P. (2017). Backward genotype-transcript-phenotype association mapping. *Methods*, *129*, 18–23

Lehner, B., & Kaneko, K. (2011). Fluctuation and response in biology. *Cellular and Molecular Life Sciences*, *68*, 1005–1010

Lenormand, T., Roze, D., & Rousset, F. (2009). Stochasticity in evolution. *Trends in Ecology & Evolution*, *24*(3), 157–165

Lestas, I., Vinnicombe, G., & Paulsson, J. (2010). Fundamental limits on the suppression of molecular fluctuations. *Nature*, *467*(7312), 174–178

Lewis, J.S., Spenkelink, L.M., Schauer, G.D., Hill, F.R., Georgescu, R.E., O'Donnell, M.E., et al. (2017). Single-molecule visualization of Saccharomyces cerevisiae leading-strand synthesis reveals dynamic interaction between MTC and the replisome. *Proceedings of the National Academy of Sciences*, *114*(40), 10630–10635

Liu, B., & Alberts, B.M. (1995). Head-on collision between a DNA replication apparatus and RNA polymerase transcription complex. *Science*, *267*, 1131–1137

Losos, J.B. (2011). Convergence, adaptation, and constraint. *Evolution*, *65*, 1827–1840

Macneil, L.T., & Walhout, A.J.M. (2011). Gene regulatory networks and the role of robustness and stochasticity in the control of gene expression. *Genome Res.*, *21*(5), 645–657

Maheshri, N., & O'Shea. E.K. (2007). Living with noisy genes: how cells function reliably with inherent variability in gene expression. *Annu. Rev. Biophys. Biomol. Struct.*, *36*, 413–434

Manceau, M., Domingues, V.S., Linnen, C.R., Rosenblum, E.B., & Hoekstra, H.E. (2010). Convergence in pigmentation at multiple levels: mutations, genes and function. *Philosophical Transactions of the Royal Society B: Biological Sciences*, *365*(1552), 2439–2450

Manhart, C.M., & Alani, E. (2017). DNA replication and mismatch repair safeguard against metabolic imbalances. *Proceedings of the National Academy of Sciences*, *114*(22), 5561–5563

McAdams, H.H., & Arkin, A. (1999). It's a noisy business! Genetic regulation at the nanomolar scale. *Trends Genet.*, *15*, 65–69

Minchin, S., & Lodge, J. (2019). Understanding biochemistry: structure and function of nucleic acids. *Essays in Biochemistry*, *63*(4), 433–456

Mori, L., & Ben Amar, M. (2023). Stochasticity and drug effects in dynamical model for cancer stem cells. *Cancers*, *15*(3), 677

Newport, J., & Kirschner, M. (1982). A major developmental transition in early Xenopus embryos: I. characterization and timing of cellular changes at the midblastula stage. *Cell*, *30*(3), 675–686

O'Farrell, P.H., Stumpff, J., & Su, T.T. (2004). Embryonic cleavage cycles: how is a mouse like a fly? *Current Biology*, *14*(1), R35–R45

Parker, J., Tsagkogeorga, G., Cotton, J.A., Liu, Y., Provero, P., Stupka, E., et al. (2013). Genome-wide signatures of convergent evolution in echolocating mammals. *Nature*, *502*, 228–231

Paulsson, J. (2004). Summing up the noise in gene networks. *Nature*, *427*(6973), 415–418

Peccoud, J., & Ycart, B. (1995). Markovian modeling of gene-product synthesis. *Theoretical Population Biology*, *48*(2), 222–234

Peterson, J.R., Cole, J.A., Fei, J., Ha, T., & Luthey-Schulten, Z.A. (2015). Effects of DNA replication on mRNA noise. *Proceedings of the National Academy of Sciences*, *112*(52), 15886–15891

Poli, J., Tsaponina, O., Crabbé, L., Keszthelyi, A., Pantesco, V., Chabes, A., … & Pasero, P. (2012). dNTP pools determine fork progression and origin usage under replication stress. *EMBO Journal*, *31*(4), 883–894

Pujadas, E., & Feinberg, A.P. (2012). Regulated noise in the epigenetic landscape of development and disease. *Cell*, *148*(6), 1123–1131

Raj, A., Peskin, C.S., Tranchina, D., Vargas, D.Y., & Tyagi, S. (2006). Stochastic mRNA synthesis in mammalian cells. *PLoS Biol.*, *4*(10), e309

Raser, J.M., & O'Shea, E.K. (2005). Noise in gene expression: origins, consequences, and control. *Science*, *309*(5743), 2010–2013

Robinson, C.V., Sali, A., & Baumeister, W. (2007). The molecular sociology of the cell. *Nature*, *450*(7172), 973–982

Robinson, P.N., & Webber, C. (2014). Phenotype ontologies and cross-species analysis for translational research. *PLoS Genetics*, *10*(4), e1004268

Rothe, M., Pehl, M., Taubert, H., & Jäckle, H. (1992). Loss of gene function through rapid mitotic cycles in the Drosophila embryo. *Nature*, *359*(6391), 156–159

Sankar, T.S., Wastuwidyaningtyas, B.D., Dong, Y., Lewis, S.A., & Wang, J.D. (2016). The nature of mutations induced by replication–transcription collisions. *Nature*, *535*(7610), 178–181

Schmidt, T.T., Reyes, G., Gries, K., Ceylan, C.Ü., Sharma, S., Meurer, M., … & Hombauer, H. (2017). Alterations in cellular metabolism triggered by URA7 or GLN3 inactivation cause imbalanced dNTP pools and increased mutagenesis. *Proceedings of the National Academy of Sciences*, *114*(22), E4442–E4451

Schultz, D., Ben Jacob, E., Onuchic, J.N., & Wolynes, P.G. (2007). Molecular level stochastic model for competence cycles in Bacillus subtilis. *Proc. Natl. Acad. Sci. USA*, *104*(45), 17582–17587

Schwander, T., Libbrecht, R., & Keller, L. (2014). Supergenes and complex phenotypes. *Current Biology*, *24*(7), R288–R294

Shahrezaei, V., & Swain, P.S. (2008). Analytical distributions for stochastic gene expression. *Proc. Natl. Acad. Sci. USA*, *105*(45), 17256–17261

Shermoen, A.W., & O'Farrell, P.H. (1991). Progression of the cell cycle through mitosis leads to abortion of nascent transcripts. *Cell*, *67*(2), 303–310

Shin, T., Kraemer, D., Pryor, J., Liu, L., Rugila, J., Howe, L., … & Westhusin, M. (2002). A cat cloned by nuclear transplantation. *Nature*, *415*(6874), 859

Skinner, D.J., & Dunkel, J. (2021). Improved bounds on entropy production in living systems. *Proceedings of the National Academy of Sciences*, *118*(18), e2024300118

Slote, C., Luu, A., George, N., and Osier, N. (2019). Ways you can protect your genes from mutations with a healthy lifestyle. *Front. Young Minds*, *7*(46). doi: 10.3389/frym.2019.00046

St Charles, J.A., Liberti, S.E., Williams, J.S., Lujan, S.A., & Kunkel, T.A. (2015). Quantifying the contributions of base selectivity, proofreading and mismatch repair to nuclear DNA replication in Saccharomyces cerevisiae. *DNA Repair*, *31*, 41–51.

Stern, D.L. (2013). The genetic causes of convergent evolution. *Nat. Rev. Genet.*, *14*, 751–764

Stonier, T. (1990). Information and entropy: the mathematical relationship. In Stonier, *Information and the internal structure of the universe: an exploration into information physics*, pp. 33–41. London: Springer

Székely Jr, T., & Burrage, K. (2014). Stochastic simulation in systems biology. *Computational and Structural Biotechnology journaJ*, *12*(20–21), 14–25

Taniguchi, Y., Choi, P.J., Li, G.W., Chen, H., Babu, M., Hearn, J., … & Xie, X.S. (2010). Quantifying E. coli proteome and transcriptome with single-molecule sensitivity in single cells. *Science*, *329*(5991), 533–538.

Thomas, P., Terradot, G., Danos, V., & Weiße, A.Y. (2018). Sources, propagation and consequences of stochasticity in cellular growth. *Nature Communications*, *9*(1), 4528

Volfson, D., Marciniak, J., Blake, W.J., Ostroff, N., Tsimring, L.S., & Hasty, J. (2006). Origins of extrinsic variability in eukaryotic gene expression. *Nature*, *439*, 861–864

Watson, J.D., & Crick, F.H. (1953). Molecular structure of nucleic acids: a structure for deoxyribose nucleic acid. *Nature*, *171*(4356), 737–738. https://doi.org/10.1038/171737a0

Wiener, N. (1948). *Cybernetics: or control and communication in the animal and the machine*. Cambridge, MA: MIT Press

Wu, K., Nan, Q., & Wu, T. (2020). Philosophical analysis of the meaning and nature of entropy and negative entropy theories. *Complexity*, *2020*, 8769060

Yousefi, R., & Rowicka, M. (2019). Stochasticity of replication forks' speeds plays a key role in the dynamics of DNA replication. *PLoS Computational Biology*, *15*(12), e1007519

4 Biological Energetics

Biological systems use energy in various forms. Specific cell-based molecular processes and interactions are behind the production of energy that is vital for life. We know biological life relies on various physicochemical processes and interactions, which happen generally at the expense of adenosine triphosphate (ATP), which is considered the currency of energy. Mitochondria, the powerhouses of the cell, play major roles in releasing energy. Production of energy and processing it (referred to as energetics) happen in physiological conditions involving major physical and chemical processes. This chapter is dedicated to explaining many of them briefly. I shall present various case studies to address the energy production and processing issues. Biophysical aspects will mainly be addressed to explain the biological phenomena involving energy transduction. Bioenergetics is highly associated with static electrical conditions and the processing of dynamical signals of biological systems. Physical techniques have been utilized to understand them. We shall learn about normal and abnormal conditions of biostructures and their functions involved in ATP production and processing within the system. Physical activities, related stresses, and nutritional states of biological systems may play important roles associated with cellular energetics. These social-aspect-related biological energetics have been parameterized too and will be presented briefly.

4.1 BIOENERGETICS ASSOCIATED WITH ELECTRICAL CONDITIONS OF BIOLOGICAL SYSTEMS

Some anaerobic archaea and bacteria (Wolfe, 1999) are known to live on substrates that do not allow the synthesis of 1.0 mol of ATP per mol of the substrate via substrate-level phosphorylation (SLP). This mentioned energy conservation is possible only by a chemiosmotic mechanism involving the generation of an electrochemical ion gradient across the cytoplasmic membrane which drives ATP synthesis via an ATP synthase (Müller and Hess, 2017). The required minimal energy for the ATP synthesis is thus dependent on the following vital conditions:

(1) the electrochemical ion gradient
(2) the phosphorylation potential in the cell
(3) the ion/ATP ratio of the ATP synthase

Anaerobic life (Wolfe, 1999) comprises a great diversity of energy-generating mechanisms (Peinemann and Gottschalk, 1992). The facultative anaerobes host a respiratory system that preferentially uses oxygen. But in cases when oxygen is absent and alternative electron acceptors, such as NO_3^-, NO_2^-, Fe(III), or others are present, alternative electron transport pathways are found to be induced. However, strict anaerobes can derive energy only by fermentation and/or by the ion

DOI: 10.1201/9781003287780-4

gradient-driven phosphorylation process. The energy yield in strict anaerobes is generally only a small fraction of the one gained by aerobes (Müller, 2001). The fermenting organisms are obligated to rely mostly on SLP, where the yield of energy ranges from 1–4 mol of ATP per mol of fermented hexose. A few organisms have also evolved to adopt additional mechanisms for increasing their ATP yields. Among them, some are found to employ electrogenic end-product efflux (Otto et al., 1980). Additionally, some of the fermentation pathways are found to comprise one or more membrane-bound reactions resulting in the generation of a transmembrane ion gradient across the membrane (Müller, 2001; Marreiros et al., 2016).

The gradients of the concentrations of Na^+ and H^+ are naturally used to drive the synthesis of ATP by membrane-bound A1A0 ATPases (Müller et al., 1999). Acetogens are found to use a similar pathway for CO_2 reduction as methanogens, the Wood-Ljungdahl pathway, but regarding the energy-conserving mechanisms they can be divided into two groups, the Na+ and the H+ organisms (Ljungdahl, 1994). There is a huge amount of information on the energetics of CO_2 reduction (homoacetogenesis) and reduction of other electron acceptors in acetogens in Müller (2003).

4.2 PHYSICS TECHNIQUES ASSESSING BIOENERGY

Bioenergies in our physiological systems originate in various organs, and their mechanisms of action are also often organ-specific (Zhang et al., 1997; Porter Jr et al., 2011; Piquereau and Ventura-Clapier, 2018). Recently, neurophysics techniques were utilized to assess muscle bioenergy (Leon-Sarmiento et al., 2019).

Both classical and quantum mechanical methods have been found applicable to address some aspects of bioenergy (Maslov, 2021). Maslov has shown that the Hamilton–Jacobi equation and the transport equation and also the theory of the mechanics of an infinitely narrow beam as a whole can be applied to some objects in bioenergy in special cases when thin organic objects such as wood splinters, straw, pellets, and so on are approximated by infinitely narrow beams.

Electric Field Driven Torque was explored recently in ATP Synthase (Miller Jr et al., 2013). Here a mechanism involving electric fields emanating from the proton entry and exit channels is proposed to act on asymmetric charge distributions in F_O-ATP synthase and drive it to rotate. F_O-ATP synthase (F_O) here is a rotary motor that converts potential energy from ions, usually protons, moving from high- to low-potential sides of a membrane into torque and rotary motion. The model predicts a scaling between time-averaged torque and proton motive force, which may be hindered by mutations. The torque, created by the ring of synthase structure, raises a rotation within the ATP-producing complex, overcoming, with the aid of thermal fluctuations, an opposing torque that rises and falls with the angular position. Using the analogy with the thermal Brownian motion of a particle in a tilted washboard potential, ATP production rates vs. proton motive force were computed. The latter was found to show a minimum, needed to drive ATP production, which scales inversely with the number of proton binding sites on the *c*-ring. Interested readers may dip into Miller Jr et al. (2013).

We see that a fundamental biological process, “bioenergetics,” rather contains many physical processes, so attracts application of physics principles. Many techniques have so far been applied to elucidate the fundamentals of bioenergetics, ATP production mechanism, etc., which will be addressed here.

4.3 BIOPHYSICS OF ATP SYNTHASE IN BIOLOGY

ATP synthesis is a chemicophysical process taking place in mitochondria (Kazakova and Markosian, 1966). ATP synthesis involves the transfer of electrons from the mitochondrial intermembrane space, through the mitochondrial inner membrane, back to the matrix. There is a mitochondrial enzyme “ATP synthase,” localized in the inner membrane, where it catalyzes the ATP synthesis from

ADP and phosphate. This process is driven by a proton flux across a gradient generated by electron transfer from the proton chemically positive to the negative side. The ATP synthase is popularly studied using biophysical techniques (Kagawa, 1999), as ATP synthesis is apparently a biophysical process (Anandakrishnan et al., 2016). All living organisms – archaea, bacteria, and eukarya – are known to use an intricate rotary molecular machine for synthesizing ATP (see Figure 4.1).

The structure of an intact mitochondrial ATP synthase dimer was recently determined by electron cryo-microscopy at near-atomic resolution (see Figure 4.2). (Klusch et al., 2017). Here a 1.6 MDa mitochondrial F1Fo ATP synthase dimer structure was revealed.

Charged and polar residues of the a-subunit stator define two aqueous channels, each spanning one-half of the membrane (see Figure 4.3). Passing through a conserved membrane-intrinsic helix hairpin, the lumenal channel protonates an acidic glutamate in the *c*-ring rotor. Upon ring rotation, the protonated glutamate encounters the matrix channel and deprotonates. An arginine between the two channels prevents proton leakage. The steep potential gradient over the sub-nm inter-channel distance exerts a force on the deprotonated glutamate, resulting in net directional rotation.

Conserved charged, polar, and hydrophobic sidechains are observed to be aligned in the channel protein interior (see Figure 4.4). The lumenal channel extends to a cluster of buried, closely spaced glutamates and histidines (aGlu172, aHis248, aHis252, aGlu288; Figure 4.4A; Figure 4.5A) that appears to serve as a local reservoir for protons to be fed to the *c*-ring glutamates. At aGlu288 about 20 Å below the lumenal membrane surface, the channel narrows to 4 by 5 Å and changes direction by 90° toward the *c*-ring (Figure 4.3D; Figure 4.4A). In the hydrophobic membrane interior, the strictly conserved polar sidechains of aAsn243 (H5) and aGln295 (H6) that face one another would stabilize the H5/H6 hairpin (Figure 4.5B). The channel passes through the hairpin at the small, conserved sidechains of aAla246, aGly247 (H5), and aAla292 (H6) (Figure 4.3C, D; Figure 4.4A; Figure 4.5B; Figure 4.2). Bulky hydrophobic sidechains close by on H5 and H6 keep the helices apart and the channel open (Figure 4.4A; Figure 4.5B). From aGlu288 the proton may jump to either of two *c*-ring glutamates. The path to cAGlu111 is 4 Å longer than the ~12 Å path to cBGlu111 (Figure 4.5C) but includes the hydrophilic sidechains of aAsn243 and cBSer112. Therefore, in our static structure, the path to cAGlu111 is more favorable for proton transfer. In a rotating *c*-ring, the estimated minimum distance from aGlu288 to cGlu111 is ~11 Å.

For details on the structural aspects, interested readers may consult the original reference, Klusch et al. (2017). This structural study explains the fundamental process known to drive ATP synthesis, which is essential for all forms of life. Therefore, this atomic model may be considered essential in

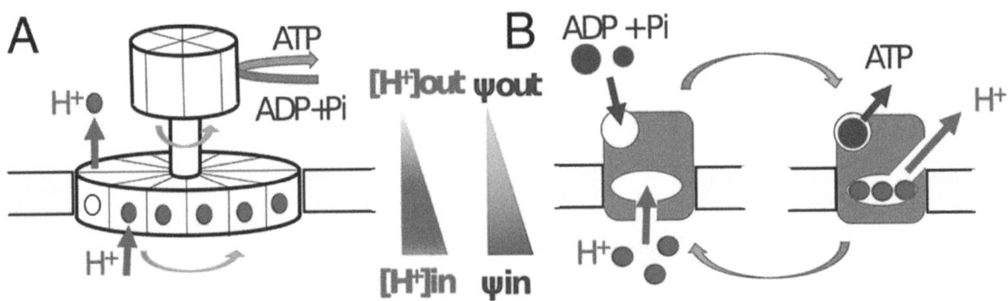

FIGURE 4.1 ATP synthesis by rotary and alternating-access mechanisms. H^+ transport across the membrane is driven by transmembrane differences in pH and electric potential ψ. (A) In the rotary mechanism, H^+ transport drives the rotation of the transmembrane ring, which induces conformational changes that catalyze phosphorylation of ADP. (B) In the alternating-access mechanism, H^+ binding allosterically produces the conformational changes necessary to catalyze phosphorylation of ADP.

***Source*: Anandakrishnan et al. (2016).**

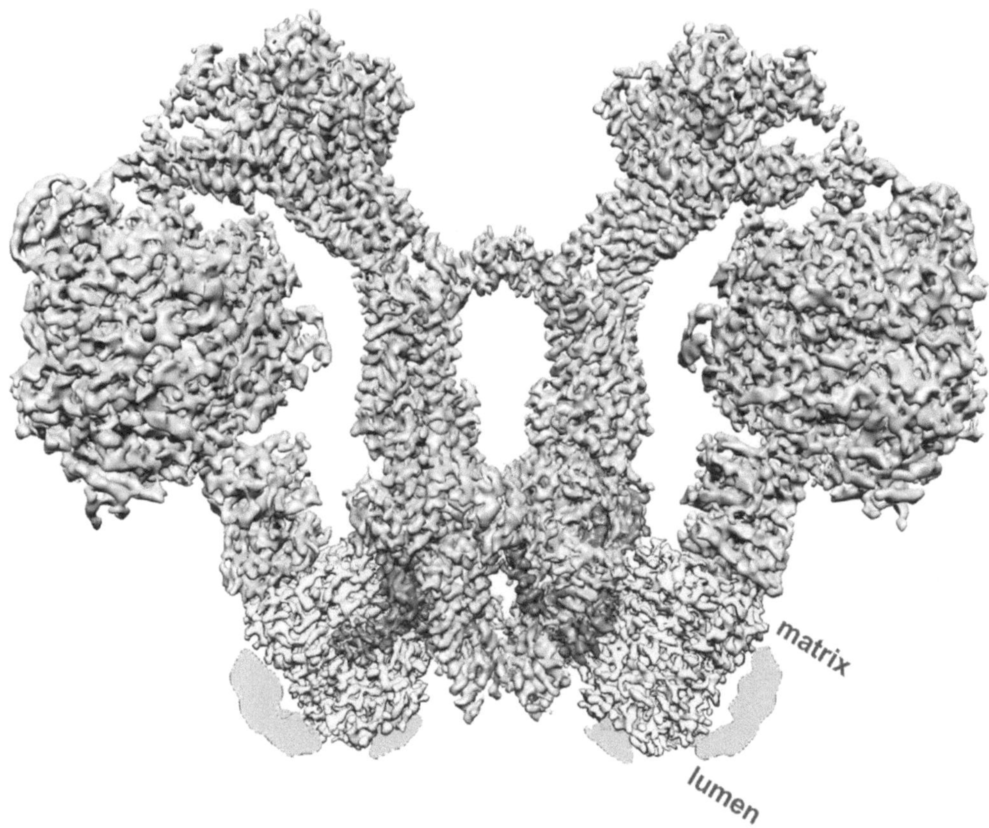

FIGURE 4.2 Cryo-EM structure of the Polytomella sp. F1 Fo ATP synthase dimer at 4.1 Å resolution. Subunit a, blue; c-ring, yellow; ASA 6, brick; other subunits, cyan; detergent micelle, grey. The color contrast in the figure here and all subsequent ones will be clear in the online version of the book.

***Source*: Klusch et al. (2017).**

evaluating the energetics of proton translocation and force generation in ATP synthases in any biological system.

4.4 REGULATION OF BIOENERGETICS DUE TO PHYSICAL FORCES

An increase in oxidative ATP synthesis rate gain and the ATP cost of contraction during all-out exercise were revealed about four decades ago (Broxterman et al. 2017). The exercise protocol utilized in this study elicited a "slow component-like" increase in intramuscular oxidative ATP synthesis rate (ATP_{OX}) gain as well as a progressive increase in the phosphate cost of contraction. Furthermore, the development of peripheral fatigue was closely related to the perturbation of specific fatigue-inducing intramuscular factors (i.e. pH and $H_2PO_4^-$ concentration). Therefore, it is not surprising to find that ATP synthesis may also be regulated by physical exercise and associated stresses (Calbet et al., 2020). The rate of mitochondrial ATP production during exercise is known to depend on the availability of O_2, carbon substrates, reducing equivalents, ADP, P_i, free creatine, and Ca^{2+}. A schematic representation of these phenomena has been provided in Figure 4.6, which provides connections of the biological energetics with various associated phenomena and parameterizations. We also know that the blood flow to contracting skeletal muscles is regulated during exercise in humans (Joyner et al., 2015). This article also addresses the fact that blood flow to the contracting muscles links oxygen in the atmosphere with

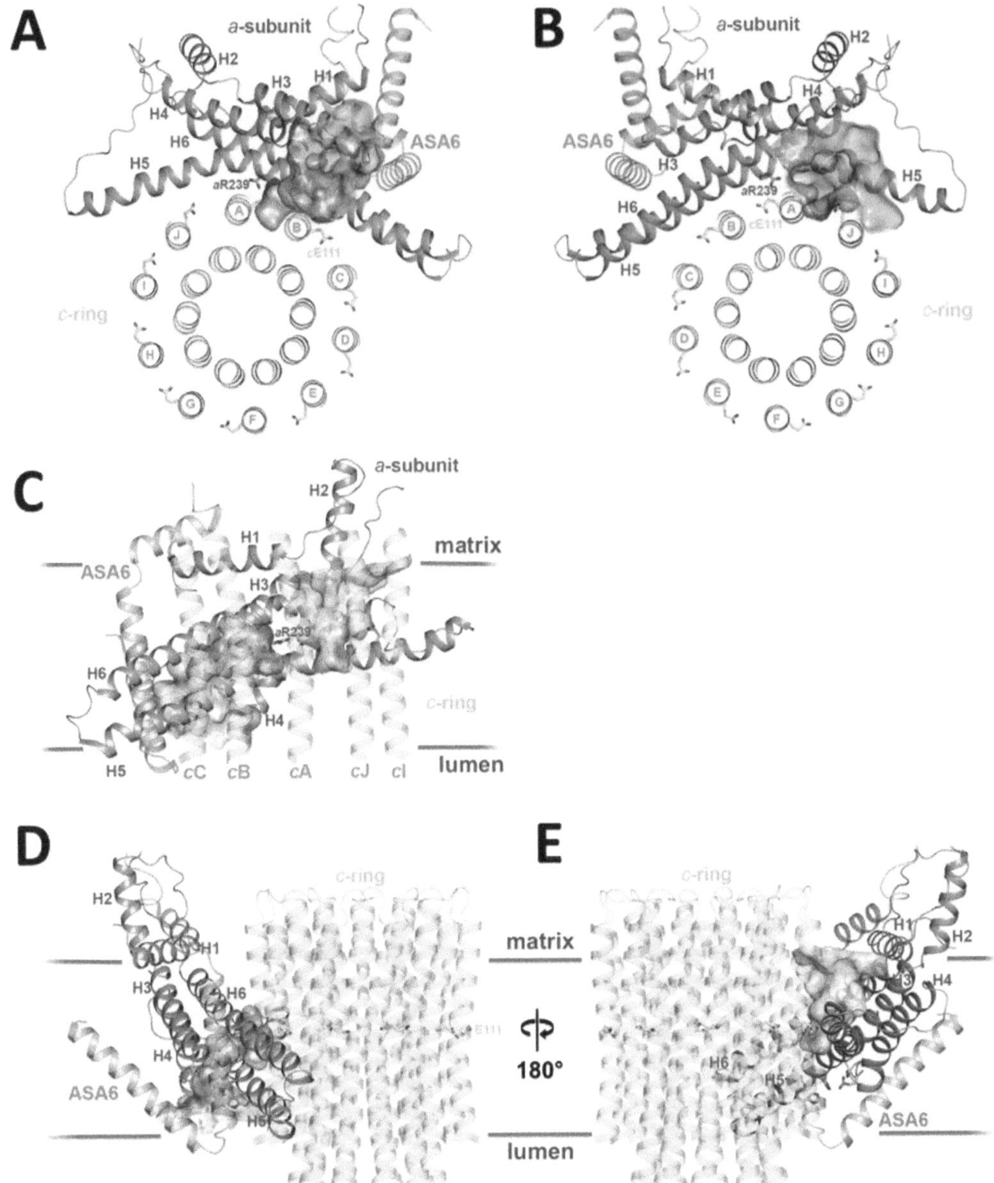

FIGURE 4.3 Two aqueous channels in F_o. (A) Lumenal channel seen from the crista lumen. (B) Matrix channel seen from the matrix. (C) Side view of both channels seen from the *c*-ring, with outer *c*-ring helices in transparent yellow. Lumenal channel, left; matrix channel, right. The strictly conserved *a*Arg239 in H5 separates the lumenal and matrix channels. (D) The lumenal channel passes through the H5/H6 hairpin at the small sidechains *a*Ala246, *a*Gly247 (H5), and *a*Ala292 (H6) (green). (E) H4, the N-terminal half of H5, and the connecting H4/H5 loop at the matrix channel. Subunit *a*, blue; c_{10}-ring, yellow; ASA 6, brick. Channels are shown as potential surfaces (red, negative; blue, positive; grey, neutral). (A) and (B) display a 5 Å slice of the c_{10}-ring at the level of the protonated *c*Glu111.

***Source*: Klusch et al. (2017).**

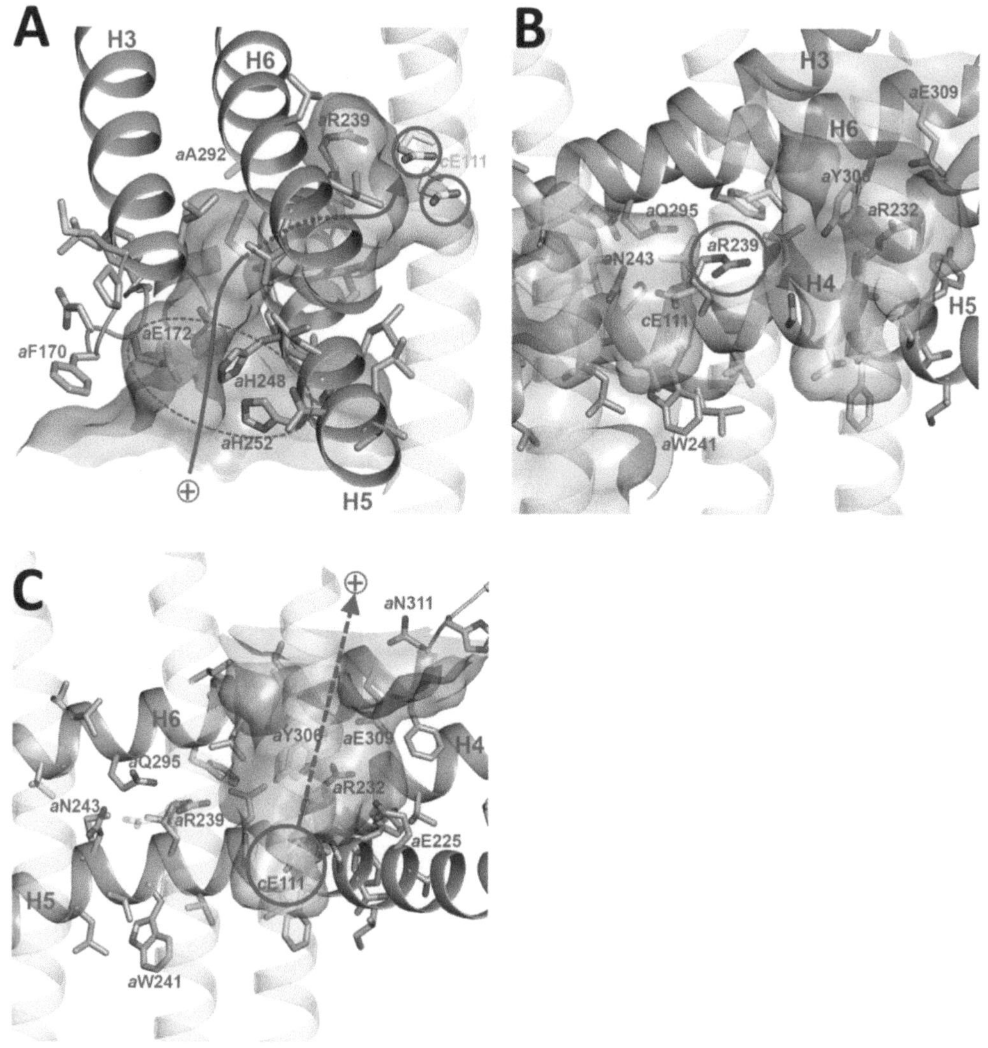

FIGURE 4.4 Proton pathway through the F_o subcomplex. (A) In the lumenal channel, protons (red arrow) pass via the local proton reservoir of *a*Glu172, *a*His248, *a*His252, and *a*Glu288 (dashed red ellipse) through the H5/H6 helix hairpin at the small sidechains of *a*Ala246, *a*Gly247 (H5), and *a*Ala292 (H6) (green) to *c*Glu111 in the rotor ring *c*-subunits (red circles). (B) *a*Arg239 (blue circle) is located halfway between the lumenal channel on the left and the matrix channel on the right, forming a seal to prevent proton leakage. *c*-ring helices (transparent yellow) with *c*Glu111 are seen in the foreground. (C) In the matrix channel, protons (dashed red arrow) can pass straight from the deprotonated *c*Glu111 to the pH 8 matrix. Subunit *a*, blue; adjacent *c*-ring helices, transparent yellow; aqueous channels, translucent grey; residues in stick representation.

***Source*: Klusch et al. (2017).**

the contracting muscles where it is consumed. We also know that mechanical forces, one of the major types of physical force, are associated with muscle development or change in the muscle's physical condition (Lemke and Schnorrer, 2017). Therefore, there are connections, via molecular mechanisms, between bioenergetics, especially the ATP synthesis process with which we are concerned in this chapter, and various physical phenomena of biological systems.

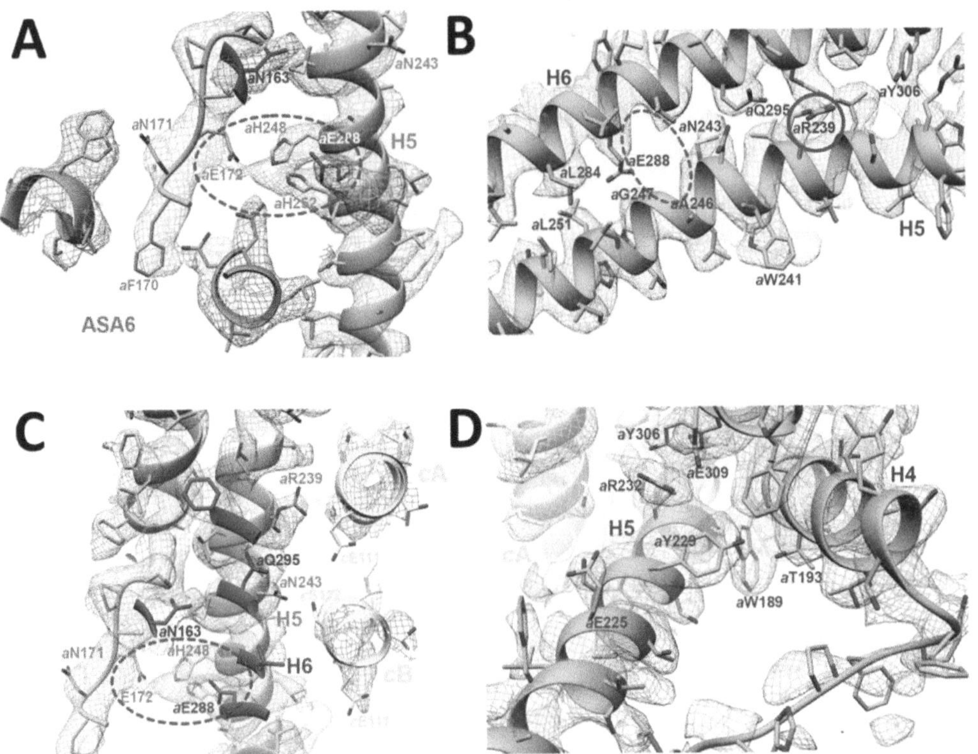

FIGURE 4.5 3.7 Å map of functionally important *a*-subunit residues with fitted atomic model. (A) Proton reservoir formed by *a*Glu172, *a*His248, *a*His252, *a*Glu288 (dashed red ellipse) in the lumenal channel; (B) Interaction of *a*Asn243 (H5) and *a*Gln295 (H6) stabilizes the H5/H6 hairpin. The space between *a*Glu288 (H6) and *a*Asn243 (H5) marks the lumenal channel (dashed red ellipse). (C) Protons in the lumenal channel can pass from *a*Glu288 to *c*Glu111 of *c*-subunit *A* near *a*Arg239 (H5) via *c*Ser112 and *a*Asn243. (D) *a*Trp189 (H4), *a*Thr193 (H4) and *a*Tyr229 (H5) act as wedges between H4 and the N-terminal end of H5, forming two sides of a triangle.

***Source*: Klusch et al. (2017).**

Like the effects of mechanical stresses, the ATP Synthase function and ATP production are also regulated due to electrical forces (Miller Jr et al., 2013). The field-driven torque model adopted here suggests a simple mechanism by which a membrane potential and ion gradient can drive a rotary motor such as ATP synthase. The ATP synthase exhibits nonlinear behavior, with a critical proton motive force needed for driving ATP production that depends strongly on the number of proton binding sites on the *c*-ring. When the *c*-ring couples to F_1 via the *γ*-subunit, the F_O-generated torque works against an opposing torque and potential due to F_1 (see Figure 4.7a), due to the energy needed to release ATP and overcome an additional energy barrier. Three ATP molecules are synthesized per rotation (Stock et al., 2000) as the *γ*-stalk rotates in 120° (2π/3 radian) steps within the *α*-*β* hexamer of F_1 (green & orange “lollipop” in Figure 4.7). The free energy ΔG required to synthesize ATP is expected to lie in the range 34–57 kJ/mol (0.35–0.59 eV/molecule) for ATP/ADP ratios ranging from 10^{-1} to 10^3 (Nicholls and Ferguson, 2002). The ATP/ADP ratio is likely to be close to *one* within the catalytically active portion of the F_1 complex. Equating ΔG, depicted in the bottom plot of Figure 4.7A, to release ATP from its catalytically active binding site in F_1 motor (Weber and Senior, 2003) to the work to drive the torque through a

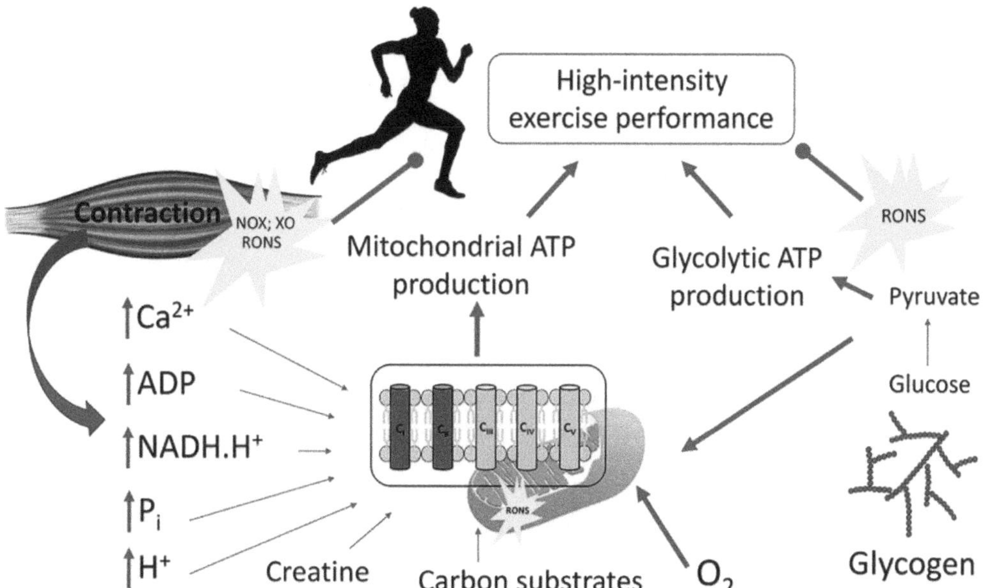

FIGURE 4.6 Association of ATP production with related biological parameters reveals regulation due to physical exercise.

***Source*: Calbet et al. (2020).**

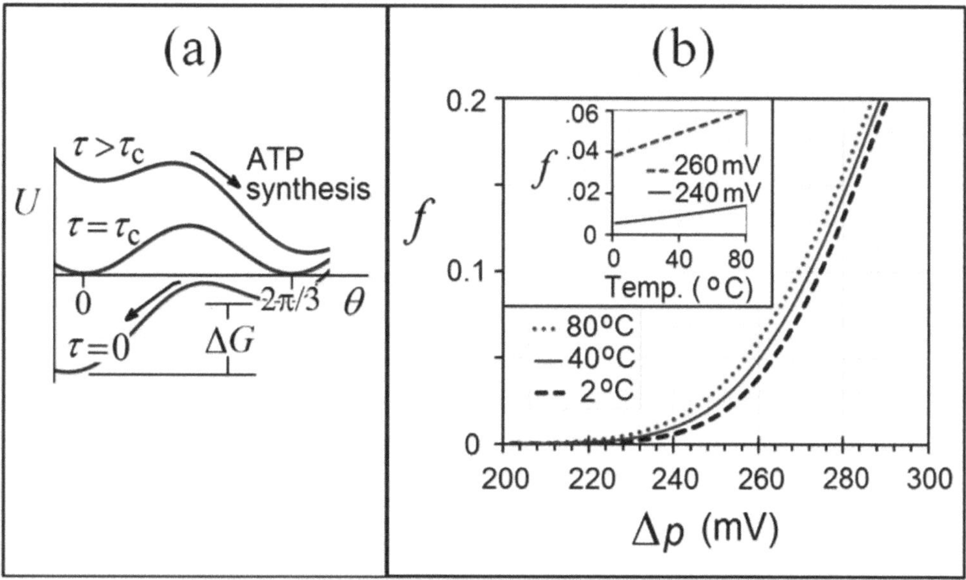

FIGURE 4.7 Torque from the *c*-ring drives rotation and ATP production by overcoming the periodic energy barrier in F_1 with the aid of thermal fluctuations.

120° (2π/3 radian) angle yields a critical torque τ_c needed to sustain ATP production in the range 27–45 pN·nm. This does not include any additional torque needed to overcome the energy barrier in F_1 (see Figure 4.7A), which is typically aided by thermal fluctuations, as will be discussed. Moreover, this barrier is much larger than the barrier for proton entry and exit from the *c*-ring and any step-like behavior of the motor is thus dominated by the 120° steps of F_1 (Palanisami and Okamoto, 2010), showing Brownian ratchet-like behavior. Figure 4.7B presents computed thermally assisted ATP production (rotation) rates f vs. pmf Δp, using the tilted washboard potential in Figure 4.7A, at various temperatures and (inset) f vs. temperature for two different pmf's, using the values $n = 8$, $\tau_c = 40$ pN·nm, and $\tau_1 = 20$ pN·nm. We avoid presenting the details here but they will be found in Miller Jr et al. (2013).

Due to the presence of so many physical processes active at the molecular level that are associated with bioenergetics, physical techniques may be utilized to measure the bioenergy (Leon-Sarmiento et al., 2019). As an example, neurophysics techniques were recently utilized to assess muscle bioenergy (Leon-Sarmiento et al., 2019). A dynamic model was recently developed in Deng et al. (2021), where an ATP reporter that tracks ATP in *E. coli* over different growth phases was explained. Our current understanding of energy dynamics and homeostasis in living cells has been limited due to the lack of easy-to-use ATP sensors and the lack of models that enable accurate estimates of energy and power consumption related to these ATP dynamics. Deng and colleagues have attempted to tackle this issue in their article.

4.5 SIMULATING CHEMICAL DYNAMICS TO ADDRESS BIOLOGICAL ENERGY CONVERSIONS

Mechanistic principles of enzymes that are responsible for biological energy conversion can be addressed using versatile simulations (Kaila, 2021). Here Kaila applied a combination of atomistic (aMD) and hybrid quantum/classical molecular dynamics (QM/MM MD) simulations with free energy (FE) sampling methods.

The catalytic biological energy conversion processes involve molecular-level elementary electron-, proton-, charge-, and energy-transfer reactions that take place in cell-based molecular machineries associated with respiration and photosynthesis processes. Understanding these crucial catalytic biological energy conversion processes using biological experimentations is quite incomplete (Alberts, 2017). The application of multiscale simulation approaches to bioenergetic enzymes with a focus on cellular respiration is progressing fast (Röpke et al., 2020; Röpke et al., 2021; Mühlbauer et al., 2020; Gamiz-Hernandez et al., 2017; Kaila, 2018). The mechanistic explorations of long-range proton-coupled electron transfer (PCET) dynamics were performed in the highly intricate respiratory chain enzyme Complex I, which functions as a redox-driven proton pump in bacterial and mitochondrial respiratory chains by catalyzing a 300 Å fully reversible PCET process (see Figure 4.8).

This PCET process is initiated by a hydride (H^-) transfer between NADH and Flavin mononucleotide (FMN), followed by long-range (>100 Å) electron transfer along a wire of 8 FeS centers leading to a quinone biding site. See Figure 4.9 for details on the simulated PCET reactions in Complex I. The reduction of the quinone to quinol initiates dissociation of the latter to a second membrane-bound binding site, and triggers proton pumping across the membrane domain of complex I, in subunits up to 200 Å away from the active site. The simulations across different size and time scales are found to be suggestive that transient charge transfer reactions lead to changes in the internal hydration state of key regions, local electric fields, and the conformation of conserved ion pairs, which in turn modulate the dynamics of functional steps along the reaction cycle. Figure 4.10 presents the schematic representation of the putative long-range PCET mechanism in Complex I.

Almost similar functional principles, which are known to operate on shorter length scales, are found in some unrelated proteins; Figure 4.11 presents a comparison. Enzymes may therefore be

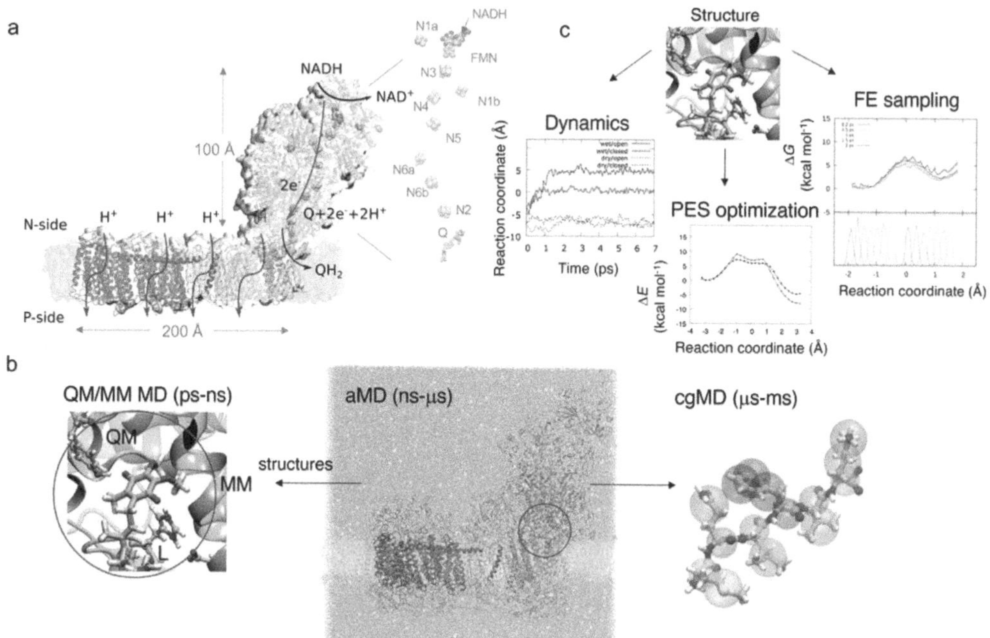

FIGURE 4.8 (a) Structure and function of Complex I. Reduced NADH donates electrons to the 100 Å long FeS chain that transfers them to quinone (Q). Q is reduced to quinol (QH_2), which triggers the transfer of four protons across the 200 Å long membrane domain. Point mutations of residues in the proton pumping subunits (shown in blue, red, yellow, gray/green) lead to inhibition of the Q oxidoreduction activity. (b) Multiscale simulation approaches can be used for probing the structure, function, and dynamics of PCET mechanisms in Complex I and other energy transducing enzymes. QM/MM models (left) allow exploring the local electronic structure, energetics, and dynamics on picosecond time scales (QM, QM region; MM, MM region; L, link atom in pink); atomistic MD (aMD) simulations (middle) enable sampling of the microsecond dynamics in a model of the biochemical environment; whereas coarse-grained models (cgMD) (right, showing a 1:4 mapping of beads:heavy atoms) allow exploring the micro- to millisecond time scale, but with loss of atomic detail. (c) The systems can be explored by unbiased MD simulations, potential energy surface (PES) scans, or free energy sampling methods at the different theory levels. The MD simulations allow probing, e.g. the dynamics of a reaction coordinate over time (here proton transfer, PT), whereas PES scan or FE sampling allows computing free energy/energy profiles along a reaction coordinate of interest (here a PT reaction).

***Source*: Kaila (2021).**

considered to employ conserved principles in the catalysis of energy transduction processes in biological systems.

4.6 MODELING CELLULAR BIOENERGETICS WITH THE USE OF A SYNTHETIC ATP REPORTER

A dynamic model that enables accurate estimates of energy and power consumption related to ATP dynamics and an ATP reporter that tracks ATP in *E. coli* over different growth phases have been reported in Deng et al. (2021). The measurement and modeling of cellular ATP dynamics are critical components in understanding fundamental physiology and its role in pathology.

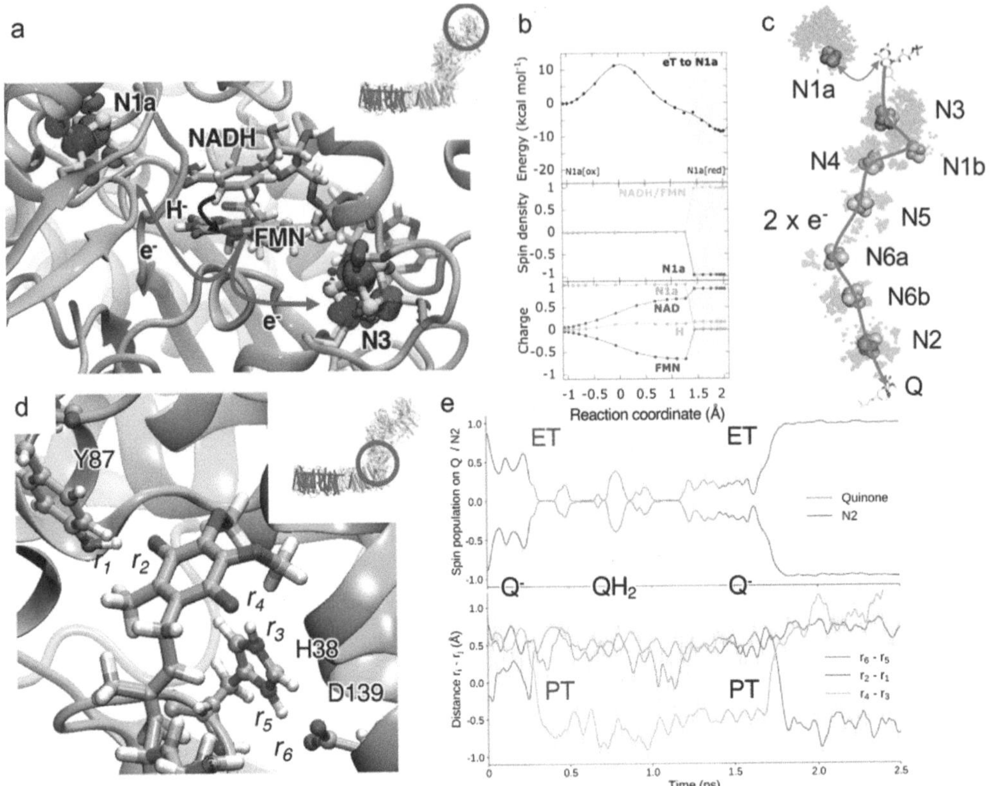

FIGURE 4.9 Multiscale simulations of PCET reactions in Complex I. (a) PCET between NADH and FMN leads to hydride (H^-) transfer between the cofactors, followed by ET to the nearby FeS centers. (b) Energy profiles, spin, and charge analysis of the cofactors along the PCET process and coupled ET (here to N1a) along a proton transfer reaction coordinate. All FeS center were initially modeled in their oxidized state. The change in redox state of N1a is indicated in the top panel. (c) ET pathway down along the 100 Å FeS chain to Q. The figure also shows water molecules surrounding the FeS clusters. (d) QM/MM model of the Q oxidoreduction process: PCET between the terminal N2 FeS and Q triggers PT from nearby Tyr and His residues. (e) QM/MM MD of the PCET process shows the reversible formation of the QH_2 by ET from the reduced N2 FeS center, and vice versa. Copied from refs. (Gamiz-Hernandez et al., 2017; Saura and Kaila, 2019; Kaila, 2021).

***Source*: Gamiz-Hernandez et al. (2017); Saura and Kaila (2019); Kaila (2021).**

Abnormal regulation of energy levels (ATP concentration) and power consumption (ATP consumption flux) in human cells is associated with numerous diseases from cancer to viral infection and immune dysfunction, while in microbes it influences their responses to drugs and other stresses. Therefore, addressing the energetics issue with regard to synthetic ATP reporters may be considered important especially in regulating ATP productions in abnormal functioning cells experiencing disease conditions.

The vacuolar ATPases (V-ATPases) may play an important role in cancer (see Figure 4.12). ATPase and ATP synthase function oppositely: ATPase is the enzyme which breaks down ATP into ADP and the free phosphate group, while ATP synthase is the enzyme which synthesizes ATP by combining ADP and a free phosphate group. The vacuolar (V-)ATPases are ATP-driven proton pumps that function to acidify intracellular compartments and transport protons across the plasma

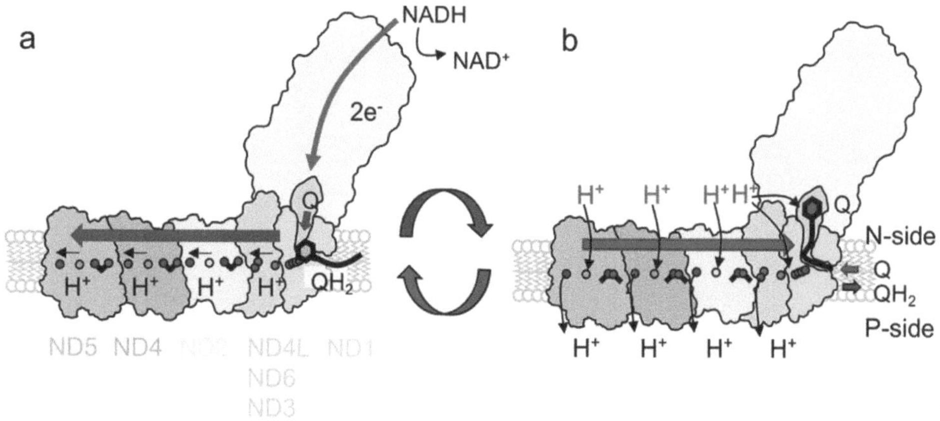

FIGURE 4.10 Schematic representation of the putative long-range PCET mechanism in Complex I. (a) NADH-driven ET along the FeS chain reduces Q to QH_2, which triggers conformational changes and motion of the QH_2 to a second binding site. The QH_2/QH^- initiate stepwise PT reactions that lead to consecutive opening/conformational changes of ion pairs, and modulate the energetics of lateral PT reactions. The signal propagates to the terminal edge of ND5 (in red). (b) Proton release across the membrane increases the pK_a of the middle Lys, leading to H^+ uptake from the bulk (H^+ in red) and closure of the IP in the last subunit (subunit in red). The closed IP destabilizes the proton stored at the ND4/ND5 interface that is ejected across the membrane. The signal propagates "backward" in a similar way via ND4 (in blue), ND2 (in yellow), and ND4L/ND1(orange/green) to the quinol, which is ejected to the membrane. The new reaction cycle is initiated by reprotonation of the Q-active site and uptake of a new Q from the membrane.

***Source*: Mühlbauer et al. (2020). Copyright American Chemical Society.**

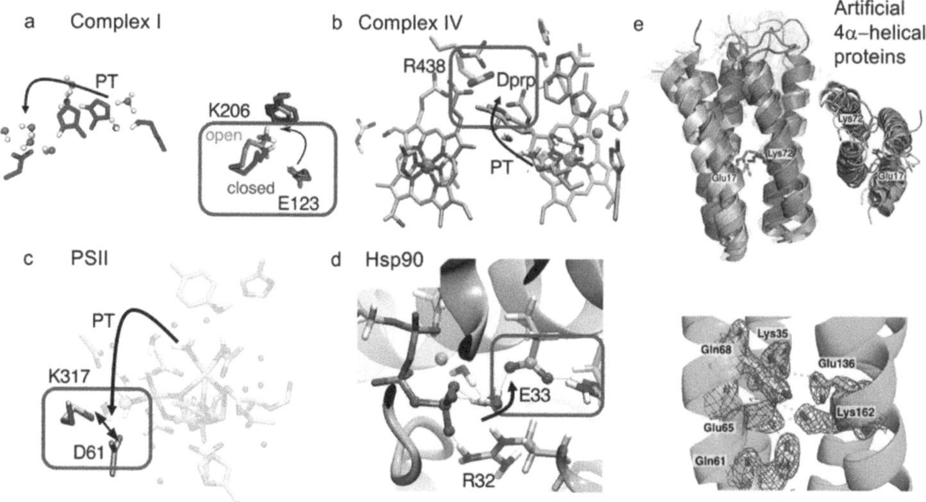

FIGURE 4.11 Comparison of functional elements involved in CT/PCET reaction in different enzymes. (a) Complex I (closeup of antiporter-like subunit ND4). (b) Cytochrome *c* oxidase (nonpolar cavity around the active site). (c) Photosystem II (vicinity of the $CaMn_4O_5$ cluster). (d) ATPase reaction in Hsp90 (active site in the N-terminal domain) (see main text). (e) Designed buried ion pairs in artificial bundle proteins, showing different ion-pair conformations, with aims to understand functional principles from a bottom-up approach. All ion pairs shown are located in buried core regions of the proteins.

***Source*: Panel (A) adapted from Röpke et al. (2020), copyright American Chemical Society; Panel (E) adapted from Baumgart et al. (2021).**

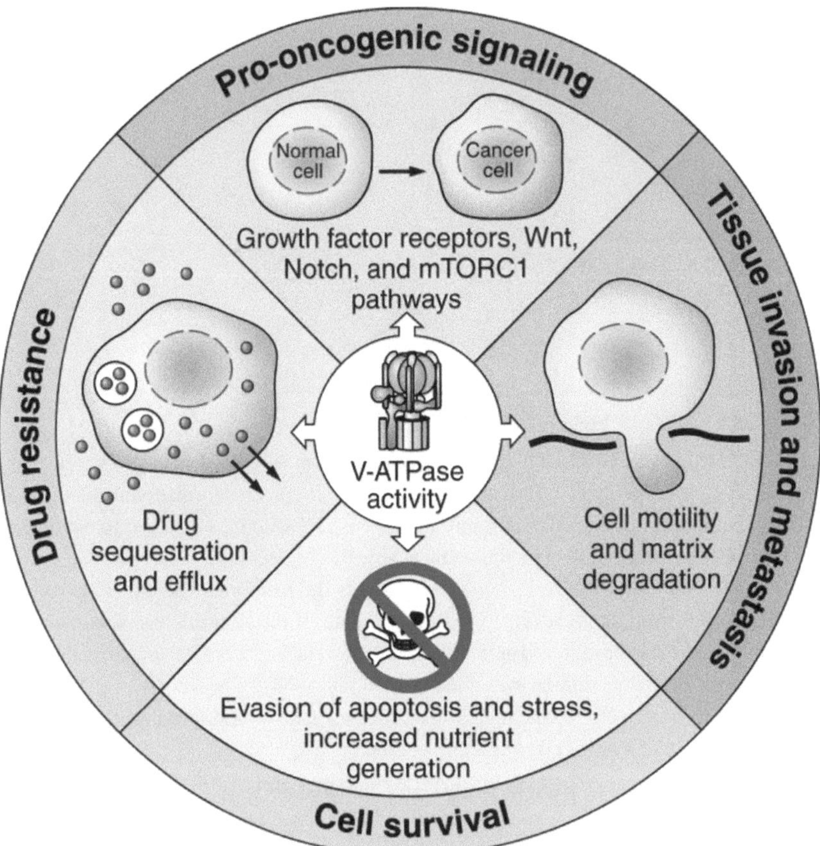

FIGURE 4.12 The role of V-ATPases in cancer. V-ATPase activity supports pro-oncogenic signaling pathways, including growth factor receptors, Wnt, Notch, and mTORC1 (top panel). V-ATPase activity allows tumor cells to evade apoptosis and various cellular stressors (right panel). V-ATPase activity promotes the ability of tumor cells to invade surrounding tissue and metastasize to secondary sites within the body by promoting cell motility and extracellular matrix degradation (bottom panel). V-ATPase activity contributes to drug resistance by promoting drug trapping in acidic cellular spaces and promoting drug efflux out of the cell (left panel).

***Source*: Stransky et al. (2016).**

membrane (Forgac, 2007). They function in various normal and disease processes, including membrane trafficking, protein degradation, bone resorption, and tumor metastasis.

V-ATPase complexes may reversibly dissociate *in vivo* into their component V_1 and V_0 domains due to stimuli, therefore shutting down ATP-dependent proton transport (see Figure 4.13a) (Forgac, 2007; Kane, 2006). This reversible dissociation occurs in yeast in response to glucose depletion (Kane, 2006) and in insects during molting (Beyenbach and Wieczorek, 2006). Increased assembly of V-ATPase complexes also occurs in dendritic cells upon activation of antigen processing (Trombetta et al., 2003), suggesting that this mechanism is quite common to mammalian systems. In both yeast and insects, dissociation represents a means of conserving cellular ATP under energy-limiting conditions because the separated V_1 and V_0 domains are silent as an ATPase or passive proton channel (Nishi and Forgac, 2002).

We see here that cellular energy conditions and ATP production issues in abnormal cells relative to normal cells may serve as biological markers associated with certain diseases. Deng

a

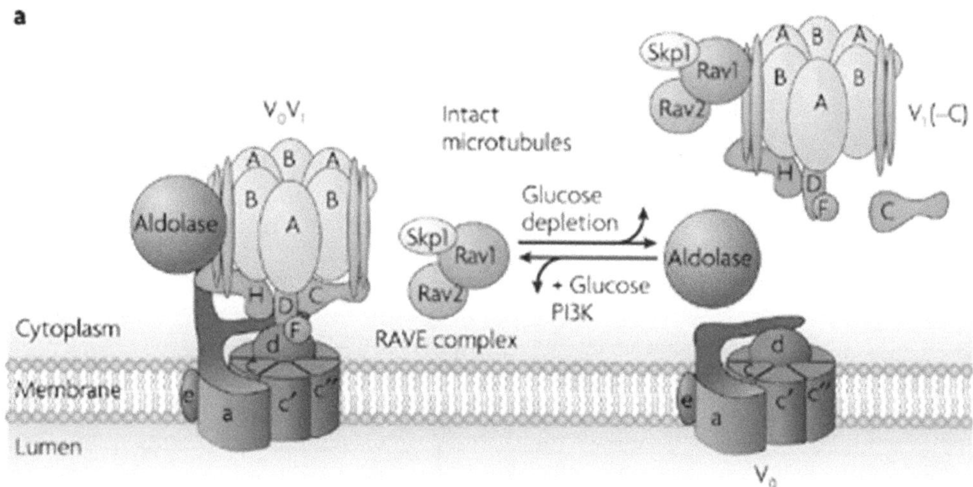

b

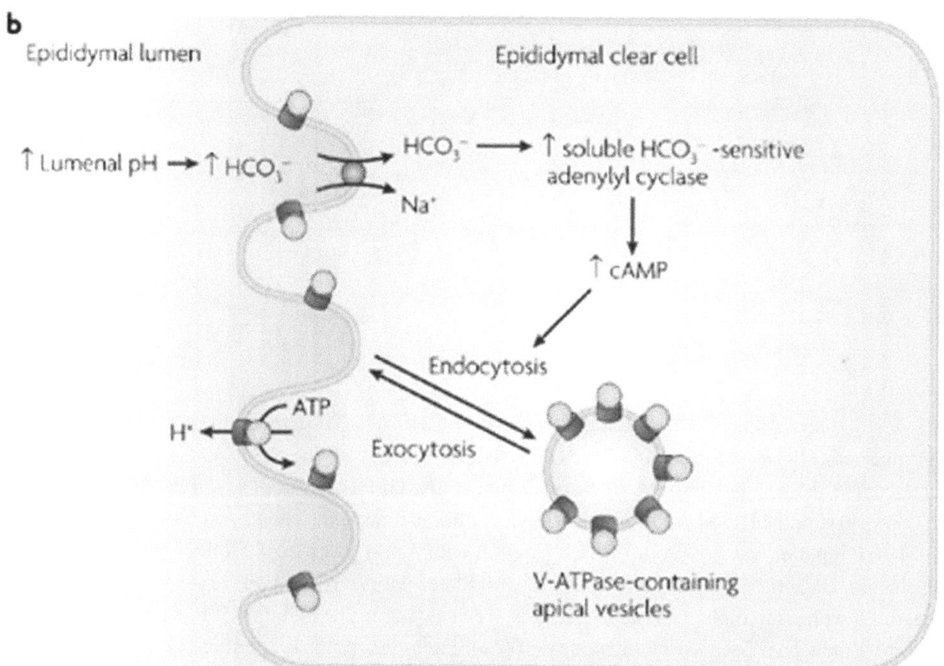

FIGURE 4.13 (a) Vacuolar (V-)ATPase complexes undergo reversible dissociation *in vivo* into a V_1 complex that lacks subunit C, free subunit C and V_0. In yeast this process happens in response to glucose depletion2 and requires an intact microtubule network. Dissociation might also be driven by altered interactions of the intact complex with aldolase or through altered binding of the non-homologous domain of subunit A with V_0. In insects, V-ATPase dissociation in the midgut occurs during molting. Assembly of the complex in yeast is promoted by the RAVE (regulator of the ATPase of vacuolar and endosomal membranes) complex. The RAVE complex, which is composed of Skp1, Rav1, and Rav2, binds to free V_1 and also appears to be involved in normal assembly of the V-ATPase. In mammalian renal cells, glucose-dependent assembly requires phosphoinositide 3-kinase (PI3K), whereas in dendritic cells assembly occurs upon activation of antigen processing. (b) Control

FIGURE 4.13 (*Continued*)

of V-ATPase activity in epididymal clear cells occurs through regulated insertion in the apical membrane. In response to elevated pH in the lumen of the epididymis, bicarbonate (HCO_3^-) entry across the apical membrane coupled to sodium (Na^+) influx (possibly via the cotransporter NBC3) stimulates a soluble bicarbonate-sensitive adenylyl cyclase. The resulting increase in cyclic (c)AMP stimulates exocytosis of V-ATPase-containing apical membrane vesicles and inhibits apical uptake of V-ATPases by endocytosis. This results in a net increase in the density of V-ATPases in the apical membrane.

***Source*: Forgac (2007).**

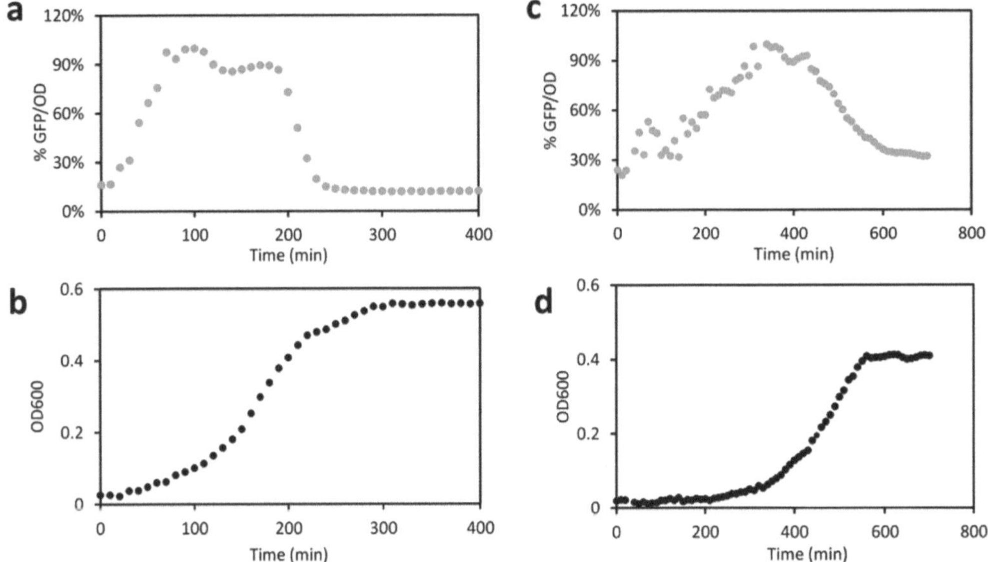

FIGURE 4.14 Fluorescence and growth characteristics of HC-M ATP reporter in rich and minimal media. (a, b) Normalized cellular GFP dynamics [% (GFP/OD)] (a) and growth (b) of *E. coli* carrying the HC-M ATP reporter grown in the rich medium. (c, d) Normalized cellular GFP dynamics [% (GFP/OD)] (c) and growth (d) of *E. coli* carrying the HC-M ATP reporter grown in minimal medium. The *E. coli* NEB 10-beta strain with the HC-M ATP reporter was grown in EZ rich medium with 5 mM glucose or MOPS minimal medium with 10 mM glucose. Bacteria were grown in black 96-well plates with shaking. GFP (ex485/em528) and OD600 were measured with a microplate reader (Molecular Devices, Inc.). The cellular GFP signals, GFP/OD, were normalized by their own peak GFP/OD values (100%). Each data point is the mean value of at least three independent experiments. The standard deviations were small (< 15 percent of the mean) and not shown.

***Source*: Deng et al. (2021).**

and colleagues (2021) made an attempt to synthesize ATP reporter to address these issues. The reporter is made by fusing an ATP-sensing rrnB P1 promoter with a fast-folding and fast-degrading green fluorescence protein (GFP). The developed and validated rrnB P1 promoter could be used as an ATP reporter. A kinetic model was developed for measuring ATP dynamics and ATP power consumption fluxes in *E. coli*. The ATP reporter was found to faithfully track cellular ATP dynamics in both the minimal and rich media across different growth phases and in different strains (Figures 4.14, 4.15, 4.16, and 4.17). The HC-M ATP reporter was found to work across different strains and conditions (Figures 4.15 and 4.16); absolute accuracy required that the ATP reporter be calibrated for different strains and conditions. For determining the dynamics and

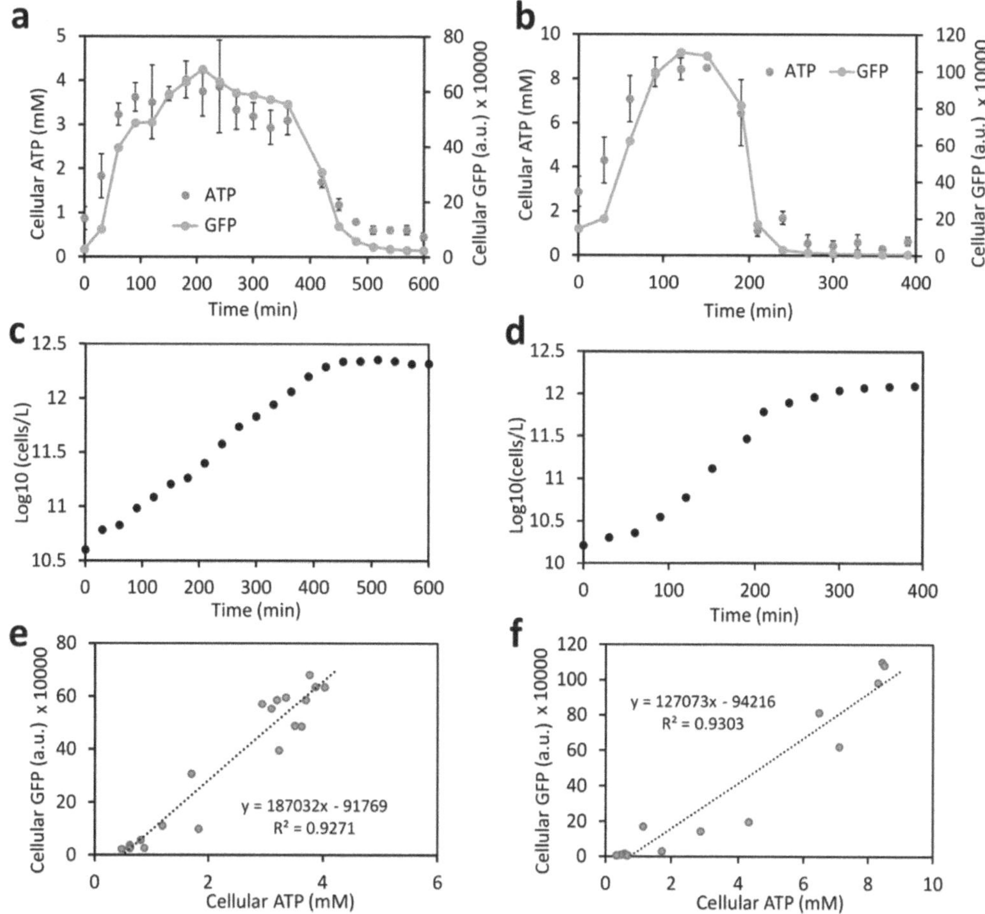

FIGURE 4.15 Correlation between the *rrnB* P1-GFP reporter and ATP. (a, b) Dynamics of the GFP reporter and cellular ATP levels in *E. coli* in minimal (a) and EZ-rich medium (b). Cellular GFP fluorescence for each sample (arbitrary fluorescence unit, a.u.) was measured by flow cytometry. Cellular ATP levels were determined by a standard luciferase assay and converted to cellular concentration in mM. Growth of bacteria is shown in minimal medium (c) and in rich medium (d). Bacterial cell counts were estimated by flow cytometry corrected with counting beads. (e, f) Linear correlations between cellular ATP concentration and cellular GFP levels in minimal (e) and rich medium (f). All experiments used the same BW25113 strain for consistency. All data points are mean values of at least three independent biological replicates with one standard deviation (SD). The SD for growth and the GFP signal were relatively small (< 15 percent) and are not shown.

***Source*: Deng et al. (2021).**

power consumption fluxes of ATP in a bacterial cell, a dynamic model was developed that was found to accurately predict bacterial growth, glucose, and acetate metabolism, dissolved oxygen dynamics, and ATP dynamics (see Figure 4.18).

4.7 ENERGY-TRANSDUCING ARTIFICIAL CELLS

Protein complexes from photosynthetic bacteria may be incorporated into artificial cells for converting light into chemical energy, in the form of a proton gradient (Allen, 2017). An artificial photosynthetic

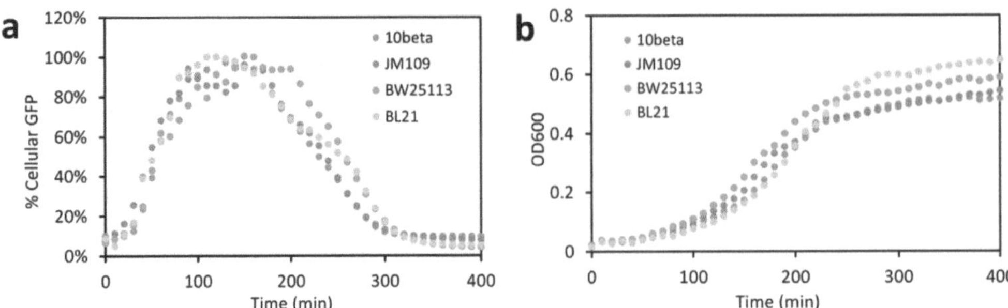

FIGURE 4.16 Robustness of the *rrnB* P1 ATP reporter in different *E. coli* strains. (a) Dynamics of cellular GFP (GFP/OD) of four *E. coli* strains (NEB 10-beta, BW25113, JM109DE3, and BL21DE3) with HC-M ATP reporter in rich medium with 6 mM glucose. (b) Corresponding bacterial growth among the four strains in the same medium. Bacterial strains were grown in black 96-well plates and GFP (ex485/em528) and OD600 were measured by a microplate reader. Cellular GFP (GFP/OD) signals of each strain were normalized to their respective peak GFP/OD values (100 percent) for comparison among different strains. Each data point represents the mean of at least three independent replicates with a standard deviation < 13 percent of the mean.

***Source*: Deng et al. (2021).**

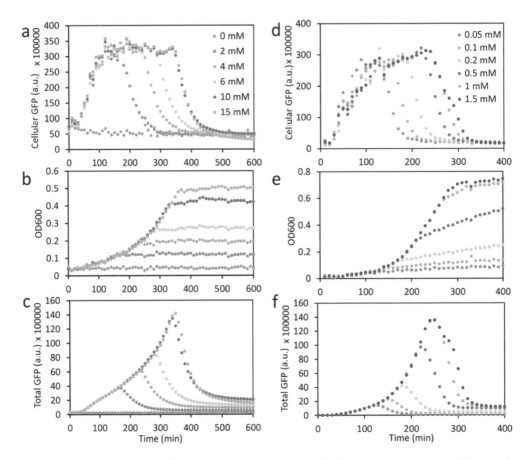

FIGURE 4.17 Cellular GFP (ATP) dynamics in response to rate-limiting nutrients using our ATP-dependent reporter. (a–c) Cellular GFP dynamics (a), bacterial growth (b), and total-population-level GFP dynamics (c),

FIGURE 4.17 *(Continued)*
in response to different amounts of glucose (0–15 mM) in minimal medium. The cellular GFP serves as a good proxy for cellular ATP and helps us monitor it in real time. (d–f) Cellular GFP dynamics (d), bacterial growth (e), and total-population-level GFP dynamics (f) in response to different amounts of phosphate (0.05–1.5 mM) in rich medium. Each data point represents the mean of four independent biological replicates with standard deviation < 15 percent of the mean. Bacteria were grown in black 96-well plates. GFP (ex485/em528) and OD600 were measured by a microplate reader. Cellular GFP is defined as the total GFP divided by OD.

***Source*: Deng et al. (2021).**

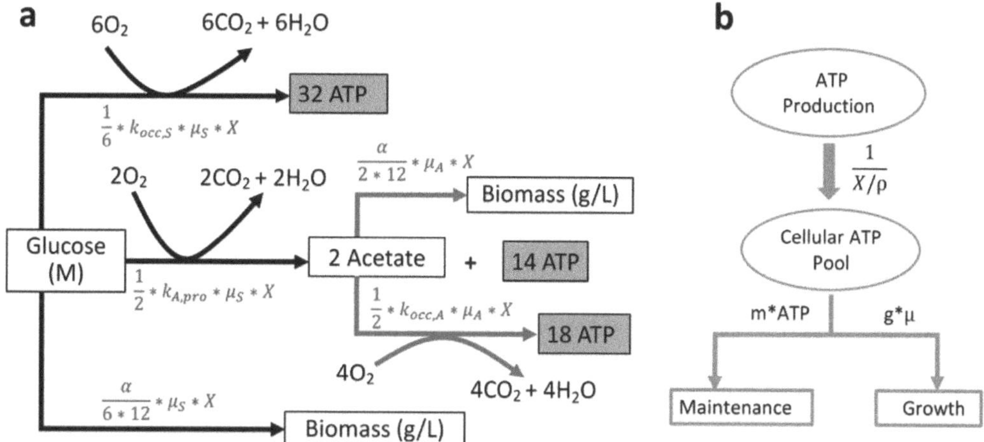

FIGURE 4.18 Workflow of kinetic model development. (a) Black arrows correspond to glucose fluxes converting glucose to biomass, carbon dioxide and water (respiration), or to the byproduct acetate, the excretion flux. Similarly, blue arrows correspond to acetate fluxes converting acetate to biomass or to carbon dioxide and water. The stoichiometry is based on general biochemical processes for glucose and acetate metabolism. Acetate utilization only occurs after glucose is depleted. The mathematical terms on each arrow describe the flux, or consumption rate, for that path in (M/s). Given the glucose or acetate consumption rate for each pathway, the total ATP production rate can then be calculated via stoichiometry (mole product per mole substrate consumed). Biomass production is based on the carbon balance between the substrate consumed and cellular carbon produced. (b) All ATP production fluxes are gathered and converted to the cellular ATP pool. This pool is simultaneously drained for bacterial growth and for cell maintenance via the mathematical flux rate terms (M/s) as indicated in the figure.

***Source*: Deng et al. (2021).**

cell system was recently proposed which has paved the way to construct an energetically independent artificial cell (Berhanu et al., 2019). A cell-free protein synthesis system was combined here with small proteoliposomes consisting of purified ATP synthase and bacteriorhodopsin, inside a giant unilamellar vesicle in order to synthesize protein by the production of ATP by light.

For developing artificial cells into the energetically independent system, it is necessary to set up a circulating energy-consumption and production system driven by an unlimited external physical or chemical energy source. A biomimetic artificial organelle producing ATP by collaborating ATP synthase and bacteriorhodopsin is applicable as a rational energy-generating system for artificial cells (Deisinger et al., 1993; Freisleben et al., 1995; Matuschka et al., 1995). Lee and colleagues (2018) recently performed ATP synthesis using a similar photosynthetic artificial organelle, where they demonstrated carbon fixation (*in vitro*) and actin polymerization within giant unilamellar

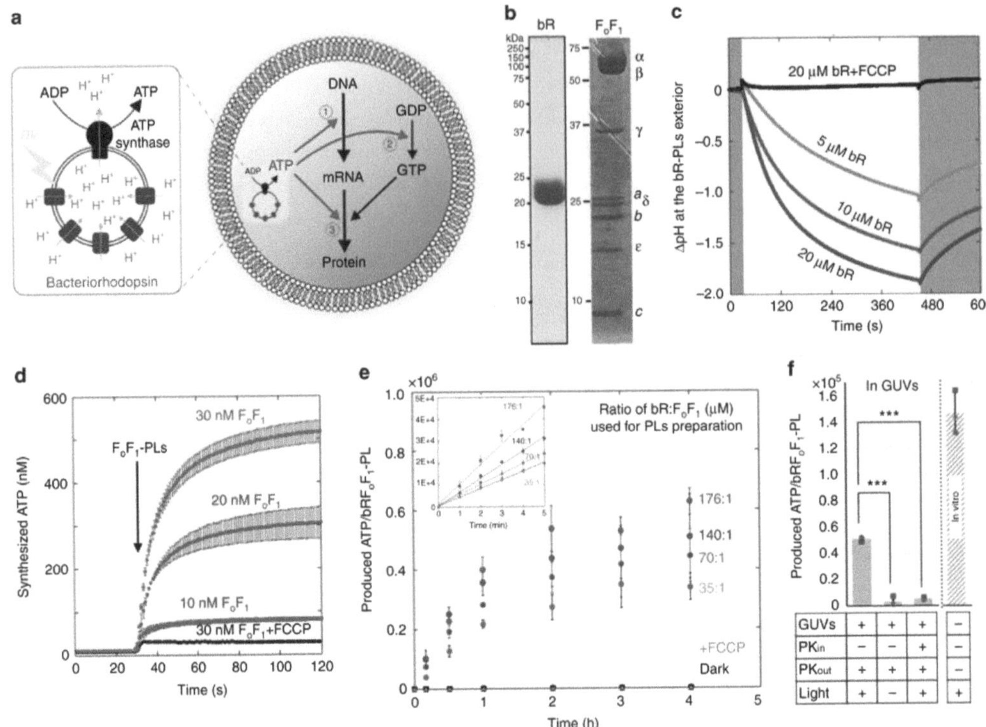

FIGURE 4.19 Light-driven ATP synthesis by artificial organelle. (a) Schematics of the artificial photosynthetic cell encapsulating artificial organelle, which consists of bacteriorhodopsin (bR) and F_0F_1-ATP synthase (F_0F_1). Synthesized ATP are consumed as substrates for messenger RNA (mRNA) (①), as energy for phosphorylation of guanosine diphosphate (GDP) (②), or as energy for aminoacylation of transfer RNA (tRNA) (③). (b) Sodium dodecyl sulfate-polyacrylamide gel electrophoresis (SDS-PAGE) analysis of purified bR and F_0F_1. The positions of molecular markers and F_0F_1 component proteins are indicated beside the gels. (c) Light-driven proton-pump activity of bR reconstituted in a proteoliposome (PL). Proton-pump activity of bR was measured by monitoring the proton concentration at the outside of bR-PLs where fluorescent proton-sensor ACMA (9-amino-6-chloro-2-methoxy acridine) was added. We defined as ΔpH = pH (original, outside) – pH (after illumination, outside). The ΔpH caused by bR activity was measured with the various bR concentrations as indicated. White and gray areas indicate light ON and OFF, respectively. An uncoupler, FCCP (carbonyl cyanide 4-(trifluoromethoxy) phenylhydrazone), was used as a control experiment. (d) ATP synthesis activity of F_0F_1 reconstituted as F_0F_1-PLs. ATP synthesis reactions were initiated by adding F_0F_1-PLs at 30 s with various F_0F_1 concentrations, as indicated. The synthesized ATP was measured by means of luciferin and luciferase (see Methods section of Berhanu et al., 2019, for the experiment details). FCCP was used for control. (e) Light-driven ATP synthesis. The amount of the photosynthesized ATP by bRF_0F_1-PLs, which was constituted in various proportions of bR against F_0F_1, were measured by luciferin and luciferase. FCCP and dark conditions were also performed as controls. The inset indicates initial rate of each PL. (f) Light-driven ATP synthesis inside giant unilamellar vesicle (GUV). bRF_0F_1-PLs were illuminated inside GUVs in the presence or absence of proteinase K (PK) that degrades the F_0F_1. The *in vitro* experiment was also performed for comparison. ***$p < 0.001$. *P* values were from two-sided *t*-test. All experiments were repeated at least three times, and their mean values and standard deviations (SD) are shown.

***Source*: Berhanu et al. (2019).**

vesicles (GUVs). The artificial organelle concept was applied to the artificial cell system, i.e. protein synthesis based on the photosynthesized ATP inside GUV (Berhanu et al., 2019). Here ATP synthesis was performed by light-driven artificial organelle inside GUV. Revising the method of arranging proteoliposomes containing bacteriorhodopsin and ATP synthase, Berhanu and colleagues succeeded in producing millimolar level ATP inside GUVs, wherein 4.6 µmol ATP per mg ATP synthase was produced after 6 h of illumination. By combining the artificial organelle and PURE system, an artificial photosynthetic cell system was constructed that produces ATP for internal protein synthesis. The produced ATP was consumed as a substrate of messenger RNA (mRNA), or as an energy for aminoacylation of transfer RNA (tRNA) and for phosphorylation of guanosine diphosphate (GDP) (see Figure 4.19). They also demonstrated photosynthesis of bacteriorhodopsin or a membrane portion of ATP synthase, an original component of the artificial organelle. This artificial cell system is found to enable the self-constitution of its own parts within a structure of the positive feedback loop.

The photo-synthesized ATP is consumed as a substrate for transcription and as an energy for translation, eventually driving the synthesis of bacteriorhodopsin or constituent proteins of ATP synthase, the original essential components of the proteoliposome. Proteoliposomes mimic lipid membranes (liposomes) into which a protein can be incorporated or inserted (Ciancaglini et al., 2012; Scalise et al., 2013). The de novo photosynthesized bacteriorhodopsin and the parts of ATP synthase integrate into the artificial photosynthetic organelle and enhance its ATP photosynthetic activity through the positive feedback of the products. Detailed analysis will be found in Berhanu et al. (2019).

REFERENCES

Alberts, B. (2017). *Molecular biology of the cell*. New York: W.W. Norton. https://doi.org/10.1201/9781315735368

Allen, J.P. (2017). Design of energy-transducing artificial cells. *Proc. Natl. Acad. Sci. USA*, *114*(15), 3790–3791. doi: 10.1073/pnas.1703163114; PMID: 28360203; PMCID: PMC5393233

Anandakrishnan, R., Zhang, Z., Donovan-Maiye, R., & Zuckerman, D.M. (2016). Biophysical comparison of ATP synthesis mechanisms shows a kinetic advantage for the rotary process. *Proceedings of the National Academy of Sciences*, *113*(40), 11220–11225

Baumgart, M., Röpke, M., Mühlbauer, M.E., Asami, S., Mader, S.L., Fredriksson, K., ... & Kaila, V.R.I. Design of buried charged networks in artificial proteins. Nature Communications, *12*, 1895. https://doi.org/10.1038/s41467-021-21909-7

Berhanu, S., Ueda, T., & Kuruma, Y. (2019). Artificial photosynthetic cell producing energy for protein synthesis. *Nat. Commun.*, *10*, 1325. https://doi.org/10.1038/s41467-019-09147-4

Beyenbach, K.W. & Wieczorek, H. (2006). The V-type H+ ATPase: molecular structure and function, physiological roles and regulation. *J. Exp. Biol.*, *209*, 577–589

Broxterman, R.M., Layec, G., Hureau, T.J., Amann, M., & Richardson, R.S. (2017). Skeletal muscle bioenergetics during all-out exercise: mechanistic insight into the oxygen uptake slow component and neuromuscular fatigue. *Journal of Applied Physiology*, *122*(5), 1208–1217

Calbet, J.A., Martín-Rodríguez, S., Martin-Rincon, M., & Morales-Alamo, D. (2020). An integrative approach to the regulation of mitochondrial respiration during exercise: focus on high-intensity exercise. *Redox Biology*, *35*, 101478

Ciancaglini, P., Simão, A.M.S., Bolean, M., Millán, J.L., Rigos, C.F., Yoneda, J.S., … & Stabeli, R.G. (2012). Proteoliposomes in nanobiotechnology. *Biophysical Reviews*, *4*, 67–81

Deisinger, B., Nawroth, T., Zwicker, K., Matuschka, S., John, G., Zimmer, G., & Freisleben, H.J. (1993). Purification of ATP synthase from beef heart mitochondria (FoF1) and co-reconstitution with monomeric bacteriorhodopsin into liposomes capable of light-driven ATP synthesis. *European Journal of Biochemistry*, *218*(2), 377–383

Deng, Y., Beahm, D.R., Ionov, S., et al. (2021). Measuring and modeling energy and power consumption in living microbial cells with a synthetic ATP reporter. *BMC Biol.*, *19*, 101. https://doi.org/10.1186/s12915-021-01023-2

Forgac, M. (2007). Vacuolar ATPases: rotary proton pumps in physiology and pathophysiology. *Nat. Rev. Mol. Cell. Biol.*, 8, 917–929. https://doi.org/10.1038/nrm2272

Freisleben, H.J., Zwicker, K., Jezek, P., John, G., Bettin-Bogutzki, A., Ring, K., & Nawroth, T. (1995). Reconstitution of bacteriorhodopsin and ATP synthase from Micrococcus luteus into liposomes of the purified main tetraether lipid from Thermoplasma acidophilum: proton conductance and light-driven ATP synthesis. *Chemistry and Physics of Lipids*, *78*(2), 137–147

Gamiz-Hernandez, A.P., Jussupow, A., Johansson, M.P., & Kaila, V.R.I. (2017). Terminal electron–proton transfer dynamics in the quinone reduction of Respiratory Complex I. *Journal of the American Chemical Society*, *139*(45), 16282–16288.

Joyner, M.J., & Casey, D.P. (2015). Regulation of increased blood flow (hyperemia) to muscles during exercise: a hierarchy of competing physiological needs. *Physiol. Rev.*, *95*(2), 549–601. doi: 10.1152/physrev.00035.2013; PMID: 25834232; PMCID: PMC4551211

Kagawa, Y. (1999). Biophysical studies on ATP synthase. *Advances in Biophysics*, *36*, 1–25

Kaila, V.R.I. (2021). Resolving chemical dynamics in biological energy conversion: long-range proton-coupled electron transfer in Respiratory Complex I. *Accounts of Chemical Research*, *54*(24), 4462–4473

Kaila, V.R.I. (2018). Long-range proton-coupled electron transfer in biological energy conversion: towards mechanistic understanding of Respiratory Complex I. *J. R. Soc., Interface*, *15* (141), 20170916, doi: 10.1098/rsif.2017.0916

Kane, P.M. (2006). The where, when and how of organelle acidification by the yeast vacuolar H+-ATPase. *Microbiol. Mol. Biol. Rev.*, *70*, 177–191

Kazakova, T.B., & Markosian, K.A. (1966). Comparison of physicochemical properties of mitochondrial and nuclear deoxyribonucleic acid from rat liver cells. *Nature*, *211*, 79–80

Klusch, N., Murphy, B.J., Mills, D.J., Yildiz, Ö., & Kühlbrandt, W. (2017). Structural basis of proton translocation and force generation in mitochondrial ATP synthase. *eLife*, *6*, e33274

Lee, K.Y., Park, S.J., Lee, K.A., Kim, S.H., Kim, H., Meroz, Y., ... & Shin, K. (2018). Photosynthetic artificial organelles sustain and control ATP-dependent reactions in a protocellular system. *Nature Biotechnology*, *36*(6), 530–535

Lemke, S.B., & Schnorrer, F. (2017). Mechanical forces during muscle development. *Mechanisms of Development*, *144*, 92–101

Leon-Sarmiento, F.E., Gonzalez-Castaño, A., Rizzo-Sierra, C.V., Aceros, J., Leon-Ariza, D.S., Leon-Ariza, J.S., ... & Wang, Z.Y. (2019). Neurophysics assessment of the muscle bioenergy generated by transcranial magnetic stimulation. *Research*, *2019*, 7109535

Ljungdahl, L.G. (1994). The acetyl-CoA pathway and the chemiosmotic generation of ATP during acetogenesis, In *Acetogenesis*, ed. H.L. Drake, pp. 63–87. New York: Chapman & Hall

Marreiros, B.C., Calisto, F., Castro, P.J., Duarte, A.M., Sena, F.V., Silva, A.F., ... & Pereira, M.M. (2016). Exploring membrane respiratory chains. *Biochimica et Biophysica Acta (BBA) – Bioenergetics*, *1857*(8), 1039–1067

Maslov, V.P. (2021). Using methods of classical and quantum physics in bioenergy. *Theor. Math. Phys.*, *206*, 391–395. https://doi.org/10.1134/S0040577921030107

Matuschka, S., Zwicker, K., Nawroth, T., & Zimmer, G. (1995). ATP synthesis by purified ATP-synthase from beef heart mitochondria after coreconstitution with bacteriorhodopsin. *Arch. Biochem. Biophys*, *322*, 135–142

Miller Jr, J.H., Rajapakshe, K.I., Infante, H.L., & Claycomb, J.R. (2013). Electric field driven torque in ATP synthase. *PLoS One*, *8*(9), e74978. doi: 10.1371/journal.pone.0074978; PMID: 24040370; PMCID: PMC3769276

Mühlbauer, M.E., Saura, P., Nuber, F., Di Luca, A., Friedrich, T., & Kaila, V.R.I. (2020). Water-gated proton transfer dynamics in Respiratory Complex I. *J. Am. Chem. Soc.*, *142*, 13718–13728

Müller, V. (2003). Energy conservation in acetogenic bacteria. *Applied and Environmental Microbiology*, *69*(11), 6345–6353

Müller, V. (2001). Bacterial fermentation. In *Encyclopedia of life sciences* [online]. London: Macmillan. www.els.net

Müller, V., & Hess, V. (2017). The minimum biological energy quantum. *Frontiers in Microbiology*, *8*, 2019.

Müller, V., Ruppert, C., & Lemker, T. (1999). Structure and function of the A1A0 ATPases from methanogenic archaea. *J. Bioenerg. Biomembr.*, *31*, 15–28

Nicholls, D.G., & Ferguson, S.J. (2002). *Bioenergetics 3*. London: Academic Press.

Nishi, T., & Forgac, M. (2002). The vacuolar (H+)-ATPases: nature's most versatile proton pumps. *Nature Rev. Mol. Cell Biol.*, *3*, 94–103

Otto, R., Sonnenberg, A.S., Veldkamp, H., and Konings, W.N. (1980). Generation of an electrochemical proton gradient in Streptococcus cremoris by lactate efflux. *Proc. Natl. Acad. Sci. USA*, *77*, 5502–5506

Palanisami, A., & Okamoto, T. (2010). Torque-induced slip of the rotary motor F1-ATPase. *Nano Letters*, *10*, 4146–4149

Peinemann, S., & Gottschalk, G. (1992). The anaerobic way of life. In *The prokaryotes*, ed. A. Balows, H.G. Trüper, M. Dworkin, W. Harder, and K.H. Schleifer, pp. 300–311. Berlin: Springer

Piquereau, J., & Ventura-Clapier, R. (2018). Maturation of cardiac energy metabolism during perinatal development. *Frontiers in Physiology*, *9*, 959

Porter Jr, G.A., Hom, J., Hoffman, D., Quintanilla, R., de Mesy, B.K, & Sheu, S.S. (2011). Bioenergetics, mitochondria, and cardiac myocyte differentiation. *Prog. Pediatr. Cardiol.*, *31*(2), 75–81. doi: 10.1016/j.ppedcard.2011.02.002; PMID: 21603067; PMCID: PMC3096664

Röpke, M., Riepl, D., Saura, P., Di Luca, A., Mühlbauer, M.E., Jussupow, A., Gamiz-Hernandez, A.P., & Kaila, V.R.I. (2021). Deactivation blocks proton pathways in the mitochondrial Complex I. *Proc. Natl. Acad. Sci. USA*, *118*, e2019498118. doi: 10.1073/pnas.2019498118

Röpke, M., Saura, P., Riepl, D., Pöverlein, M.C., & Kaila, V.R.I. (2020). Functional water wires catalyze long-range proton pumping in the mammalian Respiratory Complex I. *J. Am. Chem. Soc.*, *142*, 21758–21766. doi: 10.1021/jacs.0c09209

Saura, P., & Kaila, V.R.I. (2019). Energetics and dynamics of proton-coupled electron transfer in the NADH/FMN site of Respiratory Complex I. *J. Am. Chem. Soc.*, *141*, 5710–5719. doi: 10.1021/jacs.8b11059

Scalise, M., Pochini, L., Giangregorio, N., Tonazzi, A., & Indiveri, C. (2013). Proteoliposomes as tool for assaying membrane transporter functions and interactions with xenobiotics. *Pharmaceutics*, *5*(3), 472–497

Stock, D., Gibbons, C., Arechaga, I., Leslie, A.G., & Walker, J.E. (2000). The rotary mechanism of ATP synthase. *Curr. Opin. Struct. Biol.*, *10*, 672–679

Stransky, L., Cotter, K., & Forgac, M. (2016). The function of V-ATPases in cancer. *Physiol Rev.*, *96*(3), 1071–1091. doi: 10.1152/physrev.00035.2015; PMID: 27335445; PMCID: PMC4982037

Trombetta, E.S., Ebersold, M., Garrett, W., Pypaert, M., & Mellman, I. (2003). Activation of lysosomal function during dendritic cell maturation. *Science*, *299*, 1400–1403

Weber, J., & Senior, A.E. (2003). ATP synthesis driven by proton transport in F1F0-ATP synthase. *FEBS Letters*, *545*, 61–70

Wolfe, R.S. (1999). Anaerobic life – a centennial view. *J. Bacteriol.*, *181*(11), 3317–3320. doi: 10.1128/JB.181.11.3317-3320.1999; PMID: 10348841; PMCID: PMC93796

Zhang, J., Toher, C., Erhard, M., Zhang, Y., Ugurbil, K., Bache, R.J., ... & Homans, D.C. (1997). Relationships between myocardial bioenergetic and left ventricular function in hearts with volume-overload hypertrophy. *Circulation*, *96*(1), 334–343

5 Biochemical Energetics

Biological processes run mainly on energy. Oxidation of glucose yields a specific amount of biochemical energy. Energetic changes in fluctuating chemical structural states happen due to intra and inter chemicals interactions occurring continuously at different parts of biological systems. Chemical energetics, occurring among monomer/dimer/multimer transformations and interactions involve bond, electrostatic, and van der Waals energies. This energetic process and associated fluctuations in energies are continuously dealt with by localized structures at different active, functional, and transport-associated parts and compartments in biology. ATP synthesis is itself an energetic process, which utilizes energy originating from various catabolic mechanisms, including cellular respiration, beta-oxidation, and ketosis. Signal transduction, DNA and RNA synthesis, muscle contraction, brain function, etc. cost ATPs. Addressing biochemical energetics thus appears to be a key phenomenon for understanding biology.

5.1 FUNDAMENTALS OF BIOCHEMICAL ENERGIES

Our body gets energy through the oxidative reaction of glucose ($C_6H_{12}O_6$) as follows:

$$C_6H_{12}O_6 + 6O_2 \rightarrow 6CO_2 + 6H_2O + 670\,kcal\,/\,mol \tag{5.1}$$

The oxidation of glucose yields 670 kcal of energy for every mole of glucose oxidized. The main source of energy for general cellular metabolism is glucose, which gets catabolized in the following three successive processes in order to produce adenosine triphosphate (ATP):

- glycolysis,
- tricarboxylic acid cycle (TCA or Krebs cycle), and finally
- oxidative phosphorylation

In the first process, glucose gets converted into pyruvate and produces a low amount of ATP. Pyruvate then gets converted into acetyl coenzyme A (acetyl-CoA) and enters the tricarboxylic acid (TCA) cycle, enabling the production of NADH. Finally, the respiratory chain complexes use NADH in order to generate a proton gradient across the inner mitochondrial membrane, necessary for producing large amounts of ATP by mitochondrial ATP synthase. In addition to these explained processes, it is to be mentioned that acetyl-CoA can also be generated by the catabolism of lipids and proteins, which indirectly participate in determining cellular energy state(s) through contributing to the ATP synthesis process. The majority of ATP synthesis is known to occur in cellular respiration within the

DOI: 10.1201/9781003287780-5

mitochondrial matrix, generating approximately 29–32 ATP molecules per 1 oxidized glucose molecule (Flurkey, 2010; Dunn and Grider, 2022); some textbooks estimate the production of a lower number of ATP molecules (Nelson and Cox, 2004; Nicholson, 2001; Voet et al., 2002).

If we carefully inspect the process, we shall find that all these steps involve the strong presence of physical entity "charge," environments, and specific electrical conditions. So biophysical understanding of the processes may help us become better informed regarding the underlying molecular processes.

ATP is an excellent energy storage molecule, used as energy currency due to the phosphate groups that link through phosphodiester bonds. These bonds are high energy because of the associated electronegative charges exerting a repelling force between the phosphate groups. A significantly large quantity of energy remains stored within the phosphate-phosphate bonds. Through metabolic processes, ATP becomes hydrolyzed into ADP (Figure 5.1), or further to AMP, and free inorganic phosphate groups. The process of ATP hydrolysis to ADP is energetically favorable, yielding Gibbs-free energy of -7.3 cal/mol (Meurer et al., 2017). The general equation for ATP hydrolysis is as follows:

$$ATP + H_2O \rightarrow ADP + P_i + 7.4\ kcal/mol \tag{5.2}$$

ATP hydrolysis is an exergonic process (Equation 5.2). It produces ADP (Adenosine diphosphate), P_i (inorganic phosphate), and Gibbs-free energy. ATP must continuously undergo replenishment to fuel the ever-working cell. The routine intracellular ATP concentration is 1–10 μM (Beis and Newsholme, 1975). ATP usually reaches high concentrations within cells, in the millimolar range (Bonora et al., 2012). Due to the high rate of ATP-dependent processes, together with its low stability in water, ATP content could quickly experience depletion if it were not immediately replenished by glycolysis and oxidative phosphorylation. ATP cannot be stored easily within cells, and the storage of carbon sources for ATP production (such as triglycerides or glycogen) is the best choice for energy maintenance. Many feedback mechanisms are in place to ensure the maintenance of a consistent ATP level in the cell. The enhancement or inhibition of ATP synthase is a common regulatory mechanism. ATP inhibits phosphofructokinase-1 (PFK1) and pyruvate kinase, two key enzymes in glycolysis, effectively acting as a negative feedback loop to inhibit glucose breakdown when there is sufficient cellular ATP. A schematic presentation may help us understand how cellular ATP is maintained (see Figure 5.2).

Cellular energetics concerns both intracellular and extracellular physicochemical states associated with the cell energy currency. ATP is consumed for energy (Dunn and Grider, 2022) required for cellular processes, including ion transport, muscle contraction, nerve impulse propagation, substrate

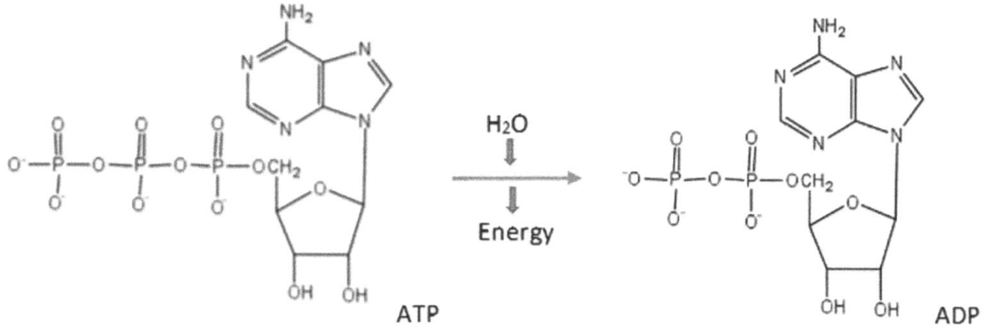

FIGURE 5.1 Hydrolysis of ATP to form ADP.

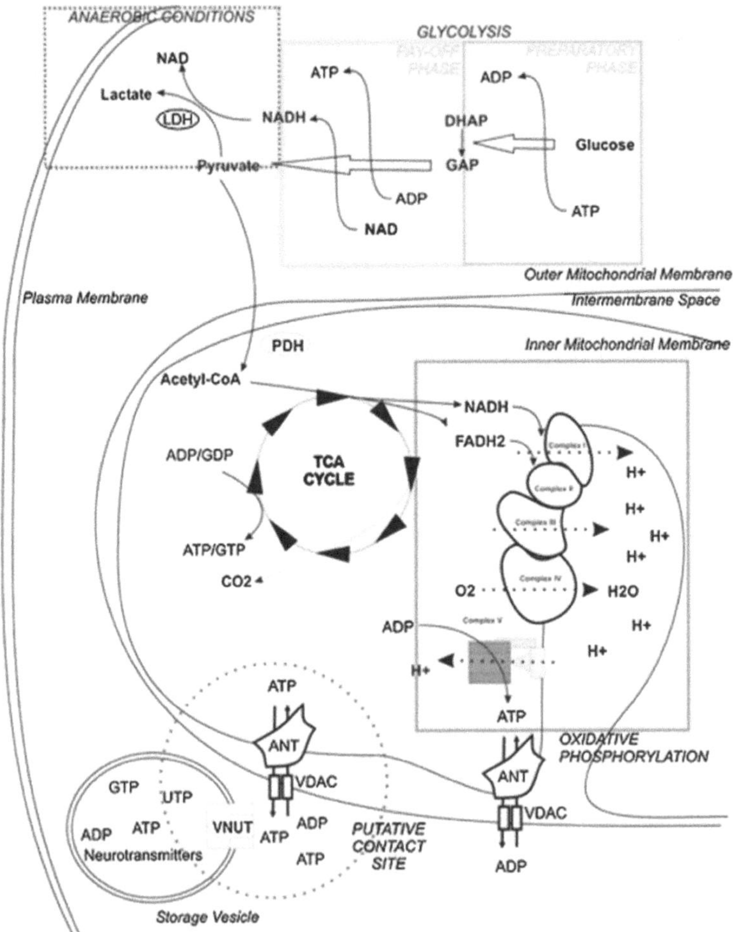

FIGURE 5.2 ATP management within the cell. Schematic representation of physicochemical processes and mechanisms of ATP synthesis and storage inside the cell. Glycolysis is represented in the yellow and blue boxes, the TCA cycle by the green circle, and oxidative phosphorylation in the orange box. Reduction of pyruvate to lactate is represented inside the red dotted rectangle. Hypothetical contacts between ATP storage vesicles and mitochondria, with preferential ATP transfer, are shown within the red dotted circle. Copied from ref.

***Source*: Bonora et al. (2012).**

phosphorylation, and chemical synthesis. These specific processes and other general ones create a high demand for ATP, with the result that cells within the human body depend upon the hydrolysis of 100–150 moles of ATP every day in order to ensure proper functioning of the body (Dunn and Grider, 2022). A quantitative analysis of ATP production, energy use, and associated environmental impacts in and around the biological cells may help make the picture of chemical energetics clearer. Recently, Mookerjee and colleagues (2017) demonstrated how extracellular fluxes quantitatively reflect the intracellular ATP turnover and the cellular bioenergetics, and provided a simple spreadsheet to calculate glycolytic and oxidative ATP production rates from raw extracellular acidification and respiration data (see Figure 5.3). For detailed on all parameters presented here the readers should consult the original article (due to space issue I have avoided presenting them all here) (Mookerjee et al., 2017).

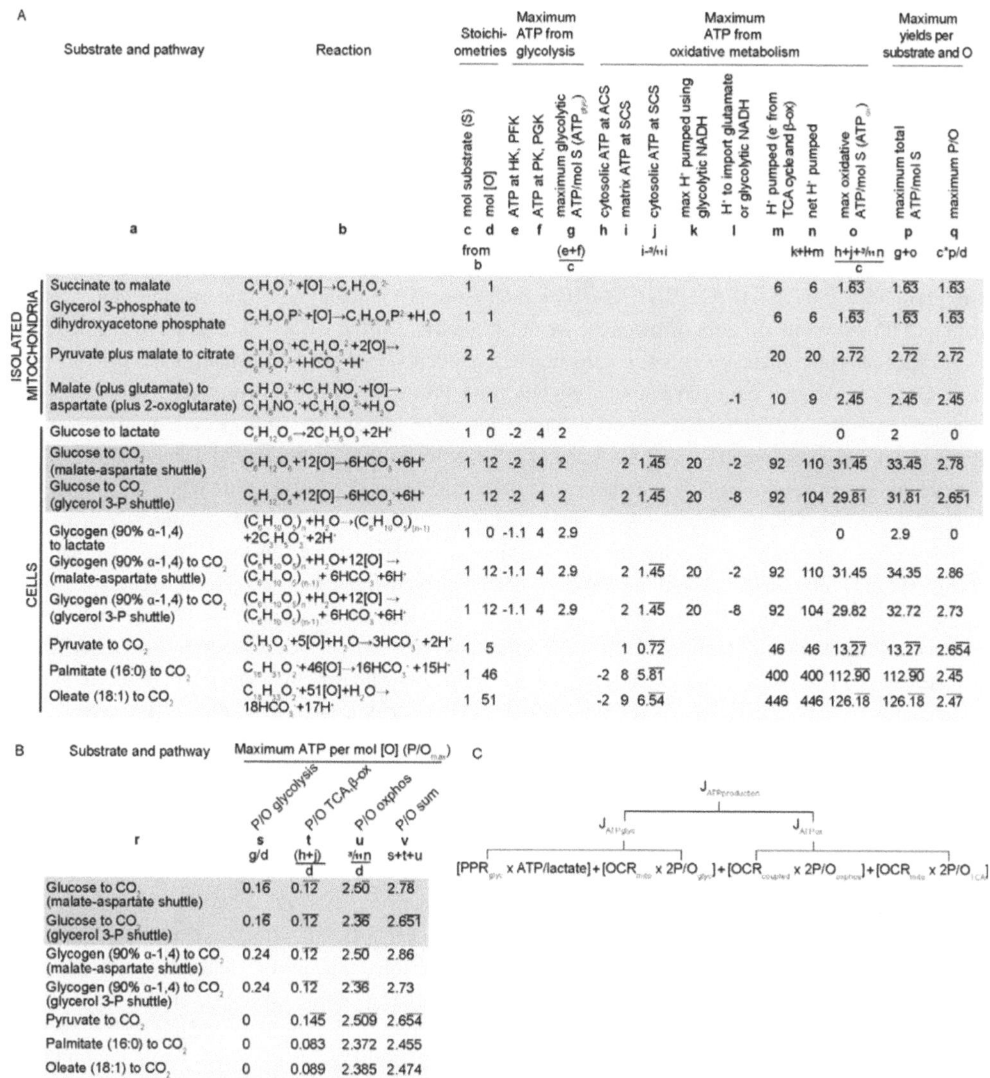

A

	Substrate and pathway	Reaction	Stoichiometries		Maximum ATP from glycolysis			Maximum ATP from oxidative metabolism								Maximum yields per substrate and O	
			mol substrate (S)	mol [O]	ATP at HK, PFK	ATP at PK, PGK	maximum glycolytic ATP/mol S (ATP_{glyc})	cytosolic ATP at ACS	matrix ATP at SCS	cytosolic ATP at SCS	max H^+ pumped using glycolytic NADH	H^+ to import glutamate or glycolytic NADH	H^+ pumped (e^- from TCA cycle and β-ox)	net H^+ pumped	max oxidative ATP/mol S (ATP_{ox})	maximum total ATP/mol S	maximum P/O
	a	b	c	d	e	f	g	h	i	j	k	l	m	n	o	p	q
			from b				(e+f)/c			i-3/11i				k+l+m	(h+j+3/11n)/c	g+o	c*p/d
ISOLATED MITOCHONDRIA	Succinate to malate	$C_4H_4O_4^{2-}+[O]\rightarrow C_4H_4O_5^{2-}$	1	1									6	6	$1.\overline{63}$	$1.\overline{63}$	$1.\overline{63}$
	Glycerol 3-phosphate to dihydroxyacetone phosphate	$C_3H_7O_6P^{2-}+[O]\rightarrow C_3H_5O_6P^{2-}+H_2O$	1	1									6	6	$1.\overline{63}$	$1.\overline{63}$	$1.\overline{63}$
	Pyruvate plus malate to citrate	$C_3H_3O_3^-+C_4H_4O_5^{2-}+2[O]\rightarrow C_6H_5O_7^{3-}+HCO_3^-+H^+$	2	2									20	20	$2.\overline{72}$	$2.\overline{72}$	$2.\overline{72}$
	Malate (plus glutamate) to aspartate (plus 2-oxoglutarate)	$C_4H_4O_5^{2-}+C_5H_8NO_4^-+[O]\rightarrow C_4H_6NO_4^-+C_5H_4O_5^{2-}+H_2O$	1	1								-1	10	9	$2.\overline{45}$	$2.\overline{45}$	$2.\overline{45}$
CELLS	Glucose to lactate	$C_6H_{12}O_6\rightarrow 2C_3H_5O_3^-+2H^+$	1	0	-2	4	2								0	2	0
	Glucose to CO_2 (malate-aspartate shuttle)	$C_6H_{12}O_6+12[O]\rightarrow 6HCO_3^-+6H^+$	1	12	-2	4	2		2	$1.\overline{45}$	20	-2	92	110	$31.\overline{45}$	$33.\overline{45}$	$2.\overline{78}$
	Glucose to CO_2 (glycerol 3-P shuttle)	$C_6H_{12}O_6+12[O]\rightarrow 6HCO_3^-+6H^+$	1	12	-2	4	2		2	$1.\overline{45}$	20	-8	92	104	$29.\overline{81}$	$31.\overline{81}$	$2.6\overline{51}$
	Glycogen (90% α-1,4) to lactate	$(C_6H_{10}O_5)_n+H_2O\rightarrow(C_6H_{10}O_5)_{(n-1)}+2C_3H_5O_3^-+2H^+$	1	0	-1.1	4	2.9								0	2.9	0
	Glycogen (90% α-1,4) to CO_2 (malate-aspartate shuttle)	$(C_6H_{10}O_5)_n+H_2O+12[O]\rightarrow(C_6H_{10}O_5)_{(n-1)}+6HCO_3^-+6H^+$	1	12	-1.1	4	2.9		2	$1.\overline{45}$	20	-2	92	110	31.45	34.35	2.86
	Glycogen (90% α-1,4) to CO_2 (glycerol 3-P shuttle)	$(C_6H_{10}O_5)_n+H_2O+12[O]\rightarrow(C_6H_{10}O_5)_{(n-1)}+6HCO_3^-+6H^+$	1	12	-1.1	4	2.9		2	$1.\overline{45}$	20	-8	92	104	29.82	32.72	2.73
	Pyruvate to CO_2	$C_3H_3O_3^-+5[O]+H_2O\rightarrow 3HCO_3^-+2H^+$	1	5					1	$0.\overline{72}$			46	46	$13.\overline{27}$	$13.\overline{27}$	$2.6\overline{54}$
	Palmitate (16:0) to CO_2	$C_{16}H_{31}O_2^-+46[O]\rightarrow 16HCO_3^-+15H^+$	1	46				-2	8	$5.\overline{81}$			400	400	$112.\overline{90}$	$112.\overline{90}$	$2.\overline{45}$
	Oleate (18:1) to CO_2	$C_{18}H_{33}O_2^-+51[O]+H_2O\rightarrow 18HCO_3^-+17H^+$	1	51				-2	9	$6.\overline{54}$			446	446	$126.\overline{18}$	$126.\overline{18}$	$2.\overline{47}$

B

Substrate and pathway	Maximum ATP per mol [O] (P/O_{max})			
	P/O glycolysis	P/O TCA,β-ox	P/O oxphos	P/O sum
r	s	t	u	v
	g/d	(h+j)/d	(3/11n)/d	s+t+u
Glucose to CO_2 (malate-aspartate shuttle)	$0.1\overline{6}$	$0.\overline{12}$	$2.5\overline{0}$	$2.\overline{78}$
Glucose to CO_2 (glycerol 3-P shuttle)	$0.1\overline{6}$	$0.\overline{12}$	$2.\overline{36}$	$2.6\overline{51}$
Glycogen (90% α-1,4) to CO_2 (malate-aspartate shuttle)	0.24	$0.\overline{12}$	2.50	2.86
Glycogen (90% α-1,4) to CO_2 (glycerol 3-P shuttle)	0.24	$0.\overline{12}$	$2.\overline{36}$	2.73
Pyruvate to CO_2	0	$0.1\overline{45}$	$2.5\overline{09}$	$2.6\overline{54}$
Palmitate (16:0) to CO_2	0	0.083	2.372	2.455
Oleate (18:1) to CO_2	0	0.089	2.385	2.474

FIGURE 5.3 Maximum extramitochondrial ATP yields and P/O ratios for the catabolism of conventional substrates by isolated mammalian mitochondria and physiological substrates by mammalian cells and calculation of the rates of ATP production from glycolysis, tricarboxylic acid cycle, β-oxidation, and oxidative phosphorylation using oxygen consumption rate. (*A*) maximum extramitochondrial ATP yields and P/O ratios. The maximum net yield of glycolytic ATP/mol of glucose or glycogen converted to pyruvate (and then to lactate or oxidized through the TCA cycle), ATP_{glyc}, is given in *column g*, after correction for ATP used to activate glucose or glycogen (calculated using *columns b* and *c* and *columns e* and *f*). *Column o* gives ATP_{ox}, the maximum oxidative yield of ATP/mol of substrate oxidized by pyruvate dehydrogenase, TCA cycle, β-oxidation, and electron transport chain, including substrate-linked phosphorylation in the TCA cycle and NADH equivalents imported from glycolytic reactions, and corrected for ATP used to activate substrates other than glucose (calculated using *columns b* and *c* and *columns h* and *n*). Values are given with *bars* (i.e. $1.\overline{63}$ = 1.636363 …) to emphasize that they are not integers or approximations but exact values arising from the arithmetic of small integers as shown (values for glycogen incorporate an assumption about branching, so they are less precise and are therefore rounded to two decimal places). *Column p* gives the maximum total yield of ATP per mol substrate (sum of *columns g* and *o*). *Column q* expresses this maximum yield of ATP per mol of oxygen atoms [O] consumed (i.e. the maximum P/O ratio for the reactions in *column b*). The values in each

FIGURE 5.3 (*Continued*)

column are calculated row-by-row as follows. During oxidation of succinate to malate by isolated mitochondria, succinate enters on the dicarboxylate carrier in exchange for malate and is oxidized to fumarate by succinate dehydrogenase, reducing Q to QH_2. QH_2 is oxidized by the electron transport chain, reducing 1 [O] to H_2O and driving 6 H^+ from the matrix to the intermembrane space. The fumarate is hydrated to malate by fumarase and exits in exchange for incoming succinate. The 6 H^+ re-enter the matrix through the ATP synthesis machinery, which translocates 8 H^+ through the ATP synthase and 3 H^+ through the phosphate and adenine nucleotide carriers for every 3 ATP generated, giving an H^+/ATP ratio of 11:3. In this way, the oxidation of succinate causes phosphorylation of 6 H^+/O × 3/11 ATP/H^+ = 1. 63—— ATP molecules/[O] reduced to water. This is the maximum P/O ratio for oxidation of succinate by mitochondria. During oxidation of glycerol 3-phosphate to dihydroxyacetone phosphate by isolated mitochondria, mitochondrial glycerol 3-phosphate dehydrogenase reduces Q to QH_2, which is oxidized as above, with the same P/O_{max} of 1. 63——. During oxidation of pyruvate plus malate by isolated mitochondria, pyruvate enters mitochondria on the pyruvate carrier (electroneutrally with a proton) and is oxidized to acetyl-CoA and CO_2 by pyruvate dehydrogenase. Malate enters in exchange for citrate and is oxidized to oxaloacetate by malate dehydrogenase. The two dehydrogenation reactions form a total of 2 NADH. Citrate synthase uses acetyl-CoA and oxaloacetate to form citrate, which exits the mitochondria with a proton (balancing the proton imported with pyruvate) on the tricarboxylate carrier in exchange for incoming malate. The 2 NADH are oxidized, driving pumping of 20 H^+ and generating 20 × 3/11 = 5. 45—— ATP with a P/O_{max} of 2. 72——. During oxidation of malate plus glutamate by isolated mitochondria, malate enters electroneutrally on the dicarboxylate carrier in exchange for 2-oxoglutarate, and glutamate enters on the glutamate-aspartate carrier electrogenically with a proton (which is therefore unavailable for ATP synthesis; *column l*) in exchange for aspartate. Malate dehydrogenase produces oxaloacetate plus 1 NADH, which is oxidized, driving pumping of 10 H^+, and then aspartate aminotransferase uses oxaloacetate and glutamate to generate 2-oxoglutarate and aspartate, which exit in exchange for incoming malate and glutamate. Overall, 9 H^+ are translocated, driving synthesis of 9 × 3/11 = 2. 45—— ATP/[O]. During glycolysis of glucose to lactate by cells, 1 ATP is consumed at hexokinase, and 1 ATP is consumed at phosphofructokinase to yield 2 trioses, each of which generates 1 ATP at phosphoglycerate kinase and 1 ATP at pyruvate kinase, for a net yield of 2 ATP/glucose. The 2 NADH generated at glyceraldehyde phosphate dehydrogenase are reoxidized during reduction of pyruvate to lactate by lactate dehydrogenase, and 2 lactates are exported from the cell accompanied by the 2 H^+ generated by the conversion of 1 uncharged glucose to 2 anionic $lactate^-$ molecules. Overall, glycolysis to lactate produces 2 ATP/glucose (1 ATP/lactate). During glycolysis of glucose to pyruvate by cells followed by oxidation of pyruvate to bicarbonate by pyruvate dehydrogenase and the TCA cycle, 2 ATP are formed by glycolytic reactions per glucose. In addition, 2 ATP are formed in the mitochondrial matrix during oxidative metabolism by the substrate-linked reaction at succinyl-CoA synthetase. Each of these ATPs is exported to the cytosol, using 1 H^+ in the process, giving a net cytosolic ATP yield from succinyl-CoA synthetase of 2 × (1 − 3/11) = 1. 45—— ATP/glucose. NADH generated at glyceraldehyde phosphate dehydrogenase enters the mitochondria on the malate-aspartate shuttle, driven by re-entry of 2 of the 20 subsequently translocated H^+ (*column l*), or on the glycerol 3-phosphate shuttle (which allows the reducing equivalents to enter the electron transport chain without passing through Complex I, so pumping 12 H^+, 8 fewer than normal for matrix NADH; *column l*). The 2 pyruvates from glycolysis are fully oxidized by pyruvate dehydrogenase and the TCA cycle, generating 8 NADH and 2 QH_2 (driving pumping of 92 H^+). The sum of 110 (or 104 if the glycerol 3-phosphate shuttle is used) translocated H^+ yields a maximum of 110 × 3/11 = 30 (or 104 × 3/11 = 28. 36—— ATP, which, together with the substrate-linked ATP production, gives a net oxidative yield of 31. 45—— (or 28. 81——) ATP/glucose. The overall ATP yield is the sum of the glycolytic and oxidative yields: 33. 45—— (or 31. 81——) ATP/glucose or a P/O_{max} of 2. 78—— (or 2. 651———). During catabolism of glycogen, the yields are the same as those for catabolism of glucose, except that less ATP is needed for the initial activation reactions. About 90 percent of the linkages in glycogen are α-1,4, which are split by the addition of phosphate, yielding glucose 1-phosphate and bypassing the consumption of ATP at hexokinase. The remainder are α-1,6, which are hydrolyzed to yield glucose, which requires activation at hexokinase. On average, activation therefore requires ~0.1 ATP at hexokinase and 1 ATP at phosphofructokinase,

FIGURE 5.3 *(Continued)*

increasing the ATP yield of glycogen catabolism by ~0.9 ATP/glucose compared with catabolism of glucose itself, giving the yields of ATP/glucose unit and P/O_{max} ratios shown. Complete oxidation of pyruvate by cells bypasses glycolysis and generates 1 ATP/pyruvate at succinyl-CoA synthetase in the matrix (0. 72—— ATP/pyruvate after export of ATP). Proton pumping yields 46 × 3/11 = 12. 54—— ATP/pyruvate, for a sum of 13. 27—— ATP/pyruvate or a P/O_{max} of 2. 654———. During complete oxidation of palmitate by cells, palmitate activation to palmitoyl-CoA generates AMP from ATP, effectively using 2 ATP/palmitate. Palmitoyl-CoA enters the matrix electroneutrally as palmitoyl carnitine on the carnitine transporter, and then β-oxidation to 8 acetyl-CoA yields 7 NADH and 7 QH_2, and oxidation of 8 acetyl-CoA in the TCA cycle yields 24 NADH, 8 QH_2, and 8 matrix ATP. The maximum overall yield is 112. 90 ATP/palmitate and a P/O_{max} of 2. 45 Oxidation of other fatty acids gives slightly different yields and P/O_{max} values; the monounsaturated oleate, whose oxidation generates 1 fewer QH_2 and 6 fewer H^+ translocated than the corresponding saturated fatty acid, is calculated out as an example. *HK*, hexokinase; *PFK*, phosphofructokinase; *PK*, pyruvate kinase; *PGK*, phosphoglycerate kinase; *ACS*, acyl- CoA synthase; *SCS*, succinyl-CoA synthetase. *Purple* highlights isolated mitochondria; *gray* highlights cells (and the *darker band* highlights complete oxidation of glucose by cells). (*B*) Calculation of yields of ATP per oxygen consumed. Non-repeating values are rounded to two decimal places for assumed estimates and to three decimal places for real numbers. *Columns s–u* divide the overall P/O ratio in *A* (*column q*) into components dependent on different subsets of the total mitochondrial oxygen consumption, to enable calculations of glycolytic and oxidative ATP yields from experimental oxygen consumption data (glycolytic ATP yields from glycolysis to lactate are calculated from extracellular acidification, but glycolytic ATP yields from glycolysis to pyruvate (subsequently oxidized) are calculated from oxygen consumption). *Column s* gives the glycolytic ATP yield from conversion of glucose or glycogen to pyruvate subsequently oxidized to bicarbonate (which depends on total mitochondrial oxygen consumption). *Column t* gives the P/O ratio for substrate-linked reactions in the TCA cycle and β-oxidation (which depends on total mitochondrial oxygen consumption), and *column u* gives the P/O ratio for oxidation of NADH and QH_2 derived from pyruvate dehydrogenase, TCA cycle, and β-oxidation (which depends on coupled mitochondrial oxygen consumption). *Column v* gives the overall sum of these partial P/O ratios, which are the same as *column q* in *A*. *β-ox*, β-oxidation; *oxphos*, oxidative phosphorylation. (C) the total rate of ATP production, $J_{ATP\ production}$, is the sum of measurable extracellular rates (PPR and OCR) multiplied by the appropriate ATP/lactate ratio from *A* or P/O ratio from *B*. For details, readers may consult Mookerjee et al. (2017).

***Source*: Mookerjee et al. (2017).**

Mookerjee and colleagues (2017) considered calculating the rates of ATP production due to glycolysis ($J_{ATPglyc}$) and oxidation (J_{ATPox}), respectively. The method introduced here was built on previous deconvolution of glycolytic and respiratory sources of acidification (detailed in Mookerjee and Brand, 2015, and Mookerjee et al., 2015), which was extended to the calculation of rates of intracellular ATP production by glycolysis and oxidative reactions from extracellular measurements of rates of acidification and oxygen consumption. The known rates of ATP production can be used to quantify and interpret classical qualitative indicators of cellular energy metabolism, the Warburg, Crabtree, and Pasteur effects, and to quantify the flexibility of substrate use and the bioenergetic capacity of cells. This method was used to characterize the bioenergetic phenotype of C2C12 myoblasts under different conditions and to assess whether different ATP-consuming reactions draw preferentially on glycolytic or oxidative ATP production (Mookerjee et al., 2017). Figure 5.4 demonstrates the pathways involved when glucose (or glycogen) is used as the substrate for cellular ATP production, and Figure 5.3 provides the associated accounting (Mookerjee et al., 2017).

Figure 5.5 presents raw extracellular flux data and the overall values and individual components of $J_{ATPglyc}$ and J_{ATPox} during a partial cell respiratory control assay (Brand and Nicholls, 2011). Figure 5.5A shows raw ECAR traces, and Figure 5.5B shows raw OCR traces during the course of

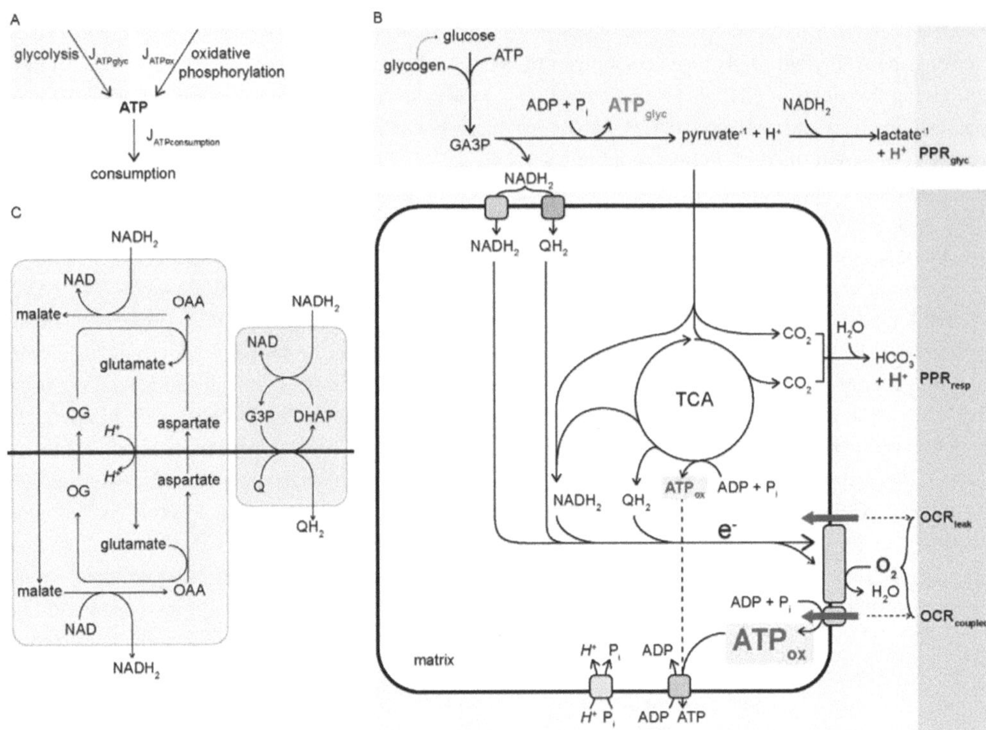

FIGURE 5.4 Substrate catabolism by glycolysis and oxidation and related extracellular measurements. (*A*) Steady-state relationship between rates of glycolytic and oxidative ATP production and rate of ATP consumption, connected by ATP (or other linked variables, such as ATP/ADP or phosphorylation potential) as an intermediate metabolite. (*B*) Pathways of glucose and glycogen catabolism giving rise to measurable extracellular rates of acid production (PPR) and oxygen consumption (OCR). The *black outline* represents the mitochondrial inner membrane enclosing the mitochondrial matrix. *Blue section*, reactions linked directly to PPR_{glyc}. Exogenous glucose that enters the cell (or endogenous glycogen) is converted to pyruvate, yielding net ATP. Pyruvate can be converted to lactate, consuming glycolytic NADH and causing extracellular acidification (PPR_{glyc}). *Red section*, reactions linked directly to PPR_{resp} and OCR. Reducing equivalents from glycolysis are delivered into the matrix by one of two shuttles (*C*). Pyruvate also enters the matrix and is fully oxidized to CO_2 by pyruvate dehydrogenase and the tricarboxylic acid cycle, yielding further reducing equivalents. Electrons (e^-) are fed into the respiratory complexes of the electron transport chain, driving oxygen consumption. OCR is shown divided into parts (*red arrows*), where OCR_{leak} is OCR due to proton leak reactions not coupled to oxidative phosphorylation ($OCR_{leak} = OCR_{oli} - OCR_{r/m}$) and $OCR_{coupled}$ is OCR coupled to oxidative phosphorylation. (*C*) Detail of shuttles for mitochondrial import of reducing equivalents shown in (*B*). *Left*, malate-aspartate shuttle; *right*, glycerol 3-phosphate shuttle. *GA3P*, glyceraldehyde 3-phosphate; *OAA*, oxaloacetate; *OG*, 2-oxoglutarate; *G3P*, glycerol 3-phosphate; *DHAP*, dihydroxyacetone phosphate. NADH + H^+ cycles in the steady state, causing no net change in H^+, and is shown as $NADH_2$ to reduce confusion over sources of PPR. For details, readers may consult Mookerjee et al. (2017). The color contrast in the figure here and all subsequent ones will be clear in the online version of the book.

***Source*: Mookerjee et al. (2017).**

a single set of experiments. Figure 5.5D presents the contributions of each component of glycolytic and oxidative ATP production to the total rates of ATP production shown in Figure 5.5C.

Describing the cellular bioenergetics in terms of the total rate of ATP production ($J_{ATP\ production}$), divided into glycolytic ($J_{ATPglyc}$) and oxidative sources (J_{ATPox}) allows quantitative and comprehensive

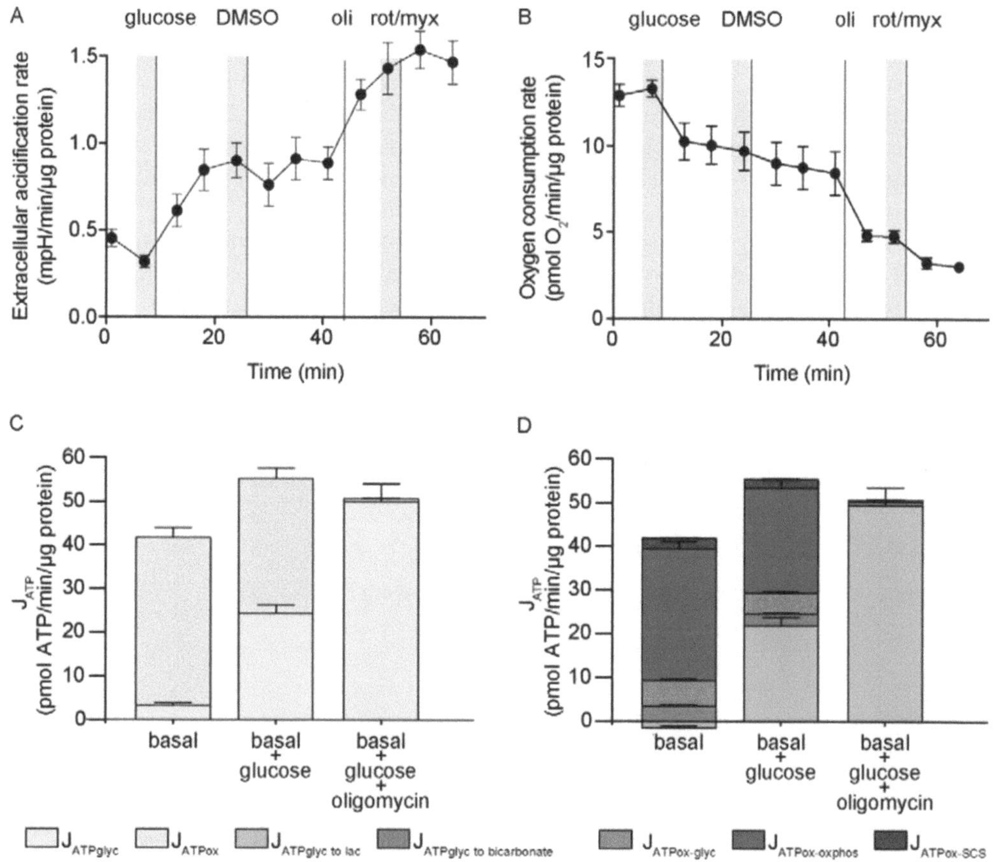

FIGURE 5.5 Raw extracellular flux data and its conversion into rates of ATP production (J_{ATP}). (A and B) raw traces of extracellular acidification and oxygen consumption by C2C12 myoblasts. A basal measurement was recorded in the absence of exogenous substrate, followed by the addition of 10 mm glucose, vehicle control (DMSO; maximum concentration <0.05% (v/v)), 2 μg/ml oligomycin (*oli*), and 1 μm rotenone with 1 μm myxothiazol (*rot/myx*). Points within *gray regions* were assumed to be at or near steady state and were part of the data set used to calculate values shown in sequential *columns* in (C) and (D). Points are means ± S.E. (*error bars*) of $n = 4$ independent experiments. The final protein content of each well was typically 10–15 μg. (C) $J_{ATP\ production}$ for each time point marked in *gray* in (A) and (B), divided into $J_{ATPglyc}$ and J_{ATPox}. Aggregate data from multiple experiments, including those in (A) and (B), gave rise to basal $n = 24$, glucose $n = 36$, and glucose + oligomycin $n = 19$. (D) Data in (C) further divided into the component reactions of $J_{ATPglyc}$ and J_{ATPox}. $J_{ATPglyc}$ is divided into ATP production during glycolytic production of pyruvate that is either converted to lactate ($J_{ATPglyc\ to\ lac}$) or imported into the matrix and oxidized ($J_{ATPglyc\ to\ bicarbonate}$). J_{ATPox} is divided into ATP production from oxidation of glycolytic NADH ($J_{ATPox\text{-}glyc}$), from oxidation of reducing equivalents generated within the mitochondria ($J_{ATPox\text{-}oxphos}$) and from succinyl-CoA synthetase ($J_{ATPox\text{-}SCS}$). For details, readers may consult Mookerjee et al. (2017).

***Source*: Mookerjee et al. (2017).**

bioenergetic analysis which is not possible using raw data. For illustrating this, a single extended data set is represented in several different ways (see Mookerjee et al., 2017).

Besides understanding ATP production modes, it is important to address the profiling of ATP consumers. The major consumers of ATP generated by oxidative phosphorylation in cells have been profiled by inferring their rates from measurements of the decrease in cellular oxygen consumption rate when individual ATP-consuming pathways are inhibited pharmacologically (Felber and Brand, 1983; Brand and Felber, 1984; Birket et al., 2011).

Both glycolysis and oxidative reactions generated cellular ATPs can be used by ATP consumption pathways (see Figure 5.6A). Combined approaches (Felber and Brand, 1983; Brand and Felber, 1984; Birket et al., 2011) were extended in Mookerjee et al. (2017) for generating a rather complete ATP consumption profile for C2C12 myoblasts in the presence of glucose, with both ATPglyc and ATPox accounted for. Figure 5.6B shows the extent of inhibition of $J_{ATPglyc}$ and J_{ATPox} by inhibitors of specific ATP-consuming pathways, allowing about 43 percent of glycolytic, 81 percent of oxidative, and 62 percent of total ATP production to be associated with specific cellular ATP-consuming processes, particularly protein synthesis and actin dynamics.

The analysis in Mookerjee et al. (2017) provides detailed descriptions and conceptual analyses of ATP production and consumption by different processes within cells. This analysis is also suggestive that vital understanding of ATP production and consumption in biological systems can be derived using simple, non-invasive, non-terminal extracellular measurements of acidification and oxygen consumption with the use of commercially available instruments.

5.2 BIOCHEMICAL INTERACTION ENERGETICS

We all in physics and chemistry know a lot about the following equation:

$$\Delta G = \Delta H - T\Delta S \tag{5.3}$$

Biochemistry deals with a lot of chemical reactions occurring in living organisms. Equation (5.3) appears a vital one in helping us describe biochemical reactions. Here G is Gibbs free energy, H is enthalpy, T is absolute temperature, and S is entropy. Biochemical reactions are associated with the change in Gibbs free energy, ΔG, change in enthalpy, ΔH, and change in entropy, ΔS in a thermodynamic condition with temperature T.

If we understand the single biomolecular structure and interaction energetics, we obtain a picture of molecular-level active mechanisms. The roles of Gibbs free energy change in protein synthesis and function (Khan et al., 2020; Razavi et al., 2022), protein–protein interaction (Golas et al., 2019), DNA structure stabilization (Lomzov et al., 2015), RNA stabilization (Dawson et al., 2021), enzyme function (Yagisawa, 1995; Bearne, 2012; Rojas-Pirela et al., 2020), etc. are evident and addressed in various studies. Relationships between Gibbs free energy changes or Gibbs free energetics and consumption of ATP (Golas et al., 2019) are understood in biochemical processes.

Several cancer cell types were recently analyzed considering the over-expression of various proteins that participate in protein–protein interaction networks and a metabolic shift from oxidative phosphorylation to glycolysis (Razavi et al., 2022). Large data sets were used here for measuring the protein–protein interaction (PPI) associated Gibbs free energy G (Equation 5.4), which revealed a strong inverse correlation between the percentage of energy production via oxidative phosphorylation and the Gibbs free energy of the protein networks.

$$G = \sum_i G_i \tag{5.4}$$

where G_i is expressed as follows:

$$G_i = c_i ln \frac{c_i}{\sum_j c_j} \tag{5.4}$$

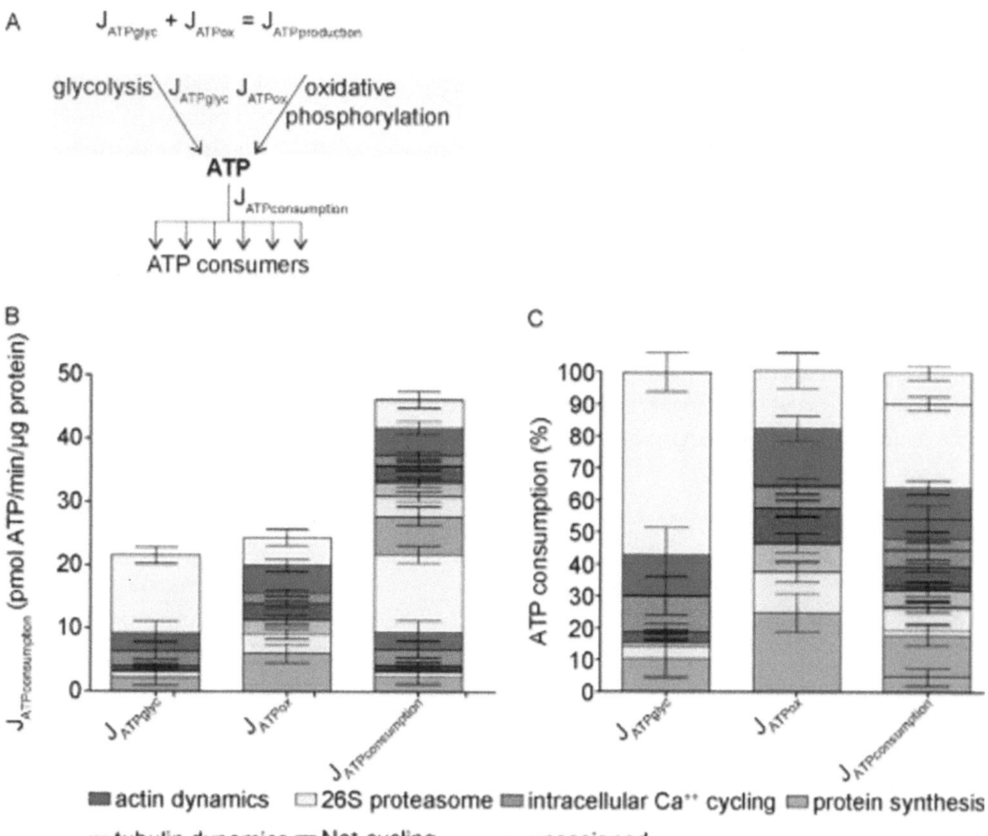

FIGURE 5.6 Rates of consumption of ATP by different pathways in C2C12 myoblasts. (*A*) Steady-state rates of glycolytic ATP production ($J_{ATPglyc}$) and oxidative ATP production (J_{ATPox}) sum to the total rate of ATP production ($J_{ATP\ production}$) and hence to the total rate of ATP consumption by multiple ATP consumers ($J_{ATP\ consumers}$). ATP production and consumption are connected by ATP (or other linked variables, such as ATP/ADP or phosphorylation potential) as the common intermediate. (*B*) Absolute $J_{ATPglyc}$ and J_{ATPox} supplying individual ATP consumers, calculated from the decrease in ECAR and OCR caused by the addition of different inhibitors of specific ATP consumption pathways (*colored bars*) or unassigned (by difference from the total $J_{ATP\ consumption}$ in the absence of inhibitors). The *third bar* shows the set of values for J_{ATPox} stacked *above* the set of values for $J_{ATPglyc}$ to display total assigned and unassigned rates of ATP consumption. (*C*) Data from (*B*) scaled to the uninhibited totals for $J_{ATPglyc}$, J_{ATPox}, and $J_{ATP\ consumption}$ as 100 percent. No significant differences were found by unpaired *t* test between percentage contribution of $J_{ATPglyc}$ and percentage contribution of J_{ATPox} for any single consumer or for "unassigned" percentage contribution of J_{ATP}. The *third bar* shows the individual values for J_{ATPox} stacked *above* the corresponding individual values for $J_{ATPglyc}$ to emphasize the contributions of each ATP consumption pathway to $J_{ATP\ consumption}$. Compounds used to define individual pathways of ATP consumption were as follows: 10 µm cycloheximide (protein synthesis), 25 µm MG132 (26S proteasome activity), 1 mm ouabain (plasma membrane Na^+ cycling), 1 µm nocodazole (tubulin dynamics), 0.5 µm thapsigargin (intracellular Ca^{2+} cycling), and 0.25 µm latrunculin A (actin dynamics). Values are means ± S.E. (*error bars*) of n = 6–8 independent experiments/compound.

***Source*: Mookerjee et al. (2017).**

where c_i is the concentration of the protein i, normalized or rescaled, to be between 0 and 1. The sum in the denominator is considered overall protein neighbors of i, and including the ith one, thus the denominator is considered a measure of the degree of entropy. Carrying out this mathematical operation essentially transforms the concentration value assigned to each protein to a Gibbs free energy G. Gibbs free energy of a protein will be $G_i \leq 0$; the magnitude of which is associated with the amount of that protein bound to interaction partners. Equation (5.4) will provide the Gibbs energy of the whole network.

Razavi and colleagues (2022) used mRNA transcriptome data as a surrogate for protein concentration, considering Kim et al. (2014), and Wilhelm et al. (2014), where an 83 percent correlation between mass spectrometry proteomic information and transcriptomic information for multiple tissue types was established. The values G for a cell can be calculated from mRNA expression, a surrogate for protein concentrations, and overlaid on an expression network (Rietman et al., 2017). The Gibbs values used in this paper were calculated using PPIs from BioGRID, a curated bioinformatics database (https://thebiogrid.org) (see Figure 5.7) for calculating G (Figure 5.8). Here gene

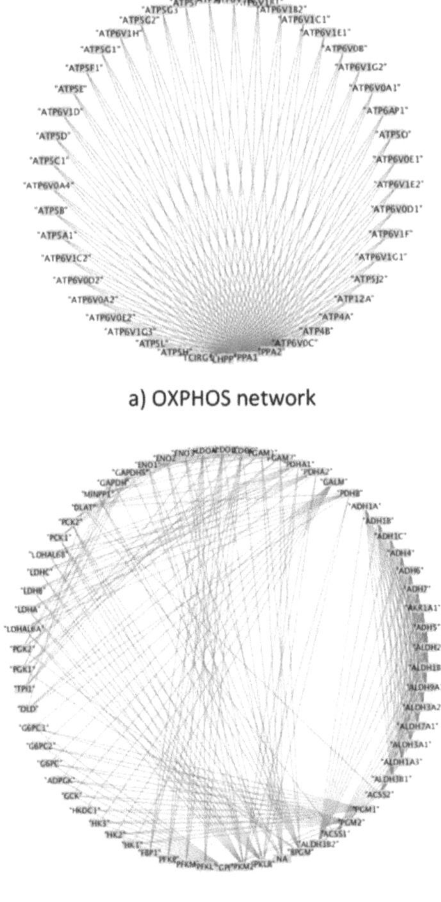

FIGURE 5.7 *Kyoto Encyclopedia of Genes and Genomes* (KEGG) networks converted to protein–protein interactions (PPI) networks with KEGGgraph and plotted in Cytoscape. For the listing of the proteins and the number of connections to each protein, see Golas et al. (2019).

***Source*: Golas et al. (2019).**

expression data for several cancers were obtained from TCGA (https://cancergenome.nih.gov/). An inverse correlation between the percentage of OXPHOS (in ATP) utilization and the Gibbs free energy of the corresponding PPI network has been reported for various cancer types (Golas et al., 2019; Rietman et al., 2017). The study suggests that the biochemical energetics and ATP processing including ATP expense rate may appear to help us understand disease (e.g. cancer) status too (Golas et al., 2019; Rietman et al., 2017).

Biomolecular energetics through the calculation of the enthalpy for DNA duplex formation is another prime example we may inspect to address the energy dependent biomolecular structural stability (see Figure 5.9 for model demonstration). (Lomzov et al., 2015).

The hybridization enthalpy, $\Delta H°$, was calculated as a difference of the oligonucleotides' total energy in the double-stranded (ds) state (E_{tot}^{ds}) and the single-stranded (ss) state (E_{tot}^{ss1} and E_{tot}^{ss2}), as follows:

$$\Delta H^{o} \approx E_{tot}^{ds} - \left(E_{tot}^{ss1} + E_{tot}^{ss2}\right) \tag{5.5}$$

This enthalpy change is related to the overall Gibbs free energy following Equation (5.3), where ΔH is used as a universal notation for the enthalpy change. The value of hybridization enthalpy was reported to be close to the total internal change because the simulation had been performed at

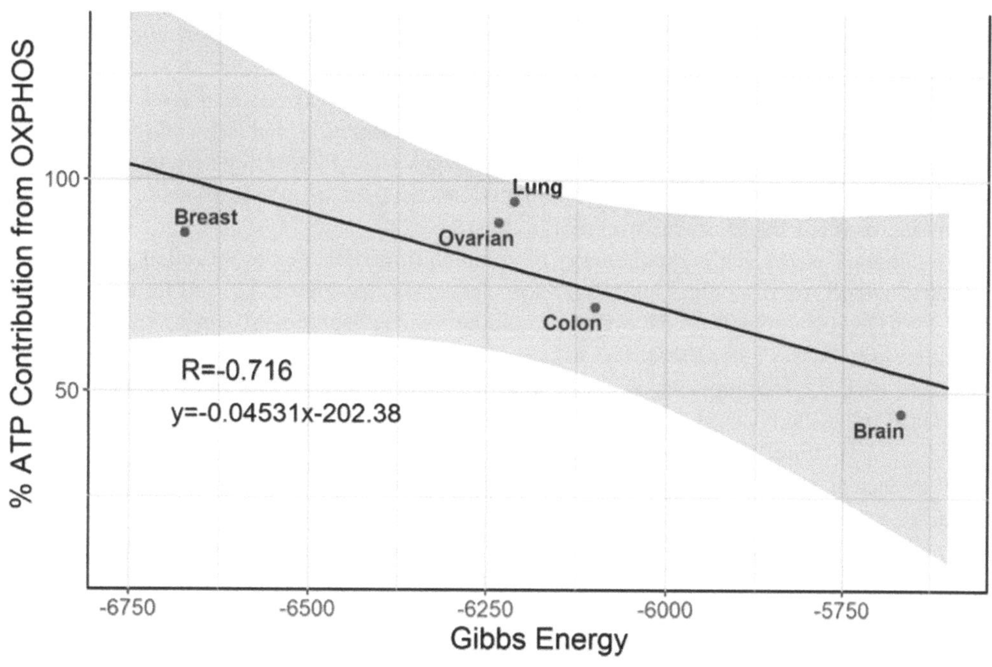

FIGURE 5.8 Cancer metabolism by type. Y-axis data refer to the overall contribution to ATP generation from oxidative phosphorylation (OXPHOS) and were determined from a number of metabolic flux analyses whose results are tabulated in Moreno-Sánchez et al. (2009). The percentages may be contrasted with percentage ATP obtained from glycolysis. "Lung" refers to lung squamous cell adenoid cancer, and "brain" refers to glioblastoma. The Gibbs energy estimates are in arbitrary units. The gray region represents the 95 percent confidence interval on the linear regression.

***Source*: Golas et al. (2019).**

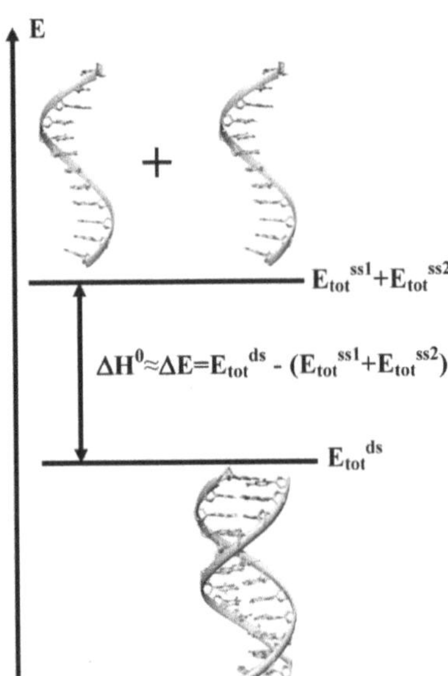

FIGURE 5.9 Scheme for the calculation of the enthalpy for DNA duplex formation.

***Source*: Lomzov et al. (2015).**

constant pressure, and the change of the volume during DNA denaturation was negligible (Lomzov et al., 2015).

The total energy of each of these three states (E_{tot}^{i}, I = ds, ss1, ss2) was taken as the average value along the molecular dynamics (MD) trajectory after an equilibration. The error of the total energy calculation was calculated as the root-mean-square deviation (RMSD) of total energy along the trajectory divided by the square root of the number of independent snapshots minus 1. For details of MD simulation methods, see Lomzov et al. (2015).

As mentioned earlier, the internal energy of complex formation was evaluated as the difference between total energies of double- and single-stranded states (see Figure 5.9, and Equation 5.5) and was close to the corresponding ΔH° value and did not depend significantly on simulation method, as detailed in Lomzov et al. (2015). The values in enthalpy change of the complex formation were reported to be 113.8 kcal/mol, which corresponds closely to values found in experimental studies ~ −116 kcal/mol (Shikiya et al., 2005) and those calculated using the nearest neighbor model (NN) (~ −95.5 ± 9.5 kcal/mol) (Lomzov et al., 2006).

The molecular system was studied by heating from 1 to 300 K linearly over 0.1–100 ns. The double DNA helix was found preserved, its topology was close to the B-form of DNA double helix, and the structures obtained by the MD calculations were nearly the same (see Figure 5.10a). Figure 5.10b shows that the equilibrium parts of molecular dynamic trajectories are close to each other in the range 0.025–10 ns, independent of the heating time of the system (0.1–100 ns). Energies of complexes in the single-stranded state and their difference (corresponding to the hybridization enthalpy change) (see Figure 5.10c) coincide with the accuracy. We have chosen 2.5 ns as a typical heating time from 1 K up to 300 K.

The double helix structural moiety did not change throughout the 1000 ns MD simulation. The analysis of the hybridization energy effect on the simulation time demonstrates that, after about 1 ns equilibration of the system, the change of enthalpy reaches a plateau and remains unchanged up to 1000 ns (see Figure 5.11). A 10 ns simulation may therefore be considered sufficient to reliably evaluate the internal energies of oligonucleotides and their complexes and the hybridization enthalpy change based on the MD trajectory.

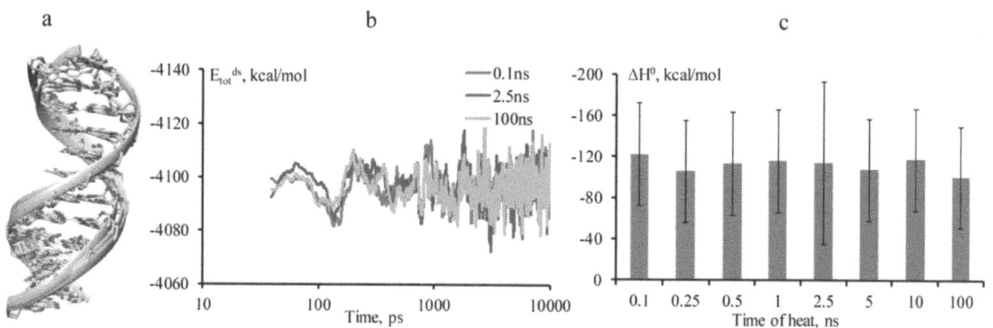

FIGURE 5.10 Analysis of 10 ns MD trajectories after equilibration. (a) The comparison of structures in double-stranded state averaged along MD trajectory in explicit (12 Å) (blue) and implicit (brown) water shell at the heating time of 2.5 ns with the structure obtained by the X-ray crystallography method (pink) (Holbrook et al., 1985). (b) Evolution of the Dickerson–Drew dodecamer duplex total energy after the equilibration at various heating times of the system in the implicit water shell. (c) The comparison of the enthalpy values of complex formation at various heating times in the implicit water shell.

***Source*: Lomzov et al. (2015).**

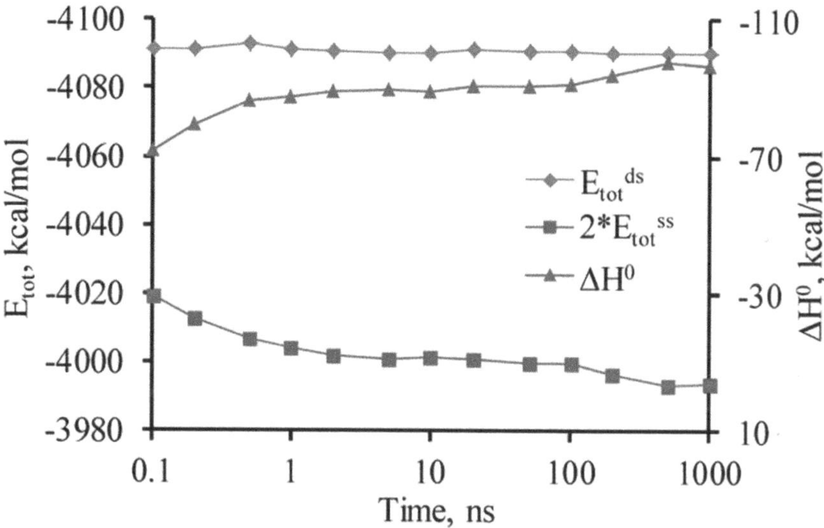

FIGURE 5.11 Dependence of full energy of single-stranded ($2*E_{tot}^{ss}$) and double-stranded (E_{tot}^{ds}) states (left axis) and the enthalpy of the self-complementary dodecamer complex formation ($\Delta H^\circ \approx E_{tot}^{ds} - 2*E_{tot}^{ss}$) (right axis) on simulation time in the implicit water shell (Andersen thermostat, 300 K, 0.1 M 1:1 salt) after heating of molecular dynamic system.

***Source*: Lomzov et al. (2015).**

Comparison between MD simulation results and those of experiments and modeling may help establish the authenticity of *in silico* data. In this regard, the correlation of the hybridization enthalpy, entropy change values, and free Gibbs energy changes has been established (details are found in Lomzov et al., 2015). As an example, I present here just the case of enthalpy values of the complex formation calculated by the MD simulations, being reported to correlate very well with those obtained in experiments and calculated using the nearest neighbor model (see Figure 5.12).

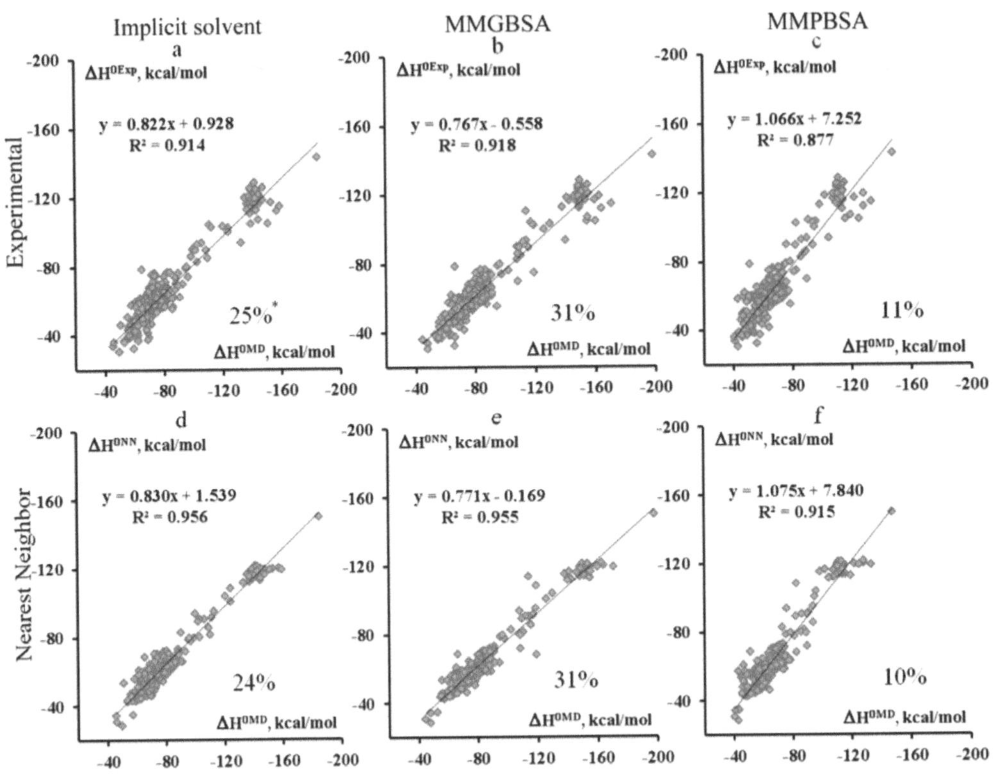

FIGURE 5.12 Correlation of the enthalpy values obtained in experiments (a, b, c) or calculated using the NN model (d, e, f) with those calculated by the MD method (along x-axis) using optimal simulation parameters in the implicit (a, d) and explicit water shells [MMGBSA calculations (b, e), MMPBSA calculations (c, f)]. Asterisk indicates mean absolute values of the calculation errors. Molecular mechanics generalized Born (or Poisson–Boltzmann) surface area (MMGB(PB)SA).

***Source*: Lomzov et al. (2015).**

5.3 CHEMICAL ENERGETICS IN LARGE BIOLOGICAL SYSTEMS

The application of the Gibbs equation to microbial growth occurred substantially around 1997, when yeast cell entropy was experimentally measured (Battley et al., 1997). Here the absolute entropy of the yeast cells was found to fall within the range of those for simple biological molecules like sugars and amino acids and more complex biopolymers like proteins. A value of 34.167 J K^{-1}·ICC-mol^{-1} for the absolute entropy of this mass of cells and of −151.46 J K^{-1}·ICC-mol^{-1} for the entropy of formation were reported by Battley and colleagues, who concluded that the thermodynamic effect of cellular organization in the dried cells was indeed negligible. Battley later examined the use of the Gibbs free energy equation (5.3) to accurately determine the change in energy accompanying the cellular growth (Battley, 2011; Battley, 2013). Here a thorough analysis of energy changes accompanying the growth of *S. cerevisiae* anaerobically on glucose and aerobically on glucose, ethanol, and acetic acid are made, and associated values of energetics including the entropy and enthalpy, Gibbs free energy changes have been listed. Readers may consult these references as vital sources (Battley, 2011; Battley 2013).

The concept of Gibbs free energy has been utilized in addressing larger systems in health-related issues, e.g. in understanding memory functions of the human brain (Rietman et al., 2020), measuring

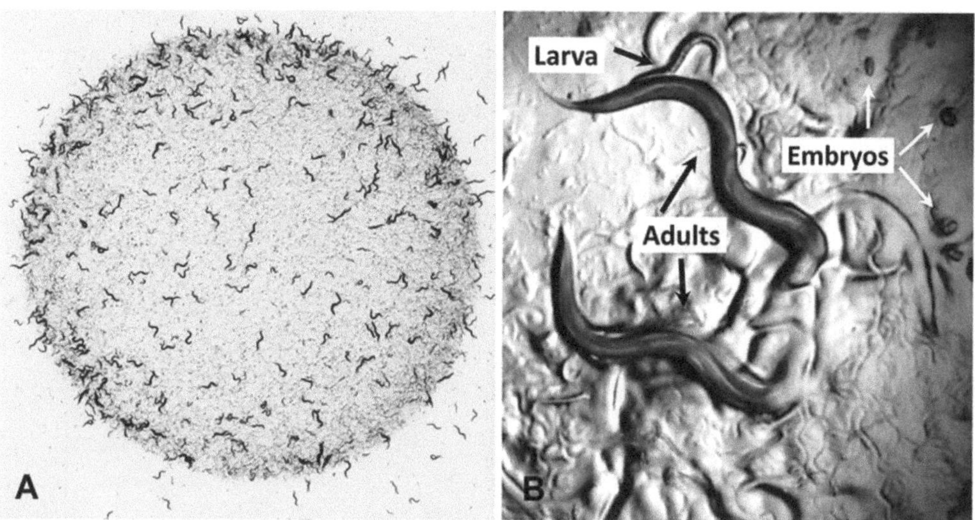

FIGURE 5.13 (A) A population of *C. elegans* feeding on a lawn of bacteria in a petri dish. Eggs and animals of all ages can be observed using sub-stage transmitted light on a stereo microscope. (B) A closeup of animals at different developmental stages feeding on a similar lawn of bacteria.

***Source*: Meneely et al. (2019). Courtesy of WormBook.org and used by permission. Due to the open access nature of the article, permission to copy the figure is automatic.**

the time-dependent rise of complexity within (C. elegans) embryonic development (McGuire et al., 2017), etc.

McGuire and colleagues demonstrated the possibility of time-dependent alterations in the Gibbs free energy in *C. elegans* development (McGuire et al., 2017). Figure 5.13 demonstrates the biological development of *C. elegans*.

The negative slope of this correlation (consider the magnitude of Gibbs free energy) (see Figure 5.14) can be partially explained by differentiating between the distinct embryogenic processes of proliferation and morphogenesis, which indicates that the cell division plays a fundamental role in the change of Gibbs free energy over the course of the development and, correspondingly, the stability of a large thermodynamic system like the studied one here, *C. elegans* (McGuire et al., 2017). This study utilized only publicly available gene expression data, raising a possible picture of developmental changes in protein concentration and, resultantly, chemical potential and free energy, so the study is a totally indirect investigation type. However, a huge amount of further associated research is required on various aspects concerning Gibbs free energy change, due to the dynamic nature of any thermodynamic system, before making such a sweeping conclusion as this study does. But one can consider the results here as a useful hint.

Recently, Gibbs free energy function was proposed as a measure of human brain development (see Figures 5.15 and 5.16; Rietman et al., 2020). The human brain development over the human lifespan, from a prenatal stage to advanced age, was inspected here (Rietman et al., 2020). The proteomic expression data with the Gibbs free energy were utilized in order to quantify the human brain's protein–protein interaction networks. The data, obtained from BioGRID, comprised tissue samples from the 16 main brain areas, at different ages, of 57 post-mortem human brains (see Kang et al., 2011; Sedmak et al., 2016). The obtained samples were from normal donors without a clinical brain pathology or signs of any serious genomic abnormalities. Different regions in the brain used in the study are briefed in Table 5.1.

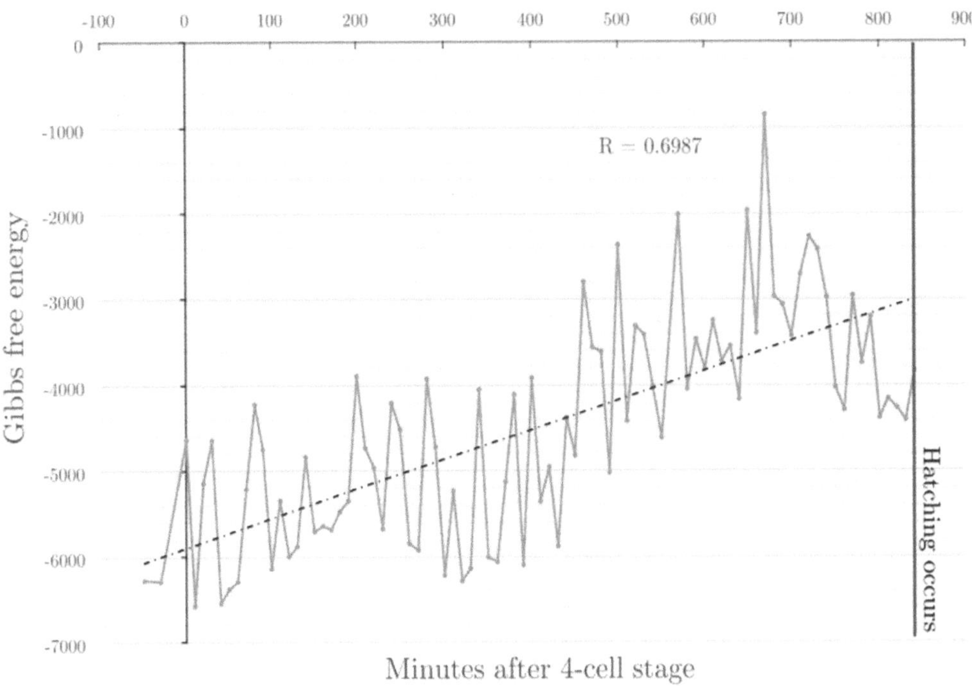

FIGURE 5.14 Gibbs free energy vs. time for embyronic development. This is computed using the methods outlined above using the expression set with accession number GSE60755, available in Levin et al. (2016), and a yeast two-hybrid *C. elegans* interactome, available at Center for Cancer Systems Biology: Worm Interactome Version 8, http://interactome.dfci.harvard.edu/C_elegans/index.php?page = download (accessed March 3, 2023). The results indicate a decrease in the Gibbs free energy magnitude across this timespan, which physically corresponds to a decrease in stability and potential work within the system, as well as an increase in disorder.

***Source*: McGuire et al. (2017).**

The gross anatomy of the brain undergoes morphological changes in both prenatal and neonatal stages. The subsequent evolution of neuronal connections takes place during the process of maturation and into adulthood correlated with learning and acquisition of skills. Reitman and colleagues (2020) have hinted that such continuous transformation taking place throughout the human lifespan is perhaps due to energetics in protein–protein interaction networks inside neurons. From the largest to the smallest hierarchies of organization of the brain, these structures are expected to undergo profound changes affecting their complex architecture and functions.

5.4 FREE ENERGY OF CHEMICAL INTERACTIONS

Figure 5.17 presents a schematic illustration demonstrating how free energy (F.E.) depends on the reaction coordinate (R.C.), when protein conformational transitions between different energy states occur (Ashrafuzzaman and Tuszynski, 2012b). E.g., the back-and-forth transitions between gramicidin A (gA) dimer (D) and monomers (M) in lipid bilayer membranes have been demonstrated here (see Figure 5.18). These states have different energy values and are separated by a potential barrier. $\Delta G_{I,II}$(Harm) and $\Delta G_{I,II}$(A.Harm) energy terms are harmonic and anharmonic energy terms, respectively, effective in the small deformation (bidirectional arrow in the picture, within a very short range from the point of the energy minimum, thanks to harmonic oscillator type behavior) and beyond small deformation region (right arrow) respectively to ensure transitions from D to M

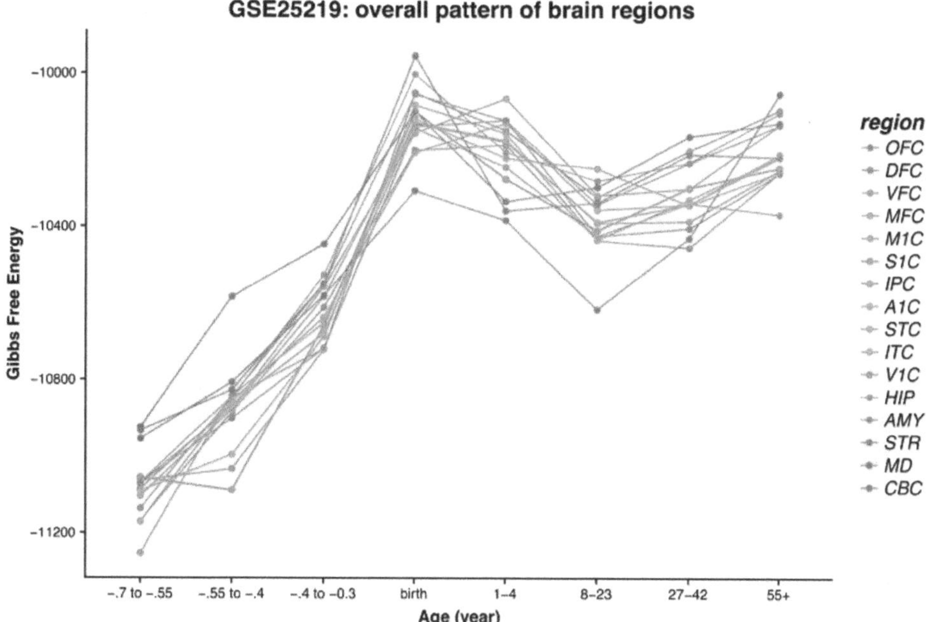

FIGURE 5.15 Plot of the Gibbs free energy for the 16 main brain regions averaged over the individual data sets and binned according to age groups.

***Source*: Rietman et al. (2020).**

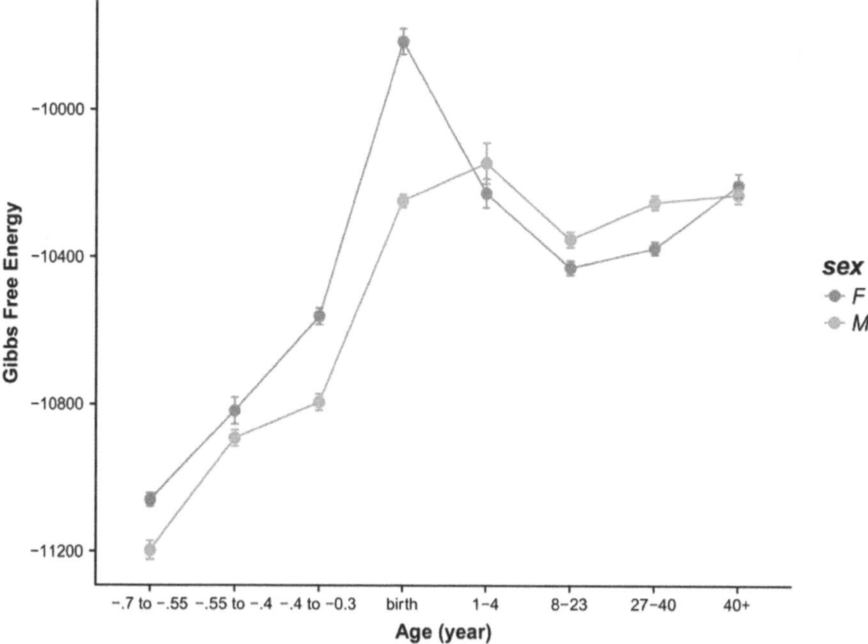

FIGURE 5.16 Plot of the Gibbs free energy values averaged over all 16 brain areas (see Figure 5.15) and presented for female and male cases separately.

***Source*: Rietman et al. (2020).**

TABLE 5.1
Summary of the abbreviations used for the brain regions with their descriptions and main functions performed

Abbreviation	Brain Region	Main Functions
OFC	orbital prefrontal cortex	Reasoning, language, decision making
DFC	dorsolateral prefrontal cortex	Working memory, motor planning, abstract reasoning
VFC	ventrolateral prefrontal cortex	Decision making, regulation of emotions, flexible behavior
MFC	medial prefrontal cortex	Sensory motor processes, cognitive and affective processes
M1C	primary motor (M1) cortex	Motor function coordination
S1C	primary somatosensory (S1) cortex	Integration of afferent somatosensory inputs
IPC	posterior inferior parietal cortex	Language, mathematical operations, body image
A1C	primary auditory (A1) cortex	Auditory system
STC	posterior superior temporal cortex	Sensation of sound, processing of speech
ITC	inferior temporal cortex	Visual object recognition
V1C	primary visual (V1) cortex	Pattern recognition
HIP	hippocampus	Learning and memory
AMY	amygdala	Processing of emotions
STR	striatum	Motor and action planning, decision making, motivation reinforcement
MD	mediodorsal nucleus of thalamus	Pain processing
CBC	cerebellar cortex	Sensory, motor, and association functions
IPC	posterior inferior parietal cortex	Language, mathematical operations, body image
A1C	primary auditory (A1) cortex	Auditory system
STC	posterior superior temporal cortex	Sensation of sound, processing of speech

Source: Rietman et al. (2020).

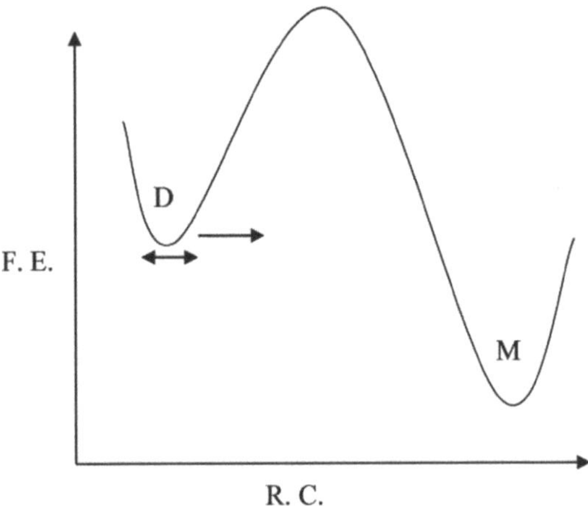

FIGURE 5.17 Free energy calculation in chemical reaction.

***Source*: Ashrafuzzaman and Tuszynski (2012b).**

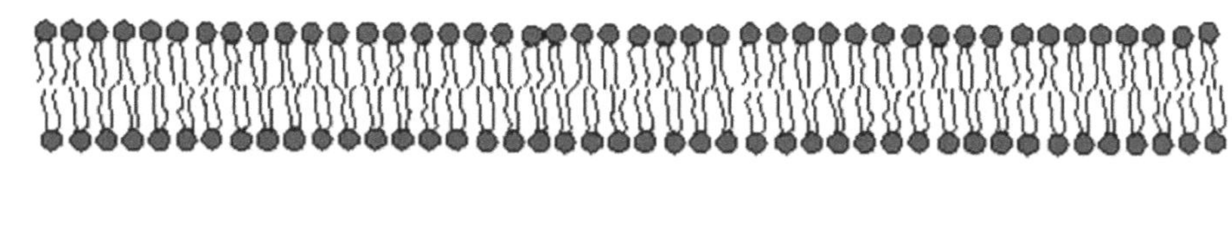

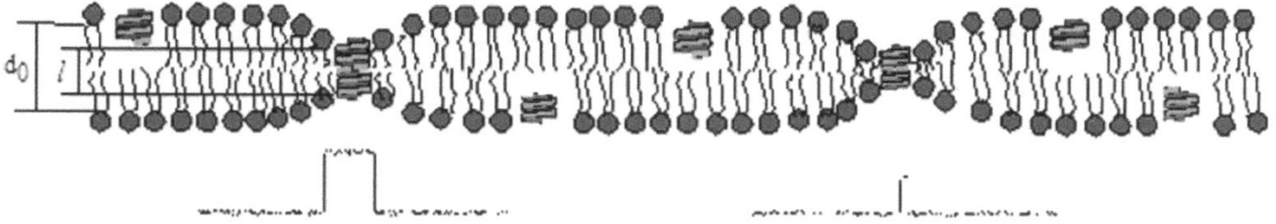

FIGURE 5.18 gA channels deform lipid bilayer's resting thickness. With channel formation bilayer conducts a current pulse with an average pulse width (τ) and height. Two types of monomers have structured two different channels.

Source**: Ashrafuzzaman and Tuszynski (2012b).**

states and vice versa of gA. Only in the limit of an extremely small bilayer deformation (d_0-l~0) the inclusion of only the harmonic term $\Delta G_{I,II}$(Harm)~$(d_0-l)^2$ may be sufficient. When d_0-l increases beyond the immediate vicinity of the free energy minima, higher order anharmonic energy terms dominate in the transition between D and M states. All such energy states appear together in our screened Coulomb interaction model calculation but many are missing in the elastic bilayer model calculation of the bilayer deformation energy as explained in the text. Moreover, at high values of d_0-l only a long-range Coulomb interaction between lipids sitting on the resting thickness of the bilayer and channel forming peptides (which is screened by intermediate lipids) can confirm the coupling between the two and in no way can the bilayer channel harmonic energy coupling, which works for a small distance due to the elastic properties of bilayers, be suitable to address the interaction between two distant species as is the case between lipids on the membrane's resting thickness and monomers in a channel.

The calculation of the energy terms in this kind of chemical association and dissociation (see Figure 5.18) is a biophysical challenge.

Electrophysiology record of gA channel current across lipid bilayer membrane helps address the kinetics of gA monomers and dimers inside the membrane. Theoretical modeling of the channel structure and channels' dimer/monomer transitions (M↔D) inside lipid bilayer was made considering a novel screened Coulomb interaction formalism, which considers the electrical coupling energetics between gA and lipid molecules in the channel–membrane coupling vicinity (for details, see Ashrafuzzaman and Tuszynski, 2012b). Here we found the free energy of associations between participating biomolecules generally takes a theoretical form as follows:

$$\Delta G_{I,II} \propto e^{d_0 - 1}$$

$$\text{so } F_{dts} = -\left(-\partial / \partial\left(d_0 - l\right) \Delta G_{I,II}\right) \propto e^{d_0 - 1} \qquad (5.6)$$

where $\Delta G_{I,II}$ and F_{dis} represent free energy of association/dissociation and the corresponding driving force acting on both gA monomers, respectively. The detailed derivation followed from a numerical

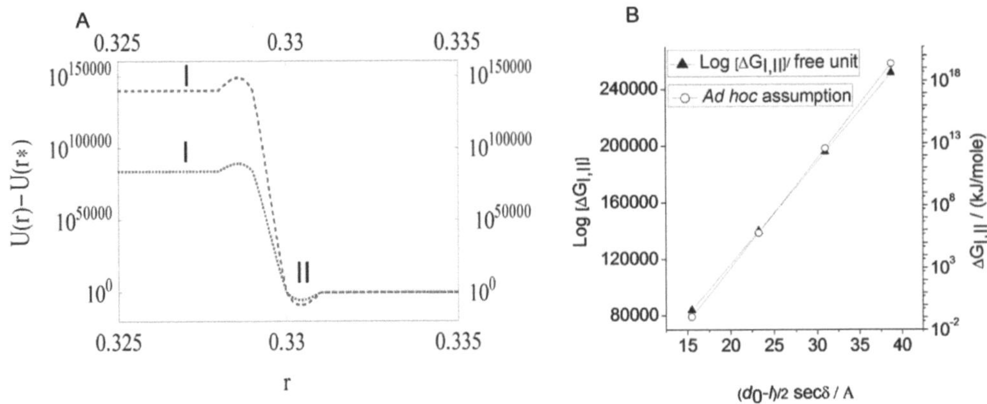

FIGURE 5.19 (A) Energy (real part) vs. reaction coordinate (r) plot for gA channels (single and double dashed curves represent 1st and 2nd order screening, respectively). $r^{*} \approx 1.5$ Å (hydrogen bond length). I and II represent energy levels G_I and G_{II}, respectively, at which gA monomers exist as free (no channel) and as gA dimmers, respectively. Although transitions (I↔II) happen at different r the range was chosen arbitrarily to better focus at a particular transition. $q_L/q_g = 0.005$, $(1/\varepsilon_0)q_L q_g \approx 1$ (for simplicity), $r_{LL} = 7.74597$ Å. Numerical integration was performed using Mathematica 7 ($-k_{max}2\pi/r_{LL}$, $k_{max}2\pi/r_{LL}$), $k_{max} = 100$, step size dr = 0.001 are judicious choices. (B) Log[$\Delta G_{I,II}$] vs d_0-l plot ($\Delta G_{I,II} = G_I$-G_{II}, free energy difference between free and dimer gA states). (d_0-l)secδ is the distance covered by lipid head groups in the deformed bilayer region near channel. δ (~constant, within 0-90^0 in all screening orders) the angle at which lipids in the deformed portion of the bilayer couples with the extension of gA channel length. Ad hoc assumptions (q_g~electron charge and other relevant parameters) give an estimate of $\Delta G_{I,II}$(○)/ (kJ/mole) which seriously depends on q_L as d_0 increases. Results are perhaps within 2nd order screening (d_0-l<40 Å).

computation using algorithms and programming in Mathematica providing us two energy states of the gA monomer (I)/dimer (II) (see Figure 5.19A). For details, see Ashrafuzzaman and Tuszynski (2012a; 2012b). Figure 5.19B shows how nice our computational and experimental values mutually associate.

Measuring free energetics of chemical interactions in a biological environment using valid theoretical analysis on experimental parameters was done, for the first time, correctly by us (Ashrafuzzaman and Tuszynski, 2012a; 2012b). While performing energetic calculations associated with addressing free energetics, we have focused here mostly on electrostatic interactions, van der Waals interactions, bond interactions, etc. that are sole physical energetic processes. These theoretical and experimental methodologies led us to create formalisms for drug discovery (Ashrafuzzaman, 2021). Interested readers may consult this patent to see how free energetics of chemical association/dissociation knowledge may even be extended toward discovering novel agents to bind with target biological structures to qualify as drugs.

REFERENCES

Ashrafuzzaman, M. (2021). Energy-based method for drug design. US Patent, US10916330B1, https://patents.google.com/patent/US10916330B1/en

Ashrafuzzaman, M., & Tuszynski, J.A. (2012a). *Membrane biophysics*. Berlin and Heidelberg: Springer. https://doi.org/10.1007/978-3-642-16105-6

Ashrafuzzaman, M., & Tuszynski, J. (2012b). Regulation of channel function due to coupling with a lipid bilayer. *Journal of Computational and Theoretical Nanoscience*, *9*(4), 564–570

Battley, E.H. (2013). A theoretical study of the thermodynamics of microbial growth using Saccharomyces cerevisiae and a different free energy equation. *Q. Rev. Biol.*, *88*(2), 69–96. doi: 10.1086/670529; PMID: 23909225

Battley, E.H. (2011). A comparison of energy changes accompanying growth processes by Saccharomyces cerevisiae . *J. Therm. Anal. Calorim.*, *104*, 193–200. https://doi.org/10.1007/s10973-011-1329-8

Battley, E.H., Putnam, R.L., & Boerio-Goates, J. (1997). Heat capacity measurements from 10 to 300 K and derived thermodynamic functions of lyophilized cells of Saccharomyces cerevisiae including the absolute entropy and the entropy of formation at 298.15 K. *Thermochimica Acta*, *298*(1–2), 37–46

Bearne, S.L. (2012). Illustrating enzyme inhibition using Gibbs energy profiles. *Journal of Chemical Education*, *89*(6), 732–737

Beis, I., & Newsholme, E.A. (1975). The contents of adenine nucleotides, phosphagens and some glycolytic intermediates in resting muscles from vertebrates and invertebrates. *Biochem. J.*, *152*(1), 23–32

Birket, M.J., Orr, A.L., Gerencser, A.A., Madden, D.T., Vitelli, C., Swistowski, A., Brand, M.D., & Zeng, X. (2011). A reduction in ATP demand and mitochondrial activity with neural differentiation of human embryonic stem cells. *J. Cell Sci.*, *124*, 348–358

Bonora, M., Patergnani, S., Rimessi, A., De Marchi, E., Suski, J.M., Bononi, A., ... & Pinton, P. (2012). ATP synthesis and storage. *Purinergic Signal*, *8*(3), 343–357. doi: 10.1007/s11302-012-9305-8; PMID: 22528680; PMCID: PMC3360099

Brand, M.D., & Felber, S.M. (1984). The intracellular calcium antagonist TMB-8 [8-(NN-diethylamino)octyl-3,4,5-trimethoxybenzoate] inhibits mitochondrial ATP production in rat thymocytes. *Biochem. J.*, *224*, 1027–1030

Brand, M.D., & Nicholls, D.G. (2011). Assessing mitochondrial dysfunction in cells. *Biochem. J.*, *435*, 297–312

Dawson, W.K., Shino, A., Kawai, G., & Morishita, E.C. (2021). Developing an updated strategy for estimating the free-energy parameters in RNA duplexes. *International Journal of Molecular Sciences*, *22*(18), 9708

Dunn, J., & Grider, M.H. (2022). Physiology, adenosine triphosphate. In *StatPearls* [Internet]. StatPearls Publishing.

Felber, S.M., & Brand, M.D. (1983). Concanavalin A causes an increase in sodium permeability and intracellular sodium content of pig lymphocytes. *Biochem. J.*, *210*, 893–897

Flurkey, W.H. (2010). Yield of ATP molecules per glucose molecule. *Journal of Chemical Education*, *87*(3), 271

Golas, S.M., Nguyen, A.N., Rietman, E.A., & Tuszynski, J.A. (2019). Gibbs free energy of protein–protein interactions correlates with ATP production in cancer cells. *J. Biol. Phys.*, *45*(4), 423–430. doi: 10.1007/s10867-019-09537-1; PMID: 31845118; PMCID: PMC6917683

Holbrook, S.R., Dickerson, R.E., & Kim, S.H. (1985). Anisotropic thermal-parameter refinement of the DNA dodecamer CGCGAATTCGCG by the segmented rigid-body method. *Acta Crystallogr., Sect. B: Struct. Sci.*, *41*, 255–262. doi: 10.1107/S0108768185002087

Kang, H.J., Kawasawa, Y.I., Cheng, F., Zhu, Y., Xu, X., Li, M., et al. (2011). Spatio-temporal transcriptome of the human brain. *Nature*, *478*, 483–489. doi: 10.1038/nature10523

Khan, M.T., Ali, S., Zeb, M.T., Kaushik, A.C., Malik, S.I., & Wei, D.Q. (2020). Gibbs free energy calculation of mutation in PncA and RpsA associated with pyrazinamide resistance. *Frontiers in Molecular Biosciences*, *7*, 52

Kim, M.S., Pinto, S.M., Getnet, D., Nirujogi, R.S., Manda, S.S., Chaerkady, R., ... & Pandey, A. (2014). A draft map of the human proteome. *Nature*, *509*(7502), 575–581

Levin, M., Anavy, L., Cole, A.G., Winter, E., Mostov, N., Khair, S., ... & Yanai, I. (2016). The mid-developmental transition and the evolution of animal body plans. *Nature*, *531*(7596), 637–641

Lomzov, A.A., Pyshnaya, I.A., Ivanova, E.M., Pyshnyi, D.V. (2006). Thermodynamic parameters for calculating the stability of complexes of bridged oligonucleotides. *Dokl. Biochem. Biophys.*, *409*, 211–215. doi: 10.1134/S1607672906040053

Lomzov, A.A., Vorobjev, Y.N., & Pyshnyi, D.V. (2015). Evaluation of the Gibbs free energy changes and melting temperatures of DNA/DNA duplexes using hybridization enthalpy calculated by molecular dynamics simulation. *Journal of Physical Chemistry B*, *119*(49), 15221–15234

McGuire, S.H., Rietman, E.A., Siegelmann, H., & Tuszynski, J.A. (2017). Gibbs free energy as a measure of complexity correlates with time within *C. elegans* embryonic development. *J. Biol. Phys.*, *43*(4), 551–563. doi: 10.1007/s10867-017-9469-0; PMID: 28929407; PMCID: PMC5696307

Meneely, P.M., Dahlberg, C.L., & Rose, J.K. (2019). Working with worms: *Caenorhabditis elegans* as a model organism. *Current Protocols: Essential Laboratory Techniques*, *19*(1), e35

Meurer, F., Do, H.T., Sadowski, G., & Held, C. (2017). Standard Gibbs energy of metabolic reactions: II. Glucose-6-phosphatase reaction and ATP hydrolysis. *Biophys Chem.*, *223*, 30–38

Mookerjee, S.A., & Brand, M.D. (2015). Measurement and analysis of extracellular acid production to determine glycolytic rate. *J. Vis. Exp.*, *10*, 3791/53464

Mookerjee, S.A., Gerencser, A.A., Nicholls, D.G., & Brand, M.D. (2017). Quantifying intracellular rates of glycolytic and oxidative ATP production and consumption using extracellular flux measurements. *Journal of Biological Chemistry*, *292*(17), 7189–7207

Mookerjee S.A., Goncalves R.L.S., Gerencser A.A., Nicholls D.G., & Brand, M.D. (2015). The contributions of respiration and glycolysis to extracellular acid production. *Biochim. Biophys. Acta*, *1847*, 171–181

Moreno-Sánchez, R., Rodríguez-Enríquez, S., Saavedra, E., Marín-Hernández, A., & Gallardo-Pérez, J.C. (2009). The bioenergetics of cancer: is glycolysis the main ATP supplier in all tumor cells? *Biofactors*, *35*(2), 209–225

Nelson, D.L., & Cox, M.M. (2004). *Lehninger principles of biochemistry*, (4th edn). New York: Freeman

Nicholson, D.E. (2001). IUBMB-Nicholson metabolic pathways charts. *Biochemistry and Molecular Biology Education*, *29*(2), 42–44.

Razavi, M., Saberi Fathi, S.M., & Tuszynski, J.A. (2022). The effect of the protein synthesis entropy reduction on the cell size regulation and division size of unicellular organisms. *Entropy*, *24*(1), 94

Rietman, E.A., Scott, J.G., Tuszynski, J.A., Klement, G.L. (2017). Personalized anticancer therapy selection using molecular landscape topology and thermodynamics. *Oncotarget*, 8, 18735. doi: 10.18632/oncotarget.12932

Rietman, E.A., Taylor, S., Siegelmann, H.T., Deriu, M.A., Cavaglia, M., & Tuszynski, J.A. (2020). Using the Gibbs function as a measure of human brain development trends from fetal stage to advanced age. *Int. J. Mol. Sci.*, *21*(3), 1116. doi: 10.3390/ijms21031116; PMID: 32046179; PMCID: PMC7037634

Rojas-Pirela, M., Andrade-Alviárez, D., Rojas, V., Kemmerling, U., Cáceres, A.J., Michels, P.A., … & Quiñones, W. (2020). Phosphoglycerate kinase: structural aspects and functions, with special emphasis on the enzyme from Kinetoplastea. *Open Biology*, *10*(11), 200302

Sedmak, G., Jovanov-Milošević, N., Puskarjov, M., Ulamec, M., Krušlin, B., Kaila, K., & Judaš, M. (2016). Developmental expression patterns of KCC2 and functionally associated molecules in the human brain. *Cereb. Cortex.*, *26*, 4574–4589. doi: 10.1093/cercor/bhv218

Shikiya, R., Li, J.S., Gold, B., & Marky, L.A. (2005). Incorporation of cationic chains in the Dickerson-Drew dodecamer: correlation of energetics, structure, and ion and water binding. *Biochemistry*, *44*, 12582–12588. doi: 10.1021/bi050897i

Voet, D.J., Voet, J.G., & Pratt, C.W. (2002). *Fundamentals of biochemistry*. New York: John Wiley and Sons. www.jove.com/science-education/11008/atp-yield

Wilhelm, M., Schlegl, J., Hahne, H., Gholami, A.M., Lieberenz, M., Savitski, M., … & Kuster, B. (2014). Mass-spectrometry-based draft of the human proteome. *Nature*, *509*(7502), 582–587.

Yagisawa, S. (1995). Enzyme kinetics based on free-energy profiles. *Biochemical Journal*, *308*(1), 305–311.

6 Biomolecular Machines

The seventeenth-century concept of "molecular machines" has taken center stage in modern biology (Piccolino, 2000; Aprahamian, 2020). Biophysics techniques have evolved to address biomolecular machines at the single-molecule level (Schliwa and Woehlke, 2003; Vale and Milligan, 2000; Boyer, 1997; Bustamante et al., 2011). These techniques are capable of detecting existing and synthesizing molecular machines in biological systems (Abendroth et al., 2015; Coskun et al., 2012; Astumian, 2001; Erbas-Cakmak et al., 2015; Kay et al., 2007; Cheng et al., 2015; Kottas et al., 2005; Sengupta et al., 2012). The application of updated biophysics and the latest engineering techniques along with systems biology approaches developed mostly in the twenty-first century has made it possible to progress fast (Gurkiewicz and Korngreen, 2007; Menon et al., 2009; Teed and Silva, 2016; George et al., 2020). The ongoing efforts are devoted to designing and constructing synthetic molecular pumps and motors that will result both in a better understanding of biomolecular machines and in the development of remarkably versatile tools for organizing complex matter in biological systems (Lehn, 2013), for developing networked nanoswitches for catalysis (Schmittel, 2015), and for harnessing the power of molecule-by-molecule assembly (Kassem et al., 2016) in biological systems. Fast-advancing bioinformatics, genomics, and various novel machine learning (ML) algorithms are emulating advanced nanotechnology in addressing various aspects of biomolecular machines, which may help us apply machine technology to deal with both naturally functioning and mutated cellular structures. Biomolecular machines are expected to be at the forefront of biomedical breakthroughs in the coming days.

6.1 BIOLOGICAL ORGAN AS BIOLOGICAL MILL: A NETWORK OF BIOMOLECULAR MACHINES

Malpighi, a prominent life scientist in the seventeenth century, was one of the first to attribute body function to an organized series of minute "organic machines" (Piccolino, 2000). Following the concept of the existence of microscopic body machines, physiology started to depend less on metaphysical theories for the interpretation of body functions, and more on investigations combining experimental study with the application of the "laws of mathematics and geometry" to body machines. In the seventeenth century, the discovery of cells by Robert Hooke using the microscopy technique aided heavily in this trend. Cells quickly appeared to attain a status as representative of the biological body, which runs on power or energy supplied by its powerhouse "mitochondria" discovered in the nineteenth century (Siekevitz, 1957). Every human cell usually contains more than 1,000 mitochondria. Mitochondria, tiny factories in our cells, may take up to 25 percent of the cell volume and are engaged in producing adenosine triphosphate (ATP) every second of every day because ATP is continuously spent by the cell and the lifetime of this currency is temporary,

DOI: 10.1201/9781003287780-6

that is, it can't be stored for future use. Mitochondria are proven to operate like both engines and factories in eukaryotes, which coordinate cellular energy production and the availability of fundamental building blocks that are required for cell proliferation (Ahn and Metallo, 2015). Having so many mitochondria (1,000–2,000), every cell appears to house a large mill containing a huge number of factories working continuously to produce fuel for biological cells, tissues, organs, and above all the whole body.

The ATP synthesis largely occurs in cellular respiration within the mitochondrial matrix, generating approximately 32 ATP molecules per molecule of oxidized glucose. Consumption of ATP happens for supplying the energy needed to run cellular machinery which works to process various cellular mechanisms such as ion transport, muscle contraction, nerve impulse propagation, substrate phosphorylation, and chemical synthesis. Human cells within the body depend upon the hydrolysis of 100 to 150 moles of ATP per day to ensure proper functioning. The intracellular ATP concentration is routinely maintained at 1–10 μM (Beis and Newsholme, 1975). From ATP production to running cellular mechanisms to helping organs to function, there are machines (small to large sizes) involved to support the body to function by continually maintaining a network among all machineries.

The human body (like other animal bodies), whose purpose is to consume, store, process (create or destroy), and release, may be conceived as acting like a mill (Corner, 1967), which contains machines of various sizes and structures with versatile objectives. To run the machinery, standard amounts of energy are required. Figure 6.1 presents a comparative analysis of the use of energy in various processes and sections of biological systems. I did some calculations (presented later) to create this figure for general comparison only, as the comparison features energy use for only specific aspects. Actual energetics is a bit different, but Figure 6.1 gives us a general idea about the energetic scales.

As we wish to learn about various biomolecular machines active in various areas and with varied dimensions within biological systems, I thought the general estimate shown in Figure 6.1 would make upfront understanding easier. The ATP synthase is itself a molecular machine, which works to produce 30–32 ATPs by breaking every glucose molecule. So our understanding of energetics in the unit of glucose-produced ATP is indeed going to help us compare among powers behind various

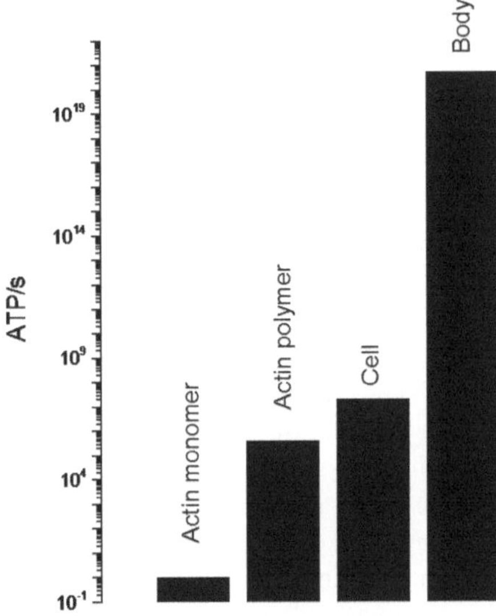

FIGURE 6.1 The change or use of ATP molecules per second in biological systems: actin monomer (~5 nm), actin polymer (~20 μm), cell, adult human body. The values here are just estimated ones.

molecular machines in biological systems in biologically obvious energy units. Most of these ATPs are used up continuously by molecular machineries working in the body.

To plot the comparable ATP conditions in Figure 6.1 we had to use information from three example cases. The size of each of the monomeric units making up an actin filament is roughly 5 nm with a molecular mass of about 40 kDa. In many eukaryotic cells, we know that the motility is driven primarily by the dynamic actin polymerization at a steady-state cost of ≈1 ATP hydrolysis per polymerizing actin monomer (Pollard and Borisy, 2003; Dominguez and Holmes, 2011). Two actin monomers are there for each ≈5 nm of filament (see Figure 6.2). The multiplications depicted in Figure 6.3 suggests that each filament must grow by ≈100 monomers/s to support the motility. This costs ~100 ATP per polymerizing filament per second. There are (2–4) × 10^6 proteins in 1 μm^3 of a cell (Milo, 2013), and the average protein is 300–400 amino acids (aa) long, yielding ≈10^9 aa/μm^3 (Flamholz et al., 2014). It requires ≈4 ATP equivalents to add an amino acid to a nascent polypeptide chain (BNID 101442). A typical goldfish keratocyte cell volume is ≈500 μm^3 (BNID 110905) and thus requires 500 μm^3 × 10^9 aa/μm^3 × 4 ATP/aa ≈ 2 × 10^{12} ATP just to synthesize its proteins from

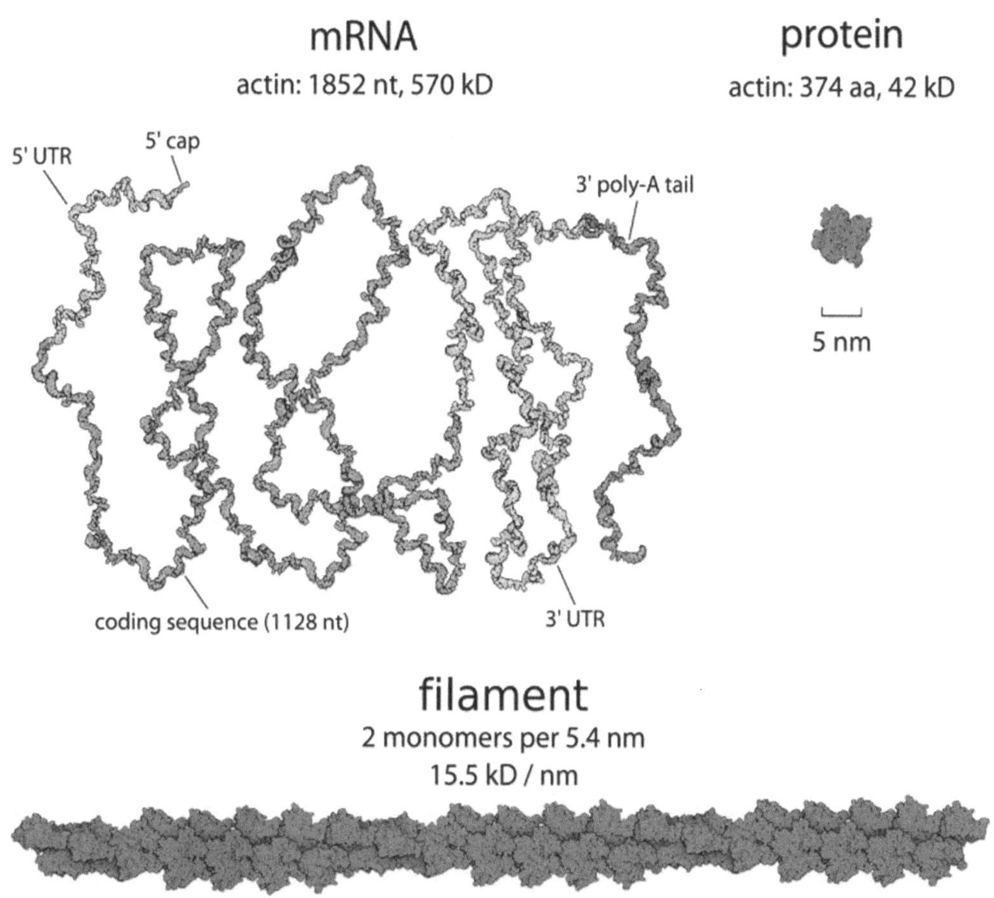

FIGURE 6.2 Which is larger, mRNA or the protein for which it codes? When we ask, most peoples' instinct is to say that proteins are larger. As seen in this figure, the opposite is overwhelmingly the case. The mRNA for actin is more massive and has a larger geometric size than the actin monomers for which it codes because the mass of a codon of mRNA is an order of magnitude greater than that of the average amino acid.

***Source*: Flamholz et al. (2014).**

How much ATP is required for actin-driven motility?

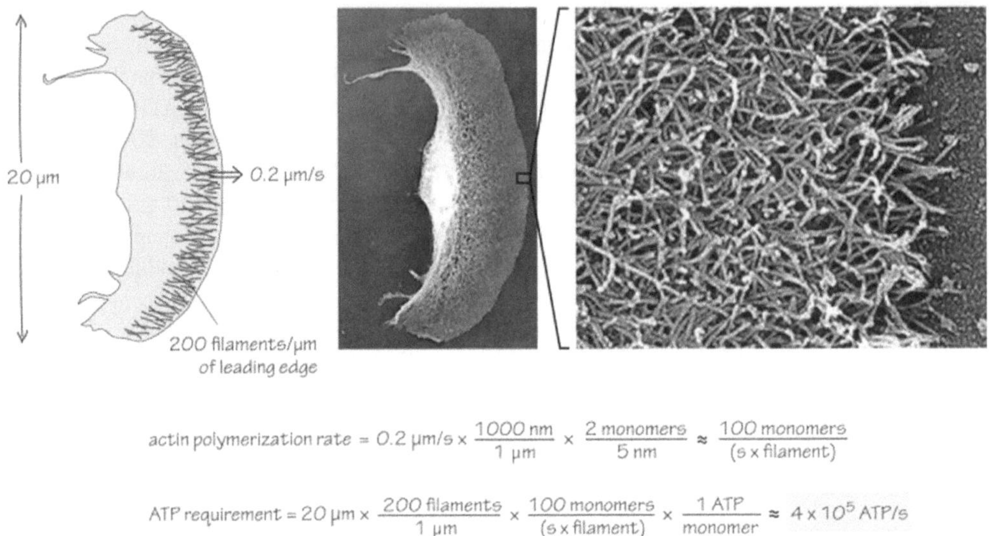

FIGURE 6.3 Back-of-the-envelope calculation of the ATP demand for motility of a cell. Actin filaments criss-cross the leading edge of a motile keratocyte, and their dynamic polymerization results in a net forward motion with a speed of 0.2 μm/s.

***Source*: Electron micrographs adapted from Svitkina et al. (1997).**

amino acids. Considering the average half-life of a protein to be about 1 day (Cambridge et al., 2011), it is found that the cell must duplicate its proteome once every 24 h ≈ 10^5 s, thereby consuming ≈2×10^7 ATP/s. If one wishes to calculate the total length of the protein, it is easy to do so taking the length of an amino acid to be 0.4 nm (Ainavarapu et al., 2007), thus the length-dependent rate of the consumption of ATP may be deducted too.

The human cell produces ~10^7–10^8 ATP/s on average (Flamholz et al. 2014) and the total cells in the adult human body ~10^{14} (Goodsell, 2009; Alberts et al., 2002). Thus the rate of ATP appearing in the whole human body is approximately 10^{21}/s. This information has been used in Figure 6.1 (see the rightmost bar/column).

ATP generation is not absolutely a constant aspect, rather may strongly depend on the environment and other parameters, including oxygen consumption. ATP generated per molecule of oxygen consumed may vary significantly both among and within individuals (Salin et al., 2015). The total quantity of ATP in an adult is approximately 0.10 mol/L. Daily 100–150 mol/L of ATP are required, which is equivalent to each ATP molecule being recycled 1,000–1,500 times per day. The human body basically turns over its weight in ATP daily. A massive amount of ATP molecules is produced in the human body, amounting to 50 kg per day in a healthy adult (Kühlbrandt, 2015). ATP consumption rate may vary due to one's activity. From the daily produced ATP, the human body uses most of it, leaving only a tiny amount as a surplus, which is of the order of the gram. The animal body stores a small quantity of ATP within its muscle cells, approximately 8 mmol/kg wet weight of muscle (Baker et al., 2010); the amount is so tiny that it can fuel only a few seconds of physical exercise. Therefore, to sustain muscle contraction during highly intense activity or exercise, ATP needs to be regenerated at a rate complementary to ATP demand. ATP regeneration has to compensate for the increased demand, which goes up as much as 1,000-fold compared to that at rest. The following three energy systems function to replenish ATP in muscle (Baker et al., 2010):

- phosphagen
- glycolytic
- mitochondrial respiration

From the analysis presented above it is clear that our body produces, uses, and recycles huge amounts of ATP continuously. A huge number of molecular machines are engaged in this energetics, independently or collectively. ATP synthesis machinery (Thomas, 2009; Kühlbrandt, 2015), protein synthesis machinery (Preiss and Hentze, 2003), gene synthesis machineries (Caruthers, 1985), lipid synthesis machineries (Caruthers, 1985), etc. work independently. But the functions and effects of their products in biological cells may not be considered exclusively independent. Cellular signaling strongly coordinates among various cellular components, including cell-based machines, to help cells function properly (Jordan et al., 2000).

6.2 BIOMOLECULAR MACHINES OF THE CELL

Structural studies of molecular machines raise actual understanding of the mechanisms of the machines. Among various physical techniques, X-ray crystallography, NMR spectroscopy, and electron microscopy are applied techniques in the field of biomolecular structural studies. These techniques provide high-resolution structural information helping us interpret various underlying mechanisms of biomolecules. They may be applied to understanding the functions of biomolecular machines. We shall brief on a few example cases here.

6.2.1 ATP Synthase Structure

ATP synthase is a rotary machine found next to the bacterial flagella motor in the biological world. This enzyme is composed of two motors, F_0 and F_1, which are connected by a common rotor shaft to exchange the energy of proton translocation and ATP synthesis/hydrolysis through mechanical rotation (see model diagram in Figure 6.4). The rotation of the ATP hydrolysis-driven isolated F_1 motor was observed with an optical microscope, and its performance was explained by Noji and colleagues (Noji and Yoshida, 2001). The motor is seen to rotate with discrete 120° steps, each driven by one ATP molecule hydrolysis. A cooperative domain bending motion of the catalytic β subunits initiated by ATP binding generates the torque. In the F_0 motor, it was proposed that torque might be generated by the large twist of one helix of F_0 *c* subunits or by the change in electrostatic forces between rigid subunits. Figure 6.5 presents the γ rotation in the F_1 motor.

Recent progress in structural studies of ATP synthase using various biophysical techniques have yielded a detailed understanding (see e.g. Morales-Rios et al., 2015; Vinothkumar et al., 2016; Montgomery et al., 2021). These studies help address the structural dynamics and functional moieties of this highly important cellular structure at high resolution.

Here we shall address the X-ray crystallography structural study of ATP synthase. X-ray crystallography at 4.0 Å resolution was applied for addressing the structure of the intact ATP synthase from the α-proteobacterium *Paracoccus denitrificans*, inhibited by its natural regulatory ζ-protein (Morales-Rios et al., 2015). The ζ-protein is bound via its N-terminal α-helix in a catalytic interface in the F_1 domain. The bacterial F_1 domain is attached to the membrane domain by peripheral and central stalks. The structure of the *P. denitrificans* F-ATPase–ζ-inhibitor complex was resolved using molecular replacement at 4.0 Å resolution. The final model (see Figure 6.6) contains the following residues (where E, TP, and DP denote the subunits comprising the empty, diphosphate-containing, and triphosphate-containing catalytic interfaces, respectively): α_E, 2–190 and 196–511; α_{TP}, 7–193, 198–405, and 411–511; α_{DP}, 28–511; β_E, 3–468; β_{TP}, 4–469; β_{DP}, 2–471; γ, 3–62, 64–73, 78–110, 115–143, 147–166, 170–199, and 212–289; δ, 5–114; ε, 9–83; subunit a, 35 residues in aH3 and aH4 modeled as poly-Ala (residues 1,001–1,035), aH5 (residues 166–198), and aH6 (residues

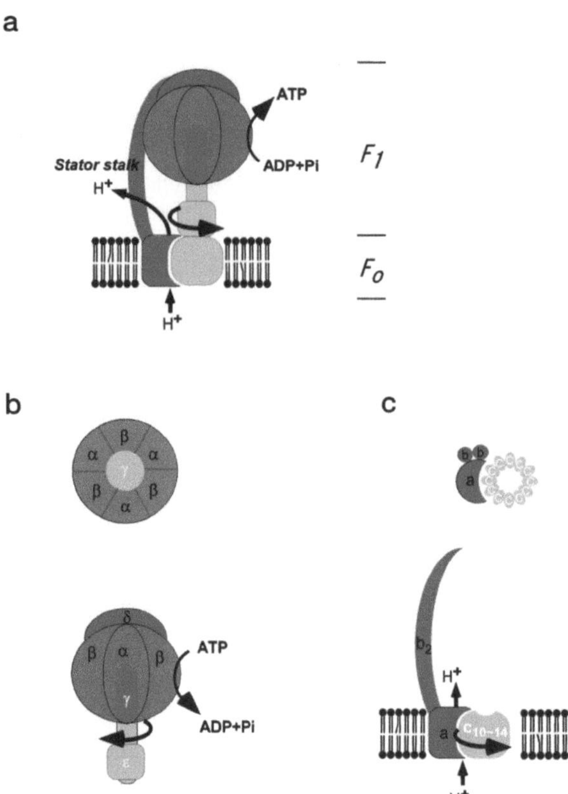

FIGURE 6.4 Schematic diagram of the ATP synthase. (*A*) Side view of the ATP synthase. ATP synthase is composed of the F_1 and F_0 motor sharing a common rotary shaft (*gray*). A stator stalk connects two motors (*red*) that do not slip. The F_0 motor generates a rotary torque powered by the proton flow-enforcing F_1 motor to synthesize ATP. The rotational direction is clockwise viewed from the membrane side. (*B*) Cross-section and side view of F_1 motor. The $\alpha_3\beta_3$ cylinder hydrolyzing ATP makes an anti-clockwise rotation of the rotor part composed of the γ and ε subunits. (*C*) Cross-section and side view of the F_0 motor. Proton flow accompanies a clockwise rotation of the ring structure made of 10–14 copies of the c subunit. The color contrast in the figure here and all subsequent ones will be clear in the online version of the book.

***Source*: Noji and Yoshida (2001).**

217–246); and each c-subunit in the c_{12}-rotor ring (3–76). It also contains five segments of secondary structure that are not assigned to any specific subunit, defined as follows: chain V, residues 1,001–1,078 (probably either subunit b or b′); W, residues 1,001–1,124 (subunit b or b′); Y, residues 1,001–1,054 (two antiparallel transmembrane α-helices); 1, residues 1,001–1,020 (subunit δ or α_{DP}); 2, residues 1,001–1,015 (subunit δ or α_{DP} or b or b′); and 3, residues 1,001–1,019 (an α-helix parallel to the plane of the membrane). The structure is seen to contain two additional α-helical segments containing residues 1–32 and 82–103 of the ζ-inhibitor. The nucleotide-binding sites in the catalytic β_{DP}- and β_{TP}-subunits and the noncatalytic α_{TP}- and α_{DP}-subunits each contain ATP-Mg, and the nucleotide-binding site in the α_E-subunit contains ADP-Mg. Neither substrates nor products are associated with the β_E-subunit.

The mode of binding of the ζ-inhibitor to the F-ATPase from *P. denitrificans* is presented in Figure 6.7. The inhibitor is bound to the F_1 domain via residues 1–19 of the N-terminal α-helix, which occupy a cleft in the lower region of the $\alpha_{DP}\beta_{DP}$-catalytic interface (see Figure 6.7A and B),

a

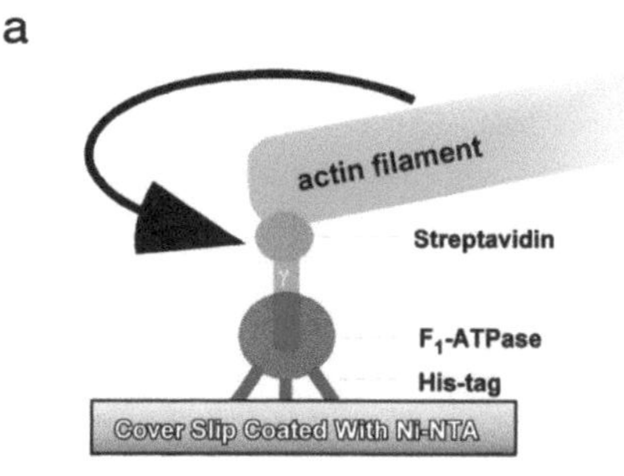

b

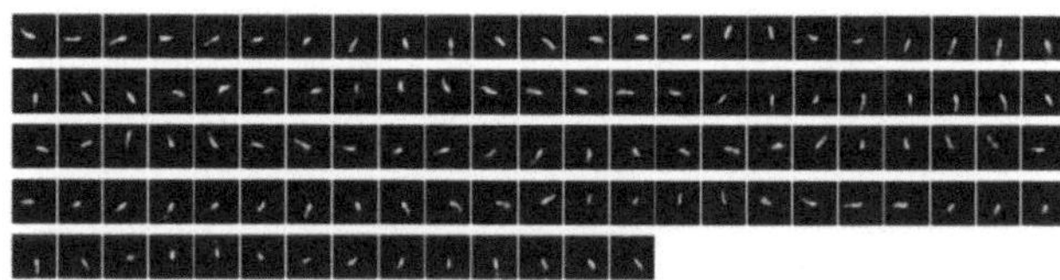

FIGURE 6.5 The direct observation of the γ rotation in the F_1 motor. (*A*) Experimental system for the observation of the γ rotation using an optical microscope. The F_1 motor tagged with 10 His residues at the N terminus of the β subunit was immobilized upside down on a coverslip coated with nickel-nitrilotriacetic acid (*Ni-NTA*). An actin filament (*green*) labeled with fluorescent dyes and biotins was attached to the biotinylated γ subunit (*gray*) through streptavidin (*blue*). (*B*) Rotary movement of an actin filament observed from the bottom, the membrane side, with an epifluorescent microscope. Length from the axis to tip, 2.6 μm; rotary rate, 0.5 revolution per s; time interval between images, 133 ms.

***Source*: Noji and Yoshida (2001).**

with the rest of the α-helix (residues 20–32) extending from the surface of the enzyme. Residues 1 and 2 are close to residue Ser13 in the N-terminal α-helix of the γ-subunit, along the central axis of the F_1 domain. Residues 3–19 probably form polar and hydrophobic interactions with other residues in α-helices in the C-terminal domains of the α_{DP}- and β_{DP}-subunits (see Figure 6.7C). α-Helix 4 (residues 82–103) is also resolved, stabilized by contacts with α-helix 1 and the α_{DP}-subunit. In solution, residues 1–18 of the 107-aa chain of the ζ-inhibitor are unstructured, with the rest of the chain folded into a four-helix bundle (residues 19–42, 46–53, 66–77, and 81–103) (Serrano et al., 2014). A complete fold of the ζ-inhibitor was constructed from the combined solution and crystal structures (see Figure 6.7D).

The δ-subunit component of the peripheral stalk binds to the N-terminal regions of two α-subunits (see Figure 6.8). The main interactions between the peripheral stalk and the F_1 domain of the enzyme involve the δ-subunit, which sits on top of the crown of the F_1 domain (see Figure. 6.8A). The N-terminal domain of the *P. denitrificans* δ-subunit is folded into a bundle of six α-helices, as in the *E. coli* δ-subunit (Wilkens et al., 2005) and the orthologous bovine oligomycin sensitivity conferral protein subunit, OSCP (Carbajo et al., 2007; Rees et al., 2009). The N-terminal regions of α-subunits project from the top of the crown, and α-helices in the N-terminal regions of the α_E- and α_{TP}-subunits (residues 2–22 and 7–22, respectively) interact with α-helices δH1 (residues 8–27) and δH5 (residues 79–91), and δH2 (residues 30–45), δH3 (residues 47–53), and δH4 (residues 61–68),

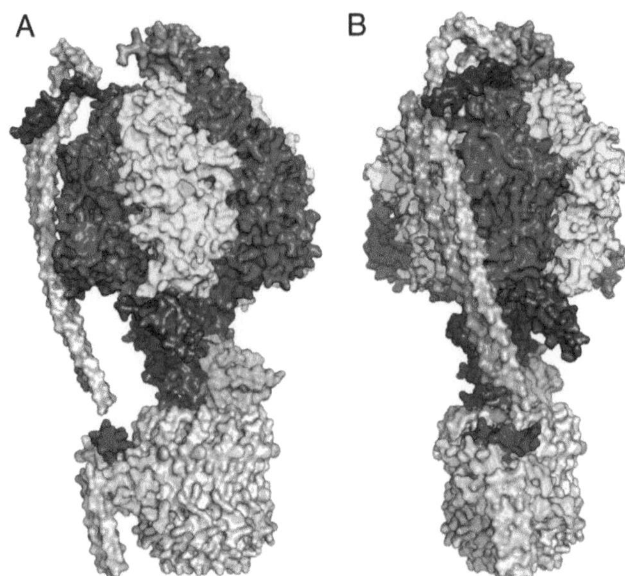

FIGURE 6.6 Structure of the complex of the F-ATPase from *P. denitrificans* with the bound ζ-inhibitor protein. (*A* and *B*) Side views of the enzyme–inhibitor complex in surface representation. (*B*) is rotated right by 90° relative to (*A*). (*Upper*) Membrane extrinsic F_1 catalytic domain (red, yellow, blue, and green corresponding to three α-subunits, three β-subunits, and single γ- and ε-subunits, respectively). In the peripheral stalk, the δ-subunit (top) is sky blue and the long and approximately parallel α-helical segments in orange and pink (chains V and W) extending down the surface of the interface between the α- and β-subunits are parts of the b- and b′-subunits (undistinguished). Unassigned α-helical segments (chains 1 and 2) in the vicinity of the junction between the δ-subunit and b- and b′-subunits are purple and light gray, respectively. Helix-1 of the ζ-inhibitor is brown. (*Lower*) In the membrane domain, the ring of 12 c-subunits is gray and a bundle of four resolved α-helices assigned to the a-subunit is lemon green. An unassigned α-helical segment in magenta (chain 3) lies approximately parallel to α-helices in subunit a, and two unassigned side-by-side transmembrane α-helices (chain Y) are colored light blue.

***Source*: Morales-Rios et al. (2015).**

respectively; only the first interaction was observed in the structure of the bovine F_1–peripheral stalk complex (Rees et al., 2009). The structure of the N-terminal region of the α_{DP}-subunit is unclear.
The membrane domain of the F-ATPase from *P. denitrificans* was resolved (see Figure 6.9A). It consists of a c_{12}-ring (gray); an associated bundle of four α-helices (green) with its axis tilted at about 30° to the plane of the membrane, with a fifth α-helix (magenta) sitting on top of the bundle close to the inner surface of the bacterial membrane; and two side-by-side α-helices (cyan) normal to the plane of the membrane. The c-ring is made of an inner ring of N-terminal α-helices and an outer ring of C-terminal α-helices, and the loops joining the helices make an extensive interface with the foot of the central stalk with a buried surface area of 522 $Å^2$. The c-ring and the central stalk together constitute the rotor of the enzyme. Based on the conservation of six hydrophobic segments in their sequences, it is likely that the a-subunit has six transmembrane α-helices (aH1–aH6), whereas it has previously been considered to have five (corresponding to aH1–aH3, aH5, and aH6) (Rees et al., 2009). The four-helix bundle in the current structure has been attributed to α-helices aH3–aH6, where aH5 and aH6 consist of 29 and 32 residues, respectively, corresponding to the unusually long hydrophobic segments 5 and 6 in the sequences of a-subunits. Segment 5 contains the absolutely conserved Arg residue Arg182, known from studies in *E. coli* to be essential for proton translocation

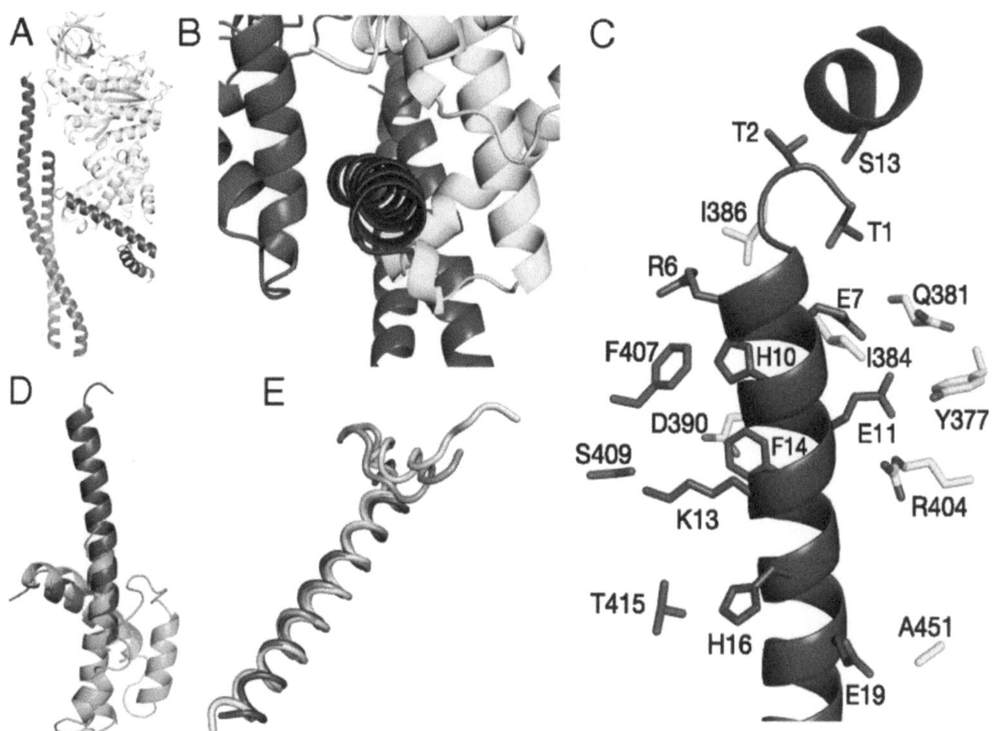

FIGURE 6.7 Mode of binding of the ζ-inhibitor to the F-ATPase from *P. denitrificans*. The inhibitor is bound in the $\alpha_{DP}\beta_{DP}$-catalytic interface of the enzyme. (*A*) Cross-sectional side view of the F_1 domain showing the interaction of the ζ-inhibitor protein (brown) with the C-terminal domain of the β_{DP}-subunit (yellow) and the coiled-coil of α-helices in the γ-subunit (blue). (*B*) View from outside the complex toward the $\alpha_{DP}\beta_{DP}$-catalytic interface with the N-terminal α-helix of the ζ-inhibitor in a cleft between the α_{DP}- and β_{DP}-subunits. (*C*) Potential interactions between side chains of the ζ-inhibitor protein with residues in the α_{DP}-, β_{DP}-, and γ-subunits. (*D*) Composite structure of the ζ-inhibitor by combination of residues 1–32 and 82–103 from the current study (brown) with residues 15–104 of the solution structure (cyan). (*E*) Superposition of the N-terminal region of the ζ-inhibitor (brown) with the corresponding inhibitory regions of IF_1 from bovine and yeast mitochondria (cyan and pink, respectively). The α-helical regions are residues 21–49, 16–36, and 3–24, respectively.

***Source*: Morales-Rios et al. (2015).**

through the membrane domain of the enzyme, and in the structure, as required, this residue is close to another essential residue Glu60 in the c-subunit (see Figure 6.9). Tilted α-helices aH3 and aH4 are shorter, as expected from the shorter hydrophobic sequence segments 3 and 4, and are packed close to aH5 and aH6. A tilted four-helix bundle has been observed also by cryo-EM, nominally at 7 Å resolution, in the membrane domain of the F-ATPase from *Polytomella*, but aH5 and aH6 in *P. denitrificans* were assigned as aH6 and aH5, respectively, in *Polytomella* (Allegretti et al., 2015). X-ray crystallography at 4.0 Å resolution study of ATP synthase from the α-proteobacterium *Paracoccus denitrificans* reveals deep insights into the workings of the extraordinary molecular machine (Morales-Rios et al., 2015). The ζ-protein is bound via its N-terminal α-helix in a catalytic interface in the F_1 domain. The bacterial F_1 domain is found attached to the membrane domain by peripheral and central stalks. The δ-subunit component of the peripheral stalk is found to bind to the N-terminal regions of two α-subunits. The stalk extends via two parallel long α-helices, one in each of the related b and b′ subunits, down a noncatalytic interface of the F_1 domain and interacts in

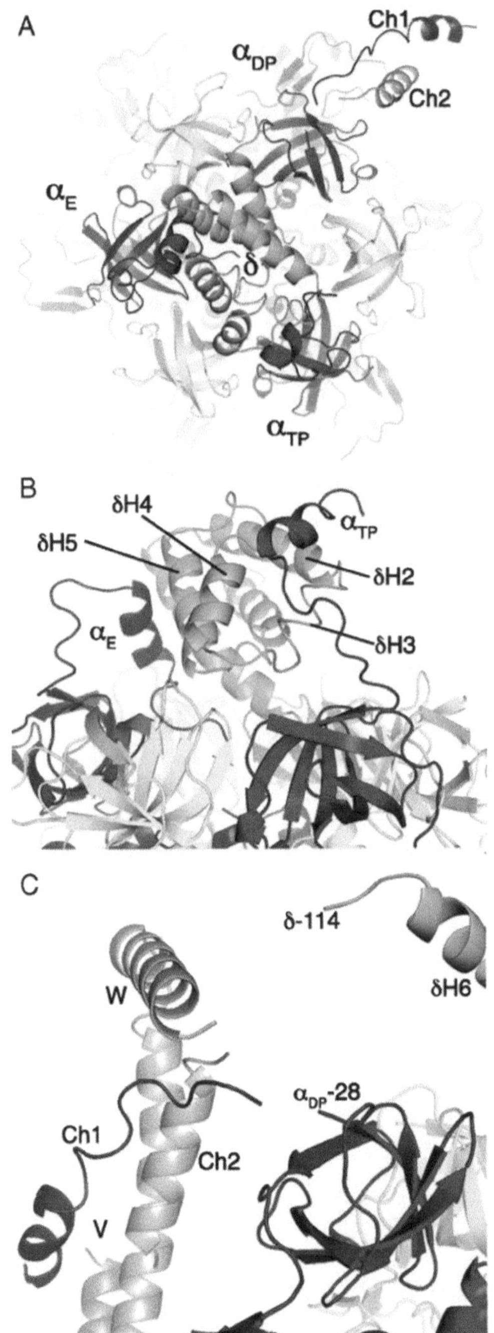

FIGURE 6.8 Interactions of the δ-subunit with N-terminal regions of α-subunits in the F-ATPase from *P. denitrificans*. (*A*) View from above the F-ATPase toward the "crown" of the F_1 domain depicting the N-terminal regions of the α_E-, α_{TP}-, and α_{DP}-subunits (red) with the δ-subunit (blue), β-subunits (yellow) and chains 1 and 2 (Ch1 and Ch2; purple and gray, respectively). (*B*) Side view of the interactions of the α_E-subunit (residues 7–22) and the α_{TP}-subunit (residues 2–22) with helices δH1 and δH5 and helices δH2, δH3, and δH4, respectively. (*C*) Side view of the region around the N-terminal part of the α_{DP}-subunit with structural elements from peripheral stalk subunits.

***Source*: Morales-Rios et al. (2015).**

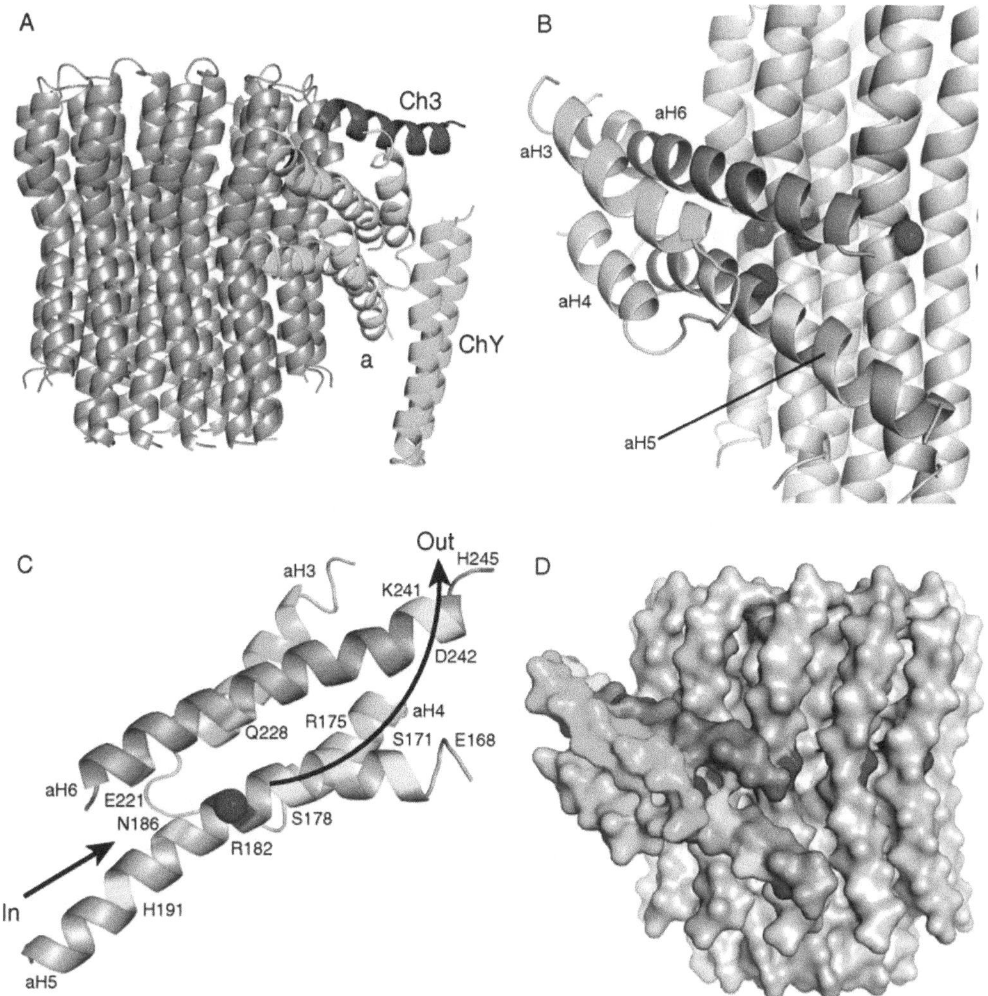

FIGURE 6.9 Topography of the membrane domain of the F-ATPase from *P. dentrificans* and a potential pathway of transmembrane proton translocation. (*A*) View of the c_{12}-rotor ring, and an associated bundle of α-helices (green), assigned to the a-subunit and named aH3–aH4 (residues 1,001–1,035), and aH5 and aH6 containing residues 166–198 and 217–246, respectively. Unassigned α-helix Ch3 and α-helical hairpin ChY are shown in magenta and blue, respectively. (*B*) View of the association of the tilted bundle of four α-helices in subunit a with the c-ring showing residue Glu60 (red) in the c-subunit in the proton transfer site and residue Arg182 (blue) in aH5. (*C*) View from the c-ring of the tilted bundle of four α-helices in subunit a showing conserved polar residues (yellow) that could provide the access path (In) for protons from the bacterial periplasm to reach the proton transfer site and the exit path (Out) for protons to be released into the bacterial cytoplasm. Residue Arg182 is colored blue. (*D*) View in solid representation of the tilted bundle of four α-helices in the a-subunit in juxtaposition with the c-ring showing the potential inlet pathway for protons (yellow) leading through the bundle to the proton transfer site containing a negatively charged Glu60 (red).

***Source*: Morales-Rios et al. (2015).**

an unspecified way with the a-subunit in the membrane domain. The a-subunit lies close to a ring of 12 c-subunits attached to the central stalk in the F_1 domain, and, together, the central stalk and c-ring form the enzyme's rotor. Rotation is driven by the transmembrane proton-motive force, by a mechanism where protons pass through the interface between the a-subunit and c-ring via two half-channels in the a-subunit. These half-channels are probably located in a bundle of four α-helices in the a-subunit that are tilted at ~30° to the plane of the membrane. Conserved polar residues in the two α-helices closest to the c-ring probably line the proton inlet path to an essential carboxyl group in the c-subunit in the proton uptake site and a proton exit path from the proton release site.

We have explained the X-ray crystallography structural study of ATP synthase (Morales-Rios et al., 2015). We may inspect whether other biophysical techniques provide similar resolution in the structural moieties that may help address the function of this important molecular machine in biological systems. Electron cryo-microscopy has been applied to the structure of the intact monomeric ATP synthase from the fungus *Pichia angusta* (see Figure 6.10) and the ATP synthase from *Mycobacterium smegmatis* (see Figure 6.11).

The structural studies of Vinothkumar and colleagues (2016) provide insights into the mechanical coupling of the transmembrane proton motive force across mitochondrial membranes in the synthesis

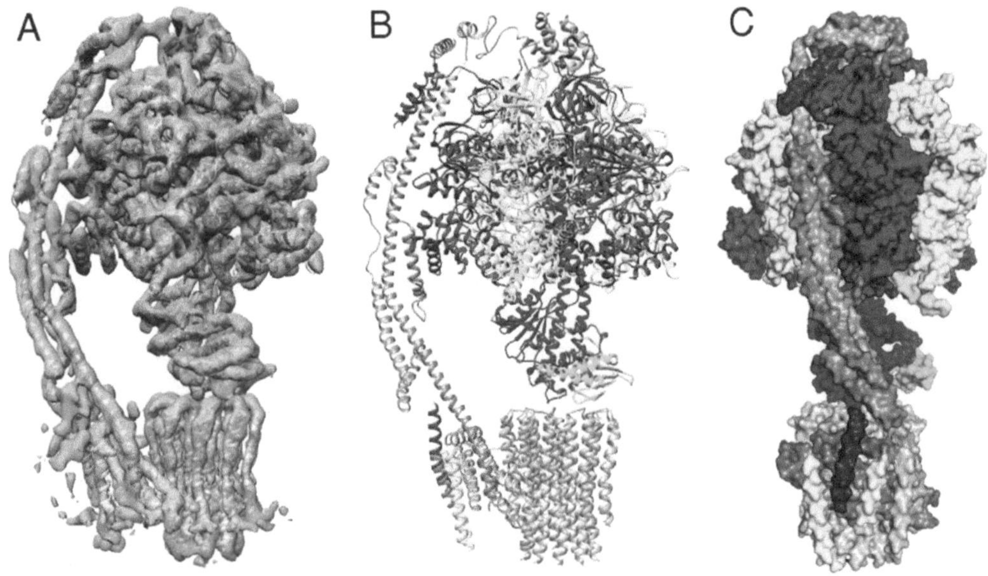

FIGURE 6.10 The structure of the F-ATPase from *P. angusta*. The enzyme was inhibited with residues 1–60 of the bovine inhibitor protein IF_1. (*A*) The cryo-EM map of state 1 and structural model viewed from the side, with the peripheral stalk on the left, the catalytic domain at the top, attached by the central and peripheral stalks to the membrane domain below. (*B* and *C*) Side views of the enzyme-inhibitor complex in cartoon and surface representation, respectively. (*C*) is rotated to the right by 90° relative to (*A*) and (*B*). The α-, β-, γ-, δ-, and ε-subunits forming the membrane extrinsic catalytic domain are red, yellow, royal blue, green, and magenta, respectively; the inhibitor protein is cyan; and the peripheral stalk subunits OSCP, b, d, and h are sea-green, pink, orange, and purple, respectively. In the membrane domain, the c_{10}-rotor is gray, the resolved region of the associated subunit a is corn-flower blue. Chains Ch1–Ch4 are pale yellow, brick-red, pale cyan, and beige, respectively, and have been assigned as transmembrane α-helices in subunit f and, in ATP8, as aH1 and bH1, respectively. The identities of subunits are placed directly on an enlarged version of (*C*).

***Source*: Vinothkumar et al. (2016).**

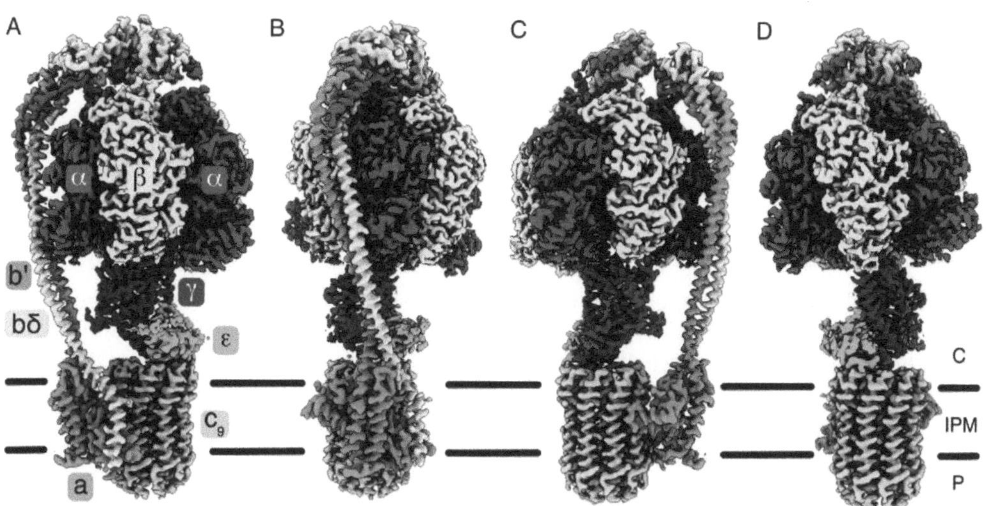

FIGURE 6.11 The structure and subunit composition of ATP synthase from *M. smegmatis*. (*A–D*) Four views of the electron density of the intact enzyme in state s1a (EMD-12377). The colors corresponding to each subunit are indicated in (*A*). In (*B–D*), the views are rotated by 90° from left to right about the vertical axis running through the central stalk and the center of the c_9-ring. The mycobacterial ATP synthase consists of three copies of the α-subunit, with an extended C terminus unique to *Mycobacteria*; three copies of the β-subunit; nine copies of the c-subunit; and one copy each of the γ-, ε-, a-, b′-, and bδ-subunits. The latter is also unique to *Mycobacteria* and comprises a canonical b-subunit fused to a canonical δ-subunit via an additional bundle of seven α-helices. The black bars represent the IPM. Protons are translocated from the periplasm (P) to the cytoplasm via the interface between a- and c-subunits.

***Source*: Montgomery et al. (2021).**

of ATP. Montgomery and colleagues (2021) recognized crucial features while understanding the mechanism and regulation of the mycobacterial enzyme. Firstly, they resolved not only the three main states in the catalytic cycle but also eight substrates that portray structural and mechanistic changes occurring during a 360° catalytic cycle. Secondly, a mechanism of auto-inhibition of ATP hydrolysis was found to involve not only the engagement of the C-terminal region of an α-subunit in a loop in the γ-subunit, as proposed before, but also a "fail-safe" mechanism involving the b′-subunit in the peripheral stalk that enhances engagement. A third unreported characteristic is that the fused bδ-subunit contains a duplicated domain in its N-terminal region where the two copies of the domain participate in similar modes of attachment of the two of three N-terminal regions of the α-subunits. The auto-inhibitory plus the associated "fail-safe" mechanisms and the modes of attachment of the α-subunits provide targets for development of drugs specifically meant for serving as antitubercular drugs. The structure also provides support for an observation made in the bovine ATP synthase that the transmembrane proton-motive force that provides the energy to drive the rotary mechanism is delivered directly and tangentially to the rotor via a Grotthuss water chain in a polar L-shaped tunnel. Readers are encouraged to go through these three studies for a comparable understanding of the structures and functions of ATP synthase, which appears to be the most important molecular machinery helping the cell to get energy supply molecularly. It is worth mentioning that biophysical techniques have emerged strongly to serve the purpose of understanding biologically important structural and functional aspects active in cellular environments that are associated with life's processes.

6.2.2 Gene Synthesis Machines

Deoxyoligonucleotides can now be synthesized rapidly and in high yield because of recent advances in nucleic acid chemistry (Caruthers, 1985). DNA polymerase is known to be responsible for synthesizing DNA, a key component in the running of biological machinery (Sengupta et al., 2014). RNA polymerase is known to serve as the vehicle of transcription (Borukhov and Nudler, 2008). Telomerase, a specialized ribonucleoprotein enzyme complex, is well known to work in maintaining telomere length at the 3′ end of chromosomes (Liu et al., 2018). Reversible primer termination, enabled by polymerase–nucleotide conjugates, has been found to provide an enzymatic method for the de novo synthesis of oligonucleotides (Tang, 2018). The oligonucleotide synthesis strategy that uses the template-independent polymerase terminal deoxynucleotidyl transferase (TdT) has recently been explained in detail (Palluk et al., 2018). Palluk and colleagues have demonstrated here that TdT–deoxyribonucleoside triphosphate (dNTP) conjugates can quantitatively extend a primer by a single nucleotide in 10–20 s. This scheme can also be iterated to extend writing any defined sequence.

A tremendous rise in recent research and commercial sectoral demands for the production of synthetic DNA, oligonucleotides (oligos), and even longer constructs such as synthetic genes and entire chromosomes (Gibson et al., 2010; Richardson et al., 2017) has made this quantitative primer production field demanding (Palluk et al., 2018). Massively parallel oligo synthesis (Kosuri and Church, 2014) has dramatically favored the commercial cost of high-throughput and genome-wide functional screens (Wang et al., 2014) and target capture for next-generation sequencing (NGS).

Palluk and colleagues (2018) conceived an approach for reversible termination wherein each polymerase molecule is labeled site-specifically with a tethered nucleoside triphosphate (see

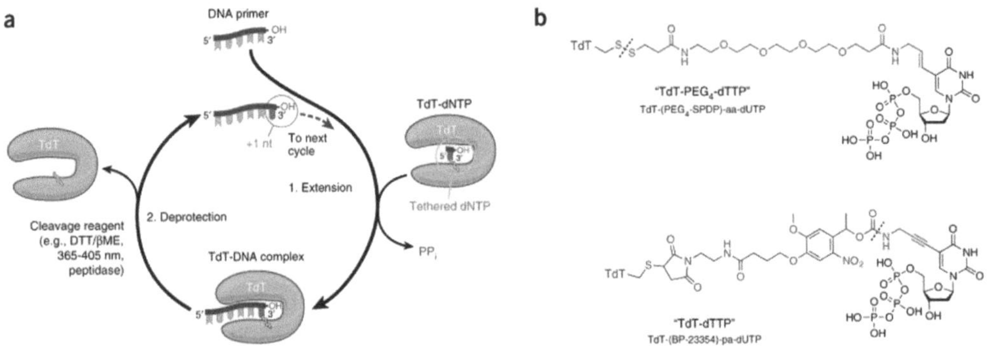

FIGURE 6.12 TdT–dNTP conjugates for reversible termination of primer elongation. (a) Scheme for two-step oligonucleotide extension using TdT–dNTP conjugates consisting of a TdT molecule site-specifically labeled with a dNTP via a cleavable linker. In the extension step, a DNA primer is exposed to an excess of TdT–dNTP conjugate. Upon incorporation of the tethered nucleotide into the 3′ end of the primer, the conjugate becomes covalently attached and prevents further extensions by other TdT–dNTP molecules. In the deprotection step, the remaining TdT–dNTP conjugates are inactivated (or removed) and the linkage between the incorporated nucleotide and TdT is cleaved by addition of the cleavage reagent (e.g. DTT, 365 nm light, peptidase), thereby releasing the primer for subsequent extension. The cycle can be iterated to extend a primer by a defined sequence. (b) Chemical structures of two types of TdT–linker-dNTP conjugates used in this study, based on the amine-to-thiol crosslinkers PEG_4-SPDP (upper, "TdT–PEG_4-dTTP") and BP-23354 (lower, "TdT–dTTP") and the dTTP analogs 5-aminoallyl dUTP (aa-dUTP) and 5-propargylamino dUTP (pa-dUTP), respectively. Upon cleavage of the linker, the atoms indicated in red remain attached to the nucleobase and are referred to as a scar. The cleavable bond is indicated with a black dotted line.

***Source*: Palluk et al. (2018).**

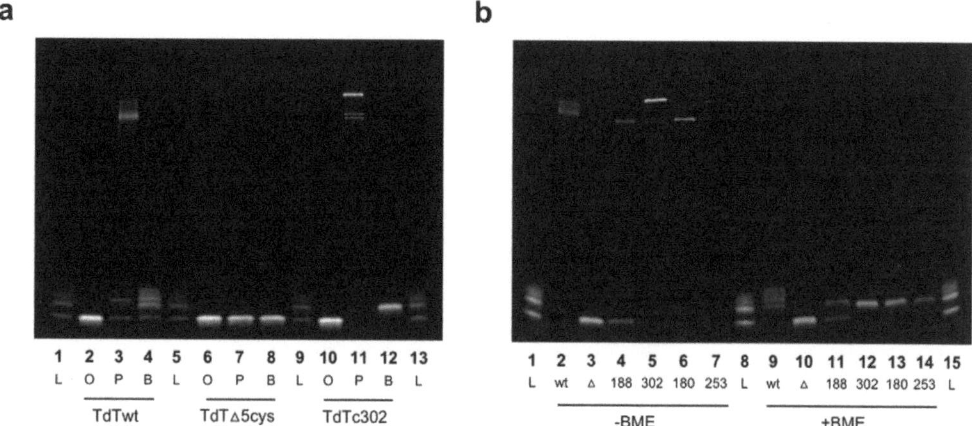

FIGURE 6.13 SDS-PAGE showing the ability of different TdT variants to incorporate a tethered nucleotide after labeling reactions with OPSS-PEG$_4$-dTTP. Both gels were imaged for the fluorescence of oligo P1 (5' FAM-dT$_{35}$), the ladder used (L) was generated by extensions of P1 by free OPSS-PEG$_4$-dTTP. (a) The oligo before the reaction (O), reaction products formed during incorporation with labeled TdT (P) and after cleavage of the linker by βME (B) are shown. TdTwt and TdTc302 form high-molecular weight complexes containing the primer, indicating tethered incorporation, while TdTΔ5cys does not. After treatment with βME, the complexes dissociate, and the migration of the released oligo indicates multiple extensions by TdTwt, a single extension by TdTc302, and no extensions by TdTΔ5cys, respectively. (b) Products formed during extension reactions (-BME) and after cleavage of the linker by βME (+BME) are shown for TdTwt (wt), TdTΔ5cys (Δ), and TdT-PEG$_4$-dTTP conjugates with a single surface exposed cysteine in positions 180, 188, 253, and 302. The formation of a fluorescent high-molecular weight complex occurs for all TdT variants except for TdTΔ5cys. Upon cleavage, the oligo shows multiple extensions for TdTwt, and a single extension for TdT variants with a single surface-exposed cysteine. These experiments were each performed twice with similar results.

***Source*: Palluk et al. (2018).**

Figure 6.12a). A two-step reaction cycle of extension and deprotection is proposed to be iterated to write a defined sequence. As a polymerase incorporates its tethered dNTP into a primer, it remains covalently attached to the 3′ end, and blocks further elongation by other polymerase-dNTP conjugates. The linker may then be cleaved to deprotect the 3′ end of the primer in order to create the subsequent extension.

Palluk and colleagues tested the feasibility of the approach by first expressing and purifying the polymerase domain of murine TdT, with five surface-accessible cysteine residues (based on the crystal structure), and then tethering the dTTP analog 5-aminoallyl-dUTP to those residues using the disulfide-forming amine-to-thiol crosslinker PEG$_4$-SPDP (see Figure 6.12B); for details see Palluk et al. (2018). When exposed to a 5′ fluorescein (FAM)-labeled DNA primer, polymerase-nucleotide conjugates formed fluorescent complexes detectable by SDS-PAGE, which indicate that one or more tethered nucleotides had been incorporated into the primer (see Figure 6.13). The addition of β-mercaptoethanol (βME), which cleaves the linkage between the tethered nucleotides and TdT, dissociated the complexes, releasing primers that had been extended by up to four nucleotides. This result helped conclude that TdT can incorporate nucleotides tethered to at least some points on its surface.

Palluk and colleagues (2018) then attempted to generate conjugates that would extend a primer by precisely one nucleotide, the key property for enabling synthesis of any defined sequences. Tethering a single nucleotide to the surface of TdT results in polymerase–nucleotide conjugates that can be used for extending a DNA primer by a single nucleotide. Figure 6.14 presents the TdT–dNTP

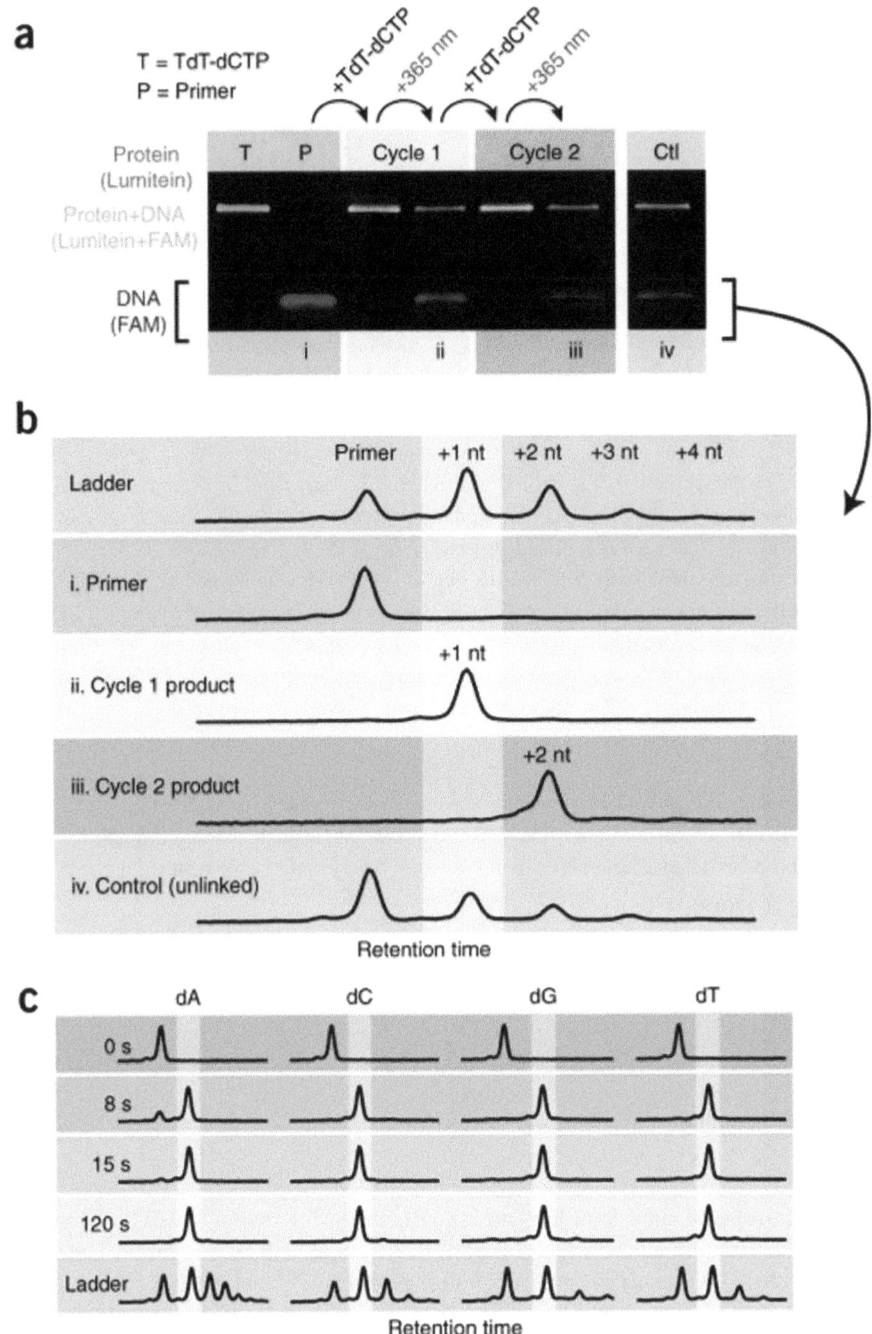

FIGURE 6.14 TdT–dNTP conjugates can extend a DNA molecule by a single nucleotide in 10–20 s, enabling stepwise DNA synthesis. (a) Monitoring of reaction cycle intermediates by SDS-PAGE with two-color fluorescence imaging. Oligonucleotides are visualized using 5′ FAM fluorescence (red channel) and protein is visualized using the fluorescent stain Lumitein (green channel). Exposing the primer (i) to an excess of the TdT–dCTP conjugate results in a new band containing both protein and DNA (yellow in composite), indicating the formation of a primer-TdT complex. Irradiation of the complex with 365 nm light recovers the distinct migration of the protein and DNA bands, indicating dissociation of the primer-TdT complex (ii). When exposed to fresh TdT–dCTP conjugate, extension products of cycle 1 again form a primer-TdT complex, which again is

FIGURE 6.14 *(Continued)*

dissociated by 365 nm irradiation (iii). In contrast, no primer-TdT complex formation is observed if the TdT–dCTP conjugates are irradiated before the incubation with the primer (iv). This experiment was repeated with conjugates of all four nucleotides with similar results. (b) Capillary electropherograms of reaction products from panel a. Each reaction cycle quantitatively extends the primer (i) by a single nucleotide (ii, iii). Exposure of the primer to pre-photolyzed conjugates results in a distribution of extension products (iv). The ladder of product standards was generated by incorporating free dNTP analogs using TdT under conditions selected to give 0–4 extensions. The electropherogram time axis is normalized using the elution times of internal standards and the intensity axis is normalized to the height of the tallest peak. (c) Capillary electropherograms of reaction time courses for the extension of a 25 nM DNA primer by 16 μM TdT–dATP, TdT–dCTP, TdT–dGTP, and TdT–dTTP conjugates, followed by photolysis. The primer is completely converted into the singly extended product in 10–20 s depending on the type of conjugate. A small amount of double extension (i.e. non-termination) is detectable in the reactions quenched at 120 s as a peak co-eluting with the +2 peaks of the ladders. The ladder of product standards was generated as described above. This experiment was performed twice with similar results.

***Source*: Palluk et al. (2018).**

conjugation scheme that could extend a DNA molecule by a single nucleotide in 10–20 s, enabling stepwise DNA synthesis.

For demonstrating the feasibility of TdT–dNTP conjugates for stepwise de novo DNA synthesis, a double-stranded DNA molecule was subjected with a 3′ overhang to ten cycles of extension and deprotection using conjugates corresponding to the sequence 5′-CTAGTCAGCT-3′. Palluk and colleagues then poly-A tailed the product using TdT with free dATP to create a (reverse) primer binding site; PCR-amplified the product (see Figure 6.15a), and analyzed the amplicon by NGS (see Figure 6.15b). By aligning the reads against the target sequence, Palluk and colleagues were able to estimate the yields of individual steps of the synthesis (see Figure 6.15c), which ranged from 99.5 percent (step 6) to 93.4 percent (step 10). The average yield of all steps was 97.7 percent, with deletions as the predominant source of errors (1.3 percent) and the remaining errors arising from insertions (1.0 percent).

We have a lot of literature containing knowledge on DNA polymerases (Loeb and Monnat, 2008), RNA polymerases (Barba-Aliaga et al., 2021), the ribosome and ATPases as familiar examples of biomolecular machines playing pivotal roles in DNA replication, gene expression, translation, and cell energy production, respectively (Ramezani and Dietz, 2020). By contrast, human-made biomolecular machines, which mimic or aim to eventually surpass the functions of their natural counterparts, remain in their infancy though they claim to have started achieving success (Palluk et al., 2018). The progress in this area has started having impacts, which over the next decades are hopefully going to be able to initiate a technological revolution. Thus we shall be able to handle issues associated with genetics and associated mutations.

6.2.3 Human Histone pre-mRNA Processing Machinery

Recently, the structure of an active human histone pre-mRNA 3′-end processing machinery was studied (Sun et al., 2020). Biochemical experiments and cryo-electron microscopy (cryo-EM) were used for determining the atomic structure of a complex assembly of these molecules. This machine plays a fundamental role in the proper activity and duplication of the cell genome. Histone proteins are known to play essential structural and functional roles in the transition between active and inactive chromatin states (Mariño-Ramírez et al., 2005). In all plants and animals histone proteins are found to form a "beads-on-a-string" arrangement where the DNA in chromosomes is wrapped around the beads of histones. This protein helps in packaging DNA efficiently and thus helps regulate which

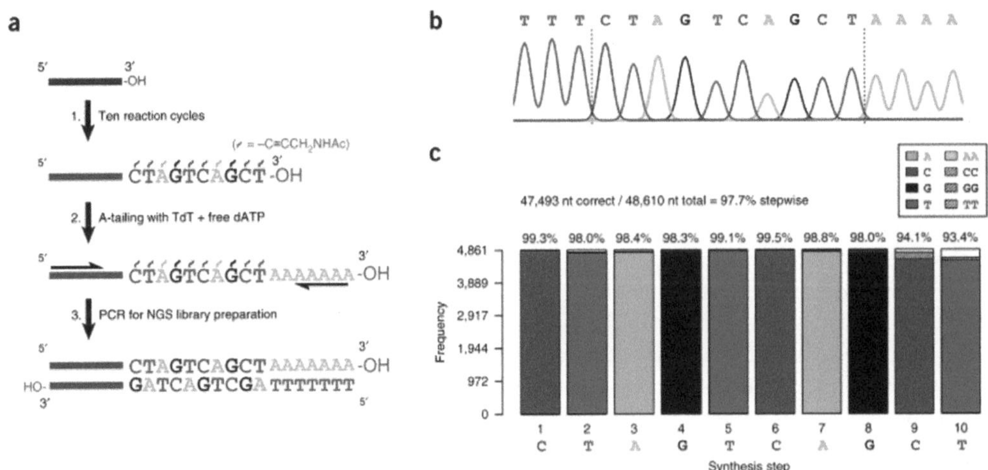

FIGURE 6.15 Sequence confirmation of 10-mer synthesis. (a) Ten-step synthesis and procedure for amplification of the products for sequencing. A double-stranded DNA molecule with a 3′ overhang is extended by the *N*-acetyl-propargylamino DNA sequence 5′-CTACTGACTG-3′ via ten cycles of extension, deprotection, and DNA purification. The products are then A-tailed using TdT with free dATP to generate a binding site for a poly-T primer. Tailing products are PCR-amplified for subsequent analysis. (b) Sanger sequencing chromatogram of the PCR-amplified synthesis products. The sequencing chromatogram indicates that the starter (ending with 5′-TTT-3′) had been extended by the intended sequence, followed by a poly-A tail. (c) Estimates of the stepwise yields of the synthesis based on an alignment of 4,861 reads (excluding singletons) against the target sequence. Solid bars indicate correct bases, hatched bars indicate insertions (non-termination), and empty bars indicate deletions. Substitutions occurred at a rate of <0.1 percent per step. This ten-step synthesis was performed once.

***Source*: Palluk et al. (2018).**

genes are turned “on” and which are kept “off,” two fundamental processes needed for proper cellular function. Genomes get packaged by complexing DNA with histone proteins, the process providing opportunities to regulate gene expression by dynamically impeding access of transcriptional regulatory proteins and RNA polymerases to DNA (Deal and Henikoff, 2011).

The 3′-end processing machinery for metazoan replication-dependent histone precursor messenger RNAs (pre-mRNAs) contains the U7 small nuclear ribonucleoprotein and shares the key cleavage module with the canonical cleavage and polyadenylation machinery. Sun and colleagues reconstituted an active human histone pre-mRNA processing machinery using thirteen recombinant proteins and two RNAs and determined its structure by cryo-electron microscopy (Sun et al., 2020). The overall structure is found to be highly asymmetrical and resembles an amphora with one long handle. Figure 6.16 presents the overall structure of the human histone pre-mRNA 3′-end processing machinery.

Sun and colleagues (2020) could capture the pre-mRNA in the active site of the endonuclease, the 73-kilodalton subunit of the cleavage and polyadenylation specificity factor, poised for cleavage. The active machinery was purified and a cryo-EM reconstruction at 3.2 Å resolution for its core was obtained (see Figure 6.16C and D), and a reconstruction at 4.1 Å resolution for the entire machinery was made. The overall machinery structure is found to resemble an amphora with one long handle (see Figure 6.16E). The machinery core is seen to constitute the body of the amphora, with the U7 snRNA 3′-end SL and the Sm ring at the base and the CTDs of CPSF73 and CPSF100 and the first few helical repeats of the symplekin CTD forming the mouth. CPSF73 and symplekin NTD are positioned opposite each other on the Sm ring (see Figure 6.16D). CPSF100 interacts with

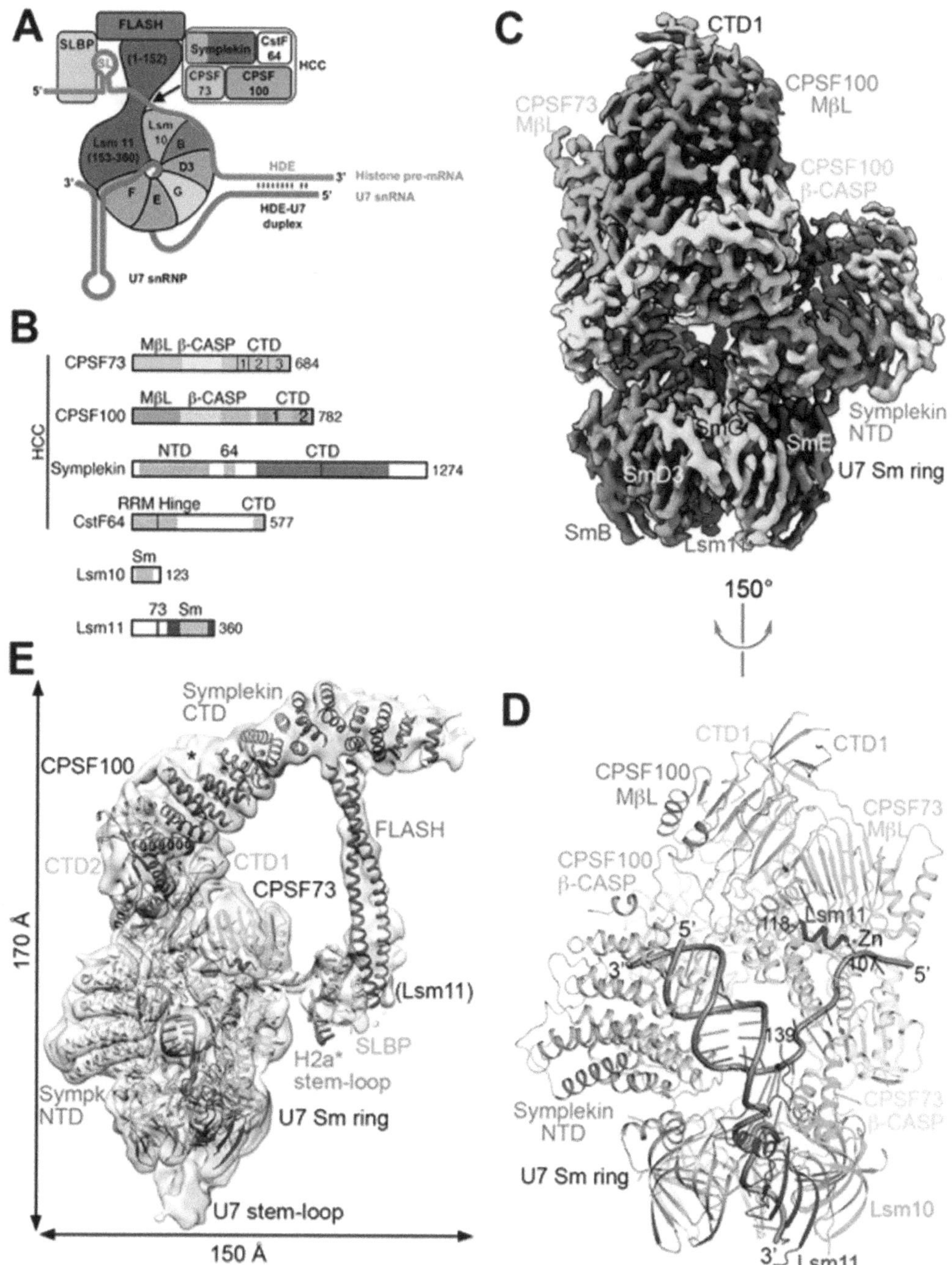

FIGURE 6.16 Overall structure of the human histone pre-mRNA 3′-end processing machinery. (A) Schematic drawing of the histone pre-mRNA 3′-end processing machinery. (B) Domain organizations of the subunits of HCC, and Lsm10 and Lsm11. The domains in CPSF100 are given slightly darker colors compared to their homologs in CPSF73. The vertical bar in the symplekin CTD marks its N-terminal half that interacts with CPSF73. (C) Cryo-EM density at 3.2 Å resolution for the core of the machinery. (D) Schematic drawing of the structure of the core of the machinery, viewed after a 150° rotation around the vertical axis from panel C. The proteins are colored as in Figs. 1A, 1B. The U7 snRNA is in dark green, and H2a* in orange. (E) Cryo-EM density for the entire machinery (gray), low-pass filtered to 8 Å resolution to show the density of FLASH and

FIGURE 6.16 *(Continued)*

SLBP. The possible density for CTD3 of CPSF73 is indicated with the asterisk. Structure figures are produced with PyMOL (www.pymol.org) unless noted otherwise. Panels C and E produced with Chimera [Pettersen, E.F., et al., UCSF Chimera – a visualization system for exploratory research and analysis, *J. Comput. Chem.* 25 (2004), 1605–1612].

***Source*: Sun et al. (2020).**

both CPSF73 and symplekin, but does not directly contact the Sm ring (see Figure 6.16C). The symplekin CTD, FLASH dimer (Aik et al., 2017), SLBP, pre-mRNA SL, and residues 20–65 of Lsm11 form the handle of the amphora (see Figure 6.16E). The FLASH dimer makes an 80 Å long connection from the symplekin CTD to the SLBP-SL complex. CstF64 was not observed in the EM density, and is not required for cleavage *in vitro*.

Twelve consecutive Watson-Crick base pairs in the HDE-U7 duplex were observed in the center of the amphora (see Figure 6.16D). The metallo-β-lactamase domain of CPSF73, the β-CASP domain of CPSF100, and the concave face of the symplekin NTD surround the duplex on three sides (see Figure 6.16D, Figure 6.17A). Although the interactions are ionic and hydrophilic in nature, they involve none of the bases in the duplex (see Figure 6.17B), which explains that base pairing rather than sequence is important for processing (Dominski and Marzluff, 2007; Bucholc et al., 2020). An extra, U-U base pair at the bottom of the duplex was revealed (see Figure 6.17C), and analysis of histone pre-mRNA sequences suggested that U-U base pairs are common in HDE-U7 duplexes. A Watson-Crick base pair between C28 and G31 of the CUAG sequence at the 3′ end of the U7 Sm site was also revealed (see Figure 6.17D).

The machinery structure suggests the ways it may get assembled in order to process (see Figure 6.18C). Figure 6.18A explains the structural differences between an active state and an inactive state, while Figure 6.18B presents schematic drawing of the CTD2 domain complex of CPSF73 (light green) and CPSF100 (darker green) and the N-terminal segment of the symplekin CTD (magenta). For details see the original article.

I have provided a few key examples of molecular machine structures revealing their roles in cellular key processes. I could continue providing additional examples, but for that, readers are encouraged to go through my other book (Ashrafuzzaman, 2018). Instead let's analyze something more general by quoting a recent article that has raised a question regarding the cell as a whole being a machine (Nicholson, 2019). This article points at four major domains of current research in which the challenges to the machine conception of the cell are particularly pronounced – cellular architecture, protein complexes, intracellular transport, and cellular behavior – through making arguments that a new theoretical understanding of the cell is emerging from the study of these phenomena, emphasizing the dynamic, self-organizing nature of its constitution, the fluidity and plasticity of its components, and the stochasticity and non-linearity of its underlying processes.

Recently, the cycle of orchestrated movements by which a cargo-carrying kinesin walks along a microtubule was simulated (see Figure 6.19). Animations, like the one in Figure 6.19, conform to what we would expect to find if the cell was indeed a machine, as these snapshots portray motor proteins as tiny robotic bipeds performing sequential cycles of precisely coordinated, mechanically powered movements along cytoskeletal tracks. Proper theoretical analysis is indeed needed to understand this machine function considering mainly their mechanical properties (Aprahamian, 2020; Moeendarbary and Harris, 2014). Cells being molecular machines or hosting versatile molecular machines that work individually or collectively requires additional theoretical and experimental analyses to help us understand the machinery from their biological function perspectives (Nicholson, 2019).

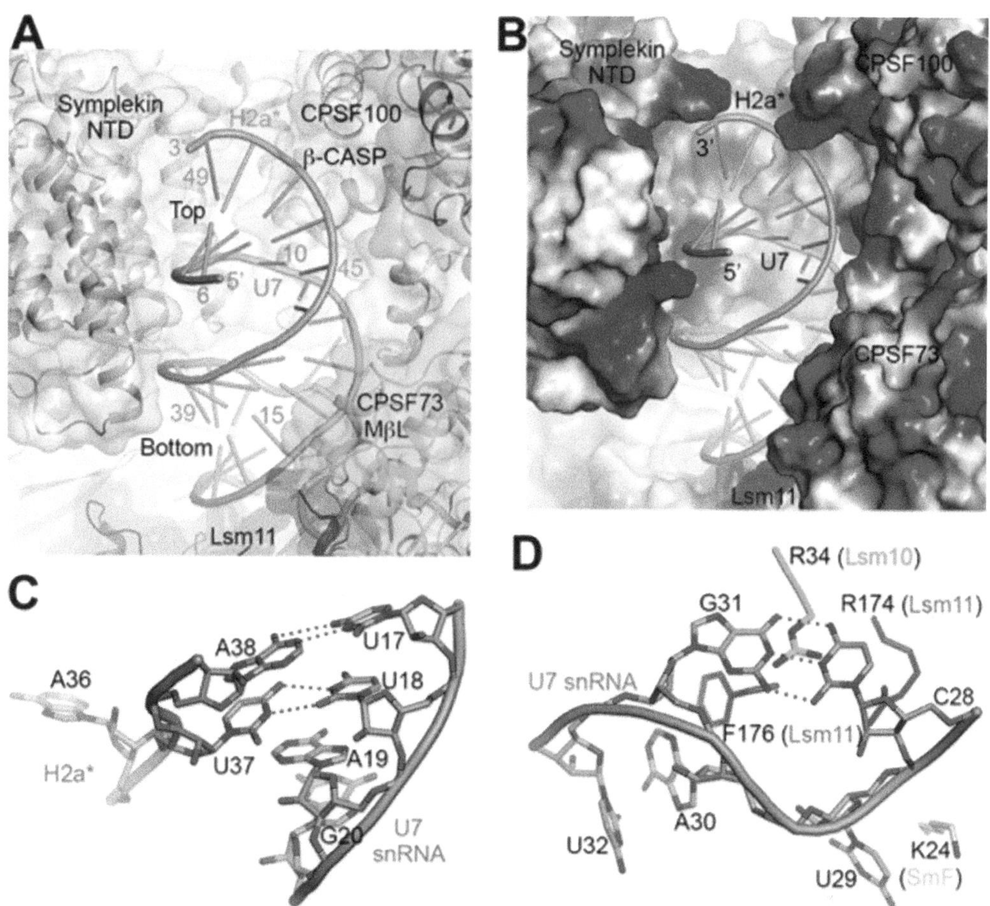

FIGURE 6.17 Recognition of the HDE-U7 duplex and the U7 Sm site. (A) The HDE-U7 duplex is surrounded by CPSF73, CPSF100 and symplekin NTD, shown as a transparent surface. Lsm11 has interactions with the bottom of the duplex. (B) Electrostatic surface of the proteins in the duplex binding site, showing charged interactions with the backbone of the duplex. (C) A U-U base pair at the bottom of the duplex, flanked on the other face by A19 of U7 snRNA. (D) A C-G base pair in the 3′ CUAG sequence of the U7 Sm site. The base pair is flanked on one side by Arg34 of Lsm10, and on the other by Arg174 of Lsm11.

***Source*: Sun et al. (2020).**

6.3 BIOMOLECULAR MACHINES OF THE TISSUE

Cell to cell communication is key to developing tissues (Rossello and Kohn, 2010). An artificial molecular motor has recently been created that may talk to living cells by gently pulling their surface with enough physical force to elicit a biochemical response. The approach could help research scientists decode the language which cells use for communicating with each other in tissues (Urquhart, 2021). Cells apply physical forces among themselves and provide signals to guide their functions (Zheng et al., 2021). This investigation revealed the potential of nanomotors for the manipulation of living cells at the molecular scale.

Mechanical stimulation of the T-cell receptor (TCR) is known to lead to T-cell activation, proven in single-molecule experiments (Liu et al., 2016; Liu et al., 2014). Zheng and colleagues (2021) tested whether the opto-mechanical actuator coupled to the T-cell receptor (TCR) was able

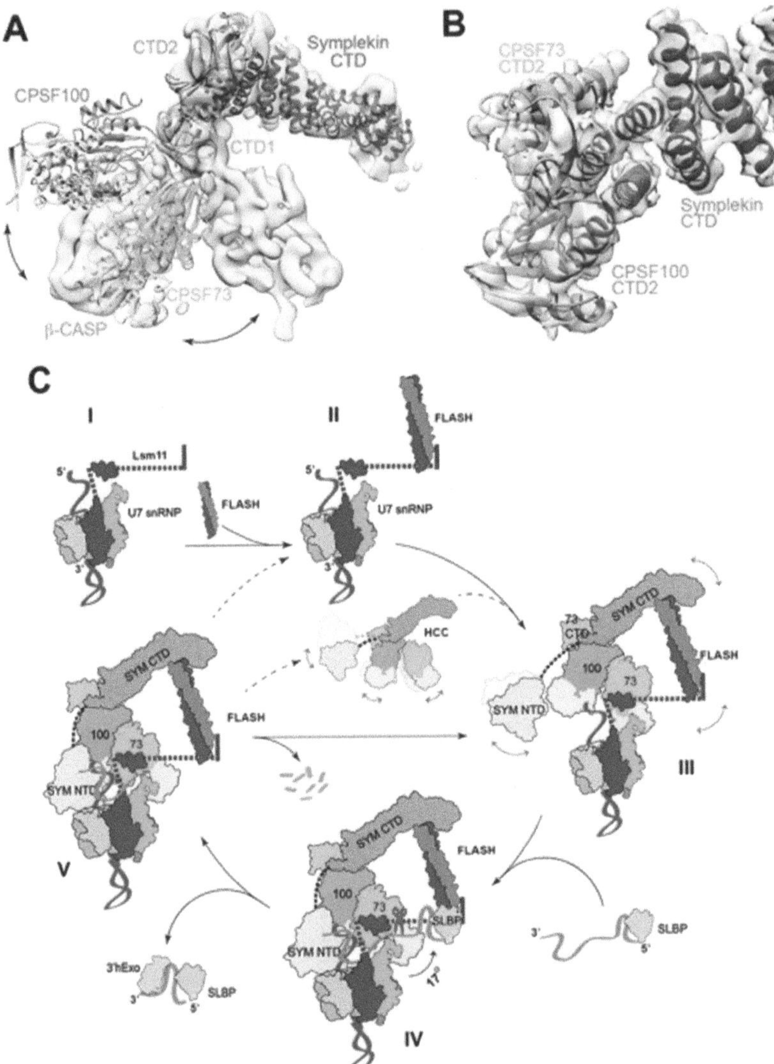

FIGURE 6.18 Schematic of histone pre-mRNA 3′-end processing cycle. (A) Significant structural differences of HCC in an active state compared to an inactive state. The structure of HCC observed here is docked into the EM density for mCF (gray surface) (Zhang, Sun, et al., 2020), using the symplekin CTD as the reference. (B) Schematic drawing of the CTD2 domain complex of CPSF73 (light green) and CPSF100 (darker green) and the N-terminal segment of the symplekin CTD (magenta). The CTD complex of IntS9 and IntS11 (Wu et al., 2017) was docked into the EM density at 4.1 Å resolution (transparent surface) using Chimera. Panels A and B produced with Chimera. (C) A putative model for histone pre-mRNA 3′-end processing cycle. The machinery is assembled from the U7 snRNP (state I) with the recruitment of FLASH (II) and HCC (III), followed by the recognition of the pre-mRNA for CPSF73 and HCC activation and pre-mRNA cleavage (IV). The machinery is likely highly dynamic before the binding of the authentic pre-mRNA, and the possible flexible regions are indicated with the curved arrows and dashed lines. After the cleavage (V), the downstream product is degraded by an exonuclease activity and the machinery can be recycled directly (solid arrow), or possibly disassembled followed by reassembly. State IV corresponds to the structure reported here, with the scissors indicating cleavage by CPSF73, and the other states are models.

***Source*: Sun et al. (2020).**

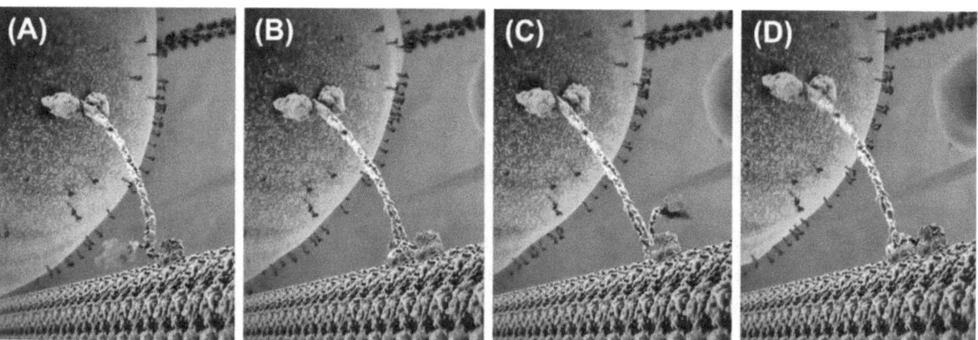

FIGURE 6.19 Cropped snapshots of the acclaimed computer animation The Inner Life of the Cell, created by XVIVO for Harvard University's Department of Molecular and Cellular Biology. The four consecutive snapshots depict the cycle of orchestrated movements by which a cargo-carrying kinesin "walks" along a microtubule. (A) ATP-binding to the motor domain of the left leg triggers a change in its conformation which generates a power-stroke in the linker region that throws the motor domain of the right leg overhead of the left leg. (B) The motor domain of the right leg reattaches to the microtubule and the products of ATP hydrolysis are released. (C) Binding of ATP to the motor domain of the right leg in turn induces a rearrangement of its structure which generates a further power-stroke in the linker region that pushes the motor domain of the left leg above the right leg. (D) The motor domain of the left leg reattaches to the microtubule and the products of ATP hydrolysis are again released, thus completing the cycle.

***Source*: www.artofthecell.com/the-inner-life-of-the-cell; © 2006 The President and Fellows of Harvard College.**

to activate T cells by light exposure. Jurkat and primary T cells were loaded with Ca^{2+} indicator Fluo-4-AM and seeded on anti-CD3/motor/PEG-modified surface (see Figure 6.20a). Ca^{2+} influx was induced in the cells when they contacted the surface, indicating recognition of the anti-CD3 antibody. For decoupling the Ca^{2+} influx induced by the motor from a possible response by simple interaction with the surface, Jurkat cells were seeded at 0 mM [Ca^{2+}] concentration, and the same volume containing 2 mM [Ca^{2+}] was added at 8 min post-seeding. At 15 min cells were exposed to pulses (1 s) of 365 nm light for 1 min. A Ca^{2+} rise was observed immediately after the first UV pulse (see Figure 6.20b and c), which decayed within 5 min. In comparison, no response was recorded when the Jurkat cells were seeded on control motor (having no rotation) surfaces exposed to UV light (see Figure 6.20b and c). This indicates that the response was not associated with UV illumination but a consequence of the applied pulling force on the TCR by the rotary motor linked to anti-CD3. Under identical conditions, primary human $CD4^+$ T cells were reported to show a similar response (see Figure 6.20d and e). Shorter light pulses (0.5 s) did not elicit an observable response (see Figure 6.20f and g), indicating that a certain threshold of force is required for force-mediated activation of T cells. Longer pulses (2 s) induced a decrease in the Ca^{2+} signal on the motor and control surfaces (Figure 6.20h and i), which is most likely related to photodamage of the cells following long UV pulses. These results define the boundaries for experimentation with this unique tool. Importantly, as negative biologically relevant control, similar experiments were performed with the CD28 receptor, whose activity has been proven to be unaffected by mechanical stimulation (Bashour et al., 2014). No response was reported, confirming that the observed force-dependent Ca^{2+} response was specific to CD3 engagement. Mechanical force applied by the motor does not change the contact area at the immunological synapse (IS). Together, these results demonstrate that synthetic molecular machines can be used to apply external forces to specific membrane receptors on T cells and study mechanotransduction events at biointerfaces. An identical motor had been recently incorporated into the membrane of living cells for killing cells by drilling pores in the membrane upon 360-nm illumination (García-López et al., 2017).

According to Del Campo of the team in Zheng et al. (2021), the approach of integrating motors in living systems will hopefully lead to better understanding of cellular function, in both cell cultures and living systems, and whole tissues. The subsequent goal may be to couple the motor to a force sensor for quantifying the force ranges at which cells will talk to each other (Urquhart, 2021). Elastic protein-based polymers may also be designed as temporary functional scaffoldings that cells can enter, attach to, spread, sense forces, and remodel, with the potential to restore natural tissue (Urry, 1999). Appropriate physical and chemical means may be explored in order to maintain cell–cell communication in biological tissues, which may be found medically important especially while dealing with disease cases like cancer.

6.4 ARTIFICIAL MOLECULAR MACHINES IN MUTATED BIOLOGICAL PROCESSES

The early progress targeting the utilization of synthetic molecular structures to perform tasks using mechanical motion addressing biomolecular-level movement is tremendous (Kay et al., 2007). The generation of sophisticated synthetic molecular machine systems, in which the controlled motion of subcomponents is used for performing complex tasks, paves the way for applications and the realization of a new era of "molecular nanotechnology" (Erbas-Cakmak et al., 2015). Artificial interventions in structural and functional disorders of molecular machinery become inevitable. The molecular machines experiencing mutations are medically dealt with using appropriate therapeutic means. (Finnigan et al., 2012). In contrast, clear defining of the interaction networks (or "interactome") formed by RNA polymerase II, a molecular machine that decodes the human genome, may help us address how these machine interaction networks naturally decode, replicate, and maintain the human genome integrity (Coulombe et al., 2004). This kind of understanding of natural processes may create a platform to attract artificial means of getting involved in resolving possible issues related to the mentioned networking of the molecular machines active in biological processes.

Molecular machines in complex living organisms are involved in performing crucial biological processes. Artificial molecular machines (AMMs) have been developed using mature synthetic technologies for exhibiting processes similar to naturally occurring ones. The use of AMMs in smart systems and materials with controllable regulations and interesting functions has been summarized, presenting the specific micro- to macroscale applications in solid surface modification, transmembrane transport, smart catalysts, liquid crystals, artificial molecular muscles, and stimuli-responsive polymers, in Shi et al. (2020). Shi and colleagues have detailed challenges of developing novel complex AMMs with intelligent functions and proposed potential solutions.

AMMs are individual molecules or self-assembled multiple-component systems undergoing controllable molecular-level motions or structural changes in response to external stimuli. AMMs are, from structural and functional perspectives, classified into three main types: (1) mechanically interlocked molecules (MIMs) (Saha and Stoddart, 2007; Zhang, Marcos, and Leigh, 2018), (2) molecular switches respond to external stimuli and thus induce reversible transitions between different states (Qu et al., 2015; Grzelczak et al., 2019), (3) molecular motors, undergoing unidirectional motion under external energy input (Koumura et al., 2002; Feringa, 2017). AMM fabrication with high complexity and controllable regulation has reached maturity (Barendt et al., 2017; Wang et al., 2017). Various smart molecular systems with practical functions have been constructed by mimicking biological systems or collecting synchronous motions in time and space (Lancia et al., 2019; Zhang et al., 2017). Shi and colleagues (2020) focused on the recent development of intelligent AMMs and summarized the contributions of converting these smart systems into practical functions. The content of this progress report is categorized according to different scales,

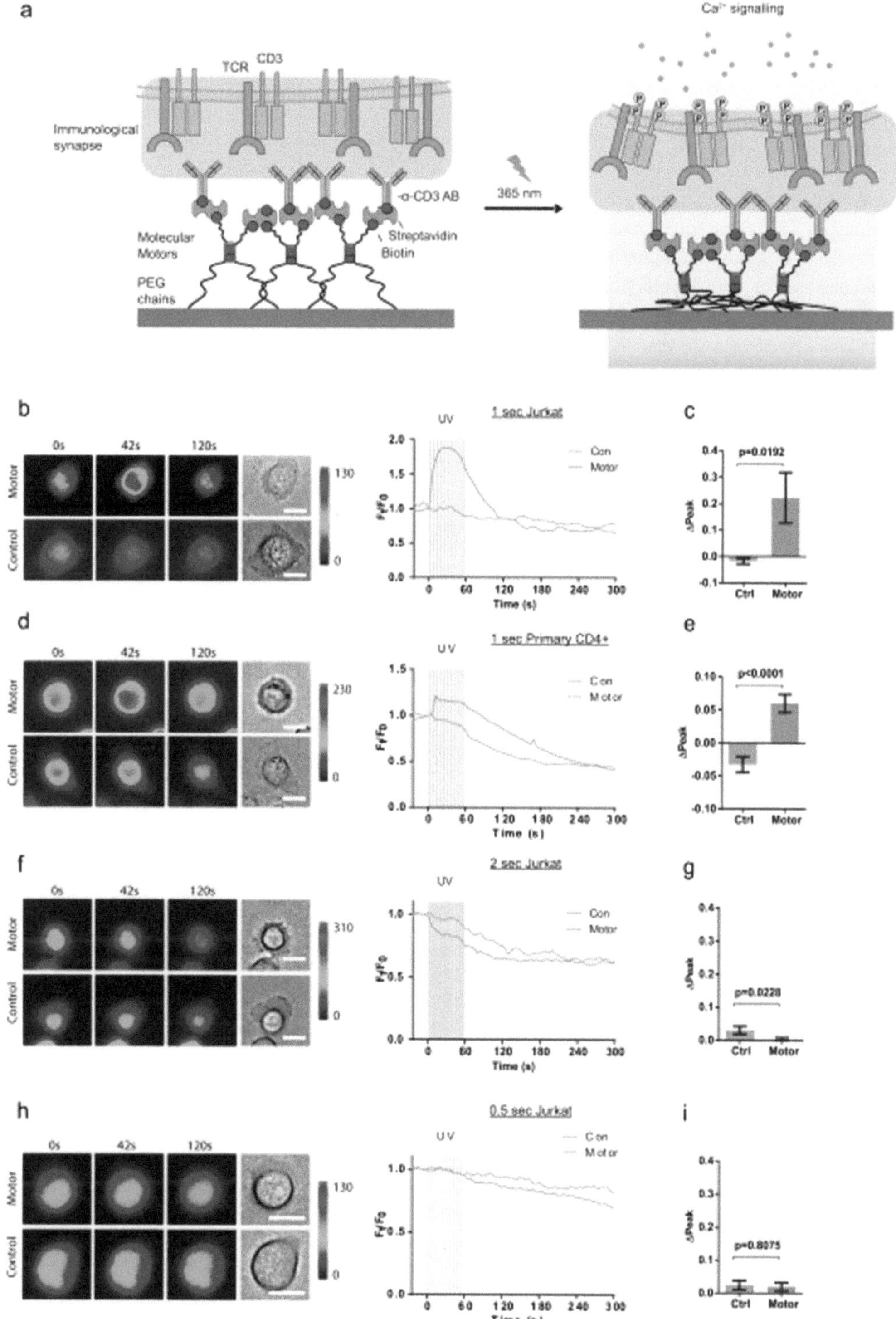

FIGURE 6.20 Mechanical force generated by rotatory motor can trigger TCR signaling. (a) Schematic of manufacture and activation of anti-CD3 antibody linked to the substrate via rotatable motor (Motor) or non-rotatable motor (Ctrl). The red box represents the motor-receptor interface at the immunological synapse.

FIGURE 6.20 *(Continued)*

Ten UV pulses were applied within 60 s to activate the motor. (b–e) Ca^{2+} influx is induced by activation of the motor. Either Jurkat T cells (b, c) or primary human $CD4^+$ T cells (d, e) loaded with Fluo-4-AM were settled on motor substrate for 20 min prior to UV illumination (starting at time 0). Duration of UV pulses was 1 s. Exemplary cells and Ca^{2+} traces are shown in (b) and (d). Ca^{2+} influx (ΔPeak) was analyzed in (c) (Ctrl, n = 137 cells from six independent experiments; Motor, n = 153 cells from nine independent experiments) and (e) (Ctrl, n = 42 cells from two independent experiments; Motor, n = 46 cells from three independent experiments). (f–i) Shorter or longer duration of UV pulses cannot induce Ca^{2+} influx. We used Jurkat cells and applied UV pulses with either shorter duration of 0.5 s (f, g Ctrl, n = 119 cells from five independent experiments; Motor, n = 201 cells from nine independent experiments) or extended duration of 2 s (h, i Ctrl, n = 101 cells from five independent experiments; Motor, n = 172 cells from nine independent experiments). LUT min and max given in a.u. to visually compare motor vs. control substrate within the same condition only. Data in (c), (g), (e), and (i) represent mean ± s.e.m. Data from motor and control motor substrates were compared using an unpaired two-tailed Student's t-test. Scale bars are 10 μm.

***Source*: Zheng et al. (2021).**

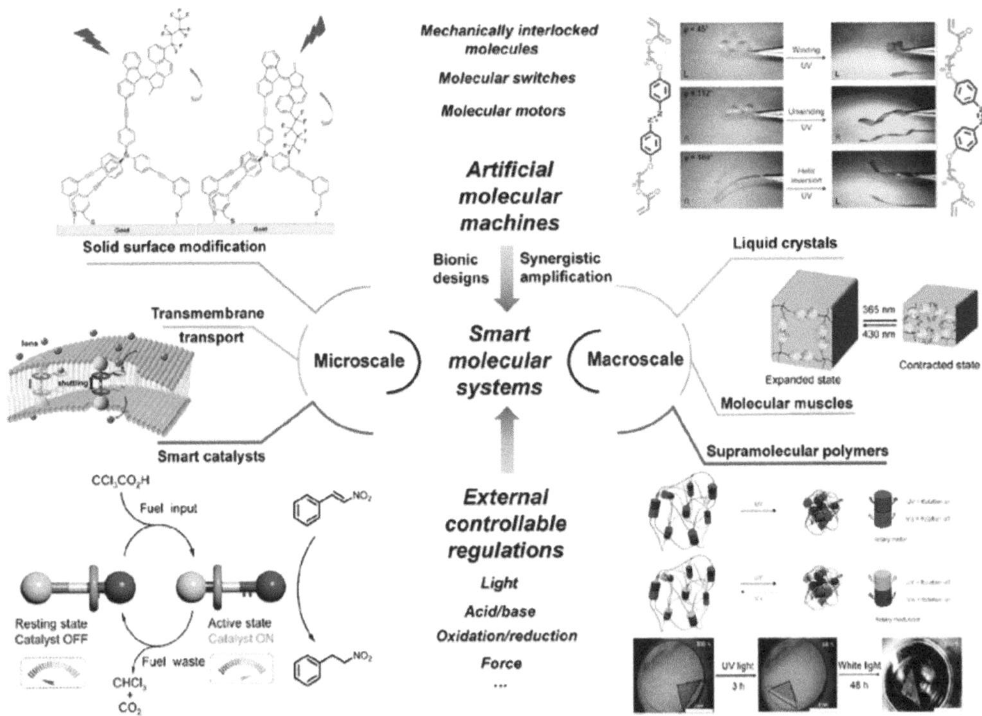

FIGURE 6.21 Schematic representation of the construction of smart molecular systems from AMMs.

***Source*: Shi et al. (2020).**

overviewing AMMs' applications on micro- and macroscales (see Figure 6.21). Applied scientists now focus on developing AMMs with smarter functionality capable of dealing with more complex systems, including mutated biological structures.

Among various biological process-associated issues, malfunction in cell membrane transport is a key one for various cellular disorders (Kuwahara and Marumo, 1996; Hatta et al., 2002); Yarwood et al., 2020). The successful construction of the switchable transport system is a vital step to the practical functioning of AMMs in biological systems and may therefore provide a new strategy for building artificial transmembrane transport systems (Shi et al., 2020), which might have the potential to intervene in cellular disorders or diseases (Didenko et al., 2004; Yu et al., 2018).

A semi-artificial molecular device, which contains a naturally occurring molecular machine – a vaccinia virus-encoded protein – linked with an artificial part was constructed by Didenko and colleagues about two decades ago (Didenko et al., 2004). The self-assembled construct makes two fluorescently labeled detector units. It was the first ever sensor designed which was capable of selectively detecting different types of DNA breaks, exemplifying a practical approach to the design of molecular devices. The molecular device was made by combining two molecules: (1) a self-complimentary dual-labeled 38-mer oligonucleotide. The oligonucleotide contains a CCCTT3′ vaccinia topoisomerase I recognition sequence located adjacent to the nick formed by folded 3′ and 5′ ends (see Figure 6.22); (2) a vaccinia DNA topoisomerase I molecule, which is a virus-encoded eukaryotic type IB topoisomerase (Shuman, 1998). The molecular device importantly operates at room temperature and physiologically neutral pH. To detect two types of breaks, the device requires the presence of ATP, Mg+2, and T4 DNA ligase; otherwise, it detects a single type of DNA break (bearing 5′OH).

Several key features keep the design of this nonmotor-driven molecular device simple yet functional in the nanoscale environment (Astumian and Derényi, 1998). Random movement is an efficient way of surveying the section or a solution for DNA breaks. The presented data show that in many instances 1 h is sufficient to complete DNA damage detection (see Figure 6.23). The construct

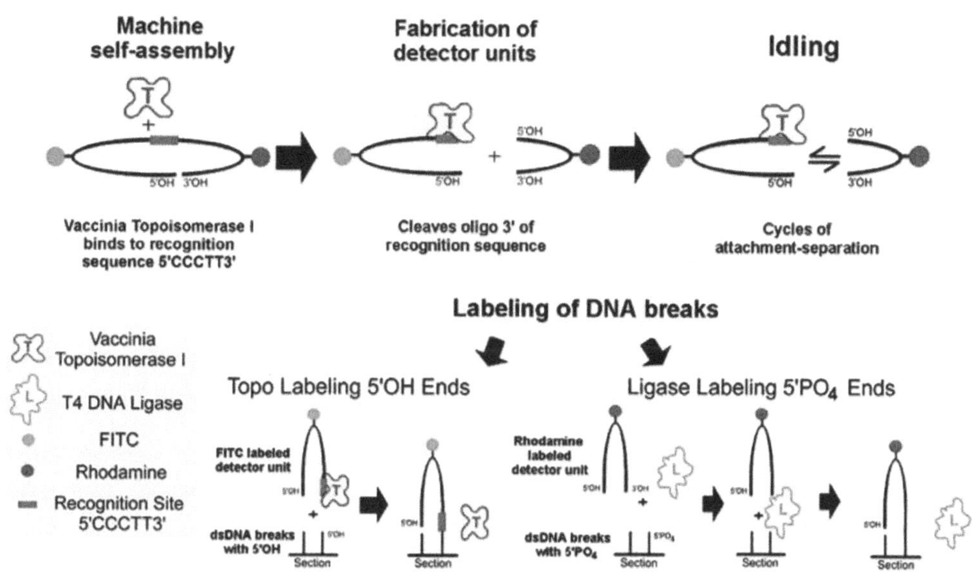

FIGURE 6.22 Light-emitting molecular machine and its usage for DNA damage detection in situ. The machine self-assembles when vaccinia topoisomerase I (TOPO) binds to the double-hairpin oligonucleotide. The machine operation starts when TOPO cleaves the oligonucleotide at the 3′ end of the recognition sequence. As a result the TOPO-activated FITC-labeled portion of the oligonucleotide continuously separates and religates back to the rhodamine-labeled part. When other acceptors, such as blunt-ended DNA breaks with 5′OH, are present in tissue section, the FITC labeled part will attach to them. If T4 DNA ligase is present in reaction solution, it will ligate the remaining rhodamine-labeled part to DNA blunt ends with 5′ PO_4. Therefore, both types of DNA breaks are detected simultaneously and directly.

***Source*: Didenko et al. (2004).**

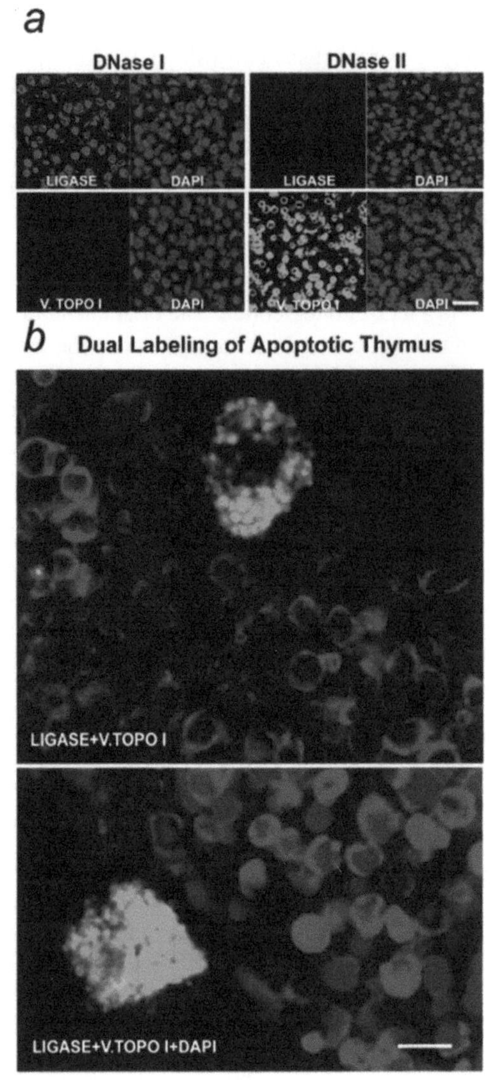

FIGURE 6.23 DNA damage detection in tissue sections using molecular machines: 5′OH blunt-ended DNA breaks detection (a) dual detection of 5′OH and 5′PO_4 blunt-ended DNA breaks (b). (a) DNase I and DNase II treated sections. Sections of the normal bovine adrenal were treated with either DNase I, producing 3′OH/5′PO_4 blunt-ended breaks, or DNase II, producing 3′ PO_4/5′ OH blunt-ended breaks. The vaccinia topoisomerase I-based molecular machines bearing single fluorophore and T4 DNA ligase selectively labeled only one type of DNA breaks each. Bar = 25 μm. (b) Dexamethasone-treated apoptotic rat thymus. Simultaneous dual detection of two types of cuts in the thymic cortical areas undergoing apoptosis using molecular machines bearing two fluorophores. Cortical macrophages with engulfed nuclear material (green fluorescence) and apoptotic thymocytes (red fluorescence) are labeled. Green cytoplasmic fluorescence reveals lysosomes containing DNA with 5′OH double-strand breaks. Surrounding thymocytes undergo apoptosis and have 5′PO_4 double-strand breaks, located at the periphery of their nuclei (red fluorescence). The lower image was additionally counterstained with DAPI to visualize cell nuclei (blue fluorescence). Red, ligase-based detection; green, topoisomerase-based detection; blue, DAPI staining of cellular DNA. Bar = 15 μm.

***Source*: Didenko et al. (2004).**

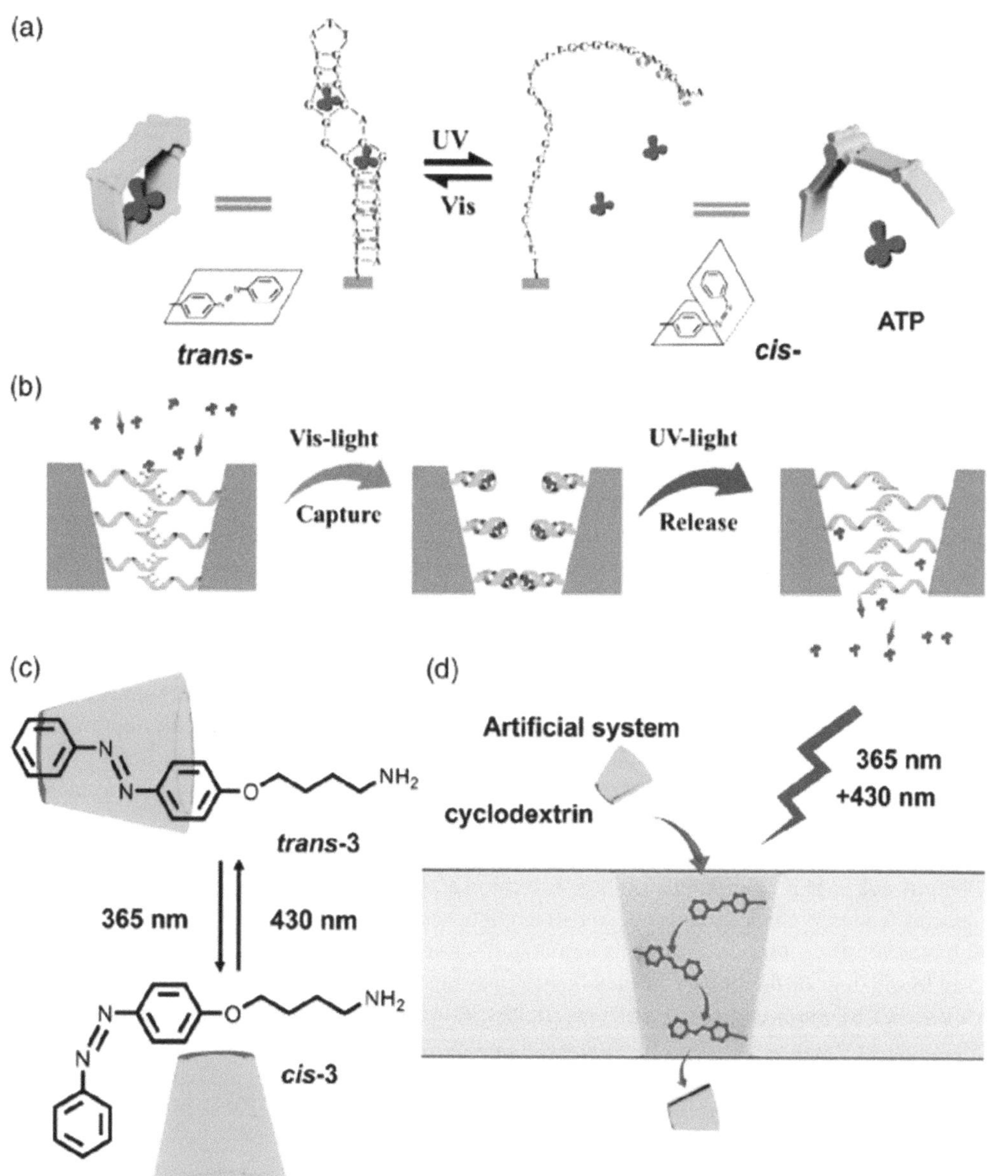

FIGURE 6.24 (a) Schematic representation of the inclusion/exclusion of ATP onto/from azo-DNA under external light. (b) Schematic illustration of the ATP transmembrane transport system. (c) Schematic representation of supramolecular photoreactions between azo group and β-CD. (d) Schematic representation of the bioinspired light-driven mass-transporting system.

***Source*: Shi et al. (2020). (A) and (B) reproduced with permission from Li et al. (2018); (C) and (D) reproduced with permission from Xie et al. (2018).**

uses both "bottom-up" and "top-down" (Balzani et al., 2002) approaches, as it first self-assembles from two molecules and then cleaves itself into smaller detector units. This "bottom-up-top-down" strategy mimics the pathways used by enzymatic systems *in vivo* (Karran and Dyer, 2001). The approach ensures both simple design and durable performance, as no external building blocks or guided assemblies are required for the fabrication, which is accomplished by the simple act of cutting. This molecular construct relies on mechanisms that are unique to the nanoenvironment. The molecular construct demonstrates that molecular devices that use mechanisms developed in the evolution of biological molecules are simpler and uniquely suitable for nanoscale environments. The described construct may be used for apoptosis and DNA damage detection in solution and in fixed cells and tissues.

Many recent developments have emerged in the area of applied scientific research where the application of AMMs in understanding diseases and finding therapeutics has been rigorous. Zhang and coworkers developed a series of light-responsive nanoscale delivery systems based on azobenzene switches (Zhang, Wen, and Jiang, 2018; Zhang, Huang, et al., 2020), as exemplified by a selective ATP transmembrane transport system prepared by immobilizing azo-DNA onto the surface of conical polyimide (PI) nanochannels (Li et al., 2018). In this system, the PI nanochannels act as templates, while the azobenzene switch renders the system light-responsive. Irradiation with visible light (450 nm) favors the planar *trans*-state of azobenzene and, hence, the formation of a hairpin structure through stacking with adjacent base pairs. Importantly, this hairpin structure can be further stabilized by capturing two ATP molecules. Upon exposure to UV light (365 nm), the azobenzene switch undergoes photoisomerization into the *cis*-structure, which induces the collapse of the hairpin structure due to the unfavorable steric hindrance and thus results in ATP release (see Figure 6.24a). The azo-DNA can be switched between folded and unfolded states by alternating irradiation with UV and visible light to realize the concomitant capture and release of ATP molecules and their transport across the membrane (see Figure 6.24b). Li and colleagues (2018) developed a bioinspired system for the selective transport of β-cyclodextrins (β-CDs) across the membrane, relying on transformations between the *trans*- and *cis*-structures of azobenzene derivatives grafted onto PI membranes. In their *trans*-state, the azobenzene units engage in strong hydrophobic supramolecular interactions with β-CD to form inclusion complexes, while *cis*-isomer formation upon irradiation with UV light results in the dissociation of these complexes because of increased steric hindrance (see Figure 6.24c). Upon simultaneous illumination with UV and visible light, the azo groups undergo persistent *trans–cis* conformation changes to result in the continuous trapping/release of β-CD molecules. This continuous rotation–inversion acts as molecular-level stirring to expel β-CD molecules from the nanochannels and thus complete transmembrane transport (see Figure 6.24d). The bionic artificial molecular systems simulate the transmembrane transport processes to deepen our understanding of the biological machinery and thus appear to have promising applications in both drug distribution and cancer therapy fields. More examples and detailed analyses of their applications in regulating various other physical processes of the biological systems may be found in Shi et al. (2020). I could continue elaborating on many such example cases but have decided to leave them for readers to explore further (Aprahamian, 2020; Baroncini et al., 2018; Pfeifer et al., 2022).

REFERENCES

Abendroth, J.M., Bushuyev, O.S., Weiss, P.S., & Barrett, C.J. (2015). Controlling motion at the nanoscale: rise of the molecular machines. *ACS Nano*, *9*(8), 7746–7768

Ahn, C.S., & Metallo, C.M. (2015). Mitochondria as biosynthetic factories for cancer proliferation. *Cancer Metab.*, *3*(1), 1. doi: 10.1186/s40170-015-0128-2; PMID: 25621173; PMCID: PMC4305394

Aik, W.S., Lin, M.H., Tan, D., Tripathy, A., Marzluff, W.F., Dominski, Z., ... & Tong, L. (2017). The N-terminal domains of FLASH and Lsm11 form a 2: 1 heterotrimer for histone pre-mRNA 3'-end processing. *PLoS One*, *12*(10), e0186034

Ainavarapu, S.R., Brujic, J., Huang, H.H., Wiita, A.P., Lu, H., Li, L., Walther, K.A., Carrion-Vazquez, M., Li, H., & Fernandez, J.M. (2007). Contour length and refolding rate of a small protein controlled by engineered disulfide bonds. *Biophys. J.*, *92*(1), 225–233. doi: 10.1529/biophysj.106.091561; PMID: 17028145; PMCID: PMC1697845

Alberts, B., Bray, D., Lewis, J., & Raff, M. (2002). *Molecular biology of the cell* (4th edn). New York: Garland Science

Allegretti, M., Klusch, N., Mills, D.J., Vonck, J., Kühlbrandt, W., & Davies, K.M. (2015). Horizontal membrane-intrinsic α-helices in the stator a-subunit of an F-type ATP synthase. *Nature*, *521*(7551), 237–240

Aprahamian, I. (2020). The future of molecular machines. *ACS Central Science*, *6*(3), 347–358

Ashrafuzzaman, M. (2018). *Nanoscale biophysics of the cell*. Cham: Springer. https://doi.org/10.1007/978-3-319-77465-7

Astumian, R.D. (2001). Making molecules into motors. *Scientific American*, *285*(1), 56–64

Astumian, R.D., & Derényi, I. (1998). Fluctuation driven transport and models of molecular motors and pumps. *European Biophysics Journal*, *27*(5), 474–489

Baker, J.S., McCormick, M.C., & Robergs, R.A. (2010). Interaction among skeletal muscle metabolic energy systems during intense exercise. *J. Nutr. Metab.*, *2010*, 905612. doi: 10.1155/2010/905612; PMID: 21188163; PMCID: PMC3005844

Balzani, V., Credi, A., & Venturi, M. (2002). The bottom-up approach to molecular-level devices and machines. *Chemistry – A European Journal*, *8*(24), 5524–5532

Barba-Aliaga, M., Alepuz, P., & Pérez-Ortín, J.E. (2021). Eukaryotic RNA polymerases: the many ways to transcribe a gene. *Frontiers in Molecular Biosciences*, *8*, 663209

Barendt, T.A., Ferreira, L., Marques, I., Felix, V., & Beer, P.D. (2017). Anion- and solvent-induced rotary dynamics and sensing in a perylene diimide [3] catenane. *Journal of the American Chemical Society*, *139*(26), 9026–9037

Baroncini, M., Casimiro, L., de Vet, C., Groppi, J., Silvi, S., & Credi, A. (2018). Making and operating molecular machines: a multidisciplinary challenge. *ChemistryOpen*, *7*(2), 169–179. doi: 10.1002/open.201700181; PMID: 29435402; PMCID: PMC5795756

Bashour, K.T., Gondarenko, A., Chen, H., Shen, K., Liu, X., Huse, M., ... & Kam, L.C. (2014). CD28 and CD3 have complementary roles in T-cell traction forces. *Proceedings of the National Academy of Sciences*, *111*(6), 2241–2246

Beis, I., & Newsholme, E.A. (1975). The contents of adenine nucleotides, phosphagens and some glycolytic intermediates in resting muscles from vertebrates and invertebrates. *Biochem. J.*, *152*(1), 23–32

Borukhov, S., & Nudler, E. (2008). RNA polymerase: the vehicle of transcription. *Trends Microbiol.*, *16*(3), 126–134. doi: 10.1016/j.tim.2007.12.006; PMID: 18280161

Boyer, P.D. (1997). The ATP synthase – a splendid molecular machine. *Annual Review of Biochemistry*, *66*(1), 717–749

Bucholc, K., Aik, W.S., Yang, X.C., Wang, K., Zhou, Z.H., Dadlez, M., ... & Dominski, Z. (2020). Composition and processing activity of a semi-recombinant holo U7 snRNP. *Nucleic Acids Research*, *48*(3), 1508–1530.

Bustamante, C., Cheng, W., & Mejia, Y.X. (2011). Revisiting the central dogma one molecule at a time. *Cell*, *144*(4), 480–497

Cambridge, S.B., Gnad, F., Nguyen, C., Bermejo, J.L., Krüger, M., & Mann, M. (2011). Systems-wide proteomic analysis in mammalian cells reveals conserved, functional protein turnover. *Journal of Proteome Research*, *10*(12), 5275–5284

Carbajo, R.J., Kellas, F.A., Yang, J.C., Runswick, M.J., Montgomery, M.G., Walker, J.E., & Neuhaus, D. (2007). How the N-terminal domain of the OSCP subunit of bovine F1Fo-ATP synthase interacts with the N-terminal region of an alpha subunit. *Journal of Molecular Biology*, *368*(2), 310–318

Caruthers, M.H. (1985). Gene synthesis machines: DNA chemistry and its uses. *Science*, *230*(4723), 281–285

Cheng, C., McGonigal, P.R., Stoddart, J.F., & Astumian, R.D. (2015). Design and synthesis of nonequilibrium systems. *ACS Nano*, *9*(9), 8672–8688

Corner, G.W. (1967). *Marcello Malpighi and the evolution of embryology* (5 vols.). Ithaca, NY: Cornell University Press.

Coskun, A., Banaszak, M., Astumian, R.D., Stoddart, J.F., & Grzybowski, B.A. (2012). Great expectations: can artificial molecular machines deliver on their promise? *Chemical Society Reviews*, *41*(1), 19–30

Coulombe, B., Jeronimo, C., Langelier, M.F., Cojocaru, M., & Bergeron, D. (2004). Interaction networks of the molecular machines that decode, replicate, and maintain the integrity of the human genome. *Molecular & Cellular Proteomics*, *3*(9), 851–856

Deal, R.B., & Henikoff, S. (2011). Histone variants and modifications in plant gene regulation. *Curr. Opin. Plant Biol.*, *14*(2), 116–122. doi: 10.1016/j.pbi.2010.11.005; PMCID: PMC3093162

Didenko, V.V., Minchew, C.L., Shuman, S., & Baskin, D.S. (2004). Semi-artificial fluorescent molecular machine for DNA damage detection. *Nano Lett.*, *4*(12), 2461–2466. doi: 10.1021/nl048357e; PMID: 17330146

Dominguez, R., & Holmes, K.C. (2011). Actin structure and function. *Annual Review of Biophysics*, *40*, 169

Dominski, Z., & Marzluff, W.F. (2007). Formation of the 3' end of histone mRNA: getting closer to the end. *Gene*, *396*, 373–390

Erbas-Cakmak, S., Leigh, D.A., McTernan, C.T., & Nussbaumer, A.L. (2015). Artificial molecular machines. *Chemical Reviews*, *115*(18), 10081–10206

Feringa, B.L. (2017). The art of building small: from molecular switches to motors (Nobel lecture). *Angewandte Chemie International Edition*, *56*(37), 11060–11078

Finnigan, G.C., Hanson-Smith, V., Stevens, T.H., & Thornton, J.W. (2012). Evolution of increased complexity in a molecular machine. *Nature*, *481*(7381), 360–364. doi: 10.1038/nature10724; PMID: 22230956; PMCID: PMC3979732

Flamholz, A., Phillips, R., & Milo, R. (2014). The quantified cell. *Mol. Biol. Cell.*, *25*(22), 3497–3500. doi: 10.1091/mbc.E14-09-1347; PMID: 25368429; PMCID: PMC4230611

García-López, V., Chen, F., Nilewski, L.G., Duret, G., Aliyan, A., Kolomeisky, A.B., ... & Tour, J.M. (2017). Molecular machines open cell membranes. *Nature*, *548*(7669), 567–572

George, A., Bisignano, P., Rosenberg, J.M., Grabe, M., & Zuckerman, D.M. (2020). A systems-biology approach to molecular machines: exploration of alternative transporter mechanisms. *PLoS Comput. Biol.*, *16*(7), e1007884. doi: 10.1371/journal.pcbi.1007884; PMID: 32614821; PMCID: PMC7331975

Gibson, D.G., Glass, J.I., Lartigue, C., Noskov, V.N., Chuang, R.Y., Algire, M.A., ... & Venter, J.C. (2010). Creation of a bacterial cell controlled by a chemically synthesized genome. *Science*, *329*(5987), 52–56

Goodsell, D.S. (2009). *The machinery of life*. New York: Springer. doi: 10.1007/978-0-387-84925-6l

Grzelczak, M., Liz-Marzán, L.M., & Klajn, R. (2019). Stimuli-responsive self-assembly of nanoparticles. *Chemical Society Reviews*, *48*(5), 1342–1361

Gurkiewicz, M., & Korngreen, A. (2007). A numerical approach to ion channel modelling using whole-cell voltage-clamp recordings and a genetic algorithm. *PLoS Computational Biology*, *3*(8), e169. https://doi.org/10.1371/journal.pcbi.0030169

Hatta, S., Sakamoto, J., & Horio, Y. (2002). Ion channels and diseases. *Med. Electron Microsc.* *35*(3), 117–126. doi: 10.1007/s007950200015; PMID: 12353132

Jordan, J.D., Landau, E.M., & Iyengar, R. (2000). Signaling networks: the origins of cellular multitasking. *Cell*, *103*(2), 193–200. doi: 10.1016/s0092-8674(00)00112-4; PMID: 11057893; PMCID: PMC3619409

Karran, L., & Dyer, M.J. (2001). Proteolytic cleavage of molecules involved in cell death or survival pathways: a role in the control of apoptosis? *Critical Reviews in Eukaryotic Gene Expression*, *11*(4), 269–277

Kassem, S., Lee, A.T., Leigh, D.A., Markevicius, A., & Solà, J. (2016). Pick-up, transport and release of a molecular cargo using a small-molecule robotic arm. *Nature Chemistry*, *8*(2), 138–143

Kay, E.R., Leigh, D.A., & Zerbetto, F. (2007). Synthetic molecular motors and mechanical machines. *Angewandte Chemie International Edition*, *46*(1–2), 72–191

Kosuri, S., & Church, G.M. (2014). Large-scale de novo DNA synthesis: technologies and applications. *Nature Methods*, *11*(5), 499–507

Kottas, G.S., Clarke, L.I., Horinek, D., & Michl, J. (2005). Artificial molecular rotors. *Chemical Reviews*, *105*(4), 1281–1376

Koumura, N., Geertsema, E.M., van Gelder, M.B., Meetsma, A., & Feringa, B.L. (2002). Second generation light-driven molecular motors: unidirectional rotation controlled by a single stereogenic center with near-perfect photoequilibria and acceleration of the speed of rotation by structural modification. *Journal of the American Chemical Society*, *124*(18), 5037–5051

Kühlbrandt, W. (2015). Structure and function of mitochondrial membrane protein complexes. *BMC Biol.*, *13*, 89. https://doi.org/10.1186/s12915-015-0201-x

Kuwahara, M., & Marumo, F. (1996). Diseases caused by disorders of membrane transport: an overview. *Nihon rinsho: Japanese Journal of Clinical Medicine*, *54*(3), 581–585.

Lancia, F., Ryabchun, A., & Katsonis, N. (2019). Life-like motion driven by artificial molecular machines. *Nature Reviews Chemistry*, *3*(9), 536–551

Lehn, J.M. (2013). Perspektiven der Chemie – Stufen zur komplexen Materie. *Angewandte Chemie*, *125*(10), 2906–2921

Li, P., Xie, G., Liu, P., Kong, X.Y., Song, Y., Wen, L., & Jiang, L. (2018). Light-driven ATP transmembrane transport controlled by DNA nanomachines. *Journal of the American Chemical Society*, *140*(47), 16048–16052

Liu, B., Chen, W., Evavold, B.D., & Zhu, C. (2014). Accumulation of dynamic catch bonds between TCR and agonist peptide-MHC triggers T cell signaling. *Cell*, *157*(2), 357–368

Liu, M.Y., Nemes, A., & Zhou, Q.G. (2018). The emerging roles for telomerase in the central nervous system. *Front. Mol. Neurosci.*, *11*, 160

Liu, Z., Liu, Y., Chang, Y., Seyf, H.R., Henry, A., Mattheyses, A.L., ... & Salaita, K. (2016). Nanoscale optomechanical actuators for controlling mechanotransduction in living cells. *Nature Methods*, *13*(2), 143–146

Loeb, L., & Monnat, R. (2008). DNA polymerases and human disease. *Nat. Rev. Genet.*, *9*, 594–604. https://doi.org/10.1038/nrg2345

Mariño-Ramírez, L., Kann, M.G., Shoemaker, B.A., & Landsman, D. (2005). Histone structure and nucleosome stability. *Expert Rev. Proteomics*, *2*(5), 719–729. doi: 10.1586/14789450.2.5.719; PMID: 16209651; PMCID: PMC1831843

Menon, V., Spruston, N., & Kath, W.L. (2009). A state-mutating genetic algorithm to design ion-channel models. *Proceedings of the National Academy of Sciences*, *106*(39), 16829–16834. doi: 10.1073/pnas.0903766106

Milo, R. (2013). What is the total number of protein molecules per cell volume? A call to rethink some published values. *Bioessays*, *35*(12), 1050–1055

Moeendarbary, E., & Harris, A.R. (2014). Cell mechanics: principles, practices, and prospects. *Wiley Interdiscip. Rev. Syst. Biol. Med.*, *6*(5), 371–388. doi: 10.1002/wsbm.1275; PMID: 25269160; PMCID: PMC4309479

Montgomery, M.G., Petri, J., Spikes, T.E., & Walker, J.E. (2021). Structure of the ATP synthase from *Mycobacterium smegmatis* provides targets for treating tuberculosis. *Proceedings of the National Academy of Sciences*, *118*(47), e2111899118

Morales-Rios, E., Montgomery, M.G., Leslie, A.G., & Walker, J.E. (2015). Structure of ATP synthase from *Paracoccus denitrificans* determined by X-ray crystallography at 4.0 Å resolution. *Proceedings of the National Academy of Sciences*, *112*(43), 13231–13236

Nicholson, D.J. (2019). Is the cell really a machine? *Journal of Theoretical Biology*, *477*, 108–126

Noji, H., & Yoshida, M. (2001). The rotary machine in the cell, ATP synthase*210. *Journal of Biological Chemistry*, *276*(3), 1665–1668

Palluk, S., Arlow, D., de Rond, T., et al. (2018). De novo DNA synthesis using polymerase-nucleotide conjugates. *Nat. Biotechnol.*, *36*, 645–650. https://doi.org/10.1038/nbt.4173

Pfeifer, L., Hoang, N.V., Crespi, S., Pshenichnikov, M.S., & Feringa, B.L. (2022). Dual-function artificial molecular motors performing rotation and photoluminescence. *Science Advances*, *8*(44), eadd0410

Piccolino, M. (2000). Biological machines: from mills to molecules. *Nature Reviews Molecular Cell Biology*, *1*(2), 149–152

Pollard, T.D., & Borisy, G.G. (2003). Cellular motility driven by assembly and disassembly of actin filaments. *Cell*, *112*(4), 453–465

Preiss, T.W., & Hentze, M. (2003). Starting the protein synthesis machine: eukaryotic translation initiation. *Bioessays*, *25*(12), 1201–1211. doi: 10.1002/bies.10362; PMID: 14635255

Qu, D.H., Wang, Q.C., Zhang, Q.W., Ma, X., & Tian, H. (2015). Photoresponsive host–guest functional systems. *Chemical Reviews*, *115*(15), 7543–7588

Ramezani, H., & Dietz, H. (2020). Building machines with DNA molecules. *Nat. Rev. Genet.*, *21*(1), 5–26. doi: 10.1038/s41576-019-0175-6; PMID: 31636414; PMCID: PMC6976304

Rees, D.M., Leslie, A.G., & Walker, J.E. (2009). The structure of the membrane extrinsic region of bovine ATP synthase. *Proceedings of the National Academy of Sciences*, *106*(51), 21597–21601

Richardson, S.M., Mitchell, L.A., Stracquadanio, G., Yang, K., Dymond, J.S., DiCarlo, J.E., ... & Bader, J.S. (2017). Design of a synthetic yeast genome. *Science*, *355*(6329), 1040–1044

Rossello, R.A., & Kohn, D.H. (2010). Cell communication and tissue engineering. *Communicative & Integrative Biology*, *3*(1), 53–56.

Saha, S., & Stoddart, J.F. (2007). Photo-driven molecular devices. *Chemical Society Reviews*, *36*(1), 77–92

Salin, K., Auer, S.K., Rey, B., Selman, C., & Metcalfe, N.B. (2015). Variation in the link between oxygen consumption and ATP production, and its relevance for animal performance. *Proc. Biol. Sci.*, *282*(1812), 20151028. doi: 10.1098/rspb.2015.1028; PMID: 26203001; PMCID: PMC4528520

Schliwa, M., & Woehlke, G. (2003). Molecular motors. *Nature*, *422*(6933), 759–765

Schmittel, M. (2015). From self-sorted coordination libraries to networking nanoswitches for catalysis. *Chemical Communications*, *51*(81), 14956–14968

Sengupta, S., Ibele, M.E., & Sen, A. (2012). Fantastic voyage: designing self-powered nanorobots. *Angewandte Chemie International Edition*, *51*(34), 8434–8445

Sengupta, S., Spiering, M.M., Dey, K.K., Duan, W., Patra, D., Butler, P.J., Astumian, R.D., Benkovic, S.J., and Sen, A. (2014). DNA polymerase as a molecular motor and pump. *ACS Nano*, *8*(3), 2410–2418

Serrano, P., Geralt, M., Mohanty, B., & Wüthrich, K. (2014). NMR structures of α-proteobacterial ATPase-regulating ζ-subunits. *Journal of Molecular Biology*, *426*(14), 2547–2553

Shi, Z.T., Zhang, Q., Tian, H., & Qu, D.H. (2020). Driving smart molecular systems by artificial molecular machines. *Advanced Intelligent Systems*, *2*(5), 1900169

Shuman, S. (1998). Vaccinia virus DNA topoisomerase: a model eukaryotic type IB enzyme. *Biochimica et Biophysica Acta – Gene Structure and Expression*, *1400*(1–3), 321–337

Siekevitz, P. (1957). Powerhouse of the cell. *Scientific American*, *197*(1), 131–144

Sun, Y., Zhang, Y., Aik, W.S., Yang, X.C., Marzluff, W.F., Walz, T., … & Tong, L. (2020). Structure of an active human histone pre-mRNA 3′-end processing machinery. *Science*, *367*(6478), 700–703

Svitkina, T.M., Verkhovsky, A.B., McQuade, K.M., & Borisy, G.G. (1997). Analysis of the actin–myosin II system in fish epidermal keratocytes: mechanism of cell body translocation. *Journal of Cell Biology*, *139*(2), 397–415

Tang, L. (2018). An enzymatic oligonucleotide synthesizer. *Nat. Methods*, *15*, 568. https://doi.org/10.1038/s41592-018-0096-x

Teed, Z.R., & Silva, J.R. (2016). A computationally efficient algorithm for fitting ion channel parameters. *MethodsX*, *3*, 577–588. doi: 10.1016/j.mex.2016.11.001

Thomas, B. (2009). ATP synthase: majestic molecular machine made by a mastermind. *Retrieved*, *1*(19), 2017

Urquhart, J. (2021). Molecular machines mechanically talk to cells. *Chemistry World*, July 1, 2021, www.chemistryworld.com/news/molecular-machines-talk-to-living-cells-for-the-first-time/4013927.article (accessed November 11, 2022)

Urry, D.W. (1999). Elastic molecular machines in metabolism and soft-tissue restoration. *Trends in Biotechnology*, *17*(6), 249–257

Vale, R.D., & Milligan, R.A. (2000). The way things move: looking under the hood of molecular motor proteins. *Science*, *288*(5463), 88–95

Vinothkumar, K.R., Montgomery, M.G., Liu, S., & Walker, J.E. (2016). Structure of the mitochondrial ATP synthase from *Pichia angusta* determined by electron cryo-microscopy. *Proceedings of the National Academy of Sciences*, *113*(45), 12709–12714

Wang, L.L., Chen, Z., Liu, W.E., Ke, H., Wang, S.H., & Jiang, W. (2017). Molecular recognition and chirality sensing of epoxides in water using endo-functionalized molecular tubes. *Journal of the American Chemical Society*, *139*(25), 8436–8439

Wang, T., Wei, J.J., Sabatini, D.M., & Lander, E.S. (2014). Genetic screens in human cells using the CRISPR-Cas9 system. *Science*, *343*(6166), 80–84

Wilkens, S., Borchardt, D., Weber, J., & Senior, A.E. (2005). Structural characterization of the interaction of the δ and α subunits of the *Escherichia coli* F1F0-ATP synthase by NMR spectroscopy. *Biochemistry*, *44*(35), 11786–11794

Wu, Y., Albrecht, T.R., Baillat, D., Wagner, E.J., & Tong, L. (2017). Molecular basis for the interaction between Integrator subunits IntS9 and IntS11 and its functional importance. *Proceedings of the National Academy of Sciences*, *114*(17), 4394–4399

Xie, G., Li, P., Zhao, Z., Kong, X.Y., Zhang, Z., Xiao, K., … & Jiang, L. (2018). Bacteriorhodopsin-inspired light-driven artificial molecule motors for transmembrane mass transportation. *Angewandte Chemie International Edition*, *57*(51), 16708–16712

Yarwood, R., Hellicar, J., Woodman, P. G., & Lowe, M. (2020). Membrane trafficking in health and disease. *Disease Models & Mechanisms*, *13*(4), dmm043448

Yu, G., Yung, B.C., Zhou, Z., Mao, Z., & Chen, X. (2018). Artificial molecular machines in nanotheranostics. *ACS Nano*, *12*(1), 7–12

Zhang, L., Marcos, V., & Leigh, D.A. (2018). Molecular machines with bio-inspired mechanisms. *Proceedings of the National Academy of Sciences*, *115*(38), 9397–9404

Zhang, Q., Wang, W.Z., Yu, J.J., Qu, D.H., & Tian, H. (2017). Dynamic self-assembly encodes a tri-stable Au–TiO2 photocatalyst. *Advanced Materials*, *29*(5), 1604948

Zhang, Y., Sun, Y., Shi, Y., Walz, T., & Tong, L. (2020). Structural insights into the human pre-mRNA 3′-end processing machinery. *Molecular cell*, *77*(4), 800–809.

Zhang, Z., Huang, X., Qian, Y., Chen, W., Wen, L., & Jiang, L. (2020). Engineering smart nanofluidic systems for artificial ion channels and ion pumps: from single-pore to multichannel membranes. *Advanced Materials*, *32*(4), 1904351

Zhang, Z., Wen, L., & Jiang, L. (2018). Bioinspired smart asymmetric nanochannel membranes. *Chemical Society Reviews*, *47*(2), 322–356

Zheng, Y., Han, M.K., Zhao, R., Blass, J., Zhang, J., Zhou, D.W., … & Del Campo, A. (2021). Optoregulated force application to cellular receptors using molecular motors. *Nature Communications*, *12*(1), 1–10

7 Forces and Motion in Biological Systems

The cellular function mainly depends on intracellular chemical signaling, the physiological condition of the environment, and especially the mechanical properties of the cells. Cells sense external mechanical inputs and transduce them into biochemical and electrical signals that influence various cellular processes, e.g. cell proliferation, adhesion, extracellular and intracellular migration, and most obviously the fate of the cells. Mechanical force-induced cellular signals also mediate organ homeostasis and drive pathophysiological processes. Mechanical forces are found to play crucial roles in vascular and skeletal organ systems (Veith et al., 2020). The formation of nucleic acid structures, e.g. the double helical structure of DNA, and their replication are both dependent on physical energetics and forces (Li et al., 2015). Likewise, the protein's structural stability and the folding mechanisms in the cellular environment rely on specific internal and external forces and energetics (Newberry and Raines, 2019). All these forces and energies, which originate from both bonded and nonbonded interactions in biological systems, cause fluctuations and motion even in a controlled aqueous environment. Bonded interactions are limited within the specific range determined by specific bond types while nonbonded interactions may extend over a wide range of distances. Both types of interactions vary in strength. As examples, covalent bonds vary in strength over 30–120 kcal mol^{-1}, hydrogen bonds vary over 2–40 kcal mol^{-1}, and electrostatic and van der Waals interactions range over a system-specific distribution of strengths (Sartori and Nascimento, 2019; Ashrafuzzaman et al., 2020). These strong and weak interaction forces contribute to constructing biochemical systems in biology. This chapter will be dedicated to addressing crucial forces and energies that are naturally active inside biological systems. The associated motion caused by applied external or internal forces and energetic fluctuations may follow general Newtonian mechanics, statistical mechanical formalisms, and various other diffusion rules, most of which will be briefly explored.

7.1 FORCES OF WATER MOLECULES IN BIOLOGY AND ASSOCIATED DYNAMICS

Biomolecular forces originate from both long- and short-range interaction energetics. From the physical behavioral perspective of biological systems, the environment in which biomolecules exist, move, and interact appears to play crucial roles in determining the interaction energetics and forces felt by the biomolecules, resulting in measurable physical movement or motion of the molecules. The water environment is quite common in this case because water occupies a healthy proportion of the body, and the fluctuations in the quantity and uptake of water by cellular systems may regulate crucial functions of cells and biomolecules thereof (King and Smythe, 2020). The behavior of water as a solvent for nonpolar and charged molecules lies partly in the stability of its caging structures

DOI: 10.1201/9781003287780-7

and the vibration or motion of the water's electric dipoles (Brini et al., 2017). One of the major roles of water is to mediate chemical and biological self-assembly (Despa and Berry, 2007; Despa et al., 2004). Water as a biomolecule influences biological processes, including the formation of the structure and determination of the function of enzymes and other biomolecules. The physics of water molecules contributes important mechanisms helping to create the biological environment.

Water–water interactions appear important in biological systems. Both source and strength of the long-range attraction between water-coated hydrophobic surfaces originate at the polarization field produced by the strong correlation and coupling of the dipoles of the water molecules at the surfaces. This polarization field is found to give rise to dipoles on the surface of the hydrophobic solutes generating long-range hydrophobic attractions. The hydrophobic aggregation thus begins with a step in which water-coated nonpolar solutes are found to approach one another due to long-range electrostatic forces. This regime occurs before the entropy increase of the water layer release (Lum et al., 1999; Huang and Chandler, 2002; ten Wolde and Chandler, 2002; Pangali et al., 1979; Wallqvist and Berne, 1995; Huang et al., 2003; Liu et al., 2005) and the short-range van der Waals attraction provides the driving force to "dry out" the contact surface. The effective force of attraction is derived from basic molecular principles, without assumptions of the structure of the hydrophobe–water interaction. Distinct from the drying-induced hydrophobic surface interaction, two hydrophobes may attract at longer distances, via the dipole-dipole and the induction-dispersion effects that are generated by the polarization fields of the water, structured at the interface (see Figure 7.1).

The strength of the effective force of water–water attraction can be measured using the atomic force microscopy (AFM) imaging of a hydrophobic molecule tethered to a surface but extending into water, and another hydrophobe attached to an atomic force probe. This phenomenon can be observed in the transverse relaxation rates in water proton magnetic resonance (see Figure 7.2). A kinetic T_2-MR imaging measurement of hydrophobic solution would allow the derivation of time-dependent ρ (η,t) maps (see detailed theoretical analysis in Despa and Berry, 2007). These maps may be used in order to extract the dynamical information on hydrophobic interactions.

The value of water molecules as life's matrix lies in both structural and dynamical characteristics of its status as a complex (Ball, 2017). Figure 7.3 shows a representative time scale that water molecular dynamics follows in biological systems in comparison to other biomolecules (Xu and Havenith, 2015). The low-frequency spectrum of the solvated biomolecule and molecular dynamics (MD) simulations provided insights into the collective hydrogen bond dynamics on the sub-ps time scale. The absorption spectrum between 1 THz and 10 THz of solvated biomolecules is found to be sensitive to changes in the fast fluctuations of the water network (Ball, 2017).

7.2 FORCES AND MOTION OF BIOMOLECULES

The water molecule with dipole moment is assumed to respond to electrical fields or forces and thus can also demonstrate dynamics. In Figure 7.3, we see that the dynamics and/or vibration, hence physical motion, associated with biomolecules including water follow distinguished time scales or regimes, ranging from high microsecond (μs) to as low as picosecond (ps) or even femtosecond (fs) (Xu and Havenith, 2015). With this fundamental physical property and other ones, the energetics of water molecules in proteins or generally in a water–protein complex is found to follow certain behavior, thus often playing important roles in proteins' functions (Clarke et al., 2001; García-Sosa et al., 2005; García-Sosa and Mancera, 2010; Kadirvelraj et al., 2008; Mancera, 2002; Michel et al., 2009; Mikol et al., 1995; Wong and Lightstone, 2010; Morozenko et al., 2014).

The physics of hydration of biomolecules is known to show interesting behavior. Based on an estimate of the average dipole moment of water molecules located in the internal cavities of the protein and their binding energies, the water dipole in the protein environment is found to show biomolecule-specific physical behavior. The water dipole in the biomolecule's core environment is found to be much different from that in the bulk water environment (Morozenko et al., 2014). This

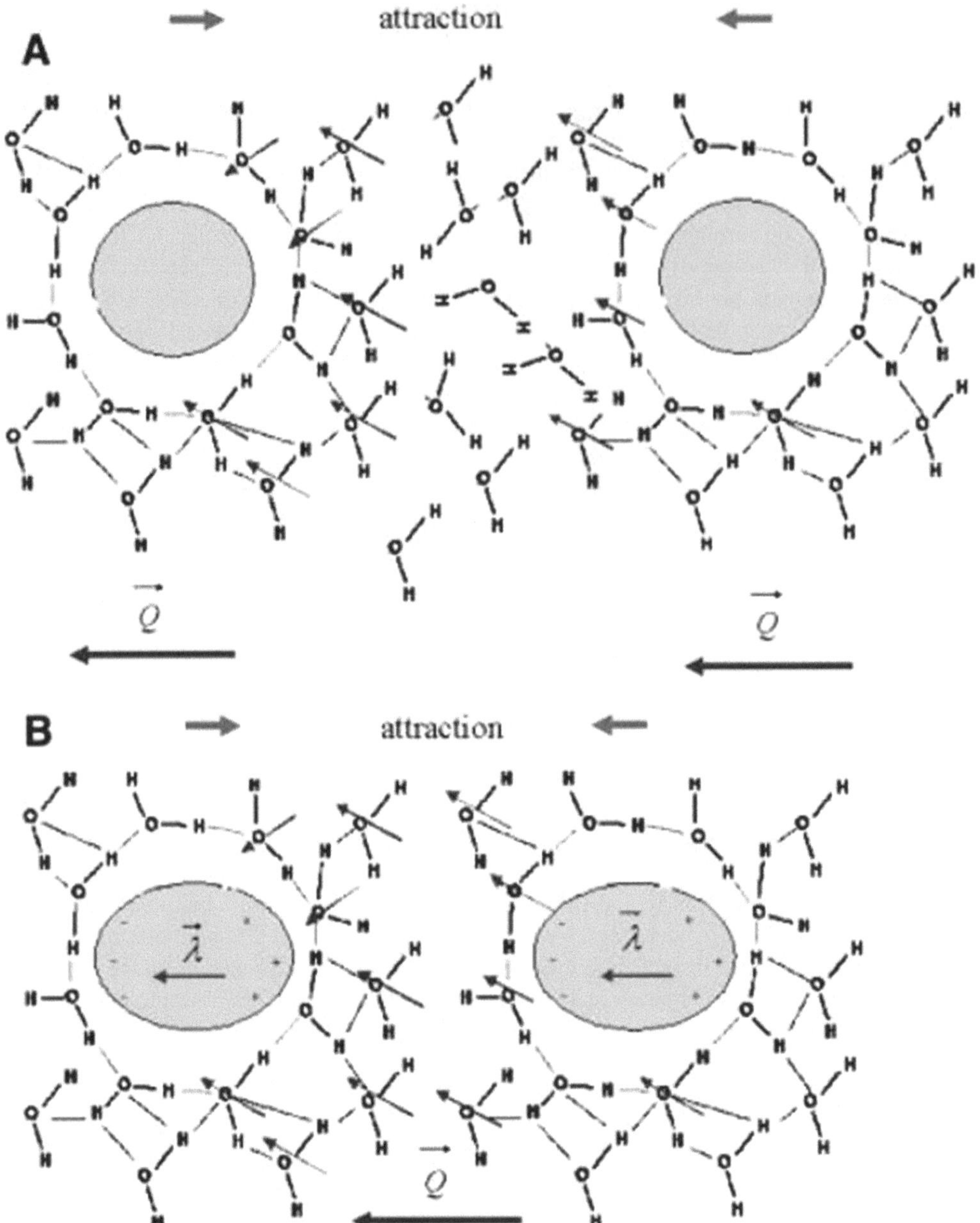

FIGURE 7.1 Schematic representation of the long-range attraction between hydrophobes initiated by the domains of polarized water (a) and by induced dipoles on the surface of the hydrophobic solutes (b).

***Source*: reproduced with permission from Despa and Berry (2007).**

suggests that as a result of the water–biomolecule association both water and hydrated biomolecules may influence various physical properties of the complex, of each other, including especially the electronic polarizability of the biomolecules.

The long-range transfer of protons of proteins is usually carried out by the Grotthuss mechanism which is a proton-jumping process by which any excess proton diffuses across the water molecules'

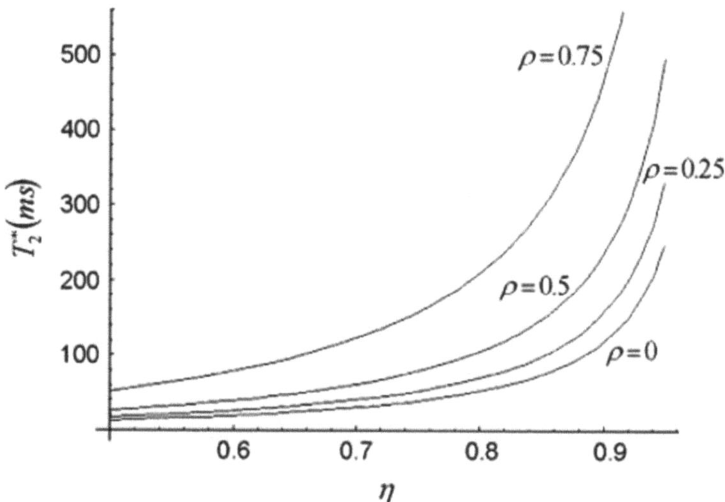

FIGURE 7.2 Predicted trends for the value of the transverse magnetic relaxation time of water protons (plotted along vertical axis) in solutions containing hydrophobic molecules that interact via long-range attraction forces. The measurement of the transverse magnetic relaxation time of water protons is performed in a solution containing hydrophobic molecules and η percent water. Here ρ is the average packing density of a hydrophobic assembly. We avoid presenting the detailed theories here, which can be found in Despa and Berry (2007).

***Source*: reproduced with permission from Despa and Berry (2007).**

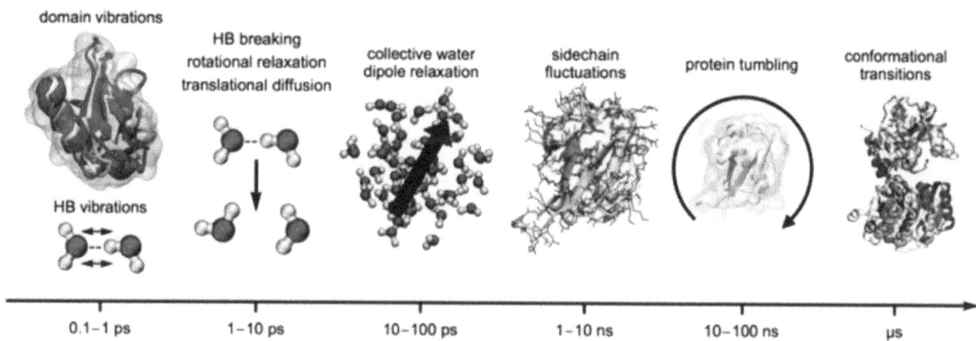

FIGURE 7.3 Hierarchy of time scales for protein motions including the water dipole relaxation time scale. From left to right: (0.1–1.0 ps/1–10 THz) intramolecular vibrations of protein domains and secondary structure elements, including skeleton motions and intermolecular vibrations of the water hydrogen bond network; (1–10 ps/0.1–1 THz) relaxation processes in the hydrogen bond network of water, i.e. hydrogen bond rearrangements, single molecule rotation, and translational diffusion; (10–100 ps/10–100 GHz) collective dipole relaxation in water; (1–10 ns/0.1–1 GHz) protein side chain fluctuations; (10–100 ns/10–100 MHz) protein rotational tumbling motions;(µs/MHz) conformational transitions in proteins, shown as a hinge motion of glycogen synthase.

***Source*: reproduced with permission from Xu and Havenith (2015).**

hydrogen bond network through the formation and cleavage of covalent bonds involving neighboring molecules (Grancha et al., 2016), This process connects among a chain of hydrogen bonds composed of internal water molecules and amino acid residues of the protein. Water molecules, as temporary proton donors/acceptors, can also facilitate the enzymes' catalytic reactions. Due to high dielectric properties, water is known to screen the electrostatic interaction of charges, leading to the stabilization of charge on ionizable residues or affecting their pK_a values. However, the dynamics of water in proteins does not allow resolution of all the molecules, but a few of the protein's water molecules are typically seen in the X-ray structure (Carugo and Bordo, 1999; Davis et al., 2003; Ernst et al., 1995). Here the existence of the optimum charge or the water molecule dipole can be qualitatively rationalized by analyzing inserted water molecules–protein atoms interactions.

The total energy of a water molecule (inserted by the water placement software Dowser), E_W^{tot} is the sum of three interactions, the 6–12 Lennard-Jones potential ($E_{L,J}$), the charge–dipole interaction of molecules with background protein charges ($U_{P,W}$), and the electrostatic dipole–dipole interaction among molecules ($U_{W,W}$), so follows Morozenko et al. (2014):

$$E_W^{tot} = E_{L,J} + U_{P,W} + U_{W,W} \tag{7.1}$$

The water dipole moment is proportional to the charges assigned to the atoms $\mu \sim q_O \sim q_H$. E_{LJ} is independent of charges (q), $U_{P,W} \sim q$, and $U_{W,W} \sim q^2$. Resultantly, E_W^{tot} is a quadratic function of q and has a stationary value at some value of $q = q^*$. For a given cutoff energy E_{cutoff}, there is an optimum value of charge q_H^* (setting the dipole μ of the water) that maximizes the ratio of the number of predicted molecules in the protein to the total number of crystallographic waters (χ). The dependence of the optimum charge q_H^* on E_{cutoff} is found to follow a relation (see Figure 7.4).

At $E_{\text{cutoff}} \sim -4$ kcal/mol, the uncertainty of the optimum charge q_H^* is lowest, and the maximum value of $\chi = \chi_{\text{ave}}(E_{\text{cutoff}}, q_H^*)$ is achieved at $q_H^* \sim 0.33$ e (indicated by a circle in Figure 7.4). Here we know that the decrease in energy reduces the number of discarded water molecules and thus χ should go up. An optimum value of charge exists at which energy is minimized; the lower the energy of a molecule, the higher the probability that the criterion $E_W^{tot} < E_{\text{cutoff}}$ is satisfied, and hence, the higher the hit ratio χ. For a given E_{cutoff}, there is an optimum value of q_H^* that maximizes the hit ratio χ. From the analysis of a set of protein structures, the optimum values of Dowser parameters μ and E_{cutoff} were determined, which essentially are the effective dipole moment and binding energy of water molecules in the protein environment. Once the correct cutoff/binding energy of −4 kcal/mol is implemented in Dowser, the amount of predicted crystallographic waters increases to its maximum value. This information may lead to the conclusion that the dipole relaxation for water as seen in Figure 7.3 may be explained in light of the energetics and charge fluctuations of water molecules associated with the host protein structures in a biological environment. This study may be extended in this direction to create an updated understanding of the role of hydration in biomolecular physics.

Besides understanding the hydration-associated fluctuations associated with specific protein structures and general physical properties, as explained earlier, we may also want to address inherited motion inside and on the surface of proteins. Almost half a century ago, translational and rotational motions were generally accepted physical properties of proteins in solution besides observing intramolecular motions in the case when the protein molecule is treated as a rigid matrix (Gurd and Rothges, 1979). Rotational reorientation of intramolecular components of proteins involves rotations about individual interatomic bonds. Concerted motions in proteins are associated with the movement of groups of side chains or segments of the backbone relative to other structures. Motions of interacting ions and small molecules also produce fluctuations in the protein structure in addition to the Brownian transfer of momentum. Small molecules or specific drugs may also regulate protein–protein interactions, suggesting also for their dragging effects on even any large and

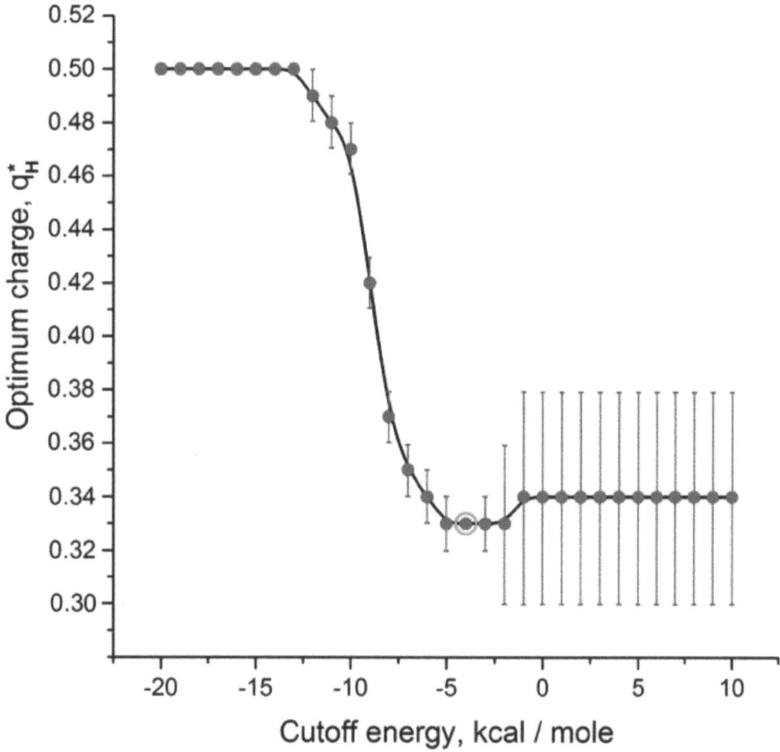

FIGURE 7.4 Optimum value of charge q_H^* versus E_{cutoff}.

***Source*: reproduced with permission from Morozenko et al. (2014).**

difficult topology causing the topology to experience altered rotation, fluctuations, and dynamics (Cesa et al., 2015). Other tuned physical processes such as modulation of the energy landscape by bound ligands were observed to funnel the dihydrofolate reductase enzyme through its reaction cycle along a preferred kinetic path, suggesting small-scale yet crucial fluctuations in the structural and functional pathways (Boehr et al., 2006). Recently, nuclear magnetic resonance spectroscopy was used to probe the dynamics of the peroxidatic cysteine (C_p)-thiolate and disulfide forms of Xanthomonas campestris peroxiredoxin Q, providing evidence that a catalytically relevant local unfolding equilibrium exists in the enzyme's C_p-thiolate form. Faster inherited motions imply an active site instability, which could promote local unfolding (Estelle et al., 2022). Here also we observe that the molecular-level fluctuations and associated motion induce effects on even chemical interactions.

Both protein–polymer interactions and the polymer adsorption of nanometer-size colloidal particles on surfaces are known to follow time-dependent kinetics (Fang et al., 2005; Wu et al., 2004). Flexible polymer molecules grafted to surfaces or interfaces impose a steric barrier that can be tuned depending upon the polymer molecular weight, surface coverage, and type of chemical structure (Milner, 1991; Halperin et al., 1992; Szleifer and Carignano, 1996). The grafted polymer layer is known to prevent the nonspecific adsorption of proteins on the surface of the biocompatible material or drug carrier (McPherson et al., 1995; Woodle, 1997). The surface contains N_g polymer molecules, with each having n_g segments and each having length l, grafted at one of their ends (see Figure 7.5). The polymer-modified surface is put, at time t = 0, in contact with a protein

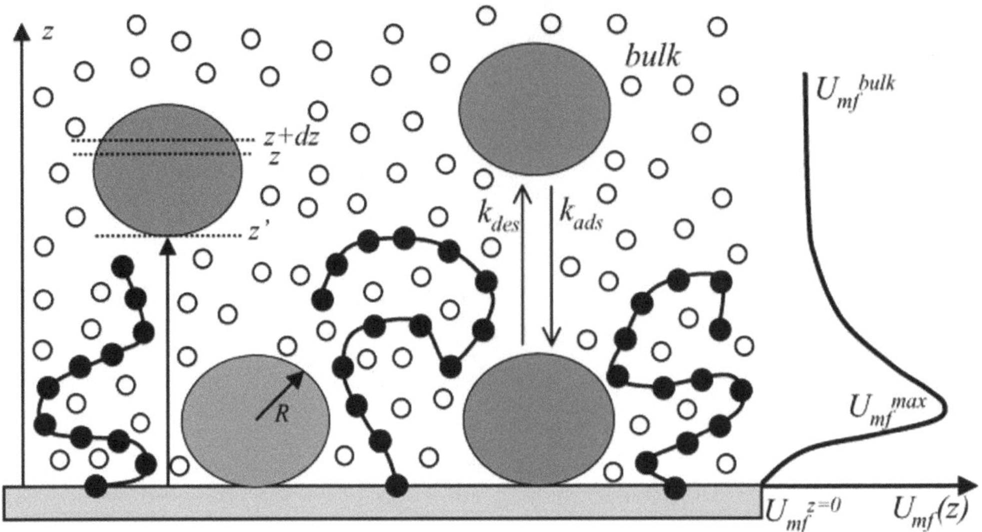

FIGURE 7.5 Schematic representation of the system containing proteins in their native conformation dissolved in a low molecular-weight solvent, and in contact with a surface with grafted polymers. The large solid circles are the proteins and the small open circles are the solvent molecules. The strings of small solid circles, tethered to the surface, represent the grafted polymers. The z direction is defined perpendicular to the surface. The position of a protein, z', refers to the lowest point of the protein, whereas the volume that a protein contributes to z refers to the volume that the protein occupies between z and $z + dz$. The two rate coefficients represent the kinetic processes involved in the adsorption of proteins onto the surface with grafted polymers. The right of the figure represents schematically the potential of mean-force felt by the adsorbing/desorbing proteins. Copied with due permission to reproduce from ref. Fang et al. (2005).

***Source*: reproduced with permission from Fang et al. (2005).**

solution in water. The bulk density $\rho_{p,bulk}$ or chemical potential $\mu_{p,bulk}$ represents the characteristics of the protein solution. When the surface is put in contact with the solution the proteins feel anisotropic interactions by the surface, raising the driving forces for the adsorption process. The protein adsorbed on surfaces with grafted polymers follows time-dependent kinetic rules (see Figure 7.6).

In Fang et al. (2005), the time-dependent adsorption for no-polymer on the surface and short-chain-length-grafted polymer showed a very fast early regime, in which the surface was found to act as a sink to the proteins due to the strong attraction between the surface and the proteins. After a certain amount of proteins adsorbed, there was a very sharp slowdown, during which the kinetic process was dominated by barrier crossing. In the case of $n_g = 50$ a kinetic barrier was reported even at the beginning of the adsorption process. The kinetics of adsorption was thus found to have a slower regime but it is dominated at all times by barrier crossing (Figure 7.6). The kinetics of the polymer adsorption of proteins perhaps originates due to the underlying physical processes associated with the change of the height of the barrier and the range of the potential of mean-force as the adsorption proceeds.

Biomolecular surface dynamics of agents rely on general aqueous physiological conditions, specific salt concentrations, certain physical properties, hydration, and energy profiles (Janc et al., 2021; Cesa et al., 2015). Protein structural stability and its surface adsorption follow nanoscale dimension fluctuations and time-dependent kinetic behavior. The biomolecular surface dynamics and dynamics of biomolecules on surfaces are both crucial for biomolecular functions.

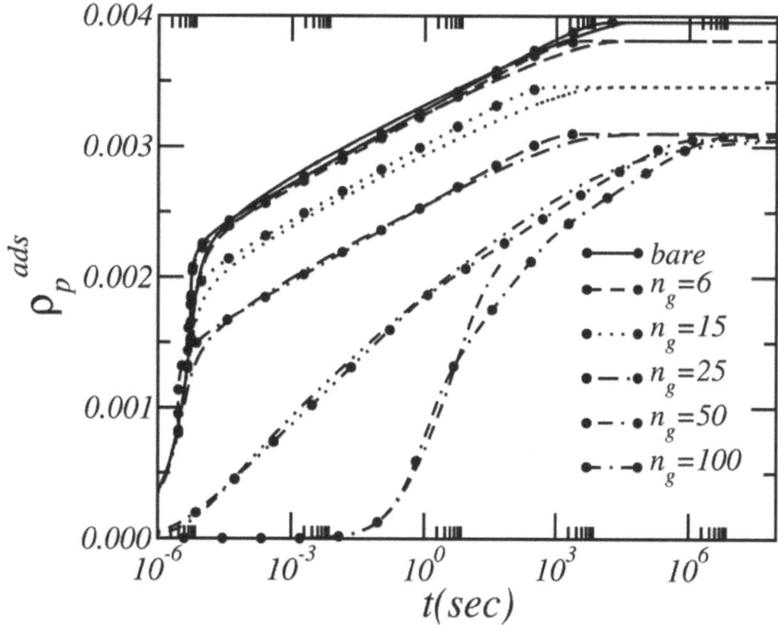

FIGURE 7.6 Variation of the amount of protein adsorbed as a function of time for surfaces with (and without) grafted polymers. The lines are the results from the generalized diffusion approach. The lines with symbols are the results from the kinetic theory approach. In all cases, σl2 = 0.01. The bulk protein volume fraction is φp, bulk = 0.001. The different chain lengths are denoted in the figure.

***Source*: reproduced with permission from Fang et al. (2005), where details of the associated kinetic theory will be found.**

7.3 FORCES AND MOTION OF THE CELL

Cell functions are associated with active physical forces (Roca-Cusachs et al., 2017). A century ago D'Arcy Thompson attempted to explain how the size and shape of living organisms could be determined by physical forces (Thompson, 1942). Cells can sense biochemical, electrical, and mechanical signals created by the environment in which cells exist, and these signals affect the differentiation and behavior of cells. Unlike specific roles of biochemical and electrical signals, mechanical signals may work differently. These physical signals can propagate without the diffusion of proteins or ions; instead, forces become transmitted through mechanically stiff structures, flowing, etc., for example, through cytoskeletal elements such as microtubules or filamentous actin (Yusko and Asbury, 2014).

In cell biology, force sensors may transduce forces into measurable physical quantities such as mechanical deformation or light. Any direct measurements of these physical quantities allow the force quantification with the known material properties of the force sensor. The techniques based on quantifying the extent to which cellular forces deform inert materials of known mechanical properties are presented in Figure 7.7, while the analogy between force sensing in the cell and human-made tools to probe physical forces is presented in Figure 7.8 (mechanotransduction).

All types of forces relevant to protein–protein interactions and enzyme catalysis in biology have summarized recently (Yusko and Asbury, 2014) and found to be on the pico-Newton (10^{-12} N) scale (see Table 7.1).

	Force range	Length scale	Measured quantity	*In vivo?*	Strengths	Limitations	References	Schematic
2D traction microscopy	$1–10^4$ Pa	$10^{-1}–10^3$ μm	Substrate displacement	No	Absolute measurement Tunability of substrate stiffness Output is a 2D map	Computationally involved High sensitivity to displacement noise	21–42	
3D traction microscopy	$10–10^4$ Pa	$10^{-1}–10^2$ μm	ECM displacement	No	Cells in 3D environment Output is a 3D map	Computationally very involved Unknown ECM material properties close to the cell Physiological ECM is non-linear	38,39	
Micropillars	$10^{-2}–10^2$ nN	$10^{-1}–1$ μm	Pillar displacement	No	Absolute measurement No reference image required Simple force calculation	Discrete rather than continuous adhesion Difficult to compare to physiological environments Small stiffness range	44–54	
Cantilevers	$10^{-2}–10^2$ nN	$10–10^3$ μm	Cantilever displacement	No	No reference image required Simple and precise force measurements in real time	Requires contact Low throughput	56–59	
Inserts	$10^{-1}–10^4$ Pa	$10–10^2$ μm	Insert deformation	Yes	*In vivo* Control of adhesion specificity Versatile	Requires microinjection No measurement of shear stress No measurement of isotropic stress	60–62	
Genetically encoded molecular sensors	1–10 pN	1–10 nm	Fluorescence signal	Yes	Measures forces per molecule Molecular specificity	No directional information Difficult calibration Low signal-to-noise ratio	64–70	
Synthesized molecular sensors	1–100 pN	1–10 nm	Fluorescence signal	No	Higher force range than genetically encoded molecular sensors Easier force calibration	Only available for extracellular ligands	72–78	
					Unknown material properties			
Monolayer stress microscopy	$1–10^3$ nN	$1–10^3$ μm	Unbalanced traction	No	Exact solution in 1D Straightforward if tractions are known	Model assumption in 2D	80–92	
Laser ablation	NA	$0.1–10^3$ μm	Wound deformation	Yes	Multiscale Relatively simple implementation High precision of perturbation	Relative measurements Invasive	93–109	
Force inference	NA	$10–10^3$ μm	Contour geometry	Yes	Non-invasive No probe required (geometry only) Largely independent of material properties	Relative measurements Sensitive to image segmentation noise	112–122	

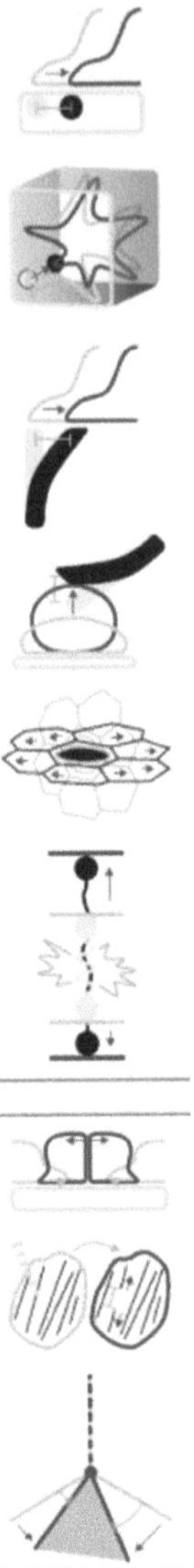

FIGURE 7.7 Techniques used for measuring cellular forces. The column "References" quotes the reference numbers (due to space constraints we do not list them here) in Roca-Cusachs et al. (2017).

***Source*: reproduced with permission from Roca-Cusachs et al. (2017).**

In Table 7.1 we see that tools for direct observation of individual protein interactions must operate with nanometer (nm) precision and on the pico-Newton (pN) force scale. Biophysical tools that are capable of dealing with this low resolution precision can be grouped into two categories, as follows: those that actively control position and therefore apply force (see Figure 7.9, A–C), and those that passively measure force (see Figure 7.9, D–F).

The cell-adhesion proteins' clustering occurs at focal adhesions and contains multiple mechanosensitive proteins that are involved in coupling transmembrane α/β-integrins to the actin network. The extracellular portion of α/β-integrins connects to fibronectin in the extracellular matrix

	Example molecule	Typical force	Length scale	Force-induced event	References	Schematic
Cytoskeleton	Actin	1 nN	1–10 μm	Cytoskeletal remodelling	46	
Molecular extension	Integrin αVβ3	10 pN	10 nm	Switch from bent to extended configuration	126	
Unfolding	Talin	1 pN	10^2 nm	Unfolding of molecular domains	128	
Domain reorientation	Filamin	Unknown	10 nm	Change in angle of dimer crosslinking, exposing integrin-binding sites	127	
Bond rupture	Integrin α5β1	10 pN	10 nm	Rupture of integrin–ECM bonds	129	
Opening of ion channels	Piezo1	10 pN	10 nm	Gating of ion channel	131	

FIGURE 7.8 The cell appearing as a force sensor. *2D traction microscopy*: a cell (pink) is laid on a hydrogel (orange) embedded with microbeads (gray). Traction forces (blue) exerted by an adherent cell (magenta) are computed from displacement (green) of the bead (black). *3D traction microscopy*: a cell (pink) is embedded on synthetic or native ECM containing microbeads (gray). Traction forces (blue) exerted by the cell (magenta) are computed from displacement (green) of the bead (black). *Micropillars*: a cell (pink) is laid on micropillars at rest (gray). Traction forces (blue) exerted by an adherent cell (magenta) are computed from the displacement (green) that they induce by bending the pillar (black). *Cantilevers*: a cell (pink) is laid on a plate (orange) underneath an AFM cantilever (gray). The deforming cell (magenta) exerts forces (blue) that can be computed through displacement (green) of the deformed cantilever (black). *Inserts*: an undeformed insert (gray) is introduced in a cell aggregate (pink). Cell forces (blue) exerted by migrating or stretched cells (magenta) are computed from deformation (green) of the insert (black). *Molecular sensors*: cell features (pink) are connected to a linker molecule (gray). Moving cell features (magenta) exert cell forces (blue) that can be computed through FRET (green) of the stretched linker molecule (black). *Monolayer stress microscopy*: interconnected cells (magenta) equilibrate cell–substrate tractions (green) through intercellular stresses (blue). *Laser ablation*: diverse features (gray) in cells (pink), specifically filaments in this illustration, may be severed through short and intense laser pulses (orange). Deformation (green) of the wounded feature (black) of these cells (magenta), displacement of a retracting fiber in this specific example, is used to compute tension (blue). *Force inference*: cell edges (pink) in equilibrium at triple junctions are displaced (magenta) by inter- and intracellular forces (blue). Angle variations (green) with respect to the equilibrium configuration are used to estimate cellular forces. *Cytoskeleton*: a cell (pink) responds to different forces (blue) transmitted to their substrate by reorganizing their cytoskeleton (gray/black). *Molecular extension/unfolding*: a molecule is in a folded/bent configuration (gray), but under force (blue) extends and exposes a binding site (black) to another molecule (green). *Domain reorientation*: a molecule (pink) changes conformation (magenta) under force (blue), altering the affinity (green) for a binding partner (black). *Bond rupture*: a bond (grey) between two molecules breaks (black) under force (blue). This bond can be intracellular, intercellular, or link cells (pink/magenta) to their surrounding ECM. *Opening of ion channels*: an ion channel on the cell membrane (pink) changes conformation under force (membrane tension, blue), altering its ability to transport ions (gray/black). The column "Reference" quotes the reference numbers (due to space constraints we do not list them here) in Roca-Cusachs et al. (2017).

***Source*: reproduced with permission from Roca-Cusachs et al. (2017).**

TABLE 7.1
Cellular Events where Forces Influence Biochemical Function

Event	Speed or lifetime	Relevant force (pN)	Note
Kinesin movement on a microtubule	800 nm/s	5–7.5	Maximum load on motor before it stalls
Dynein movement on a microtubule	85 nm/s	7–10	Maximum load on motor before it stalls
Myosin movement on an actin filament[a]	0.03 s at 6 pN	10 and 80	Rupture forces at ramp rates ~5 and 1000 pN/s
Activation of titin kinase by removal of inhibitory peptide	—	30	Equivalent to the activity of ~5 or 6 myosin units
VWF tethering platelets to endothelial cells[a]	0.2 s at 20 pN	5–80	Force required to reveal protease cleavage site
One kinetochore complex binding to one microtubule[a]	50 min at 5 pN	9	Rupture force at a ramp rate of 0.25 pN/s
FimH-mannose bond[a]	—	~150	Rupture force at a ramp rate of 250 pN/s
Single integrin in vitro[a]	10 s at 30 pN	13–50	Rupture force at ramp rates of 50–100 pN/s
Unfolding of talin to reveal vinculin binding sites		5	Force at a ramp rate of 5 pN/s
Single kinesin transporting a 30-nm quantum dot	570 ± 20 nm/s	0.6	Estimated drag force on quantum dot during transport
Chromosome segregation in anaphase	100 nm/s	0.1–10	Estimated force to move chromosome in vivo during anaphase
Force to stop chromosome movement during anaphase	—	700; 50	Force per chromosome; force per kinetochore microtubule
Single integrin in cells to RGD on surface	—	1–5	FRET sensor in ECM
Single vinculin connecting talin to F-actin in cells	Minutes	2.5 ± 1.0 up to 10	FRET sensor in cells
Activation of Notch during cell adhesion	5–15 min	<12	Based on tension-gauge-tether sensor
Contractile forces through focal adhesion complexes	Minutes to hours	100–165	Estimate for a complex of 3–5 integrins

[a] Event involving protein with catch-bond behavior.

Source: Yusko and Asbury (2014).

(ECM) and exhibits classic catch-bond behavior, likely through an allosteric pathway (Kong et al., 2009). Contractile forces generated in the actomyosin network are transmitted through talin to the integrins and the ECM. Talin contains a C-terminal rod-like structure consisting of 13 α-helical bundles. Under ~5 pN of tension, several α-helical bundles unfold, revealing binding sites for vinculin (see Figure 7.10) (del Rio et al., 2009). A recent report suggests that the mechanoregulation occurring at integrins in the cell membrane may route forces through the cytoplasmic cytoskeleton directly to the nuclear cortex, ultimately to the transcription factors in the nucleus (see Figure 7.10) (Swift et al., 2013). All these observations suggest that the mechanisms for mechanical signals' transmitting, routing, and sensing in mechanobiology are associated with creating physical signaling active behind vital chemical processes taking place in cells.

Besides versatile intracellular force-based signaling, as explained here, vital forces are also active that participate in biochemical association among cells. These inter-cellular active forces contribute considerably to underlying molecular mechanisms active behind cellular mortality (Zhang et al.,

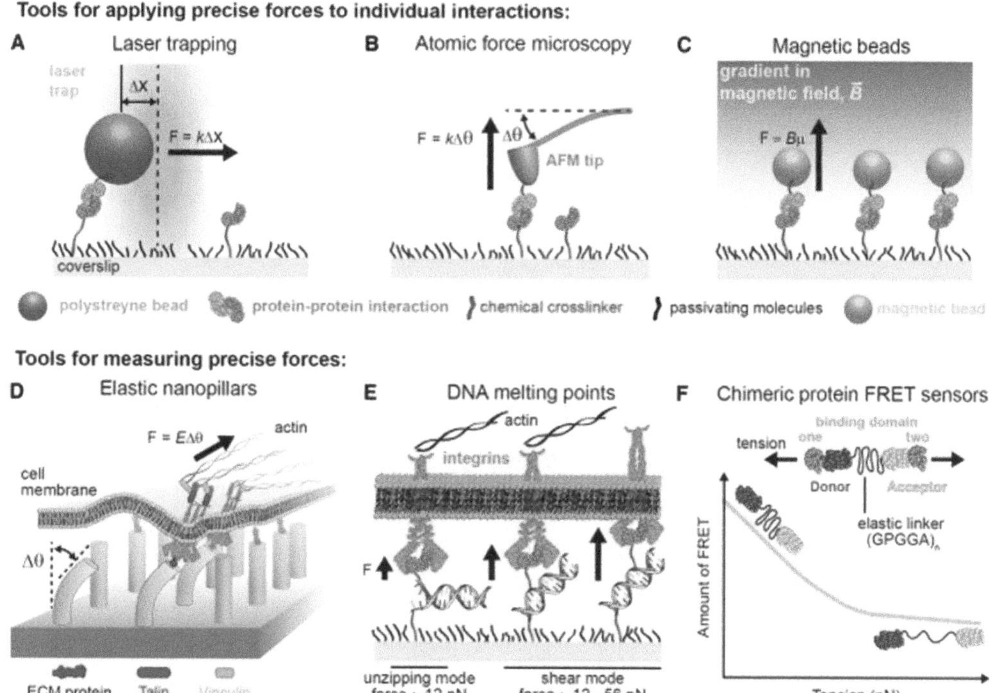

FIGURE 7.9 Methods for applying and measuring precise forces to single molecules and molecular complexes. (A–C) Instruments often used to apply precise forces to individual macromolecules or complexes. (A) In laser trapping, a focused laser beam behaves roughly like a Hookean spring, pulling a submicrometer bead toward its center with a force proportional to the stiffness of the laser trap, k, multiplied with the displacement of the bead from the trap center, Δx; beads are often decorated with a protein or receptor of interest and can be controlled by manipulating the position of the laser beam relative to the microscope slide. (B) Atomic force microscopes employ a micrometer-width cantilever, at the tip of which is a nanometer-sized pointer that can be decorated with proteins or receptors; once these proteins bind their receptors on the surface of a glass slide, the cantilever is retracted, causing it to deflect. (C) Magnetic tweezers employ magnetic beads with a magnetic moment, μ; when subjected to a magnetic field, the force on the beads is proportional to the magnetic field strength multiplied by μ. Up to several hundred magnetic beads can be pulled at the same time. (D–F) Techniques for measuring forces precisely between and within molecules. (D) Pillars with diameters and lengths on the nanometer to submicrometer scale can be formed from elastic polymers and decorated with extracellular matrix proteins, such that cultured cells adhere and form focal adhesions; the deflection of each nanopillar from its resting position reveals the contractile forces exerted at the corresponding focal adhesion. (E) Hybridized dsDNA molecules for which one strand is tethered to a surface and the complementary strand is tethered to a protein or receptor can act as a "tension-gauge-tether" by which the number of base pairs within the dsDNA that support the load dictates a well-defined force at which the dsDNA will unzip or melt; unzipping of the dsDNA can be observed using fluorescent tags on the DNA molecules or by cell phenotypes, allowing estimation of the range of forces to which a protein–ligand interaction might be subjected during a cellular event such as early stages of cell adhesion. (F) An intramolecular strain sensor based on FRET can be used to determine the forces exerted through a protein by engineering the probe into the protein structure and monitoring the level of FRET.

***Source*: reproduced with permission from Yusko and Asbury (2014).**

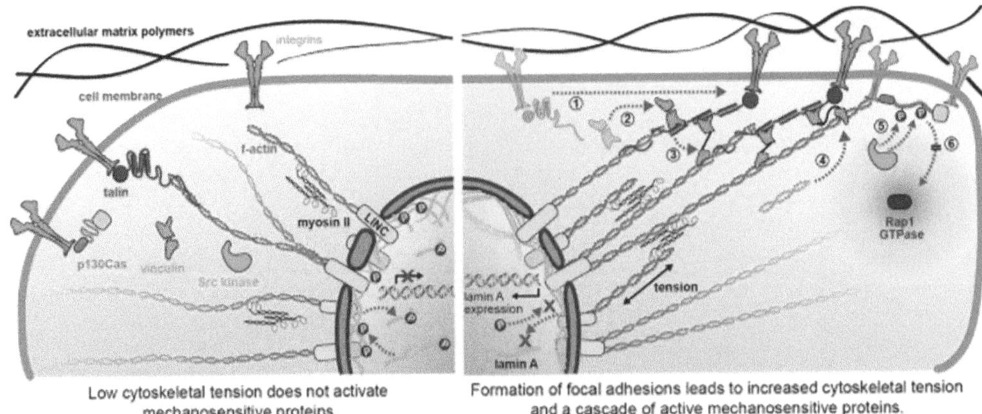

FIGURE 7.10 Forces in cells are routed, transmitted, and transduced by mechanically sensitive proteins and have cell-wide implications, influencing biochemical signaling in the cytoplasm and gene expression in the nucleus. Forces generated in the actin cytoskeleton by myosin II cross-bridges are transmitted several micrometers between adhesion proteins in cell membrane and LINC complexes in the nuclear cortex. Tension-dependent unfolding of talin (step 1) reveals substrates for vinculin binding (step 2), which in turn recruits additional actin filaments (steps 3 and 4) as part of focal adhesion development. Tension-dependent unfolding of p130Cas reveals phosphorylation sites for Src kinase as part of integrin signaling and ultimately generates the active form of a diffusible GTPase Rap1(steps 5 and 6). Concomitantly, tension in the actin cytoskeleton is transmitted through LINC complexes to the nuclear cortex. Lamin A, an intermediate filament of the nuclear cortex, mechanically couples the nuclear cortex to LINC complexes and therefore the cytoplasmic cytoskeleton; it affects DNA transcription of the gene for lamin A and the transcription of stress fiber genes (Swift et al., 2013). Increased cytoskeletal tension on LINC complexes correlates with decreasing phosphorylation of lamin A, decreasing turnover of lamin A in the nuclear cortex, increasing stiffness of the nuclear cortex, and ultimately, through the retnonic acid pathway, increasing levels of lamin A. Note that many intermediate proteins are not shown, for simplicity.

2020). The confluent cell sheets appear with a collective behavior which is strongly influenced by polar forces, arising through cytoskeletal propulsion, and active inter-cellular forces, being mediated by interactions across cell–cell junctions. The distribution of forces within cell monolayers was experimentally measured using traction force and monolayer stress microscopy (MSM) (Tambe et al., 2011; Kim et al., 2013; Tambe et al., 2013). MSM helped measure the local stress within a monolayer (see Figure 7.11). For details on techniques, see Tambe et al. (2011). Here, using inverted optical microscopes, measurements are made on cell-generated displacements of fluorescent markers that are embedded near the surface of a collagen-coated polyacrylamide gel substrate on which the cells are adherent. A map is constructed representing the traction forces (T) that are exerted by the monolayer upon the gel (Trepat et al., 2009). These traction forces at the interface between the cell and its substrate are used to obtain the distribution of the mechanical line forces everywhere within the cell sheet (see Figure 7.11A); the line forces (force/length) are then converted to the intercellular stresses (force/area) (Figure 7.11B). At each point within the sheet the coordinate system (see Figure 7.11C) may be rotated in the plane of the cells which eventually help define two principal stress components (σ_{max} and σ_{min}) and the two corresponding, mutually perpendicular, principal orientations (see Figure 7.11D).

The image representing the average local normal stress, $\bar{\sigma} = (\sigma_{max} + \sigma_{min})/2$, of an advancing monolayer of rat pulmonary microvascular endothelial (RPME) cells shows remarkable behavior (see Figure 7.12A). The underlying distribution of the local normal stress is severely heterogeneous – normal stresses are mostly positive (tensile) with values exceeding 300 Pa in regions spanning tens

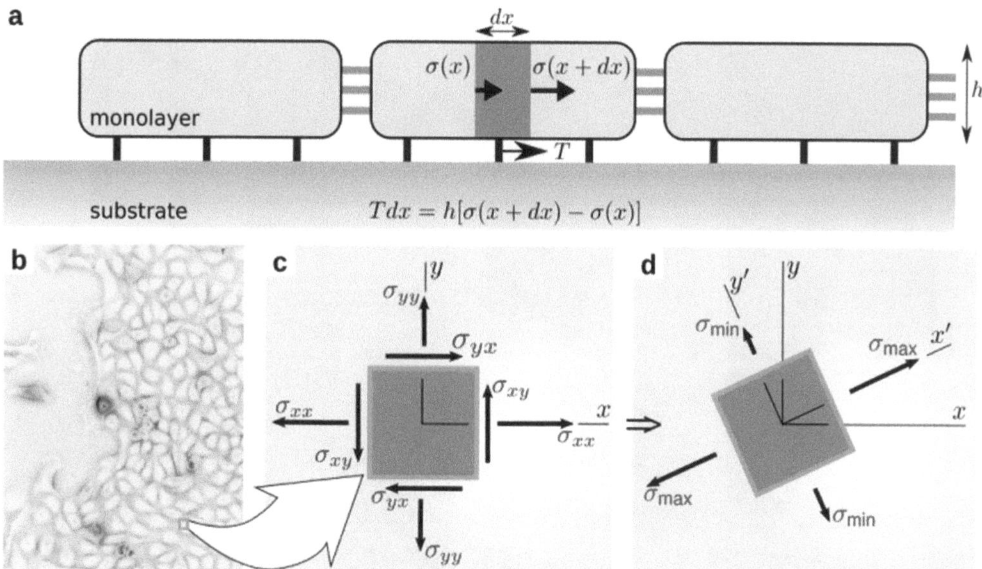

FIGURE 7.11 (a) Simplified representation of the physical relationship between cell-substrate tractions, T, which have been reported previously (Trepat et al., 2009), and intercellular stresses, σ, which are reported for the first time here. Intercellular stresses arise from the accumulation of unbalanced cell-substrate tractions. At any point within the monolayer (b), the intercellular stresses, defined in laboratory frame (x, y), (c), have shear (σ_{xy}, and σ_{yx}) and normal (σ_{xx}, and σ_{yy}) components. This frame can be rotated locally to obtain the principal frame (x', y'), (d), where shear stresses vanish and the resulting normal stresses are called principal stresses (σ_{max} and σ_{min}). The corresponding axes are called maximum, aligned with x', and minimum, aligned with y', principal orientations.

***Source*: reproduced with permission from Tambe et al. (2011).**

of cells. These regions of predominantly tensile stresses alternate with regions of weakly negative (compressive) stresses (Figure 7.12B). These fluctuations span over multiple cell widths and define a stress landscape that is rugged (Figure 7.12B, K). Like the normal stress, the shear stress (DePaola et al., 1992) at a point in a material varies with orientation and attains the maximal value, $\mu = (\sigma_{max} - \sigma_{min})/2$, at 45° from the principal orientations (Figure 7.12D). The local maximal shear stress, which appeared systematically smaller than the local normal stress, was also characterized by a rugged landscape (Figure 7.12C). The dependence of local stresses upon orientation signifies stress anisotropy (see Figure 7.12E).

Since the phase-contrast images and the stress maps are mutually independent measurements, the coincidence between orientation of the cell body versus orientation of the maximal principal stress is striking (see Figure 7.12E). The cell–cell junction, and the cell body, are found to support high normal stresses, accounting for being overwhelmingly tensile, but minimal shear stresses. The actin structures spanning the RPME cells might align with maximal principal orientations (see Figure 7.12E). Cells have been found to not only align with the maximal principal orientation, but also migrate along that orientation. Regions of higher stress anisotropy were predicted to exhibit stronger alignment between the direction of local maximal principal stress and that of local cellular migration velocity. The validity of this hypothesis was proven by measuring the alignment angle ϕ between the orientation of the local maximal principal stress and the orientation of the local cellular migration velocity vector (Figure 7.12F, inset). The greater the local shear stress, the narrower was the distribution of ϕ (Figure 7.12F–H). The cumulative probability distribution function, $\bar{P}(\phi)$, was then constructed, which reasoned that if there were perfect alignment between the orientation of

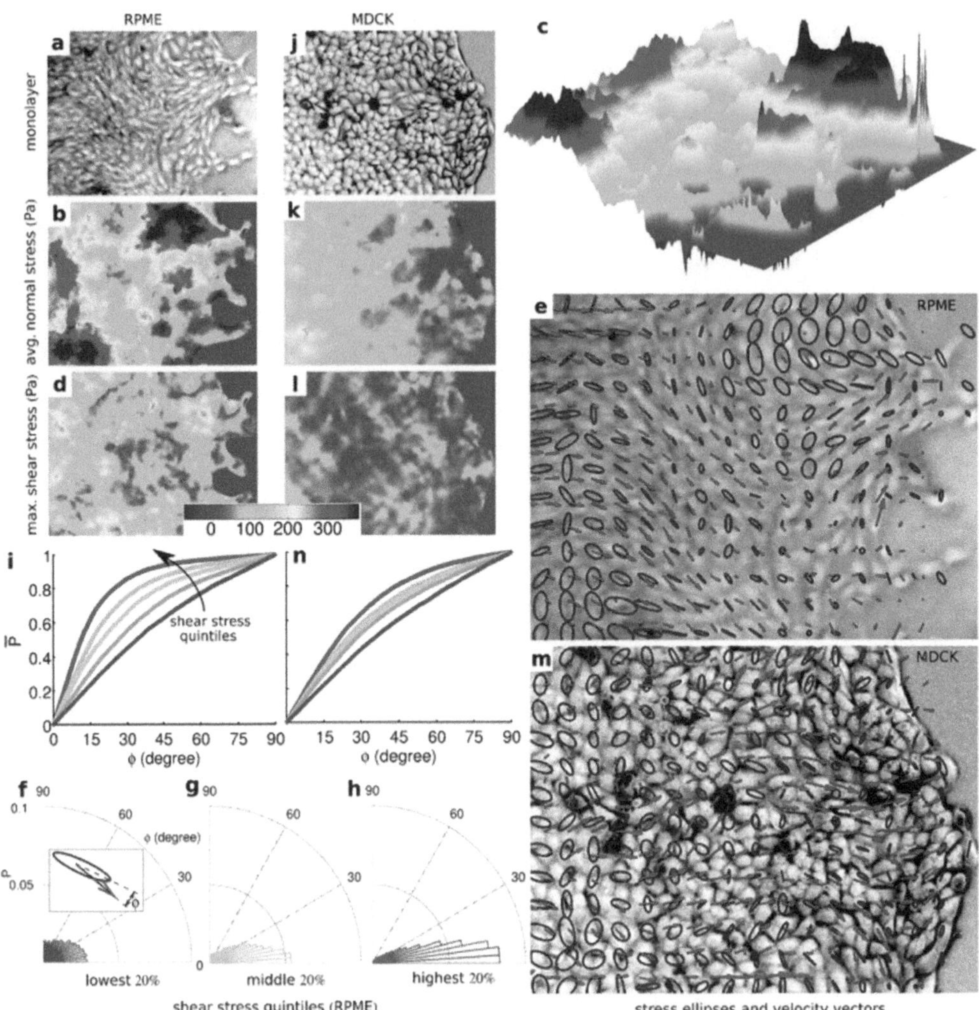

FIGURE 7.12 Intercellular stress maps and mechanical guidance of collectively migrating monolayers. (a) Transmitted light image of rat pulmonary microvascular endothelial (RPME) cell monolayer. Corresponding to this image are the maps of average normal stress (b), which is predominately tensile but forms a rugged stress landscape (c), the maximum shear stress (d), principal stress ellipses (blue) and cell velocity vectors (red) (e). The alignment angle, ϕ, between major axis of the principal stress ellipse and direction of the cellular motion (f, inset) shows that the greater the local shear stress the narrower is the distribution of ϕ (f, g, h). The cumulative probability distribution $\overline{P}(\phi)$ varied strongly and systematically with stress anisotropy (i); curves from blue, to red are in the order of higher quintiles. Comparable maps are found for the Madin-Darby canine kidney (MDCK) cell monolayer (j–n). Note that the average tensile stress (k) increased systematically with increasing distance from the advancing front, thus contributing to the state of global tug-of-war (Trepat et al., 2009). Vertical size of the images of cell monolayer: RPME – 545 μm, MDCK – 410 μm. Each curve in (i) and (n) and distributions in (f), (g), and (h) have more than 8,000 observations. For color contrast see the online version of the book.

***Source*: reproduced with permission from Tambe et al. (2011).**

local cellular migration velocity and that of local maximal principal stress, then all angles ϕ would be 0° and the cumulative probability distribution would be a step function from probability 0 to probability 1 occurring at 0°. The stress anisotropy was found to be stronger as the overall degree of alignment became greater.

The generality of this finding was assessed by examining monolayers comprising Madin-Darby canine kidney (MDCK) epithelial cells (see Figure 7.12J), which were rounded in the plane, not spindle-shaped as are RPME cells. Despite these differences in cell type and cell morphology, the stresses were found to be dramatically heterogeneous (see Figure 7.12K, L) and the local orientation of cellular migration was also found to follow the local orientation of maximal principal stress (see Figure 7.12M, N). The generality of this finding was further assessed examining the behavior of monolayers of well-established breast-cancer model systems MCF10A cells. The details of the result are not presented here but will be found in Tambe et al. (2011).

The importance of cell–cell adhesion was assessed by weakening cell–cell contacts of MCF10A vector cells by calcium chelation (see Figure 7.13G, I). As expected, alignment between orientations of local stress and orientation of local cellular motions was lessened (Figure 7.13, magenta), which was restored upon returning to normal growth medium (Figure 7.13I, S, blue). The reversibility was blocked in the presence of E-cadherin antibodies (Figure 7.13R, S, red). The transmission of mechanical stresses from cell-to-cell across many cells was thus found necessary for plithotaxis, i.e., for each individual cell to follow the local orientation of the maximal principal stress.

Following the findings explained above, an article appeared recently that used a phase-field model to explore the interplay between the polar forces of cytoskeletal propulsion and the active inter-cellular forces being mediated by cell–cell junction interactions (Zhang et al., 2020). The phase-field approach used here allows for verification of the following single-cell properties:

1. cell deformability or cell polarity
2. inter-cellular interactions

Focus has been applied to the following two active forces: the polar force on each cell and the active inter-cellular forces. Varying them independently allows us to understand the comparable conditions that lead to three different states: jamming, liquid, and flocking. Hence, this model of a cell sheet helps unify a wide range of collective cell dynamics based on different forcing in a single coherent description. Thus the mechanisms behind the collective motion of cells may become clearer here. Considerable theoretical understanding has been recently achieved using two-dimensional simplified model systems of confluent cell layers (Ladoux and Mège, 2017; Alert and Trepat, 2020).

The role of active forces in collective cell dynamics can be addressed by analyzing inter-cellular active stresses, which may be defined based on the cell deformations, and the active polar forces, proportional to the polarization of each cell (Zhang et al., 2020). The strength of the former is controlled by a parameter ζ, while another parameter α controls the latter. Here ζ measures the strength of the forcing associated with the stress field due to all the cells and taken $\zeta>0$ here in the simulation. The simulations were started with randomly oriented velocities of magnitude α.

The simulation was performed on the dynamics of 233 cells of radius $R = 4$ in a periodic domain of size 108 × 108 in simulation units. For details on theories and other steps adopted in simulations, readers are encouraged to read the original article (Zhang et al., 2020). The unjamming trajectories of cells and the roles of these two parameters ζ and α in the simulation have been presented in Figure 7.14. It is clear from the rearrangement rates plotted in Figure 7.14 that very similar behavior is seen for all choices of the polarity alignment (see values of the associated parameters presented in the index of the figure).

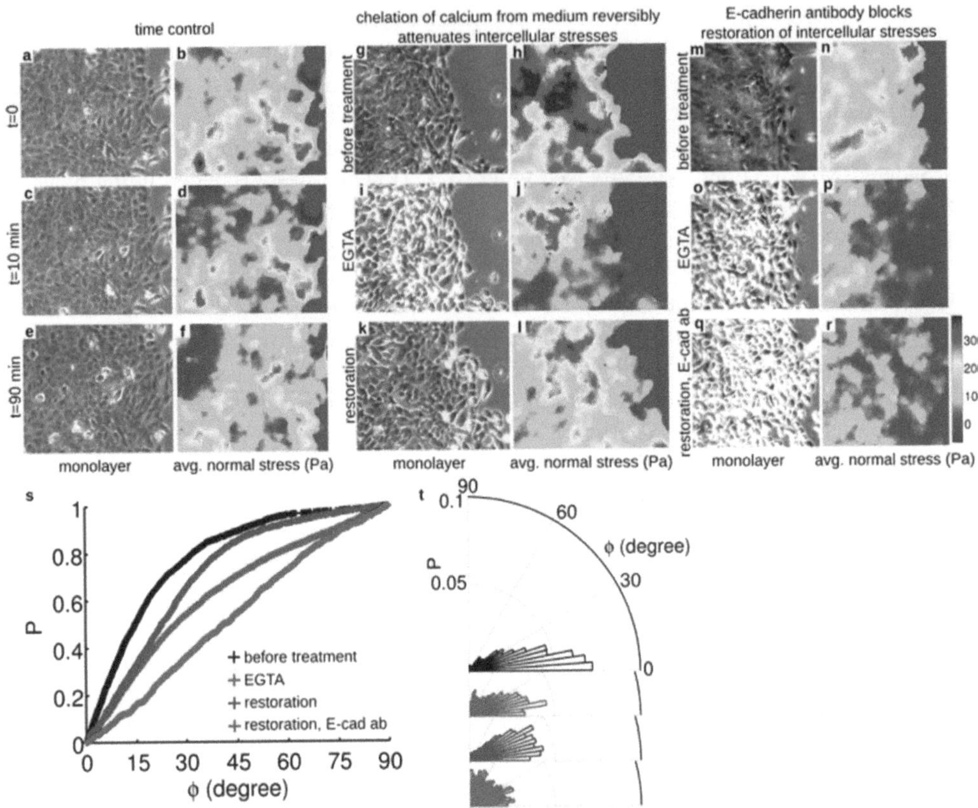

FIGURE 7.13 Local cell guidance requires force transmission from cell-to-cell. Time-controls of intercellular stress maps of MCF10A-vector cell monolayers (a–f). The stress patterns do not change appreciably over a period of 80 minutes. After 10 minutes in presence of the calcium chelator EGTA (4mM), however, cells lose contacts with their neighbors (g, i and m, o). These changes lead to attenuation of intercellular average normal stress (h, j and n, p). After returning to normal growth medium for 80 minutes, the stresses and the cell–cell contacts are largely restored (k, l), but if the growth medium is supplemented with E-cadherin antibody (7 μg/ml), recovery of stresses and cell–cell contact are blocked (q, r). EGTA treatment widens the distribution of angle (ϕ) between local cellular velocity and local maximum principal orientation corresponding to highest of the maximum shear stress quintiles (s, t). The distribution of ϕ is narrowed if calcium is restored (s and t, blue), but widened further if the restoration medium is supplemented with E-cadherin antibody (s and t, red). Together, these data show that local cell guidance along the orientation of maximal principal stress (plithotaxis) requires force transmission across cell–cell junctions. These preferred orientations correspond to those engendering minimal intercellular shear stresses. Increased intensity at cell boundaries in phase contrast images (panels i, o, and q) reveals disruption of cell–cell junctions. Vertical size of the images of monlayer: 410 μm. Each data set in (s and t) has more than 1,500 observations. For color contrast see the online version of the book.

***Source*: reproduced with permission from Tambe et al. (2011).**

The liquid state properties of the cell cluster were then investigated in the simulation (Zhang et al., 2020). Figure 7.15A shows a snapshot of the velocity field being characterized by localized bursts of higher velocities. Figure 7.15B shows the vorticity field with vortices on the scale of a few cells. The other panels in the figure display this phenomenon more quantitatively by showing the vorticity–vorticity and velocity–velocity correlation functions for layers driven by inter-cellular driving (Figure 7.15C, D) or polar forcing (Figure 7.15E, F). Correlations are seen to persist over

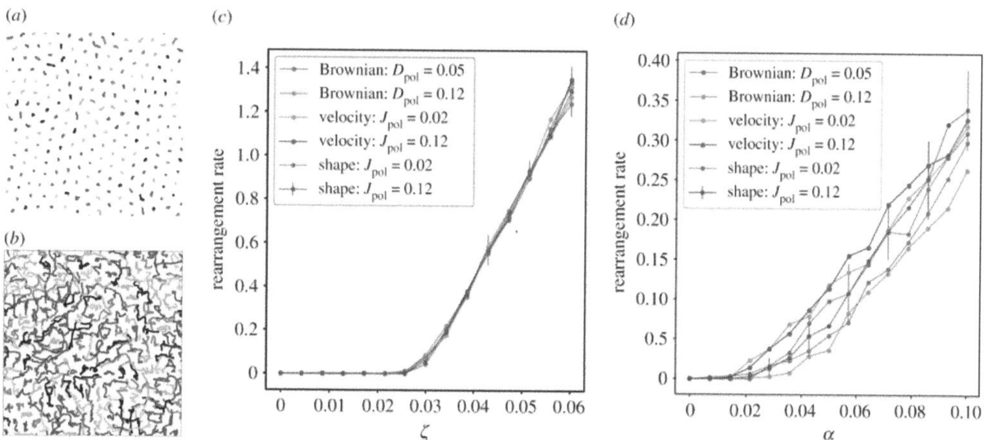

FIGURE 7.14 *Unjamming:* trajectories of cells in (*a*) the jammed state with $\omega = 0.005$, $\alpha = 0.02$, $\zeta = 0.00$ and (*b*) the liquid state with $\omega = 0.005$, $\alpha = 0.00$, $\zeta = 0.04$. Trajectories are followed for 2100 time steps and in both (*a*) and (*b*) the polarization of a cell aligns to the long axis. Rearrangement rate as a function of the strength of (*c*) the inter-cellular force ζ for $\alpha = 0.01$, (*d*) the polar force α for $\zeta = 0.06$, comparing different definitions of the polarization alignment. $\omega = 0.01$ in both (*c*) and (*d*). ω is associated with energy scale in the simulation.

***Source*: reproduced with permission from Zhang et al. (2020).**

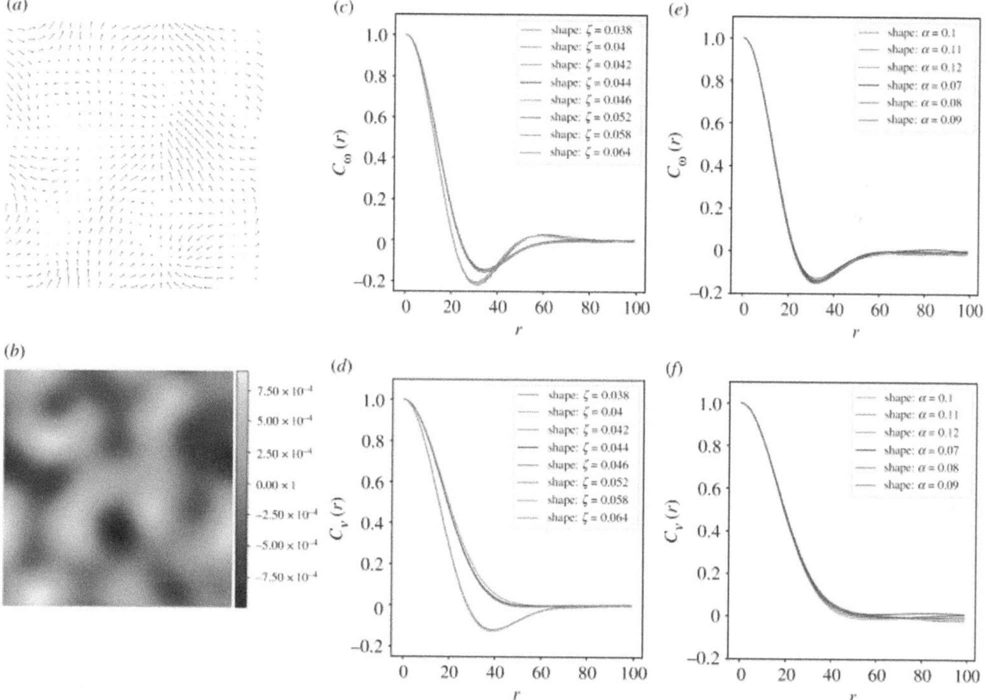

FIGURE 7.15 *Liquid state:* snapshot showing a typical (*a*) velocity field, (*b*) vorticity field in the liquid state. The color bar in (*b*) indicates the magnitude of the vorticity field. (*c*) Vorticity–vorticity and (*d*) velocity–velocity correlation functions for $\alpha = 0$ and different values of the intercellular driving ζ. (*e*) Vorticity–vorticity and (*f*) velocity–velocity correlation functions for $\zeta = 0$ and different values of the polar driving α. $\omega = 0.005$.

***Source*: reproduced with permission from Zhang et al. (2020).**

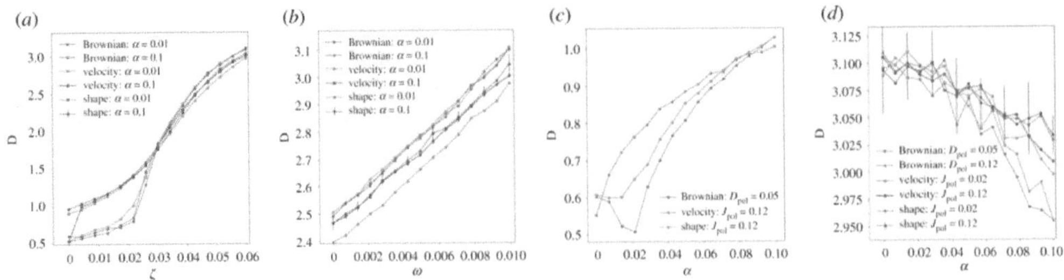

FIGURE 7.16 *Cell shape:* average cell deformation D as a function of (a) the active inter-cellular force ζ for $\omega = 0.01$; (b) the passive adhesion force ω for $\zeta = 0.06$, and the polar force α for (c) small active, inter-cellular forces $\zeta = 0.005$, (d) large active, inter-cellular forces $\zeta = 0.06$. $J_{pol} = 0.12$ and $D_{pol} = 0.12$ in (a) and (b). $\omega = 0.01$ in (c) and (d).

***Source*: reproduced with permission from Zhang et al. (2020).**

approximately four cell diameters and are largely independent of the details or strength of the active forces, suggesting that the range of the correlations is set by the passive forces. Polar forces are found to result in larger velocities and there is a tendency of inter-cellular active forces to elongate the cells. Therefore, the dependence of cell deformations was measured on the active driving (see Figure 7.16).

In the absence of active forcing, the ground state of the cells is a honeycomb lattice with the cells taking a hexagonal lattice shape. Here investigations were performed to understand how they are stretched by the activity by measuring the average cell deformation (D) (Figure 7.16), for details on the definition of D and associated xx and xy components of the deformation tensor, see Zhang et al. (2020). $D = 0.5$ corresponds to isotropic cells, while $D = 3$ to cells with an aspect ratio ~ 1.75.

The cell monolayer force distribution has been measured experimentally using traction force microscopy and monolayer stress microscopy (Tambe et al., 2011; Kim et al., 2013; Park et al., 2015). But there is no clear understanding of how the forces should be divided into passive or active, or into polar and inter-cellular contributions. The inter-cellular stresses may be regulated by mechanosensors (Saw et al., 2017). Comparing the force density in different situations experimentally and in models (Zhang et al., 2020) is expected to provide a framework helping to interpret current and future measurements of dynamical parameters characterizing the forces and motion in cell distribution/migration (Trepat et al., 2009), hence helping us understand both cell movement and tissue mechanics.

7.4 FORCES AND MOTION OF THE TISSUE

The distribution of forces associated with the development of tissues participates in driving cell behavior to shape organs. In growing tissues, as cells grow and divide, they exert pressure on their neighbors. Mechanical stress due to the internal distribution of forces within cells and tissues makes cells move, change shape, exchange neighbors, etc. Specific mechanisms active behind signaling pathways that pattern the embryo impact on the distribution of mechanical stress, and determine cell behavior, have been discussed in various studies (e.g. Heisenberg and Bellaïche, 2013; Rauzi et al., 2008; Zallen and Wieschaus, 2004). Cell intercalation (Walck-Shannon and Hardin, 2014) is reported to drive the extension of the neural tube in vertebrates (Nishimura et al., 2012). All this background information suggests that tissue structures consist of forces leading to specific slow movement that originate mainly due to physical properties active behind cell organization, inter-cell association, etc.

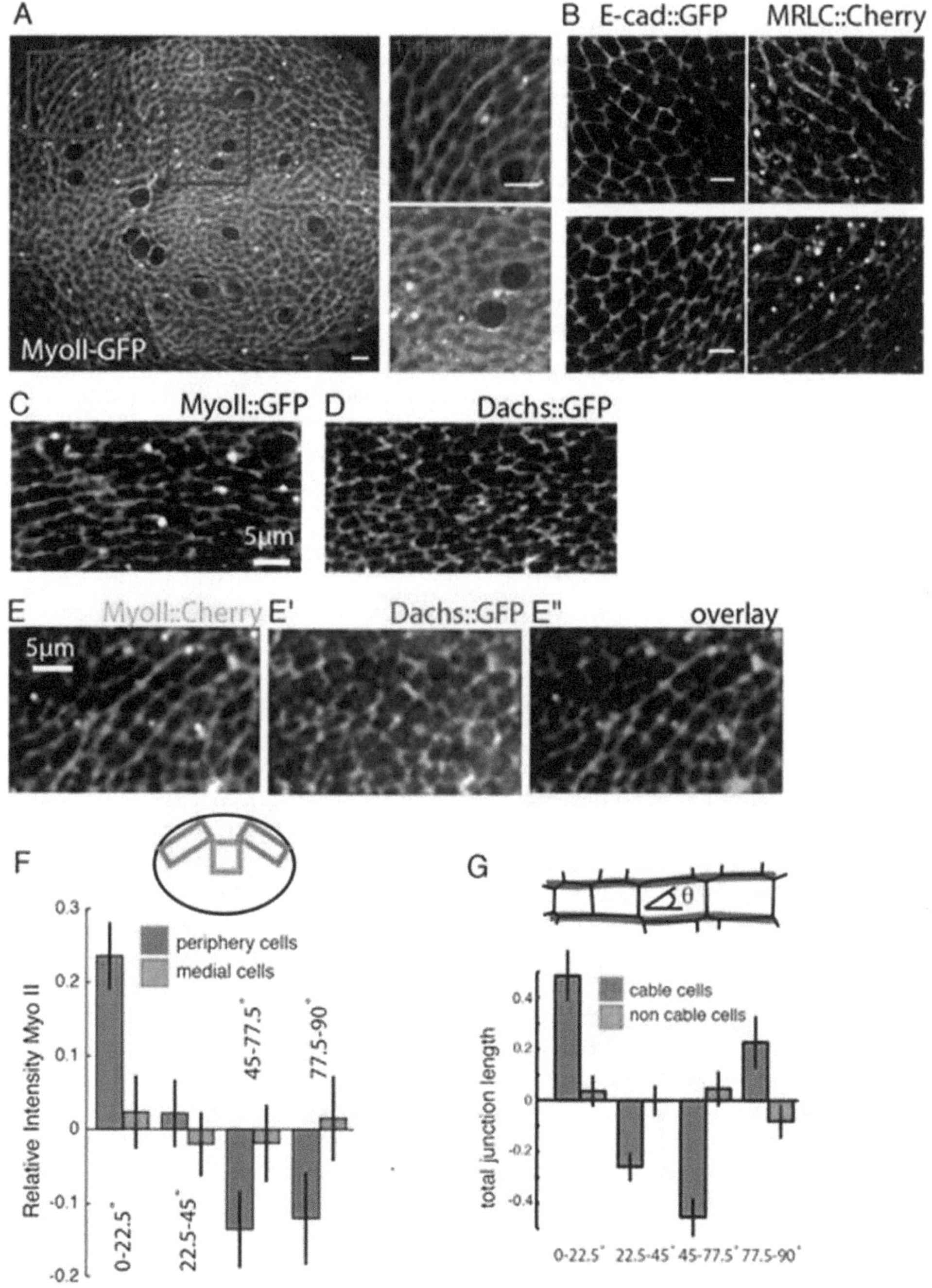

FIGURE 7.17 MyoII enrichment in stretched regions. (A) MyoII-GFP in the pouch. Insets: Polarized enrichment in periphery but not in medial regions. (B) E-cad-GFP and MyoII-Cherry co-visualization shows absence of E-cad polarity in cables (arrowheads). (C, D) MyoII polarity contrasts with Dachs absence of polarity even in the periphery (D, same as Fig. 3D). (E–E') Co-visualization of Dachs::GFP and MyoII::Cherry confirms MyoII polarity in the absence of Dachs polarity. (F) Angular distribution of junctional MyoII shows a polarity in the periphery (magenta) but not in the medial region (green). Angles are measured with respect to the average tissue orientation. Relative intensity is normalized by the mean $I^r_i = (I_i - I_{mean})/I_{mean}$. (G) Summed length of junctions as a function of the angle made with the axis of the cables. Junctions perpendicular or

FIGURE 7.17 (*Continued*)

tangent to the cables dominate the distribution, thus giving cells rectangular shapes. Summed length of junctions is expressed in relative normalized length $L^r_i = (L_i - L_{mean})/L_{mean}$. For control (green), strings of cells were arbitrarily handpicked in the medial region. Error bars represent s.e.m. Scale bars: 5 µm. See LeGoff et al. (2013) for details.

***Source*: reproduced with permission from LeGoff et al. (2013).**

Legoff and colleagues (2013) recently addressed the distribution of forces in developing tissues in the growing wing imaginal disc of Drosophila larvae, the precursor of the adult wing. As an example demonstration, we may brief some of their important findings here. The interplay between tissue mechanics and morphogenesis was investigated at the tissue level in the precursor of the Drosophila wing: the wing disc. A global pattern of mechanical stress that impacts tissue morphogenesis was found. The mechanics of the developing precursor of the wing disc was first addressed, then its impact on tissue morphogenesis, and finally the clonal perturbation of growth rates was found to affect the mechanics of the tissue non-autonomously.

The precursor of the Drosophila wing was found to experience mechanical stress far beyond the compartment boundaries in the course of its development. The peripheral region of the wing pouch was found to be stretched tangentially. The growing wing disc context raises a new question: how do the constant cell rearrangements provided by cell divisions affect the distribution of stress in the tissue? Although the theoretical hypothesis suggested that divisions dissipate mechanical stress, maintaining the tissue in a stress-free, liquid-like state (Ranft et al., 2010), the observations of Legoff and colleagues (2013) show that this is not the case in the wing pouch. Although stress dissipation by divisions is likely to be at play in the pouch, notably in the form of cell divisions oriented along the axis of stretch, it is not sufficient to completely dissipate the stress. The apparent absence of exchange of neighbors in this tissue during the development of larvae (Gibson et al., 2006) is also expected to contribute to a low dissipation of stress. Preventing the tissue from dissipating mechanical stress might be a common way to generate the shape. The existence of stress in the wing pouch might play a role in polarizing mechanics for subsequent changes, especially in shapes.

At a local scale, the tissue's cells have been found to respond to the stretch by polarizing their cytoskeleton (see Figures 7.17 and 7.18), and orienting their divisions (see Figure 7.19). The fact that cells respond to stretch by polarizing their cytoskeleton has the following two implications on the cellular lattice of the tissue: Firstly, it serves as a homeostatic mechanism: by stiffening or contracting their cortex along the axis of stretch, cells reduce the deformation they undergo (Fig. 7.18 D,E). Secondly, it leads to the emergence of higher-order structures in which cells assemble linearly along MyoII cables, which might work to limit cell mixing in the bulk of the tissue or participate in orienting cell divisions there. Overall, the presence of mechanical stress, and the active response mediated by MyoII polarity, gives a representation of the cellular lattice that contrasts with the classic view inherited from Lewis, where the shape of cells is considered to be hexagonal (Gibson et al., 2006; Lewis, 1926): in the periphery of the pouch, the shape of cells is driven by stress fields and not by topology.

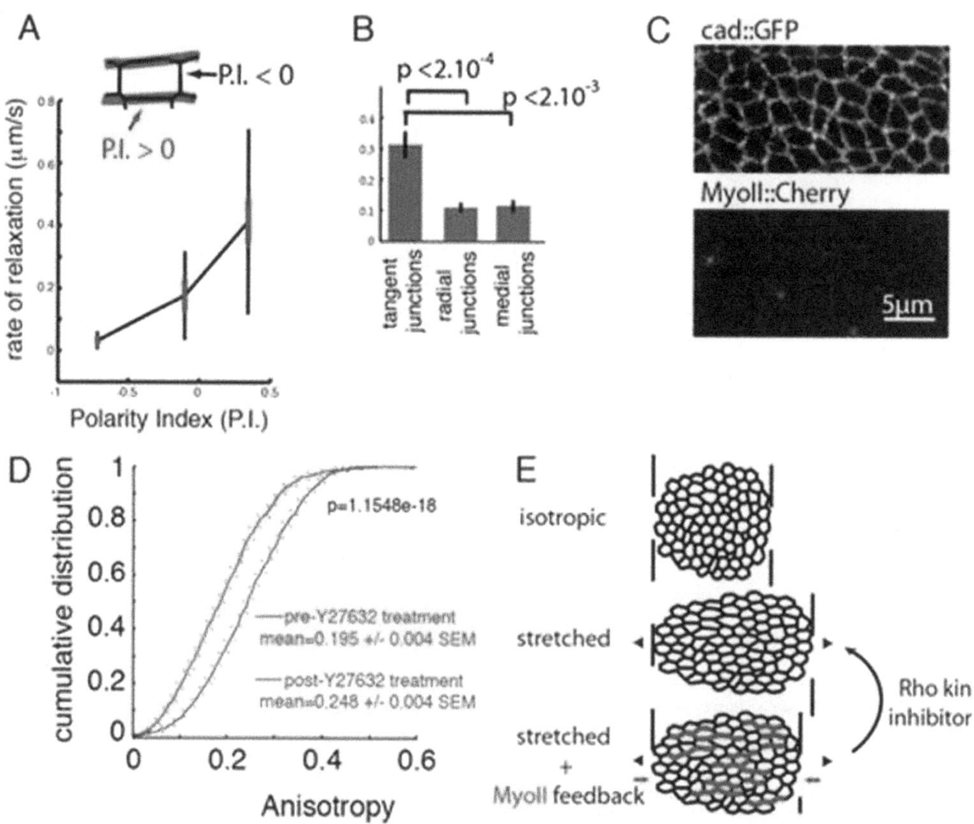

FIGURE 7.18 Linking MyoII enrichment and mechanical stress. (A) Junctional recoil velocity after ablation versus polarity index shows that MyoII enrichment correlates with cortical tension. Black error bars represent s.d.; red error bars represent s.e.m.; *P*-values are from KS test. (B) Recoil velocity after Y27632 treatment. The polarity in the periphery is still observed as indicated by the different recoil for radial and tangent junction. Medial junctions are also still significantly less tensed than tangent peripheral ones. Error bars represent s.e.m.; *P*-values are from KS-test. (C) Region of interest in a disc treated with Y27632 at the level of the adherence plane: MyoII has been removed from junctions. (D) Tissue anisotropy in peripheral regions before and after Y27632 treatment. The cumulative distribution shows a small but significant increase in anisotropy after treatment. (E) Model for MyoII polarity: the stretch at the periphery polarizes MyoII (red lines), resulting in a small reduction of anisotropy (red arrows). The tissue goes back to its basal mechanical stretch after MyoII inhibition.

***Source*: reproduced with permission from LeGoff et al. (2013).**

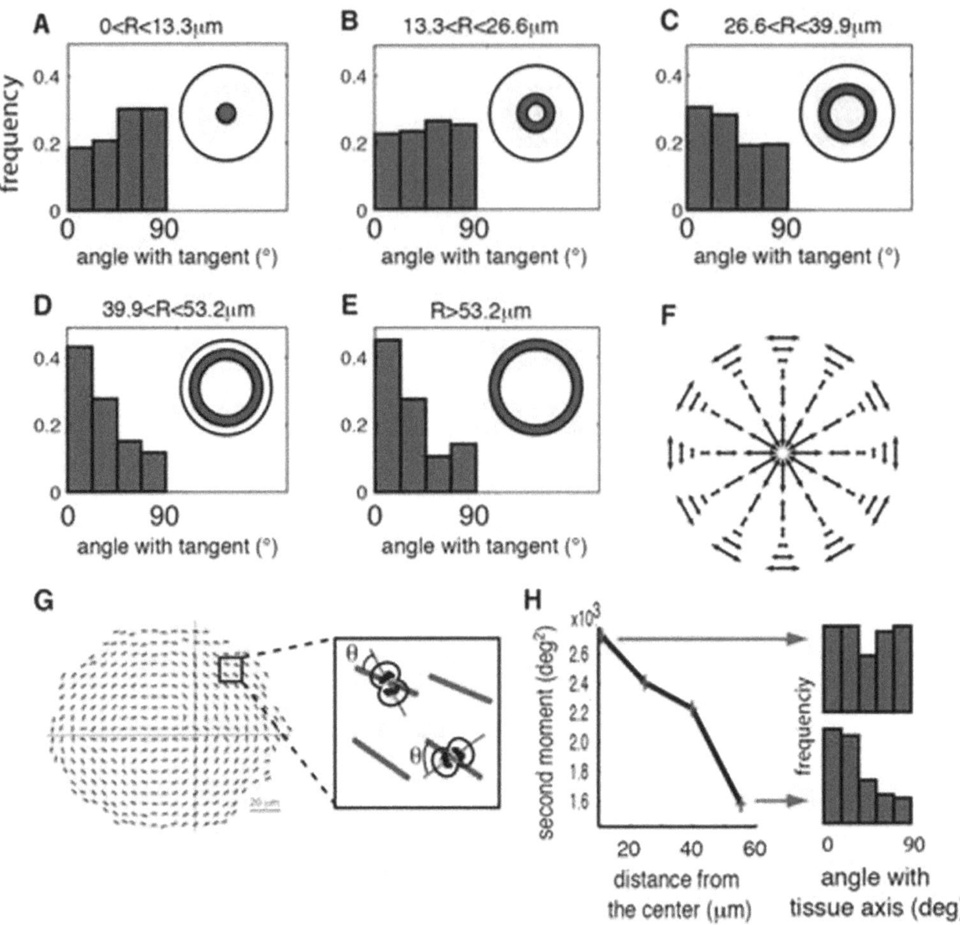

FIGURE 7.19 Orientation of cell divisions in the wing pouch. (A–E) Orientation of cell divisions (with respect to tangent) in concentric circles spanning the wing pouch. Divisions become gradually tangent in the periphery. Histograms represent the normalized angular distribution. (F) Illustration of the pattern divisions. Divisions are radial in the center and tangent in the periphery. Length of arrows reflects the degree of polarity. (G, H) Direct comparison of the orientation of divisions and the local tissue axis. The second moment of the angular distribution is plotted against the distance to the center of the tissue. Two distributions (center versus periphery) are represented on the right.

***Source*: reproduced with permission from LeGoff et al. (2013).**

REFERENCES

Alert, R., & Trepat, X. 2020. Physical models of collective cell migration. *Annu. Rev. Condens. Matter Phys.*, 11, 77–101. doi:10.1146/annurev-conmatphys-031218-013516

Ashrafuzzaman, M., Tseng, C.Y., & Tuszynski, J.A. (2020). Charge-based interactions of antimicrobial peptides and general drugs with lipid bilayers. *Journal of Molecular Graphics and Modelling*, *95*, 107502. doi: 10.1016/j.jmgm.2019.107502; PMID: 31805474

Ball, P. (2017). Water is an active matrix of life for cell and molecular biology. *Proceedings of the National Academy of Sciences*, *114*(51), 13327–13335

Boehr, D.D., McElheny, D., Dyson, H.J., & Wright, P.E. (2006). The dynamic energy landscape of dihydrofolate reductase catalysis. *Science*, *313*(5793), 1638–1642

Brini, E., Fennell, C.J., Fernandez-Serra, M., Hribar-Lee, B., Luksic, M., & Dill, K.A. (2017). How water's properties are encoded in its molecular structure and energies. *Chemical Reviews*, *117*(19), 12385–12414

Carugo, O., & Bordo, D. 1999. How many water molecules can be detected by protein crystallography? *Acta Crystallogr., Sect. D: Biol. Crystallogr.*, *55* (2), 479–483

Cesa, L.C., Mapp, A.K., & Gestwicki, J.E. (2015). Direct and propagated effects of small molecules on protein–protein interaction networks. *Front. Bioeng. Biotechnol.*, *3*, 119. doi: 10.3389/fbioe.2015.00119

Clarke, C., Woods, R.J., Gluska, J., Cooper, A., Nutley, M.A., & Boons, G.-J. (2001). Involvement of water in carbohydrate–protein binding. *J. Am. Chem. Soc.*, *123*(49), 12238–12247

Davis, A.M., Teague, S.J., & Kleywegt, G.J. (2003). Application and limitations of X-ray crystallographic data in structure-based ligand and drug design. *Angew. Chem., Int. Ed.*, *42*(24), 2718–2736

del Rio, A., Perez-Jimenez, R., Liu, R.C., Roca-Cusachs, P., Fernandez, J.M., & Sheetz, M.P. (2009). Stretching single talin rod molecules activates vinculin binding. *Science*, *323*, 638–641

DePaola, N., Gimbrone Jr, M.A., Davies, P.F., & Dewey Jr, C.F. (1992). Vascular endothelium responds to fluid shear stress gradients. *Arteriosclerosis and Thrombosis: A Journal of Vascular Biology*, *12*(11), 1254–1257

Despa, F., & Berry, R.S. (2007). The origin of long-range attraction between hydrophobes in water. *Biophys J.*, *92*(2), 373–378. doi: 10.1529/biophysj.106.087023

Despa, F., Fernández, A., & Berry. R.S. (2004). Dielectric modulation of biological water. *Phys. Rev. Lett.*, *93*, 228104

Ernst, J., Clubb, R., Zhou, H., Gronenborn, A., & Clore, G. (1995). Demonstration of positionally disordered water within a protein hydrophobic cavity by NMR. *Science*, *267*(5205), 1813–1817

Estelle, A.B., Reardon, P.N., Pinckney, S.H., Poole, L.B., Barbar, E., & Karplus, P.A. (2022). Native state fluctuations in a peroxiredoxin active site match motions needed for catalysis. *Structure*, *30*(2), 278–288

Fang, F., Satulovsky, J., & Szleifer, I. (2005). Kinetics of protein adsorption and desorption on surfaces with grafted polymers. *Biophys. J.*, *89*(3), 1516–1533. doi: 10.1529/biophysj.104.055079

García-Sosa, A.T., Firth-Clark, S., & Mancera, R.L. (2005). Including tightly-bound water molecules in de novo drug design: exemplification through the in silico generation of poly(ADP-ribose)polymerase ligands. *J. Chem. Inf. Model.*, *45*(3), 624–633

García-Sosa, A.T., & Mancera, R.L. (2010). Free energy calculations of mutations involving a tightly bound water molecule and ligand substitutions in a ligand–protein complex. *Mol. Inf.*, *29*(8–9), 589–600

Gibson, M.C., Patel, A.B., Nagpal, R., & Perrimon, N. (2006). The emergence of geometric order in proliferating metazoan epithelia. *Nature*, *442*, 1038–1041

Grancha, T., Ferrando-Soria, J., Cano, J., Amorós, P., Seoane, B., Gascon, J., … & Pardo, E. (2016). Insights into the dynamics of Grotthuss mechanism in a proton-conducting chiral bio MOF. *Chemistry of Materials*, *28*(13), 4608–4615

Gurd, F.R., & Rothges, T.M. (1979). Motions in proteins. *Advances in Protein Chemistry*, *33*, 73–165

Halperin, A., Tirrell, M., and Lodge, T. (1992). Tethered chains in polymer microstructures. *Adv. Polym. Sci.*, *100*, 31–71

Heisenberg, C.P., & Bellaïche, Y. (2013). Forces in tissue morphogenesis and patterning. *Cell*, *143*, 948–962

Huang, D.M., and Chandler, D. (2002). The hydrophobic effect and the influence of solute-solvent attractions. *J. Phys. Chem. B.*, *106*, 2047–2053

Huang, X., Margulis, C.J., and Berne, B.J. (2003). Dewetting-induced collapse of hydrophobic particles. *Proc. Natl. Acad. Sci. USA*, *100*, 11953–11958

Janc, T., Korb, J.P., Luksic, M., Vlachy, V., Bryant, R.G., Mériguet, G., … & Rollet, A.L. (2021). Multiscale water dynamics on protein surfaces: protein-specific response to surface ions. *Journal of Physical Chemistry B*, *125*(31), 8673–8681

Kadirvelraj, R., Foley, B.L., Dyekjær, J.D., & Woods, R.J. (2008). Involvement of water in carbohydrate–protein binding: concanavalin A revisited. *J. Am. Chem. Soc.*, *130*(50), 16933–16942

Kim, J.H., Serra-Picamal, X., Tambe, D.T., Zhou, E.H., Park, C.Y., Sadati, M., … & Fredberg, J.J. (2013). Propulsion and navigation within the advancing monolayer sheet. *Nature Materials*, *12*(9), 856–863

King, J.S., & Smythe, E. (2020). Water loss regulates cell and vesicle volume. *Science*, *367*(6475), 246–247

Kong, F., Garcia, A.J., Mould, A.P., Humphries, M.J., & Zhu, C. (2009). Demonstration of catch bonds between an integrin and its ligand. *J. Cell Biol.*, *185*, 1275–1284

Ladoux, B., & Mège, R.-M. (2017). Mechanobiology of collective cell behaviours. *Nat. Rev. Mol. Cell Biol.*, *18*, 743–757. doi: 10.1038/nrm.2017.98

LeGoff, L., Rouault, H., & Lecuit, T. (2013). A global pattern of mechanical stress polarizes cell divisions and cell shape in the growing Drosophila wing disc. *Development*, *140*(19), 4051–4059. https://doi.org/10.1242/dev.090878

Lewis, F.T. (1926). The effect of cell division on the shape and size of hexagonal cells. *Anat. Rec.*, *33*, 331–355

Li, J., Wijeratne, S.S., Qiu, X., & Kiang, C.H. (2015). DNA under force: mechanics, electrostatics, and hydration. *Nanomaterials*, *5*(1), 246–267

Liu, P., Huang, X., Zhou, R., and Berne, B.J. (2005). Observation of a dewetting transition in the collapse of the melittin tetramer. *Nature*, *437*, 159–162

Lum, K., Chandler, D., and Weeks, J. D. (1999). Hydrophobicity at small and large length scales. *J. Phys. Chem. B.*, *103*, 4570–4577

Mancera, R. (2002). De novo ligand design with explicit water molecules: an application to bacterial neuraminidase. *J. Comput.-Aided Mol. Des.*, *16*(7), 479–499

McPherson, T.B., Lee, S.J., & Park, K. (1995). Analysis of the prevention of protein adsorption by steric repulsion theory. *ACS Symp. Series*, *602*, 395–404

Michel, J., Tirado-Rives, J., &Jorgensen, W.L. (2009). Energetics of displacing water molecules from protein binding sites: consequences for ligand optimization. *J. Am. Chem. Soc.*, *131*(42), 15403–15411

Mikol, V., Papageorgiou, C., & Borer, X. (1995). The role of water molecules in the structure-based design of (5-hydroxynorvaline)-2-cyclosporin: synthesis, biological activity, and crystallographic analysis with cyclophilin A. *J. Med. Chem.*, *38*(17), 3361–3367

Milner, S.T. (1991). Polymer brushes. *Science*, *251*, 905–914

Morozenko, A., Leontyev, I.V., & Stuchebrukhov, A.A. (2014). Dipole moment and binding energy of water in proteins from crystallographic analysis. *Journal of Chemical Theory and Computation*, *10*(10), 4618–4623

Newberry, R.W., & Raines, R.T. (2019). Secondary forces in protein folding. *ACS Chemical Biology*, *14*(8), 1677–1686

Nishimura, T., Honda, H., & Takeichi, M. (2012). Planar cell polarity links axes of spatial dynamics in neural-tube closure. *Cell*, *149*, 1084–1097

Pangali, C., Rao, M., and Berne, B.J. (1979). Monte-Carlo simulation of the hydrophobic interaction. *J. Chem. Phys.*, *71*, 2975–2980

Park, J.A., Kim, J.H., Bi, D., Mitchel, J.A., Qazvini, N.T., Tantisira, K., ... & Fredberg, J.J. (2015). Unjamming and cell shape in the asthmatic airway epithelium. *Nature Materials*, *14*(10), 1040–1048

Ranft, J., Basan, M., Elgeti, J., Joanny, J.F., Prost, J., & Jülicher, F. (2010). Fluidization of tissues by cell division and apoptosis. *Proc. Natl. Acad. Sci. USA*, *107*(49), 20863–20868. doi: 10.1073/pnas.1011086107; PMID: 21078958; PMCID: PMC3000289

Rauzi, M., Verant, P., Lecuit, T., & Lenne, P.-F. (2008). Nature and anisotropy of cortical forces orienting Drosophila tissue morphogenesis. *Nat. Cell Biol.*, *10*, 1401–1410

Roca-Cusachs, P., Conte, V., & Trepat, X. (2017). Quantifying forces in cell biology. *Nat. Cell Biol.*, *19*, 742–751. https://doi.org/10.1038/ncb3564

Sartori, G.R., & Nascimento, A.S. (2019). Comparative analysis of electrostatic models for ligand docking. *Frontiers in Molecular Biosciences*, *6*, 52

Saw, T.B., Doostmohammadi, A., Nier, V., Kocgozlu, L., Thampi, S., Toyama, Y., ... & Ladoux, B. (2017). Topological defects in epithelia govern cell death and extrusion. *Nature*, *544*(7649), 212–216

Swift, J., Ivanovska, I.L., Buxboim, A., Harada, T., Dingal, P., Pinter, J., ... & Tewari, M. (2013). Nuclear lamin-A scales with tissue stiffness and enhances matrix-directed differentiation. *Science*, *341*, 1240104

Szleifer, I., and Carignano, M.A. (1996). Tethered polymer layers. *Adv. Chem. Phys.*, *44*, 165–260

Tambe, D.T., Croutelle, U., Trepat, X., Park, C.Y., Kim, J.H., Millet, E., Butler, J.P., & Fredberg, J.J. (2013). Monolayer stress microscopy: limitations, artifacts, and accuracy of recovered intercellular stresses. *PLoS ONE*, *8*, e55172. doi: 10.1371/journal.pone.0055172

Tambe, D.T., Hardin, C.C., Angelini, T.E., Rajendran, K., Park, C.Y., Serra-Picamal, X., ... & Trepat, X. (2011). Collective cell guidance by cooperative intercellular forces. *Nature Materials*, *10*(6), 469–475. https://doi.org/10.1038/nmat3025

ten Wolde, P.R., and Chandler, D. (2002). Drying-induced hydrophobic polymer collapse. *Proc. Natl. Acad. Sci. USA*, *99*, 6539–6543

Thompson, D.A.W. (1942). *On growth and form* (2nd edn). Cambridge: Cambridge University Press

Trepat, X., Wasserman, M.R., Angelini, T.E., Millet, E., Weitz, D.A., Butler, J.P., & Fredberg, J.J. (2009). Physical forces during collective cell migration. *Nature Physics*, *5*(6), 426–430

Veith, A., Conway, D., Mei, L., Eskin, S.G., McIntire, L.V., & Baker, A.B. (2020). Effects of mechanical forces on cells and tissues. In *Biomaterials science*, ed. W.R. Wagner, pp. 717–733. London: Academic Press

Walck-Shannon, E., & Hardin, J. (2014). Cell intercalation from top to bottom. *Nat. Rev. Mol. Cell Biol.*, *15*, 34–48. https://doi.org/10.1038/nrm3723

Wallqvist, A., and Berne, B.J. (1995). Computer-simulation of hydrophobic hydration forces on stacked plates at short-range. *J. Phys. Chem.*, *99*, 2893–2899

Wong, S.E., & Lightstone, F.C. (2010). Accounting for water molecules in drug design. *Expert Opin. Drug Discov.*, *6*(1), 65–74

Woodle, M.C. (1997). Poly(ethylene glycol)-grafted liposome therapeutics. *ACS Symp. Series*, *680*, 60–81

Wu, T., Genzer, J., Gong, P., Szleifer, I., Vlček, P., and Subr. V. (2004). Behavior of surface-anchored poly(acrylic acid) brushes with grafting density gradients on solid substrates. In *Polymer brushes*, ed. R. Advincula, W. Brittain, K. Caster, and J. Rüen, pp. 287–315. Weinheim: Wiley-VCH

Xu, Y., & Havenith, M. (2015). Perspective: watching low-frequency vibrations of water in biomolecular recognition by THz spectroscopy. *J. Chem. Phys.*, *143*, 170901

Yusko, E.C., & Asbury, C.L. (2014). Force is a signal that cells cannot ignore. *Molecular Biology of the Cell*, *25*(23), 3717–3725. https://doi.org/10.1091/mbc.E13-12-0707

Zallen, J.A., & Wieschaus, E. (2004). Patterned gene expression directs bipolar planar polarity in Drosophila. *Dev. Cell*, *6*, 343–355

Zhang, G., Mueller, R., Doostmohammadi, A., & Yeomans, J.M. (2020). Active inter-cellular forces in collective cell motility. *Journal of the Royal Society Interface*, *17*(169), 20200312

8 Surface Biophysics

The surfaces of biological systems such as tissues, cells, large or small biomolecular structural complexes, and single biomolecules appear with measurable mechanical stiffness (Handorf et al., 2015; Thomas et al., 2013) and analyzable electrical charge distributions (Krishnamurthy and Soundararajan, 1966; Yamamoto et al., 2021). To understand the general biological properties and specific physiological roles of any biosystem, certain physical properties are popularly targeted and utilized in research. Generally, surface movement regulates friction, and the surface interaction regulates geometrical fluctuations and surface tension. The viscosity of biological fluids plays a crucial role in setting up the physical states of any biochemical in fluids. The hydrophobic and hydrophilic properties of biosystem-contained substances help construct distinguishable compartments. Therefore, within any biosystem, the naturally constructed boundaries demarcating hydrophobic and hydrophilic regions also appear as intra-system surfaces besides the systems' global surfaces. These intra-system physical boundaries or surfaces are found to regulate vital transports of the systems (Ashrafuzzaman, 2015) that often participate in versatile signaling processes. The roles of membranes in biological cells are prime examples. Fundamental physics principles active on the biological systems' surfaces certainly play crucial roles in helping biosystems to become structured and work properly. Some of the associated principles and fundamental physical processes may become compromised in disease conditions when biosystems experience mutations. Surface physics is therefore also associated with diseases or biological disorders originating at the biomolecular structures. Students and faculty will find a substantial amount of basic materials to help them understand the surface biophysics of biosystems in this chapter. Researchers will also find a crucial understanding that will help them plan projects toward developing tools to strengthen our applied knowledge that can be translated into medical sciences discoveries.

8.1 SURFACE PHYSICS OF BIOLOGICAL FLUIDS

The most important physical property of the surface of a liquid is surface tension. Every liquid in a capillary tube experiences a capillary rise against the downward gravitational pull. The surface tension of a liquid surface helps it act as if it were a stretched elastic membrane. The surface tension helps a tissue to take shape and experience growth kinetics (Ehrig et al., 2019). Here a piece of quantitative evidence has been provided that the osteoid-like tissues can make use of the physics of fluids to generate complex three-dimensional (3D) patterns, corresponding to equilibrium shapes with constant mean curvature.

The surface tension can be measured in the form of either energy per unit area or force per unit length active on the surface of the liquid. The presence of surface active substances in biological or body fluids influences the value of the surface tension (Mottaghy and Hahn, 1981). Alterations of

DOI: 10.1201/9781003287780-8

the biological fluids' surface tension compared to those of the normal values may indicate a pathophysiological status (Azarbayjani and Jouyban, 2015). The measurement of the surface tension of biological fluids may help diagnose the physicochemical status, including the rise of disease conditions (Azarbayjani and Jouyban, 2015).

The measurement of dynamic surface tension of human biological fluids has been found to depend on various parameters such as sex, age, and changes during pregnancy or certain diseases (Azarbayjani and Jouyban, 2015). These determinants are either physiological or pathophysiological conditions that affect the values of this physical parameter, the dynamic surface tension of fluids in biological systems. Here we need to know the difference between dynamic and static surface tensions. Static surface tension or interfacial tension is the value of the surface tension in a thermodynamic equilibrium which is independent of time, while dynamic surface tension is referred to a particular surface age that is it is time dependent.

The pharmaceutical formulation (both scientific and technological aspects of it) draws quite a considerable amount of impact from the surface tension of the liquid in which general chemicals or specific drugs are introduced (Khoubnasabjafari et al., 2013; Azarbayjani et al., 2011; Azarbayjani, Jouyban, and Chan, 2009; Azarbayjani et al., 2010; Jouyban and Azarbayjani, 2012; Azarbayjani, Tan, et al., 2009; Jouyban et al., 2004). But the surface tensiometry of biological fluids is not yet properly considered while quantifying the impacts of medicines, perhaps because surface rheology is yet to be introduced to medicine, and theoretical studies and characteristics of all kinds of biological fluids are not well established (Kazakov et al., 1998; Kazakov et al., 2000; McGuire, 2002; Depalma, 1967). However, the measurement of physical properties such as surface tension, viscosity, etc. of biological fluids like water, plasma, and other related ones has been considered over the past century or more (Harkins and Harkins, 1929; Pelofsky, 1966; Boda et al., 1997; Azarbayjani and Jouyban, 2015).

The general surface physics properties of biological fluids are interrelated to some extent. For example, an increase in temperature lowers the net intermolecular interaction forces of a fluid and hence decreases its surface tension. As a liquid is heated up, its molecules become excited due to the addition of excess thermal fluctuations and begin to experience increased movement. The increased energy of this movement helps overcome the forces that bind the molecules together, allowing the liquid to become more fluid and decreasing its viscosity. The physiological condition of biological fluids is therefore very much environment sensitive, and so are the fluid's physical properties like surface tension and viscosity.

Biological tissues are viscoelastic materials and the cells in a tissue behave like single particles or molecules in a biological fluid. The cells can move inside the fluidic tissue. The geometry (shape and size) of a cell is the result of a balance of intracellular and extrinsic forces exerted on it. This behavior is defined through surface tension which tends to minimize the exposed area of the cell aggregate and maximize the cohesive forces. The intracellular forces on the membrane are a result of the cytoskeleton reorganization. The energy at the cellular level is usually measured through physical properties such as cell adhesion, viscosity, and cortical tension (Azarbayjani, Jouyban, and Chan, 2009; Wozniak et al., 2004). Cell adhesion can be expressed using the following equation (Azarbayjani, Jouyban, and Chan, 2009):

$$\Delta F^{adh} = \gamma_{cs} - \gamma_{cl} - \gamma_{sl} \tag{8.1}$$

Here ΔF^{adh} is the free energy of adhesion, γ_{cs}, γ_{cl}, and γ_{sl} are the cell-substrate interfacial tensions, cell-fluid interfacial tension, and substrate-fluid interfacial tension, respectively.

Both intracellular adhesion and repulsion between various cell types are known to cause mobility within cells. Cell coupling, cell morphology, and tissue rearrangement are associated with cell forces (Steinberg, 1963; Davis et al., 1997; Kalantarian et al., 2009). The balance of intracellular and extracellular forces determines the setting of size, shape, and structure and leads to the rounding of the

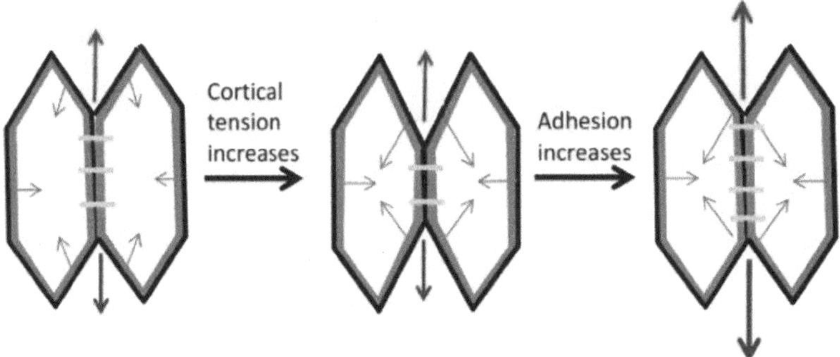

FIGURE 8.1 Schematic image of the interaction between cortical tension and cellular adhesion during the formation of cell–cell contacts. Directions of the forces are depicted with arrows, the red line represents cortical cytoskeleton, black lines are plasma membrane, and the purple rods depict adhesion sites. For color contrast in this figure and other ones presented later, see the online version of the book.

***Source*: Mombach et al. (2005).**

tissue, helping to minimize its surface tension and to maximize intracellular adhesion. Figure 8.1 is the schematic representation of the interaction between cortical tension and cellular adhesion during the formation of cell–cell contact.

The adhesion contact among the neighboring cells helps construct the tissue to have stable organization. In this regard, forces originating from the following three physical properties are mechanically coupled to each other:

- cortical tension
- cellular adhesion
- cellular viscoelasticity

The combined effects of these mechanical properties help govern cell shape and cell arrangement in the tissue. These forces regulate cellular behaviors such as tissue development, including helping to determine the tissue surface morphology, embryogenesis, protein expression, cell proliferation and differentiation, and will help in studying cell behavior in normal and pathological conditions and invasiveness of malignant tumors (Wozniak et al., 2004; Davis et al., 1997; Kalantarian et al., 2009).

Forces generated by surface tension are found to play a leading role in the position rearrangement of cells and to act in minimizing the exposed area of the aggregate in a tissue. Specific adhesion molecules, e.g. cadherins, help maintain the cell–cell adhesion. Different types of adhesion molecules are expressed on the cell surface and they are cell type-specific. The intensity of adhesion among cells is also adhesion molecule type-specific and it varies markedly (Walker et al., 2005; Mombach et al., 2005).

The varying cell surface adhesion causes cell migration. The regulation of the adhesion molecules in different cell groups determines cell shape and helps to gain information about cellular aggregates for deciding on the next migration step. When cells with varied adhesive forces are put together, the less adhesive cells migrate gradually and surround the more adhesive cells (see Figure 8.2) (Thiessen and Man, 1999; Balaban et al., 2001; Voß-Böhme and Deutsch, 2010). We see here that the surface tension-induced interstitial dynamics and redistribution among cells in a tissue largely contribute to constructing the tissue's mechanical properties, including especially the surface morphology.

Active cellular forces are essential for generating sufficient surface stresses for ensuring the liquid-like behavior and growth of the tissue, suggesting that the mechanical signaling between cells

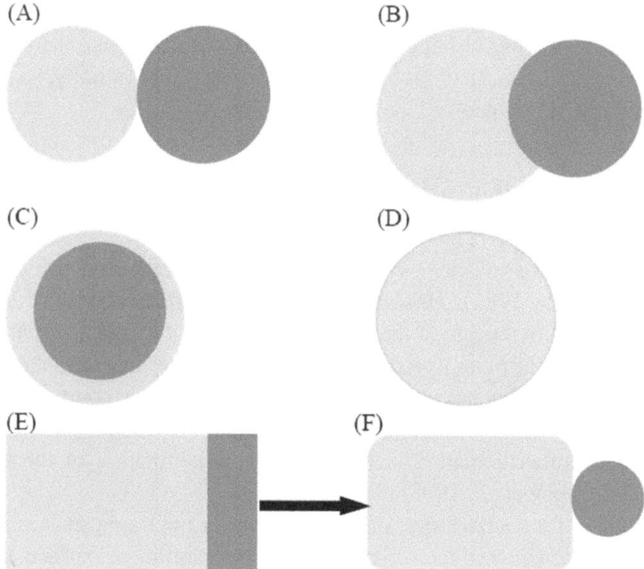

FIGURE 8.2 Configuration of two different cell populations with varying adhesion forces. The dark sphere represents cells with high self-adhesion whereas the white sphere represents cells with lower self-adhesion. (A) no cross-adhesion, (B) relatively weak crossadhesion, (C) intermediate cross-adhesion, (D) preferential cross-adhesion between the cells, and (E) differential adhesion and morphological changes during somatogenesis.

***Source*: adapted from Thiessen and Man (1999).**

and their physical environment, along with the continuous reorganization of cells and matrix, is a key principle for the emergence of tissue shape (Ehrig et al., 2019). This mechanical behavior of the tissue surface was confirmed during investigation of the shape developed by a growing osteoid-like tissue, grown on the polydimethylsiloxane scaffolds using a preosteoblastic cell line (MC3T3-E1), which has been demonstrated to synthesize a collagen-rich extracellular matrix (ECM) (Rumpler et al., 2008; Bidan et al., 2012).

The interstitial tissue fluids are relatively slow dynamic entities experiencing modest to negligible fluctuations affecting the local organization of tissue components. This interstitial tissue fluid makes a solution that bathes and surrounds the cells of multicellular animals. The interstitial biological fluid is present between the cells and tissues, comprising a similar composition to that of the plasma. The composition and function of the transcellular fluid vary inside the human body. One specific example is that the fluid at joints serves lubrication. The function of the interstitial fluids that are almost static is dependent mainly on the collective actions of forces and energies active between cells, as depicted in Figure 8.1. We may consider another rather highly dynamic fluid, the blood in the vessel, and see how the surface of this dynamic fluid behaves (Harkins and Harkins, 1929). The dynamic surface tension of water is another important physical parameter in biological systems that helps address the physiological condition (Hauner et al., 2017).

The surface tension of the undiluted normal blood serum was found to be 52 and 48 dynes per centimeter (cm) at 20° C and 37° C, respectively (Harkins and Harkins, 1929). The surface tension of a freshly created water surface is ~90 mN/m or 90 dynes per cm while under equilibrium conditions ~72 mN/m or 72 dynes per cm with a relaxation process occurring on a long time scale (~1 ms) (Hauner et al., 2017). The tissue surface tension has been measured for the embryonic F9 cell aggregates, where the non angle corrected Circular Arc (CA) method gives a much lower surface tension value ~3.3±2.1 mN/m than the angle independent methods value ~5.3±0.7 and

5.6±0.8mN/m for Local Polynomial Fit method and circular arc approximation (CAcm), respectively between 20° and 30° C (Mgharbel et al., 2009). It is clear here that the values of the surface tension among various fluid surfaces in biological systems vary substantially, so the surface physics is expected to naturally be fluid-specific.

8.2 SURFACE PHYSICS OF BIOLOGICAL FLUIDS IS REGULATED IN PATHOLOGICAL CONDITIONS

The presence of surface active substances in biological fluids influences the value of the surface tension (Mottaghy and Hahn, 1981). This article also demonstrates how the values of the surface tension change (decrease) with the age of the systems. The time-dependent dynamic surface tension may be a quantitative physical parameter used for understanding the physical nature of fluids in pathological conditions (Azarbayjani and Jouyban, 2015). Here a comparison has been made between surface tension values of biological fluids in healthy (Table 8.1) and the following disease conditions (Table 8.2): glomerulonephritis, which is the inflammation of the tiny filters in your kidneys (glomeruli) (Kazakov et al., 1998).

Here γ_1 surface tension for $t = 0.01$ s; γ_2 surface tension for $t = 1$ s; γ_3 surface tension $t \to \infty$.

Many diseases have been investigated regarding the regulation of surface tension of the biological fluids due to the rise of pathological conditions. A few such are presented in Table 8.3 for diseases specifically of the nervous systems and in Table 8.4 for various other diseases associated with serum and urine (Kazakov et al., 2000).

TABLE 8.1
Dynamic Surface Tension Characteristics in Urine of Healthy Volunteers

		Surface tension parameters		
Sex	**Age**	$\gamma_{1\,(mNm^{-1})}$	$\gamma_{2\,(mNm^{-1})}$	$\gamma_{3\,(mNm^{-1})}$
Males	>20	72–73	69–73	61–63
	20–35	70–72	68–70	59–62
	36–50	70–72	68–70	57–60
	≤50	70–73	67–71	56–61
Females	>20	72–73	69–71	65–68
	20–35	70–72	68–70	64–66
	36–50	70–72	68–70	61–65
	≤50	70–73	67–71	60–65

TABLE 8.2
Surface Tension of Blood and Urine for Patients with Various Forms of Glomerulonephritis

	Serum				Urine			
Types of glomerolonephritis	γ_1	γ_2	γ_3	λ	γ_1	γ_2	γ_3	λ
Acute	+	+			+			
Chronic				+	+	-	-	-
Lupus	+	+	+	+	+			
Genoch	+	+			+	+	+	

+ and – represent statistically significant increase and decrease, respectively, of parameters in disease condition compared to the healthy status.

TABLE 8.3
Surface Tension of Serum and Cerebrospinal Fluid during Various Types of Nervous System Disease

	Surface tension Serum				Urine			
Disease of the nervous system	γ_1	γ_2	γ_3	λ	γ_1	γ_2	γ_3	λ
Infection		-	-			+	+	+
Vascular					+	+	+	+
Spondylogenic			-	+	+	+		+
Neoplasm		-	-					+
Trauma		-	-			+	+	+

TABLE 8.4
Surface Tension Changes During Various Rheumatic Diseases

	Surface tension Serum				Urine			
Diseases	γ_1	γ_2	γ_3	λ	γ_1	γ_2	γ_3	λ
Rheumatism		+	+					-
Systemic lupus erythematosus	+			+		-		-
Scleroderma systematica	+	+	+	+	+	-		
Haemorrhagic vasculitis	+	+			+	+		
Rheumatoid arthritis	+	+	+					-
Bechterew's disease				-			+	
Reiter's disease	+			-	+		+	
Psoriatic arthropathy						+	+	-
Gout				-	+	+	+	
Osteoarthrosis					+	+	+	-

8.3 BIOPHYSICS OF THE TISSUE SURFACE

Physical properties of the tissue surface contribute to maintaining both the shape and growth of tissues (Ehrig et al., 2019). The fluidity of tissue largely depends on the collective self-organization of cells resulting in three-dimensional (3D) structures. That is to say, the tissue grows three-dimensionally and fluidity helps it maintain the growth and create a certain physical shape. Recently, it has been reported that tissues, while growing on curved surfaces, develop shapes with outer boundaries of constant mean curvature (Ehrig et al., 2019). This resembles identical physical properties observed on the wet liquid surface with some kind of energy minimization.

The tissue growth process is dynamic though slow on the biological time scale (Bodenstein, 1986). While growing geometrically, the biological tissue follows important physics principles, some of which will be addressed here. The cells participating in the formation of the tissue are interactive and motile (Friedl and Gilmour, 2009), which helps in the rise of physical properties, such as viscosity. This kind of physical behavior has been found in epithelial monolayers during embryogenesis (Mongera et al., 2018; He et al., 2014) and in cell agglomerates (Schötz et al., 2013) with a measurable physical parameter surface tension (Foty and Steinberg, 2005). The tissue also holds elastic properties as the mechanical integrity of tissues is provided by extracellular matrices (ECMs) inducing solidification into tissues (Ophir et al., 1991). However, it has also been shown that even osteoid-like tissue with large amounts of ECM grows according to rules reminiscent of the fluid behavior (Rumpler et al., 2008; Bidan et al., 2012).

Recently, a study has shown quantitatively, by constraining growing tissues to surfaces of the controlled mean curvature, that osteoid-like tissues develop physical shapes (Ehrig et al., 2019) similar to the equilibrium shapes of fluids (Langbein, 2002). The tissue growing on the curved surface was pinned to the flat edges of the bridge (see Figure 8.3A, red circle). The tissue growing in between the capillary bridge (CB) and the sample holder is seen to form an almost constant angle with the flat surfaces. Both observations are reminiscent of the liquid wetting. For a constant CB height, the average tissue thickness formed after one month of the culture decreased with increasing neck radius of the CB (Figure 8.3B, C, F–I, and J–L). The tissue surfaces were also seen to be rotational symmetric, observed in 3D light-sheet fluorescence microscopy (LSFM) imaging of fixed tissues (see Figure 8.3B–E). This rotational symmetry allowed estimation of tissue volume and mean curvature from 2D images.

To explore the behavior of the growing tissue, physical parameters such as mean curvatures and surface areas of the tissue were plotted as a function of total volume (CB volume plus tissue) and compared to the theoretical predictions of the Laplace-Young law for a liquid drop adhering to the scaffold (see Figure 8.4A and B) (Brakke, 1992). A surface energy minimization scheme was utilized to predict the shape of liquid interfaces that are subject to the surface tension. The liquid-like behavior was demonstrated for CBs with smaller end spacings (see Figure 8.4C and D) that cover a larger range of the mean curvature. The experimental results match the theoretical predictions perfectly (see the shaded area, Figure 8.4A–D). The growing osteoid-like tissues are conclusively found to behave like fluids at time scales relevant for the tissue growth (Latorre and Humphrey, 2018).

An important physical property of the fluidic tissue surface is to adapt to sustained changes in mechanical loads. These slow macroscale adaptations, resulting from the mechanobiological cellular responses, are important determinants of various physiologically relevant behaviors with clinical importance. Relative roles of the rates of tissue responses and external loading have been modeled to understand the evolution of soft tissue growth in Latorre and Humphrey (2018), where the physics demonstrating the following two cases are presented:

A. long-term, steady-state, tissue maintenance solution
B. instantaneously adapted, quasi-equilibrium, slow evolution

The time scales inherent to the rates of mechanical loading and growth and remodeling (G&R) responses (Fung, 1995) have been used for determining conditions under which a rate-independent ("pseudoelastic") theory can hold throughout G&R (Latorre and Humphrey, 2018). For purposes of illustration and application, a theoretical approach has been developed (Latorre and Humphrey, 2018) to simulate arterial responses to altered pressure, flow, and axial stretch. Instead of presenting the theoretical details, we explain here key results. Figure 8.5 presents the numerical results for the case of the "long-term, steady-state, tissue maintenance solution." As said earlier, for details on methods readers may consult the original article (Latorre and Humphrey, 2018). The goal of this first example was to compute exactly the long-term response of a thin-walled bilayered artery when subjected to multiple external loads that were sustained for long periods. Consider, therefore, two different combinations of loading consisting of 1.15-fold increases in the applied pressure P and flowrate Q, each sustained following initial transients. To delineate better the responses to different stimuli, the loads were applied sequentially in the orders $P^{(1)} \rightarrow Q^{(2)}$ (first case) and $Q^{(1)} \rightarrow P^{(2)}$ (second case), each taking 21 days overall to reach steady values and time-shifted from the other by 14 days (see Figure 8.5a). Figure 8.5b shows the resulting/evolving stimulus function (Υ^c) for (both medial M and adventitial A) collagen c (an example). Figure 8.5c–f present the case-specific evolving responses predicted for the different combinations of loads shifted over time. Fold

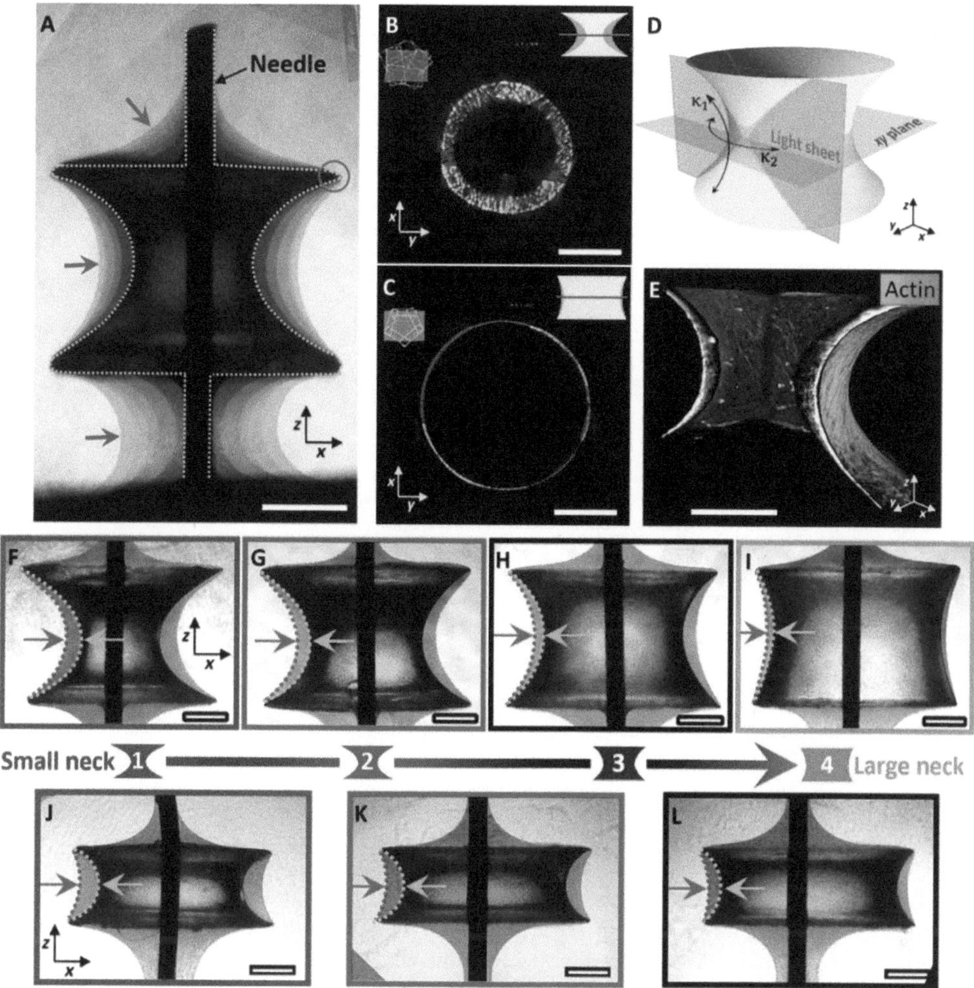

FIGURE 8.3 CB geometry as a means to control 3D tissue growth. (A) Composition of phase-contrast images of tissues grown on a CB taken after 4, 7, 21, 32, 39, and 47 days. The tissue is pinned at the edges of the CB (red circle) and shows a moving contact line in between the scaffold and the Teflon holder, reminiscent of a liquid. Dashed line indicates the CB surface, and red arrows point toward the tissue–medium interface. Scale bar, 500 μm. (B and C) Radial slices at the neck for two different CB sizes with initial volumes of 1.1 μl (B) and 2.8 μl (C) obtained with LSFM from five different views. (D) Sample geometry and orientation of the light sheet. $\kappa_1$and κ_2 are minimum and maximum principal curvatures. (E) 3D rendering of actin fibers on the sample shown in (F) color-coded according to fluorescence intensity. (F to I) Phase-contrast images of tissues grown on four different CB surfaces with initial volumes of 1.1 μl (F), 1.6 μl (G), 2.2 μl (H), and 2.8 μl (I). (J to L) CB surfaces with initial volumes of 1.1 μl (J), 1.3 μl (K), and 1.5 μl (L). Sample neck size increases from left to right. Green arrow indicates the interface of the initial shape, and red arrow indicates the position of the tissue–medium interface after 32 days. Scale bars, 400 μm. For the color contrast see the online version of the book.

***Source*: reproduced with permission from Ehrig et al. (2019).**

differences in shear, circumferential, and axial stresses from homeostatic are shown separately in Figure 8.5g–i.

Figure 8.6 shows the numerical results for the case "instantaneously adapted, quasi-equilibrium, slow evolution." For details on methods, interested readers may consult the original article (Latorre

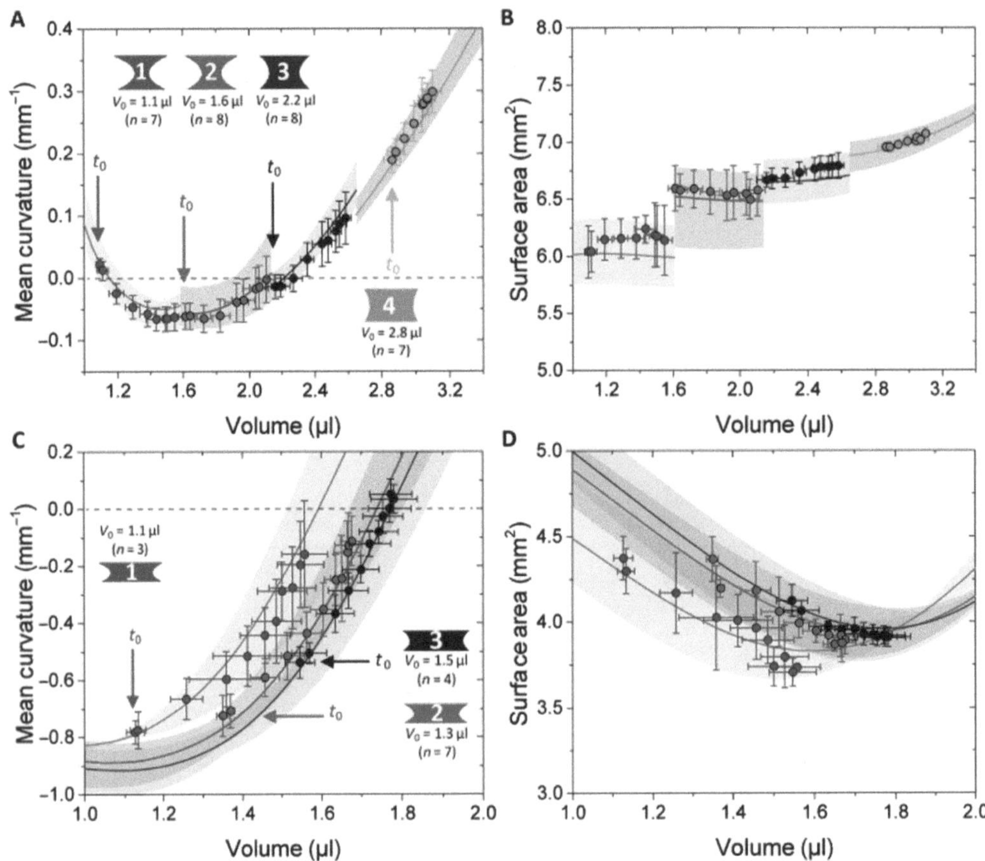

FIGURE 8.4 Tissues behave like liquids when constrained by curved surfaces. (A to D) Evolution of the surface mean curvature (A and C) and surface area (B and D) of the tissue interfaces as a function of total volume (CB volume V_0 plus tissue volume) for the two different CB shapes shown in Figure 8.3. (A and B) Tissues grown on CB surfaces with a height of ~1.2 mm and top/bottom radius of ~1 mm with initial volumes of 1.1 μl (size 1; blue), 1.6 μl (size 2; red), 2.2 μl (size 3; black), and 2.8 μl (size 4; orange). Colored areas delineate the theoretical predictions of liquid interfaces of the same dimension based on the scaffold geometries obtained from the experiment. Colored curves are the corresponding mean values. (C and D) Tissues grown on CB surfaces with a height of ~0.7 mm and top/bottom radius of ~1 mm with initial volumes of 1.1 μl (size 1; blue), 1.3 μl (size 2; red), and 1.5 μl (size 3; black). Curves are color-coded according to the corresponding theoretical predictions of the liquid interface; for details see Ehrig et al. (2019). For color contrast see the online version of the book.

***Source*: reproduced with permission from Ehrig et al. (2019).**

and Humphrey, 2018), where a huge theoretical analysis is presented. I have decided not to present these details here to keep things simpler.

In this section we have addressed a few fundamental physical properties of the tissue surface that help tissue to become constructed, expand three-dimensionally, and show associated fluidity. The mechanical properties of the tissue surface may also be regulated using engineering techniques (Xu et al., 2015; Denchai et al., 2018). It may be noted that the surface mechanics of the tissue may result due to specific responses originated at the underlying organization, orientation, and association among the cells of the tissue.

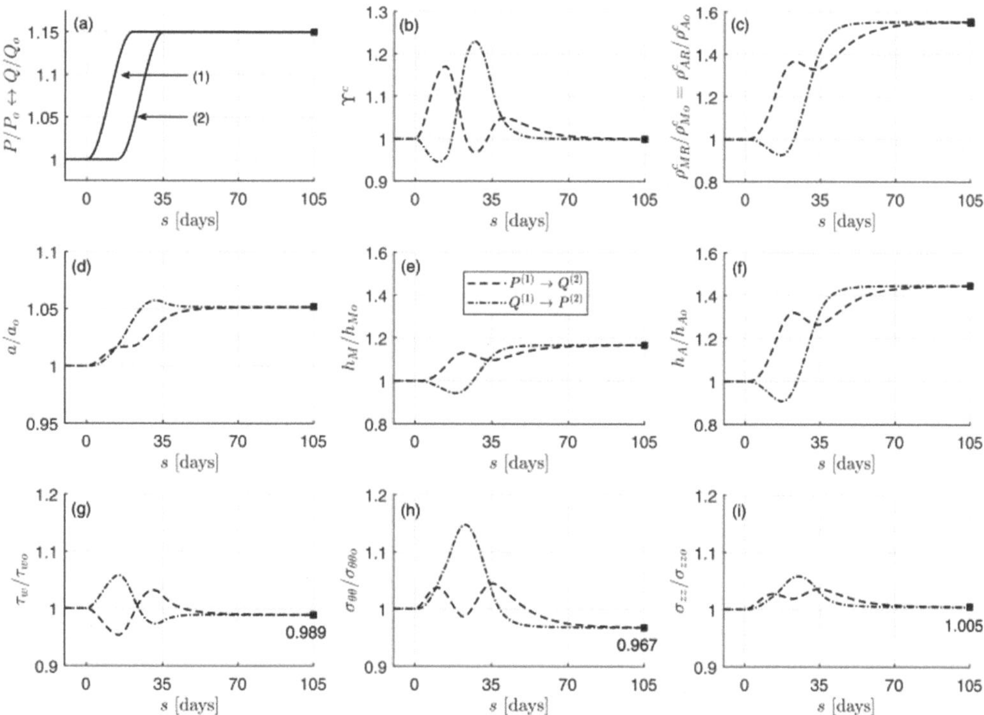

FIGURE 8.5 Predictions by the full constrained mixture model (first case, dashed; second case, dash-dotted) for the evolution of (c) medial and adventitial collagen (referential) mass density, (d) inner radius, (e) medial thickness, and (f) adventitial thickness, each normalized to original values, which result from two different cases of perturbations in loading [(a), solid lines] that cause (b) evolving stimulus functions (i.e. stresses different from homeostatic target values). Panels (g)–(i) show, separately, associated deviations in stress components from original homeostatic values. Note the two different time-delayed combinations of changes in pressure and flow (a). Shown, too, is the long-term, mechanobiologically equilibrated solution (solid square), which is the same for each of the two (same final) loading conditions. Note the perfect correspondence of the long-term steady-state solution computed with the present time-independent formulation and the full (hereditary) constrained mixture model.

***Source*: reproduced with permission from Latorre and Humphrey (2018).**

8.4 BIOPHYSICS OF THE CELL SURFACE

In the book *Nanoscale Biophysics of the Cell*, I outlined crucial physics principles active on the surface of the cell (Ashrafuzzaman, 2018). Cell surface morphology is usually addressed using versatile imaging techniques, including atomic force microscopy (AFM) (Ashrafuzzaman et al., 2021), confocal microscopy (Chidambaram et al., 2019; Jia et al., 2019), and various others (Denchai et al., 2018; Xu et al., 2015). The imaging of a single cell or collection of cells helps us address cell surface physics principles and properties (Ashrafuzzaman et al., 2021; Ashrafuzzaman et al. 2022; Denchai et al., 2018; Xu et al., 2015).

Figures 8.7 and 8.8 present AFM images demonstrating cell (human colorectal adenocarcinoma cells, LoVo) surface morphology (Ashrafuzzaman et al., 2021). We can see that the control cell surface looks relatively smooth while the drug (colchicine) treated cell surface shows the presence of physical bumps (drug clusters), which statistically appeared at random locations. The cluster heights

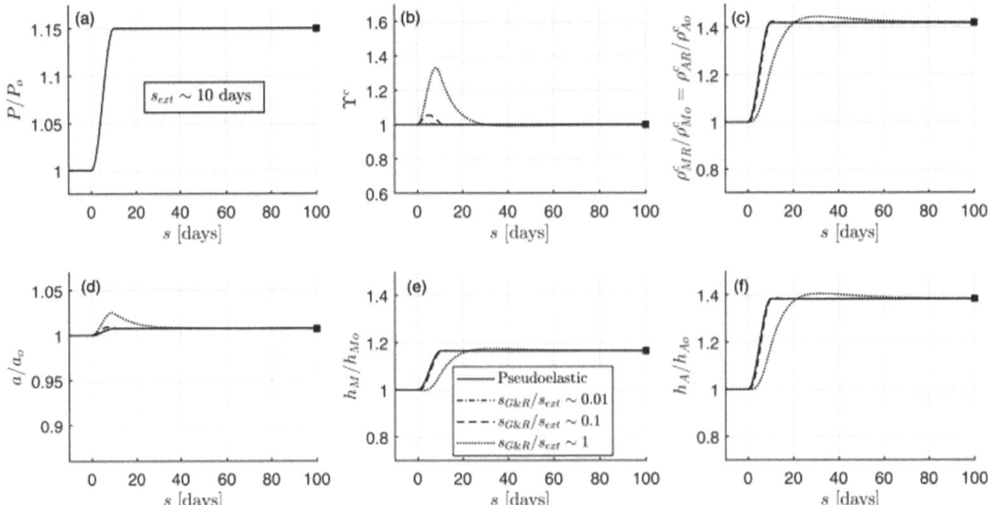

FIGURE 8.6 Rate-independent (solid line) and rate-dependent (non-solid lines) evolutions computed, respectively, with the pseudoelastic and the full constrained mixture models, the latter with different characteristic G&R times $s_{G\&R} = \{0.1, 1, 10\}$ days, for an isolated increase in pressure with $s_{ext} \sim 10$ days. Shown are (a) prescribed load P/P_o from 1 to 1.15, (b) mechano-stimulus function Υ^c, (c) referential mass densities of collagen, (d) relative inner radius a/a_o, (e) relative medial thickness h_M/h_{Mo}, and (f) relative adventitial thickness h_A/h_{Ao}. The final total wall thickness is $h = 0.0494$ mm (with 67 percent due to medial thickening and 33 percent due to adventitial thickening). Finally, shown too is the long-term mechanobiologically equilibrated solution (solid square), which reveals perfect correspondence of all three methods at the final adapted state. Copied from ref. Latorre and Humphrey (2018).

***Source*: Latorre and Humphrey (2018).**

generally increase with increasing cell-treating drug concentrations (Figure 8.7, right panels). Within the same cell surface, the distribution of drug cluster heights follows a Gaussian distribution. The number of drug clusters and cluster-contained drugs appears to increase substantially with the increase of drug concentration following some power law (see Figure 8.9) (Ashrafuzzaman et al., 2021).

Some crucial aspects related to cell surface physics may be concluded from Figures 8.7–8.9, as follows:

1. Cell surface is generally non-smooth, that is, the three-dimensional (3D) geometry of the cell surface appears with fluctuating surface morphology.
2. Cell surface structural fluctuations vary depending on the presence of cell surface accumulated agents such as cell targeted drugs, nanoparticles, proteins, etc. The change in surface morphology may be very much cell surface or cell targeted agent specific.
3. The cell surface accumulation, distribution, clustering, and adsorption of cell surface or cell targeted agents follow standard statistical mechanical formalism.

Based on the above-mentioned changes in the cell surface morphology due to the adsorption of chemotherapy drug (CD) colchicine, we attempted to understand the adhesive forces of the cell surface by measuring the adhesive energy (see Figure 8.10). Quantitative adhesion force (F_{ads}) between probe and surface (SiO_2, drug, and cell membrane) was extracted following techniques detailed in

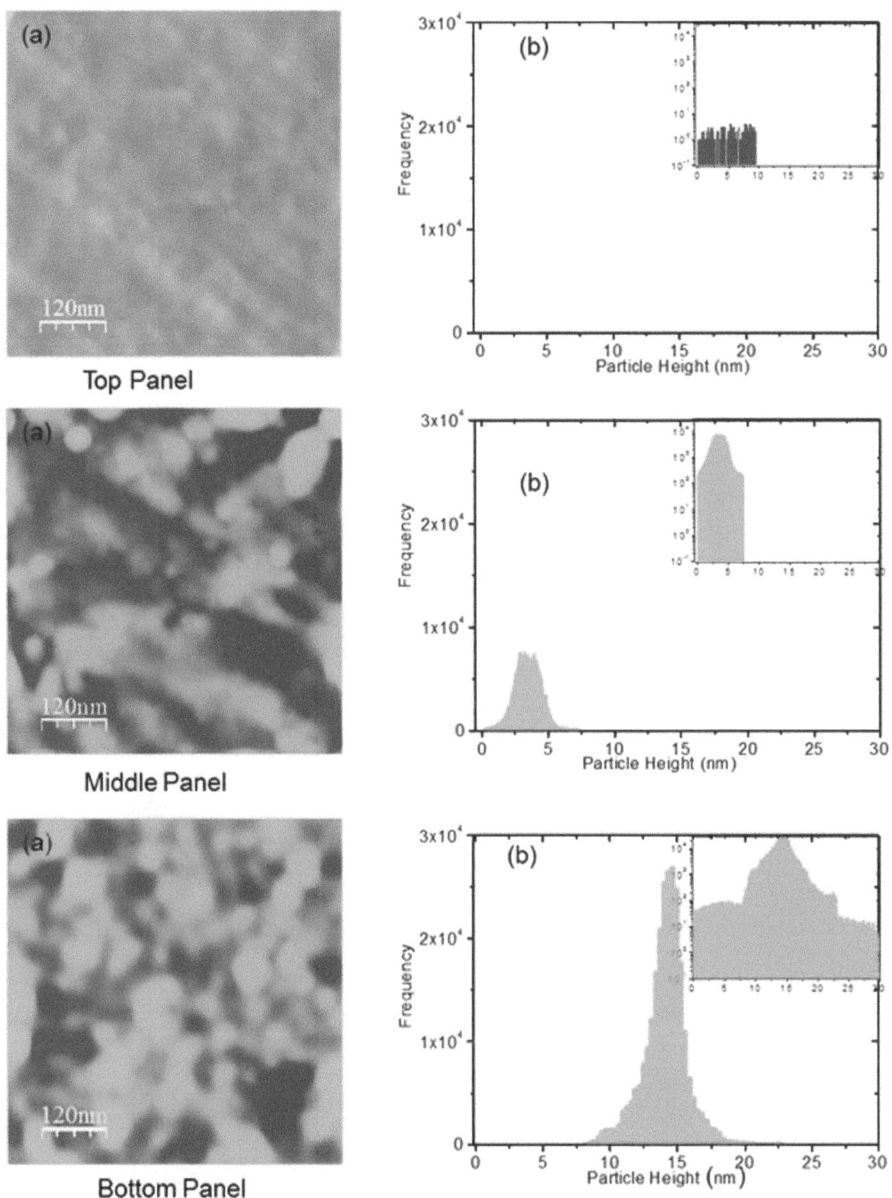

FIGURE 8.7 Top-left panel (a), high resolution TM-AFM image of a control (colchicine untreated) cell surface. Top-right panel (b), histograms presented here are based on counted clusters on the section of cell surface shown in the left panel (a). Each histogram represents the number of counted clusters (called frequency, let's assume it as f_i, plotted along y-axis. i = 1, 2, 3, …, etc.) versus a height profile (known as particle height, let's assume it as h_i, plotted in nm scale along x-axis) shown in b, top-right panel. As mentioned in Figure 8.8, the height profiles are recorded at the centers of the corresponding clusters (see the left panel, a). Middle and bottom panels present data for cells treated with 1 and 100 μM colchicine, respectively. The higher presence of clusters (in green color) on blue background is visible. We repeated the experiments three times but inspected no substantial changes in the distribution patterns presented in the right panels. Origin 9.1 was used to plot histograms after detecting the numbers using program WSxM. The origin of the detected few events (visible in log plots, see in insets of b) for untreated/control cell surface is nothing but noise.

***Source*: reproduced with permission from Ashrafuzzaman et al. (2021).**

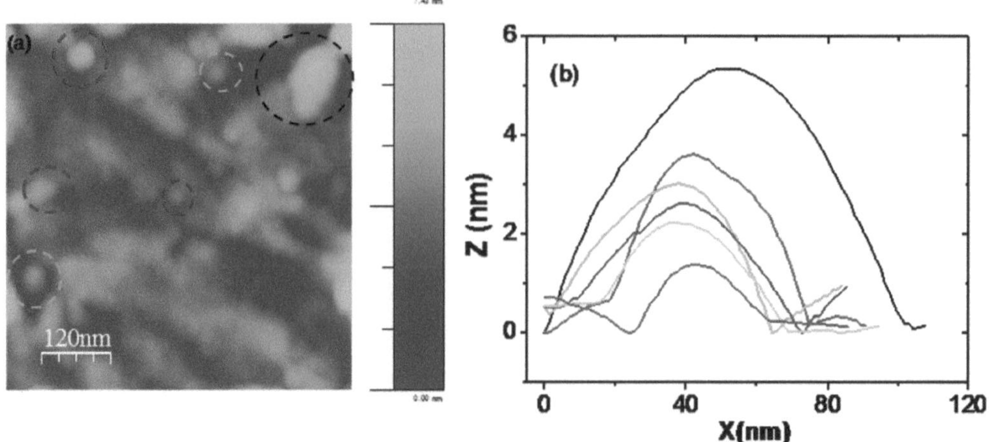

FIGURE 8.8 (a) A high resolution AFM height image of a 1 μM colchicine treated cell wall bound to SiO_2 substrate surface. The height of the cell wall is presented in color scale 0 (dark fields) – 7.43 nm (light fields). (b) A number of height profiles were recorded through the center of the circles. Here we marked the blobs (or clusters of colchicine molecules) of different sizes by circles of various colors. It can be concluded that the colchicine molecules bind to the cell membrane either in single or cluster form.

***Source*: reproduced with permission from Ashrafuzzaman et al. (2021).**

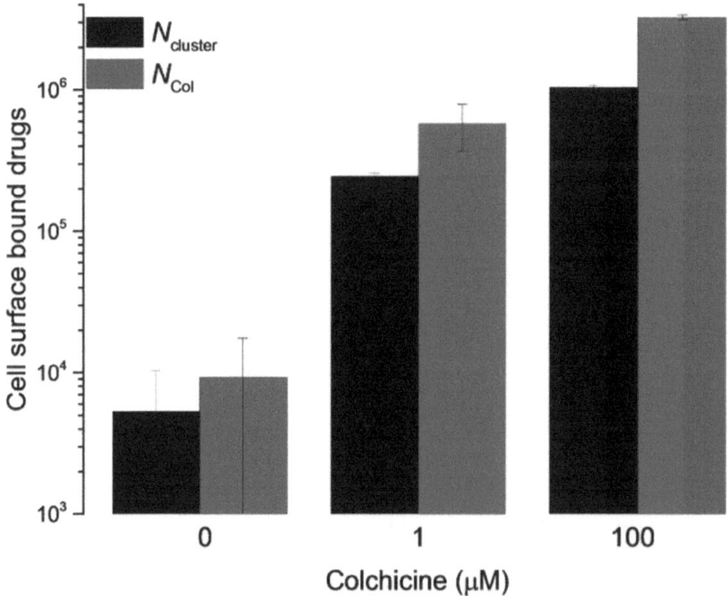

FIGURE 8.9 Cell surface bound/adsorbed colchicine molecules (N_{Col}) and cell surface clusters of the colchicine molecules ($N_{cluster}$) have been plotted (y-axis) as histograms at the cell treating 0 (control), 1 and 100 μM colchicine concentrations. Both sets of data represent the mean ± SD, n = 3. The number each histogram represents is the counted number of clusters against a particle height profile (see Figure 8.7b and supplementary Figure 8.8b). The height profiles are recorded at the centers of the corresponding clusters in supplementary Figure 8.8a.

***Source*: reproduced with permission from Ashrafuzzaman et al. (2021).**

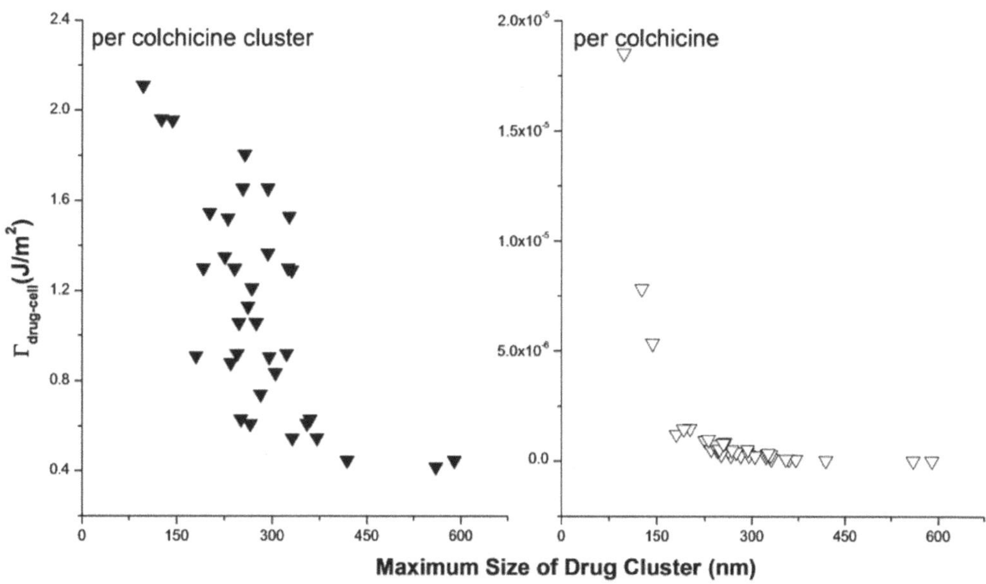

FIGURE 8.10 Adhesion energy between drug and cell surface is plotted against drug cluster size (see clusters in Figures 8.7 and 8.8). Left panel: energy per drug (colchicine) cluster. Right panel: energy per drug molecule in a cluster.

***Source*: reproduced with permission from Ashrafuzzaman et al. (2021).**

Ashrafuzzaman et al. (2021). This adhesion force was then converted into adhesion energy (Γ_{ads}) using the following equation (Horcas et al., 2007):

$$\Gamma_{ads} = \frac{F_{ads}}{2\pi R} \tag{8.2}$$

Here R is the radius of the AFM tip, approximately 8 nm. The adhesion energy between drug and cell membrane ($\Gamma_{drug-cell}$) has been calculated by using the following equation (Grierson et al., 2005):

$$\Gamma_{drug-cell} = 2\left(\gamma_{drug} \times \gamma_{cell}\right)^{1/2} \tag{8.3}$$

Here γ is the surface energy. Using the same equation, we calculated the surface energy of AFM tip, $\gamma_{tip} = 0.252$ J.m^{-2} where surface energy of SiO_2 surface was used, $\gamma_{SiO2} = 0.15$ J.m^{-2} (Grierson et al., 2005). The adhesive energy $\Gamma_{drug-cell}$ between colchicine clusters and cell membranes has been calculated from quantitative adhesive force measurements by AFM on 35 clusters over 97-590 nm size distribution, chosen randomly from 5 different cells. $\Gamma_{drug-cell}$ is plotted against cluster size in Figure 8.10. $\Gamma_{drug-cell}$ is observed to decrease with increasing cluster size. We then calculated $\Gamma_{drug-cell}$ per single colchicine drug and found that the adhesion energy per colchicine decreased with increasing cluster size before gradually saturating beyond 225 nm cluster size (see Figure 8.10).

The adhesive energy plot as a function of drug cluster size (Figure 8.10) suggests that the energetics is dependent on the phenomenon of drug clustering and shows no constant values; instead it suggests that the energetics may lead to a different phenomenon as the cluster size changes. To understand this I modelled the drug clustering on the cell membrane surface, as shown in Figure 8.11 (Ashrafuzzaman et al., 2021).

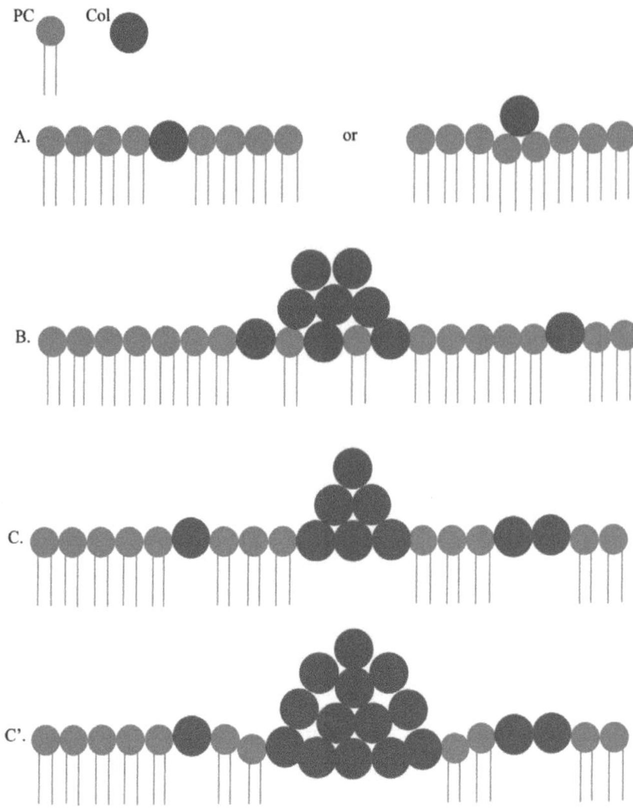

FIGURE 8.11 The distribution of drugs (Col: colchicine) on the cell surface has been modeled. For simplicity, we have assumed that the cell membrane is constructed using just phosphatidylcholine (PC). Top panel (A) demonstrates the single drug adsorption on the cell surface, which we brand as the drug in the lipid cluster (right side confirmation is also supported due to lipid membrane elasticity, Huang, 1986, which ensures modest lipid monolayer bending before actually breaking or rearranging the structure as shown on the left side). B, C, and C' demonstrate the creation of drug clusters on the cell surface. B shows that the drugs and lipids may coexist inside a cluster in its 2D base (cross-sectional view over one lipid layer is shown here) aligned with the surface of the membrane monolayer, whereas C suggests that the drug cluster expels lipids totally, the scenario we have branded as drugs in drug clusters. Drug clusters in both distinguishable conditions presented in B, and in C and C', may push the monolayer a bit toward the hydrophobic membrane core, depending on the amount of downward acting gravity, which is predicted (not calculated here) to be higher for larger clusters (see C'). C and C' represent identical conditions, except for the latter being a little bigger than the former cluster. The different types of drug clustering may also concomitantly exist on different locations of the cell surface. As both drugs and lipids are continuously dynamic with the time scale of molecular movement of the order of milliseconds (ms) or less (maybe considered equivalent to the ion pore stability as detected in various studies, including e.g. Ashrafuzzaman and Tuszynski, 2012; Ashrafuzzaman et al. 2013), the whole process follows principles of statistical mechanics. For simplicity, we have just sketched a lipid monolayer, although cell membrane is a lipid bilayer. But it is also true that while penetrating into the cell through lipid membrane from the cell surface, drugs face cell surface lipid monolayer, as we have modeled here. The chemotherapy drug (CD)-induced pore formation certainly involves both monolayers (models are presented in Ashrafuzzaman et al., 2011; Ashrafuzzaman et al., 2012).

***Source*: reproduced with permission from Ashrafuzzaman et al. (2021).**

Following the modeling of the drug clustering on the cell surface, I utilized a powerful theoretical tool (Ashrafuzzaman and Tuszynski, 2012; Ashrafuzzaman et al. 2013), the screened Coulomb interaction (SCI) among participating agents through the charges they hold, to calculate the energetics of drug insertion into the clusters. This SCI takes the following form:

$$V_{sc}(\vec{r}) = \int d^3k Exp\left\{i\vec{k}.\vec{r}\right\} V_{sc}\left(\vec{k}\right) \tag{8.4}$$

which is in Fourier space:

$$V_{sc}\left(\vec{k}\right) = \frac{V\left(\vec{k}\right)}{1 + \frac{V\left(\vec{k}\right)}{2\pi k_B T} n} \tag{8.5}$$

where $V(k) = (1/\varepsilon_0\ \varepsilon_r) q_D q_L / k^2$ is the direct Coulomb interaction between drug (D) (charge q_D) in a drug/lipid cluster and the nearest-neighbor drug/lipid (lipid charge q_L), k is wave number, T is the absolute temperature of the aqueous phase, n is the density of particles, k_B is Boltzmann's constant. ε_0 is the dielectric constant in vacuum, ε_r is the relative dielectric constant where interaction takes place.

The Fourier transformation of Equation (8.4) was performed using Mathematica 9.1 program-based NCs (for detailed methodology, see Ashrafuzzaman and Tuszynski, 2012; Ashrafuzzaman et al., 2013) to detect values of drug binding energy δG between drug–drug/lipid in drug or lipid clusters. δG is 1st, 2nd, etc. order terms in the expansion of $V_{sc}(r)$ for the hydrophobic mismatch to be filled by single, double, etc. particles representing 1st, 2nd, etc. order screening, respectively, in the cluster. In our current investigation q_D and q_L are charges of Col (q_{Col}) and PC (q_{PC}), respectively, in the aqueous phase. $K \approx 2\pi/a$, a is the average Col-Col or Col-PC distance in Col or PC clusters while both appear as nearest neighbors to each other. The values could be assumed as average Col dimension (~1 nm) or PC headgroup dimension (~0.77 nm), respectively. In reality, we might also need to take into consideration many other parameters related to e.g. variations in the membrane's electrical conditions, the presence of hydrocarbons within, etc. n, the particle (drug or lipid) density, was chosen ~1/60 Å^2 (for a~0.77 nm). $k_B T \approx 1.38 \times 10^{-23}$ Joule/K (300 K), ε_0 is the dielectric constant in vacuum, ε_r (~2) is the relative dielectric constant inside the membrane (Parsegian, 1969).

A series of NCs have been performed considering the drug distributions in Col and PC domains on the cell surface (see Figure 8.11). The approximate consideration of 2D treatment for a Col cluster is possible when the cluster size is small (the height of the cluster approaches the lowest possible value). Numerically computed energy versus reaction coordinate plots of Equation (8.4) have been presented in Figures 8.12 and 8.13. All curves have two energy states, namely low energy state (G_I) and high energy state (G_{II}) (see Figure 8.12 where we have presented single transitions for all four SCI orders). The curves undergo transitions (accounting for drug–drug/lipid association ↔ dissociation) between these two energy states at various values of r (see Figure 8.13). $\delta G = |G_I\text{-}G_{II}|$. As values of G_Is for all SCI orders are almost the same, the difference in δG arises from increasing (on a geometric order) values of G_{II}s with increasing SCI order.

δGs for all four SCI orders have been deducted and their logarithmic values are plotted against corresponding inter-particle separations in Figure 8.14 for different particle charge (relative charges of Col and PC) and lattice constant (a) (Col and PC dimensions) conditions. δG is observed to increase exponentially with increasing SCI order (or corresponding increasing inter-particle separation d) for drug interactions in both drug and lipid clusters. Values of δG for drug clusters are higher than that for lipid clusters. Another important observation is made that as the value of lattice constant (a) decreases (e.g. from 1 nm to 0.77 nm), the value of δG increases for corresponding values

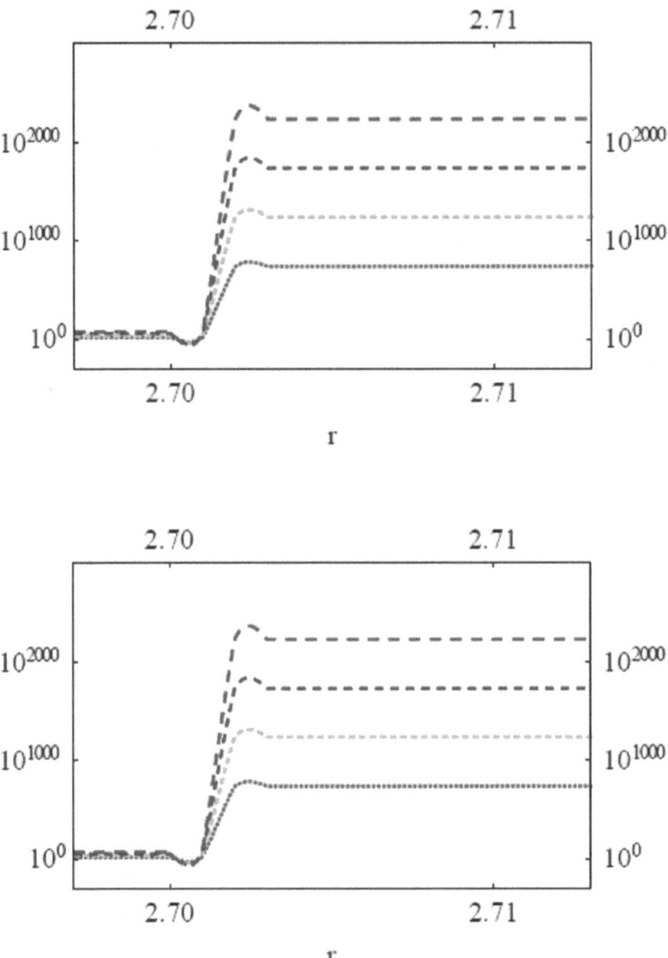

FIGURE 8.12 Numerically computed scale free energy versus reaction coordinate r plots of Equation (8.4). Col in Col cluster (top panel) and Col in lipid (e.g. PC) cluster (bottom panel). In both panel figures there are four curves representing energies of four SCI orders, 1st, 2nd, 3rd, 4th (bottom to top). Here a = 1 nm, $q_{Col}q_{PC} = 10^{-2}e^2$, as PC is nearly charge neutral, but membrane active agents (MAAs) are negatively charged in physiological aqueous phase. *e* is the electron charge.

***Source*: reproduced with permission from Ashrafuzzaman and Tuszynski (2012).**

of *d*, due to perhaps closer packing of particles (checked for $q_{Col}q_{PC} = 10^{-16}e$). The formation of bigger drug clusters accounting for the requirement of higher SCI orders in drug–drug interactions is harder (Ashrafuzzaman and Tuszynski, 2012). Concomitantly, the formation of smaller drug clusters is easier and is expected to happen with a higher probability than that of bigger drug clusters. Similarly, binding of a drug with distant lipids requiring higher SCI orders in drug–lipid interactions is less likely, so the probability of existing single drug with the minimum (low order SCI) cell-surface interactions is higher. Figure 8.15 presents the values of δG in kJ/mole for Col-PC interactions (for a = 1nm and $q_{Col}q_{PC} = 10^{-2}e$). It seems that the required free energy change for inter-particle association↔dissociation goes fast beyond the physiologically relevant energies (~kJ/mole in single-digit [22]) as *d* increases. δG is found to be in order that was detected in MD simulations for a drug–lipid pair interaction where $d \leq 2$ nm (Ashrafuzzaman et al., 2012). The binding energetics of

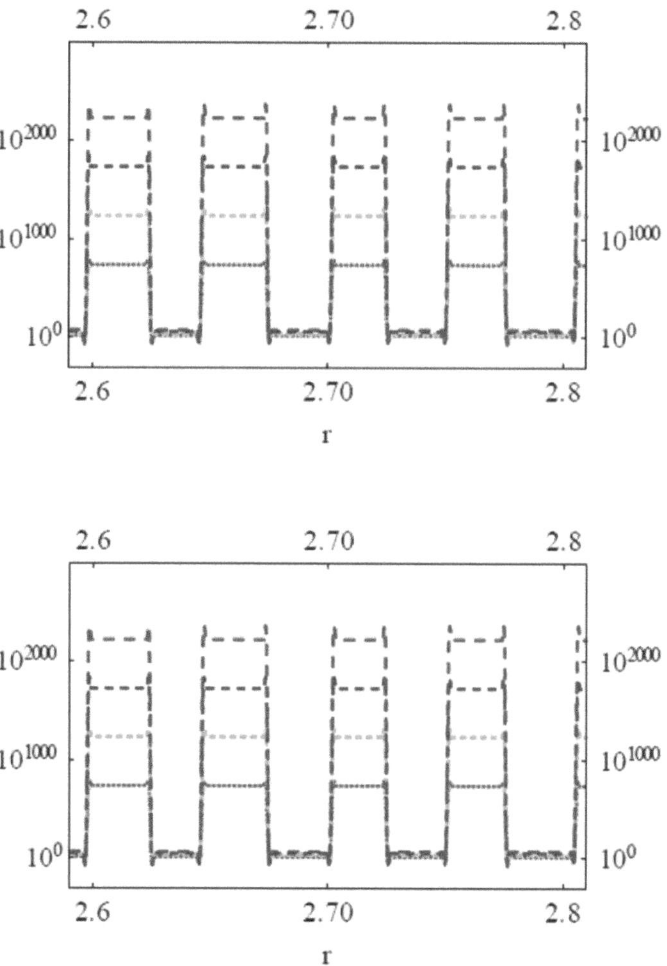

FIGURE 8.13 Numerically computed energy versus reaction coordinate plots of Equation (8.4). Col in Col cluster (top panel) and Col in lipid cluster (bottom panel). In both panel figures there are four curves representing energies of four SCI orders, 1st, 2nd, 3rd, 4th (bottom to top). Here a = 1 nm, $q_{Col}q_{PC} = 10^{-2}e^2$. The back-and-forth transitions for any SCI order shown here happen between two energy states; the pattern of the transition is shown (from lower to higher energy state transition) in expanded scale plots in Figure 8.12.

***Source*: reproduced with permission from Ashrafuzzaman and Tuszynski (2012).**

drugs and lipids on the cell surface demonstrates that the probability of observing single colchicine on the cell surface is highest, and the probability of finding drug clusters decreases as cluster size increases. This molecular-level energetics supports the AFM data (see Figure 8.7).

The plot of Log($\delta G_{Col\text{-}Col}/\delta G_{Col\text{-}PC}$) addresses the relative energetics any drug experiences in a drug cluster in comparison to that in a lipid cluster (see Figure 8.16). Values of Log($\delta G_{Col\text{-}Col}/\ \delta G_{Col\text{-}PC}$) increase proportionally with *d*, suggesting heavier energetic costs for the transition into drug cluster while drugs interact over higher-order screening. We may hypothesize that the single drug on the lipid membrane causes the initiation of drug cluster which is the smallest theoretical size limit of a cluster. Adding more and more drugs into the cluster requires exponentially increasing higher-order energetics. The first collapse of the adhesion energy (accounting for stronger binding) between drug and cell surface with increasing cluster size (Figure 8.10, right panel) hints at the requirement

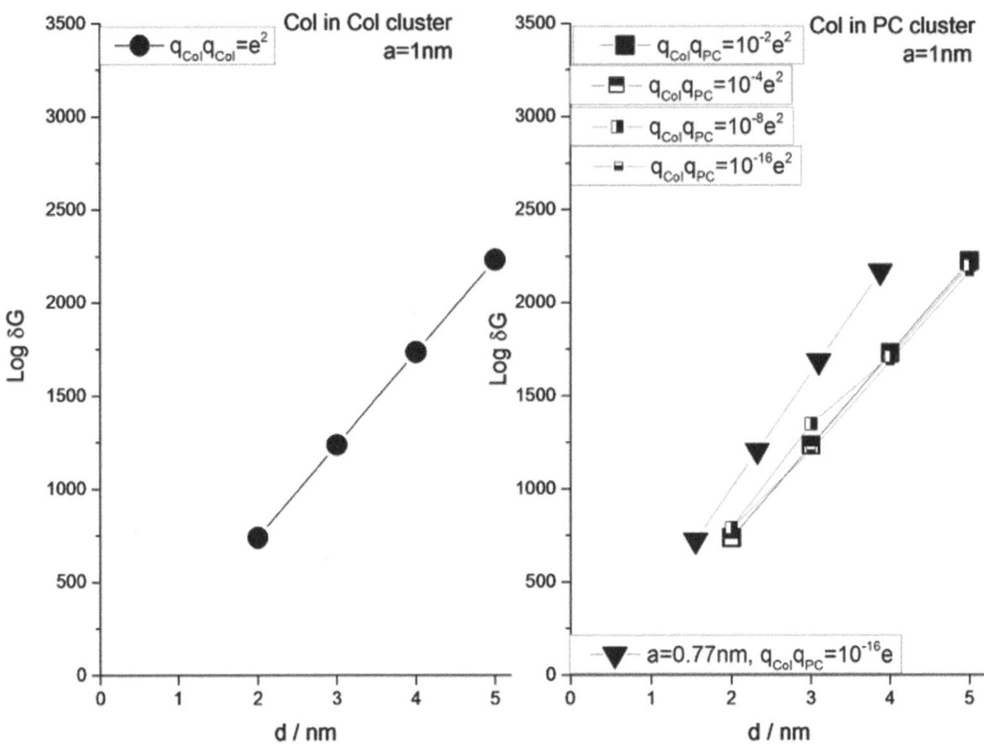

FIGURE 8.14 Energy for inter-particle distance for 4 SCI orders. The inter-particle separation *d* is deducted for each SCI order considering a close drug–drug (left panel) or drug–lipid (right panel) packing conformation.

***Source*: reproduced with permission from Ashrafuzzaman and Tuszynski (2012).**

of higher-order change of energetics for drugs' association↔dissociation, an experimental finding explained in the trend found in numerical investigations of SCI model calculations (Figure 8.16).

The above-presented explanations have already provided enough evidences of active cell surface energetics that may regulate the cell surface morphology. We have provided experimental evidence and theoretical analysis on the understanding of the cell surface physics. We can also address the surface physics, e.g. specifically energetics of cell surface binding phenomena using various *in silico* computed parameters. My group recently performed several case studies where molecular dynamics (MD) simulation has been performed to address dynamical behavior among agents present on the cell surface. As a concrete example, we may present here the results of the study on interactions of cells approaching various drugs with cell membrane lipids, equivalent to the case modelled in Figure 8.11, where we have induced dynamics (Ashrafuzzaman et al., 2020). Here only pairwise interactions and dynamics have been presented to understand the molecular-level active interaction energetics contributing to their mutual dynamics.

Figure 8.17 summarizes the MD results through a probability function (Ashrafuzzaman et al., 2020). MD simulations for PS aptamer- and CD-lipid were conducted earlier (Ashrafuzzaman et al., 2012; Tseng et al., 2011; Ashrafuzzaman et al., 2013). Similar MD simulations have been carried out for gA- and Alm-lipid pairs in Ashrafuzzaman et al. (2020). Three physical quantities – (a) the separation distance between the centers of mass of the agent and lipid, $d_{agent\text{-}lipid}$, (b) van der Waals (vdW) and (c) electrostatic (ES) energies – were used to analyze simulation results. The solvent accessible surface area (SASA) was calculated using Amber Tools 11 (Case et al., 2010) when both drug and lipid molecules were completely separated. In order to investigate features of physical

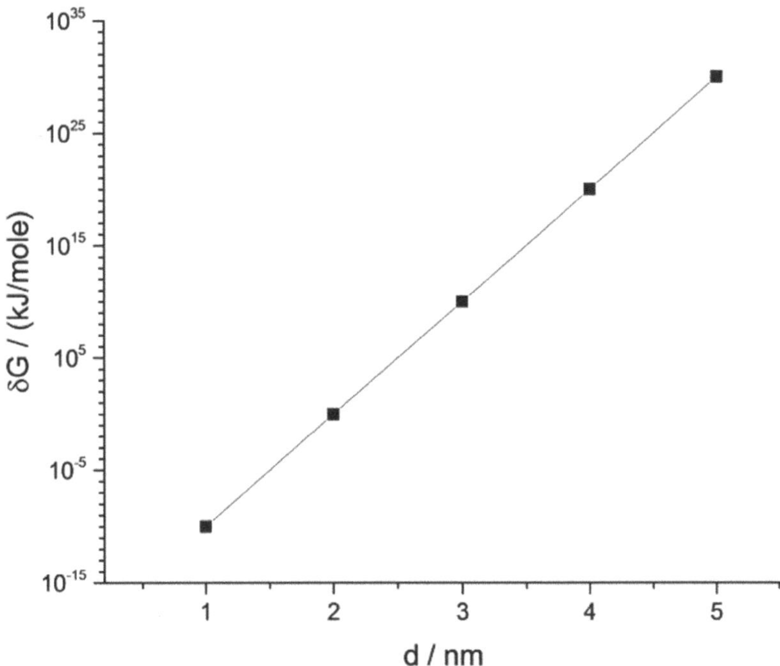

FIGURE 8.15 Energy in kJ/mole for inter particle distance, following ad hoc assumptions (see "Materials and methods: NCs considering relevant physiological parameters," in Ashrafuzzaman and Tuszynski, 2012). Values of δG are found in order detected in MD simulations for a drug–lipid pair interaction where d ≤ 2 nm.

***Source*: reproduced with permission from Ashrafuzzaman and Tuszynski (2012).**

interactions of all pairs of lipids and agents from MD results, a probabilistic description is proposed (Ashrafuzzaman et al., 2020). We first evaluate the probability of observing a pair within $d_{agent\text{-}lipid}$ as $P(d_{agent\text{-}lipid}) = \Delta t(d_{agent\text{-}lipid})/T_{sim}$, where $\Delta t(d_{agent\text{-}lipid})$ is the time duration the agent-lipid pair stays within $d_{agent\text{-}lipid}$ and T_{sim} is the total simulation time. Second, the probability of having either vdW or ES energy of a lipid and an agent staying at distance $d_{agent\text{-}lipid}$ is given by following Boltzmann distribution:

$$P\left(E\left(d_{agent-lipid}\right)\right) = \exp - \beta E\left(d_{agent-lipid}\right) / Z \tag{8.6}$$

Here the partition function is $Z = \sum_{d_{agent-lipid}} \exp - \beta E\left(d_{agent-lipid}\right)$ and $\beta = 1/k_B T$ with T = 300 K.

Universal footprint revealed by MD results. Figure 8.17 shows plots of $P(E(d_{agent\text{-}lipid}))$ against $P(d_{agent\text{-}lipid})$, and the corresponding $d_{agent\text{-}lipid}$ values are represented by symbol size, which is illustrated in the bottom panel. Considering all three variables together, Figure 8.17 shows a corresponding three-dimensional plot. The upper row shows the case with PC and lower row for PS and the bottom row shows the size of symbols, which denotes the distance between an agent and lipid. The left, middle, and right columns represent vdW, ES, and the sum of vdW and ES energies, respectively. This figure reveals several interesting features. First, it shows similar trends for all three categories of agents against either PC or PS from the vdW interactions' point of view. Probabilities of having a pair within $d_{agent\text{-}lipid}$ and vdW energy $E_{vdW}(d_{agent\text{-}lipid})$ gradually decrease when the separation distance between the lipid and an agent $d_{agent\text{-}lipid}$ increases. Namely, the vdW force is likely to play a crucial

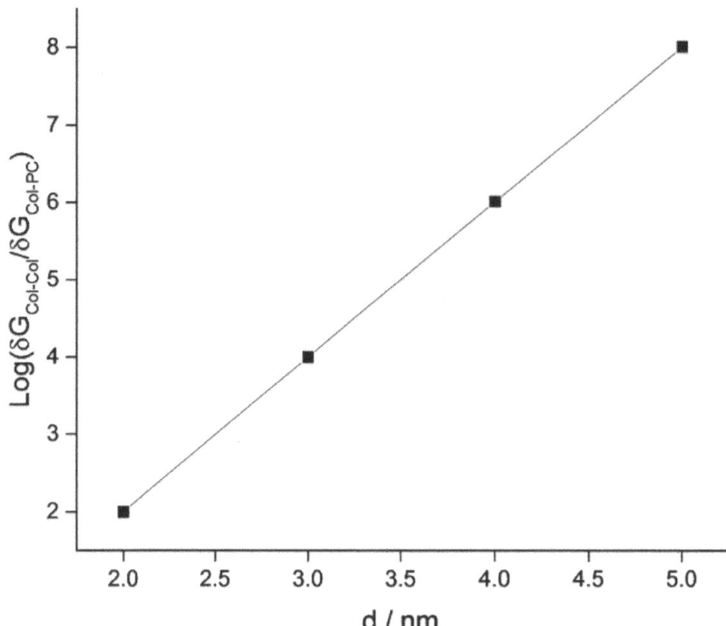

FIGURE 8.16 Plot of Log($\delta G_{Col\text{-}Col}/\ \delta G_{Col\text{-}PC}$) addresses the relative energetic cost a drug experiences in falling into a drug cluster from lipid bound state. $\delta G_{Col\text{-}Col}$ and $\delta G_{Col\text{-}PC}$ are the values of δG for Col in Col cluster and Col in PC cluster, respectively (plotted independently). A = 1 nm.

***Source*: reproduced with permission from Ashrafuzzaman and Tuszynski (2012).**

role in all types of agents binding with lipids (short $d_{agent\text{-}lipid}$ range). Second, from the ES energy point of view, two chemotherapy drugs (CDs) (TCC and TXL) are the only type of agents to show similar trends, namely the larger the separation distance $d_{agent\text{-}lipid}$ is, the lower the probabilities $P(d_{agent\text{-}lipid})$ and $P(E(d_{agent\text{-}lipid}))$ are, for both PC and PS cases. It suggests that similarly to the vdW force, the ES force is also likely a mechanism for the binding process of CDs. Yet, the ES force is likely to play only a minor role in the binding of lipids and agents such as peptides and aptamers. This is probably due to the polarities of charges on participating agents and thus the ES force can play a role either to favor or to disfavor the binding. PC is nearly neutral, and PS and all MAAs except gA are negatively charged (Ashrafuzzaman et al., 2020). The negligible lipid binding for aptamer II is naturally valid as the sequence has been designed as a negative control not to bind with the target lipids (Tseng et al., 2011; Ashrafuzzaman et al., 2013). Despite having different membrane effects, all MAAs bind directly to lipids with considerable binding stability and the vdW and ES interactions represent a universal molecular mechanism producing the necessary driving forces to bring charges together. We view this novel mechanism as a very important finding in membrane science. The most important message here is that if we can determine all the important molecular-level interactions, we can correctly predict the binding phenomena. In this regard, detection of energies in a single MAA-lipid complex (Figure 8.17) provides better molecular-level understanding than in a usually expected MAA-lipid complex in a membrane. Consequently, we can detect the primary (not the collective) energy values. The overall energies in a MAA-lipid complex in a membrane are simply a combination of these individually detected contributing energies. Therefore, our MD simulation strategy and the detected energy-based discovery of the two associated probabilities appear as strong functions correlating the molecular-level information with the phenomenological observations of MAA-lipid complexes in lipid membranes as predicted from various *in vitro* experiments. This new

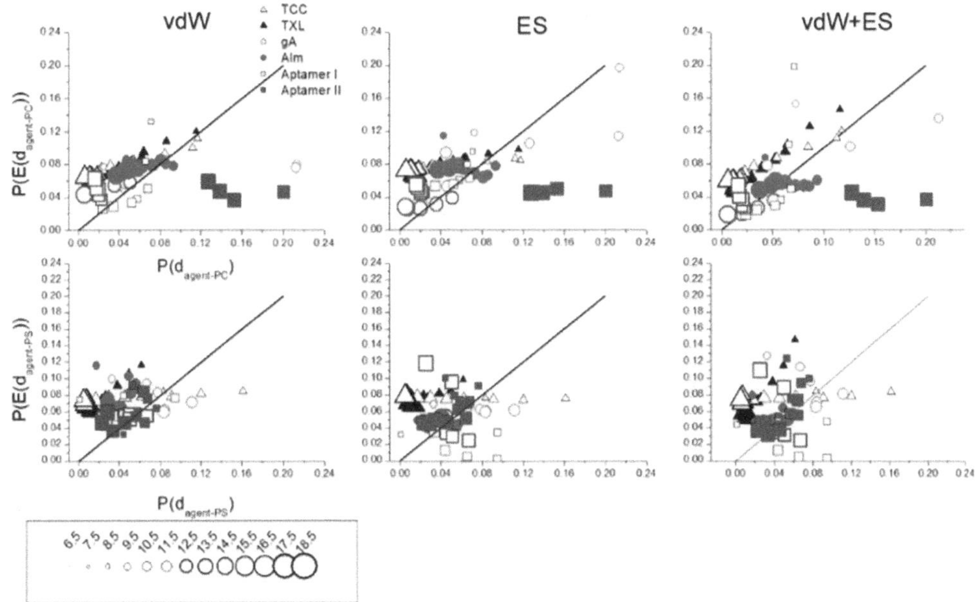

FIGURE 8.17 Analysis of MD results of all lipid-agent pairs. Values of $d_{\text{agent-lipid}}$ are represented by symbol sizes (for simplicity only circle sizes are shown) (Ashrafuzzaman, 2018). Lipid binding agents utilized in MD simulations are TCC: thiocolchicoside, TXL: taxol, gA: gramicidin A, Alm: alamethicin, Aptamer I: 5'-AAAAGA-3', Aptamer II: 5'-AAAGAC-3'.

***Source*: Ashrafuzzaman et al. (2020).**

approach closely correlates information between *in silico* and *in vitro* experiments addressing the cell surface active physics.

8.5 BIOPHYSICS OF THE BIOMOLECULAR SURFACE

Proteins in aqueous solution of biological systems change conformations by folding, wrapping, and winding. These conformational changes are influenced by the surrounding water molecules which solvate proteins (see Figure 8.18) (Metzler, 2018). The presence of proteins also influences water environment as water molecules may be restricted in their movement while navigating around the complex protein surface (Pal et al., 2002).

A femtosecond (fs) resolution dynamical solvation study addressing the physical behavior of water molecules on protein surfaces provided crucial understanding (Pal et al., 2002). The protein surface hydration is a dynamical process showing two distinct trajectories, as follows:

1. Fast solvation. This results from the weak interactions with the selected surface site, giving rise to bulk-type solvation with solvation time ~1 picosecond (ps)
2. Slow solvation. This results from a stronger interaction, capable of defining a rigid water structure, with a solvation time ~38 ps

Beyond 7 Å from the protein surface, essentially all trajectories appear as bulk-type (see Figure 8.19).

The bulk-like water environment in the close vicinity of the protein surface is expected to enhance the interaction with the substrate. Concomitantly, the structured water molecules with specific time

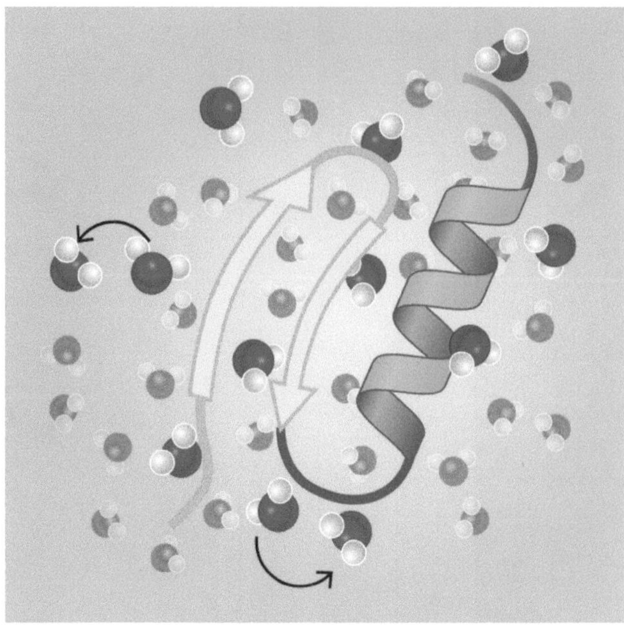

FIGURE 8.18 Schematic of a single protein molecule surrounded by a surface water layer. Some individual water molecules of this layer are depicted with small arrows indicating in-layer jumps as well as jumps to higher water layers or the surrounding bulk water.

***Source*: reproduced with permission from Metzler (2018).**

scale for hydration and exchange dynamics surrounding the protein surface are chemically important and lead to maintaining the three-dimensional (3D) protein structure, as seen in Figure 8.20.

A recent study of the water dynamics around proteins combining experiments, simulations, and modeling has provided a picture of the hopping behavior of the protein surface water molecules (Tan et al., 2018). Neutron scattering experiments, molecular dynamics (MD) simulations, and related analytic modeling on the hydrated perdeuterated protein powders altogether report water molecules to jump randomly between protein surface trapping sites, whose waiting times are found to obey a broad distribution, resulting in subdiffusion. The diffusive dynamics of the surface water on hydrated perdeuterated green fluorescent protein (GFP) and cytochrome P450 (CYP) at physiological conditions has been provided. Figure 8.21 presents the neutron susceptibility spectra $\chi''(q,\nu)$, which is expressed as

$$\chi''(q,\nu) = {S(q,\nu)}\Big/{n_B(\nu)} \tag{8.7}$$

where $S(q,\nu)$ is the dynamic structure factor, presenting the distribution of the dynamic modes in the sample over frequency at a given wave vector q. $n_B(\nu)$ is the Bose factor, expressed as

$$n_B(\nu) = \frac{1}{\left(\exp\left(\frac{h\nu}{kT} - 1\right)\right)} \tag{8.8}$$

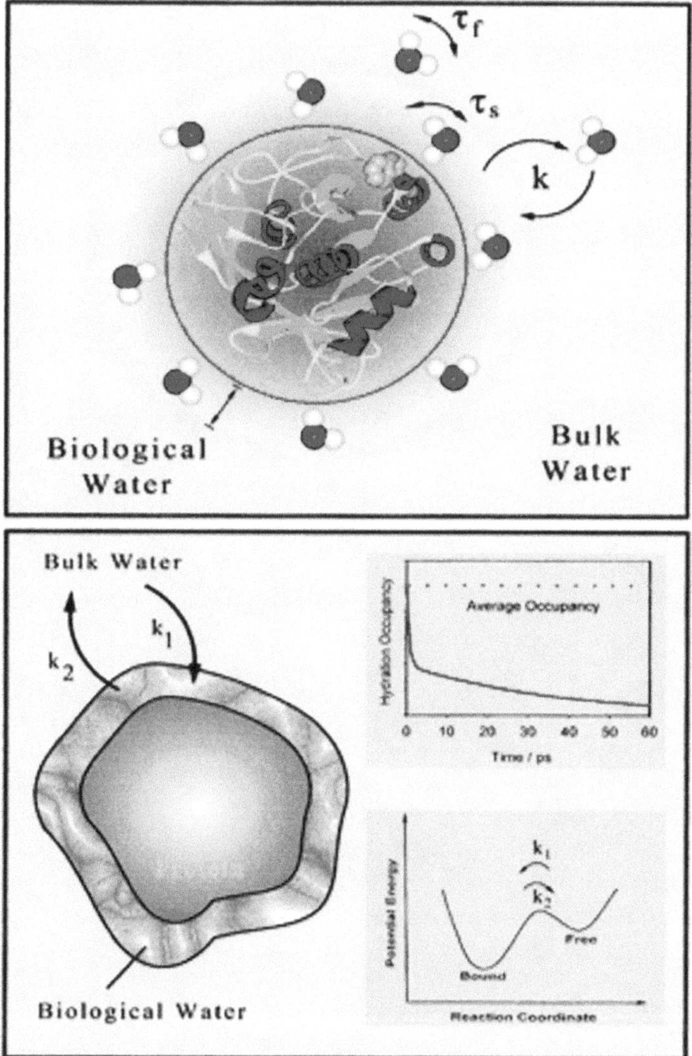

FIGURE 8.19 A representation of the dynamic model discussed in the text for the solvation of proteins. (*Upper*) A water structure is shown, but it should not be viewed as a static layer. Instead, the distribution in residence time at the surface site defines two types of water motions: those that are as fast as bulk water, τ_f (weak interactions), and those much slower, τ_s (strong interactions). The exchange of surface and bulk water is displayed with two rate constants, $k = k_1$ or k_2. (*Lower*) We correlate the solvation behavior to a hydration occupancy, but the time scale shown is only for illustration. The free energy change is shown (*Inset*).

***Source*: reproduced with permission from Pal et al. (2002).**

The characteristic relaxation time, t_c at a given q, is roughly the time for water molecules to diffuse a distance of ~$2\pi/q$; see Figure 8.21c for the q dependence of t_c). Both simulation and experimental results (the trend of the q dependence of t_c) appear with considerable mutual agreement.

Figure 8.22a presents a scatter plot to project the MD trajectory of the oxygen atom of one selected water molecule on CYP recorded at every 100 ps continuously over 100 ns. These trajectories on the ps-to-100-ns time scales suggest clearly that the water dynamics consists of two

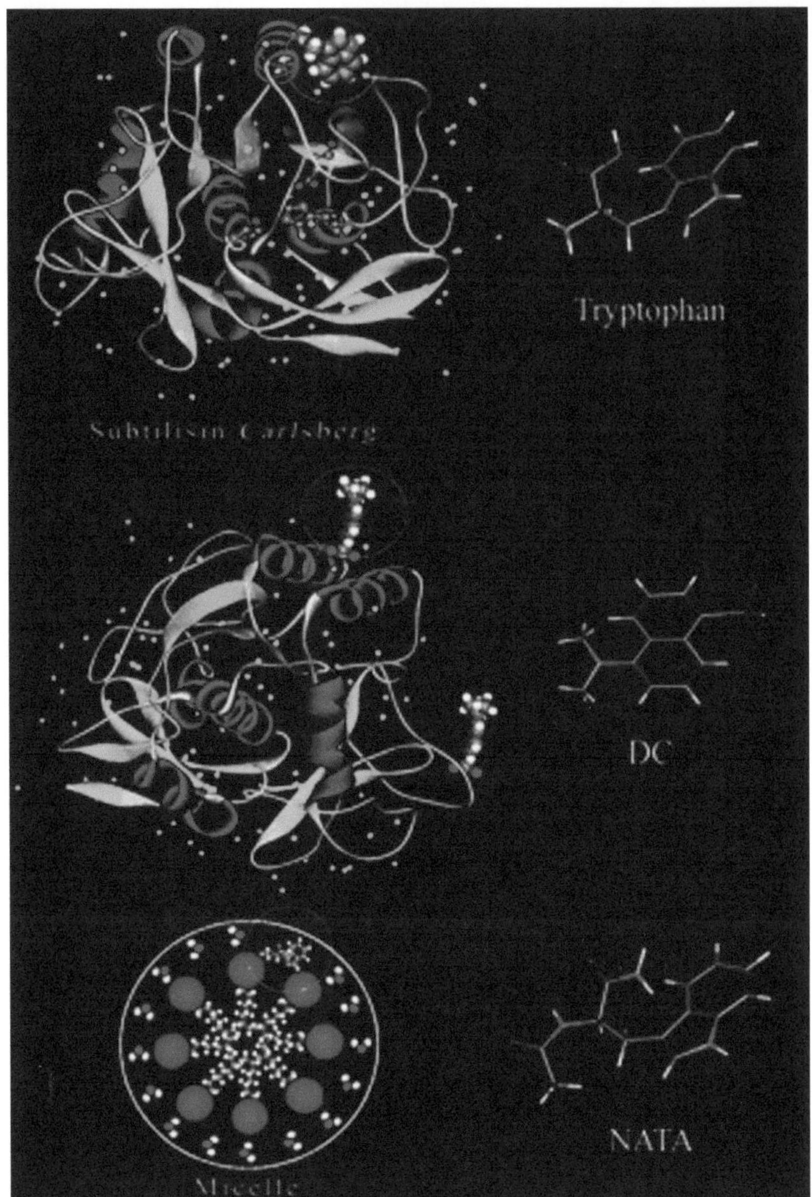

FIGURE 8.20 High-resolution X-ray structure of the protein subtilisin *Carlsberg* (SC). This structure was downloaded from the Protein Data Bank and processed with WEBLAB-VIEWERLITE, Accelrys, San Diego, CA. (*Top*) the position of the single Trp residue of the protein. Note the bound water molecules around this residue. (*Middle*) Two of the nine potential binding sites for DC labeling are shown. (*Bottom*) Illustration of a micelle with a NATA molecule included. Molecular structures of the probes are presented at the right of each illustration.

***Source*: reproduced with permission from Pal et al. (2002).**

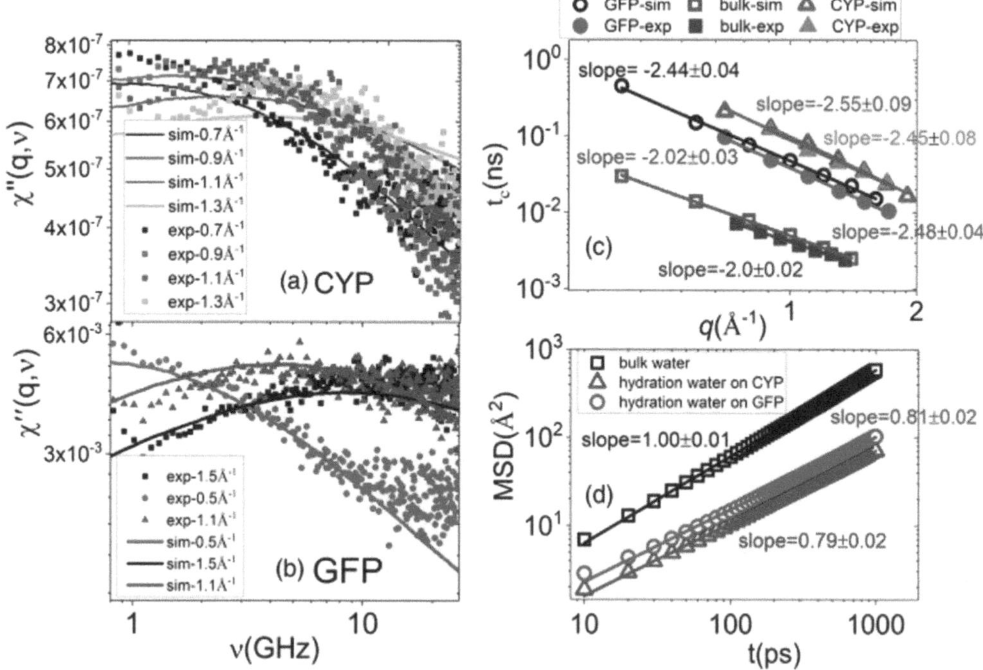

FIGURE 8.21 Neutron susceptibility spectra $\chi''(q,\nu)$ for hydration water derived from experiment and from MD simulation on perdeuterated (a) CYP and (b) GFP (Nickels et al., 2012) at various q. (c) q dependence of the characteristic relaxation time t_c in nanosecond (ns) of hydration water and bulk water derived by fitting the χ'' spectra to the Cole-Cole function (see Tan et al., 2018). Solid symbols represent experimental values while open ones denote the MD-derived values. Spheres denote hydration water on GFP, triangles correspond to hydration water on CYP, and squares represent bulk water. (d) The MD-derived mean square atomic displacement (MSD) for both hydration and bulk water in the time window from 10 ps to 1 ns, black (bulk water), red (water on GFP), and blue (water on CYP). Solid lines are power-law fits, with the exponents displayed accordingly. Copied with due permission to reproduce from ref. Tan et al. (2018).

***Source*: reproduced with permission from Tan et al. (2018).**

modes: rattling within one trapping site (basin) at short time scales and jumping over to neighboring traps at longer time scales. The hopping distance is about 2–3 Å, with possibility to obtain larger jumps. This supports the scenario of temporal disorder. The typical size of trapping basins is about 1–2 Å. It is unlikely that two water molecules occupy the same basin at the same time, as the interwater molecule distance is about 3 Å, which is larger than the size of the traps. The step-step auto correlation function C(k) doesn't change from 0 with the progressing values of k.

The many-body volume-exclusion effect makes water molecules jump preferentially among shallow sites, and thus effectively diffuse faster. The resulting greater mobility in water can eventually be delivered to the enclosed protein molecule to gain sufficient flexibility required for its function. This might be the reason why certain hydration, ~ 20 percent in weight, is required for enzymes to present appreciable anharmonic dynamics and bioactivity (Rupley and Careri, 1991), as such many-body effect will be insufficient when the hydration is considerably low.

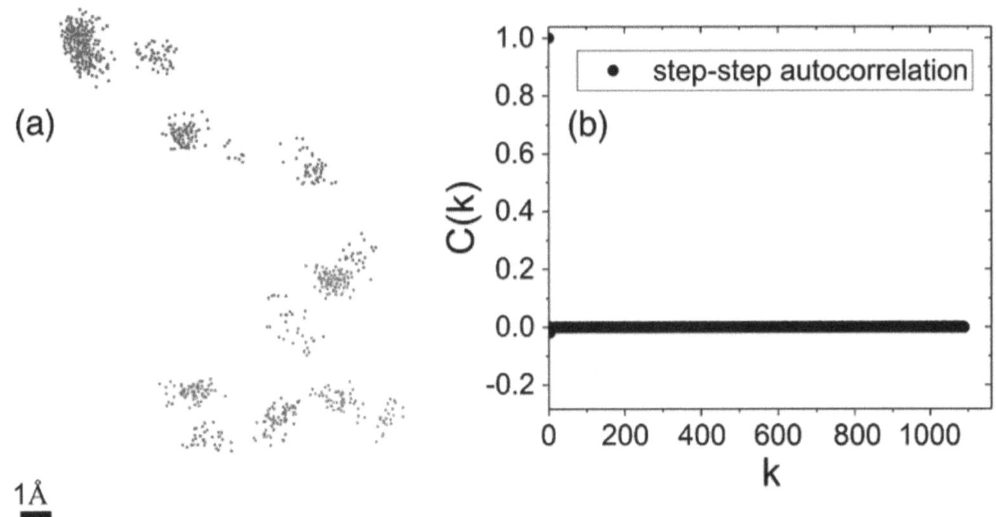

FIGURE 8.22 (a) The scatter plot obtained by projection of the MD trajectory of a selected water molecule on CYP. The water molecule's oxygen atom position is projected at every 100 ps for a continuous trajectory of 100 ns. (b) The step-step auto correlation function C(k) of jump displacements between centers of trap sites, ensemble averaged over all the hydration water molecules on CYP, has been plotted against k.

***Source*: reproduced with permission from Tan et al. (2018).**

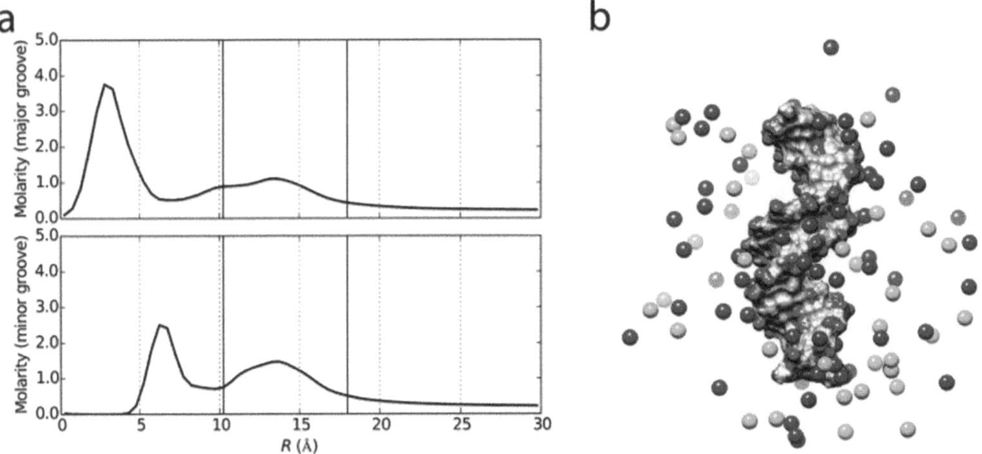

FIGURE 8.23 (a) Radial distribution of K^+ ions along the distance to the DNA helical axis, averaged over a series of microsecond simulations of different tetranucleotides at physiological ion concentration, in the major (upper panel) and minor (lower panel) grooves. (b) Distribution of K^+ (blue) and Cl^- (green) ions around DNA, where the oxygen and phosphorus atoms are highlighted in red and purple.

***Source*: reproduced with permission from Pasi et al. (2015) and Laage et al. (2017).**

Like hydrated proteins, hydrated nucleic acids show important surface physics properties. Molecular dynamics (MD) simulations of the hydrated B-DNA (Korolev et al., 2003; Lavery et al., 2014) provide detailed insight into the time-averaged location of counterions around DNA and their spatial distribution functions. A recent microsecond (μs) MD simulation included a physiological aqueous condition (150 mM K^+Cl^- ion concentration) around the double-stranded DNA oligomers consisting of 18 base pairs, 36 phosphate groups, and their Na^+ counterions (see Figure 8.23) (Pasi et al., 2015). We observe an enhanced distribution of positive ions in the DNA–water interface with ~75 percent of these ions residing at radial distances from the helical axis of less than 20 Å. Two layers of ions are formed, with maxima in the radial distribution function at $r \approx 5$–7 and 12 Å. Sodium and potassium counterions are found to reside in the minor groove, modifiying the chain of tightly bound water molecules. The ion occupancy is seen to change with the DNA sequence. The minor groove ion occupancy is less than 10 percent and ions affect only the few molecules in their hydration shell and affect the overall DNA first hydration shell dynamics just modestly. The K^+ ion atmosphere is seen to extend to a radial distance of up to about 25 Å (see Figure 8.24). Counterion

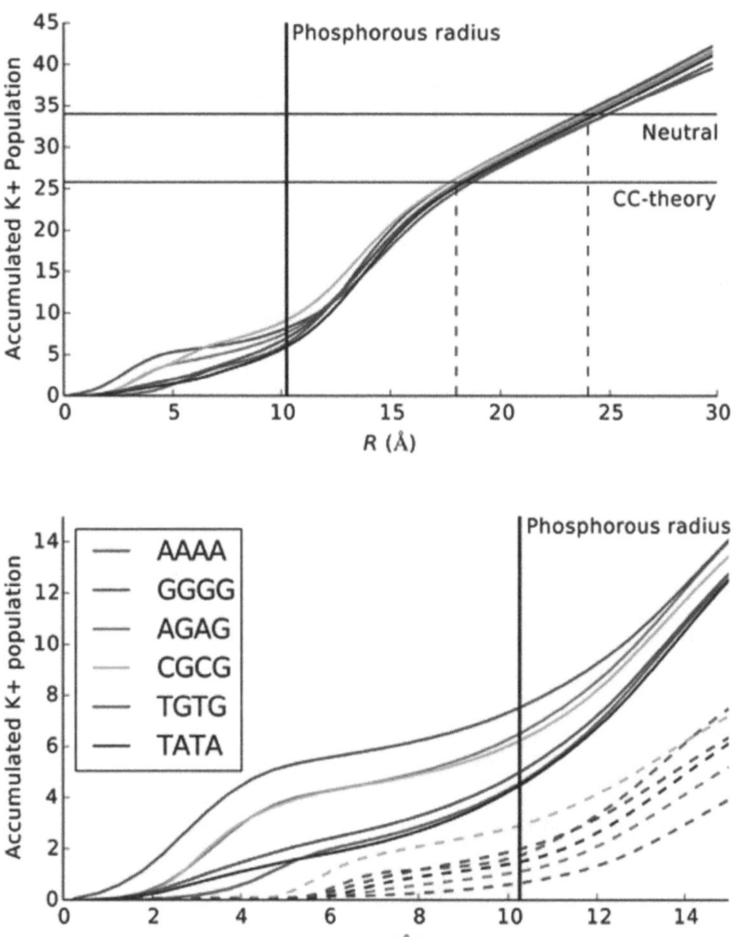

FIGURE 8.24 Accumulated K^+ populations. Cumulative K^+ ion population as a function of distance from the helical axis (R, in Å). Populations are integrated along the full length of six 18-mers (see legend of bottom panel) and either over the full angular range (top panel), or discriminating the major and minor grooves (bottom panel, solid and dashed lines, respectively). The solid vertical line indicates the phosphorous radius at R = 10.25 Å. Horizontal lines in the top panel show the accumulated ion population required to fully neutralize the DNA (one negative charge on each of the 34 phosphates of an 18-mer), or to achieve 76 percent neutralization (CC-theory). Vertical dashed lines indicate the average distances (24 and 18 Å, respectively) at which these conditions are achieved.

***Source*: reproduced with permission from Pasi et al. (2015).**

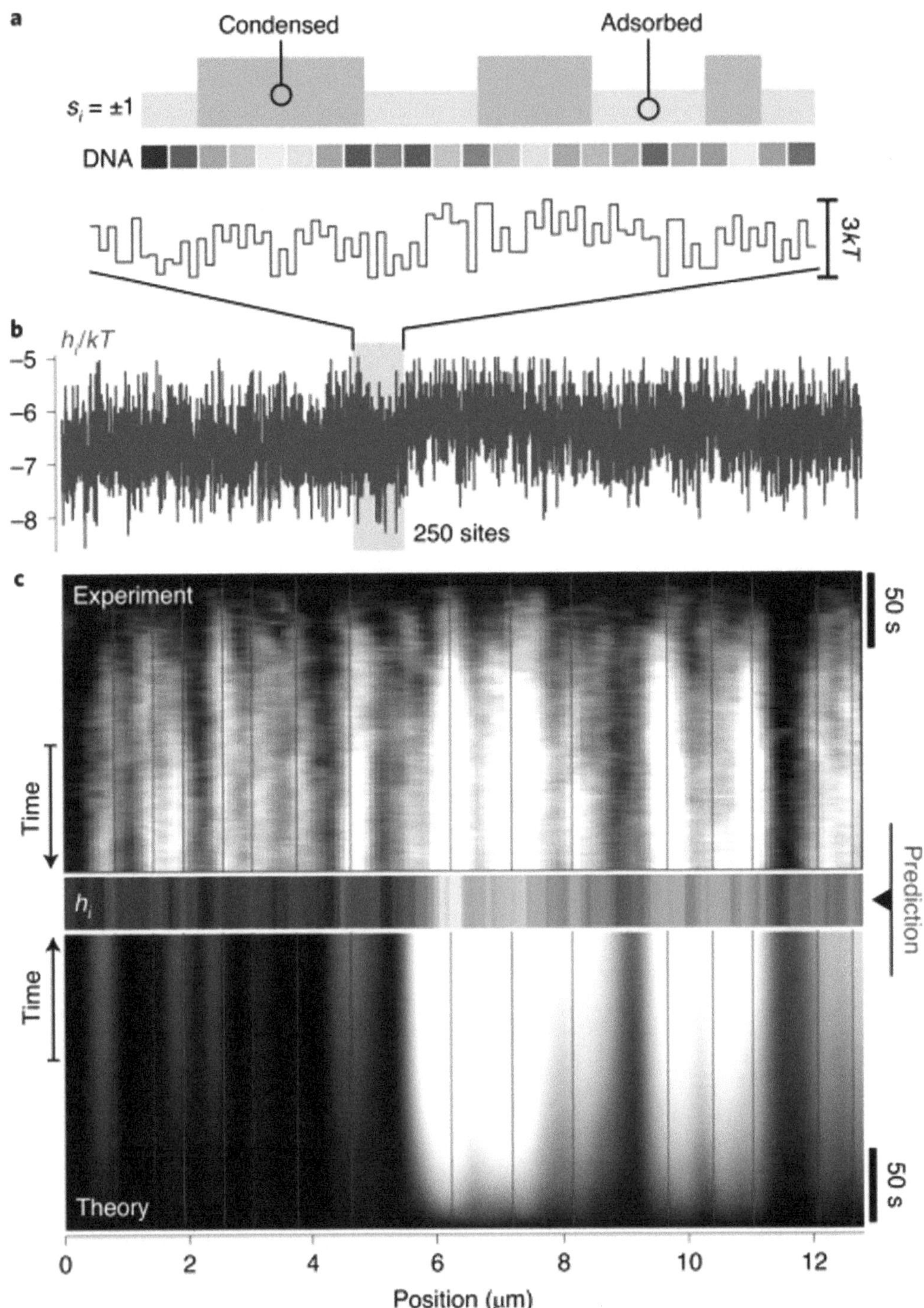

FIGURE 8.25 (a) Schematic of the two-state heterogeneous model of prewetting. The DNA is considered as a one-dimensional lattice of sites that can be in either one of the two states, adsorbed state ($s_i = -1$) or condensed state ($s_i = +1$). (b) DNA sites have an inhomogeneous propensity for condensation (denoted by h_i) given by the binding energy landscape inferred from the position weight matrix. (c) Average experimental kymograph ([Klf4-GFP]: 105–281 nM, $N = 34$; top) shows a spatial localization pattern consistent with the coarse-grained

FIGURE 8.25 *(Continued)*

binding energy landscape (middle, color map representation of h_i). The average kymograph obtained from the model ($N = 400$; bottom) captures this pattern. The spatial dimension of the model kymograph was convolved with a Gaussian function with a full-width at half-maximum corresponding to that of the point spread function of the microscopy system. For additional details, see Morin et al. (2022), where from this figure is copied with due permission to reproduce.

***Source*: reproduced with permission from Morin et al. (2022).**

residence times around phosphates are on the order of 100 ps and range from 100 ps to more than 10 ns in the grooves, with limited diffusive motion along the grooves (Várnai and Zakrzewska, 2004; Pérez et al., 2007).

Surface physics of biomolecules appears as an important determinant for the biomolecules to function. In the example case explained above, I have provided evidence using the physics on the hydrated protein and DNA surfaces. But other physics such as surface fluidity, surface charge profiles (Leung et al., 2009), etc. are also active that may dictate in biomolecular functions. Transcriptional condensates on the DNA surface appeared to contain thousands of molecules. Although the physics behind the formation of small condensates on DNA surfaces is still poorly known, a recent investigation to address the nature of transcription factor condensates using the pioneer transcription factor Krüppel-like factor 4 (Klf4) revealed that Klf4 can phase separate on its own at high concentrations, but at low concentrations, Klf4 only forms condensates on DNA surfaces (Morin et al., 2022). A prewetting transition of pioneer transcription factors on DNA surfaces underlies the formation and positioning of transcriptional condensates and provides robustness to transcriptional regulation. A two-state heterogeneous model of prewetting on DNA captures the sequence-dependent switching of Klf4 to a condensed state (see Figure 8.25). This suggests the presence of important physics active on the DNA surface. Like the protein surface and DNA surface, other biomolecular surfaces always hold important physical properties and active physical laws. But the field is still new, so not enough information is available. Over time we shall acquire more detail in this field.

REFERENCES

Ashrafuzzaman, M. (2018). *Nanoscale biophysics of the cell.* Cham: Springer. https://doi.org/10.1007/978-3-319-77465-7

Ashrafuzzaman, M. (2015). Phenomenology and energetics of diffusion across cell phase states. *Saudi Journal of Biological Sciences*, *22*(6), 666–673

Ashrafuzzaman, M., AlMansour, H.A., AlOtaibi, M.A., Khan, Z., & Shaik, G.M. (2022). Lipid specific membrane interaction of aptamers and cytotoxicity. *Membranes*, *12*(1), 37

Ashrafuzzaman, M., Duszyk, M., & Tuszynski, J. (2011). Chemotherapy drug molecules thiocochicoside and taxol permeabilize lipid bilayer membranes by forming ion channels. *J. Phys. Conf. Ser.*, *329*, 012029

Ashrafuzzaman, M., Khan, Z., Alqarni, A., Alanazi, M., & Alam, M.S. (2021). Cell surface binding and lipid interactions behind chemotherapy-drug-induced ion pore formation in membranes. *Membranes*, *11*, 501

Ashrafuzzaman, M., Tseng, C.-Y., Duszyk, M., & Tuszynski, J.A. (2012). Chemotherapy drugs form ion pores in membranes due to physical interactions with lipids. *Chem. Biol. Drug Des.*, *80*, 992–1002. doi: 10.1111/cbdd.12060

Ashrafuzzaman, M., Tseng, C.Y., Kapty, J., Mercer, J.R., & Tuszynski, J.A. (2013). Computationally designed DNA aptamer template with specific binding to phosphatidylserine. *Nucl. Acid. Ther.*, *223*, 418–426

Ashrafuzzaman, M., Tseng, C.-Y., & Tuszynski, J.A. (2020). Charge-based interactions of antimicrobial peptides and general drugs with lipid bilayers. *J. Mol. Graph. Model.*, *95*, 107502

Ashrafuzzaman, M., & Tuszynski, J.A. (2012). Regulation of channel function due to coupling with a lipid bilayer. *J. Comput. Theor. Nanosci.*, *9*(4,) 564–570. doi:10.1166/jctn.2012.2062

Azarbayjani, A.F., & Jouyban, A. (2015). Surface tension in human pathophysiology and its application as a medical diagnostic tool. *BioImpacts*, *5*(1), 29

Azarbayjani, A.F., Jouyban, A., & Chan, S.Y. (2009). Impact of surface tension in pharmaceutical sciences. *Journal of Pharmacy and Pharmaceutical Sciences*, *12*(2), 218–228

Azarbayjani, A.F., Khu, J.V., Chan, Y.W., & Chan, S.Y. (2011). Development and characterization of skin permeation retardants and enhancers: a comparative study of levothyroxine-loaded PNIPAM, PLA, PLGA and EC microparticles. *Biopharmaceutics & Drug Disposition*, *32*(7), 380–388

Azarbayjani, A.F., Lin, H., Yap, C.W., Chan, Y.W., & Chan, S.Y. (2010). Surface tension and wettability in transdermal delivery: a study on the in-vitro permeation of haloperidol with cyclodextrin across human epidermis. *Journal of Pharmacy and Pharmacology*, *62*(6), 770–778

Azarbayjani, A.F., Tan, E.H., Chan, Y.W., & Chan, S.Y. (2009). Transdermal delivery of haloperidol by proniosomal formulations with non-ionic surfactants. *Biological and Pharmaceutical Bulletin*, *32*(8), 1453–1458

Balaban, N.Q., Schwarz, U.S., & Riveline, D. (2001). Force and focal adhesion assembly: a close relationship studied using elastic micropatterned substrates. *Nat. Cell Biol.*, *13*, 466–472

Bidan, C.M., Kommareddy, K.P., Rumpler, M., Kollmannsberger, P., Bréchet, Y.J.M., Fratzl, P., & Dunlop, J.W.C. (2012). How linear tension converts to curvature: geometric control of bone tissue growth. *PLoS One*, *7*, e36336

Boda, D., Eck, E., & Boda, K. (1997). Measurement of surface tension in biological fluids by a pulsating capillary technique. *J. Perinat. Med.*, *25*(2),146–152. doi: 10.1515/jpme.1997.25.2.146; PMID: 9189834

Bodenstein, L. (1986). A dynamic simulation model of tissue growth and cell patterning. *Cell Differ.*, *19*(1), 19–33. doi: 10.1016/0045-6039(86)90022-9; PMID: 375538

Brakke, K.A. (1992). The surface evolver. *Exp. Math.*, *1*, 141–165

Case, D.A., Darden, T.A., Cheatham, T.E., et al. (2010). AMBER 11. University of California, San Francisco

Chidambaram, J.D., Prajna, N.V., Palepu, S., Lanjewar, S., Shah, M., Elakkiya, S., … & Burton, M.J. (2019). Cellular morphological changes detected by laser scanning in vivo confocal microscopy associated with clinical outcome in fungal keratitis. *Scientific Reports*, *9*(1), 8334

Davis, G.S., Phillips, H.M., & Steinberg, M.S. (1997). Germ-layer surface tensions and "issue affinities" in rana pipiens gastrulae: quantitative measurements. *Dev. Biol.*, *192*, 630–644. doi: 10.1006/dbio.1997.8741

Denchai, A., Tartarini, D., & Mele, E. (2018). Cellular response to surface morphology: electrospinning and computational modeling. *Frontiers in Bioengineering and Biotechnology*, *6*, 155

Depalma, R.G. (1967). Surface forces in biological material: measurement of surface tension by drop volume. *J. Surg. Res.*, *7*, 317–322

Ehrig, S., Schamberger, B., Bidan, C.M., West, A., Jacobi, C., Lam, K., … & Dunlop, J.W. (2019). Surface tension determines tissue shape and growth kinetics. *Science Advances*, *5*(9), eaav9394

Foty, R.A., & Steinberg, M.S. (2005). The differential adhesion hypothesis: a direct evaluation. *Dev. Biol.*, *278*, 255–263

Friedl, P., & Gilmour, D. (2009). Collective cell migration in morphogenesis, regeneration and cancer. *Nat. Rev. Mol. Cell Biol.*, *10*, 445–457

Fung, Y.C. (1995). Stress, strain, growth, and remodeling of living organisms. In *Theoretical, experimental, and numerical contributions to the mechanics of fluids and solids*, ed. J. Casey and M.J. Crochet, pp. 469–482. Basel: Birkhäuser

Grierson, D.S., Flater, E.E., & Carpick, R.W. (2005). Accounting for the JKR-DMT transition in adhesion and friction measurements with atomic force microscopy. *J. Adhes. Sci. Technol.*, *19*, 291–311

Handorf, A.M., Zhou, Y., Halanski, M.A., & Li, W.J. (2015). Tissue stiffness dictates development, homeostasis, and disease progression. *Organogenesis*, *11*(1), 1–15

Harkins, H.N., & Harkins, W.D. (1929). The surface tension of blood serum, and the determination of the surface tension of biological fluids. *Journal of Clinical Investigation*, *7*(2), 263–281

Hauner, I.M., Deblais, A., Beattie, J.K., Kellay, H., & Bonn, D. (2017). The dynamic surface tension of water. *Journal of Physical Chemistry Letters*, *8*(7), 1599–1603

He, B., Doubrovinski, K., Polyakov, O., & Wieschaus, E. (2014). Apical constriction drives tissue-scale hydrodynamic flow to mediate cell elongation. *Nature*, *508*(7496), 392–396

Horcas, I., Fernández, R., Gómez-Rodríguez, J.M., Colchero, J., Gómez-Herrero, J., & Baro, A.M. (2007). WSXM: a software for scanning probe microscopy and a tool for nanotechnology. *Rev. Sci. Instrum.*, *78*, 013705

Huang, H. (1986). Deformation free energy of bilayer membrane and its effect on gramicidin channel lifetime. *Biophys. J.*, *50*, 1061–1070

Jia, H.R., Zhu, Y.X., Xu, K.F., Pan, G.Y., Liu, X., Qiao, Y., & Wu, F.G. (2019). Efficient cell surface labelling of live zebrafish embryos: wash-free fluorescence imaging for cellular dynamics tracking and nanotoxicity evaluation. *Chemical Science*, *10*(14), 4062–4068

Jouyban, A., & Azarbayjani, A.F. (2012). Experimental and computational methods pertaining to surface tension of pharmaceutical. In *Toxicity and drug testing*, ed. W. Acree, pp. 47–70. Croatia: InTechOpen

Jouyban, A., Azarbayjani, A.F., & Acree, W.E. (2004). Surface tension calculation of mixed solvents with respect to solvent composition and temperature by using Jouyban–Acree model. *Chemical and Pharmaceutical Bulletin*, *52*(10), 1219–1222

Kalantarian, A., Ninomiya, H., Saad, S.M.I., David, R., Winklbauer, R., & Neumann, A.W. (2009). Axisymmetric drop shape analysis for estimating the surface tension of cell aggregates by centrifugation. *Biophys. J.*, *96*, 1606–1616. doi: 10.1016/j.bpj.2008.10.064

Kazakov, V.N., Sinyachenko, O.V., Trukhin, D.V., & Pison, U. (1998). Dynamic interfacial tensiometry of biologic liquids – does it have an impact on medicine. *Colloids and Surfaces A: Physicochemical and Engineering Aspects*, *143*(2–3), 441–459

Kazakov, V.N., Vozianov, A.F., Sinyachenko, O.V., Trukhin, D.V., Kovalchuk, V.I., & Pison, U. (2000). Studies on the application of dynamic surface tensiometry of serum and cerebrospinal liquid for diagnostics and monitoring of treatment in patients who have rheumatic, neurological or oncological diseases. *Advances in Colloid and Interface Science*, *86*(1–2), 1–38

Khoubnasabjafari, M., Jouyban, V., Azarbayjani, A., & Jouyban, A. Application of Abraham solvation parameters for surface tension prediction of mono-solvents and solvent mixtures at various temperatures. *J. Mol. Liq.*, *178*, 44–56. doi: 10.1016/j.molliq.2012.11.010

Korolev, N., Lyubartsev, A.P., Laaksonen, A., & Nordenskiöld, L. A molecular dynamics simulation study of oriented DNA with polyamine and sodium counterions: diffusion and averaged binding of water and cations. *Nucl. Acids Res.*, *31*, 5971–5981. doi: 10.1093/nar/gkg802

Krishnamurthy, V.N., & Soundararajan, S. (1966). Charge distribution, electric moments, and molecular structure of thiols and thio ethers. *Journal of Organic Chemistry*, *31*(12), 4300–4301

Laage, D., Elsaesser, T., & Hynes, J.T. (2017). Water dynamics in the hydration shells of biomolecules. *Chemical Reviews*, *117*(16), 10694–10725

Langbein, D. (ed.). (2002). *Capillary surfaces: shape – stability – dynamics, in particular under weightlesness.* Berlin: Springer

Latorre, M., and Humphrey, J.D. (2018). Critical roles of time-scales in soft tissue growth and remodeling. *APL Bioeng.*, *2*(2), 026108. doi: 10.1063/1.5017842; PMID: 31069305; PMCID: PMC6324203

Lavery, R., Maddocks, J.H., Pasi, M., & Zakrzewska, K. (2014). Analyzing ion distributions around DNA. *Nucl. Acids Res.*, *42*, 8138–8149. doi: 10.1093/nar/gku504

Leung, C., Kinns, H., Hoogenboom, B.W., Howorka, S., & Mesquida, P. (2009). Imaging surface charges of individual biomolecules. *Nano Letters*, *9*(7), 2769–2773

McGuire, J.F. (2002). Surfactant in the middle ear and eustachian tube: a review. *Int. J. Ped. Otorhinolaryngol.*, *66*, 1–15. doi: 10.1016/S0165-5876(02)00203-3

Metzler, R. (2018). The dance of water molecules around proteins. *Physics*, *11*, 59

Mgharbel, A., Delanoë-Ayari, H., & Rieu, J.P. (2009). Measuring accurately liquid and tissue surface tension with a compression plate tensiometer. *HFSP Journal*, *3*(3), 213–221. doi: 10.2976/1.3116822

Mombach, J.C., Robert, D., Graner, F., Gillet, G., Thomas, G.L., Idiart, M., & Rieu, J.P. (2005). Rounding of aggregates of biological cells: experiments and simulations. *Physica A: Statistical Mechanics and its Applications*, *352*(2–4), 525–534

Mongera, A., Rowghanian, P., Gustafson, H.J., Shelton, E., Kealhofer, D.A., Carn, E.K., ... & Campàs, O. (2018). A fluid-to-solid jamming transition underlies vertebrate body axis elongation. *Nature*, *561*(7723), 401–405

Morin, J.A., Wittmann, S., Choubey, S., Klosin, A., Golfier, S., Hyman, A.A., ... & Grill, S.W. (2022). Sequence-dependent surface condensation of a pioneer transcription factor on DNA. *Nature Physics*, *18*(3), 271–276

Mottaghy, K., & Hahn, A. (1981). Interfacial tension of some biological fluids: a comparative study. *J. Clin. Chem. Clin. Biochem.*, *19*, 267–271

Nickels, J.D., O'Neill, H., Hong, L., Tyagi, M., Ehlers, G., Weiss, K.L., … & Sokolov, A.P. (2012). Dynamics of protein and its hydration water: neutron scattering studies on fully deuterated GFP. *Biophysical Journal, 103*(7), 1566–1575

Ophir, J., Cespedes, I., Ponnekanti, H., Yazdi, Y., & Li, X. (1991). Elastography: a quantitative method for imaging the elasticity of biological tissues. *Ultrasonic Imaging, 13*(2), 111–134

Pal, S.K., Peon, J., & Zewail, A.H. (2002). Biological water at the protein surface: dynamical solvation probed directly with femtosecond resolution. *Proceedings of the National Academy of Sciences, 99*(4), 1763–1768

Parsegian, A. (1969). Energy of an ion crossing a low dielectric membrane: solutions to four relevant electrostatic problems. *Nature, 221*, 844–846

Pasi, M,; Maddocks, J.H., & Lavery, R. (2015). Analyzing ion distributions around DNA: sequence-dependence of potassium ion distributions from microsecond molecular dynamics. *Nucl. Acids Res., 43*, 2412–2423. doi: 10.1093/nar/gkv080

Pelofsky, A.H. (1966). Surface tension-viscosity relation for liquids. *Journal of Chemical and Engineering Data, 11*(3), 394–397

Pérez, A., Luque, F.J., & Orozco, M. (2007). Dynamics of B-DNA on the microsecond time scale. *J. Am. Chem. Soc., 129*, 14739–14745. doi: 10.1021/ja0753546

Rumpler, M., Woesz, A., Dunlop, J.W., van Dongen, J.T., & Fratzl, P. (2008). The effect of geometry on three-dimensional tissue growth. *J. R. Soc. Interface, 5*, 1173–1180

Rupley, J.A., & Careri, G. (1991). Protein hydration and function. *Advances in Protein Chemistry, 41*, 37–172

Schötz, E.M., Lanio, M., Talbot, J.A., & Manning, M.L. (2013). Glassy dynamics in three-dimensional embryonic tissues. *Journal of the Royal Society Interface, 10*(89), 20130726

Steinberg, M.S. (1963). Reconstruction of tissues by dissociated cells. *Science, 141*, 401–408

Tan, P., Liang, Y., Xu, Q., Mamontov, E., Li, J., Xing, X., & Hong, L. (2018). Gradual crossover from subdiffusion to normal diffusion: a many-body effect in protein surface water. *Physical Review Letters, 120*(24), 248101

Thiessen, D.B., & Man, K.F. (1999). *Surface tension measurements*. Boca Raton: CRC Press

Thomas, G., Burnham, N.A., Camesano, T.A., & Wen, Q. (2013). Measuring the mechanical properties of living cells using atomic force microscopy. *Journal of Visualized Experiments, 76*, e50497

Tseng, C.Y., Ashrafuzzaman, M., Mane, J., Kapty, J., Mercer, J., & Tuszynski, J.A. (2011). Entropic fragment based approach to aptamer design. *Chem. Biol. Drug Des., 78*, 1–13

Várnai, P., & Zakrzewska, K. (2004). DNA and its counterions: a molecular dynamics study. *Nucl. Acids Res., 32*, 4269–4280. doi: 10.1093/nar/gkh765;

Voß-Böhme, A., & Deutsch, A. (2010). The cellular basis of cell sorting kinetics. *J. Theor Biol., 263*, 419–436. doi: 10.1016/j.jtbi.2009.12.011

Walker, J.L., Fournier, A.K., & Assoian, R.K. (2005). Regulation of growth factor signaling and cell cycle progression by cell adhesion and adhesion-dependent changes in cellular tension. *Cytokine & Growth Factor Rev., 16*, 395–405. doi: 10.1016/j.cytogfr.2005.03.003

Wozniak, M.A., Modzelewska, K., Kwong, L., & Keely, P.J. (2004). Focal adhesion regulation of cell behavior. *Biochimica et Biophysica Acta (BBA) – Molecular Cell Research, 1692*(2–3), 103–119

Xu, H., Li, H., Ke, Q., & Chang, J. (2015). An anisotropically and heterogeneously aligned patterned electrospun scaffold with tailored mechanical property and improved bioactivity for vascular tissue engineering. *ACS Applied Materials & Interfaces, 7*(16), 8706–8718

Yamamoto, Y., Kominami, H., Kobayashi, K., & Yamada, H. (2021). Surface charge density measurement of a single protein molecule with a controlled orientation by AFM. *Biophysical Journal, 120*(12), 2490–2497

9 Biochemical Thermodynamics

Chemical and biochemical thermodynamics require specialized formalisms (Sabatini et al., 2019; Sabatini et al., 2021). Both chemical and biochemical reaction equations have corresponding equilibrium constants, K and K', respectively (Alberty, 1994a, 1994b). Chemical reaction equations deal generally with species while biochemical reaction equations need to deal with reactants at specified pH and concentrations of free metal ions that are bound by reactant species. At specified pH one needs to deal with new thermodynamic properties <transformed thermodynamic properties>, and new values, which are different, especially for the standard transformed Gibbs energy. It has been considered important to distinguish between the standard thermodynamic properties calculated from K and its temperature coefficient and the standard transformed thermodynamic properties calculated from K' and its temperature coefficient. Recently, however, a generalization approach has been proposed to unify conventional thermodynamics and transformed thermodynamics (Sabatini et al., 2021). This chapter will focus on these specialized areas, considering a few example cases to let readers understand dip insights into thermodynamic treatments of chemical and biochemical reactions.

9.1 CONVENTIONAL CHEMICAL THERMODYNAMICS AND TRANSFORMED BIOCHEMICAL THERMODYNAMICS

Chemical thermodynamics employs conventional thermodynamic potentials to deal with chemical reactions (Ewing et al., 1994; IUPAC, 1982, 1988). Biochemical thermodynamics employs transformed thermodynamic quantities to deal with biochemical reactions (Alberty, 1993, 1994a, 1994b, 1996; Alberty et al., 2011; Vinnakota et al., 2009; Moss, 1994; Liebecq, 1997).

All species involved in a chemical reaction are explicitly considered and their atoms and charges are balanced. At equilibrium condition, the thermodynamic or standard equilibrium constant, K, for the chemical reaction is expressed by the following relation:

$$K = \prod_i \left(\frac{\gamma_i c_i}{c_i^0} \right)_{eq}^{\nu_i} \tag{9.1}$$

Here γ_i and c_i are activity coefficient and concentration of the ith chemical species, which is the reactant or product, respectively. c_i^0 is the standard concentration of any species i. Species concentration is considered in unit of mol/dm^3 for the "chemical" convention, while it differs in biochemical convention where the concentration values are ~ 10^{-7} mol/dm^3, e.g. it is the standard concentration

DOI: 10.1201/9781003287780-9

for species H^+ (Sabatini et al., 2021). v_i is the corresponding stoichiometric coefficient (positive for products and negative for reactants) and eq stands for equilibrium. K is dimensionless, which is defined by the following relation (Sabatini et al., 2021; Ewing et al., 1994; IUPAC, 1982, 1988; Alberty, 1994b; Alberty, 1996; Vinnakota et al., 2009; Moss, 1994):

$$\Delta_r G^0 = -RT \ln K \tag{9.2}$$

A biochemical reaction is known to involve many species and simultaneous ancillary reactions. The determination of the equilibrium composition requires many equilibrium expressions and many conservation equations. Under steady state conditions, networks of biochemical reactions take place in living organisms. This steady state condition of living systems typically allows work to be done by chemical reactions at maximal efficiency, i.e. to increase entropy at the minimal rate (Katchalsky and Curran, 2013).

A typical feature of living systems is homeostasis, which keeps the conditions at which biochemical reactions occur within a narrow range. Due to this feature, in a biochemical reaction, the concentration of certain ancillary chemical species, e.g. H^+, Mg^{2+} ions, etc., remains essentially constant. Biochemical reactions are known to require the use of an apparent equilibrium constant or conditional equilibrium constant and in such biochemical reaction, reagents and products are written in terms of the sum of species instead of specific species (Alberty, 1994b, revised by IUBMB in Alberty, 1996; Alberty, 1993; Vinnakota et al., 2009; Moss, 1994). It is valid at a given condition with certain pH, temperature, and ionic strength (Sabatini et al., 2021).

Let us consider a biochemical reaction as follows:

$$aA + bB = cC + dD \tag{9.3}$$

Here A, B, C, and D are biochemical reagents. Then, the conditional equilibrium constant K' is expressed as follows (Sabatini et al., 2021):

$$K' = \frac{\left([C]/c^0\right)^c \left([D]/c^0\right)^d}{\left([A]/c^0\right)^a \left([B]/c^0\right)^b} \tag{9.4}$$

Here c^0 is the standard concentration considered equal for all species, which is, according to the chemical convention, 1 mol/dm^3 (Alberty, 1993, 1994b; Vinnakota et al., 2009; Moss, 1994). The value of K', the conditional equilibrium constant, depends, besides temperature T, pressure p, and ionic strength I of the solution, also on pH and pMg (Ewing et al., 1994; IUPAC, 1982) (according to IUPAC convention). In Sabatini et al. (2021), pH and pMg are defined as $-\log 10 aH^+$ and $-\log 10 aMg^{2+}$, respectively. In *in vivo* systems, biochemical reactions occur at constant pH and pMg (for details, see Alberty, 1992).

The chemical species of the reactants A and B bound amount of H^+ and Mg^{2+} is different from that of these ions bound to the chemical species of the products C and D (Equation 9.3). H^+ and Mg^{2+} ions are thus produced or consumed during the reaction. Therefore, chemical and biochemical reactions require different thermodynamic formulations, since chemical equations are written in terms of specific ionic and elemental species and balance elements and charge, while biochemical equations are written using biochemical reactants consisting of species in equilibrium with each other and do not balance elements assumed to be fixed, such as H^+ and Mg^{2+}. Thus, while pH and pMg are specified, K' for a biochemical reaction is written in terms of sums of species which can be used for calculating a standard Gibbs energy of reaction $\Delta_r G'^0$. Here the prime symbol (′) indicates that the concentrations of H^+ and Mg^{2+} ions are constant, but not at the standard value. Similar to equation (9.2), $\Delta_r G'^0$ is defined as

$$\Delta_r G'^0 = -RT \ln K' \tag{9.5}$$

We may now consider an example case (Sabatini et al., 2021). In cytosolic solutions, the biochemical reactant adenosine 5′-triphosphate (ATP) is composed of the chemical species ATP^{4-}, $HATP^{3-}$, $MgATP^{2-}$, $MgHATP^-$, H_2ATP^{2-}, and Mg_2ATP. For the hydrolysis of ATP, the biochemical reaction is written as:

$$ATP + H_2O = ADP + P_i \tag{9.6}$$

Here adenosine 5′-diphosphate (ADP) is understood to be composed of the chemical species ADP^{3-}, $HADP^{2-}$, H_2ADP^-, $MgADP^-$, and MgHADP. P_i (inorganic phosphate) is similarly understood to be composed of the chemical species PO_4^{3-}, HPO_4^{2-}, $H_2PO_4^-$, and $MgHPO_4$.

According to the definition of Alberty, the pseudoisomers are all the chemical species that form a biochemical reagent (Alberty, 1993, 1994b, 1996; Alberty et al., 2011; Vinnakota et al., 2009; Moss, 1994). The number of pseudoisomers of ATP are six, but of ADP are five, and of P_i are four. The complex species are adducts of these Lewis bases with the Lewis acids H^+ and Mg^{2+}. The equilibria between the pseudoisomers of ATP, ADP, and P_i are shown in Figure 9.1.

$MgATP^{2-}$ is the active species in the binding of enzyme in the cellular active transport and the form responsible for energy production and muscular contraction (Iotti et al., 2005). The biologically relevant ATP pseudoisomer is consequently $MgATP^{2-}$ and the chemical reaction related to ATP hydrolysis is:

$$MgATP^{2-} + H_2O = MgADP^- + H_2PO_4^- \tag{9.6}$$

Considering the equilibrium $H_2PO_4^- \leftrightarrows H^+ + HPO_4^{2-}$, the equilibrium constant for the chemical reaction can be written as:

$$K = \frac{\left(\left[MgADP^-\right]/c^0\right)\left(\left[HPO_4^{2-}\right]/c^0\right)\left(\left[H^+\right]/c^0\right)}{\left[MgATP^{2-}\right]/c^0} \tag{9.7}$$

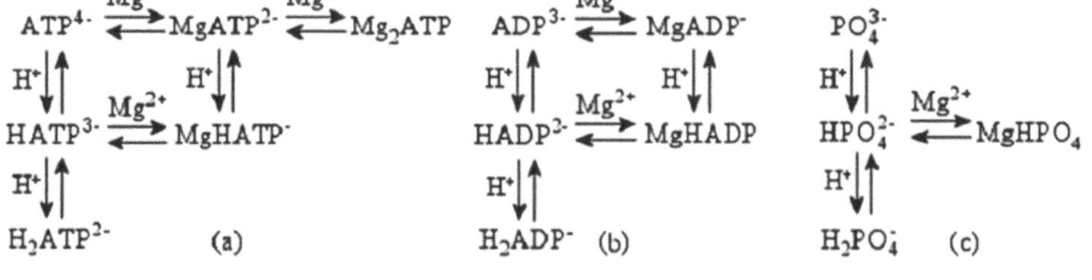

FIGURE 9.1 Equilibria between pseudoisomers of ATP (a), ADP (b) and P_i (c).

***Source*: Reproduced with permission from Iotti et al. (2017) and Sabatini et al. (2021).**

Although the value of K depends on temperature T, pressure p, and ionic strength I of the solution, the conditional equilibrium constant K' depends on these parameters as well as on pH and pMg, and for the biochemical reaction, K' is expressed as:

$$K' = \frac{\left([ADP]/c^0\right)\left([P_i]/c^0\right)}{[ATP]/c^0} \tag{9.8}$$

From the values of K and K', the corresponding values of the standard Gibbs energy of reaction can be obtained using Equations (9.2) and (9.5).

Two distinguishable thermodynamics, chemical thermodynamics and biochemical thermodynamics, based on different concepts and different formalisms, are now generally considered. Chemical thermodynamics uses the conventional thermodynamic properties and is only suitable in dealing with chemical reactions. Biochemical thermodynamics uses transformed thermodynamic properties and is only suitable in dealing with biochemical reactions.

The biochemical reaction complexities entail conceptual and experimental problems in order to determine experimentally the energy released by a specific enzymatic reaction (often considered as free energetics of interactions), and to assess the associated thermodynamic properties.

The biochemical reactions occurring at constant pH and pMg can be conveniently described by conventional thermodynamics (Iotti et al., 2017). Chemical and biochemical thermodynamics may be viewed as being within the same consistent thermodynamic framework.

While addressing chemical and/or biochemical reactions we are used to dealing with chemical equilibria occurring in a solution at constant pH and having reagents that are the "sum of species." The corresponding equilibrium constant, K', is referred to as *conditional* (Ringbom, 1958; Burgot, 2012), whereas $\Delta_r G'$, $\Delta_r H'$, and $\Delta_r S'$ have been so far termed, according to Alberty, transformed thermodynamics properties Gibbs energy, enthalpy, and entropy, respectively (Alberty, 1994b, revised by IUBMB in Alberty, 1996; Alberty, 1993). Sabatini and colleagues (2021) propose the unification of the terminology using the appellation conditional for both K′ and $\Delta_r G'$, $\Delta_r H'$, and $\Delta_r S'$. The term *transformed* thermodynamics is proposed here to be abolished and *conditional* thermodynamics used instead.

9.2 THERMODYNAMICS OF THE ATP SYNTHESIS

An organic compound and hydrotrope *adenosine triphosphate* (*ATP*) provides energy to drive most of the cellular processes. ATP is the principal biomolecule for storing and transferring energy in cells (Zhang et al., 2018). The chemical synthesis of ATP happens in mitochondria through oxidative phosphorylation (Song et al., 2018). Mitochondrial synthesis of ATP involves tricarboxylic acid cycle enzymes, electron transport chain complexes, and ATP synthase, in which acetyl-coenzyme A derived from food molecules is oxidized to produce ATP. During bioenergetic reactions, mitochondria produce several physiologically important molecules such as reactive oxygen species (see Figure 9.2) (Zhang et al., 2018; Nunnari and Suomalainen, 2012).

The thermodynamic efficiency of ATP synthesis in the mitochondrial energy transduction oxidative phosphorylation process has been found to lie around 40–41 percent from the following four different approaches as explained in Nath (2016):

(a) estimation using structural and biochemical data
(b) fundamental nonequilibrium thermodynamic analysis
(c) novel insights arising from Nath's torsional mechanism of energy transduction and ATP synthesis
(d) the overall balance of cellular energetics

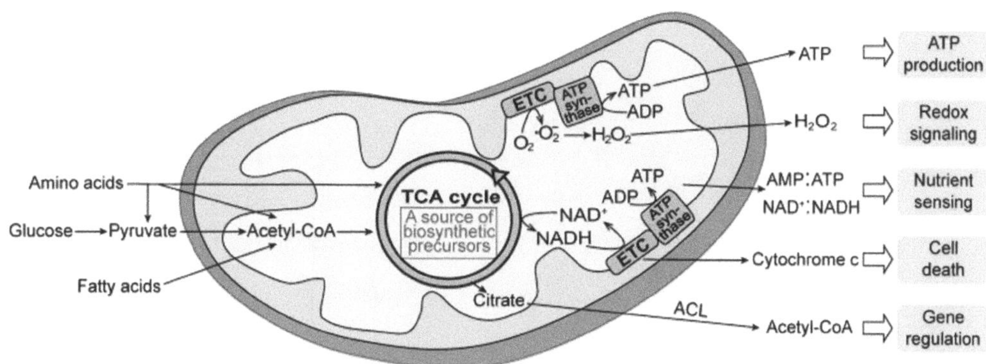

FIGURE 9.2 The mitochondrion is a multifunctional organelle. During mitochondrial bioenergetics, acetyl-coenzyme A converted from food molecules is oxidized to produce ATP. In addition to ATP production, mitochondria play essential roles in cell signaling, including nutrient sensing, redox signaling, cell death, and many other functions. ADP, adenosine diphosphate; AMP, adenosine monophosphate; CoA, coenzyme A; ETC, electron transport chain; TCA, tricarboxylic acid.

The coupled chemical reactions culminating in ATP synthesis by oxidative phosphorylation have been the focus of key studies in biothermodynamics (Jou and Ferrer, 1985; Stucki, 1980; Dewar et al., 2006). The degree of coupling between the exergonic and endergonic reactions is central to a quantitative description of the coupled nonequilibrium process, including the chemical reactions associated with ATP production and ATP/adenosine diphosphate (ADP) conversions (Roels, 1983; Caplan and Essig, 1999; Hansen et al., 2005; Lebon et al., 2008). Recently, the entropy production and its application to the coupled nonequilibrium ATP Synthesis process has been rigorously analyzed, where a few important things are clearly demonstrated (Nath, 2019).

Quantifying experimental data, the rate of entropy production in oxidative phosphorylation and its parsing into its various elementary reaction and transport steps have been determined. Then, the thermodynamic efficiency of the coupled nonequilibrium process is calculated utilizing the knowledge of the rate of entropy production. This is tested by a linear nonequilibrium thermodynamic analysis of ATP synthesis and applied to test the thermodynamic efficacy. Then the degree of coupling between oxidation and ATP synthesis may be measured from the data using nonequilibrium thermodynamics.

Two decades ago, H^+/ATP ratio of proton transport-coupled ATP synthesis was inspected (Turina et al., 2003). The thermodynamic H^+/ATP ratio of the H^+–ATP synthase from chloroplasts was measured in proteoliposomes after energization of the membrane by an acid base transition (Turina et al., 2003). The standard Gibbs free energy of ATP synthesis was determined with the help of a chemiosmotic model system (see the scheme in Figure 9.3). The purified H^+-translocating ATP synthase from chloroplasts was reconstituted into phosphatidylcholine/phosphatidic acid liposomes. The Gibbs free energies of ATP synthesis were determined to be 37 ± 2 kJ/mol at pH 8.45 and 36 ± 3 kJ/mol at pH 8.05, respectively. In a subsequent article the same group reported the thermodynamic H^+/ATP ratio in the chloroplast ATP synthase to be 4.0 ± 0.1 (Turina et al., 2016).

As per the chemiosmotic theory, the ATP synthesis/hydrolysis by the H^+–ATP synthase is coupled with the transport of $n = H^+$/ATP protons across the membrane between the internal and the external aqueous phases of a vesicle (Figure 9.3), i.e.

$$\mathrm{ADP} + \mathrm{P_i} + n\mathrm{H}^+_{\mathrm{in}} \rightleftharpoons \mathrm{ATP} + \mathrm{H_2O} + n\mathrm{H}^+_{\mathrm{out}} \tag{9.9}$$

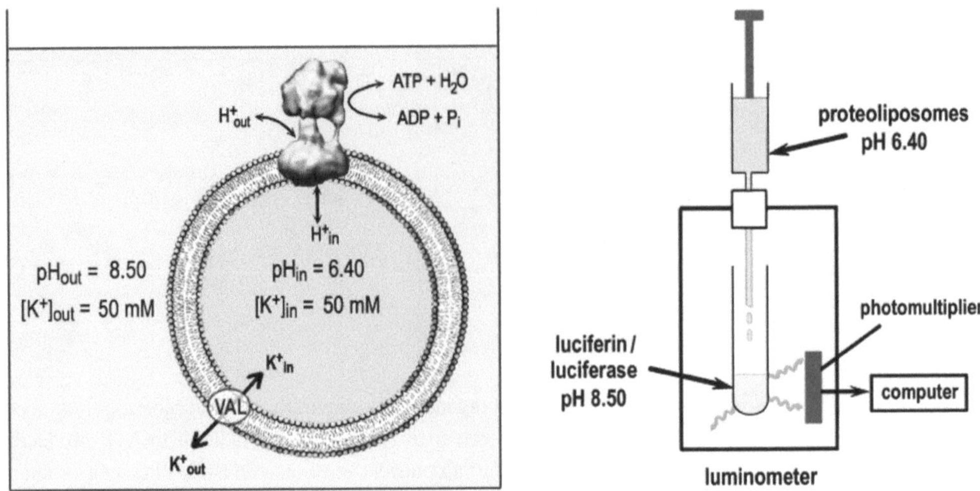

FIGURE 9.3 Scheme of the chemiosmotic system and of the experimental setup. Left: proteoliposomes with a mean diameter of 150 nm are made from phosphatidylcholine/phosphatidic acid and contain 1 CF_0F_1 molecule. Basic structural model of rotary CF_0F_1-ATP synthase may be found in Pänke and Rumberg (1999). The basic buffer is depicted in blue, the acidic buffer in red. Right: scheme of the experimental setup for measurements of ATP synthesis and ATP hydrolysis after generation of a transmembrane ΔpH (± Δφ) by an acid–base transition. Proteoliposomes in the acidic reconstitution medium (red) are injected into a cuvette placed in front of a photomultiplier. The cuvette contains the basic medium and the luciferin/luciferase assay (blue). The luminescence intensity is recorded continuously before and after injection of the proteoliposomes.

***Source*: Reproduced with permission from Turina et al. (2016).**

At equilibrium the ratio of the chemical reactants ATP/(ADP P_i) is equal to the ratio of the transport protons $([H^+]_{in}/[H^+]_{out})^n$. Therefore, the number of protons n that are thermodynamically coupled with the production of one ATP can be determined by measuring the equilibrium between different pre-established ATP/(ADP P_i) ratios and the $([H^+]_{in}/[H^+]_{out})^n$ ratio (Mitchell, 1961, 1966; Pänke and Rumberg, 1997). The H^+/ATP ratio obtained from measuring the chemical equilibrium of the ATP synthesis/hydrolysis reaction in response to ΔpH is referred to as the "thermodynamic H^+/ATP ratio." The transmembrane electric potential difference, Δφ, is a second factor contributing to shift the equilibrium. Such thermodynamic considerations do not require a detailed knowledge of the structure and mechanism of the enzyme.

The Gibbs free energy (ΔG) of the coupled reactions (see Equation 9.10) is given in Turina et al. (2016) (detailed derivations will be found there):

$$\Delta G = \Delta G_{obs} + RT\ln Q - n\Delta\mu_{H^+} \tag{9.10}$$

Here $\Delta\mu_{H^+}$ is the transmembrane electrochemical potential difference of protons. It is usually expressed in terms of the transmembrane pH-difference, $\Delta pH = pH_{out} - pH_{in}$ and of the transmembrane electric potential difference $\Delta\varphi = \varphi_{in} - \varphi_{out}$, as follows (details in Turina et al., 2016):

$$\Delta\mu_{H^+} = 2.3RT\Delta pH + F\Delta\varphi s \tag{9.11}$$

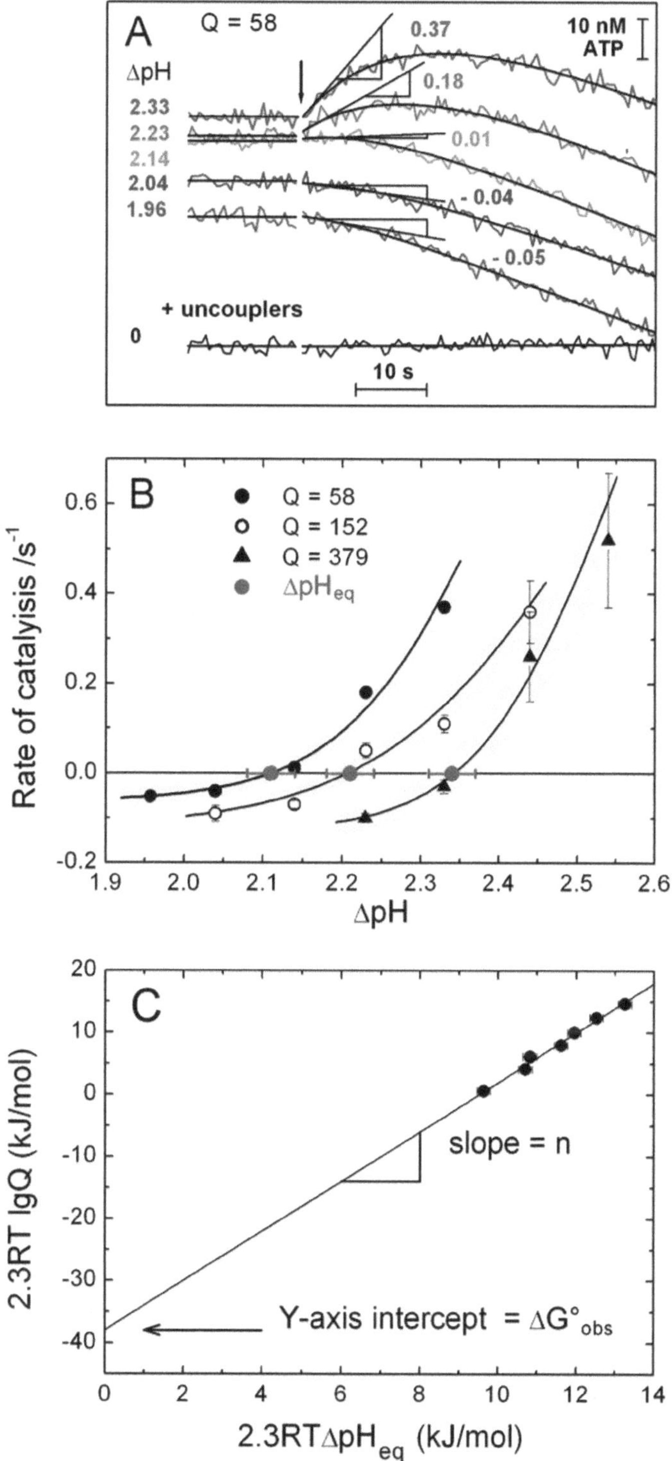

FIGURE 9.4 Determination of H^+/ATP ratio from the pH_{in} dependence of the equilibrium. (A) Luminescence intensity time traces obtained at different ΔpH values. The cuvette with the basic medium (pH_{out} = 8.47, Q = 58 with ADP = 3.4 μM, ATP = 245 nM, P_i = 1.25 mM and luciferin, luciferase) is placed in the luminometer.

FIGURE 9.4 (*Continued*)

Proteoliposomes in the acidic buffer (pH_{in}) are injected into the basic buffer, thereby generating the transmembrane ΔpH (black arrow). An increase in luminescence indicates ATP synthesis; a decrease indicates ATP hydrolysis. The initial rates at t = 0 are calculated from a fit of these traces (black lines) and are given in turnover per second. In the presence of an uncoupler, neither synthesis nor hydrolysis is observed. The luminescence is calibrated by addition of known amounts of ATP after the reaction. (B) Initial rates of ATP synthesis (positive values) and ATP hydrolysis (negative values) are depicted as a function of ΔpH values. Black circles show data from (A), open circles and triangles refer to different reaction products, Q. Each point is the average of three measurements, and black error bars are the corresponding standard deviations. The equilibrium ΔpH (ΔpH_{eq}, red circles) is obtained by interpolation to zero rate with a sigmoidal function. Red error bars of ΔpH_{eq} resulted from the interpolation with the same function of the upper or lower error limits of the rates (shaded area). (C) The reaction product (in units of 2.3RTlgQ) is shown as a function of ΔpH_{eq} (in units of $2.3RT\Delta pH_{eq}$). The pH_{out} = 8.47 is constant and pH_{in} is varied. The slope gives the $n = H^+/ATP$ ratio, the y-axis intercept gives $-\Delta G°_{obs}$ (in this case $H^+/ATP = 4.0 \pm 0.2$ and $\Delta G°_{obs} = (38 \pm 3)$ kJ/mol). Color contrast will be visible in the online version of the book.

***Source*: Reproduced with permission from Turina et al. (2016).**

Hence, we get from Equations (9.10) and (9.11), at equilibrium (ΔG = 0):

$$2.3RTlgQ = -\Delta G_{obs} + n2.3RT\Delta pH_{eq} + nF\Delta\varphi_{eq} \tag{9.12}$$

Equation (9.12) can be used to determine $n = H^+/ATP$ from a plot of 2.3RTlgQ versus $2.3RT\Delta pH_{eq}$ at constant $\Delta\varphi_{eq}$. An exemplification of these measurements is shown in Figure 9.4.

Avoiding presentation of the details, we will just quickly jump to concluding remarks on the thermodynamics of the ATP synthesis (though readers may learn more from the original reference, Turina et al., 2016), as follows:

1. The standard free energy of the reference reaction of the ATP synthesis results $\Delta G°_{ref} = 33.8 \pm 1.6$ kJ/mol.
2. The thermodynamic H^+/ATP ratio is 4.0 ± 0.1.
3. The difference between the thermodynamic (4.0) and the structural (4.7) H^+/ATP ratio is larger than the error limits of the measured values.
4. The thermodynamic H^+/ATP ratio reflects the number of protons necessary for shifting the chemical equilibrium, while the structural H^+/ATP ratio reflects the number of protons translocated during a 360° rotation of the c-ring to the number of ATP generated during a 360° rotation of the γ-subunit. Regarding understanding the γ-subunit rotation in FoF1-ATP synthase, one may read Pu and Karplus (2008). It implies that 2 out of 14 transported protons have to be involved in processes within the enzyme which do not directly shift the chemical equilibrium. Therefore, the ratio 12/14 = 0.85 is assumed to represent the overall efficiency of CF_0F_1 during catalysis.

REFERENCES

Alberty, R.A. (1996). Recommendations for nomenclature and tables in biochemical thermodynamics: recommendations 1994. *European Journal of Biochemistry*, *240*(1), 1–14

Alberty, R.A. (1994a). Biochemical thermodynamics. *Biochimica et Biophysica Acta (BBA) – Protein Structure and Molecular Enzymology*, *1207*(1), 1–11

Alberty, R. A. (1994b). Recommendations for nomenclature and tables in biochemical thermodynamics (IUPAC recommendations 1994). *Pure and Applied Chemistry*, *66*(8), 1641–1666

Alberty, R.A. (1993). Levels of thermodynamic treatment of biochemical reaction systems. *Biophysical Journal*, *65*(3), 1243–1254

Alberty, R.A. (1992). Degrees of freedom in biochemical reaction systems at specified pH and pMg. *Journal of Physical Chemistry*, *96*(24), 9614–9621

Alberty, R.A., Cornish-Bowden, A., Goldberg, R.N., Hammes, G.G., Tipton, K., & Westerhoff, H.V. (2011). Recommendations for terminology and databases for biochemical thermodynamics. *Biophysical Chemistry*, *155*(2–3), 89–103

Burgot, J.L. (2012). *Ionic equilibria in analytical chemistry*, New York: Springer. https://doi.org/10.1007/978-1-4419-8382-4

Caplan, S.R., & Essig, A. (1999). *Bioenergetics and linear nonequilibrium thermodynamics: the steady state*. Cambridge, MA: Harvard University Press

Dewar, R.C., Juretić, D., & Županović, P. (2006). The functional design of the rotary enzyme ATP synthase is consistent with maximum entropy production. *Chem. Phys. Lett.*, *430*, 177–182. doi: 10.1016/j.cplett.2006.08.095

Ewing, M.B., Lilley, T.H., Olofsson, G.M., Ratzsch, M.T., & Somsen, G. (1994). Standard quantities in chemical thermodynamics: fugacities, activities and equilibrium constants for pure and mixed phases (IUPAC recommendations 1994). *Pure and Applied Chemistry*, *66*(3), 533–552

Hansen, L.D., Criddle, R.S., Smith, B.N., Macfarlane, C., Church, J.N., Thygerson, T., Jovanovic, T., & Booth, T. (2005). Thermodynamic law for adaptation of plants to environmental temperatures. *Pure Appl. Chem.*, *77*, 1425–1444

Iotti, S., Frassineti, C., Sabatini, A., Vacca, A., & Barbiroli, B. (2005). Quantitative mathematical expressions for accurate in vivo assessment of cytosolic [ADP] and ΔG of ATP hydrolysis in the human brain and skeletal muscle. *Biochimica et Biophysica Acta (BBA) – Bioenergetics*, *1708*(2), 164–177

Iotti, S., Raff, L., & Sabatini, A. (2017). Chemical and biochemical thermodynamics: is it time for a reunification? *Biophysical Chemistry*, *221*, 49–57

IUPAC Green Book – Physical Chemistry Division. (1988). *Quantities, Units and Symbols in Physical Chemistry*. Oxford: Blackwell Scientific Publications (2nd edn 1993)

IUPAC Physical Chemistry Division. (1982). Manual of symbols and terminology for physicochemical quantities and units. Appendix IV. *Pure Appl. Chem.*, *54*, 1239–1250

Jou, D., & Ferrer, F. (1985). A simple nonequilibrium thermodynamic description of some inhibitors of oxidative phosphorylation. *J. Theor. Biol.*, *117*, 471–488. doi: 10.1016/S0022-5193(85)80155-7

Katchalsky, A., & Curran, P.F. (2013). *Nonequilibrium thermodynamics in biophysics*. Cambridge, MA: Harvard University Press

Lebon, G., Jou, D., &Casas-Vázquez, J. (2008*). Understanding non-equilibrium thermodynamics: foundations, applications, frontiers*. Berlin: Springer

Liebecq, C. (1997). IUPAC-IUBMB Joint Commission on Biochemical Nomenclature (JCBN) and Nomenclature Committee of IUBMB (NC-IUBMB). Newsletter 1996. *European Journal of Biochemistry*, *247*(2), 733–739

Mitchell, P. (1966). Chemiosmotic coupling in oxidative and photosynthetic phosphorylation. *Biol. Rev.*, 41, 445–502

Mitchell, P. (1961). Coupling of phosphorylation to electron and hydrogen transfer by a chemiosmotic type of mechanism. *Nature*, *191*, 144–148

Moss, G.P. (1994). IUBMB-IUPAC Joint Commission on Biochemical Nomenclature (JCBN): Recommendations for nomenclature and tables in biochemical thermodynamics. https://iubmb.qmul.ac.uk/thermod

Nath, S. (2019). Entropy production and its application to the coupled nonequilibrium processes of ATP synthesis. *Entropy*, *21*(8), 746. https://doi.org/10.3390/e21080746

Nath, S. (2016). The thermodynamic efficiency of ATP synthesis in oxidative phosphorylation. *Biophys. Chem.*, *219*, 69–74. doi: 10.1016/j.bpc.2016.10.002; PMID: 27770651

Nunnari, J., & Suomalainen, A. (2012). Mitochondria: in sickness and in health. Cell, *148*(6), 1145–1159. PMID: 22424226

Pänke, O., & Rumberg, B. (1999). Kinetic modeling of rotary CF0F1-ATP synthase: storage of elastic energy during energy transduction. *Biochimica et Biophysica Acta (BBA) – Bioenergetics*, *1412*(2), 118–128

Pänke, O., & Rumberg, B. (1997). Energy and entropy balance of ATP synthesis. *Biochim. Biophys. Acta*, *1322*, 183–194

Pu, J., & Karplus, M. (2008). How subunit coupling produces the γ-subunit rotary motion in F1-ATPase. *Proceedings of the National Academy of Sciences*, *105*(4), 1192–1197

Ringbom, A. (1958). The analyst and the inconstant constants. *Journal of Chemical Education*, *35*(6), 282

Roels, J.A. (1983). *Energetics and Kinetics in Biotechnology*. Amsterdam: Elsevier Biomedical Press

Sabatini, A., Borsari, M., Moss, G.P., & Iotti, S. (2021). Chemical and biochemical thermodynamics reunification (IUPAC Technical Report). *Pure and Applied Chemistry*, *93*(2), 243–252

Sabatini, A., Borsari, M., Raff, L.M., Cannon, W.R., & Iotti, S. (2019). Chemical and biochemical thermodynamics reunification. *Chemistry International*, *41*(2), 34

Song, J., Pfanner, N., & Becker, T. (2018). Assembling the mitochondrial ATP synthase. *Proceedings of the National Academy of Sciences*, *115*(12), 2850–2852

Stucki, J.W. (1980). The optimal efficiency and economic degrees of coupling of oxidative phosphorylation. *Eur. J. Biochem.*, *109*, 269–283. doi: 10.1111/j.1432-1033.1980.tb04792.x

Turina, P., Petersen, J., & Gräber, P. (2016). Thermodynamics of proton transport coupled ATP synthesis. *Biochimica et Biophysica Acta (BBA) – Bioenergetics*, *1857*(6), 653–664

Turina, P., Samoray, D., & Gräber, P. (2003). H+/ATP ratio of proton transport-coupled ATP synthesis and hydrolysis catalysed by CF0F1-liposomes. *EMBO J.*, *22*(3), 418–426

Vinnakota, K.C., Wu, F., Kushmerick, M.J., & Beard, D.A. (2009). Multiple ion binding equilibria, reaction kinetics, and thermodynamics in dynamic models of biochemical pathways. *Methods in Enzymology*, *454*, 29–68

Zhang, J., Han, X., & Lin, Y. (2018). Dissecting the regulation and function of ATP at the single-cell level. *PLoS Biol.*, *16*(12), e3000095. https://doi.org/10.1371/journal.pbio.3000095

10 Nucleic Acid Structure and Stability

Nucleic acids' structure and function are keys to fundamental biological processes related to defining aspects of life. Therefore, I wish to write an independent chapter for the book focusing only on nucleic acids. Both deoxyribonucleic acid (DNA) and ribonucleic acid (RNA) have been explored quite rigorously since their discoveries over a century ago. Following the DNA double helical structure model discovery by Watson and Crick in 1953, nucleic acid research came to the forefront of attention and importance in biomedical science research. Besides understanding primary sequences, the nucleic acid secondary structures and their other alternative ones have been inspected using versatile molecular biology techniques, including biophysical and biochemical analysis and theoretical modelings. Energetics, bonding, base pairing, bending, mutations, etc. are key molecular-level active processes we wish to especially focus on here. Both DNA and RNA, and their respective building blocks, will be individually treated to address their structures, functions, and participation in modern means of medical interventions, including developing vaccines.

10.1 DISCOVERY OF NUCLEIC ACIDS

Molecular biologist James Watson and biophysicist Francis Crick discovered the double helical structure of deoxyribonucleic acid (DNA) in 1953. But the discovery of nucleic acids dates back to 1868 when Friedrich Miescher discovered a substance containing both phosphorus and nitrogen in the nuclei of white blood cells found in pus. Watson and Crick explained their DNA structural findings using a single statement: "This structure has two helical chains each coiled round the same axis" (Watson and Crick, 1953; Figure 10.1). This DNA structure is popularly known as the "DNA double helix model." In their chemical assumptions, Watson and Crick meant each chain to consist of phosphate di-ester groups joining p-D-deoxy-ribofuranose residues with 3', and 5' linkages. Both chains are related by a dyad perpendicular to the fiber axis, both chains following right-handed helices, but owing to the dyad the sequences of the atoms in the two chains running in opposite directions. The bases are on the inside of the helix and the phosphates are on the outside.

There is a residue on each chain every 3.4 Å in the z-direction. An angle of 36 degrees between adjacent residues in the same chain was assumed, so that the structure repeats after 10 residues on each chain, that is, after 34 Å. The distance of a phosphorus atom from the fiber axis was found to be 10 Å. The two chains were found to be held together by the purine and pyrimidine bases. The planes of the bases were predicted to be perpendicular to the fiber axis. They were found joined together in pairs, a single base from one chain being hydrogen-bonded to a single base from the other chain so that the two lie side by side with identical z-coordinates. The pairing occurs between a purine and a pyrimidine to ensure bonding between two strands.

DOI: 10.1201/9781003287780-10

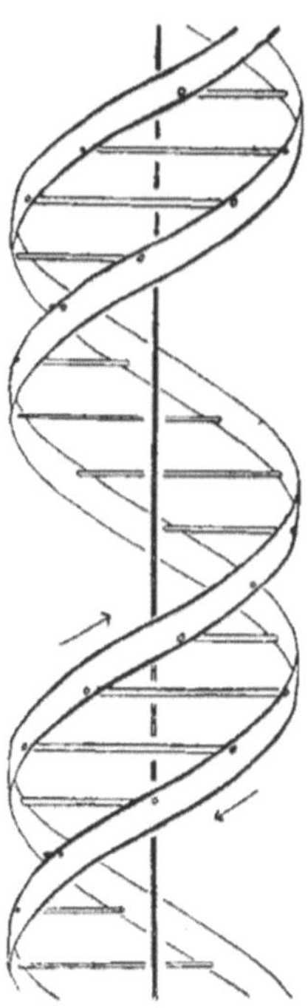

FIGURE 10.1 The DNA double helix structure, as modeled by Watson and Crick in 1953 (Watson and Crick, 1953).

Before the 1953 discovery of the DNA double helix model, there were other understandings of DNA structures, e.g. by Pauling and Corey (1953), and by Fraser as mentioned in Watson and Crick (1953). The model proposed by Pauling and Corey consists of three intertwined chains, with the phosphates near the fiber axis, and the bases on the outside. In the three-chain structure model of Fraser, the phosphates were proposed to be on the outside and the bases on the inside, linked together by hydrogen bonds. Both of these models were declared unsatisfactory by Watson and Crick (1953), due mainly to a lack of valid physical reasoning . In 1962, the Nobel Prize in Physiology or Medicine was awarded to James Watson, Francis Crick, and Maurice Wilkins for their discoveries concerning the molecular structure of nucleic acids and their significance for information transfer in living material. This successful X-ray crystallography data-based understanding of biomolecular structures led in the field of structural biology and has since been helping scientists to discover the structures of all kinds of new biochemicals in living systems.

Although the DNA structure discovery appeared as a single groundbreaking scientific event, the discoveries of ribonucleic acid (RNA) molecules appeared with a complicated series of discoveries by a huge number of scientists (Cobb, 2015; Kresge et al., 2005). During the 1950s Paul Zamecnik, Mahlon Hoagland, and Mary Stephenson discovered transfer RNA (tRNA) (Hoagland et al., 1958). The discovery of messenger RNA (mRNA) was the result of a combination of research efforts during the 1940s and 1950s, and was the product of years of work by a community of researchers such as André Boivin, Raymond Jeener, Al Hershey's group, Volkin and Astrachan, the Brenner-Crick-Meselson group, Jacob and Monod (who named mRNA), Watson's team, and Nirenberg and Matthaei (Cobb, 2015). Severo Ochoa and Arthur Kornberg were awarded the Nobel Prize in Physiology or Medicine in 1959 for their discovery of the mechanisms in the biological synthesis of ribonucleic acid and deoxyribonucleic acid. Thomas Cech and Sidney Altman jointly received the Chemistry Nobel Prize in 1989 for discovering that certain RNAs, now known as ribozymes, showed enzymatic activity. rRNAs are known to combine with proteins and enzymes in the cytoplasm to form ribosomes, which act as the site of protein synthesis. Briefly, this is the history of the discoveries of various aspects of three RNAs: mRNA, tRNA, and rRNA.

10.2 CURRENT STATUS OF NUCLEIC ACID PHYSICAL STRUCTURES

There have been tremendous developments in our understanding of both DNA and RNA structures since their original discoveries. Here I wish to brief on various DNA and RNA structures. Both of the nucleic acids DNA and RNA have subclasses with various structural moieties. We shall outline them here.

10.2.1 DNA Structures

DNA appears with various secondary structures (Mitchell, 1998; Lu et al., 2000) aside from the canonical B-form (Vargason et al., 2001). The sequences that are capable of forming these alternative structures *in vitro* are often sites of genomic instability *in vivo* (Bochman et al., 2012). The B-form DNA (B-DNA) forms the canonical right-handed double helical secondary structure assumed by bulk DNA *in vivo*.

The right-handed double helical B-DNA structure has been known since the discovery of the DNA double helix (Watson and Crick, 1953). DNA can also adopt alternative conformations based on particular sequence motifs and interactions with various proteins. These non-B-form secondary structures include G-quadruplex structures (G4 structures) (Figure 10.2), Z-DNA, cruciform, and triplexes (Figure 10.3), characterized *in vitro* using biophysical techniques (Kypr et al., 2009).

Torsion angles were used by Svozil and colleagues (2008) to describe the structural variability of the sugar–phosphate backbone of DNA and to identify the main DNA conformers. Here 7,739 dinucleotide units from a large number of crystal structures of naked (noncomplexed) DNA structures (187) and complexed DNA structures (260) were analyzed, which confirmed that most conformational variation is covered by several major conformational families (BI, BII, AI, AII). The DNA duplex is not uniform and the dominant conformers appear with many variants having specific roles.

The double helical structure of DNA's A-, B-, and Z-forms has been detailed in Dickerson (1992) and Neidle (2002). DNA is known to adopt the following other forms: triple (Felsenfeld et al., 1957) and quadruple helices (Morgan, 1970), junction (cruciform) structures (Beerman and Lebowitz, 1973), and parallel helices (Van De Sande et al., 1988). According to Svozil and colleagues (2008), the architecture of some of these DNA forms may be in full analogy to the double-helical DNA, almost completely based on the self-assembly of two or more DNA strands and not forming complicated folds analogous to RNA.

Computing DNA structures is a modern technique utilized to understand various structural and functional sites of DNA molecules (Jonoska et al., 1999). A single DNA molecule is capable of providing both the input data and all of the necessary fuel for a molecular automaton (Benenson et al., 2003). The computer modeling of natural sequences was reported more than three decades ago to be a viable approach to the study of the biological implications of alternative DNA structures (Eckahl and Anderson, 1987). Rigid DNA structures can be designed that will serve as scaffolds for the organization of matter at the molecular scale and can be built into simple DNA-computing devices, diagnostic machines, and DNA motors, which are quite important advancements in nanotechnology and medical applications. These integrated biological and engineering advances offer great potential for therapeutic and diagnostic applications, and for nanoscale electronic engineering (Condon, 2006).

The axes of the DNA double helices are unbranched lines. Joining DNA molecules by sticky ends can yield longer lines with specific components in a particular linear or cyclic order in one dimension (1D). The chromosomes inside cells exist as packed 1D arrays. To produce DNA-based

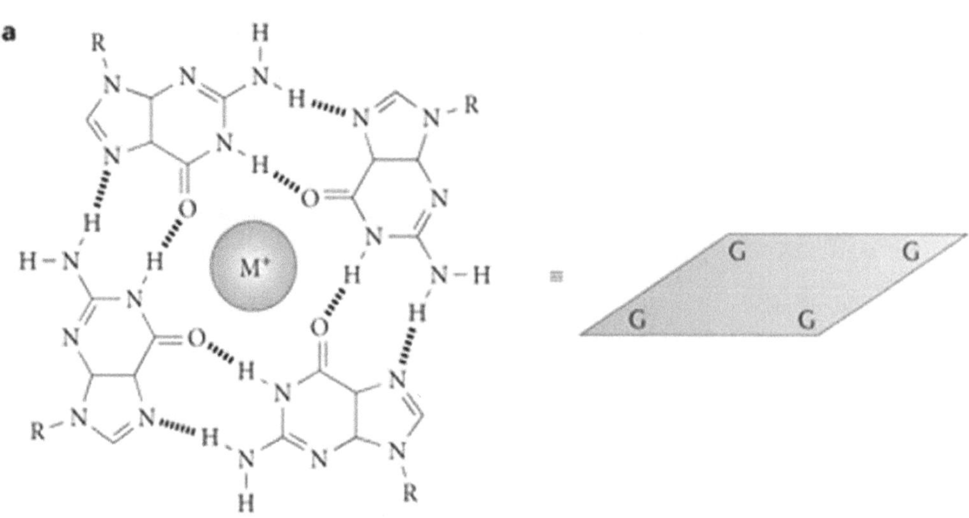

FIGURE 10.2 (a) An illustration of the interactions in a G-quartet. This quartet is represented schematically as a square in the other panels of this figure. M^+ denotes a monovalent cation. (b) Schematic diagrams of intramolecular (left) and intermolecular (right) G-quadruplex (G4) DNA structures. The arrowheads indicate the direction of the DNA strands. The intermolecular structures shown have two (upper) or four (lower) strands.

***Source*: reproduced with permission from Bochman et al. (2012).**

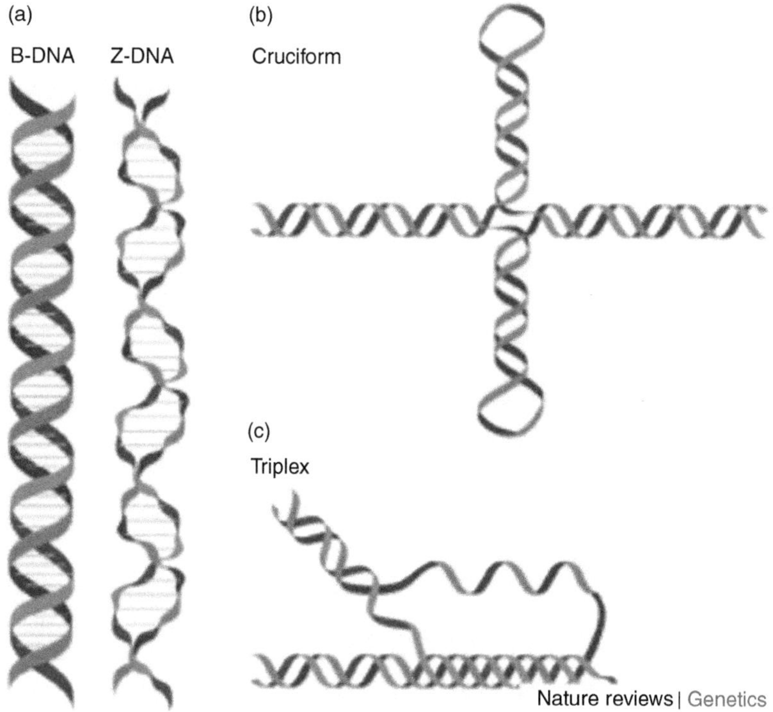

FIGURE 10.3 Various DNA secondary structures. (a) In contrast to B-DNA, Z-DNA is a left-handed helix (Gessner et al., 1989). (b) Negative supercoiling can also cause B-DNA to adopt a four-armed, cruciform secondary structure resembling a Holliday junction (Palecek, 1991). (c) Three-stranded triplex DNA occurs when single-stranded DNA forms Hoogsteen hydrogen bonds in the major groove of purine-rich double-stranded B-DNA (Jain et al., 2008).

***Source*: reproduced with permission from Bochman et al. (2012).**

materials, synthesis is required in multiple dimensions. For this purpose, branched DNA is required (Seeman, 2003; see Figure 10.4).

Understanding the flexibility in DNA structures may help us determine how this molecule rearranges itself structurally in biological systems. Recently, molecular dynamics (MD) simulations have been performed where it is demonstrated how the local environment to which the DNA molecules have to adapt influences the DNA structural alterations, including bending and deformability (Liebl and Zacharias, 2021; see Figure 10.5). The DNA deformability at the base level is typically described by nearest-neighbor-coupled harmonic oscillators (see Figure 10.6, which models the deformability of a DNA polymer). DNA is also capable of adopting different transiently stable substates that also depend on neighbors. A model that combines a harmonically coupled oscillator description with an Ising model to include also all possible substates of a DNA duplex has been proposed in Liebl and Zacharias (2021). It claims not only to give an improved prediction of the equilibrium distribution of DNA helical variables but also to predict indirect readout contributions to protein–DNA binding in better agreement with the experiment than a unimodal harmonic model.

10.2.2 RNA Structures

The secondary RNA structures have been described in several classes of RNAs, including noncoding RNAs such as ribosomal RNA (rRNA), transfer RNA (tRNA), Ribonuclease P (RNase P),

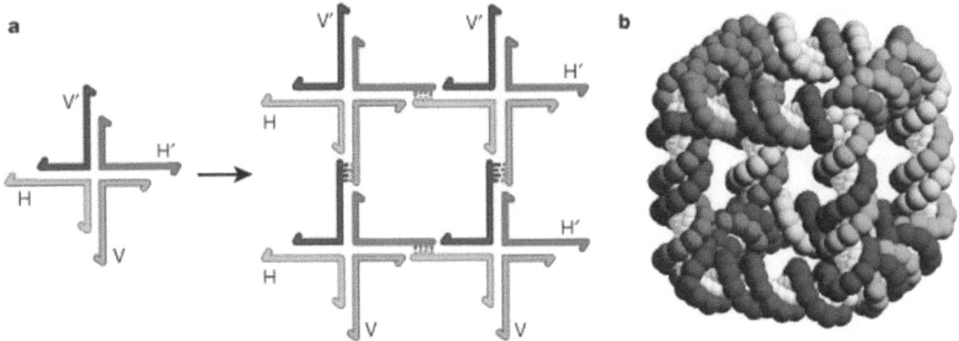

FIGURE 10.4 (a) Self-assembly of branched DNA molecules into a two-dimensional crystal. A DNA branched junction forms from four DNA strands; those strands colored green and blue have complementary sticky-end overhangs labelled H and H′, respectively, whereas those coloured pink and red have complementary overhangs V and V′, respectively. A number of DNA branched junctions cohere based on the orientation of their complementary sticky ends, forming a square-like unit with unpaired sticky ends on the outside, so more units could be added to produce a two-dimensional crystal. (b) Ligated DNA molecules form interconnected rings to create a cube-like structure. The structure consists of six cyclic interlocked single strands, each linked twice to its four neighbors, because each edge contains two turns of the DNA double helix. For example, the front red strand is linked to the green strand on the right, the light blue strand on the top, the magenta strand on the left, and the dark blue strand on the bottom. It is linked only indirectly to the yellow strand at the rear. The color contrast in the figure here and all subsequent ones will be clear in the online version of the book.

***Source*: reproduced with permission from Seeman (2003).**

and signal recognition particle (SRP), as well as cellular and viral messenger RNA (mRNA). A few example mRNA structures include the iron-responsive element (IRE) located in the 5′- or 3′-untranslated regions (UTRs) of mRNAs involved in iron metabolism and transport (Theil, 1998; Kim et al., 1996), stem–loops in the 3′-UTRs of histones and vimentin (Son, 1993; Zehner et al., 1997), and IRES elements in the 5′-UTRs of picornaviruses, pestiviruses, and flaviviruses (Le et al., 1996). Despite substantial sequence variation, in all known RNA structures the secondary structure remains conserved during evolution.

RNA biology is complex and diverse. The cellular RNA functional classes include rRNA, tRNA, small nuclear RNA (snRNA), microRNA, and other noncoding RNAs (Noller, 1984; Rich & RajBhandary, 1976; Allmang et al., 1999; Geisler and Coller, 2013).

The cellular RNAs are typically single-stranded, the fold of which is largely determined by nucleotide base pairing, including canonical base pairing (A–U, C–G, and non-Watson–Crick pairing G–U), and non-canonical base pairing (Fallmann et al., 2017; Westhof and Fritsch, 2000). These base-paired structures are often referred to as the RNA secondary structure (Fox and Woese, 1975; see Figure 10.7). More than half a century ago attempts were made to deduce the secondary structure of 5S RNA (Jordan et al., 1974; Kearns and Wong, 1974; Lewis and Doty, 1970; Cantor, 1968). Various structures were proposed, but each turned out to be inconsistent with either physical, chemical, biochemical, or comparative evidence. The functional 5S RNA molecule was found to be an integral part of the 50S ribosomal subunit, whereas tRNA was found to be a transient inhabitant of the ribosome milieu.

The secondary structure is essential for the correct RNA functions, and the role of secondary structure in many cases appears to be a more important determinant than that of primary structure. Many homologous RNA species are found to demonstrate conserved secondary structures, despite exhibiting diverse sequences (Mathews et al., 2010; see Figures 10.8 and 10.9).

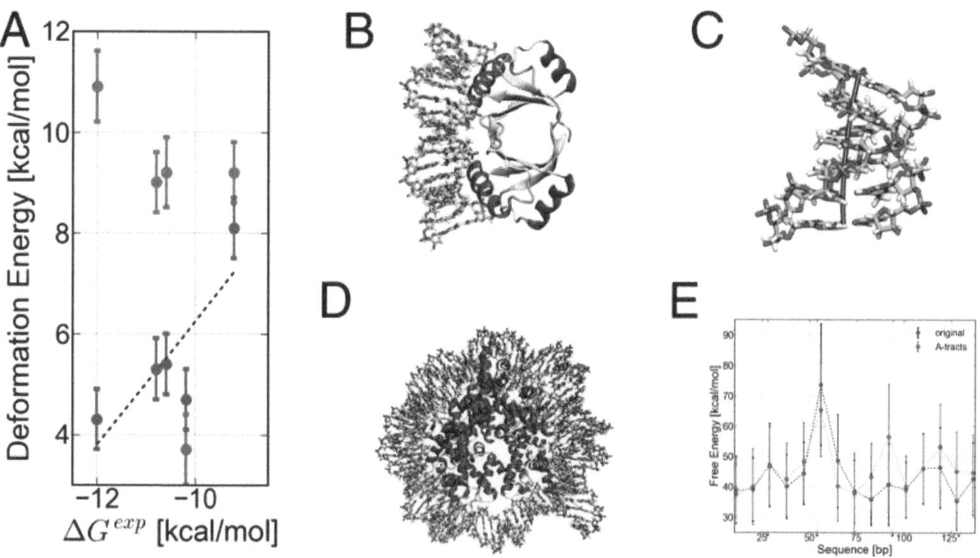

FIGURE 10.5 Free-energy calculations on the papillomavirus E2 and nucleosome system. (*A*) Correlation of computed deformation energies (red, unimodal harmonic; blue, Ising model) with experimental binding affinities for the spacer sequences AATT, AAAT, AAAA, TTAA, and ACGT. (*B* and *C*) Snapshot of the E2 protein in complex with the DNA (*B*) and bending of the central spacer region (*C*). (*D*) Snapshot of the human nucleosome core particle (denoted by "original"). (*E*) Deformation energies for the original DNA sequence and the A-tracts, including sequence as computed with our Ising model. Deformation energies have been calculated over nine-base-pair-long DNA sequences and averaged over the last 10 ns of the MD trajectory. Positions of the A-tracts are labeled by asterisks; error bars represent SDs. A, T, G, C stand for DNA nucleotides adenine, thymine, guanine, and cytosine, respectively.

***Source*: reproduced with permission from Liebl and Zacharias (2021).**

RNA secondary structures can be determined using atomic coordinates obtained from X-ray crystallography, nuclear magnetic resonance (NMR), or cryogenic electron microscopy (Fürtig et al., 2003; Cheong et al., 2004; Fica and Nagai, 2017). These experimental methods have low throughput and the structures of only a tiny fraction of RNAs have been thus determined. To overcome this limitation, other experimental methods have found inferring base pairing by using probes based on enzymes, chemicals, and cross-linking techniques that are coupled with high throughput sequencing (Ehresmann et al., 1987; Knapp, 1989; Bevilacqua et al., 2016; Underwood et al., 2010). These promising methods are still under development, so unable to provide precise base pairing at a single nucleotide solution.

The understanding of biological functions of RNA molecules certainly requires an understanding of structure and dynamics at the atomistic level. Molecular dynamics (MD) simulations presented strong and diversified approaches in this regard (McDowell et al., 2007; Sponer et al., 2018). A schematic representation of the time scales that are relevant for RNA structural dynamics is presented by Sponer et al. (2018; see Figure 10.10). Figure 10.11 shows example MD simulation case studies over 90-3300 nanosec (ns), where simulations of preQ1 riboswitch aptamer start from an X-ray structure upon removal of the ligand, showing the most severe artifacts that may occur in RNA molecules (Banáš et al., 2012; Šponer et al., 2018).

Recently, algorithm-based machine learning and deep learning methods have been popularly utilized to predict nucleic acid secondary structures, e.g. RNA structures have been predicted recently (Fu et al., 2022; see Figure 10.12). A deep learning-based method, called UFold, has been

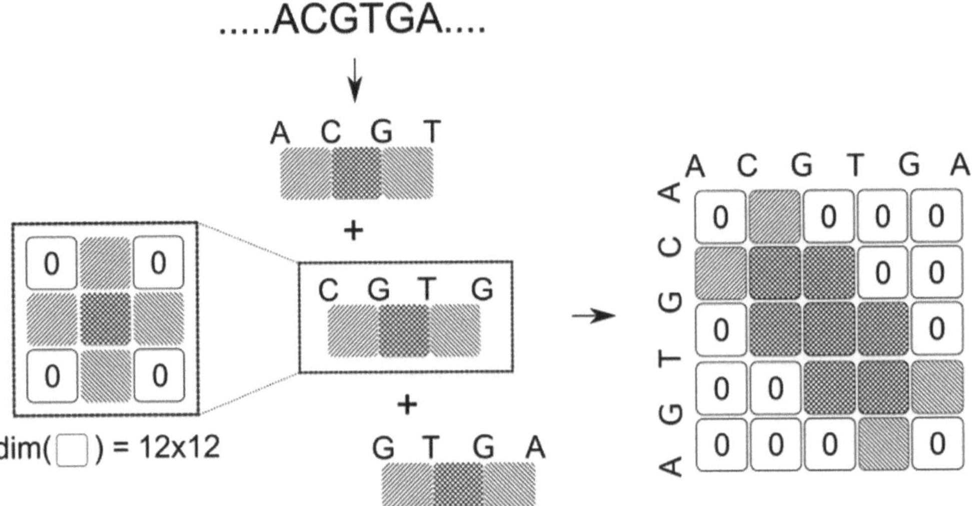

FIGURE 10.6 Modeling the deformability of a DNA polymer: DNA's sequence is split up into tetranucleotides. For every tetranucleotide, the deformability of the central step is fully retained (cross-hatched), whereas nearest-neighbor coupling is scaled by a factor of 1/2 (linear-hatched). Self-terms of the flanking sites as well as non-nearest-neighbor couplings are set to zero (indicated by 0-panels). Deformability for the entire polymer is then obtained by connecting the tetranucleotide descriptions additively.

***Source*: reproduced with permission from Liebl and Zacharias (2021).**

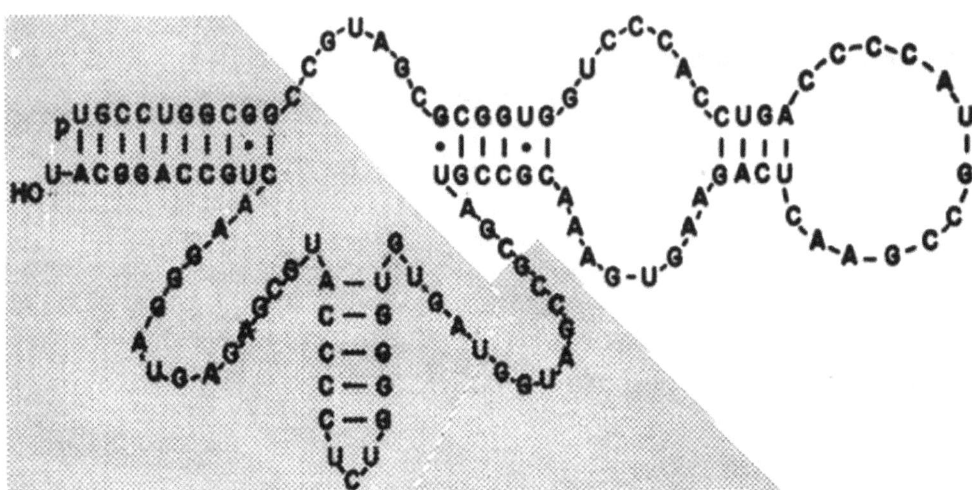

FIGURE 10.7 Schematic representation of 5S RNA secondary structure. Normal base pairs are connected by vertical lines, whereas G–U pairs are indicated by dots. The shaded portion of the molecule is that which has been identified as being involved in an interaction with specific 50S subunit proteins. The structure is drawn with helices aligned along the long axis in accordance with the suggestion of Connors and Beeman (1972).

***Source*: reproduced with permission from Fox and Woese (1975).**

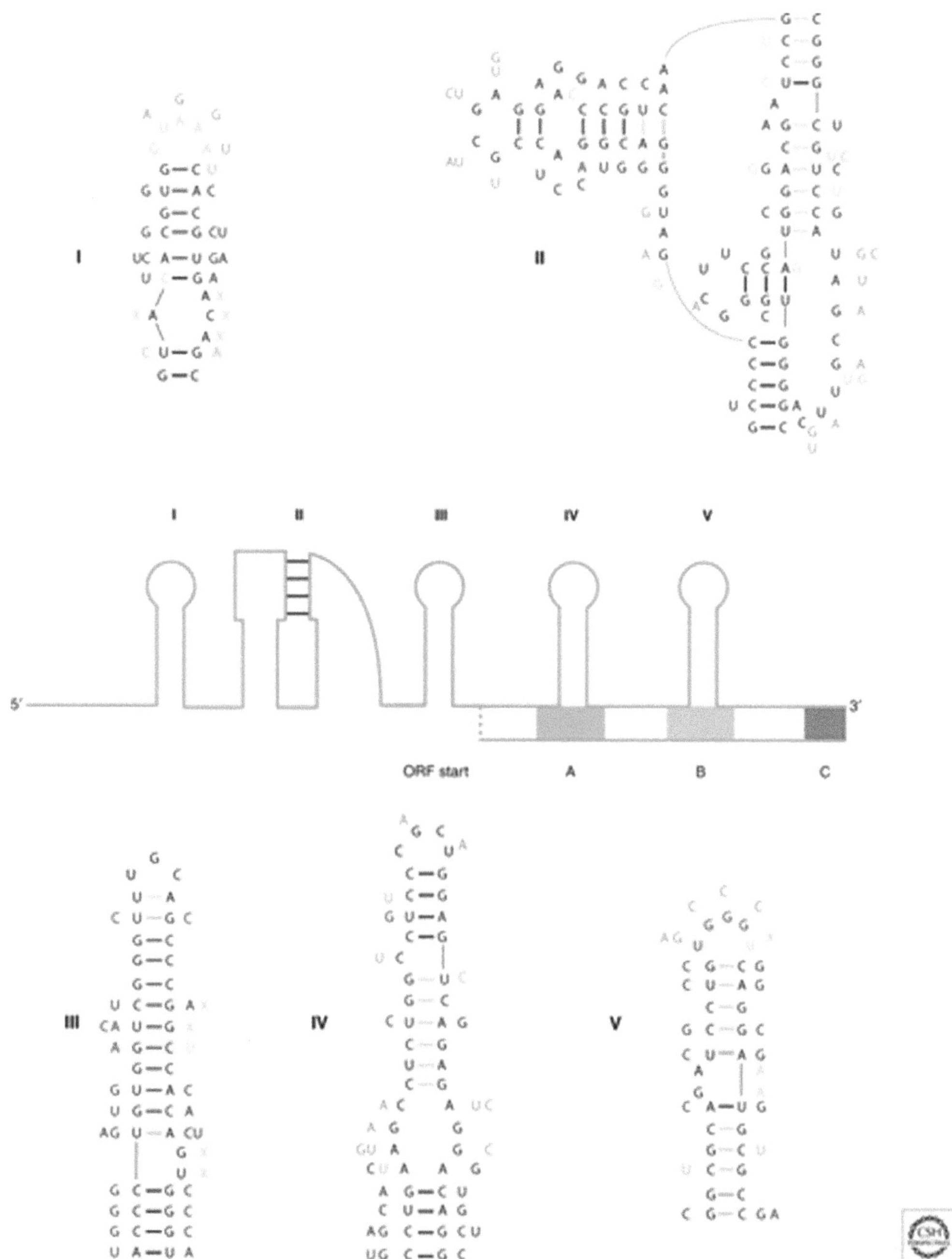

FIGURE 10.8 Determination of structured regions of an RNA. A cartoon of the 5′ region of the silk moth R2 retrotransposon is shown. The conserved structure is organized into four hairpin loops (labeled I and III–V) and a pseudoknot (labeled II). Also shown are three conserved coding regions (A–C) and a putative open reading frame (ORF) start site. The five conserved structures are detailed with data that went into the structural modeling. The sequences shown are for *B. mori* whereas mutations are those that occur in four other moth species. Mutational data appear next to the main sequence and are color annotated: dark blue are double mutations that maintain base pairing (compensatory), light blue are single point mutations that maintain pairing (consistent), gray are mutations in loops, red disrupt canonical base pairs (inconsistent), green are insertions (green X represents a deletion). Experimental mapping is color annotated on the backbone sequence: red are

FIGURE 10.8 *(Continued)*

NMIA only modifications and orange are modifications by both traditional mapping agents (DMS or CMCT) and NMIA. Base pairs are indicated with dashes between nucleotides and are color annotated for probability from partition function calculation: Red, probability (P) ≥99%; Orange, 99% > P ≥ 95%; Yellow, 95% > P ≥ 90%; Dark Green, 90% > P ≥ 80%; Light Green, 80% > P ≥ 70%; Light Blue, 70% > P ≥ 60%; Dark Blue, 60% > P ≥ 50%; Black <50%. Many base pairs in the pseudoknot have low probability because the RNA structure program does not allow pseudoknots, and thus undercounts them in the partition function.

***Source*: reproduced with permission from Mathews et al. (2010).**

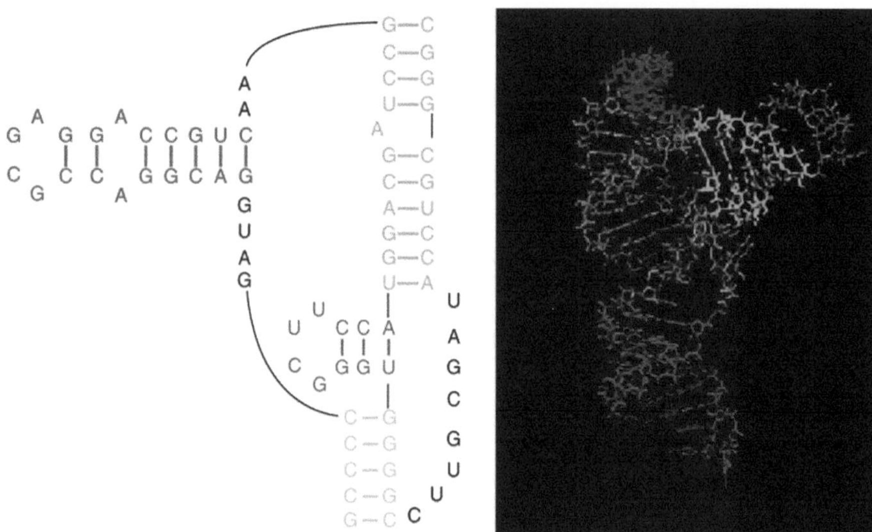

FIGURE 10.9 Experiment and sequence comparison are used to model 3D structure. Homology with known structures was used to propose 3D folds for the B. mori R2 element pseudoknot from MC-Sym (Parisien and Major, 2008) which were then screened with respect to experimental data (e.g. solvent accessibility to chemical reagents (Kierzek, 2009) and helix stacking from NMR (Hart et al., 2008)). Helical motifs in the 3D model are color coded to match the secondary structural model.

***Source*: reproduced with permission from Mathews et al. (2010).**

used here for RNA secondary structure prediction, trained directly on annotated data and base-pairing rules. UFold proposes a novel image-like representation of RNA sequences, which can be efficiently processed by Fully Convolutional Networks (FCNs). The performance of UFold has been benchmarked on both within- and cross-family RNA datasets. We avoid detailing the deep learning method here but encourage interested readers to learn from Fu et al. (2022).

10.3 NUCLEIC ACID ENERGETICS ASSOCIATED WITH STRUCTURAL ALTERATIONS

Nucleic acid structural and functional diversity is associated with corresponding energetics at the intramolecular level, inter-strands and inter-molecular levels, and inter-nucleic acid building block levels. These energetics contribute to causing physical nucleic acid molecular structural organizations, structural alterations, structural transitions, translations, etc. Here we wish to brief on this matter for both DNA and RNA molecules in cellular systems.

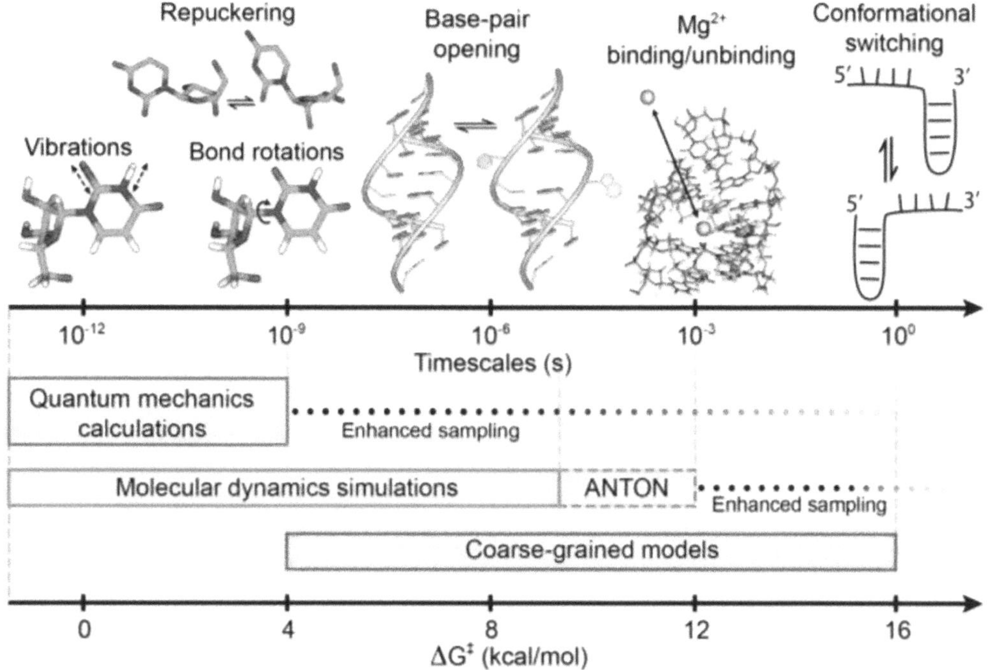

FIGURE 10.10 Schematic representation of the time scales that are relevant for RNA structural dynamics. In the upper part, sample structural changes are depicted. In the lower part, the typical time scales nowadays accessible by the techniques discussed in this review are shown, including quantum mechanical calculations, atomistic explicit-solvent MD simulations, and coarse-grained models. Hardware and software improvements have led to an order of magnitude gain every few years in the past. Dedicated hardware such as Anton (Shaw et al., 2008) allows for significantly longer time scales to be accessed in atomistic MD, and massively parallel approaches like the folding@home infrastructure (Shirts and Pande, 2000; Pande et al., 2003) allow one to gather large cumulated simulation times (although typically composed of a large number of short trajectories). However, complex conformational changes still remain out of reach of MD simulations. Enhanced sampling techniques might be used to probe those time scales. The bottom arrow shows typical free-energy barriers involved in these processes. Complex and biologically relevant molecular machines such as the ribosome utilize an endless spectrum of dynamical processes, extending from movements of single nucleotides up to large-scale movements of their whole subunits on a wide range of time scales.

***Source*: reproduced with permission from Sponer et al. (2018).**

10.3.1 DNA Energetics

The DNA's supercoiled configurations have been investigated using deterministic techniques (Schlick and Olson, 1992). While stochastic methods, e.g. simulated annealing and Metropolis-Monte Carlo sampling, are successful at generating a large number of configurations and estimating thermodynamic properties of topoisomer ensembles, deterministic methods are found to offer an accurate characterization of the minima and a systematic following of their dynamics. Circular and interwound energy minima were reported to be rapidly identified for small DNA rings for a series of imposed linking-number differences. The energetic exchange of stability between states was also determined. The 100 femtosecond (fs) time step MD simulation trajectories revealed the rapid folding of the unstable circular state into supercoiled forms. Significant motion for bending and twisting modes of structures was observed. This information may be useful to understand transition states along the folding pathway and the role of enzymes that regulate supercoiling.

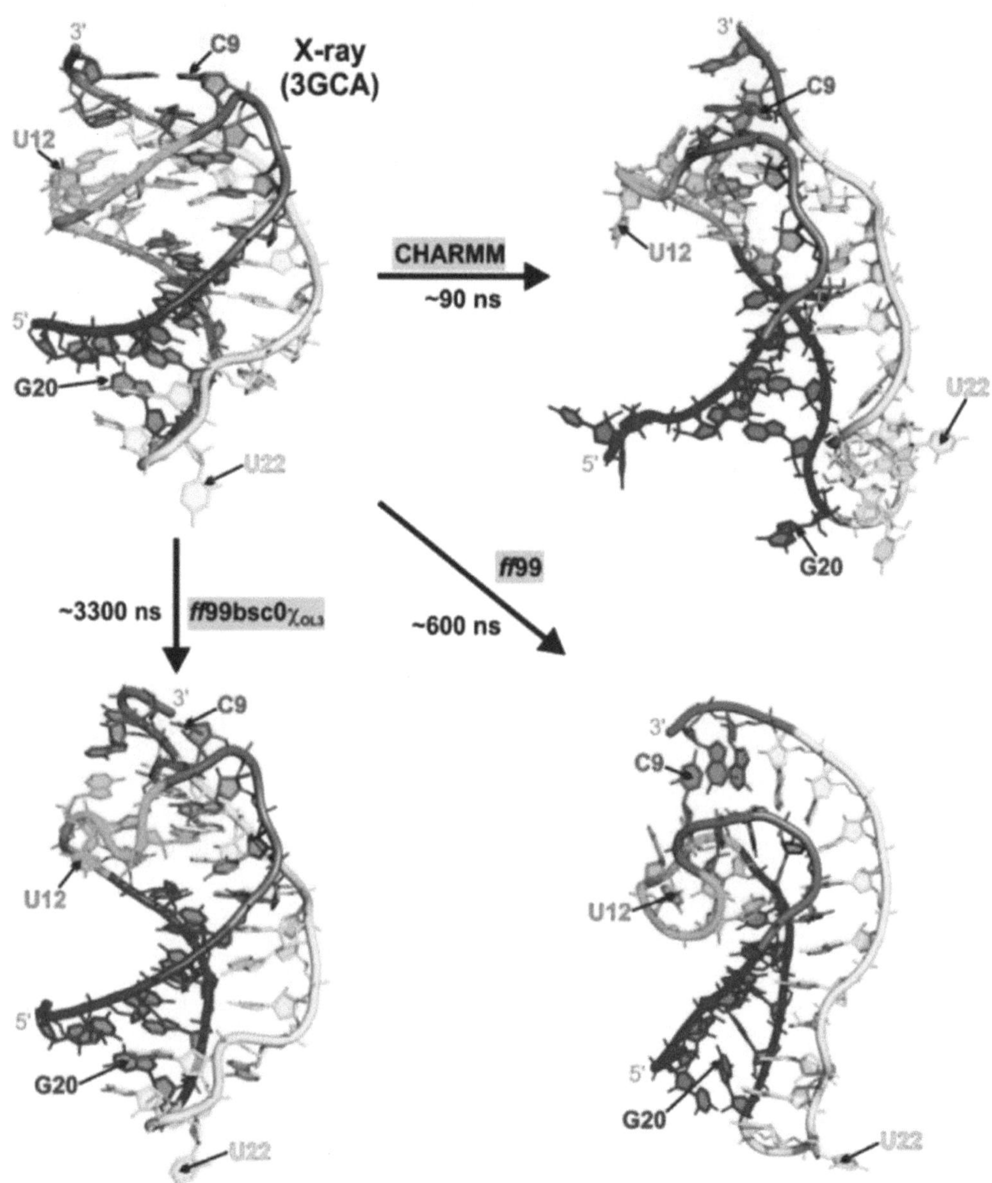

FIGURE 10.11 Simulations of preQ_1 riboswitch aptamer starting from an X-ray structure upon removal of the ligand, showing the most severe artifacts that may occur in RNA MD simulations. While the χ_{OL3} force field shows a stable trajectory with just local dynamics, the ff99 simulation collapses to the ladder-like structure (the bsc0 correction would not improve this behavior) and the CHARMM27 simulation experiences extended fraying; the latter behavior was only partly corrected by CHARMM36. The difference between the χ_{OL3} simulation and the experimental structure reflects an intricate mixture of genuine dynamics of the system, effects due to removal of crystal packing deformations, and contributions from more subtle force-field imbalances.

***Source*: reproduced with permission from Banáš et al. (2012). © 2012 American Chemical Society.**

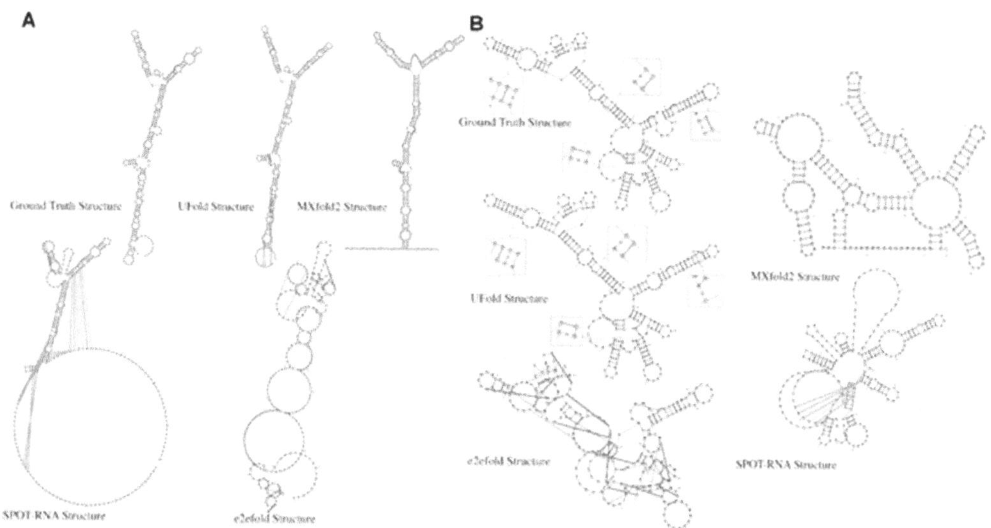

FIGURE 10.12 Visualization of two example UFold RNA secondary structure predictions. From top to bottom: ground truth, UFold prediction, and E2Efold prediction. Two RNA sequences are (A) *Aspergillus fumigatus* species, RNA ID GSP-41122, as recorded in SRPDB database; and (B) Alphaproteobacteria subfamily 16S rRNA sequence whose database ID is U13162, as recorded in RNAStralign database (http://rna.urmc.rochester.edu). Non-canonical base pairs are colored light green. In both cases, UFold produces predictions more aligned with the ground-truth.

***Source*: reproduced with permission from Fu et al. (2022).**

A recent review has addressed major theoretical models used to calculate free energy changes associated with common, torsion-induced conformational changes in DNA (Xu et al., 2021). This review has mainly addressed the following issues:

1. The energy change of the negative supercoiling-induced B- to L-DNA transition.
2. A rigorous discussion of the energetics associated with the transition to Z-form DNA.
3. It describes the energy changes associated with the formation of DNA curls and plectonemes. This energetics can regulate DNA-protein interactions and promotes cross-talk between distant DNA elements, respectively.

Studies of the folding/unfolding mechanism of the model telomeric human DNA, 5′-AGGGTTAGGGTTAGGGTTAGGG-3′ (Tel22), indicated recently that in the presence of potassium (K^+) ions, Tel22 folds into two hybrid G-quadruplex structures characterized by one double and two reversal TTA loops arranged differently (Bončina et al., 2012). A new unfolding pathway from the initial mixture of hybrid G-quadruplexes via the corresponding intermediate triplex structures into the final, fully unfolded state was predicted. The unfolding of Tel22 is proposed to be a monomolecular equilibrium three-state process that involves thermodynamically distinguishable folded (F), intermediate (I), and unfolded (U) states. The estimated number of K^+ ions released upon each unfolding step in the thermodynamic model applied here agrees well with the corresponding values predicted by the theoretical model, and the observed changes in enthalpy, entropy, and heat capacity accompanying the $F \rightarrow I$ and $I \rightarrow U$ transitions can be reasonably explained only if the intermediate state I is considered to be a triplex structural conformation.

DNA polymerases need to operate under kinetic control to achieve the high fidelity that defines successful replication. It is yet to be correctly understood whether DNA base-pairing thermodynamics plays an important role in the high error rates of less accurate polymerases. The question arises: what is the free energy source enabling high-fidelity DNA polymerases (pols) to favor incorporation of correct over incorrect base pairs by 10^3- to 10^4-fold, corresponding to free energy differences of $\Delta\Delta G_{inc}$ ~ 5.5–7 kcal/mol (Oertell et al., 2016)? The standard free energy differences $\Delta\Delta G°$ values (~0.3 kcal/mol), calculated using melting temperature measurements comparing matched vs. mismatched base pairs at duplex DNA termini, are far too low to explain the pol accuracy. The pol active-site steric constraints were earlier found to amplify DNA free energy differences at the transition state (kinetic selection). Vent pol was used to catalyze incorporations in the presence of inorganic pyrophosphate intended to equilibrate forward (polymerization) and backward (pyrophosphorolysis) reactions (Olson et al., 2013). In a steady state and with long reaction times, the equilibrium between polymerization and pyrophosphorolysis was achieved, yielding apparent $\Delta G° = -RT \ln K_{eq}$, indicating $\Delta\Delta G°$ of 3.5–7 kcal/mol, which is sufficient to account for pol accuracy without the need of kinetic selection. Experiments have been performed in order to measure and account for pyrophosphorolysis explicitly (Oertell et al., 2016). This study shows that forward and reverse reactions attain steady states far from equilibrium for wrong incorporations such as G opposite T. Therefore, free energy differences ($\Delta\Delta G°_{inc}$) values obtained from such steady-state evaluations of K_{eq} are not dependent on DNA properties alone but depend largely on constraints imposed on right and wrong substrates in the polymerase active site.

Olson and colleagues (2013) made the contrary argument that amplification is superfluous if polymerase $\Delta\Delta G°_{inc}$ values measured from equilibrium constants for nucleotide incorporation vs. pyrophosphorolysis are significantly larger than those obtained from DNA thermal melting studies. By using high $[PP_i]/[dNTP]$ levels to cause a leveling off in the right (R) and wrong (W) nucleotide incorporation profiles for Vent pol, an average $\Delta\Delta G°_{inc} = +5.2 \pm 1.34$ kcal/mol was found, apparently sufficient for >1,000-fold discrimination (Olson et al., 2013). It was tacitly assumed that pyrophosphorolysis was responsible for balancing the forward and reverse reaction rates so that observed steady-state levels represented equilibrium for W incorporation, e.g., G opposite T (G•T), as well as for R incorporation (A•T). If so, then incorporation and pyrophosphorolysis reactions of DNA polymerase should reach the same $[DNA_{n+1}]/[DNA_n]$ level. That same level should occur when starting with DNA_n and incorporating G opposite T to form DNA_{n+1} or starting with DNA_{n+1} and phosphorolytically removing G opposite T in DNA_{n+1} to form DNA_n under identical reaction conditions. Thus, to determine comparable values of $\Delta G°_{inc}$ Oertell and colleagues performed G•T incorporation and pyrophosphorolysis reactions, using Vent pol to extend a 21-mer substrate to a 22-mer product (Figure 10.13) (for detailed theoretical analysis on conditions required to achieve equilibrium for right and wrong incorporation reactions, for both polymerization and pyrophosphorolysis, see Oertell et al., 2016). Although the rate of extension diminishes with time, a steady state has not been attained after 48 h, at any of the five dGTP concentrations (Figure 10.13B). Taking the extension levels determined at 48 h to obtain Q for comparison with Olson et al. (2013), we find apparent $\Delta G°_{inc}$ (G•T) = −1.19 kcal/mol, which differs substantially from the published Vent result, $\Delta G°_{inc}$ (G•T) = −0.32 kcal/mol. An empirical asymptotic extrapolation of the misincorporation data results in an apparent $\Delta G°_{inc}$ (G•T) = −1.27, which is fourfold greater in magnitude than the earlier estimate. Definitively, however, when starting with product DNA_{22} ending in G•T (at the primer 3′ terminus) and using the same $[PP_i]/[dGTP]$, we were not able to detect pyrophosphorolysis (producing DNA_{21}) in reactions carried out to 2 d at 37 °C or 72 °C (Figure 10.13C), in the presence or absence of dGTP, so, in fact, K_{eq}' cannot be measured. Because pyrophosphorolysis at G•T was not observed even at millimolar PP_i concentrations, the reduction in misincorporation with reaction time is likely caused by some PP_i inhibitory effects on pol activity, such as inhibition of dGTP binding opposite template T in DNA_{21} bound to polymerase.

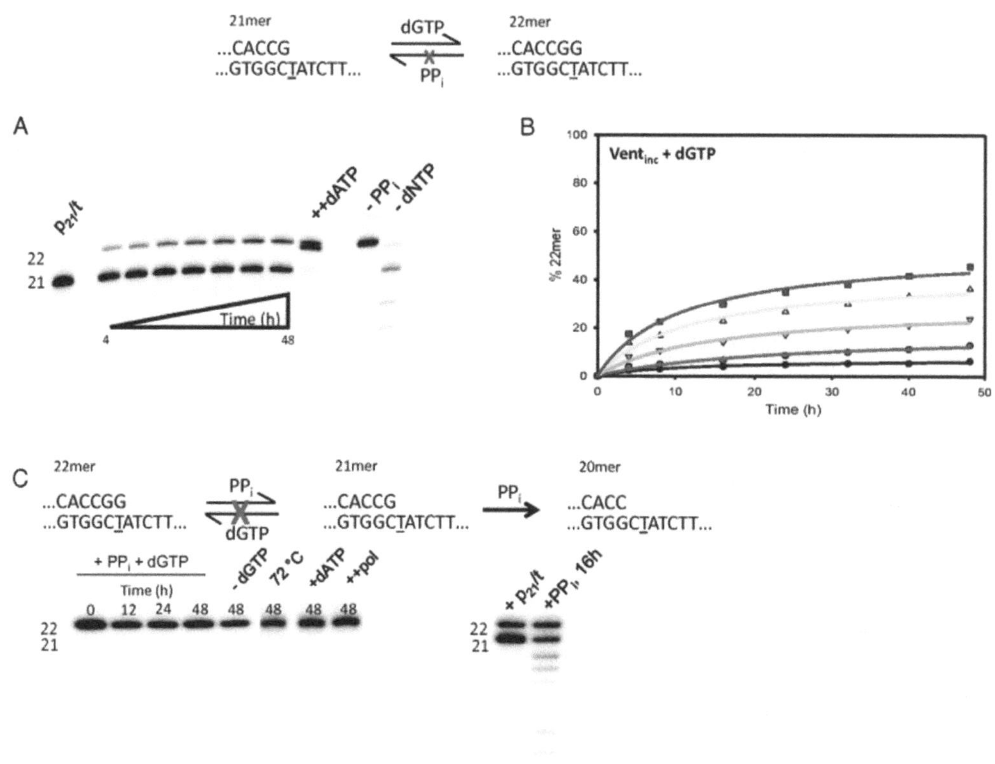

FIGURE 10.13 Forward and reverse reactions for misincorporation of dGTP opposite T by Vent pol. (*A*) Gel data obtained for a representative reaction with 1.6 mM dGTP and 11 mM PP_i. The sketch at the top illustrates the reaction proceeding from primer 21mer to 22mer, by incorporation of dGTP opposite the underlined templating T, but the reverse reaction is not detected. The amount of DNA_{22} (G•T) increases with time. At the end of 48 h incubation, 1 mM dATP is added and allowed to react for 20 min. (++dATP) to show the enzyme is still active at the end of reaction. Note that DNA_{22} (G•T) and DNA_{22} (A•T) are clearly separated as bands in the gel. The final two lanes show two controls, one containing only dGTP (−PP_i) and one only PP_i (−dNTP), demonstrating that the enzyme can fully extend the 21mer to 22mer with dGTP and no PP_i and can perform extensive pyrophosphorolysis with PP_i and no dNTP, respectively. (*B*) Amount of extended primer (% 22mer) as a function of reaction time, at various dGTP concentrations: 100 µM (black circles), 200 µM (red circles), 400 µM (green triangles), 800 µM (yellow triangles), and 1,600 µM (blue squares). By the end of 48-h reactions, the amount of extended primer DNA_{22} (G•T) is still increasing slightly. (*C*) Starting from DNA_{22} (G•T), Vent pol was unable to perform pyrophosphorolysis. The first lane shows the starting 22mer before any reaction (*t* = 0). The subsequent lanes are quenched at *t* = 12 h, 24 h, and 48 h incubation with 200 µM dGTP and 11 mM PP_i. (*C, Left*) The lane marked −dGTP shows that no pyrophosphorolysis is occurring without the inclusion of dGTP in reaction mixture, or at higher temperature (72 °C), or with inclusion of dATP in the reaction mixture (+dATP), or with excess enzyme (++pol). (*C, Right*) After 48 h, 100 nM of DNA_{21} (G•C) was added to reveal the enzyme was still active (lane 1), and then an aliquot was quenched 16 h later (lane 2).

***Source*: reproduced with permission from Oertell et al. (2016).**

One possible explanation for the lack of reaction in the reverse direction when faced with a mismatch at the 3′ primer terminus (Figure 10.13) is that the pol is simply unable to bind to the substrate. However, the complete extension of the G•T mismatch eliminates that explanation. Although we cannot rule out the possibility that Vent pol can perform incorrect pyrophosphorolysis under synthesis conditions, we can say that the activation energy must be very high. Thus, the reverse reaction, if it occurs at all when the pol is faced with an existing mismatch as opposed to having just performed the misincorporation, must be on a time scale that is longer than the residence time of the pol on the mismatched DNA.

In contrast to the absence of an equilibrium for the mismatched base pair, the corresponding incorporation and pyrophosphorolysis reactions for a correctly matched base pair appear to reach an approximate equilibrium (Figure 10.14). The A•T incorporation leveled off within 1–2 h (Figure 10.14B), yielding a ΔG°_{inc}(A•T) = −5.22 kcal/mol in agreement with that reported previously, ΔG°_{inc} (A•T) = −5.12 kcal/mol (Olson et al., 2013). The pyrophosphorolysis reaction, initiated from an A•T base pair at DNA_{22}, also leveled off within a similar 1- to 2-h time frame (Figure 10.14D), indicating an apparent ΔG°_{inc} = +5.51 kcal/mol and corresponding to a predicted apparent ΔG°_{inc} = − ΔG°_{inc} = −5.51 kcal/mol. The small difference between the observed −5.22 kcal/mol and predicted −5.51 kcal/mol suggests that equilibrium between polymerization and pyrophosphorolysis has been approximately reached for A•T reactions. However, because pyrophosphorolysis was not detected in the G•T reactions, the resultant $\Delta\Delta G^{\circ}_{inc}$ (A•T vs. G•T) = +4.03 kcal/mol and the average $\Delta\Delta G^{\circ}_{inc}$ = +5.2 ± 1.34 kcal/mol obtained for a variety of base pairs and mispairs reported previously (Olson et al., 2013) are clearly not measures of standard free energy differences for R and W incorporation intrinsic to DNA itself.

Pfu polymerase also readily incorporates A opposite T, reaching a steady-state level within 2.5–3 h (Figure 10.15C), yielding apparent ΔG°_{inc}(A•T) = −5.41 kcal/mol. We also find that *Pfu* pol catalyzes pyrophosphorolysis at A•T primer termini, reaching an approximate steady state in about 3 h (Fig. 10.15 D), yielding apparent ΔG°_{pyro} (A•T) = +6.03 kcal/mol = −ΔG°_{inc} (A•T). The difference of 0.62 kcal/mol between observed and predicted ΔG°_{inc} is about twofold greater than that observed for Vent pol. However, unlike Vent pol, *Pfu* pol does show some weak pyrophosphorolysis at mismatched G•T primer termini (Figure 10.15B), giving an apparent ΔG°_{pyro} (G•T) = +5.89 kcal/mol. Notably, however, the resultant prediction of ΔG°_{inc} = −ΔG°_{pyro} = −5.89 kcal/mol is very different from the apparent ΔG°_{inc} = −0.21 kcal/mol observed for G•T incorporation by *Pfu* pol. Such a huge discrepancy, 5.68 kcal/mol, between observed apparent ΔGoinc and predicted ΔG°_{inc} = −ΔG°_{pyro} for G•T clearly shows that observed steady-state incorporation of G opposite T is far from the equilibrium condition (ΔGoinc = −ΔGopyro) expected for polymerization balanced by pyrophosphorolysis. We do not wish to elaborate on this here, but for further details, interested readers may consult Oertell et al. (2016).

DNA interactions with the environmental components often cause DNA stability to become regulated. Due to the absorption of ultraviolet (UV) light in DNA, genetic materials appear to be vulnerable. This photodamage of the DNA may lead to mutations. The UV light-induced chemical reactions may result in carcinogenic mutations (Cadet and Vigny, 1990; Ravanat et al., 2001; Dumaz ct al., 1997). Theoretically, we have shown evidence of such DNA damage (especially in the telomere structure levels) due to its interaction with light, especially UV rays (Ashrafuzzaman and Shafee, 2005).

Femtosecond (fs) time-resolved broadband spectroscopy helped trace the electronic excitation in both time and space along the base stack in a series of single-stranded and double-stranded DNA oligonucleotides (Buchvarov et al., 2007). It demonstrated the presence of delocalized electronic domains (excitons) as a result of UV light absorption and thus revealed the spatial extent of the excitons.

Fs broadband pump-probe spectroscopy enabled monitoring of the temporal evolution of the excited-state absorption (ESA), including the absorption of "dark states" (i.e. states with negligible

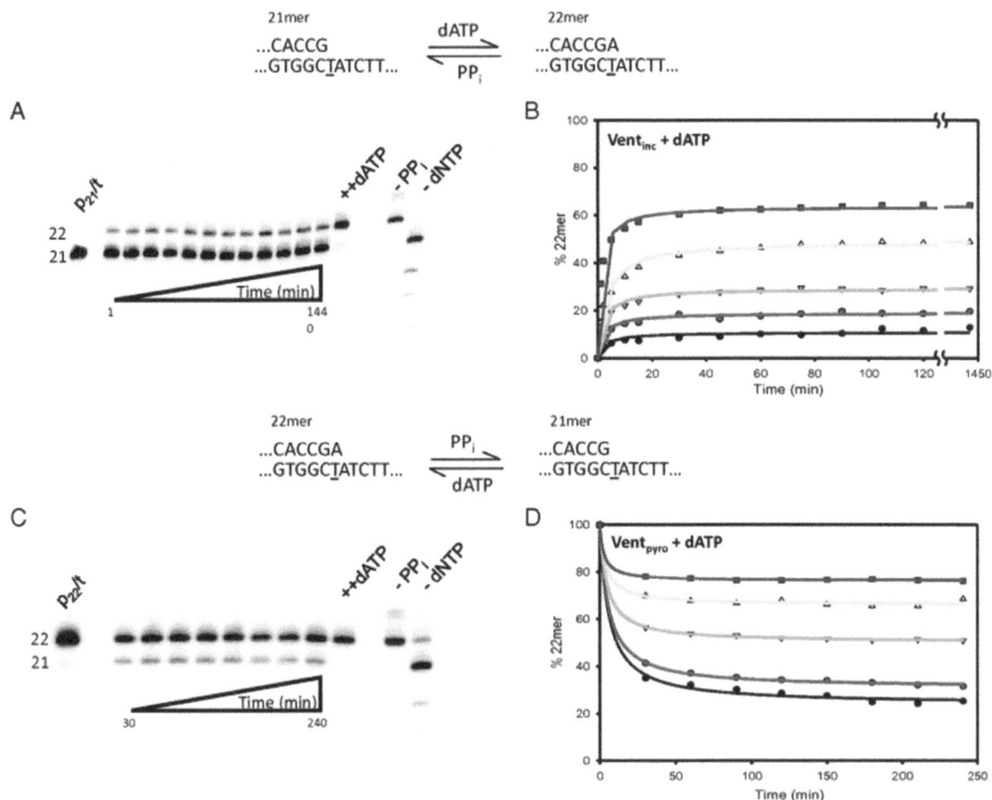

FIGURE 10.14 Forward and reverse reactions for correct incorporation of dATP opposite T by Vent pol. (*A*) Gel data showing the forward reaction for correct (A•T) incorporation. The sketch at the top shows the starting primer (21mer) being extended by the correct incorporation of dATP. The gel bands show results obtained at 8 μM concentration of dATP. The amount of DNA_{22} (A•T) increases with time until an apparent steady state is reached, within 20 min., and stays nearly constant for 24 h. Upon addition of 1 mM dATP after the 24-h incubation, the primer becomes fully extended, showing the enzyme is active throughout the reaction. (*A*, *Right*) The two controls show the reaction without PP_i and dNTP. (*B*) Amount of DNA_{22} (A•T) observed as a function of time at various dATP concentrations: 0.5 μM (black circles), 1 μM (red circles), 2 μM (green triangles), 4 μM (yellow triangles), and 8 μM (blue squares). In each case dGTP is present at 50 μM in the reaction to prevent pyrophosphorolysis continuing beyond the 20mer. (*C*) The sketch at the top describes the pyrophosphorolysis reaction converting 22mer primer to 21mer. The gel bands show results of reaction with 8 μM dATP and 15 mM PP_i starting with 22mer as initial primer. As the reaction is carried out, aliquots are quenched at various times to show the increase in amount of 21mer produced [DNA_{21} (G•C)]. At the end of the 4-h reaction, a large amount of dATP is added (++dATP) to show the enzyme is still active. The final two lanes show the reaction products when only dNTPs are included (–PP_i) and when only PP_i is included (–dNTP). (*D*) Reverse reactions starting from a correctly matched p/t [DNA_{22} (A•T)] are shown at various dATP concentrations: 0.5 μM (black circles), 1 μM (red circles), 2 μM (green triangles), 4 μM (yellow triangles), and 8 μM (blue squares). dGTP is also present at 50 μM in each case to prevent pyrophosphorolysis continuing beyond the 20mer.

***Source*: reproduced with permission from Oertell et al. (2016).**

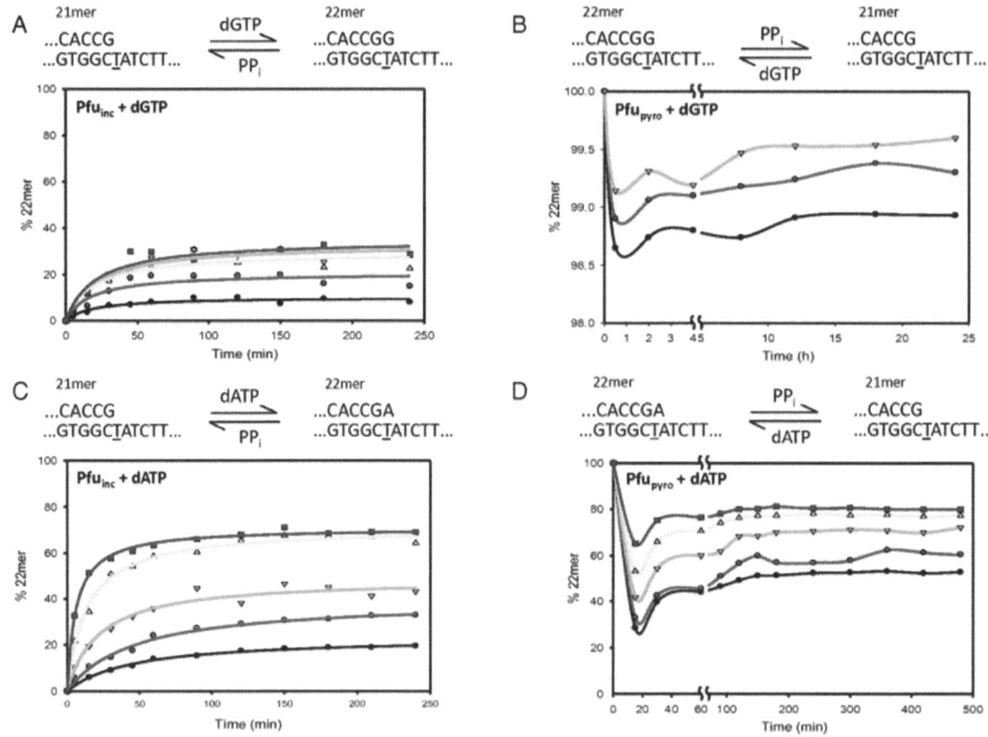

FIGURE 10.15 Incorrect and correct forward and reverse reactions with D473G *Pfu* pol. At the top of each plot, a sketch is shown to describe the reaction. All plots are shown as percentage of 22mer vs. time. (*A*) Incorrect incorporation of dGTP opposite T, at various dGTP concentrations: 200 μM (black circles), 400 μM (red circles), 600 μM (green triangles), 700 μM (yellow triangles), and 800 μM (blue squares). Reactions in the presence of 2.5 mM PP_i were carried out for 4 h. (*B*) Pyrophosphorolysis reaction starting with the incorrectly matched [DNA_{22} (G•T)] at various dGTP concentrations, in the presence of 5 mM PP_i: 25 μM (black circles), 50 μM (red circles), and 100 μM (green triangles). Pyrophosphorolysis is the primary reaction occurring initially, followed by incorporation until a steady state is reached. Very little DNA_{21} (G•C) results from this reverse reaction (note scale on *y* axis). (*C*) Correct incorporation of dATP by D473G *Pfu* pol at various dATP concentrations: 156 nM (black circles), 312 nM (red circles), 625 nM (green triangles), 1,250 nM (yellow triangles), and 2,500 nM (blue squares). In all cases, dGTP is present at 50 μM in the reaction to prevent pyrophosphorolysis continuing beyond the 20mer primer. Reactions were carried out for 4 h. (*D*) Pyrophosphorolysis reaction starting from DNA_{22} (A•T) at various dATP concentrations: 156 nM (black circles), 312 nM (red circles), 625 nM (green triangles), 1,250 nM (yellow triangles), and 2,500 nM (blue squares). Also, dGTP is present at 50 μM in the reaction to prevent pyrophosphorolysis continuing beyond the 20mer primer. Reactions were carried out for 8 h. Reactions with D473G *Pfu* pol apparently begin with a large decrease in the amount of initial 22mer, and then there is a slow increase again with incorporation taking over for pyrophosphorolysis until the reactions reach a steady state.

***Source*: reproduced with permission from Oertell et al. (2016).**

fluorescence quantum yields), which are dominant in excited DNA. The combination of broadband spectral probing (300–700 nm) with a time resolution of <100 fs allowed monitoring of a variety of spectroscopic transitions simultaneously. Within the first 100 fs after optical excitation, the ESA spectra of all nucleic acids were found to change drastically. The initially excited state of AMP decays to a vibrationally hot ground state with a time constant of <200 fs (Peon and Zewail, 2001;

Pecourt et al., 2001; Markovitsi et al., 2003). The hot ground-state absorption manifests itself as a broad spectral tail (Sension et al., 1990) with monotonically rising intensity toward the short-wavelength probing limit at 300 nm (Buchvarov et al., 2007; see Figure 10.16A). Figure 10.16B displays the early-time spectral dynamics of $(dA)_6$ where the initial band at ≈380 nm decays and a new band rises at ≈330 nm. Here $(dA)_6$ was chosen representatively for all $(dA)_n$ systems that show qualitatively identical spectral dynamics. The time scale for the formation of the ESA band ≈330 nm is sub-1 ps in all $(dA)_n$ systems studied. Dynamic quenching, i.e. the shortening of this ultrafast time component in $(dA)_n$ oligomers (with respect to AMP) caused by excimer formation, has not been observed. After 1.5 ps the spectral position of the maximum of the ESA spectrum remains constant at 330 nm throughout the excited-state lifetime of several hundred ps (see Figure 10.16C). The observed spectral dynamics can be illustrated in a simple energy level diagram (see Figure 10.17).

The temporal evolution of the ESA spectra (Figure 10.16) leads us to some brief concluding remarks (for details read the original article, Buchvarov et al., 2007). The electronic coupling between stacked bases leads to the formation of delocalized exciton states upon UV absorption. In single-stranded homoadenine sequence, the typical 1/e delocalization length is 3–4 bases. However, given the conformational inhomogeneiety of these flexible biopolymers, more extended delocalization is likely to be present in some molecules. Ensembles of A·T duplexes have a larger fraction of more extended delocalized domains. This is evidenced by the significant increase in the intensity ratio ρ_{435} from $(dA)_{12}\cdot(dT)_{12}$ to $(dA)_{18}\cdot(dT)_{18}$. The observed temporal decay of ρ_{435} is a clear indication that the electronic exciton structure is dynamic, and that changes in the delocalization length occur during the lifetime of the excited base stack. Because of the observed time scale, these changes must be induced by nuclear motions within the base stack and/or in the immediate surroundings of the DNA. Despite favorable stacking and apparent strong electronic coupling, there is a substantial fraction of excited DNA molecules found to undergo ultrafast IC to the hot ground state, similar to single bases (Crespo-Hernández et al., 2005). It is reasonable to assume that the optical excitation in this molecular subensemble remains localized due to static and dynamic disorder in the stack. Random DNA sequences containing both A·T and G·C base pairs are even more likely to yield localized "monomer-type" electronic states. The effective competition of the monomer-type photophysical pathway with the exciton formation is critical from an evolutionary viewpoint because excess energy is funneled to the ground state in times shorter than needed to make and break chemical bonds. The amount of disorder in the genome will define what fraction of stacked bases can avoid irreversible photodamage by eliminating electronic excess energy in the same fashion as single DNA bases.

10.3.2 RNA Energetics

RNA folding is one of the most essential processes underlying RNA functions (Zemora and Waldsich, 2010). RNA folding energy landscapes were studied using a statistical mechanical treatment by Chen and Dill (2000). The thermodynamic folding intermediates were predicted here. Unfolding is a simple unzipping process for one hairpin sequence. But for another sequence, the unfolding process is more complex. It involves multiple stable intermediates and a rezipping into a completely non-native conformation before unfolding. The principle that emerges is that although the protein folding tends to involve highly cooperative two-state thermodynamic transitions, without detectable intermediates, the folding of RNA secondary structures may involve rugged landscapes, often with more complex intermediate states.

The energy landscape model used in this investigation to address the RNA folding phenomena is detailed in Chen and Dill (1995, 1998, 2000). The theoretical method/model has been tested using the following basic considerations:

- A good folding algorithm has been developed.
- It should predict the point of global minimum, the native structure, on the energy landscape.

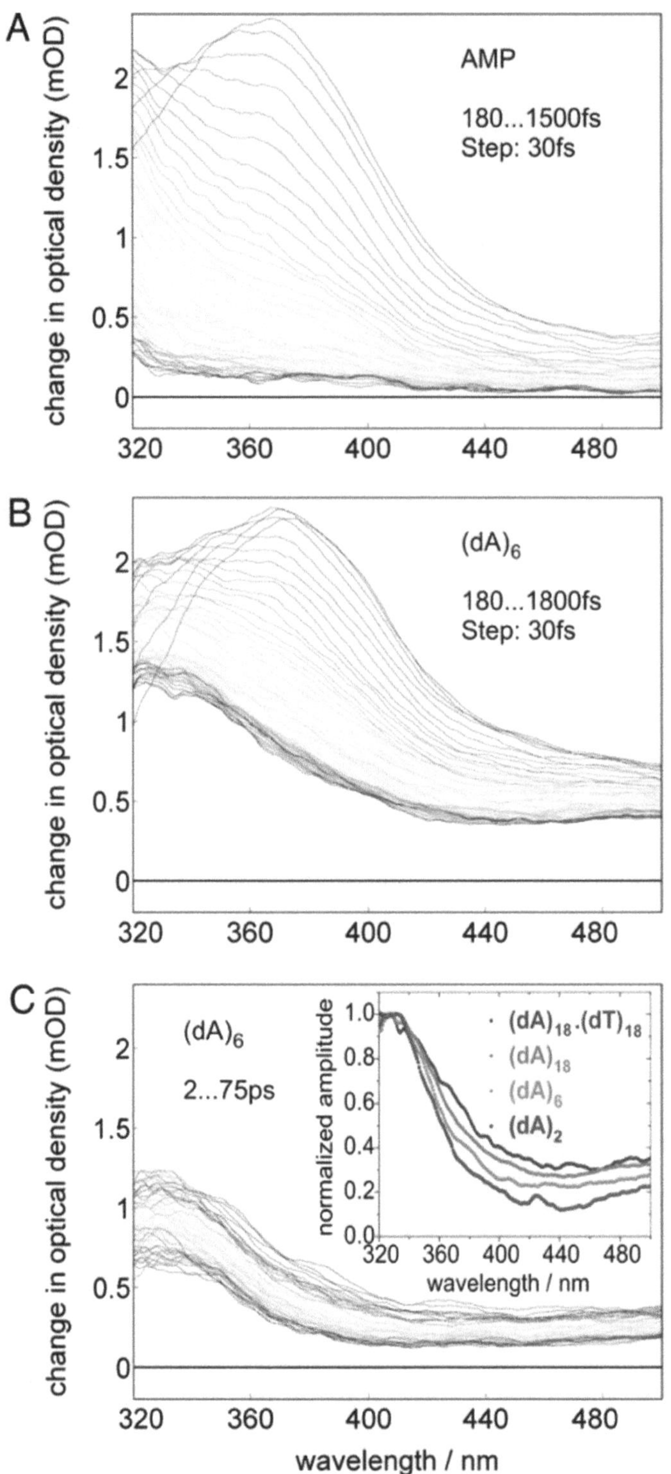

FIGURE 10.16 Temporal evolution of the ESA spectra of AMP (*A*) and $(dA)_6$ (*B* and *C*). Early spectra are shown in blue/green, and late spectra are in orange/red. (*C Inset*) The averaged normalized spectra between 5 and 9 ps of $(dA)_2$, $(dA)_6$, $(dA)_{18}$, and $(dA)_{18}\cdot(dT)_{18}$.

***Source*: reproduced with permission from Buchvarov et al. (2007).**

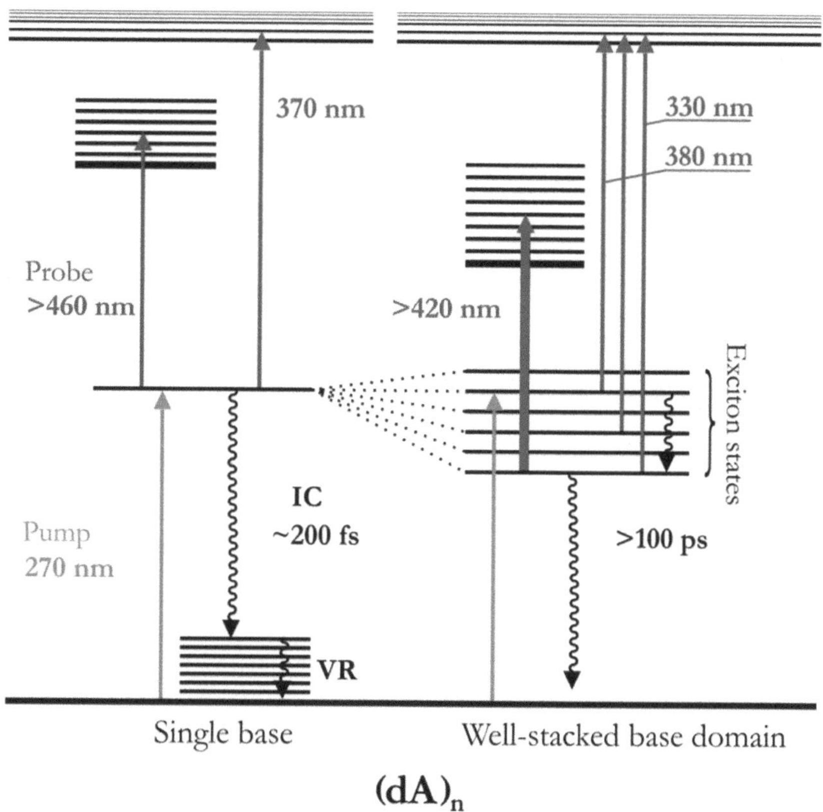

FIGURE 10.17 Energy-level diagram for the energy dissipation pathways in homoadenine sequences $(dA)_n$. In addition to poorly stacked single bases that show very similar behavior to AMP, there are well stacked base domains that can be excited cooperatively by UV light, forming delocalized exciton states. Energy dissipation in these delocalized states involves IC and can be traced spectroscopically.

***Source*: reproduced with permission from Buchvarov et al. (2007).**

In this regard, the RNA molecules were tested by predicting native RNA secondary structures (see Figure 10.18), which has about the same level of accuracy as existing methods for predicting native RNA secondary structures (Gluick and Draper, 1994; Laing et al., 1994).

The model's aim was to relate microscopic conformations to macroscopic properties, such as the melting thermodynamics. The structures predicted for a series of equilibrium melting experiments were described (Chen and Dill, 2000). The results were described in terms of energy landscapes. Because a full high-dimensional energy landscape was neither visualizable nor illuminating, Figure 10.19 shows instead a "reduced" energy landscape, the free energy $F(n,nn) = -kT\ln Q(n,nn,T)$, where $Q(n,nn,T)$ is the partition function, the count of all of the conformations that contain n native contacts and nn non-native contacts. $F(n,nn)$ can be treated as a projection of the full landscape on the (n,nn) plane, assuming all of the other degrees of freedom are in thermal equilibrium and hence can be averaged out. A contact is called "native" if that particular hydrogen-bonded base pair exists in the native structure, and "non-native" otherwise.

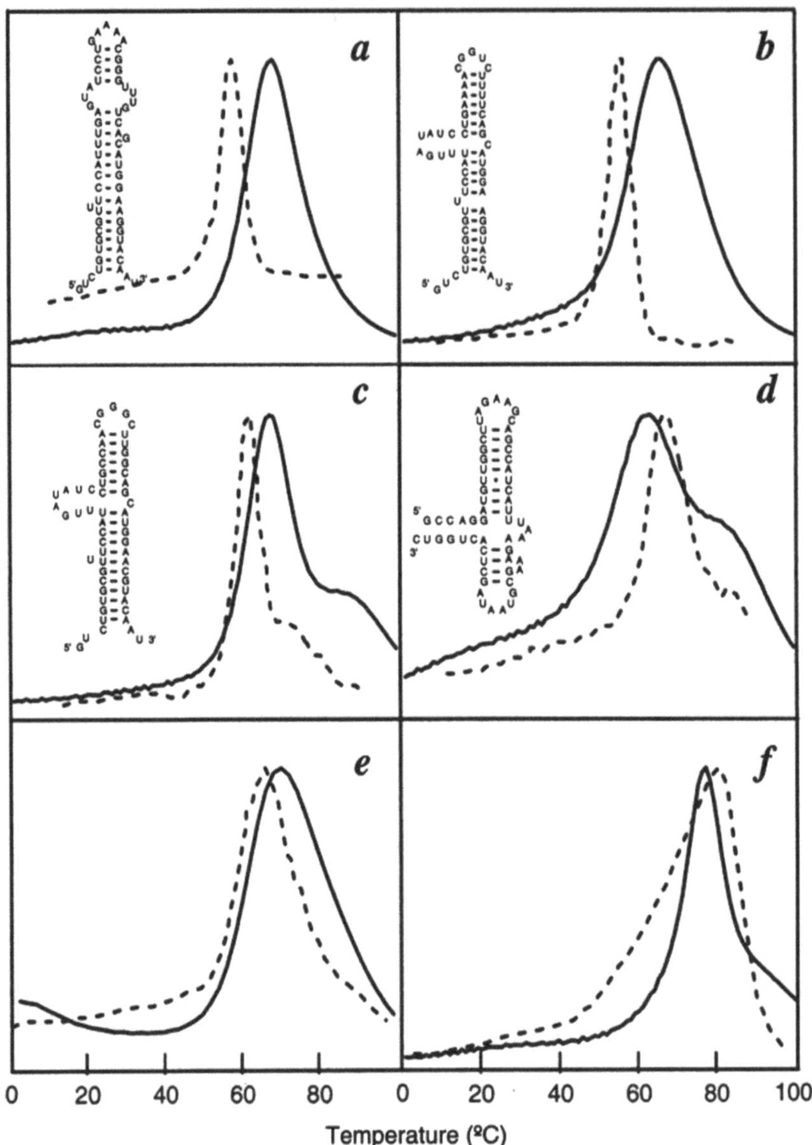

FIGURE 10.18 Predicted (continuous lines) and experimental differential melting curves (dashed lines). *a*, *b*, *c*, and *e* show the results for α operon mRNA fragment and the variants in 100 mM *(a, b*, and *c)* or 1,000 mM *(e)* KCl with 10 mM Mops (Gluick and Draper, 1994): G16-A72 *(a)*; G36→U *(b)*; AA44→CC, UU55→GG *(c)*; and G16-U127 *(e)*, with added nucleotides at the 3′ end: AU *(a, b*, and *c)*. *d* and *f* show denaturation profiles for *E. coli* 23S rRNA fragment G1051-C1109 in 1 M KCl with 10 mM Mops (Laing et al., 1994) and *E. coli* 5S RNA in 1 M NaCl with 5 mM sodium phosphate and 1 mM EDTA, respectively. *a* and *f* are from calorimetric experiments, and *b*, *c*, and *e* are from UV absorbance experiments. Because our model energies are from Turner's rules, taken in 1 M NaCl, our model should in general overestimate the melting temperatures (T_m) for experiments in lower salt conditions. Because our model accounts only for the secondary structures, peaks in the experimental curves for the disruption of the tertiary structure *(f)* were removed. The denaturation curve shows the heat capacity as a function of temperature. The heat capacity $C(T)$ is obtained from the partition function $Q(T)$ by using $C(T) = \partial/\partial T[kT^2(\partial/\partial T)\ln Q(T)]$. The computer time scales with the chain length L as L^4: for each temperature, the calculation of $Q(T)$ and $C(T)$ took 3 sec. on an Intel Pentium Pro 200 computer for the 59-mer

FIGURE 10.18 *(Continued)*

sequences (*a–d*), 30 sec. for the 100-mer sequence, and 54-sec. for the 120-mer sequence (*f*). We chose $\mu = e^{-2}$ to give a best fit to the experimental curve in *c*, and used this value for all other calculations. Small variations in μ (e.g., $\mu = e^{-2.5}$) cause only minor changes of the positions and shapes of the main and satellite peaks, but larger changes in μ lead to significant changes of the melting curves. Although the theory gives good predictions, given the simple approach, it is conceivable that the theory may differ from experimental denaturation profiles in some details. Improvements can come from accounting for tertiary interactions, treating unpaired terminal nucleotides, and other refinements in the theory.

***Source*: reproduced with permission from Chen and Dill (2000).**

All-atom modeling of RNA energetics was recently analyzed by Smith and colleagues (2017). This review presents a summary of the current state of molecular dynamics as applied to RNA. Figure 10.20 presents an easily visualizable free energy landscape to reliably predict structure and dynamics.

Understanding the RNA energetics and interaction-based structural stability may also require us to consider interactions involving the nucleotide base, sugar and phosphate moieties. The structured RNA molecules form three-dimensional (3D) complex architectures stabilized by multiple interactions involving these nucleotide base, sugar and phosphate moieties. A significant proportion of the bases in the structured RNA molecules in the Protein Data Bank (PDB) hydrogen-bond with the phosphates of other nucleotides. By extracting and superimposing base-phosphate (BPh) interactions from a reduced-redundancy subset of 3D structures from the PDB, Zirbel and colleagues identified recurrent phosphate-binding sites on the RNA bases (Zirbel et al., 2009).

Identification of BPh interactions in 3D structures was made, where Zirbel and colleagues (2009) searched the reduced-redundancy set of RNA 3D structures to locate all phosphate groups within H-bonding distance of some nucleobase (see Stombaugh et al., 2009). For visualizing the potential interactions, they superposed the interacting bases for each potential BPh interaction and then plotted the positions of the phosphorus and the phosphate oxygen atoms that are closest to the corresponding H-bond donor atom of the base. A set of projections of these data onto the plane of the donor base is shown in Figure 10.21, color-coded according to the distance between the H-bond donor and the nearest phosphate oxygen atom.

The calculations also show that the most stable phosphate-binding sites occur on the Watson–Crick edge of guanine and the Hoogsteen edge of cytosine. We modified the "Find RNA 3D" (FR3D) software suite to automatically find and classify BPh interactions. Comparison of the 3D structures of the 16S and 23S rRNAs of *Escherichia coli* and *Thermus thermophilus* revealed that most BPh interactions are phylogenetically conserved and they occur primarily in hairpin, internal, or junction loops or as part of tertiary interactions. Bases that form BPh interactions, which are conserved in the rRNA 3D structures, are also conserved in homologous rRNA sequence alignments.

BPh interactions and their corresponding category labels have been shown in Figure 10.22. Each of the bases presents distinct combinations of H-bond donors and acceptors, but certain bases have electropositive functional groups at equivalent positions, therefore can form equivalent BPh interactions (Zirbel et al., 2009). The BPh interactions, formed by more than one base, include 0BPh, 5BPh, 6BPh, 7BPh and 9BPh. All of the bases can form 0BPh interactions using the purine H8 or pyrimidine H6 proton as H-bond donor. Both A (Figure 10.22, upper-left panel) and C (upper-right panel) can form 6BPh and 7BPh interactions using equivalent exocyclic amino groups at the A(N6) and C(N4) positions. Similarly, G (lower-left panel) and U (lower-right panel) can form equivalent 5BPh interactions using the G(N1) or U(N3) imino groups on their WC edges. The 3BPh, 4BPh, and 5BPh interactions all occur on the WC edge of G and are possibly interchangeable during conformational changes or thermal fluctuations of RNA molecules. Likewise, the 7BPh, 8BPh, and

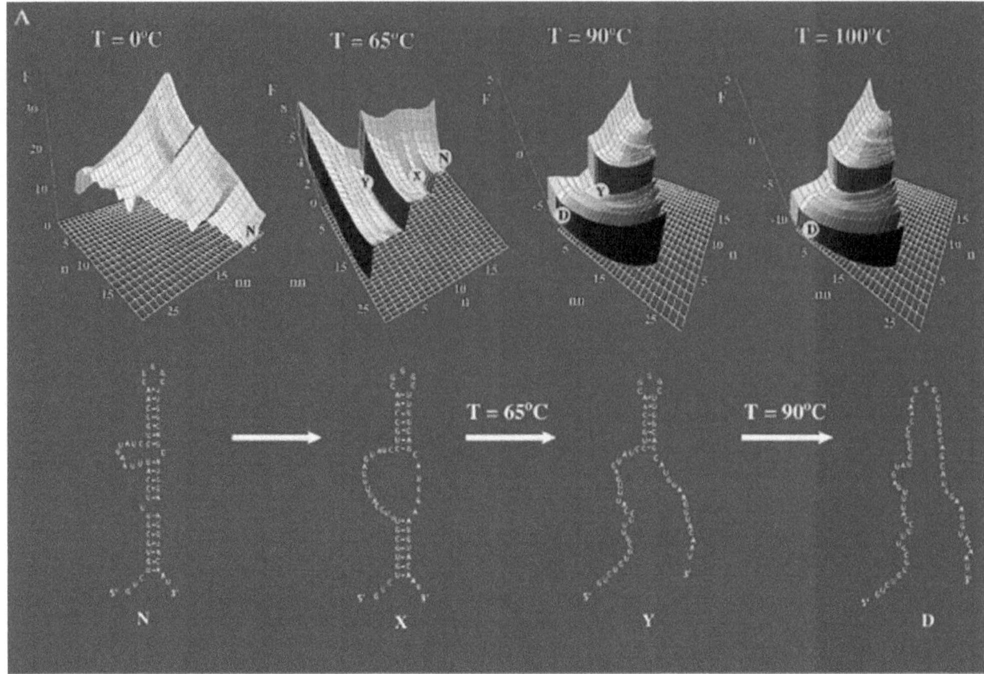

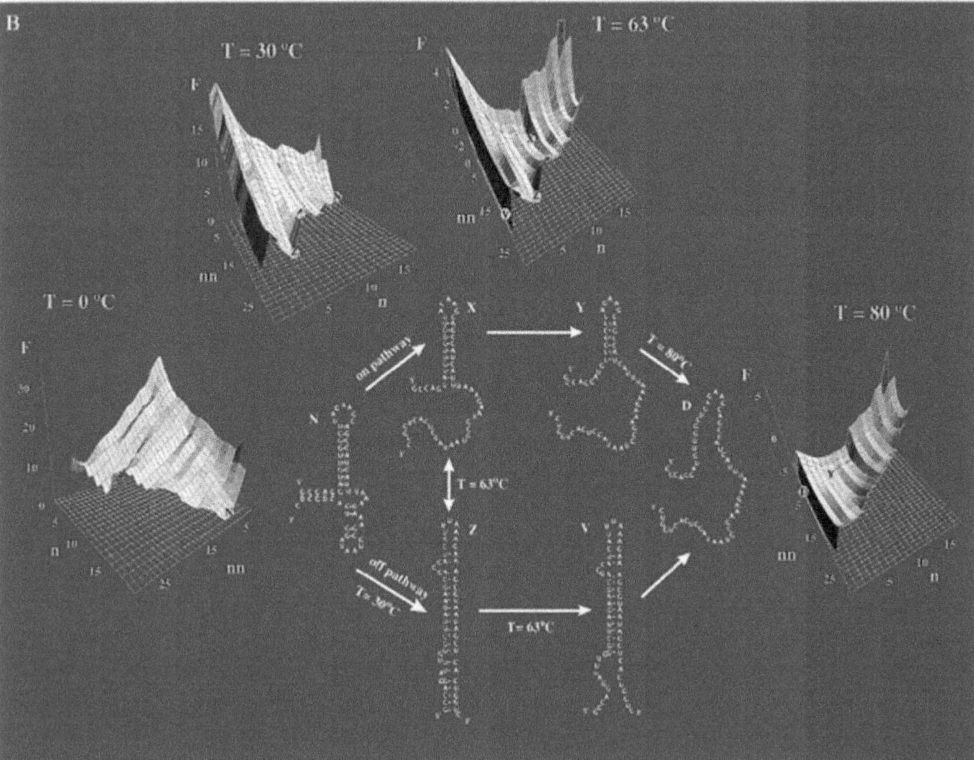

FIGURE 10.19 The energy landscapes $F(n,nn)$ vs. Temperature T (in °C) for (A) mutant α mRNA fragment (see Figure 10.18c) and (B) *E. coli* 23S rRNA fragment (see Figure 10.18d). $F(n,nn)$ is the free energy for a state with n native contacts and nn non-native contacts, where the native and non-native contacts are defined

FIGURE 10.19 (*Continued*)

according to the native structure N. The free energies (in kcal/mol) are relative to the native states. Under native conditions, the number of non-native contacts need not equal zero, because loops can bump into the chain in ways that involve no stable contact. Stable states are valleys, highlighted by the symbols on the energy landscapes.

***Source*: reproduced with permission from Chen and Dill (2000).**

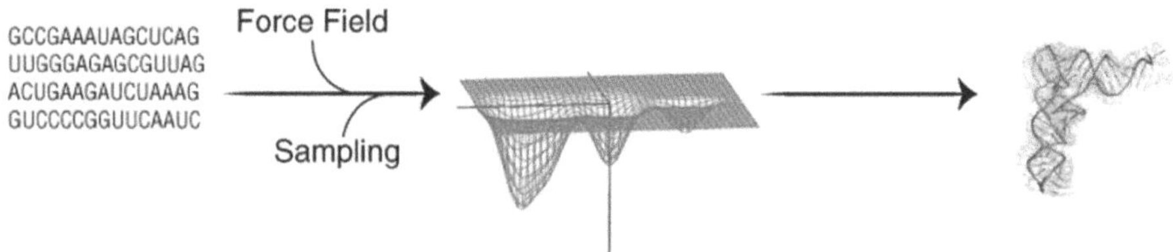

FIGURE 10.20 A perfect RNA force field and convergent sampling would generate a free energy landscape to reliably predict structure and dynamics.

***Source*: reproduced with permission from Smith et al. (2017).**

9BPh interactions all occur on the Hoogsteen edge of C and may also be interchangeable. The other interactions are specific to individual bases: 1BPh, 3BPh, and 4BPh are specific to G, 2BPh is specific to A, and 8BPh is specific to C. BPh interactions that occur on the same edge of the same base are referred as neighboring interactions. When comparing homologous 3D structures, it is expected that the same or possibly neighboring BPh interactions will be observed at homologous locations where the base forming the BPh interaction is conserved. When a base substitution occurs, it is expected that for the BPh interaction to be maintained without a large conformational change or disruption of the structure, an equivalent BPh interaction must form. This suggests BPh interactions that are crucial for RNA folding, stability, and function constrain the sequence variation of homologous RNA molecules.

Applying the quantum chemical calculations on model systems representing each BPh interaction, Zirbel and colleagues (2009) showed that the centers of each cluster obtained from the structure superpositions correspond to energy minima on the potential energy hypersurface. Here the quantum calculations were carried out for each of the identified 17 distinct potential BPh interactions. The model geometries were constructed for each interaction formed by each base using the empirical structures as a guide. The geometries were optimized quantum mechanically and then interaction energies were calculated (PDB files with the optimized geometries are provided at http://rna.bgsu.edu/FR3D/BasePhosphates). The calculated interaction energies and optimal distances are presented in Table 10.1.

The interaction energy, ΔE^{BPh} (represented by E^{INT} for simplicity) (Table 10.1) is expressed by the following equation (Zirbel et al., 2009):

$$E^{INT} = \Delta E^{BPh} = \Delta E^{BSSE}_{COSMO} \tag{10.1}$$

The "basis set superposition error" BSSE (for details, see Boys and Bernardi, 1970) corrected interaction energy ΔE^{BSSE}_{COSMO} was computed with the COSMO dielectric continuum model that adds the mean-field effect of solvent screening of the electrostatic forces (Barone and Cossi, 1998; Cossi et al., 2003).

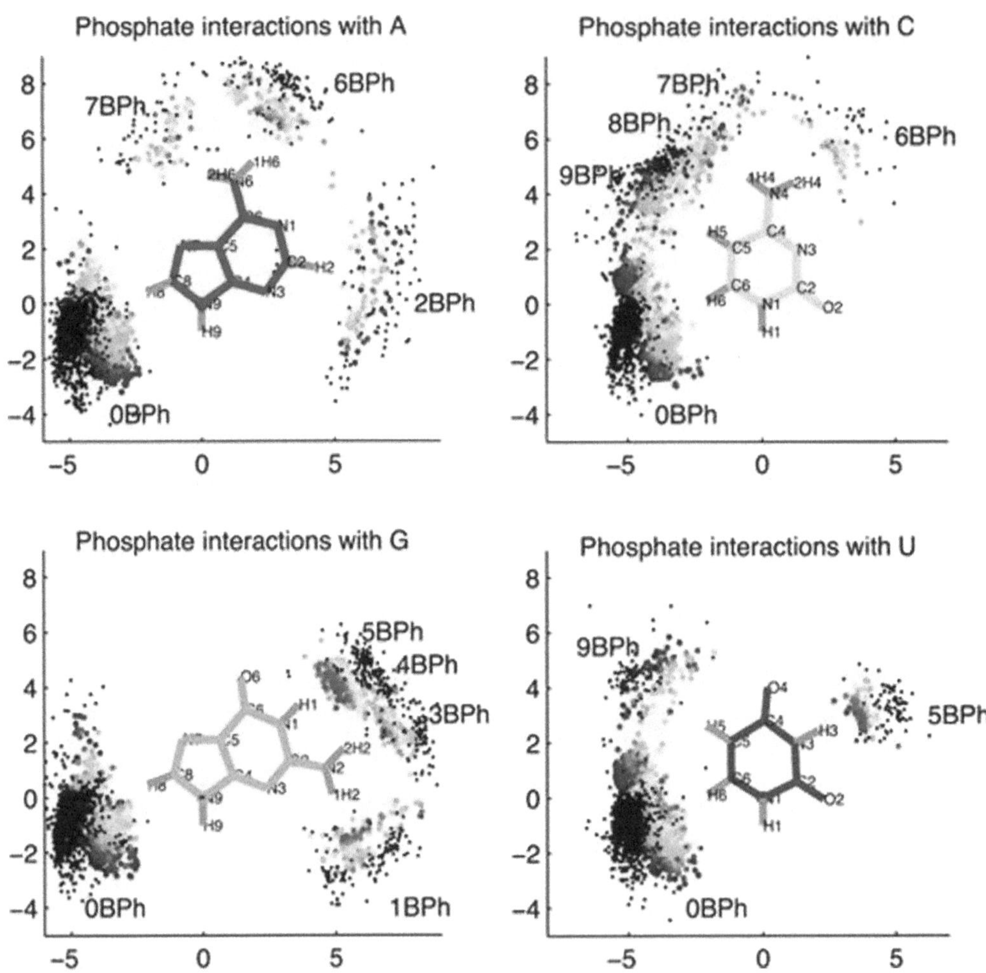

FIGURE 10.21 Potential BPh interactions extracted from the reduced-redundancy dataset of RNA 3D structures having resolution 2.5 Å or better. Bases from each instance were superposed and relative locations of phosphorus (black dots) and nearest phosphate oxygen atoms (colored dots) are plotted. Dots representing oxygen atoms are colored to indicate distance to the nearest base H-bond donor. These range from 2.5 Å or less (red) to 4.5 Å (dark blue). Only those phosphate oxygens within 4.5 Å of a base H-bond donor and with bond angle >110° are included.

***Source*: reproduced with permission from Zirbel et al. (2009).**

The analysis of the 3D structures of *E. coli* and *T. thermophilus* 16S and 23S rRNAs by Zirbel and colleagues (2009) indicates that a significant fraction of bases in structured RNA molecules (about 13 percent) form conserved inter-nucleotide BPh interactions. Moreover, they are widespread in hairpin, internal, and multi-helix junction loops. About ~87 percent of hairpin loops, ~80 percent of junction loops, and ~71 percent of internal loops of *E. coli* 16S and 23S rRNA were found to contain one or more BPh interactions involving nucleotides of the same loop. As many of these motifs are recurrent, BPh interactions are also likely to be found widely in other structured RNAs. A significant number of the conserved BPh interactions are long-range (143 out of 496), which suggests that BPh interactions also play significant roles stabilizing the tertiary structures of structured RNA molecules.

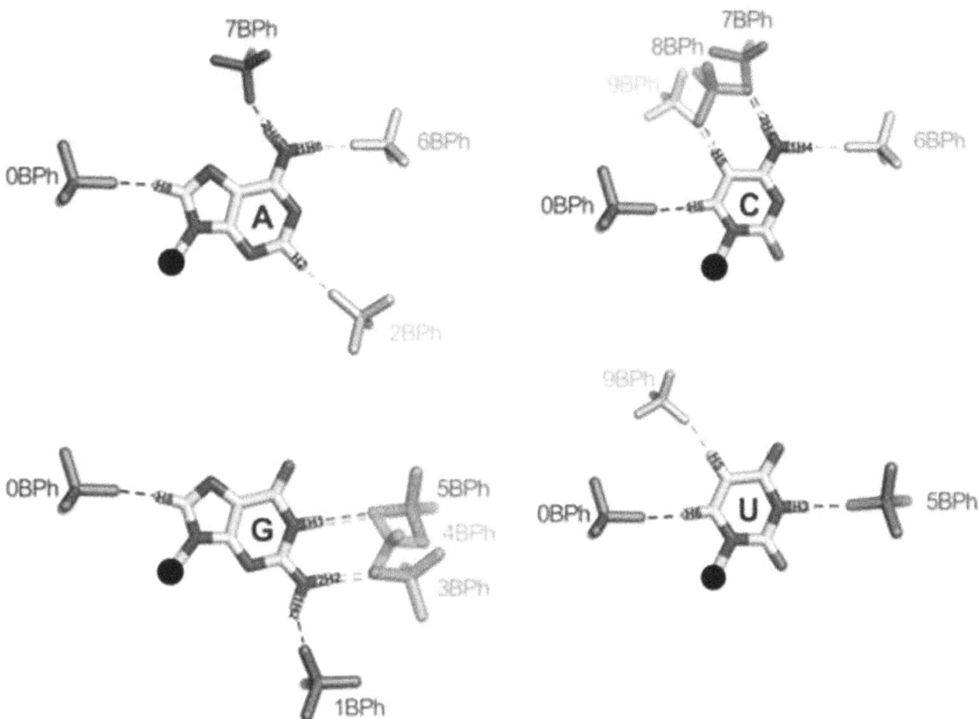

FIGURE 10.22 Proposed nomenclature for BPh interactions and superpositions of idealized BPh interactions observed in RNA 3D crystal structures for each base. H-bonds are indicated with dashed lines. BPh categories are numbered 0–9, starting at the H6 (pyrimidine) or H8 (purine) base positions. BPh interactions that involve equivalent functional groups on different bases are grouped together, i.e. 0BPh (A, C, G, U), 5BPh (G, U), 6BPh (A, C), 7BPh (A, C), and 9BPh (C, U).

***Source*: reproduced with permission from Zirbel et al. (2009).**

Detailed analysis revealed 17 unique phosphate-binding sites on the standard RNA bases (A, C, G, and U) (Zirbel et al., 2009). Based on the reported quantum chemical calculations and consideration of the locations of these binding sites along the base perimeters (edges), Zirbel and colleagues classified them into 10 families. This groups together, under the same designation, phosphate-binding sites that occur at equivalent sites on different bases. Thus, 5BPh designates the imino nitrogen H-bond donors on the WC edges of G and U with phosphate oxygen. Other BPh interactions that are shared by more than one base are 6BPh, on the WC edges of A and C, 7BPh, on the Hoogsteen edges of A and C, 9BPh, on the Hoogsteen edges of U and C, and the non-specific 0BPh, also on the Hoogsteen edge. The other BPh interactions are base-specific. BPh interactions that occur on the same base edge, such as the 3BPh, 4BPh, and 5BPh interactions of G, are called "neighboring interactions" and are very conserved.

There is a large amount of data that cannot be accommodated here due to space, so I encourage more focused readers to refer to the original article (Zirbel et al., 2009). In conclusion, by combining empirical and quantum chemical methods, they have identified optimal phosphate-binding geometries for each RNA base and implemented software in FR3D to find and classify these interactions in 3D structures (see http://rna.bgsu.edu/FR3D/BasePhosphates). Considering energetics, they have shown that BPh interactions are highly conserved in homologous rRNA molecules. BPh interactions have been found to strongly constrain the sequences of homologous structured RNAs. These

TABLE 10.1
Energies and H-bond Distances for BPh Interactions (0BPh-9BPh) by Nucleotide (A, C, G, U)

	A		C		G		U	
	E^{INT} (kcal/mol)	H-Bond Distance (Å)	E^{INT} (kcal/mol)	H-Bond Distance (Å)	E^{INT} (kcal/mol)	H-Bond Distance (Å)	E^{INT} (kcal/mol)	H-Bond Distance (Å)
1BPh					−4.3	2.85 (N2)		
2BPh	−0.1	3.41 (C2)						
3BPh					−5.4	2.80 (N2)		
4BPh					−10.1	2.85 (N2) (1)		
						2.89 (N1) (2)		
5BPh					−4.0	2.92 (N1)	−4.2	2.81 (N3)
6BPh	−3.1	2.90 (N6)	−3.5	2.89 (N4)				
7BPh	−2.8	2.77 (N6)	−4.8	2.85 (N4)				
8BPh			−5.6	2.86 (N4) (1)				
				3.53 (C5) (2)				
9BPh			−0.6	3.35 (C5)			−0.6	3.34 (C5)
0BPh	−1.1	3.22 (C8)	−1.0	3.21 (C6)	−1.1	3.25 (C8)	−1.1	3.23 (C6)

Each nucleotide is represented by two columns:
(1) The calculated interaction energy (kcal/mol);
2 Distance (in Å) from H-bond donor to acceptor. The H-bond donor site is given in parentheses.

Source: reproduced with permission from Zirbel et al. (2009).

conclusions have been made based on studies of rRNA, but the methods have been predicted to be applicable for understanding other homologous structured RNAs.

10.5 NUCLEIC ACID STRUCTURE-BASED VACCINES

Knowledge of RNA structural complexity or simplicity helps scientists to create complexes mixing RNA molecules with other biomolecules to accomplish medical breakthroughs. During the late twentieth century, Robert Malone was successful in mixing strands of mRNA with droplets of fat, to create a kind of molecular stew. Human cells bathed in this genetic gumbo were found to absorb the mRNA, and began producing proteins from it (Malone, 1989).

According to Malone, if cells could create proteins using mRNA delivered into them, he thought that it might be possible to "treat RNA as a drug." Malone's subsequent experiments showed that frog embryos absorbed such mRNA (Malone, 1989), thus it appeared to be the first-ever successful demonstration using fatty droplets to ease mRNA's passage into a living organism. These experiments are now considered stepping stones toward the discovery of two of the most important and profitable vaccines in history (Dolgin, 2021; Ball, 2021; see Figure 10.23): the mRNA-based COVID-19 vaccines given to hundreds of millions of people around the world which helped the companies to earn hundreds of billions of dollars (Ball, 2021).

Both RNA- and DNA-involved nucleic acid-based vaccines have achieved quite considerable successes (Leitner et al., 1999). DNA vaccines were introduced more than two decades ago and have

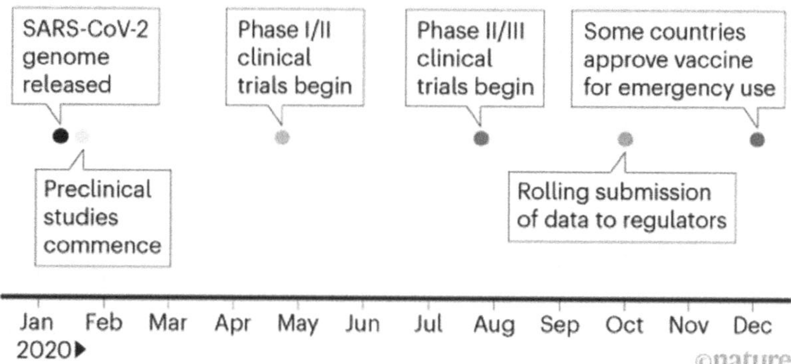

FIGURE 10.23 Vaccine discovery phases using mRNA.

***Source*: BioNTech/Pfizer; Nature analysis. Reproduced with permission from Ball (2021).**

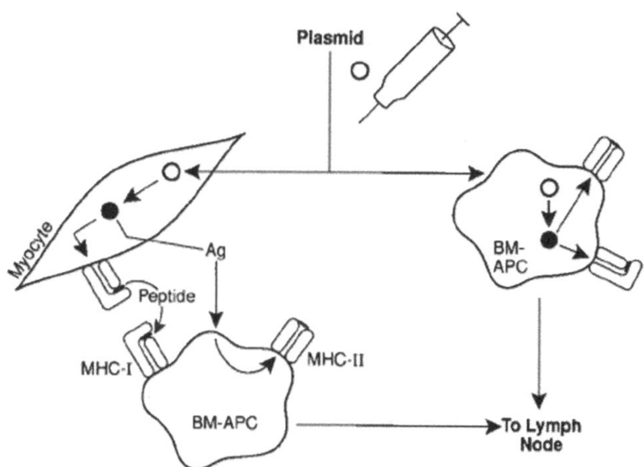

FIGURE 10.24 Transfection of host cells with plasmid DNA. Plasmid (◯) is actively or passively taken up by host cells. Antigen (v) produced by transfected myocytes can be taken up by bone marrow (BM)-derived antigen-presenting cells (APCs). Alternatively, BM-APC can be transfected directly. Antigen-bearing APC then can process and present peptides complexed with MHC-molecules to the immune system after migrating to lymphoid tissue.

***Source*: reproduced with permission from Leitner et al. (1999).**

already been applied to a wide range of infectious and malignant diseases. DNA vaccination has become the fastest-growing field in vaccine technology since the early nineties when plasmid DNA was reported to induce an immune response to the plasmid-encoded antigen (Wolff et al., 1990; Tang et al., 1992). Although DNA vaccination is often hampered by poor efficacy, this method is generally considered by many to be one of the most important discoveries in the history of vaccinology (Mor, 1998; Whalen, 1996).

The genetic vaccines consist of DNA (as plasmids) or RNA (as mRNA), which is taken up by cells and then translated into protein. The plasmid DNA is precipitated onto an inert particle (generally gold beads) and forced into the cells with a helium blast. The transfected cells then express the antigen encoded on the plasmid, resulting in an immune response (see Figure 10.24). Like live or attenuated viruses, DNA vaccines effectively engage both MHC-I and MHC-II pathways, allowing for the induction of CD8+ and CD4+ T cells (Wang et al., 1998), whereas antigen present in soluble form, such as recombinant protein, generally induces only antibody responses.

REFERENCES

Allmang, C., Kufel, J., Chanfreau, G., Mitchell, P., Petfalski, E., & Tollervey, D. (1999). Functions of the exosome in rRNA, snoRNA and snRNA synthesis. *EMBO J.*, *18*, 5399–5410

Ashrafuzzaman, M., & Shafee, A. (2005). Possible role of telomeres in DNA in aging process. In *Trends in chemical physics research*, ed. A.N. Linke, pp. 281–296. Hauppauge, NY: Nova Science

Ball, P. (2021). The lightning-fast quest for COVID vaccines – and what it means for other diseases. *Nature*, *589*, 16–18. https://doi.org/10.1038/d41586-020-03626-1

Banáš, P., Sklenovsky, P., Wedekind, J.E., Šponer, J., & Otyepka, M. (2012). Molecular mechanism of preQ1 riboswitch action: a molecular dynamics study. *Journal of Physical Chemistry B*, *116*(42), 12721–12734. doi: 10.1021/jp309230v

Barone, V., & Cossi, M. (1998). Quantum calculation of molecular energies and energy gradients in solution by a conductor solvent model. *Journal of Physical Chemistry A*, *102*(11), 1995–2001

Beerman, T.A., & Lebowitz, J. (1973). Further analysis of the altered secondary structure of superhelical DNA: sensitivity to methylmercuric hydroxide a chemical probe for unpaired bases. *Journal of Molecular Biology*, *79*(3), 451–470

Benenson, Y., Adar, R., Paz-Elizur, T., Livneh, Z., & Shapiro, E. (2003). DNA molecule provides a computing machine with both data and fuel. *Proceedings of the National Academy of Sciences*, *100*(5), 2191–2196

Bevilacqua, P.C., Ritchey, L.E., Su, Z., & Assmann, S.M. (2016). Genome-wide analysis of RNA secondary structure. *Annual Review of Genetics*, *50*, 235–266

Bochman, M., Paeschke, K., & Zakian, V. (2012). DNA secondary structures: stability and function of G-quadruplex structures. *Nat. Rev. Genet.*, *13*, 770–780. https://doi.org/10.1038/nrg3296

Bončina, M., Lah, J., Prislan, I., & Vesnaver, G. (2012). Energetic basis of human telomeric DNA folding into G-quadruplex structures. *Journal of the American Chemical Society*, *134*(23), 9657–9663

Boys, S.F., & Bernardi, F.J.M.P. (1970). The calculation of small molecular interactions by the differences of separate total energies: some procedures with reduced errors. *Molecular Physics*, *19*(4), 553–566

Buchvarov, I., Wang, Q., Raytchev, M., Trifonov, A., & Fiebig, T. (2007). Electronic energy delocalization and dissipation in single-and double-stranded DNA. *Proceedings of the National Academy of Sciences*, *104*(12), 4794–4797

Cadet, J., & Vigny, P. (1990). The photochemistry of nucleic acids. In *Bioorganic photochemistry* (vol. 1), ed. H. Morrison. pp. 1–272. New York: Wiley

Cantor, C.R. (1968). The extent of base pairing in 5S ribosomal RNA. *Proceedings of the National Academy of Sciences*, *59*(2), 478–483

Chen, S.J., & Dill, K.A. (2000). RNA folding energy landscapes. *Proceedings of the National Academy of Sciences*, *97*(2), 646–651

Chen, S.J., & Dill, K.A. (1998). Theory for the conformational changes of double-stranded chain molecules. *Journal of Chemical Physics*, *109*(11), 4602–4616

Chen, S.J., & Dill, K.A. (1995). Statistical thermodynamics of double-stranded polymer molecules. *Journal of Chemical Physics*, *103*(13), 5802–5813

Cheong, H.-K., Hwang, E., Lee, C., Choi, B.-S., & Cheong, C. (2004). Rapid preparation of RNA samples for NMR spectroscopy and X-ray crystallography. *Nucleic Acids Res.*, *32*, e84

Cobb, M. (2015). Who discovered messenger RNA? *Current Biology*, *25*(13), R526–R532

Condon, A. (2006). Designed DNA molecules: principles and applications of molecular nanotechnology. *Nature Reviews Genetics*, *7*(7), 565–575. doi: 10.1038/nrg1892

Connors, P.G., & Beeman, W.W. (1972). Size and shape of 5 S ribosomal RNA. *Journal of Molecular Biology, 71*(1), 31–37

Cossi, M., Rega, N., Scalmani, G., & Barone, V. (2003). Energies, structures, and electronic properties of molecules in solution with the C-PCM solvation model. *Journal of Computational Chemistry, 24*(6), 669–681

Crespo-Hernández, C., Cohen, B. & Kohler, B. (2005). Base stacking controls excited-state dynamics in A·T DNA. *Nature, 436*, 1141–1144. https://doi.org/10.1038/nature03933

Dickerson, R.E. (1992). DNA structure from A to Z: how do you tell if a structure is right by reading the paper? *Methods Enzymol., 211*, 67–111

Dolgin, E. (2021). The tangled history of mRNA vaccines. *Nature, 597*(7876), 318–324. https://doi.org/10.1038/d41586-021-02483-w

Dumaz, N., van Kranen, H.J., de Vries, A., Berg, R.J., Wester, P.W., van Kreijl, C.F., ... & de Gruijl, F.R. (1997). The role of UV-B light in skin carcinogenesis through the analysis of p53 mutations in squamous cell carcinomas of hairless mice. *Carcinogenesis, 18*(5), 897–904

Eckahl, T.T., & Anderson, J.N. (1987). Computer modelling of DNA structures involved in chromosome maintenance. *Nucleic Acids Research, 15*(20), 8531–8545

Ehresmann, C., Baudin, F., Mougel, M., Romby, P., Ebel, J.P., & Ehresmann, B. (1987). Probing the structure of RNAs in solution. *Nucleic Acids Research, 15*(22), 9109–9128

Fallmann, J., Will, S., Engelhardt, J., Grüning, B., Backofen, R., & Stadler, P.F. (2017). Recent advances in RNA folding. *J. Biotechnol., 261*, 97–104

Felsenfeld, G., Davies, D., & Rich, A. (1957). Formation of a three-stranded polynucleotide molecule. *J. Am. Chem. Soc., 79*, 2023–2024

Fica, S.M., & Nagai, K. (2017). Cryo-electron microscopy snapshots of the spliceosome: structural insights into a dynamic ribonucleoprotein machine. *Nat. Struct. Mol. Biol., 24*, 791 –799

Fox, G.E., & Woese, C.R. (1975). 5S RNA secondary structure. *Nature, 256*, 505–507

Fu, L., Cao, Y., Wu, J., Peng, Q., Nie, Q., & Xie, X. (2022). UFold: fast and accurate RNA secondary structure prediction with deep learning. *Nucleic Acids Research, 50*(3), e14. https://doi.org/10.1093/nar/gkab1074

Fürtig, B., Richter, C., Wöhnert, J., & Schwalbe, H. (2003). NMR spectroscopy of RNA. *ChemBioChem, 4*, 936–962

Geisler, S., & Coller, J. (2013). RNA in unexpected places: long non-coding RNA functions in diverse cellular contexts. *Nat. Rev. Mol. Cell Biol., 14*, 699–712

Gessner, R.V., Frederick, C.A., Quigley, G.J., Rich, A. & Wang, A.H. (1989). The molecular structure of the left-handed Z-DNA double helix at 1.0-Å atomic resolution: geometry, conformation, and ionic interactions of d(CGCGCG). *J. Biol. Chem., 264*, 7921–7935

Gluick, T.C., & Draper, D.E. (1994). Thermodynamics of folding a pseudoknotted mRNA fragment. *Journal of Molecular Biology, 241*(2), 246–262

Hart, J.M., Kennedy, S.D., Mathews, D.H., & Turner, D.H. (2008). NMR-assisted prediction of RNA secondary structure: identification of a probable pseudoknot in the coding region of an R2 retrotransposon. *J. Am. Chem. Soc., 130*, 10233–10239

Hoagland, M.B., Stephenson, M.L., Scott, J.F., Hecht, L.I., & Zamecnik, P.C. (1958). A soluble ribonucleic acid intermediate in protein synthesis. *J. Biol. Chem, 231*(1), 241–257

Jain, A., Wang, G., & Vasquez, K.M. (2008). DNA triple helices: biological consequences and therapeutic potential. *Biochimie, 90*, 1117–1130

Jonoska, N., Karl, S.A., & Saito, M. (1999). Three dimensional DNA structures in computing. *BioSystems, 52*(1–3), 143–153

Jordan, B.R., Galling, G., & Jourdan, R. (1974). Sequence and conformation of 5 S RNA from Chlorella cytoplasmic ribosomes: comparison with other 5 S RNA molecules. *Journal of Molecular Biology, 87*(2), 205–225

Kearns, D.R., & Wong, Y.P. (1974). Investigation of the secondary structure of *Escherichia coli* 5 S RNA by high-resolution nuclear magnetic resonance. *Journal of Molecular Biology, 87*(4), 755–774

Kierzek, K. (2009). Binding of short oligonucleotides to RNA: studies of the binding of common RNA structural motifs to isoenergetic microarrays. *Biochemistry, 48*, 11344–11356

Kim, H.Y., LaVaute, T., Iwai, K., Klausner, R.D., and Rouault, T.A. (1996). Identification of a conserved and functional iron-responsive element in the 5′-untranslated region of mammalian mitochondrial aconitase. *J. Biol. Chem.*, *271*, 24226–24230

Knapp, G. (1989). Enzymatic approaches to probing of RNA secondary and tertiary structure. In *Methods in enzymology* (vol. 180), ed. J.E. Dahlberg & J.N. Abelson, pp. 192–212. New York: Academic Press

Kresge, N., Simoni, R.D., & Hill, R.L. (2005). The discovery of tRNA by Paul C. Zamecnik. *Journal of Biological Chemistry*, *280*(40), e37–e39

Kypr, J., Kejnovska, I., Renciuk, D., & Vorlickova, M. (2009). Circular dichroism and conformational polymorphism of DNA. *Nucleic Acids Res.*, *37*, 1713–1725

Laing, L.G., Gluick, T.C., & Draper, D.E. (1994). Stabilization of RNA structure by Mg ions: specific and nonspecific effects. *Journal of Molecular Biology*, *237*(5), 577–587

Le, S.Y., Siddiqui, A., and Maizel Jr., J.V. (1996). A common structural core in the internal ribosome entry sites of picornavirus, hepatitis C virus and pestivirus. *Virus Genes*, *12*, 135–147

Leitner, W.W., Ying, H., & Restifo, N.P. (1999). DNA and RNA-based vaccines: principles, progress and prospects. *Vaccine*, *18*(9–10), 765–777. doi: 10.1016/s0264-410x(99)00271-6

Lewis, J.B., & Doty, P. (1970). Derivation of the secondary structure of 5S RNA from its binding of complementary oligonucleotides. *Nature*, *225*, 510–512

Liebl, K., & Zacharias, M. (2021). Accurate modeling of DNA conformational flexibility by a multivariate Ising model. *Proceedings of the National Academy of Sciences*, *118*(15), e2021263118

Lu, X.J., Shakked, Z., & Olson, W.K. (2000). A-form conformational motifs in ligand-bound DNA structures. *Journal of Molecular Biology*, *300*(4), 819–840

Malone, R.W. (1989). mRNA transfection of cultured eukaryotic cells and embryos using cationic liposomes. *Focus*, *11*(4), 61–66

Markovitsi, D., Sharonov, A., Onidas, D., & Gustavsson, T. (2003). The effect of molecular organisation in DNA oligomers studied by femtosecond fluorescence spectroscopy. *ChemPhysChem*, *4*(3), 303–305

Mathews, D.H., Moss, W.N., & Turner, D.H. (2010). Folding and finding RNA secondary structure. *Cold Spring Harb. Perspect. Biol.*, 2, a003665

McDowell, S.E., Špačková, N.A., Šponer, J., & Walter, N.G. (2007). Molecular dynamics simulations of RNA: an *in silico* single molecule approach. *Biopolymers: Original Research on Biomolecules*, *85*(2), 169–184. doi: 10.1002/bip.20620

Mitchell, A. (1998). The A to Z of DNA. *Nature*, *396*, 524. https://doi.org/10.1038/25014

Mor, G. (1998). Plasmid DNA: a new era in vaccinology. *Biochem. Pharmacol.*, *55*,1151–1153

Morgan, A.R. (1970). Model for DNA replication by Kornberg's DNA polymerase. *Nature*, *227*, 1310–1313

Neidle S. (2002). *Nucleic acid structure and recognition*. Oxford: Oxford University Press

Noller, H.F. (1984). Structure of ribosomal RNA. *Annu. Rev. Biochem*, *53*, 119–162

Oertell, K., Harcourt, E.M., Mohsen, M.G., Petruska, J., Kool, E.T., & Goodman, M.F. (2016). Kinetic selection vs. free energy of DNA base pairing in control of polymerase fidelity. *Proceedings of the National Academy of Sciences*, *113*(16), E2277–E2285

Olson, A.C., Patro, J.N., Urban, M., & Kuchta, R.D. (2013). The energetic difference between synthesis of correct and incorrect base pairs accounts for highly accurate DNA replication. *J. Am. Chem. Soc.*, *135*, 1205–1208

Palecek, E. (1991). Local supercoil-stabilized DNA structures. *Crit. Rev. Biochem. Mol. Biol.*, *26*, 151–226

Pande, V.S., Baker, I., Chapman, J., Elmer, S.P., Khaliq, S., Larson, S.M., … & Zagrovic, B. (2003). Atomistic protein folding simulations on the submillisecond time scale using worldwide distributed computing. *Biopolymers: Original Research on Biomolecules*, *68*(1), 91–109. doi: 10.1002/bip.10219

Parisien, M., & Major, F. (2008). The MC-Fold and MC-Sym pipeline infers RNA structure from sequence data. *Nature*, *452*, 51–55

Pauling, L., & Corey, R.B. (1953). Compound helical configurations of polypeptide chains: structure of proteins of the α-keratin type. *Nature*, *171*(4341), 59–61

Pecourt, J.M.L., Peon, J., & Kohler, B. (2001). DNA excited-state dynamics: ultrafast internal conversion and vibrational cooling in a series of nucleosides. *Journal of the American Chemical Society*, *123*(42), 10370–10378

Peon, J., & Zewail, A.H. (2001). DNA/RNA nucleotides and nucleosides: direct measurement of excited-state lifetimes by femtosecond fluorescence up-conversion. *Chemical Physics Letters*, *348*(3–4), 255–262

Ravanat, J.L., Douki, T., & Cadet, J. (2001). Direct and indirect effects of UV radiation on DNA and its components. *Journal of Photochemistry and Photobiology B: Biology*, *63*(1–3), 88–102

Rich, A., & RajBhandary, U. (1976). Transfer RNA: molecular structure, sequence, and properties. *Annu. Rev. Biochem.*, *45*, 805–860

Schlick, T., & Olson, W.K. (1992). Supercoiled DNA energetics and dynamics by computer simulation. *Journal of Molecular Biology*, *223*(4), 1089–1119

Seeman, N. (2003). DNA in a material world. *Nature*, *421*, 427–431. https://doi.org/10.1038/nature01406

Sension, R.J., Repinec, S.T., & Hochstrasser, R.M. (1990). Femtosecond laser study of energy disposal in the solution phase isomerization of stilbene. *Journal of Chemical Physics*, *93*(12), 9185–9188

Shaw, D.E., Deneroff, M.M., Dror, R.O., Kuskin, J.S., Larson, R.H., Salmon, J.K., … & Wang, S.C. (2008). Anton, a special-purpose machine for molecular dynamics simulation. *Communications of the ACM*, *51*(7), 91–97. doi: 10.1145/1364782.1364802

Shirts, M., & Pande, V.S. (2000). Screen savers of the world unite! *Science*, *290*(5498), 1903–1904. doi: 10.1126/science.290.5498.1903

Smith, L.G., Zhao, J., Mathews, D.H., & Turner, D.H. (2017). Physics-based all-atom modeling of RNA energetics and structure. *Wiley Interdisciplinary Reviews: RNA*, *8*(5), e1422

Son, S.Y. (1993). The structure and regulation of histone genes. *Saenghwahak Nyusu*, *13*, 64–70

Sponer, J., Bussi, G., Krepl, M., Banáš, P., Bottaro, S., Cunha, R.A., … & Otyepka, M. (2018). RNA structural dynamics as captured by molecular simulations: a comprehensive overview. *Chemical Reviews*, *118*(8), 4177–4338. doi: 10.1021/acs.chemrev.7b00427; PMID: 29297679; PMCID: PMC5920944

Stombaugh, J., Zirbel, C.L., Westhof, E., & Leontis, N.B. (2009). Frequency and isostericity of RNA base pairs. *Nucleic Acids Research*, *37*(7), 2294–2312

Svozil, D., Kalina, J., Omelka, M., & Schneider, B. (2008). DNA conformations and their sequence preferences. *Nucleic Acids Research*, *36*(11), 3690–3706. doi: 10.1093/nar/gkn260

Tang, D.C., DeVit, M., & Johnston, S.A. (1992). Genetic immunization is a simple method for eliciting an immune response. *Nature*, *356*(6365), 152–154

Theil, E.C. (1998). The iron responsive element (IRE) family of mRNA regulators: regulation of iron transport and uptake compared in animals, plants and microorganisms. *Met. Ions Biol. Syst.*, *35*, 403–434

Underwood, J.G., Uzilov, A.V., Katzman, S., Onodera, C.S., Mainzer, J.E., Mathews, D.H., … & Haussler, D. (2010). FragSeq: transcriptome-wide RNA structure probing using high-throughput sequencing. *Nature Methods*, *7*(12), 995–1001

Van De Sande, J.H., Ramsing, N.B., Germann, M.W., Elhorst, W., Kalisch, B.W., v. Kitzing, E., … & Jovin, T.M. (1988). Parallel stranded DNA. *Science*, *241*(4865), 551–557

Vargason, J.M., Henderson, K., & Ho, P.S. (2001). A crystallographic map of the transition from B-DNA to A-DNA. *Proceedings of the National Academy of Sciences*, *98*(13), 7265–7270

Wang, B., Godillot, A.P., Madaio, M.P., Weiner, D.B., & Williams, W.V. (1998). Vaccination against pathogenic cells by DNA inoculation. *Curr. Top. Microbiol. Immunol.*, *226*, 21–35

Watson, J.D., & Crick, F.H. (1953). Molecular structure of nucleic acids: a structure for deoxyribose nucleic acid. *Nature*, *171*(4356), 737–738. https://doi.org/10.1038/171737a0

Westhof, E., & Fritsch, V. (2000). RNA folding: beyond Watson–Crick pairs. *Structure*, *8*, R55–R65

Whalen, R.G. (1996). DNA vaccines, cyberspace, and self-help programs. *Intervirology*, *39*,120–125

Wolff, J.A., Malone, R.W., Williams, P., Chong, W., Acsadi, G., Jani, A., & Felgner, P.L. (1990). Direct gene transfer into mouse muscle in vivo. *Science*, *247*(4949), 1465–1468

Xu, W., Dunlap, D., & Finzi, L. (2021). Energetics of twisted DNA topologies. *Biophysical Journal*, *120*(16), 3242–3252. https://doi.org/10.1016/j.bpj.2021.05.002

Zehner, Z.E., Shepherd, R.K., Gabryszuk, J., Fu, T.-F., Al-Ali, M., & Holmes, W.M. (1997). RNA–protein interactions within the 3′ untranslated region of vimentin mRNA. *Nucleic Acids Res.*, *25*, 3362–3370

Zemora, G., & Waldsich, C. (2010). RNA folding in living cells. *RNA Biol.*, *7*(6), 634–641. doi: 10.4161/rna.7.6.13554

Zirbel, C.L., Šponer, J.E., Šponer, J., Stombaugh, J., & Leontis, N.B. (2009). Classification and energetics of the base-phosphate interactions in RNA. *Nucleic Acids Research*, *37*(15), 4898–4918. https://doi.org/10.1093/nar/gkp468

11 Biophysics of Proteins and Proteomics

Proteins are central cellular machines. Understanding their evolutionary origins poses crucial questions in modern biology. Their structural integrity and functions in host cells are partly governed by physical laws. Biophysical techniques have been developed for centuries to address physics and biology of proteins in physiological environments. This chapter will present pictures of our understanding of versatile aspects of proteins in biological systems. Starting with the history of proteins, I shall accumulate most of the relevant information about our current knowledge and predictions of proteins' structural integrities and functional roles in important cellular events. The chapter will also brief the current state of proteomics, and thus will highlight how proteins' independent and collective physical actions help determine the versatile physiological roles of biological cells.

11.1 A BRIEF HISTORY OF PROTEINS' BEGINNING

The last universal common ancestor (LUCA), a microbe, lived around 4 billion years ago. This microbe is considered to be the beginning of a long lineage encapsulating all forms of life on Earth (https://astrobiology.nasa.gov/news/looking-for-luca-the-last-universal-common-ancestor/, accessed March 19, 2023). For a long time, we thought that the tree of life formed the following three main branches, or domains, with LUCA at the base:

- eukarya
- bacteria
- archaea

However, a new picture has emerged that places eukarya as an offshoot of bacteria and archaea (Figure 11.1). More light has been shed on the evolution of eukaryotes in a recent publication (Spang et al., 2015). The results provided here support hypotheses in which the eukaryotic host evolved from a bona fide archaeon and appear to demonstrate that many components that underpin eukaryote-specific features were already present in that ancestor. This is hypothesized to have provided the host with a rich genomic "starter kit" to support the increase in cellular and genomic complexity that is characteristic of eukaryotes.

The modern ribosome is hypothesized to have been formed at the time of the LUCA (Fox, 2010). Ribosomes are made up of proteins and ribosomal RNA (rRNA). Ribosomes in prokaryotes are roughly 40 percent proteins and 60 percent rRNAs, while in eukaryotes, the ratio of proteins and rRNAs is 1:1. The history of protein is linked to that of ribosomes. In the earliest stages of ribosome

DOI: 10.1201/9781003287780-11

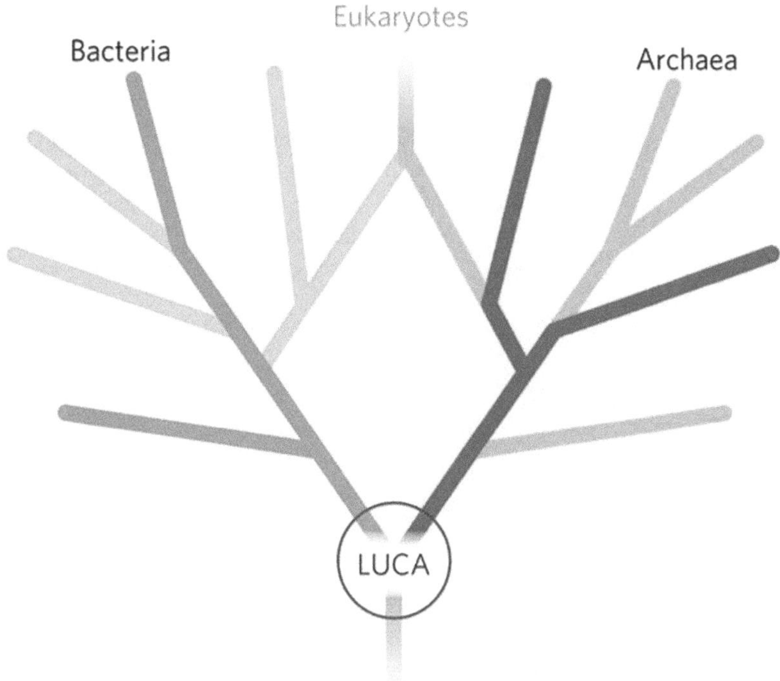

FIGURE 11.1 A schematic of the two-domain tree, with eukaryotes evolving from endosymbiosis between members of the two original trunks of the tree, archaea and bacteria.

***Source*: reproduced with permission from https://astrobiology.nasa.gov/news/looking-for-luca-the-last-universal-common-ancestor/. Image credit: Weiss et al./*Nature Microbiology*.**

evolution, the cellular entities carrying "protoribosomes" would have lacked a genetic code and the complex dynamic systems of the modern ribosome. The protoribosome, a universal internal small RNA pocket-like segment, embedded in the contemporary ribosome, is a vestige of the primordial ribosome, and the protoribosome may be the missing link between the RNA-dominated world and contemporary nucleic acids/proteins life (Bose et al., 2022).

The chemical world before LUCA (Fox, 2010) and associated pre-LUCA events was perhaps in the "throes of evolving the genotype-phenotype relationship" and could properly be considered to be a progenote (Woese and Fox, 1977a). By the time of LUCA, the ribosome is assumed to have developed to essentially its modern form. This strongly suggests that the ribosome reached a critical stage of development that facilitated the final transition from the RNA world to the RNA/protein world. Considering the phases of ribosomal evolution, ribosomal protein (rProtein) segments reveal an atomic-level history of protein folding (Kovacs et al., 2017). Here a model on aboriginal oligomers evolving into globular proteins has been proposed that considers a hierarchical step-wise process. The complexity of assembly and the folding of polypeptide are found to increase incrementally in concert with the expansion of rRNA. The phases are as follows (Kovacs et al., 2017):

1. Short random coil proto-peptides bound to rRNA, and
2. lengthened over time and coalesced into β-β secondary elements. These secondary elements
3. accreted and collapsed, primarily into β-domains. The domains
4. accumulated and gained complex super-secondary structures composed of mixtures of α-helices and β-strands.

The translation of mRNA to protein underpins the macromolecular partnership dominating the biological earth for nearly 4 billion years. It provides a blueprint of the common origin and interrelatedness of living systems (Woese and Fox, 1977b). As mentioned earlier, rRNAs and rProteins are molecular fossils from the pre-LUCA era of life (Lecompte et al., 2002; Söding and Lupas, 2003; Fox and Naik, 2004). In the Accretion Model of ribosomal evolution, rRNA was predicted to recursively accrete and freeze, experiencing a time-dependent mass increase over a long time (Bokov and Steinberg, 2009; Hsiao et al., 2009). The ribosome is thought to have sequentially acquired capabilities for RNA folding, noncoded condensation of amino acids to form peptides, subunit association, correlated subunit evolution and decoding, and energy transduction (Petrov et al., 2015). The rRNA growth is partitioned into six phases in prokaryotes (see Figure 11.2A) with two additional eukaryotic phases. The most ancient rRNA is drawn from the first phase while the final phase contains the most recent one. Phases 1–6 of ribosomal evolution are the extension and the elaboration of the exit tunnel.

The rProteins from the Large Ribosomal Subunit (LSU) were recently incorporated into the Accretion Model (Figure 11.2B) of ribosomal evolution through the establishment of temporal correlations between the acquisition of rRNA elements and acquisition of rProtein segments (Kovacs et al., 2017). The age of a given segment of rProtein is assumed here to be the same as that of the rRNA with which it interacts, thus the results should provide a test of the Accretion Model. The Accretion Model phases are a series of coarse-grained states that incorporate the highly detailed temporal information provided by insertion fingerprints, A-minor interactions, and other ribosomal elements (Petrov et al., 2014). Here the rProteins were computationally segmented (cleaved) and the associated segments were partitioned into phases corresponding to those of the rRNA in the Accretion Model. Each segment phase is determined by the phase of the rRNA with which it interacts. Figure 11.3 illustrates the segmenting of rProteins uL22 and uL13,

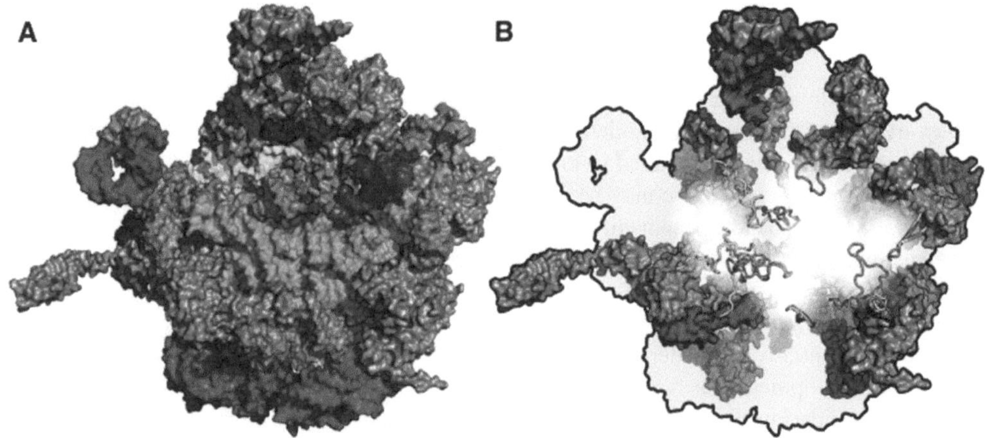

FIGURE 11.2 (A) The rRNA of the large subunit of the *T. thermophilus* ribosome colored by relative age. Phase 1, the most ancient phase, is dark blue. Phase 2 is light blue. Phase 3 is green. Phase 4 is yellow. Phase 5 is orange. Phase 6, the most recent prokaryotic phase, is red. rProteins are grey. (B) The orientation is maintained but rRNA is colored in light grey, universal rProteins are colored by evolutionary phase, and bacterial rProteins are colored dark grey. Phases 3 (green) and 4 (yellow) are shown in cartoon representation. Phases 5 (orange) and 6 (red) are shown in surface representation. From PDB entry 1VY4. The color contrast in the figure here and all subsequent ones will be clear in the online version of the book.

Source: **reproduced with permission from Kovacs et al. (2017).**

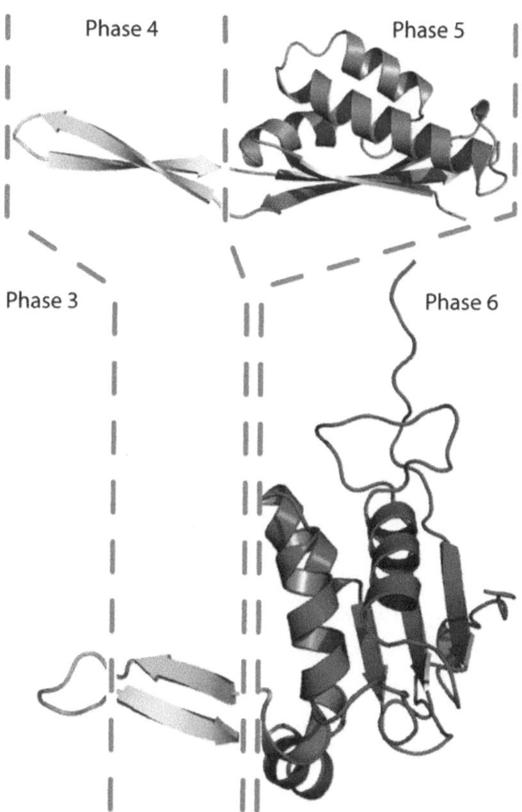

FIGURE 11.3 The history of protein folding illustrated by LSU rProteins uL22 (top) and uL13 (bottom). rProtein segments are colored by their phase, in accordance with rRNA and rProtein phases in Figure 1. Segment boundaries are indicated by dashed lines. rProtein uL22 has segments in Phases 4 and 5. uL13 has segments in Phases 3, 4, and 6. The Phase 3 segment of uL13 is random coil. Phase 4 segments of uL22 and uL13 contain isolated $\beta - \beta$ structures. The Phase 5 and 6 segments of uL22 and uL13 contain globular domains with extensive intramolecular hydrogen bonds and reduced solvent accessible surface area. These domains contain hydrophobic cores and hydrophilic surfaces. rProtein segments in lower numbered phases are more ancient than those in higher numbered phases. Structures are extracted from the *T. thermophilus* ribosome.

***Source*: reproduced with permission from Kovacs et al. (2017).**

and their assignment into phases. The age of any phase is unknown except that the beginning of Phase 6 may correspond to the era of LUCA, around 3.8 billion years ago, and Phase 1 was at or near the beginning of life.

Kovacs and colleagues focused on universal rProteins, for maximizing the universality of the results. The observed patterns appear to be general and robust in that the trends and the structural characteristics for each phase are the same for universal and for all LSU rProteins of ribosomes of two bacteria, *Thermus thermophilus* or *Escherichia coli*, or an archaean *Pyrococcus furiosus* (Kovacs et al., 2017). All 15 universal rProteins in the LSU of the *T. thermophilus* ribosome were segmented and partitioned into phases of rRNA evolution (see Figure 11.4).

With increasing phase, rProtein segments show increasing extent and folding complexity. The rProtein folding change with phase highlights the directionality of time of the protein evolution. It also reveals the evolutionary transitions of the unstructured → simple → complex structures.

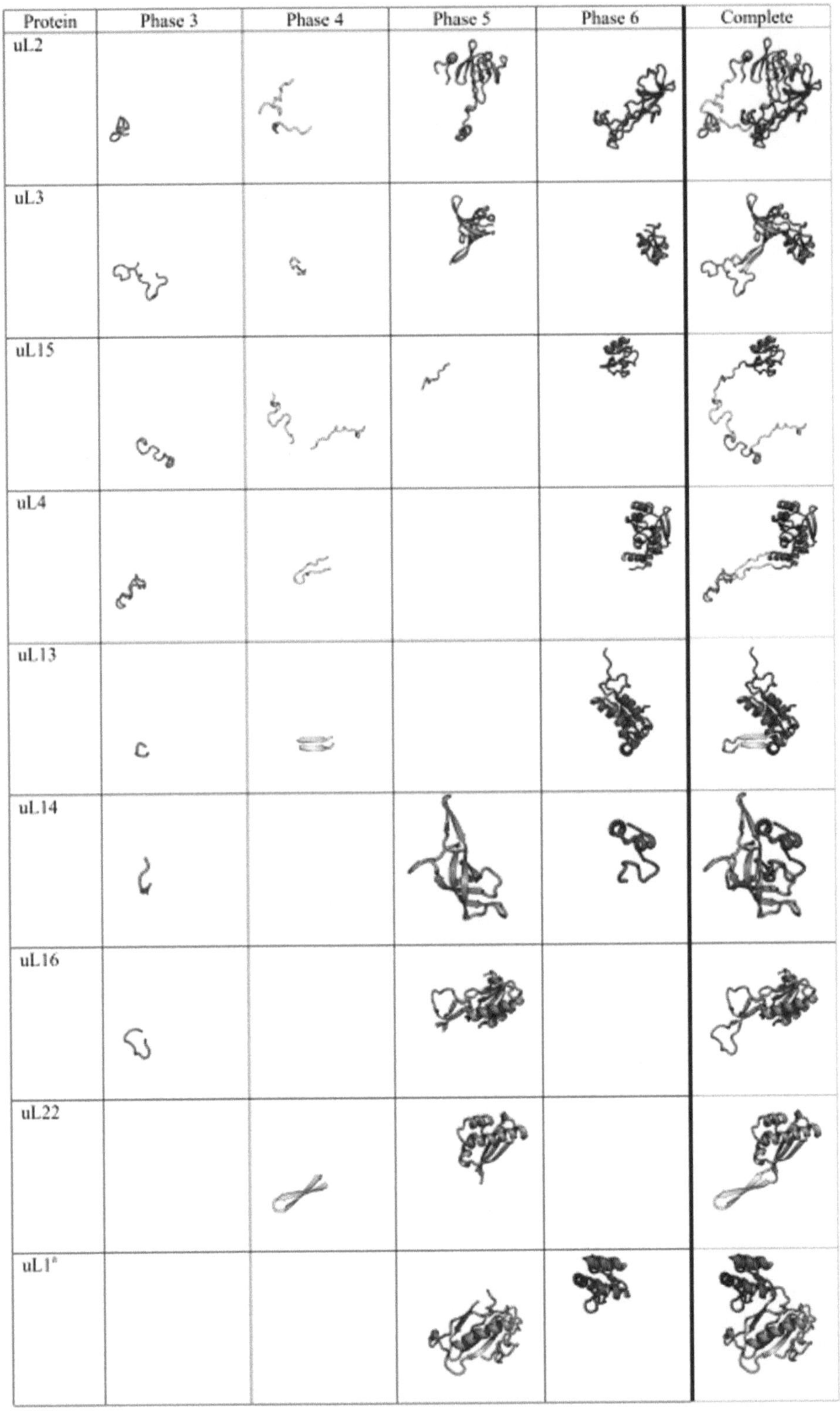

Protein	Phase 3	Phase 4	Phase 5	Phase 6	Complete
uL2					
uL3					
uL15					
uL4					
uL13					
uL14					
uL16					
uL22					
uL1[a]					

FIGURE 11.4 The evolution of protein folding has been revealed from segmentation of universal rProteins from *Thermus thermophilus*. All rProtein (from the *T. thermophilus* crystal structure) segments are colored by phases following the coloring scheme used in Figure 11.2. Empty cells contain no rProtein segment in that

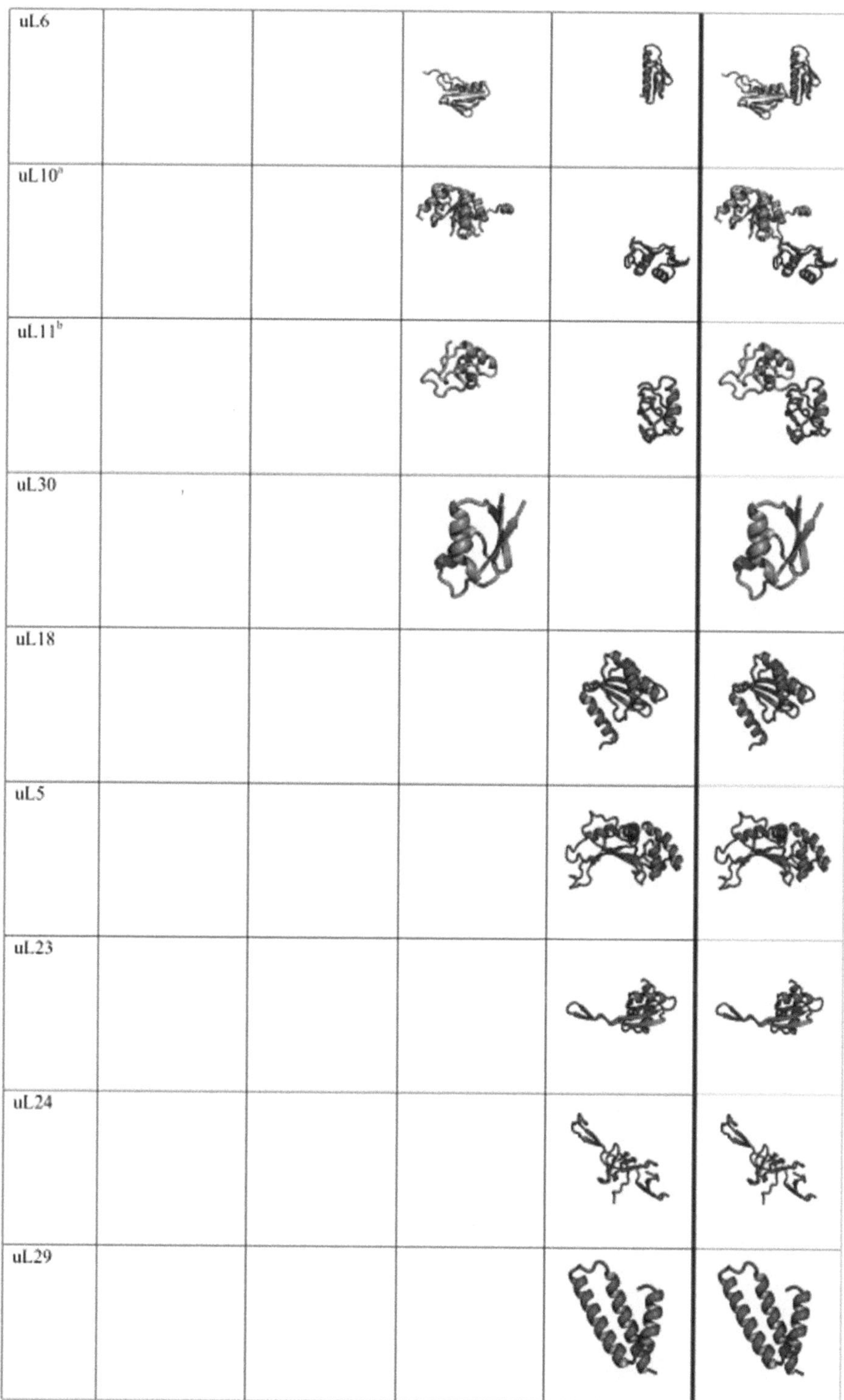

FIGURE 11.4 *(Continued)*

phase for that protein. Due to their absence in the *T. thermophilus* structure, [a]these rProteins are from the *P. furiosus* ribosomal structure and [b]this rProtein is from *E. coli* ribosomal structure.

***Source*: reproduced with permission from Kovacs et al. (2017).**

11.2 PHYSICOCHEMICAL ASPECTS OF PROTEIN STRUCTURES

Proteins get structured due to physical processes such as the condensation of amino acids and the formation of peptide bonds. All four forms of protein structures, primary, secondary, tertiary, and quaternary structure, are maintained in biological systems due to physical processes. The primary structure of protein provides the amino acid sequence(s). For the secondary structure, the dihedral angles of the peptide bonds, and for the tertiary structure the folding of protein chains in space appear as determinants. The association of folded polypeptide molecules with complex functional proteins leads to the quaternary structure.

The sequence of amino acids in a protein is fixed, as discovered by the Nobel Prize-winning biochemist Frederick Sanger in 1951 while determining the primary structure of bovine insulin. The existence of proteins in biological systems was discovered more than a century prior to Sanger's discovery of the primary structure of proteins, by the Dutch chemist Gerardus Johannes Mulder, and named by the Swedish chemist Jöns Jacob Berzelius in 1838. Kaj Ulrik Linderstrøm-Lang introduced the protein secondary structure concept at Stanford in 1952. The α-helix and the β-sheet, proposed in 1951 by Linus Pauling, Robert Corey, and Herman Branson, are now known to form the backbones of thousands of proteins (Pauling et al., 1951). Two hydrogen-bonded helical structures for a polypeptide chain were found in which the residues were stereochemically equivalent, the interatomic distances and bond angles appeared with values found in amino acids, peptides, and other simple substances related to proteins, and the conjugated amide system was discovered to be planar. The Pauling-Corey-Branson models were astoundingly correct, including bond lengths that were not surpassed in accuracy for over 40 years (Eisenberg, 2003). The Pauling-Corey-Branson trio did not, however, consider the hand of the helix or the possibility of bent sheets. As the Pauling-Corey-Branson trio constructed helices with planar amide groups, with the precise bond dimensions they had observed in crystal structures, and with linear hydrogen bonds of length 2.72 Å, they found only two possibilities, which they called the helix with 3.7 residues per turn and the helix with 5.1 residues per turn (see Figure 11.5), soon to become familiar as the α-helix and the γ-helix.

In 1965, β-sheets and single-stranded β-ribbons were first observed in globular proteins in the structure of egg white lysozyme (Blake et al., 1965). Both strands and sheets were surprisingly found twisted, unlike the straight strands and pleated sheets of Pauling and Corey. In 1989, Pauling recalled that as soon as he saw the structure of lysozyme with its twisted sheet he realized he should have incorporated the twist in the original model. Later there were thorough analyses of twist and shear in β-structures (Chothia, 1973). It is suggestive here that β-pleated sheets with a right-hand twist when viewed along the polypeptide chain direction have a lower free energy than sheets that are straight or have a left-hand twist.

All of these fundamentals and associated developments have been presented in many textbooks. Therefore, I avoid elaborating on them here. Instead, I shall focus on the latest breakthroughs helping to enrich our understanding of the physical aspects of general protein structures and associated advanced proteomics.

11.3 DISTRIBUTION OF TORSION ANGLES IN PROTEIN STRUCTURES

Three-dimensional (3D) protein structures contain regions of local order, called secondary structures, such as α-helices and β-sheets. The secondary protein structure is characterized by the local rotational state of the protein backbone, quantified by two dihedral angles called ϕ and ψ. Particular secondary structures can generally be described by a single (diffuse) location on a two-dimensional (2D) plot drawn in the space of the angles ϕ and ψ, called a Ramachandran plot (Ramachandran, 1962). Here ϕ represents the dihedral angle between N(i-1)-C(i)-CA(i)-N(i) and ψ is the backbone dihedral angle between C(i)-CA(i)-N(i)-C(i+1). ϕ and ψ values are plotted to obtain the conformation of the peptide. The angular spectrum lies between −180 and +180° on the x-axis and y-axis.

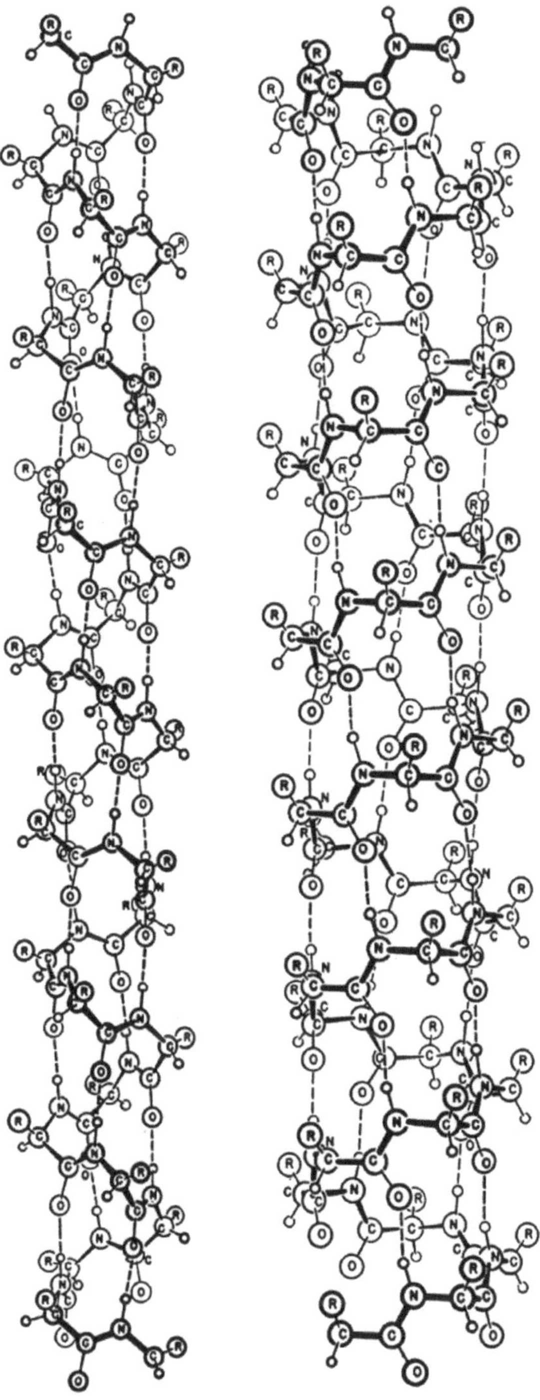

FIGURE 11.5 The α-helix (left panel) and the γ-helix (right panel), as depicted in the 1951 paper by Pauling, Corey, and Branson (1951). The CO groups of the α-helix point in the direction of its C terminus, whereas those of the γ-helix point toward its N terminus, and, further, the α-helix shown is left-handed and made up of D-amino acids. Copied from ref. Eisenberg (2003).

***Source*: reproduced with permission from Eisenberg (2003).**

Ramachandran outlier is a representation of those amino acids which lie in the unfavorable regions of the plot.

Following the discovery of the protein helical structures by Pauling, Corey, and Branson (1951) various types of chain configurations were proposed for proteins and polypeptides. Due to a lack of any analytical methods for writing down these configurations, Ramachandran and colleagues worked out a convenient notation of this type, now well known as the famous Ramachandran plot (Ramachandran and Sasisekharan, 1961a, 1961b; Ramachandran, 1962; Ramachandran et al., 1962; Ramachandran et al., 1963). The Ramachandran plot can be used to evaluate the accuracy of predicted protein structures. Figure 11.6 shows the original plot Ramachandran and colleagues published in 1963 (Ramachandran et al., 1963).

The secondary structure is characterized by the local rotational state of the protein backbone, quantified dihedral angles ϕ and ψ. Particular types of secondary structures can generally be described by a single (diffuse) location on a 2D Ramachandran plot drawn in the space angles ϕ and ψ. By contrast, a recently discovered nanomaterial made from peptoids, structural isomers of peptides, displays a secondary-structure motif corresponding to two regions on the Ramachandran plot (Mannige et al., 2015). As shown in Figure 11.7a–d, regular chiral protein structures (e.g. the α-helix and β-sheet) correspond roughly to a single location, *i*, on the Ramachandran plot. In contrast, stable nanosheets are composed of peptoids. Their backbone states occupy two specific regions of the Ramachandran plot (see Figure 11.7f), labelled *i* and *j*. The chains whose adjacent residues

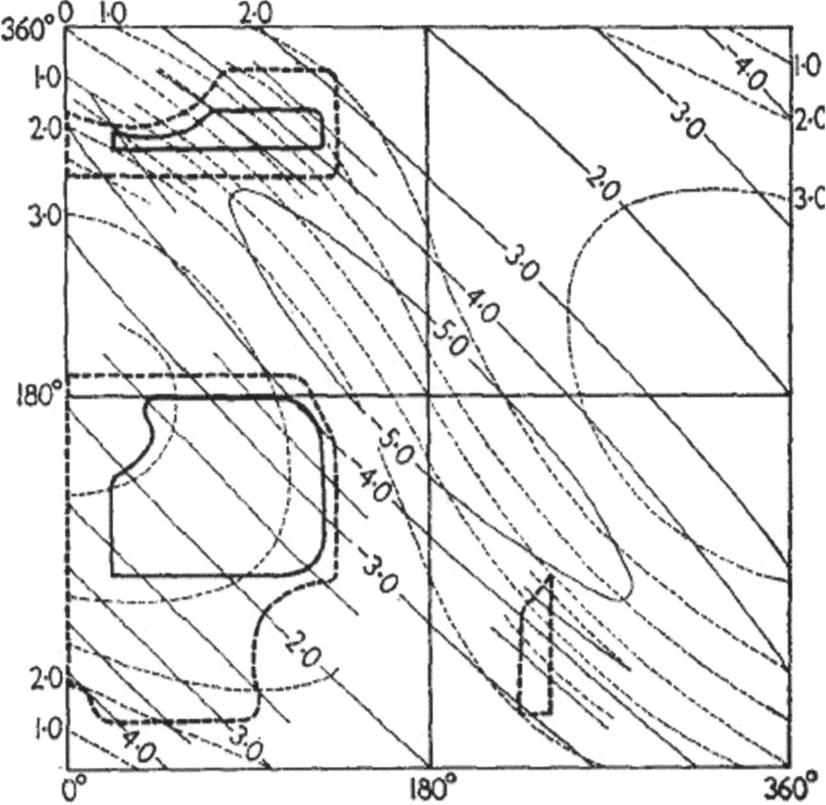

FIGURE 11.6 The fully allowed (---) and outer limit (- – -) regions of (ϕ, ψ) for angle N-αC-C' = 110° along with the configurations of various known di-, tri- and polypeptide and protein structures.

Source: **reproduced with permission from Ramachandran et al. (1963).**

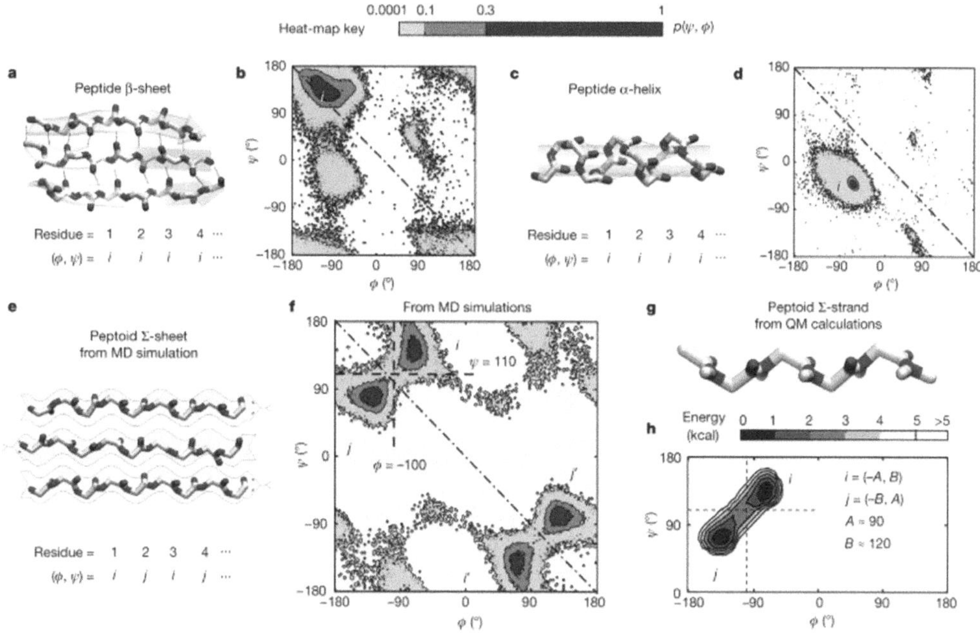

FIGURE 11.7 (a–d) Typical secondary structures found in proteins, such as the β-sheet (a) and α-helix (c), are described by specific pairs of backbone dihedral angles, ϕ and ψ, whose position is labelled by i in their respective Ramachandran plots (b, d). The heat map is a two-dimensional histogram showing the normalized probability $p(\psi, \phi)$ that a residue in one of the indicated structures will adopt particular combinations of the dihedral angles ϕ and ψ. (e, f) In contrast, the secondary structure observed in the peptoid nanosheet (e) consists of two characteristic (ϕ, ψ) positions, i and j (f). (The other occupied regions in (f), that is, the pale blue locations, indicate polymer ends and linker regions. Owing to the achirality of the peptoid backbone's α-carbon, i' and j' are equivalent to i and j, respectively.) Positions i and j are visibly equidistant from the achiral diagonal of the Ramachandran plot, and so adopting the two states in an alternating fashion allows the peptoid backbone to remain linear and untwisted. This motif, the Σ-strand, allows the formation of extended planar structures that have not been made previously using protein-like building motifs. (g) Quantum-mechanical (QM) calculations of a range of possible Σ-type strands show energy minima (h) that match the high-occupancy regions of (f), indicating that the rotational tendencies of isolated polymers are preserved by the side chain interactions established within the nanosheet. (MD, molecular dynamics. The alternating pattern is defined by the numbers A and B.)

***Source*: reproduced with permission from Mannige et al. (2015).**

alternate between these two states remain linear (see Figure 11.7e for a snapshot of three backbone segments emphasizing this alternating motif), named here the Σ-strand, because its linear, twist-free nature derives from the combination, or sum ('Σ'), of its two rotational states (and because the resulting polymer "snakes" back and forth). One could have a Σ-strand built from any two opposed rotational states. The density functional theory calculations show that the particular rotational states observed in atomistic simulations are the lowest-energy Σ-type arrangement for isolated polymers (see e.g. Figure 11.7g and supplementary figures thereof). This confirms that the basic rotational tendency of an isolated peptoid backbone (Mirijanian et al., 2014; see Figure 11.7h) is preserved by the side chain–side chain interactions established within the nanosheet (see Figure 11.7f).

In order to describe "higher-order" secondary structure deduced in Mannige et al. (2015) in a compact way, Mannige and colleagues introduced means of describing regions on the Ramachandran plot in terms of a single Ramachandran number, R, which is found to be a structurally meaningful

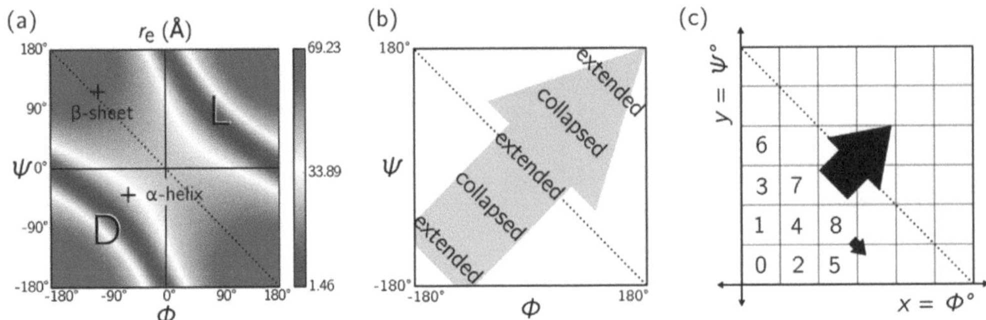

FIGURE 11.8 Physical trends within the Ramachandran plot suggest a way of describing regions of it with a single number. (a) First, the sense of residue twist changes from right-handed ("D") to left-handed ("L") as one moves from the bottom left of the Ramachandran plot to the top right. Second, contours (colored) of end-to-end polymer distance R_e (here calculated for a 20-residue glycine) have a negative slope, resulting in the general trend shown in panel (b). Panel (c) indicates one method of indexing the Ramachandran plot so as to move from the region of right-handed twist to the region of left-handed twist with R_e changing as slowly as possible. This method provides the basis for the construction of the Ramachandran number, *R*.

***Source*: reproduced with permission from Mannige et al. (2016).**

combination of ϕ and ψ (Mannige et al., 2016). Given a way of describing regions of the Ramachandran plot in terms of one number instead of two as done traditionally, one can then draw diagrams that give insight into protein geometry that is difficult to obtain by other means. The residue twist implicit in the Ramachandran plot changes sign as one crosses the negative-sloping diagonal (Figure 11.8a). The structures with the backbones occupying the bottom left-hand triangle of the Ramachandran plot appear to have a right-handed (dextrorotatory or D) sense of twist, while structures in the top right-hand triangle appear to have a left-handed (levorotatory or L) sense of twist (Berg et al., 2012; Mannige et al., 2015). This observation is suggestive of an indexing system that proceeds from the bottom left of the plot to the top right area. To gain insight into how this should be done, a protein backbone was built with dihedral angles chosen from designated regions of the Ramachandran plot (Mannige et al., 2016). The behavior of the end-to-end distance R_e of polymers built was examined. The polymer radius of gyration behaves similarly, shown in Figure 11.8a and b. The shapes of the contours of R_e are suggestive of an indexing system that proceeds in a sweeping fashion across the Ramachandran plot (see Figure 11.8c), so that R_e changes as slowly as possible. Proceeding in this manner one moves from structures having one sense of twist to structures having the opposite sense of twist, with the degree of compactness of the backbone varying only in a gradual fashion. Thus the indexing system suggested graphically in Figure 11.8c is sensitive both to the twist state and the degree of compactness of the polymer backbone, allowing one to distinguish, for example, compact α-helices from extended β-sheets, or nearly twist-free β-sheets from twisted loop regions (Mannige et al., 2016).

An indexing system has been constructed by Mannige and colleagues (2016) taking the Ramachandran plot axes to have the range $[-\lambda/2, \lambda/2)$ where $\lambda = 360°$ (Berg et al., 2012; Alberts et al., 2002). The plot was divided into a square grid of $(360° \sigma)^2$ sites. Here σ is a scaling factor that is measured in reciprocal degrees. It is straightforward to make σ large enough that the error incurred upon converting angles from structures in the protein databank to Ramachandran numbers and back again is less than the characteristic error associated with the coordinates of structures in that database. A theoretical derivation is provided in Mannige et al. (2016) to define the integer-valued Ramachandran number $R_Z(\phi, \psi)$ as follows:

$$R_Z(\phi,\ \psi) \equiv [\phi'] + \lambda'[\psi'] \tag{11.1}$$

Here the coordinates ϕ and ψ are given by the following relations:

$$\phi' \equiv \frac{(\phi - \psi + \lambda)\sigma}{\sqrt{2}} \tag{11.2}$$

$$\psi' \equiv \frac{(\phi + \psi + \lambda)\sigma}{\sqrt{2}} \tag{11.3}$$

And they correspond to a clockwise rotation by 45 degrees, which is a shift, and a rescaling of the Ramachandran coordinates ϕ and ψ. λ' is defined as

$$\lambda' \equiv \left[\sqrt{2}\lambda\sigma\right] \tag{11.4}$$

Considering these, rescaling $R_Z(\phi, \psi)$ is expressed as

$$R_Z(\phi, \psi) \equiv \left[\frac{(\phi - \psi + \lambda)\sigma}{\sqrt{2}}\right] + \left[\sqrt{2}\lambda\sigma\right]\left[\frac{(\phi + \psi + \lambda)\sigma}{\sqrt{2}}\right] \tag{11.5}$$

It is helpful for the real-valued, normalized Ramachandran number to be plotted easily, so it is defined as follows:

$$R(\phi, \psi) \equiv \frac{R_Z(\phi, \psi) - R_{Z,min}}{R_{Z,max} - R_{Z,min}} \tag{11.6}$$

$R(\phi, \psi)$ is a value which is practically invariant of σ. Given that $\phi, \psi \in [-\lambda/2, \lambda/2)$ or $[-180, 180)$, the ranges of R_Z and R are $[R_{Z,min}, R_{Z,max})$ and $[0,1)$, respectively. Further detailed analysis on deduction of the Ramachandran number is found in Mannige et al. (2016).

By "slicing" across the Ramachandran plot, Mannige and colleagues grouped together structures that might be relatively distant in dihedral angle space, the more so as they approach the negative-sloping diagonal (near $R = 0.5$). One consequence of this grouping is that the set of structures described by a small interval displays a distribution of properties, such as end-to-end distance, as shown in Figure11.9a. The mean of this distribution gives rise to a smoothly varying trend, but the variance of this distribution is nonzero, and is largest near $R = 0.5$. Some unavoidable structural coarse-graining therefore occurs upon going from the Ramachrandran plot to the Ramachandran number. Despite this drawback, we could show that R can function as an order parameter for protein geometry, in large part because the Ramachandran plot is in general relatively sparsely occupied: many hypothetical structures that possess distinctly different structural properties but that would be assigned similar Ramachandran numbers simply do not arise in the protein world. Consequently, R can resolve the major classes of protein secondary structure, such as the α and β motifs; see Figures 11.9b and 11.10.

R can be used both for assessing, in a compact manner, the geometric content of protein structures, and for drawing diagrams that reveal, at a glance, the frequency of occurrence of regular secondary structures and disordered regions in large protein datasets. R may also be useful in the analysis of intrinsically disordered proteins, whose 3D structures are less well understood than those of globular proteins (Dunker et al., 2001; Dunker et al., 2008; Dunker et al., 2013).

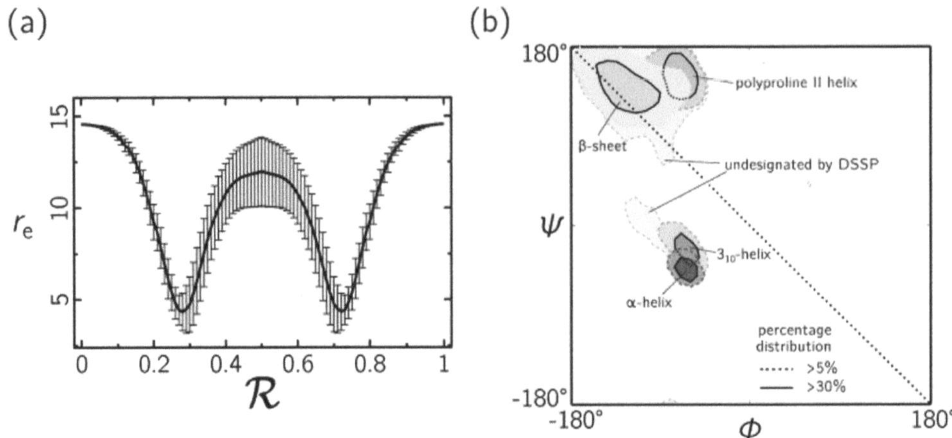

FIGURE 11.9 Potential pathologies of R are avoided by the sparse occupancy of the Ramachandran plot. (a) We construct R by slicing across the Ramachandran plot, which can cause points distant in dihedral angle space to be grouped together, the more so as we approach the negative-sloping diagonal (near $R = 0.5$). This grouping can be inferred by superposing the standard deviation (error bars) in polymer end-to-end distance on top of the mean value (smooth line) for hypothetical structures built from the relevant part of the Ramachandran diagram. (b) However, many structures distant in dihedral angle space but close in R do not arise in proteins; the Ramachandran diagram is in general relatively sparsely occupied. Consequently, R can resolve the major types of protein secondary structure, which can be inferred from the fact that lines parallel to the negative-sloping diagonal (marked), along which R varies only slowly, can touch each region of known secondary structure (colored) individually. This sensitivity allows R to function as an order parameter for protein geometry. Data in (a) were calculated for a 5-residue peptoid; R values are shown at discrete intervals of 0.01.

Source: **reproduced with permission from Mannige et al. (2016).**

11.4 SOLVING PROTEIN STRUCTURES

Following the discovery of Wilhelm Röntgen's first "medical" X-ray of his wife's hand taken in late 1895, application of X-rays as a biophysics technique experienced fast growth, especially in crystallography (Waseda et al., 2011). X-ray crystallography has so far been the best technique for the determination of three-dimensional (3D) crystalline structures at the atomic scale. As example, we may recall the discovery of DNA structures, the double helical model. DNA was first identified in 1869 by Swiss physiological chemist Friedrich Miescher (Pray, 2008). But it took until 1952 for the first X-ray picture of DNA to be successfully taken by Rosalind Franklin, which led to the discovery of its correct molecular structure by Watson and Crick (1953). X-ray crystallography helps determine the 3D atomic arrangement of a crystalline solid. There is another technique, nuclear magnetic resonance (NMR), developed during the first half of the last century by 1944 Nobel physics laureate Isidor Isaac Rabi. He developed a resonance method for recording the magnetic properties of atomic nuclei. In 1946, a technique to perform NMR in condensed matter state was discovered simultaneously by 1952 Nobel physics laureates Edward Purcell and Felix Bloch. Both X-ray crystallography (Muirhead and Perutz, 1963; Kendrew et al., 1958) and NMR (Anil-Kumar et al., 1980; Wüthrich, 2001; Williamson et al., 1985) were applied later to discover protein structures. Both these techniques helped us understand physics of the protein, which is a key component of the network of life (Banavar and Maritan, 2007).

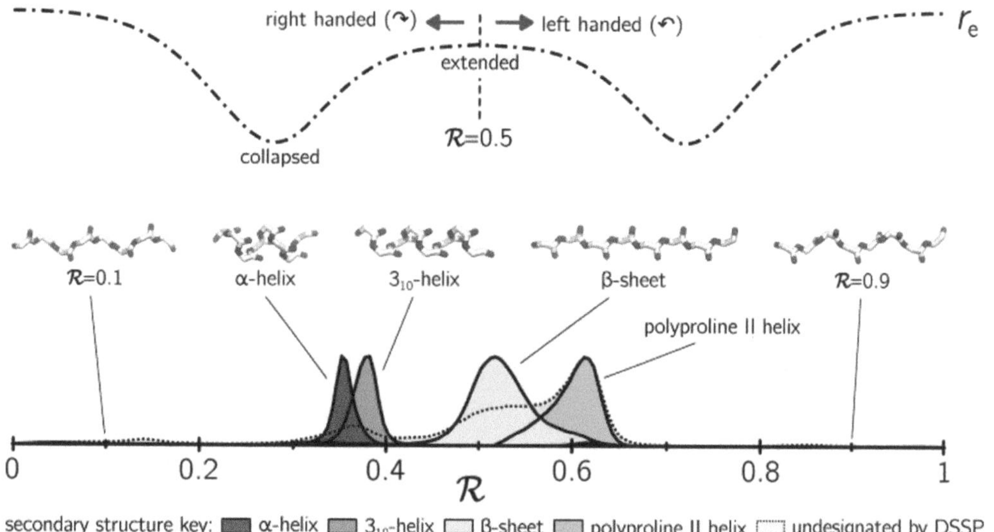

FIGURE 11.10 The indexing system presented here (Equations 11.6 and 11.1) collapses the Ramachandran plot into a single line, the Ramachandran number R. This number can act as an order parameter to distinguish secondary structures of different geometry, as shown (the overlap between distributions exists in the original Ramachandran plot representation; see Figure 12.8b). Top: R interpolates between regions of right-handed and left-handed twist, with polymer extension R_e varying smoothly throughout.

***Source*: reproduced with permission from Mannige et al. (2016).**

11.4.1 Protein Crystallography Using X-rays

Within years of the remarkable discovery of the most important genetic structure (Watson and Crick, 1953) two pioneers of protein crystallography, Max Perutz and John Kendrew, emerged to take the lead in solving the first ever X-ray crystal structures of proteins hemoglobin (Muirhead and Perutz, 1963) and myoglobin (Kendrew et al., 1958). Figure 11.11 shows the myoglobin model as it first appeared in Kendrew et al. Myoglobin, a heme-containing globular protein, is found in abundance in myocyte cells of heart and skeletal muscle. In 1958 the structure of sperm whale myoglobin was resolved by X-ray diffraction at 6 Å resolution, which was refined to 2 Å in 1960 (Perutz et al., 1960). This is the first 3D structure obtained of any protein.

Both myoglobin and its haem-free derivative apomyoglobin have been studied using various techniques (Ajaj et al., 2009). Recently, autonomous sequences (Figure 11.12) in myoglobin (Figure 11.13) emerged from X-ray structure of holomyoglobin (Narita et al., 2019).

Two autonomous sequences emerged in myoglobin (see Figure 11.13). The autonomous sequences encode tertiary structure information for semifolds. The sequences fold autonomously into small sets of continuous folding structure units in order to grow separate semifolds on each separate framework. Thus the autonomous sequence may be defined as the local sequence assigned to the small set of continuous folding structure units. The sequences create the discrete hydrophobic region in a semifold by assembly of their hydrophobic regions. The significant feature of semifolds in apomyoglobin was that they could be verified by the X-ray structure of holomyoglobin regardless of the instability of folds characteristic of autonomous sequence fragments.

X-ray crystallography has been applied to discover many protein structures (Drenth, 2007). A very important example is the structure of the potassium channel (Doyle et al., 1998), for which Roderick MacKinnon was co-awarded the 2003 Nobel Prize in chemistry. In 1998, MacKinnon

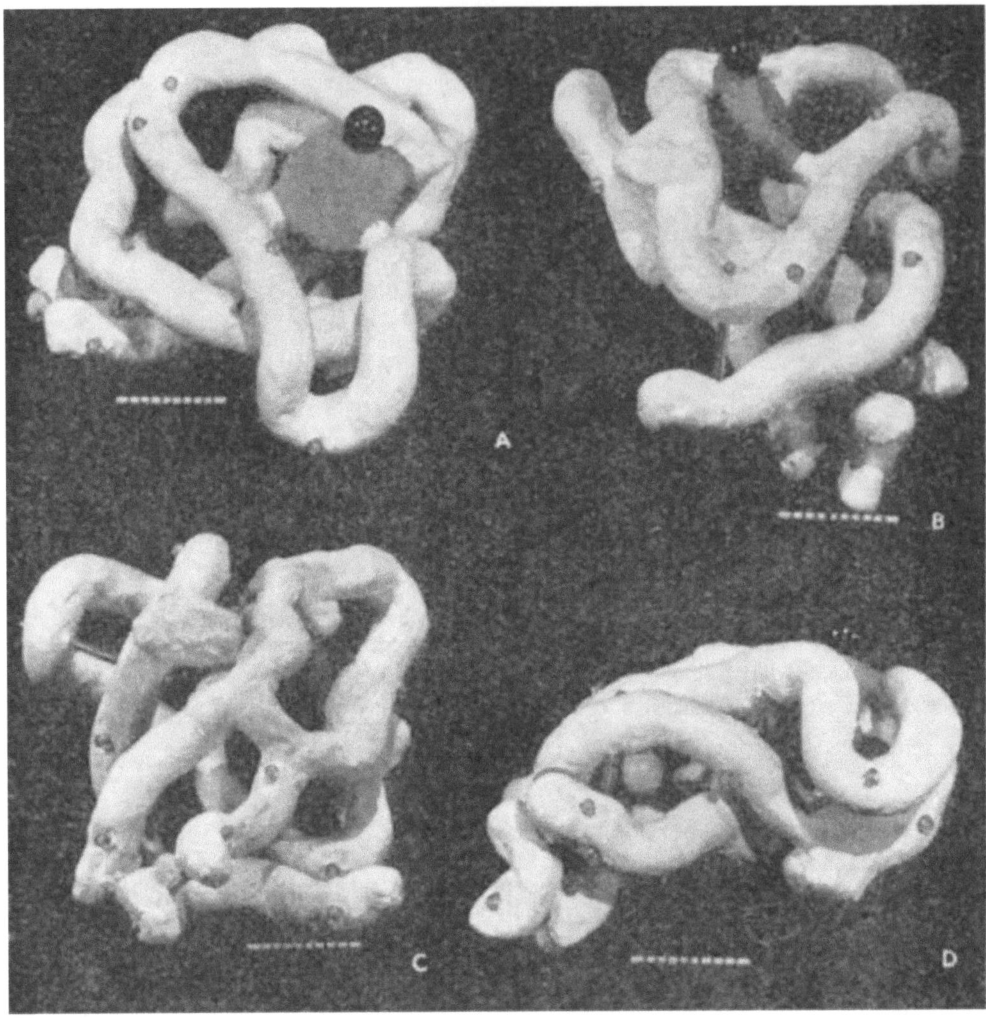

FIGURE 11.11 Photographs of a model of the myoglobin molecule. Polypeptide chains are white; the grey disk is the hsem group. The three spheres show positions at which heavy atoms were attached to the molecule (black: Hg of p-chloro-mercuri-benzene-sulphonate; dark grey: Hg of mercury diammine; light grey: Au of auri-chloride). The marks on the scale are 1 Å apart.

***Source*: reproduced with permission from Kendrew et al. (1958).**

led a group of scholarly scientists who succeeded in determining the 3D molecular structure of a potassium channel from an actinobacterium, *Streptomyces lividans*, utilizing X-ray crystallography. With this structure and other biochemical experiments, MacKinnon and colleagues were able to explain the exact mechanism by which potassium channel selectivity occurs (see Figure 11.14). X-ray analysis with 3.2 angstroms resolution shows data revealing that four identical subunits create an inverted teepee, or cone, cradling the selectivity filter of the pore in its outer end. The narrow selectivity filter is only 12 angstroms long, whereas the remainder of the pore is wider and lined with hydrophobic amino acids. A large water-filled cavity and helix dipoles are positioned so as to overcome electrostatic destabilization of an ion in the pore at the center of the bilayer. Main chain carbonyl oxygen atoms from the K^+ channel signature sequence line the selectivity filter, which is held open by structural constraints to coordinate K^+ ions but not smaller Na^+ ions.

$V_1\ L_2\ S_3\ E_4\ G_5\ E_6\ W_7\ Q_8\ L_9\ V_{10}\ L_{11}\ H_{12}\ V_{13}\ W_{14}\ A_{15}\ K_{16}\ V_{17}\ E_{18}\ A_{19}\ D_{20}\ V_{21}\ A_{22}\ G_{23}\ H_{24}\ G_{25}\ Q_{26}\ D_{27}\ I_{28}\ L_{29}\ I_{30}\ R_{31}$

a` a b c d e e e e e e e e e e f g h i i`

H20(L_2-V_{21})(29) A helix

z z4 z4'

NH3(V_1-S_3)(1.9)

a` a b c d e e e e e e e e

H19(A_{19}-P_{37})(14) B helix

$L_{32}\ F_{33}\ K_{34}\ S_{35}\ H_{36}\ P_{37}\ E_{38}\ T_{39}\ L_{40}\ E_{41}\ K_{42}\ F_{43}\ D_{44}\ R_{45}\ F_{46}\ K_{47}\ H_{48}\ L_{49}\ K_{50}\ T_{51}\ E_{52}\ A_{53}\ E_{54}\ M_{55}\ K_{56}\ A_{57}\ S_{58}\ E_{59}\ D_{60}\ L_{61}\ K_{62}$

a` a b c d f g h i i`

H10(S_{35}-D_{44})(45) C helix

a' a b c d f g h i i'

H10(K_{50}-E_{59})(10) D helix

e f g h i i`

H19(A_{19}-P_{37})(14) B helix

a` a b c k g h i i`

H9(K_{42}-K_{50})(1.0) C' helix

z2` z24 z4`

HH3(L_{49}-T_{51})(1.6)

a` a b c d e

H24(A_{57}-G_{80})(6.1) E helix

$K_{63}\ H_{64}\ G_{65}\ V_{66}\ T_{67}\ V_{68}\ L_{69}\ T_{70}\ A_{71}\ L_{72}\ G_{73}\ A_{74}\ I_{75}\ L_{76}\ K_{77}\ K_{78}\ K_{79}\ G_{80}\ H_{81}\ H_{82}\ E_{83}\ A_{84}\ E_{85}\ L_{86}\ K_{87}\ P_{88}\ L_{89}\ A_{90}\ Q_{91}\ S_{92}\ H_{93}$

a' a b c d e e e e e e e e e

H20(G_{80}-I_{99})(3.0) F helix

e e e e e e e e e e e e e f g h i i'

H24(A_{57}-G_{80})(6.1) E helix

HH22(K_{79}-P_{100})(0.41)

z2 z24 z4'

HH3(K_{79}-H_{81})(1.2)

$A_{94}\ T_{95}\ K_{96}\ H_{97}\ K_{98}\ I_{99}\ P_{100}\ I_{101}\ K_{102}\ Y_{103}\ L_{104}\ E_{105}\ F_{106}\ I_{107}\ S_{108}\ E_{109}\ A_{110}\ I_{111}\ I_{112}\ H_{113}\ V_{114}\ L_{115}\ H_{116}\ S_{117}\ R_{118}\ H_{119}\ P_{120}\ G_{121}\ D_{122}\ F_{123}\ G_{124}$

e f g h i i'

H20(G_{80}-I_{99})(3.0) F helix

a` a b j h i i`

T7(R_{118}-G_{124})(2.8)α-turn

HH22(K_{79}-P_{100})(0.41)

a' a b c d e e e e e e e e e e e e f g h i i'

a' a

z2' z24 z4'

HH3(K_{98}-P_{100})(1.3)

H22(I_{99}-P_{120})(35) G helix

H29

$A_{125}\ D_{126}\ A_{127}\ Q_{128}\ G_{129}\ A_{130}\ M_{131}\ N_{132}\ K_{133}\ A_{134}\ L_{135}\ E_{136}\ L_{137}\ F_{138}\ R_{139}\ K_{140}\ D_{141}\ I_{142}\ A_{143}\ A_{144}\ K_{145}\ Y_{146}\ K_{147}\ E_{148}\ L_{149}\ G_{150}\ Y_{151}\ Q_{152}\ G_{153}$

b c d e e e e e e e e e e e e e e e e e e e f g h i i'

H29(F_{123}-Y_{151})(378)H helix

z2' z2 z z

HC4(G_{150}-G_{153})(5.3)

Autonomous sequence 1:

Autonomous sequence 2:

α-helix :

α-turn :

interconnecting folding structure :

FIGURE 11.12 The folding structural units in myoglobin.

Source: reproduced with permission from Narita et al. (2019).

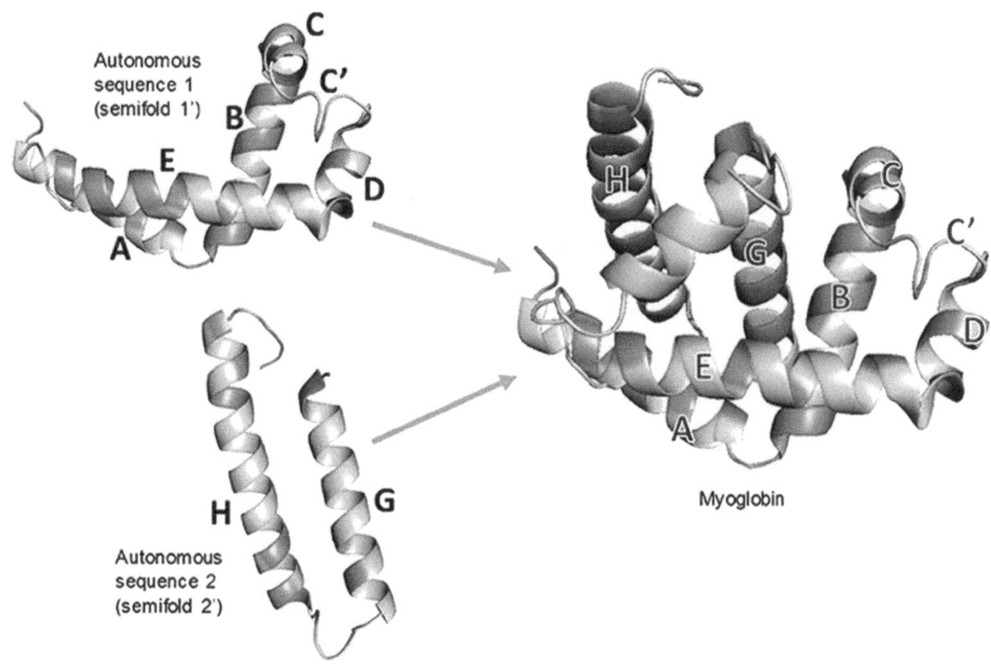

FIGURE 11.13 3D structure of myoglobin.

Source: reproduced with permission from Narita et al. (2019).

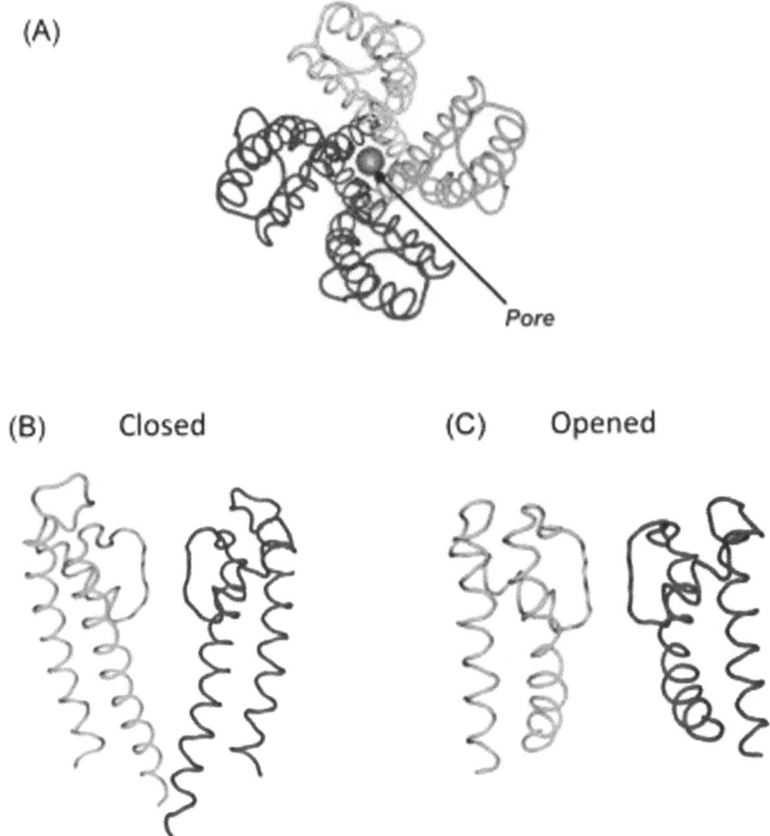

FIGURE 11.14 The bacterial potassium channel structure as discovered by the MacKinnon group (Doyle et al., 1998). Molecular models of the bacterial ion channels, KcsA and MthK, are presented based upon the crystal structures and depicting the movement of the intracellular molecular gate upon opening. (A) presents the KcsA structure as seen from the extracellular side; a potassium ion is modeled in the pore. All four subunits are depicted. (B) presents the side view of the KcsA structure with the intracellular gate shown closed. For clarity, only two of the four subunits are shown. (C) shows the structure of MthK. The gate is open. Only two of the four subunits are shown and the RCK domains have been omitted.

***Source*: reproduced with permission from Herrington and Arey (2014).**

The MacKinnon group continued their prestigious research by exploring the mutated structures of the channels (Zhou and MacKinnon, 2004). They mutated threonine 75 to cysteine (site 4 mutation) and then solved the structure in the presence of 200 mM KCl and refined to a resolution of 2.2 Å. The cysteine was found to replace threonine without significantly altering the side chain volume. See the electron density in the selectivity filter in Figure 11.15A, compared to the wild-type channel structure determined at 2.0 Å resolution under similar conditions (Figure 11.15B). In the mutant channel the sulfur atom of the cysteine side chain takes the position of the γ-carbon of the wild-type threonine side chain. The result is an almost perfect removal of the threonine hydroxyl with little further perturbation of the channel structure (RMSD for the wild-type and mutant channels from residue 71 to 80 is 0.12 Å) (see Figure 11.15C). Site 4 is rendered incomplete in the mutant channel: where K^+ fits snugly into a cubic cage of eight oxygen atoms surrounding site 4 in the wild-type channel, it experiences a wider, almost vestibule-like opening in the mutant channel.

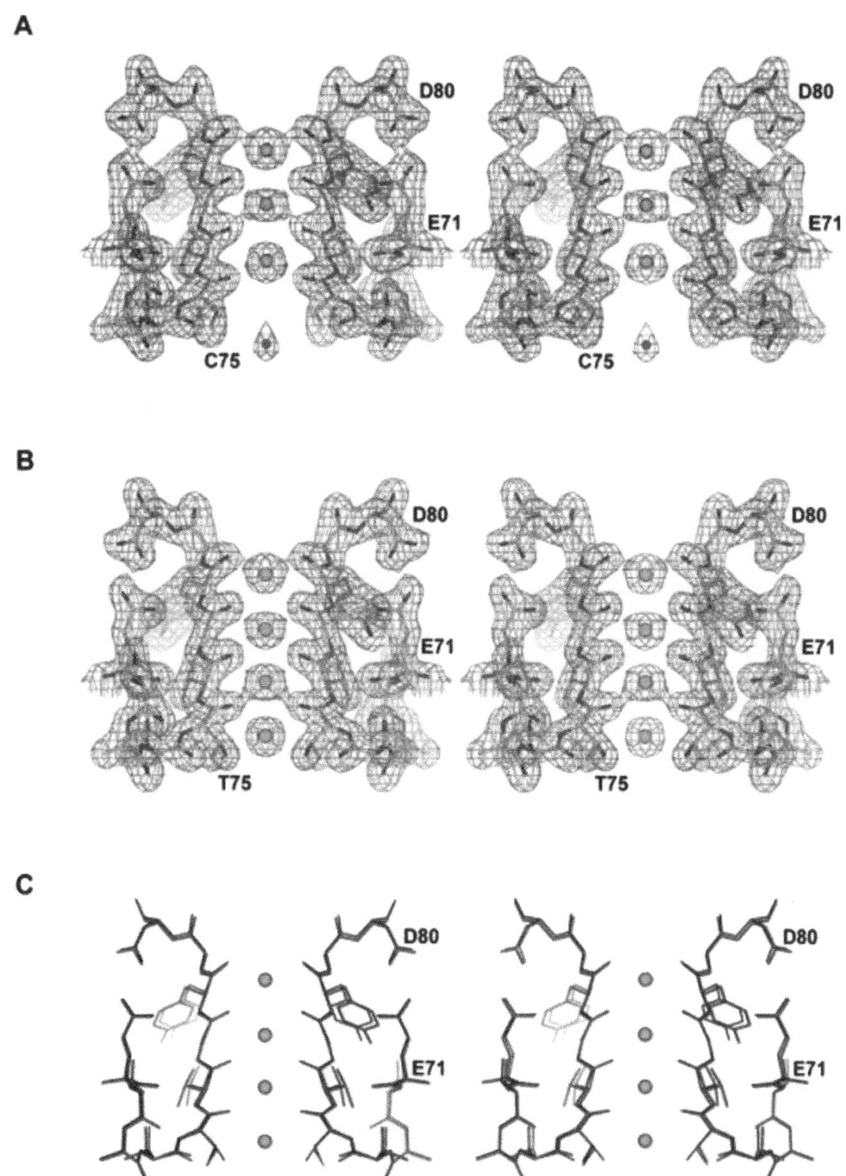

FIGURE 11.15 (A) Stereo view of the selectivity filter region for the T75C mutant KcsA channel. The structure, solved in 200 mM K^+, is shown with residues E71 to D80 (from two diagonally opposed subunits) rendered in a ball-and-stick representation. Green spheres represent K^+ binding sites in the filter, and the magenta sphere represents a site with residual electron density. The $2F_o–F_c$ electron density map (contoured at 2σ) validating the structure is shown as a blue mesh. (B) Stereo view of the selectivity filter region for the wild-type KcsA. The structure (PDB code 1K4C), solved in 200 mM K^+, is represented in the same way as in (A). (C) Stereo view of the selectivity filter region of the T75C (red) superimposed on that of the wild-type (blue).

***Source*: reproduced with permission from Zhou and MacKinnon (2004).**

One-dimensional (1D) electron density profiles were presented, which shows the wild-type (Figure 11.16A) and mutant (Figure 11.16B) channel structures. The areas have been converted into approximate occupancy (Zhou and MacKinnon, 2003). In the wild-type channel, the occupancies of sites 1 and 4 are about 0.55 and of sites 2 and 3 are about 0.45. Interpreted in the context of the two ion filter model (Zhou and MacKinnon, 2003) it is supposed that the 1,3 and 2,4 configurations each represent about 45 percent of the channels in the crystal while the remainder (~10 percent) correspond to a 1,4 configuration (K^+ at sites 1 and 4) (Morais-Cabral et al., 2001). In the mutant channel, the electron density profile is altered (see Figure 11.16). Site 4 has a broader and shifted peak with a reduced area consistent with lowered occupancy. The occupancy of site 2 is somewhat reduced. The apparent "action at a distance," that is, alteration of occupancy at site 2 as a consequence of a mutation at site 4, is easy to understand if ions tend to reside in specific configurations, because an energetic alteration of site 4 will affect the occupancy of the 2,4 configuration. The site 4 mutation

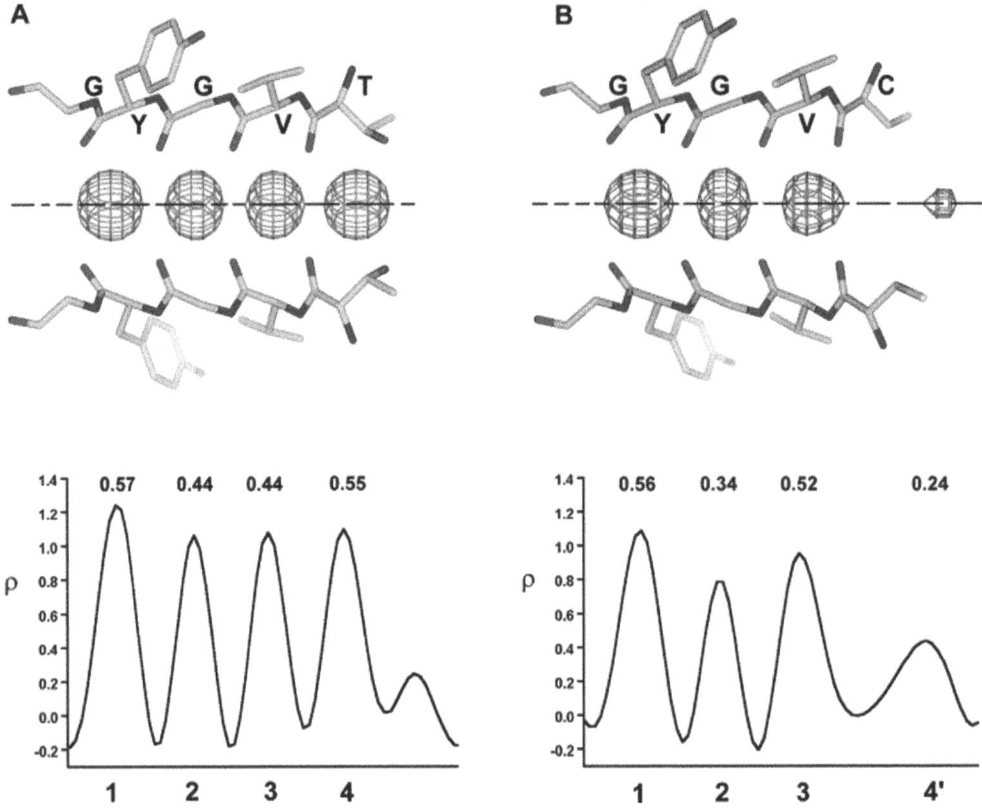

FIGURE 11.16 K^+ distributions in the selectivity filter of the wild-type (A) and the T75C (B) KcsA channels. Upper panel: the amino acid residues of the selectivity filter signature sequence (T/CVGYG) are shown in a ball-and-stick representation. F_o–F_c electron density omit maps (contoured at 3σ) validating the four K^+ binding sites are shown as a blue mesh. The left-most site corresponds to site 1. The black broken line shows the central axis of the selectivity filter, along which the difference Fourier omit map was sampled to obtain the one-dimensional electron density (ρ). Lower panel: one-dimensional electron density profiles corresponding to different K^+ binding sites in the selectivity filter. The numbers along the *x*-axis identify the four binding sites. The number on top of each peak indicates the fractional K^+ occupancy.

***Source*: reproduced with permission from Zhou and MacKinnon (2004).**

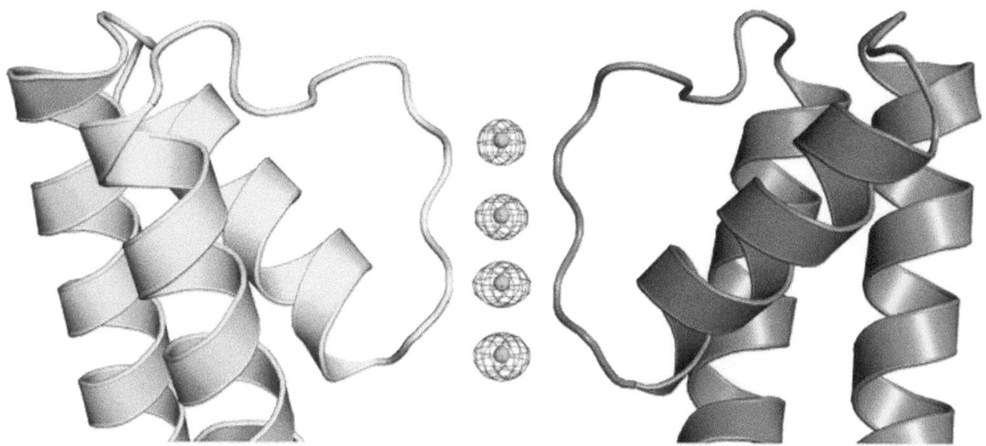

FIGURE 11.17 Anomalous difference map. The anomalous difference Fourier map contoured at 8σ is shown as a magenta mesh for subunit A. Strong anomalous difference peaks corresponding to the K^+ ions (cyan spheres) are present within the selectivity filter of the ion channel. For clarity, only two of the four subunits that make up the ion channel are shown.

***Source*: reproduced with permission from Langan et al. (2018).**

influences K^+ conduction in KcsA K+ channels, which was reconstituted into planar lipid bilayers. See Zhou and MacKinnon (2004) for details on channel conductance phenomena.

In a recent study, single wavelength anomalous dispersion (SAD) X-ray diffraction data were collected near the potassium absorption edge to show experimentally that all ion binding sites within the selectivity filter (SF) are fully occupied by K^+ ions (Langan et al., 2018). Here a 2.26 Å resolution SAD dataset was collected using X-rays with wavelength 3.35 Å. They conducted SAD studies of K^+ selective NaK2K (Derebe et al., 2011; Sauer et al., 2013) (NaK D66Y and N68D) at a K^+ ion concentration of 100 mM. This selective NaK mutant contains the same selectivity filter as KcsA (TVGYG), and therefore four equivalent K^+ binding sites (Derebe et al., 2011).

In order to determine the K^+ occupancy in NaK2K, the structure could be directly solved by K-SAD using the Shelx program suite (Sheldrick, 2010), and Anode (Thorn and Sheldrick, 2011) identified strong anomalous difference peaks between 28 and 39 sigma corresponding to the K^+ ions within the ion channel (see Figure 11.17).

On the top and bottom of the K^+ ions in the SF are bound water molecules. As atomic displacement parameters (ADPs) and occupancy are closely coupled, a careful analysis was made of the refined occupancies and ADPs of the K^+ atoms and the oxygen atoms that are in contact with them. All four K^+ ions in the channel refine to occupancy values close to the maximum possible value (0.25) with ADPs that are almost identical to the oxygen atoms located at the sides of the ion channel with which they are interacting. Finally, Langan and colleagues used phenix.refine to conduct occupancy refinements on the K^+ ions using 100 starting models with random occupancy and ADP values. Phenix.refine refines occupancies and ADP (or *B-factors*) separately at all times (Afonine et al., 2012). Following the refinement, the occupancy values for all the K^+ ions in the channel clustered around 0.25, suggesting that all four K^+ binding sites in the NaK2K SF are fully occupied with K^+ ions (see Figure 11.18).

A homo-FRET approach was recently found, which is assumed to provide complementary information to X-ray crystallography in which the protein conformational dynamics is usually compromised (Renart et al., 2019). Renart and colleagues investigated the SF dynamics of a single Trp mutant of the potassium channel KcsA (W67) using polarized time-resolved fluorescence measurements. For the first time, an analytical framework was reported to analyze the homo-Förster

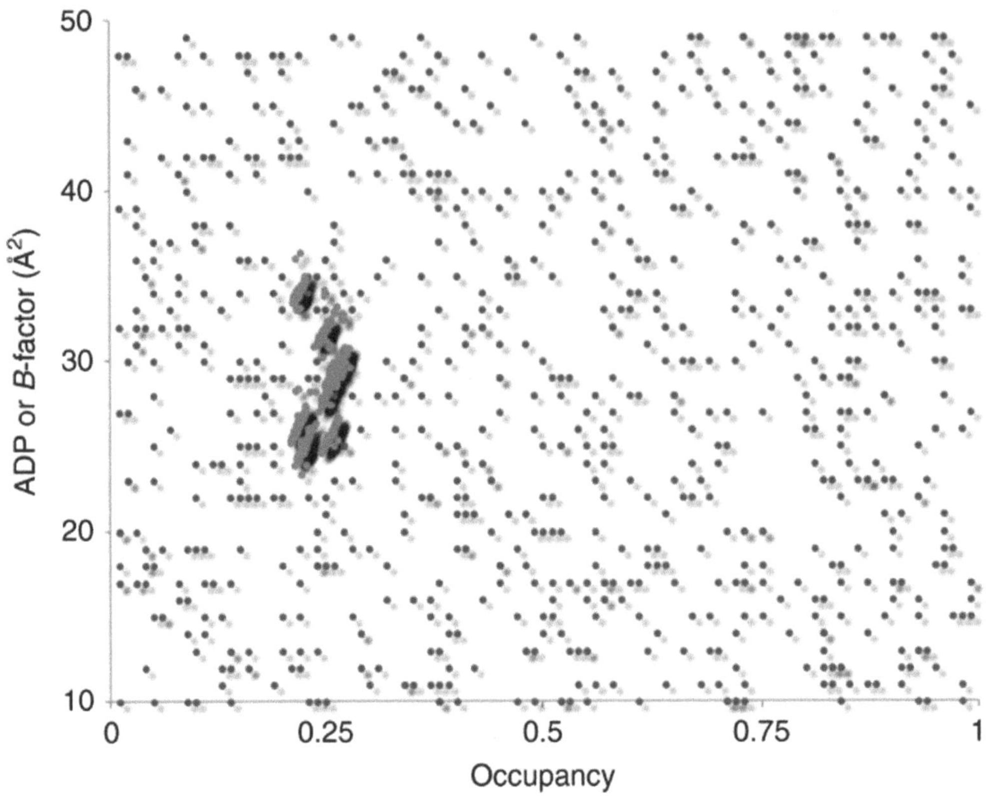

FIGURE 11.18 Results of occupancy refinement with differing starting points. One hundred starting models with initial occupancies and ADP values for K^+ ions randomly drawn from (0–1) to (10–50) ranges for each of eight K+ ions per structure (gray dots). The refined K+ occupancy values for each K+ ion (blue dots) cluster around 0.25.

***Source*: reproduced with permission from Langan et al. (2018).**

resonance energy transfer (homo-FRET) within a symmetric tetrameric protein with a square geometry. They reported that in the closed state at pH 7, the W67-W67 intersubunit distances become shorter as the average ion occupancy of the SF increases according to cation type and concentration. The hypothesis that the inactivated SF at pH 4 is structurally similar to its collapsed state, detected at low K+, pH 7, was ruled out, emphasizing the critical role played by the S2 binding site in the inactivation process of KcsA.

My last book, *Biophysics and Nanotechnology of Ion Channels*, detailed various aspects (structures and functions) of diversified ion channels and channel proteins (Ashrafuzzaman, 2021). I therefore do not wish to repeat this topic here. Interested readers are invited to consult this book to enhance their knowledge.

11.4.2 NMR for Understanding Protein Structures in Condensed Matter State

NMR can be used to solve protein structures in solution (Anil-Kumar et al., 1980; Wüthrich, 2001; Williamson et al., 1985). The first protein structure using NMR came to light as shown in Figures 11.19 and 11.20 (Wüthrich, 2001; Williamson et al., 1985; Bolognesi et al., 1982). Two-dimensional (2D) nuclear Overhauser enhancement spectroscopy (NOESY) was used in order to

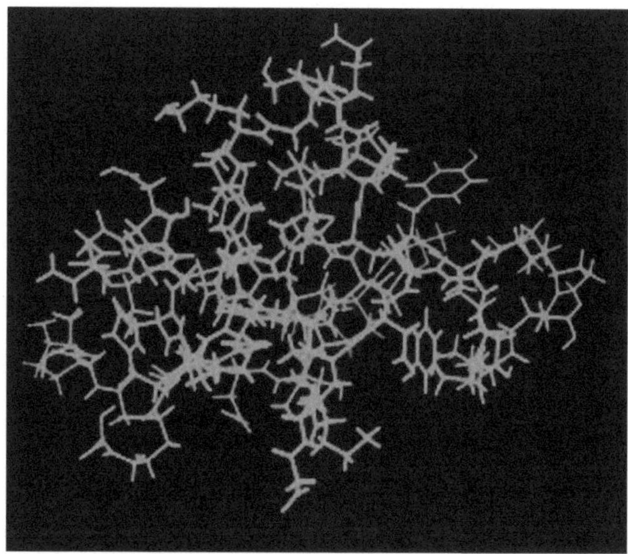

FIGURE 11.19 The first protein structure determined by NMR. All heavy-atom presentation of the NMR structure of the proteinase inhibitor IIA from bull seminal plasma (BUSI IIA).

***Source*: reproduced with permission from Wüthrich (2001), Williamson et al. (1985), and Bolognesi et al. (1982).**

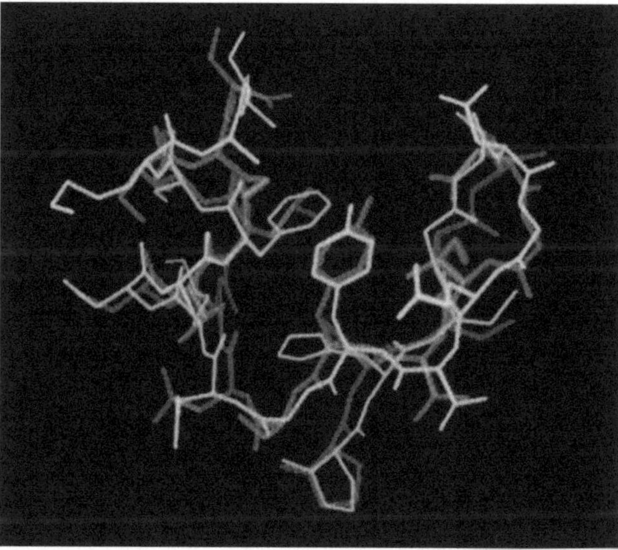

FIGURE 11.20 Superposition of the core region of residues 23–42 in the NMR structure of BUSI IIA (green) with the corresponding polypeptide segment in the X-ray crystal structure of the homologous porcine pancreatic secretory trypsin inhibitor (PSTI) (blue).

***Source*: reproduced with permission from Wüthrich (2001), Williamson et al. (1985), and Bolognesi et al. (1982).**

obtain a list of 202 distance constraints between individually assigned hydrogen atoms of the polypeptide chain, for identifying the positions of the three disulfide bridges, and for locating the single cis peptide bond. Wüthrich and colleagues derived supplementary geometric constraints from the vicinal spin-spin couplings and the locations of certain hydrogen bonds, as determined by NMR. With the use of a distance geometry program (DISGEO), capable of computing all-atom structures for proteins, the size of BUSI IIA, five conformers were computed from the NOE distance constraints alone, and another five were computed with the supplementary constraints included.

Within a few years of the successful NMR determination of protein structures, the leading investigator Wüthrich himself wrote a book, *NMR of Proteins and Nucleic Acids*, which provided an introduction to underlying principles and experimental procedures using the newest strategies and techniques for obtaining extensive NMR assignments in biopolymers based on NMR data and the primary structure available in those early days (Wüthrich, 1991). Now the field has matured (Cavalli et al., 2007; Purslow et al., 2020). Further details on NMR methods may be found in other books and chapters (e.g. Rule and Hitchens, 2006; Lian and Roberts, 2011; Delsuc et al., 2020; Chandra et al., 2022).

The determination of protein structures using NMR chemical shift was demonstrated in detail in Cavalli et al. (2007). Chemical shifts are readily and accurately measurable NMR parameters, as they reflect with great specificity conformations of native and nonnative states of proteins. Cavalli and colleagues showed, using 11 examples of proteins that are representative of the major structural classes and contain up to 123 residues, that it is possible to use chemical shifts as structural restraints in combination with a conventional molecular mechanics force field for determining the protein conformations at a resolution of 2 Å or beyond toward higher resolutions (Cavalli et al., 2007).

Chemical shifts are often used in structural biology for predicting regions of the secondary structure in proteins' native and nonnative states (Cornilescu et al., 1999; Wishart et al., 1992) to aid in the refinement of complex structures (Clore and Gronenborn, 1998), and to characterize conformational changes associated with partial unfolding (Korzhnev et al., 2004) or binding (Eisenmesser et al., 2005). Chemical shifts are also known to aid in the determination of proteins' tertiary structures when used in combination with other NMR probes that report on interproton distances (NOEs) and the relative orientations of the different nuclei in a protein structure (residual dipolar couplings, RDC) (Wishart and Case, 2002).

Chemical shifts are sometimes found to be the only NMR parameters that can be obtained on a given state of a protein with any degree of completeness (Korzhnev et al., 2004; Eisenmesser et al., 2005), helping to determine high-resolution structures. Chemical shifts in NMR protein spectra inherently carry sufficient information for determining the low molecular weight protein structures at high resolution (Sanders et al., 1993). The chemical shifts contain structural information that is different in nature from that provided by NOEs. The latter report on pairwise distances between specific protons, and thus provide unequivocal information about the relative spatial locations of different protein sequence residues (Wüthrich, 1986). The chemical shift associated with a specific atom, by contrast, is a summation of many contributing factors (Meiler, 2003), thus the reliable identification of interaction partners is very difficult, even though they may be substantially influenced by contacts between residues, such as hydrogen bonding and proximity to aromatic rings, that are at very different locations in the protein sequence.

Cavalli and colleagues (2007) described how NMR chemical shifts could be used to define the structures of the native states of proteins at high resolution. They present a procedure, termed CHESHIRE (protein structure determination with CHEmical SHIft Restraints), which exploits the availability of fast empirical methods that have recently been developed in order to enable the chemical shifts to be calculated approximately but very rapidly for a given structure (Xu and Case, 2001; Neal et al., 2003; Meiler, 2003). In their computational strategy, they used the molecular fragment replacement approach in ab initio structure prediction (Rosetta) (Simons et al., 1997; Bradley et al., 2005; Schueler-Furman et al., 2005) and in the analysis of RDC (Rohl and Baker, 2002; Delaglio

TABLE 11.1
Quality of the Structures Determined using Chemical Shift Restraints Compared with Conventional Methods from Reference PDB Structures

Protein name	%α*	%β†	N_{res}‡	N_{CS}§	$\%_{SS}$¶	$\%_{da}$ǀ	$RMSD_{bb}$**	$RMSD_{aa}$††	N_{pp}‡‡	Q_{RDC}§§
Ubiquitin	25	32	76	281	74	93	1.33	2.13	3	0.55
FF domain	77	0	54	214	90	86	1.46	2.30	9	0.48
Calbindin	60	0	74	286	85	95	1.47	2.16	5	0.54
HPr	37	29	85	331	87	86	1.83	2.59	10	0.60
Sda	60	0	46	181	89	86	1.37	2.19	0	0.43
MrR5	0	51	70	264	80	75	1.58	2.61	5	0.89
PhS018	21	50	92	350	83	91	1.21	2.17	4	0.47
Bet v 4	64	4	84	325	92	96	1.64	2.35	4	0.57
Δ27-GG	0	65	106	402	83	77	1.46	2.59	8	0.60
TM1442	44	20	110	428	80	90	1.32	2.26	12	0.63
Sen15	32	29	123	478	83	91	1.72	2.47	3	0.62

* Percentage of α-helical structure in the native state.
† Percentage of β-sheet structure in the native state.
‡ Number of residues in the protein.
§ Number of chemical shifts used as restraints.
¶ Percentage of residues with the same predicted secondary structures (α, β, coil) as in the reference conformations, as determined in the 3PRED phase.
ǀ Percentage of predicted backbone dihedral angles within 60° of those in the reference conformations, as determined in the TOPOS phase.
** RMSD for all backbone and C_β atoms. Residues before the first secondary structure element and after the last one are excluded from the calculations.
†† RMSD on all atoms.
‡‡ Number of interproton distances <5.5 Å in the reference structures but >6.5 Å in the structures determined here.
§§ Estimated Q factors (see text) for the HN-N, CA-HA, CA-C, CA-CB residual dipolar couplings.

Reproduced with permission from Cavalli et al. (2007).

et al., 2000), and sparse data of NMR (Bowers et al., 2000), including the unassigned chemical shifts (Meiler and Baker, 2003).

Firstly, in the CHESHIRE procedure, the chemical shifts are used to predict the secondary structure of the protein. The method that was developed, termed 3PRED, uses Bayesian inference in order to predict the secondary structure of amino acids from the known chemical shifts in combination with the intrinsic secondary structure propensity of amino acids triplets (Cavalli et al., 2007). In the first phase of 3PRED, the experimental chemical shifts were used for predicting the secondary structure of protein fragments of three and of nine residues (Cornilescu et al., 1999). For ≈85 percent of the residues, correct prediction was made regarding the presence of α, β or coil conformation (see Table 11.1).

Then in the second phase TOPOS, a library of trial conformations was generated for each protein fragment by screening a database to search for those of similar sequence, secondary structure, and chemical-shift patterns (see Methods in Cavalli et al., 2007). Thus values of backbone dihedral angles that are in ≈95 percent of the cases within 60° are predicted from those in the high-resolution structures determined by conventional methods (see Figure 11.4). In the third phase, the fragments are assembled (see Figure 11.21), and the structures in the resulting ensembles are refined with the use of a scoring function defined by a combination of chemical shifts and a force field similar to the standard ones used in classical molecular dynamics simulations. The tertiary information contained

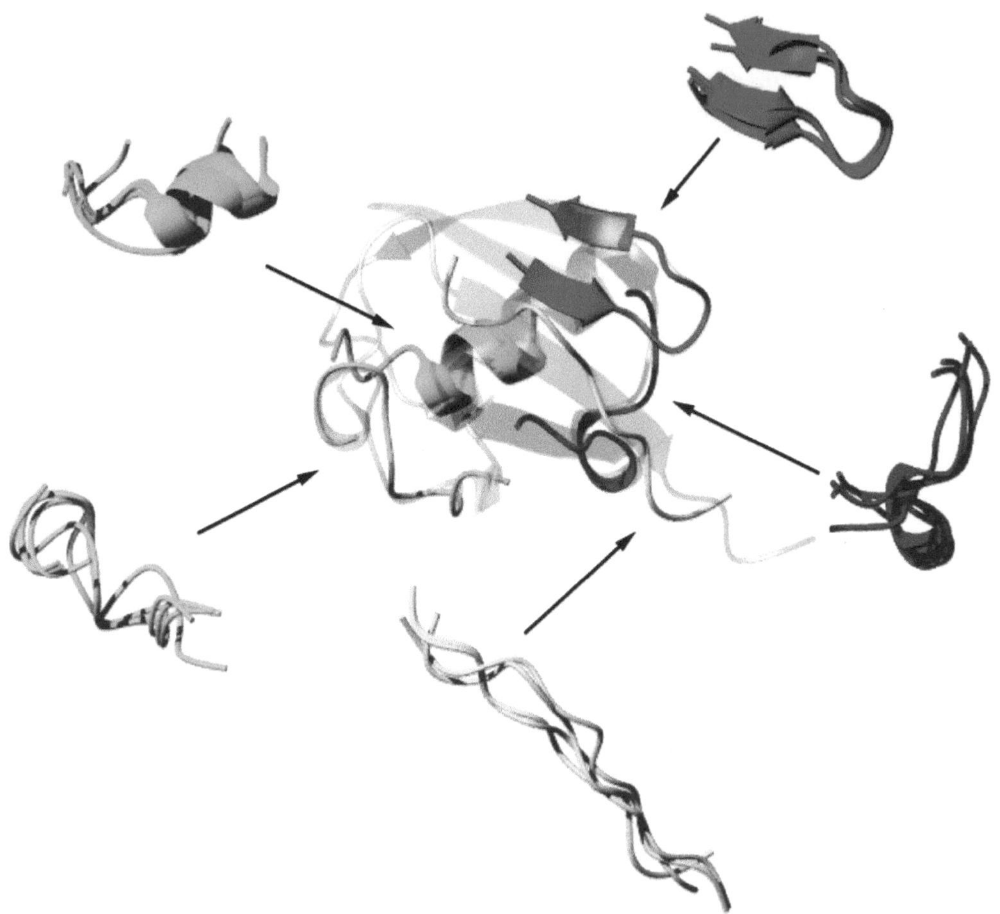

FIGURE 11.21 Schematic illustration of the molecular fragment replacement procedure implemented for chemical shifts in the CHESHIRE procedure. The protein shown is ubiquitin, and fragments are generated with main-chain dihedral angles compatible with the information contained in the chemical shifts. The fragments are then assembled in a combinatorial manner to produce an ensemble of trial structures that are subsequently refined by exploiting the information about tertiary structure contained in the chemical shifts.

***Source*: reproduced with permission from Cavalli et al. (2007).**

in the experimental chemical shifts, including orientations of aromatic rings and hydrogen bonds, were effectively exploited here (Neal et al., 2003).

Analysis of the quality of the structures (CHESHIRE procedure) was applied to 11 proteins chosen from the literature to be representative of different structural classes and to have a well defined set (^{1}H, $^{13}C_\alpha$, $^{13}C_\beta$, and ^{15}N) of chemical shifts for the main-chain atoms (Figure 11.4; see the example cases in Figure 11.22). In all cases, the overall RMSD between the structures (emerging from the use of the CHESHIRE procedure and the corresponding previously determined high-resolution X-ray or NMR structures, which are used as reference conformations), was between 1.21 Å and 1.83 Å for the backbone atoms, and between 2.13 Å and 2.61 Å for all atoms (see Figure 11.4). The average pairwise backbone RMSDs between the 10 lowest-energy structures determined for each protein range from 1.0 Å to 1.5 Å. Comparable similarities in structural predictions from both methods have been observed. It is therefore concluded that the determination of high-resolution structures of protein molecules by using NMR chemical shifts as the only source of experimental information is quite possible.

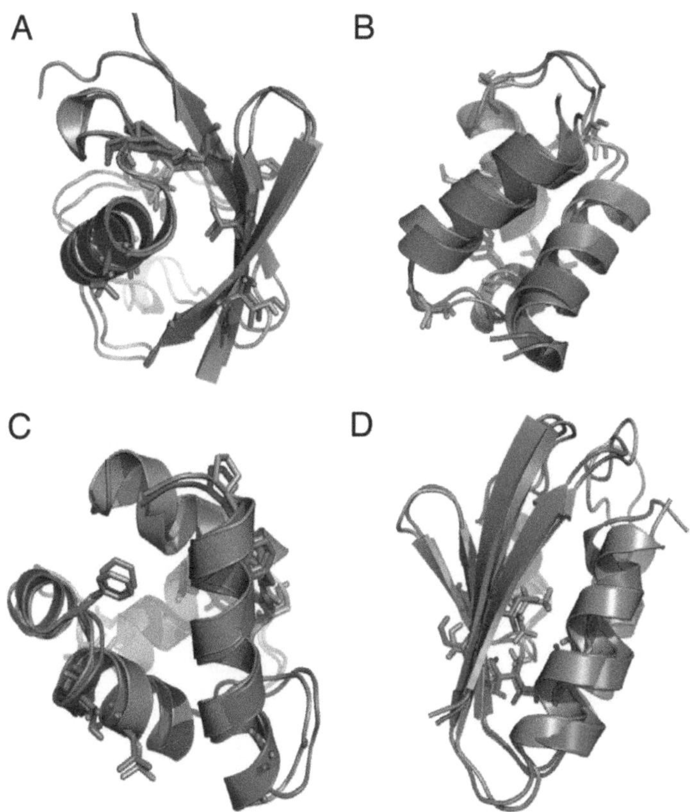

FIGURE 11.22 Comparison of the structures, also showing side chains in the hydrophobic cores, determined from chemical-shift information using the CHESHIRE procedure and those determined by standard X-ray or NMR methods. (A) Ubiquitin (blue) and PDB entry 1UBQ (pink). (B) FF domain (blue) and PDB entry 1UZC (pink). (C) Calbindin (blue) and PDB entry 4ICB (pink). (D) HPr (blue) and PDB entry 1POH (pink).

***Source*: reproduced with permission from Cavalli et al. (2007).**

Both X-ray crystallography and NMR techniques are now highly utilized to predict various protein structures. Both are often used even to address the structure of the same protein. Comparison between NMR and X-ray crystallography data-based potassium channel structures has also been performed (Chill et al., 2006; see Figure 11.23). A comparison of the secondary chemical shift, amide proton NOE connectivities, and hydrogen-exchange NMR data (see Chill et al., 2006) with the crystal structure reported for residues 23–119 of KcsA (Doyle et al., 1998) reveals excellent agreement. The three previously identified structural elements of KcsA, the TM1, pore, and TM2 helices, are located in the crystal structure at residues 27–51, 62–74, and 86–112. NMR C^{α} secondary chemical shifts identify these structural elements at residues 31–52, 62–73, and 86–115, respectively. The R27AAGA31 segment preceding the TM1 helix shows slow HX rates and strong sequential amide NOEs, suggestive of a helical conformation for this region too. Ala and Gly residues within the transmembrane helices tend to have smaller 13C secondary chemical shifts than other residues, possibly accounting for the modest secondary chemical shifts seen in this region. For details on the comparison, see Chill et al. (2006). Readers may also consult Chill et al. (2019), which has addressed the challenges and opportunities of NMR studies of membrane proteins. This article also reviews a case study focusing on the C-terminal domain of the bacterial potassium channel KcsA, describing how improvements in membrane-mimicking conditions eventually enabled us to present a structural view of the pH-dependent behavior of this cytoplasmic channel domain. Both

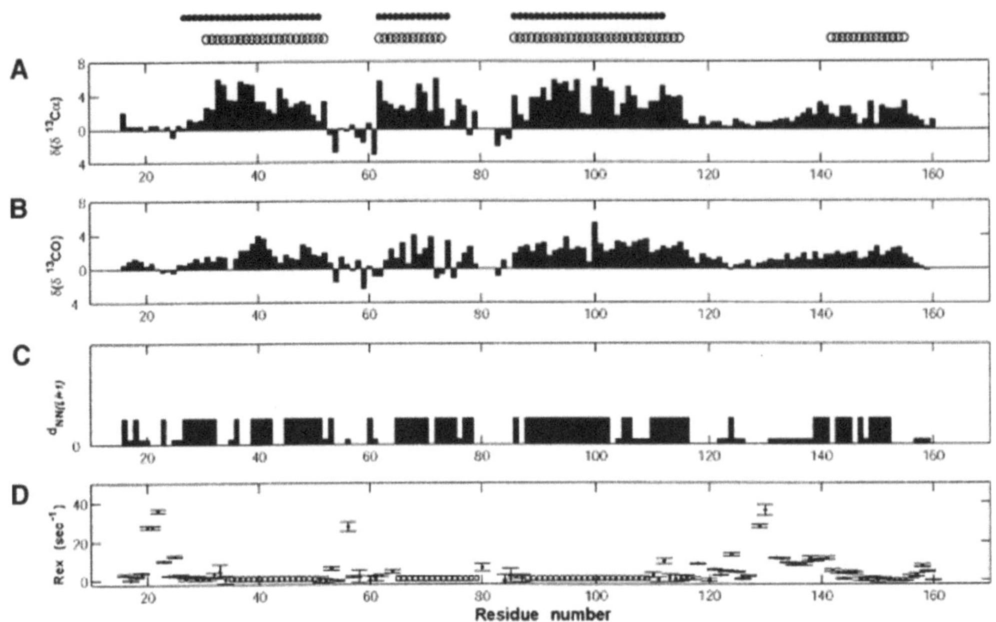

FIGURE 11.23 Backbone NMR data predict the secondary structure of KcsA. Deviations of chemical shifts from random coil values (based on Zhang et al., 2003) are shown for $^{13}C^{\alpha}$ (*A*) and $^{13}C'$ (*B*) nuclei along the KcsA backbone. Downfield values (positive deviations) of the ^{13}C resonance indicate a backbone helical conformation. (*C*) $^{1}H^{N}(i)$-$^{1}H^{N}(i+1)$ NOE connectivities observed in ^{15}N-separated NOE-HMQC spectra of KcsA. Data from several spectra were combined by measuring ratios between cross-peak and diagonal peak intensities, and classifying these in two categories, corresponding to strong and weak $^{1}H^{N}(i)$-$^{1}H^{N}(i + 1)$ interactions. (*D*) HX rates of KcsA amide protons. Exchange rates were measured by comparing the intensity of a given tr-HNCO peak acquired at 600 MHz with and without a preceding water-inversion peak. KcsA^{TM} protons are considered nonexchangeable on this scale, as designated by the open squares along the axis. No effort was made to exclude the effect of exchangeable-proton-mediated magnetization transfer, which may increase apparent exchange rates. For example, the elevated rate observed for T112 is likely the result of fast exchange of its β-OH proton with solvent, followed by NOE magnetization transfer to its backbone amide. The secondary structure elements in the crystal structure (Doyle et al., 1998) and as determined by NMR are shown above in filled and open circles, respectively.

Source: **reproduced with permission from Chill et al. (2006).**

NMR and X-ray crystallography have been actively serving structural biologists to understand versatile protein structures in different physiological conditions and environments.

11.5 PROTEIN STRUCTURE PREDICTIONS USING COMPUTATIONAL AND ARTIFICIAL INTELLIGENCE TECHNIQUES

The structures of around 100,000 proteins have been determined (wwPDB Consortium, 2018) using various experimental explorations (see e.g. Thompson et al., 2020; Bai et al., 2015; Jaskolski et al., 2014; Wüthrich, 2001). The number of known protein structures is negligible compared to billions of known protein sequences (Mitchell et al., 2020; Steinegger et al., 2019). The protein structural predictions started some seventy years ago when Pauling and Corey predicted helical and sheet conformations for protein polypeptide backbone (Pauling et al., 1951; Pauling and Corey, 1951) even before the first protein structure was determined (Kendrew et al., 1958) (for a detailed history, see Yang et al., 2018). Due to slow progress in experimental understanding of protein structures, alternative methods are explored. Computational approaches may help predict protein structures (Floudas, 2007; Majumder, 2020; Yang et al., 2020). Over many decades homology modeling has been utilized to predict the tertiary structure of a protein if this protein has sequence similarity to a protein with known atomic structure (Blundell et al., 1987; Blundell et al., 1978). The methods are based on the observation that the structures are more conserved than the sequences of amino acids. Therefore, an almost accurate molecular model of a protein may be constructed by assigning a conformation that is based on sequence alignment, followed by model building and energy minimization.

Recent progress (Senior et al., 2020; Wang et al., 2017; Zheng et al., 2019; Abriata et al., 2019; Pearce and Zhang, 2021), especially in the AlphaFold predictions (Jumper et al., 2021) of protein structures, is remarkable. Overcoming existing methods' inability to achieve atomic accuracy, especially when no homologous structure is available, Jumper and colleagues (2021) have provided the first computational method that can predict protein structures with atomic accuracy even when no similar structure is known. DeepMind's program AlphaFold is known to create improved computational methods for accurately predicting protein structures (Callaway, 2020). AlphaFold's structure predictions are occasionally found indistinguishable from those determined using standard experimental methods, such as X-ray crystallography and cryo-electron microscopy (cryo-EM).

A redesigned version of the neural network-based model, AlphaFold, was validated (Moult et al., 2020), demonstrating accuracy competitive with experimental structures in a majority of cases and greatly outperforming other methods. AlphaFold is a novel machine-learning approach that incorporates physical and biological knowledge about protein structure, leveraging multi-sequence alignments, into the design of the deep learning algorithm (Jumper et al., 2021). In CASP14, AlphaFold structures appeared more accurate than competing methods. These structures had a median backbone accuracy of 0.96 Å r.m.s.d.$_{95}$ (Cα root-mean-square deviation at 95 percent residue coverage) (95 percent confidence interval = 0.85–1.16 Å), but the next best performing method had a median backbone accuracy of 2.8 Å r.m.s.d.$_{95}$ (95 percent confidence interval = 2.7–4.0 Å) (measured on CASP domains; see Figure 11.24a for backbone accuracy, and Supplementary materials in Jumper et al., 2021, for all-atom accuracy). Besides the accurate domain structures (see Figure 11.24b), AlphaFold is found to be able to produce highly accurate side chains (see Figure 11.24c) when the backbone is highly accurate, and considerably improves over template-based methods even when strong templates are available. The all-atom accuracy of AlphaFold was 1.5 Å r.m.s.d.$_{95}$ (95 percent confidence interval = 1.2–1.6 Å) compared with the 3.5 Å r.m.s.d.$_{95}$ (95 percent confidence interval = 3.1–4.2 Å) of the best alternative method. Our methods are scalable to very long proteins with accurate domains and domain-packing (see Figure 11.24d for the prediction of a 2,180-residue protein with no structural homologues). Finally, the model is able to provide precise, per-residue estimates of its reliability that should enable the confident use of these predictions.

Figure 11.25a shows the high accuracy that AlphaFold demonstrated in CASP14 extending to a large sample of recently released PDB structures (Jumper et al., 2021). Jumper and colleagues also observed high side-chain accuracy when the backbone prediction is accurate (see Figure 11.25b) and the predicted local-distance difference test (pLDDT) reliably predicts the Cα local-distance difference test (lDDT-Cα) accuracy of the corresponding prediction (see Figure 11.25c). The global superposition metric template modeling was found to yield a score (TM-score) (Zhang and Skolnick, 2004) that can be accurately estimated (see Figure 11.25d). Overall, these analyses validate that the high accuracy and reliability of AlphaFold on CASP14 proteins also transfer to an uncurated collection of recent PDB submissions (see supplementary materials in Jumper et al., 2021).

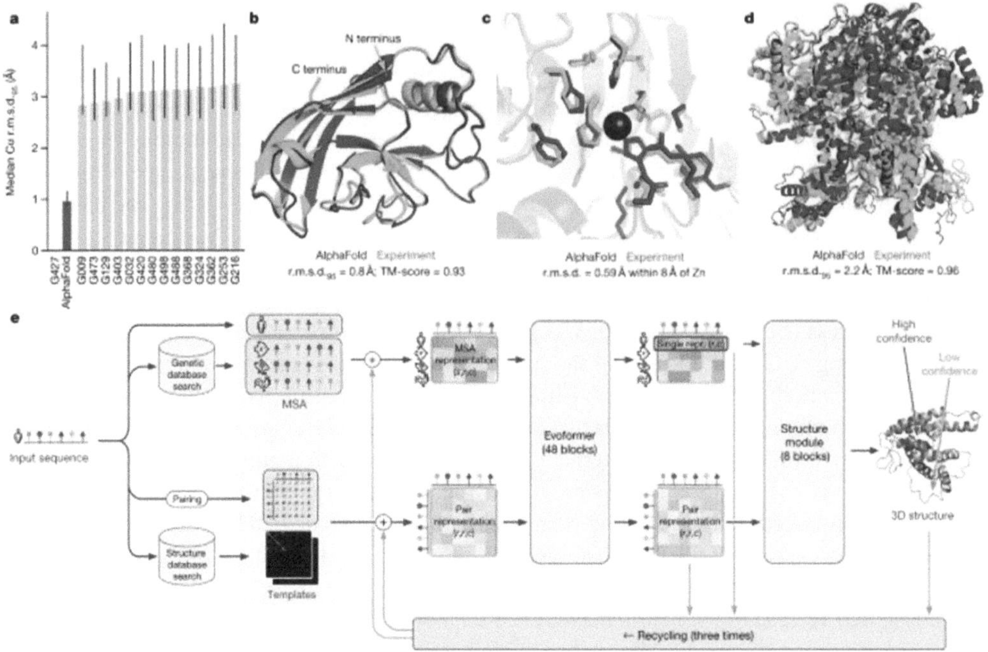

FIGURE 11.24 (a) The performance of AlphaFold on the CASP14 dataset (n = 87 protein domains) relative to the top-15 entries (out of 146 entries); group numbers correspond to the numbers assigned to entrants by CASP. Data are median and the 95 percent confidence interval of the median, estimated from 10,000 bootstrap samples. (b) The prediction of CASP14 target T1049 (PDB 6Y4F, blue) compared with the true (experimental) structure (green). Four residues in the C terminus of the crystal structure are B-factor outliers and are not depicted. (c) CASP14 target T1056 (PDB 6YJ1). An example of a well-predicted zinc-binding site (AlphaFold has accurate side chains even though it does not explicitly predict the zinc ion). (d) CASP target T1044 (PDB 6VR4) – a 2,180-residue single chain – was predicted with correct domain packing (the prediction was made after CASP using AlphaFold without intervention). (e) Model architecture. Arrows show the information flow among the various components described in this paper. Array shapes are shown in parentheses with s, number of sequences (Nseq in the main text); r, number of residues (Nres in the main text); c, number of channels.

***Source*: reproduced with permission from Jumper et al. (2021).**

11.6 PROTEOMICS AND THREE-DIMENSIONAL MAPPING OF THE CELL

Biological cells, organisms, and various organelles within biological complexes host various specific and nonspecific proteins. Proteins are found to work both independently and with participants in a complex (Donev and Karabencheva-Christova, 2021). A set of proteins produced in an organism, system, or biological context is referred to as a proteome. In a protein complex, a group of polypeptide chains are linked by noncovalent protein–protein interactions. Protein complexes are found to play important roles in biological systems and perform various functions, e.g. DNA transcription, mRNA translation, and signal transduction (Graves and Haystead, 2002).

The proteomics concept first emerged as a popular topic in 1995, defined then as the large-scale characterization of the entire protein complement of a cell line, tissue, or organism (Anderson and Anderson, 1996; Wasinger et al., 1995; Wilkins et al., 1996). Later, two definitions for proteomics appeared, as follows: first, the more classical definition, restricting the large-scale analysis of gene products to studies involving only proteins; second, this inclusive definition combines protein studies with analyses that have a genetic readout such as mRNA analysis, genomics, and the yeast

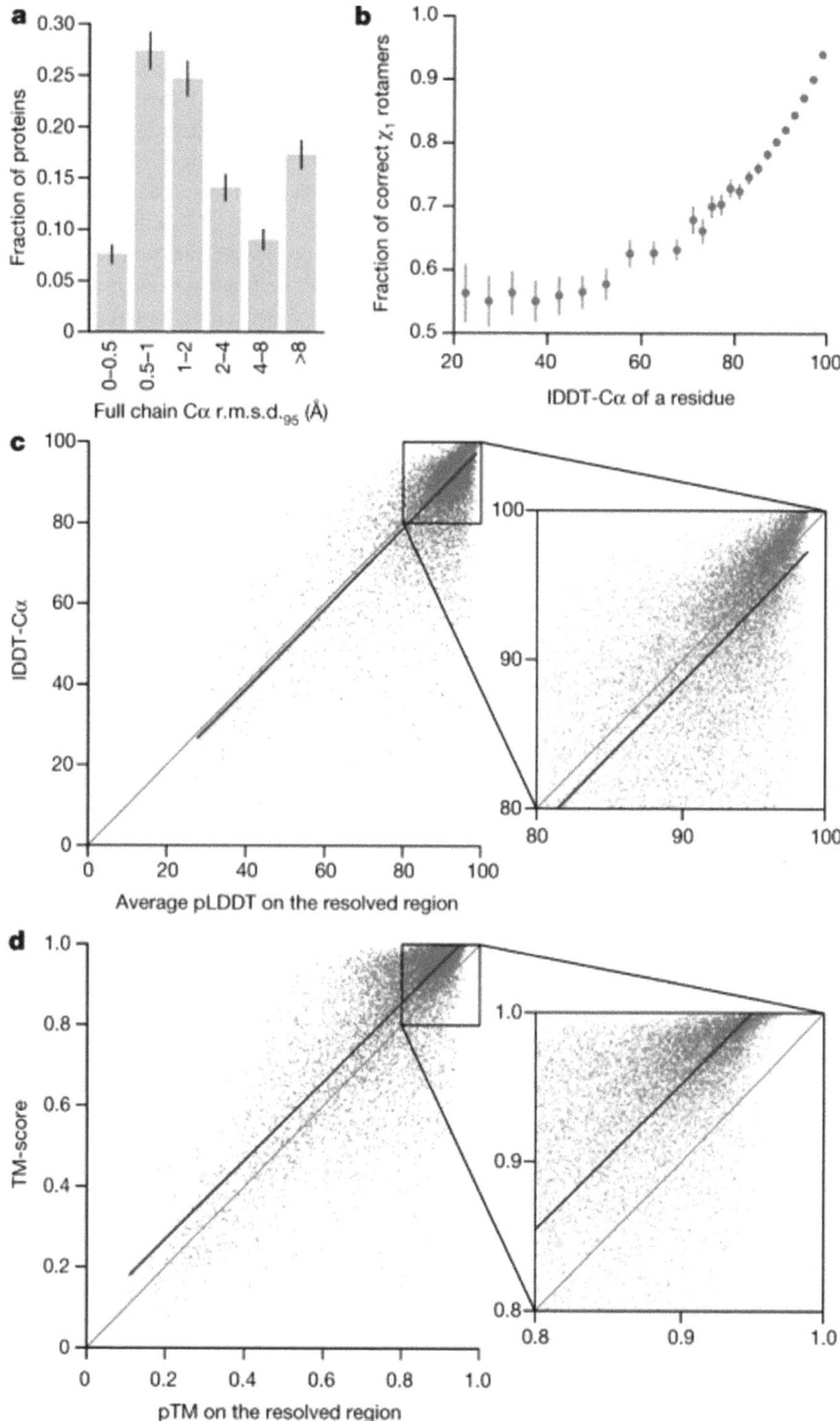

FIGURE 11.25 The analysed structures are newer than any structure in the training set. Further filtering is applied to reduce redundancy. (a) Histogram of backbone r.m.s.d. for full chains (Cα r.m.s.d. at 95 percent coverage). Error bars are 95 percent confidence intervals (Poisson). This dataset excludes proteins with a

FIGURE 11.25 *(continued)*

template (identified by hmmsearch) from the training set with more than 40 percent sequence identity covering more than 1 percent of the chain (n = 3,144 protein chains). The overall median is 1.46 Å (95 percent confidence interval = 1.40–1.56 Å). Note that this measure will be highly sensitive to domain packing and domain accuracy; a high r.m.s.d. is expected for some chains with uncertain packing or packing errors. (b) Correlation between backbone accuracy and side-chain accuracy. Filtered to structures with any observed side chains and resolution better than 2.5 Å (n = 5,317 protein chains); side chains were further filtered to B-factor <30 Å^2. A rotamer is classified as correct if the predicted torsion angle is within 40°. Each point aggregates a range of lDDT-Cα, with a bin size of 2 units above 70 lDDT-Cα and 5 units otherwise. Points correspond to the mean accuracy; error bars are 95 percent confidence intervals (Student t-test) of the mean on a per-residue basis. (c) Confidence score compared to the true accuracy on chains. Least-squares linear fit lDDT-Cα = 0.997 × pLDDT − 1.17 (Pearson's r = 0.76). n = 10,795 protein chains. The shaded region of the linear fit represents a 95 percent confidence interval estimated from 10,000 bootstrap samples. In the companion paper (Tunyasuvunakool et al., 2021), additional quantification of the reliability of pLDDT as a confidence measure is provided. (d) Correlation between pTM and full chain TM-score. Least-squares linear fit TM-score = 0.98 × pTM + 0.07 (Pearson's r = 0.85). n = 10,795 protein chains. The shaded region of the linear fit represents a 95 percent confidence interval estimated from 10,000 bootstrap samples.

***Source*: reproduced with permission from Jumper et al. (2021).**

two-hybrid analysis (Pandey and Mann, 2000). However, the goal of proteomics remains the same, i.e. to obtain a more global and integrated view of biology by studying all the proteins of a cell rather than each one individually.

The majority of proteins in eukaryotic cells are involved in complex formation at some point in the life of the cells and each protein may have on average six to eight interacting partners (Tong et al., 2004). Network connectivity was here predictive of functions, because interactions often occurred among the functionally related genes, and similar patterns of interactions tended to identify components of the same pathway. The genetic network exhibited dense local neighborhoods. Therefore, the position of a gene on a partially mapped network was predictive of other genetic interactions. Because digenic interactions are common in yeast, identical networks may underlie the complex genetics associated with inherited phenotypes in other organisms. A recent study described the construction and analysis of a comprehensive genetic interaction network for a eukaryotic cell (Costanzo et al., 2016). A global genetic interaction network is expected to highlight the functional organization of a cell and provide a resource for predicting gene and pathway function (see Figure 11.26). This network emphasizes the prevalence of genetic interactions and their potential to compound phenotypes associated with single mutations. Negative genetic interactions may connect functionally related genes and thus may be predicted using alternative functional information. Less functionally informative, positive interactions may provide insights into general mechanisms of genetic suppression or resiliency. Costanzo and colleagues anticipated that the ordered topology of the global genetic network, in which genetic interactions connect coherently within and between protein complexes and pathways, may be exploited to decipher genotype-to-phenotype relationships (see details in Costanzo et al., 2016).

Costanzo and colleagues (2021) recently extended their understanding of the global network of genetic interaction by studying the effects this networking may draw from the environment. A phenotype may be affected by genes interacting with other genes, the environment, or both other genes and the environment (a differential interaction). To understand better how these interactions function in yeast, Costanzo et al. mapped the gene–gene interactions using single- and

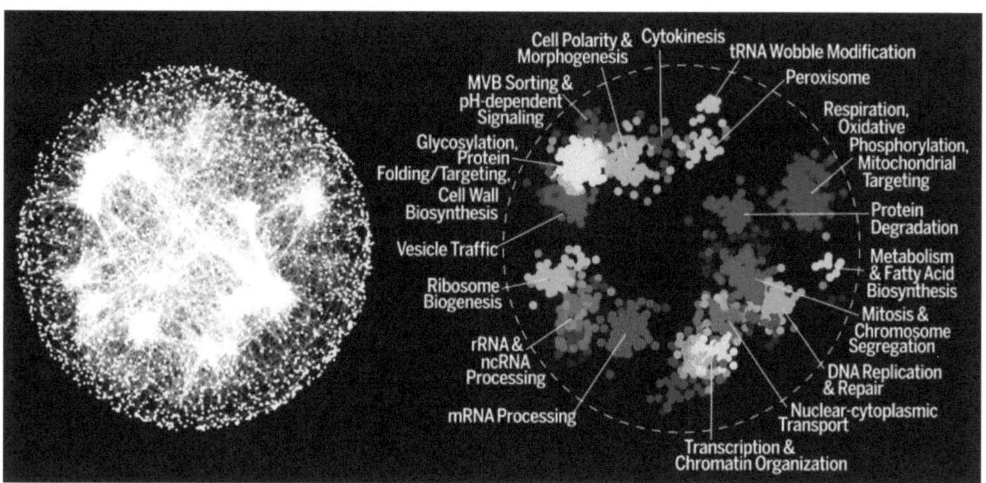

FIGURE 11.26 A global network of genetic interaction profile similarities. (Left) Genes with similar genetic interaction profiles are connected in a global network, such that genes exhibiting more similar profiles are located closer to each other, whereas genes with less similar profiles are positioned farther apart. (Right) Spatial analysis of functional enrichment was used to identify and color network regions enriched for similar gene ontology bioprocess terms.

***Source*: reproduced with permission from Costanzo et al. (2016).**

double-mutant deletions and temperature-sensitive alleles under 14 environmental conditions. Many deleted or temperature-sensitive nonessential genes were found to affect the yeast fitness both positively and negatively under at least one of the tested environmental conditions. Up to 24 percent of yeast genes were found to be affected. A minority of these differential interactions point to previously unknown genetic connections across functional networks, showing how genetic architecture responds to environmental variation. The results have been summarized in Figure 11.27. The conclusions have been made based on results that came out of their tests on ~4,000 yeast single mutants for GxE interactions across 14 diverse environments, which included an alternative carbon source, osmotic and genotoxic stress, and treatment with 11 bioactive compounds that targeted distinct yeast bioprocesses. To quantify GxGxE interactions, they constructed ~30,000 different double mutants, involving genes annotated to all major yeast bioprocesses, and the scoring was made for genetic interactions.

It is generally accepted now that proteins are the central cellular machines, and the actions they perform happen through interactions among them and with other cell-based biomolecules. Large-scale efforts in genomics, proteomics, lipidomics, and metabolomics are attempting to produce complete lists of the molecules in a cell as well as in different subcellular compartments, including the membrane. These initiatives are slowly allowing us to construct a cellular three-dimensional proteome map (Elofsson, 2021) to address the cell functions in which proteins contribute their roles individually and collectively. To have any accurate picture on this matter we still need to wait perhaps decades and acquire more research data.

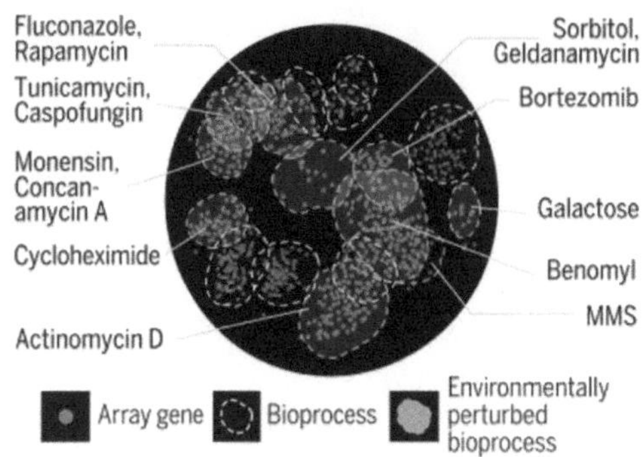

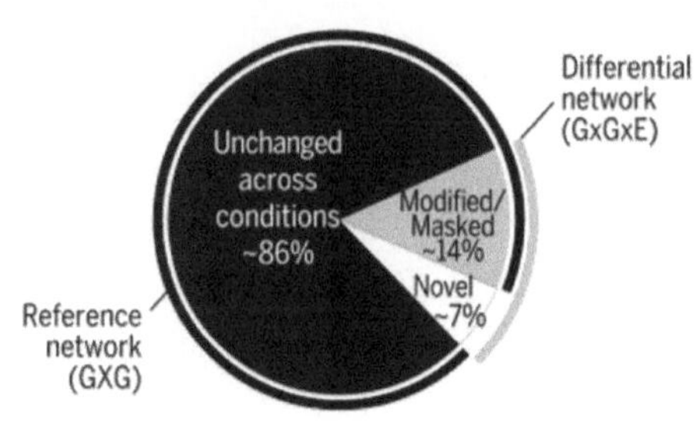

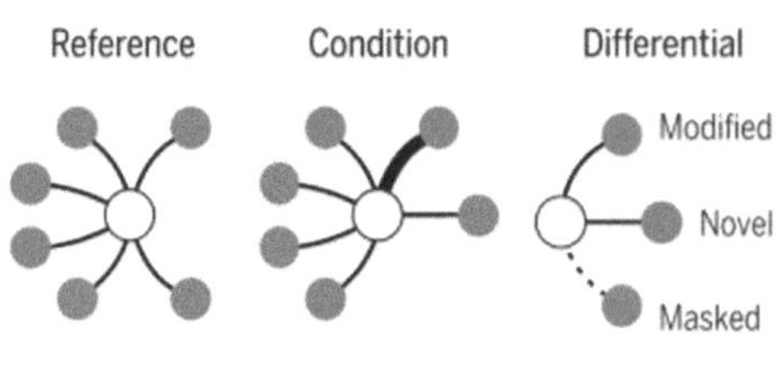

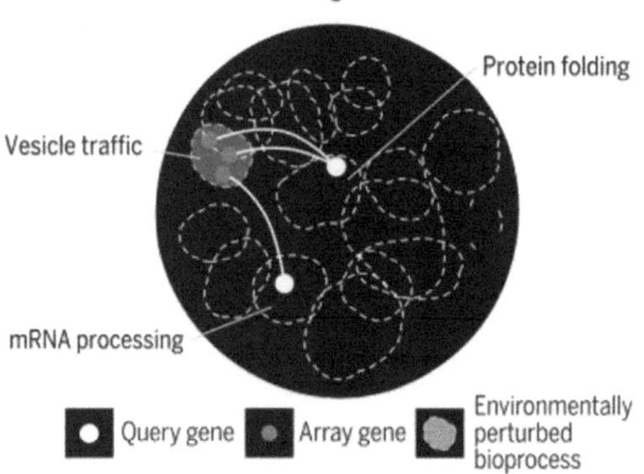

FIGURE 11.27 Systematic analysis of environmental impact on the global yeast genetic interaction network. (Top left) Mapping GxE interactions and (bottom left) GxGxE differential interactions reveals (top right) the environmental robustness of the global yeast genetic interaction network, (bottom right) highlighting new and distant functional connections associated with novel differential interactions.

***Source*: reproduced with permission from Costanzo et al. (2021).**

REFERENCES

Abriata, L.A., Tamò, G.E., & Dal Peraro, M. (2019). A further leap of improvement in tertiary structure prediction in CASP13 prompts new routes for future assessments. *Proteins*, *87*, 1100–1112

Afonine, P.V., Grosse-Kunstleve, R.W., Echols, N., Headd, J.J., Moriarty, N.W., Mustyakimov, M., … & Adams, P.D. (2012). Towards automated crystallographic structure refinement with phenix.refine. *Acta Crystallographica Section D: Biological Crystallography*, *68*(4), 352–367

Ajaj, Y., Wehner, M., & Weingärtner, H. (2009). Myoglobin and apomyoglobin in their native, molten globule and acid-denaturated states: a dielectric relaxation study. *Zeitschrift für physikalische Chemie*, *223*(9), 1105–1118. https://doi.org/10.1524/zpch.2009.6061

Alberts, B., Johnson, A., Lewis, J., Raff, M., Roberts, K., & Walter, P. (2002). *Molecular biology of the cell*. New York: Garland Science

Anderson, N.G., and Anderson, N.L. (1996). Twenty years of two-dimensional electrophoresis: past, present and future. *Electrophoresis*, *17*, 443–453

Anil-Kumar, E.R., Ernst, R.R., & Wüthrich, K. (1980). A two-dimensional nuclear Overhauser enhancement (2D NOE) experiment for the elucidation of complete proton–proton cross-relaxation networks in biological macromolecules. *Biochem. Biophys. Res. Commun.*, *64*, 2229–2246

Ashrafuzzaman, M. (2021). *Biophysics and nanotechnology of ion channels*, Boca Raton, FL: CRC Press

Bai, X.-C., McMullan, G., & Scheres, S.H.W. (2015). How cryo-EM is revolutionizing structural biology. *Trends Biochem. Sci.*, *40*, 49–57

Banavar, J.R., & Maritan, A. (2007). Physics of proteins. *Annu. Rev. Biophys. Biomol. Struct.*, *36*, 261–280. doi: 10.1146/annurev.biophys.36.040306.132808; PMID: 17477839

Berg, J.M., Tymoczko, J.L., & Stryer, L. (2012). *Biochemistry* (7th edn). New York: W.H. Freeman.

Blake, C.C.F., Koenig, D.F., Mair, G.A., North, A.C.T., Phillips, D.C., & Sarma, V.R. (1965). Structure of hen egg-white lysozyme: a three-dimensional Fourier synthesis at 2 Å resolution. *Nature*, *206*, 757–761. https://doi.org/10.1038/206757a0

Blundell, T.L., Bedarkar, S., Rinderknecht, E., & Humbel, R.E. (1978). Insulin-like growth factor: a model for tertiary structure accounting for immunoreactivity and receptor binding. *Proc. Nat. Acad. Sci., USA*, *75*, 180–184

Blundell, T.L., Sibanda, B.L., Sternberg, M.J.E., & Thornton, J.M. (1987). Knowledge-based prediction of protein structures and the design of novel molecules. *Nature*, *326*, 347–352

Bokov, K., & Steinberg, S.V. (2009). A hierarchical model for evolution of 23S ribosomal RNA. *Nature*, 457, 977–980

Bolognesi, M., Gatti, G., Menegatti, E., Guarneri, M., Marquart, M., Papamokos, E., & Huber, R. (1982). Three-dimensional structure of the complex between pancreatic secretory trypsin inhibitor (Kazal type) and trypsinogen at 1.8 Å resolution: structure solution, crystallographic refinement and preliminary structural interpretation. *Journal of Molecular Biology*, *162*(4), 839–868

Bose, T., Fridkin, G., Davidovich, C., Krupkin, M., Dinger, N., Falkovich, A H., … & Yonath, A. (2022). Origin of life: protoribosome forms peptide bonds and links RNA and protein dominated worlds. *Nucleic Acids Research*, *50*(4), 1815–1828. https://doi.org/10.1093/nar/gkac052

Bowers, P.M., Strauss, C.E., & Baker, D. (2000). De novo protein structure determination using sparse NMR data. *Journal of Biomolecular NMR*, *18*(4), 311–318

Bradley, P., Misura, K.M., & Baker, D. (2005). Toward high-resolution de novo structure prediction for small proteins. *Science*, *309*(5742), 1868–1871

Callaway, E. (2020). 'It will change everything': DeepMind's AI makes gigantic leap in solving protein structures. *Nature*, *588*(7837), 203–205

Cavalli, A., Salvatella, X., Dobson, C.M., & Vendruscolo, M. (2007). Protein structure determination from NMR chemical shifts. *Proceedings of the National Academy of Sciences*, *104*(23), 9615–9620

Chandra, K., Emwas, A.H., Al-Harthi, S., Al-Talla, Z., Hajjar, D., Makki, A.A., … & Jaremko, M. (2022). Theory and applications of NMR spectroscopy in biomolecular structures and dynamics of proteins. In NMR spectroscopy for probing functional dynamics at biological interfaces, ed. A. Bhunia, H.S. Atreya, & N. Sinha, pp. 1–28. Cambridge: Royal Society of Chemistry

Chill, J.H., Louis, J.M., Miller, C., & Bax, A. (2006). NMR study of the tetrameric KcsA potassium channel in detergent micelles. *Protein Science*, *15*(4), 684–698. doi: 10.1110/ps.051954706; PMID: 16522799; PMCID: PMC2242490

Chill, J.H., Qasim, A., Sher, I., & Gross, R. (2019). NMR perspectives of the KcsA potassium channel in the membrane environment. *Israel Journal of Chemistry*, *59*(11–12), 1001–1013

Chothia, C. (1973). Conformation of twisted β-pleated sheets in proteins. *Journal of Molecular Biology*, *75*(2), 295–302

Clore, G.M., & Gronenborn, A.M. (1998). New methods of structure refinement for macromolecular structure determination by NMR. *Proceedings of the National Academy of Sciences*, *95*(11), 5891–5898

Cornilescu, G., Delaglio, F., & Bax, A. (1999). Protein backbone angle restraints from searching a database for chemical shift and sequence homology. *Journal of Biomolecular NMR*, *13*, 289–302

Costanzo, M., Hou, J., Messier, V., Nelson, J., Rahman, M., VanderSluis, B., … & Andrews, B. (2021). Environmental robustness of the global yeast genetic interaction network. *Science*, *372*(6542), eabf8424

Costanzo, M., VanderSluis, B., Koch, E.N., Baryshnikova, A., Pons, C., Tan, G., … & Boone, C. (2016). A global genetic interaction network maps a wiring diagram of cellular function. *Science*, *353*(6306), aaf1420

Delaglio, F., Kontaxis, G., & Bax, A. (2000). Protein structure determination using molecular fragment replacement and NMR dipolar couplings. *Journal of the American Chemical Society*, *122*(9), 2142–2143

Delsuc, M.A., Vitorino, M., & Kieffer, B. (2020). Determination of protein structure and dynamics by NMR: state of the art and application to the characterization of biotherapeutics. *Structural Biology in*

Drug Discovery: Methods, Techniques, and Practices, ed. J.-P. Renaud, pp. 295–323. Hoboken, NJ: John Wiley & Sons. https://doi.org/10.1002/9781118681121.ch13

Derebe, M.G., Sauer, D.B., Zeng, W., Alam, A., Shi, N., & Jiang, Y. (2011). Tuning the ion selectivity of tetrameric cation channels by changing the number of ion binding sites. *Proceedings of the National Academy of Sciences*, *108*(2), 598–602

Donev, R., & Karabencheva-Christova, T. (eds.). (2021). *Proteomics and Systems Biology*. Advances in protein chemistry and structural biology 127. Cambridge, MA: Academic Press

Doyle, D.A., Cabral, J.M., Pfuetzner, R.A., Kuo, A., Gulbis, J.M., Cohen, S.L., … & MacKinnon, R. (1998). The structure of the potassium channel: molecular basis of K+ conduction and selectivity. *Science*, *280*(5360), 69–77. doi: 10.1126/science.280.5360.69

Drenth, J. (2007). *Principles of protein X-ray crystallography*. New York: Springer Science & Business Media. https://doi.org/10.1007/0-387-33746-6

Dunker, A.K., Babu, M., Barbar, E., Blackledge, M., Bondos, S.E., Dosztányi, Z., et al. (2013). What's in a name? Why these proteins are intrinsically disordered? *Intrinsically Disordered Proteins*, *1*, e24157

Dunker, A.K., Lawson, J.D., Brown, C.J., Williams, RM., Romero, P., Oh, J.S., … & Obradovic, Z. (2001). Intrinsically disordered protein. *Journal of Molecular Graphics and Modelling*, *19*(1), 26–59

Dunker, A.K., Silman, I., Uversky, V.N., & Sussman, J.L. (2008). Function and structure of inherently disordered proteins. *Current Opinion in Structural Biology*, *18*(6), 756–764

Eisenberg, D. (2003). The discovery of the α-helix and β-sheet, the principal structural features of proteins. *Proceedings of the National Academy of Sciences*, *100*(20), 11207–11210

Eisenmesser, E.Z., Millet, O., Labeikovsky, W., Korzhnev, D.M., Wolf-Watz, M., Bosco, D.A., … & Kern, D. (2005). Intrinsic dynamics of an enzyme underlies catalysis. *Nature*, *438*(7064), 117–121. doi: 10.1038/nature04105; PMID: 16267559

Elofsson, A. (2021). Toward characterising the cellular 3D-proteome. *Frontiers in Bioinformatics*, *1*, 598878

Floudas, C.A. (2007). Computational methods in protein structure prediction. *Biotechnology and Bioengineering*, *97*(2), 207–213

Fox, G.E. (2010). Origin and evolution of the ribosome. *Cold Spring Harbor Perspectives in Biology*, *2*(9), a003483. doi: 10.1101/cshperspect.a003483

Fox, G.E., & Naik, A.K. (2004). The evolutionary history of the translation machinery. In *The genetic code and the origin of life*, ed. L.R. de Pouplana, pp. 92–105. New York: Kluwer Academic/Plenum Publishers

Graves, P.R., & Haystead, T.A. (2002). Molecular biologist's guide to proteomics. *Microbiology and Molecular Biology Reviews*, *66*(1), 39–63

Herrington, J., & Arey, B.J. (2014). Conformational mechanisms of signaling bias of ion channels. In *Biased Signaling in Physiology, Pharmacology and Therapeutics*, ed. B.J. Arey, pp. 173–207. San Diego, CA: Academic Press

Hsiao, C., Mohan, S., Kalahar, B.K., & Williams, L.D. (2009). Peeling the onion: ribosomes are ancient molecular fossils. *Mol. Biol. Evol.*, *26*, 2415–2425

Jaskolski, M., Dauter, Z., & Wlodawer, A. (2014). A brief history of macromolecular crystallography, illustrated by a family tree and its Nobel fruits. *FEBS J.*, *281*, 3985–4009

Jumper, J., Evans, R., Pritzel, A., et al. (2021). Highly accurate protein structure prediction with AlphaFold. *Nature*, *596*, 583–589. https://doi.org/10.1038/s41586-021-03819-2

Kendrew, J.C., Bodo, G., Dintzis, H., et al. (1958). A three-dimensional model of the myoglobin molecule obtained by X-ray analysis. *Nature*, *181*, 662–666. https://doi.org/10.1038/181662a0

Korzhnev, D.M., Salvatella, X., Vendruscolo, M., Di Nardo, A.A., Davidson, A.R., Dobson, C.M., & Kay, L.E. (2004). Low-populated folding intermediates of Fyn SH3 characterized by relaxation dispersion NMR. *Nature*, *430*(6999), 586–590

Kovacs, N.A., Petrov, A.S., Lanier, K.A., & Williams, L.D. (2017). Frozen in time: the history of proteins. *Molecular Biology and Evolution*, *34*(5), 1252–1260. doi: 10.1093/molbev/msx086; PMID: 28201543; PMCID: PMC5400399

Langan, P.S., Vandavasi, V.G., Weiss, K.L., et al. (2018). Anomalous X-ray diffraction studies of ion transport in K+ channels. *Nat. Commun.*, 9, 4540. https://doi.org/10.1038/s41467-018-06957-w

Lecompte, O., Ripp, R., Thierry, J.C., Moras, D., & Poch, O. (2002). Comparative analysis of ribosomal proteins in complete genomes: an example of reductive evolution at the domain scale. *Nucleic Acids Research*, *30*, 5382–5390

Lian, L.Y., & Roberts, G. (eds.). (2011). *Protein NMR spectroscopy: practical techniques and applications.* Hoboken, NY: John Wiley & Sons. doi:10.1002/9781119972006

Majumder, P. (2020). Computational methods used in prediction of protein structure. In *Statistical modelling and machine learning principles for bioinformatics techniques, tools, and applications: algorithms for intelligent systems*, ed. K. Srinivasa, G. Siddesh, & S. Manisekhar, pp. 119–133. Singapore: Springer.

Mannige, R., Haxton, T., Proulx, C., et al. (2015). Peptoid nanosheets exhibit a new secondary-structure motif. *Nature*, *526*, 415–420. https://doi.org/10.1038/nature15363

Mannige, R.V., Kundu, J., & Whitelam, S. (2016). The Ramachandran number: an order parameter for protein geometry. *PLoS ONE*, *11*(8), e0160023. https://doi.org/10.1371/journal.pone.0160023

Meiler, J. (2003) PROSHIFT: protein chemical shift prediction using artificial neural networks. *Journal of Biomolecular NMR*, *26*(1), 25–37

Meiler, J., & Baker, D. (2003). Rapid protein fold determination using unassigned NMR data. *Proceedings of the National Academy of Sciences*, *100*(26), 15404–15409

Mirijanian, D.T., Mannige, R.V., Zuckermann, R.N. & Whitelam, S. (2014). Development and use of an atomistic CHARMM-based forcefield for peptoid simulation. *J. Comput. Chem.*, *35*, 360–370

Mitchell, A.L., Almeida, A., Beracochea, M., Boland, M., Burgin, J., Cochrane, G., ... & Finn, R.D. (2020). MGnify: the microbiome analysis resource in 2020. *Nucleic Acids Research*, *48*(D1), D570–D578

Morais-Cabral, J.H., Zhou, Y., & MacKinnon, R. (2001). Energetic optimization of ion conduction rate by the K+ selectivity filter. *Nature*, *414*(6859), 37–42

Moult, J., Fidelis, K., Kryshtafovych, A., Schwede, T., & Topf, M. (2020). Critical assessment of techniques for protein structure prediction, fourteenth round. CASP 14 Abstract Book, www.predictioncenter.org/casp14/doc/CASP14_Abstracts.pdf

Muirhead, H., & Perutz, M.F. (1963). Structure of hæemoglobin: a three-dimensional Fourier synthesis of reduced human haemoglobin at 5.5 Å resolution. *Nature*, *199*, 633–638. https://doi.org/10.1038/199633a0

Narita, M., Narita, M., Itsuno, Y., & Itsuno, S. (2019). Autonomous sequences in myoglobin emerging from X-ray structure of holomyoglobin. *ACS Omega*, *4*(1), 992–999. https://doi.org/10.1021/acsomega.8b03218

Neal, S., Nip, A.M., Zhang, H., & Wishart, D.S. (2003). Rapid and accurate calculation of protein 1H, 13C and 15N chemical shifts. *Journal of Biomolecular NMR*, *26*(3), 215–240

Pandey, A., and Mann, M. (2000). Proteomics to study genes and genomes. *Nature*, *405*, 837–846

Pauling, L., & Corey, R.B. (1951). Configurations of polypeptide chains with favored orientations around single bonds: two new pleated sheets. *Proceedings of the National Academy of Sciences*, *37*(11), 729–740

Pauling, L., Corey, R.B., & Branson, H.R. (1951). The structure of proteins: two hydrogen-bonded helical configurations of the polypeptide chain. *Proceedings of the National Academy of Sciences*, *37*(4), 205–211. doi: 10.1073/pnas.37.4.205; PMID: 14816373; PMCID: PMC1063337

Pearce, R., & Zhang, Y. (2021). Deep learning techniques have significantly impacted protein structure prediction and protein design. *Curr. Opin. Struct. Biol.*, *68*, 194–207

Perutz, M.F., Rossmann, M.G., Cullis, A..F., Muirhead, H., Will, G., & North, A.C.T. (1960). Structure of hæmoglobin: a three-dimensional Fourier synthesis at 5.5-Å. resolution, obtained by X-ray analysis. *Nature*, *185*, 416–422. https://doi.org/10.1038/185416a0

Petrov, A.S., Bernier, C.R., Hsiao, C., Norris, A.M., Kovacs, N.A., Waterbury, C.C., ... & Williams, L.D. (2014). Evolution of the ribosome at atomic resolution. *Proceedings of the National Academy of Sciences*, *111*(28), 10251–10256

Petrov, A.S., Gulen, B., Norris, A.M., Kovacs, N.A., Bernier, C.R., Lanier, K.A., ... & Williams, L.D. (2015). History of the ribosome and the origin of translation. *Proc. Natl. Acad. Sci. USA*, *112*, 15396–15401

Pray, L. (2008). Discovery of DNA structure and function: Watson and Crick. *Nature Education*, *1*(1), 100

Purslow, J.A., Khatiwada, B., Bayro, M.J., & Venditti, V. (2020). NMR methods for structural characterization of protein–protein complexes. *Frontiers in Molecular Biosciences*, *7*, 9. doi: 10.3389/fmolb.2020.00009

Ramachandran, G.N. (1962). The triple helical structure of collagen. In *Collagen: proceedings of a symposium*, ed. N. Ramanathan, p. 3. New York: John Wiley

Ramachandran, G.N., Ramakrishnan, C., & Sasisekharan, V. (1963). Stereochemistry of polypeptide chain configurations. *J. Mol. Biol.*, *7*, 95–99. doi: 10.1016/s0022-2836(63)80023-6; PMID: 13990617

Ramachandran, G.N., & Sasisekharan, V. (1961a). Structure of collagen. *Nature*, *190*, 1004–1005

Ramachandran, G.N., & Sasisekharan, V. (1961b). Structure of collagen. *Curr. Sci.*, *30*, 127–130

Ramachandran, G. N., Saaisekharan, V., & Thathaehari, Y.T. (1962). Structure of collagen. In *Collagen: proceedings of a symposium*, ed. N. Ramanathan, p. 81. New York: John Wiley

Renart, M.L., Giudici, A.M., Poveda, J.A., et al. (2019). Conformational plasticity in the KcsA potassium channel pore helix revealed by homo-FRET studies. *Sci. Rep.*, *9*, 6215. https://doi.org/10.1038/s41598-019-42405-5

Rohl, C.A., & Baker, D. (2002). De novo determination of protein backbone structure from residual dipolar couplings using Rosetta. *Journal of the American Chemical Society*, *124*(11), 2723–2729

Rule, G.S., & Hitchens, T.K. (2006). *Fundamentals of protein NMR spectroscopy* (vol. 5). Dordrecht: Springer. https://doi.org/10.1007/1-4020-3500-4

Sanders, J.K., Constable, E.C., & Hunter, B.K. (1993). *Modern NMR spectroscopy: a workbook of chemical problems* (vol. 2). Oxford: Oxford University Press

Sauer, D.B., Zeng, W., Canty, J., Lam, Y., & Jiang, Y. (2013). Sodium and potassium competition in potassium-selective and non-selective channels. *Nature Communications*, *4*(1), 2721

Schueler-Furman, O., Wang, C., Bradley, P., Misura, K., & Baker, D. (2005). Progress in modeling of protein structures and interactions. *Science*, *310*(5748), 638–642

Senior, A.W., Evans, R., Jumper, J., Kirkpatrick, J., Sifre, L., Green, T., … & Hassabis, D. (2020). Improved protein structure prediction using potentials from deep learning. *Nature*, *577*(7792), 706–710

Sheldrick, G.M. (2010). Experimental phasing with SHELXC/D/E: combining chain tracing with density modification. *Acta Crystallogr. D.*, *66*, 479–485

Simons, K.T., Kooperberg, C., Huang, E., & Baker, D. (1997). Assembly of protein tertiary structures from fragments with similar local sequences using simulated annealing and Bayesian scoring functions. *Journal of Molecular Biology*, *268*(1), 209–225

Söding, J., & Lupas, A.N. (2003). More than the sum of their parts: on the evolution of proteins from peptides. Bioessays, *25*(9), 837–846

Spang, A., Saw, J.H., Jørgensen, S.L., Zaremba-Niedzwiedzka, K., Martijn, J., Lind, A.E., … & Ettema, T.J. (2015). Complex archaea that bridge the gap between prokaryotes and eukaryotes. *Nature*, *521*(7551), 173–179

Steinegger, M., Mirdita, M., & Söding, J. (2019). Protein-level assembly increases protein sequence recovery from metagenomic samples manyfold. *Nature Methods*, *16*, 603–606

Thompson, M.C., Yeates, T.O., & Rodriguez, J.A. (2020). Advances in methods for atomic resolution macromolecular structure determination. *F1000Res.*, *9*, 667

Thorn, A., & Sheldrick, G.M. (2011). ANODE: anomalous and heavy-atom density calculation. *J. Appl. Crystallogr.*, *44*, 1285–1287

Tong, A.H.Y., Lesage, G., Bader, G.D., Ding, H., Xu, H., Xin, X., … & Chang, M. (2004). Global mapping of the yeast genetic interaction network. *Science*, *303*, 808–813

Tunyasuvunakool, K., Adler, J., Wu, Z., Green, T., Zielinski, M., Žídek, A., … & Hassabis, D. (2021). Highly accurate protein structure prediction for the human proteome. *Nature*, *596*(7873), 590–596

Wang, S., Sun, S., Li, Z., Zhang, R., & Xu, J. (2017). Accurate de novo prediction of protein contact map by ultra-deep learning model. *PLoS Comput. Biol.*, *13*, e1005324

Waseda, Y., Matsubara, E., & Shinoda, K. (2011). *X-ray diffraction crystallography: introduction, examples and solved problems*. Berlin and Heidelberg: Springer. https://doi.org/10.1007/978-3-642-16635-8

Wasinger, V.C., Cordwell, S.J., Cerpa-Poljak, A., Yan, J.X., Gooley, A.A., Wilkins, M. R., … & Humphery-Smith, I. (1995). Progress with gene-product mapping of the Mollicutes: Mycoplasma genitalium. *Electrophoresis*, *16*(1), 1090–1094

Watson, J.D., & Crick, F.H. (1953). A structure for deoxyribose nucleic acid. *Nature*, *171*, 737–738

Wilkins, M.R., Sanchez, J.C., Gooley, A.A., Appel, RD., Humphery-Smith, I., Hochstrasser, D.F., & Williams, K.L. (1996). Progress with proteome projects: why all proteins expressed by a genome should be identified and how to do it. *Biotechnology and Genetic Engineering Reviews*, *13*(1), 19–50

Williamson, M.P., Havel, T.F., & Wüthrich, K. (1985). Solution conformation of proteinase inhibitor IIA from bull seminal plasma by 1H nuclear magnetic resonance and distance geometry. *Journal of Molecular Biology*, *182*(2), 295–315. doi: 10.1016/0022-2836(85)90347-x; PMID: 3839023

Wishart, D.S., & Case, D.A. (2002). Use of chemical shifts in macromolecular structure determination. *Methods in Enzymology*, *338*, 3–34

Wishart, D.S., Sykes, B.D., & Richards, F.M. (1992). The chemical shift index: a fast and simple method for the assignment of protein secondary structure through NMR spectroscopy. *Biochemistry*, *31*(6), 1647–1651

Woese, C.R., & Fox, G.E. (1977a). The concept of cellular evolution. *Journal of Molecular Evolution*, *10*(1), 1–6

Woese, C.R., & Fox, G.E. (1977b). Phylogenetic structure of the prokaryotic domain: the primary kingdoms. *Proceedings of the National Academy of Sciences*, *74*(11), 5088–5090

Wüthrich, K. (2001). The way to NMR structures of proteins. *Nature Structural Biology*, *8*(11), 923–925

Wüthrich, K. (1991). *NMR of proteins and nucleic acids*. New York: John Wiley & Sons

Wüthrich, K. (1986). NMR with proteins and nucleic acids. *Europhysics News*, *17*(1), 11–13

wwPDB Consortium. (2018). Protein Data Bank: the single global archive for 3D macromolecular structure data. *Nucleic Acids Research*, *47*, D520–D528

Xu, X.P., & Case, D.A. (2001). Automated prediction of 15N, 13Cα, 13Cβ and 13C′ chemical shifts in proteins using a density functional database. *Journal of Biomolecular NMR*, *21*(4), 321–333

Yang, J., Anishchenko, I., Park, H., Peng, Z., Ovchinnikov, S., & Baker, D. (2020). Improved protein structure prediction using predicted interresidue orientations. *Proceedings of the National Academy of Sciences*, *117*(3), 1496–1503

Yang, Y., Gao, J., Wang, J., Heffernan, R., Hanson, J., Paliwal, K., & Zhou, Y. (2018). Sixty-five years of the long march in protein secondary structure prediction: the final stretch? *Briefings in Bioinformatics*, *19*(3), 482–494. https://doi.org/10.1093/bib/bbw129

Zhang, H., Neal S., & Wishart D.S. (2003). RefDB: a database of uniformly referenced protein chemical shifts. *J. Biomol. NMR*, *25*, 173–195

Zhang, Y., & Skolnick, J. (2004). Scoring function for automated assessment of protein structure template quality. *Proteins*, *57*, 702–710

Zheng, W., Li, Y., Zhang, C., Pearce, R., Mortuza, S.M., & Zhang, Y. (2019). Deep-learning contact-map guided protein structure prediction in CASP13. *Proteins: Structure, Function, and Bioinformatics*, *87*(12), 1149–1164

Zhou, M., & MacKinnon, R. (2004). A mutant KcsA K+ channel with altered conduction properties and selectivity filter ion distribution. *Journal of Molecular Biology*, *338*(4), 839–846

Zhou, Y., & MacKinnon, R. (2003). The occupancy of ions in the K+ selectivity filter: charge balance and coupling of ion binding to a protein conformational change underlie high conduction rates. *Journal of Molecular Biology*, *333*(5), 965–975

12 Physics Tools and Techniques in Bioimaging

Physicist Robert Hooke discovered cells in 1665 (Hooke, 1665). This was the first ever successful imaging of any biological system. Hooke improved the existing compound microscope using three lenses and a stage light, which illuminated and enlarged the specimens. He saw a piece of cork through the newly developed microscope and succeeded in discovering cells. Within less than a decade of Hooke's discovery, Antonie van Leeuwenhoek saw bacteria and protozoa through his microscopes. These two great microscopic discoveries initiated the ever-expanding field of biological imaging. This is one of a few important fields in biology which relies uniquely on utilizing the principles of light (National Geographic, 2023). The practice of the utilization of physics tools and techniques in bioimaging has been advancing fast since the microscopic discovery of cells almost four centuries ago. Medically purposeful imaging was initiated with the discovery of the X-ray in 1895 by Wilhelm Röntgen (Kaye, 1934). X-ray crystallography later helped nucleic acid DNA be imaged, which led to the discovery of its double helical structure in 1953 (Watson and Crick, 1953) and thus helped revolutionize the field of "genetics." Going beyond individual biological structural imaging, confocal microscopy emerged about a century ago for imaging fluorescently labeled specimens with significant three-dimensional (3D) structures. The confocal microscopy application in the biomedical sciences includes the imaging of the spatial distribution of macromolecules in either fixed or living cells, the automated collection of 3D data, the imaging of multiple labeled specimens, and the measurement of physiological events in living cells (Paddock and Eliceiri, 2014). Intense laser pulses are utilized to image the dynamics of biomolecules (Corkum, 2005). A recently developed method has been found capable of producing high-resolution images of individual biomolecules without requiring crystallization, which could even allow zoomed-in imaging of specific sites within the molecules (MIT News, 2017). The method "quantum interference" was utilized for achieving high-fidelity interpolation of the quantum dynamics between hardware-allowed time samplings, thus allowing high-resolution sensing (Ajoy et al., 2017). The nanoscale (nanometer (nm) resolution) magnetic resonance imaging enabled by quantum sensors appears to be a promising path toward the outstanding goal of determining the structure of single biomolecules at room temperature. Recently, the generation of the polarization-entangled two-photon state through spontaneous four-wave mixing in enhanced green fluorescent proteins was addressed (Shi et al., 2017). The quantumness, including classical and quantum correlations, of the state in the decoherent environment was also explored using a method for photonic entanglement generation, which is expected to have the potential for developing quantum spectroscopic techniques. I shall briefly address all this remarkable progress in this chapter. This enriching knowledge helps us understand gross and specific biological structures, pinpointed interaction sites, and many biophysical processes active in biological systems. I shall retain a special focus on understanding the technological developments occurring over centuries that have helped to accomplish progress in biomolecular imaging.

DOI: 10.1201/9781003287780-12

12.1 BIOMEDICAL IMAGING UTILIZING OPTICS PRINCIPLES

Biomedical optical imaging, a subdivision of optical imaging, helps us understand the anatomy, physiology, and associated functions of biological life. The biomedical optical imaging system forms an image by manipulating the excitation light and detecting the signals originating from light and tissue interactions. The sixteenth-century discovery of compound microscopes by a Dutch father and son team named Hans and Zacharias Janssen was further developed by Robert Hooke, who demonstrated the first-ever successful biological imaging by detecting cells in 1665 (Hooke, 1665). Within the subsequent decade, Antonie van Leeuwenhoek discovered bacteria and protozoa using his developed microscopes. These two optics-based discoveries were the first considerable bioimaging breakthroughs history has recorded.

Biomedical optical imaging technologies have been steadily evolving to enable faster, deeper, and higher-resolution imaging. This optical technological uplifting has led to a more comprehensive understanding of life at the macro-, micro-, and nanoscales and has helped improve clinical diagnosis and treatment (Wang and Xia, 2019). Besides basic biomedical imaging, optics-based microscopy has found a huge application in understanding physiological conditions and various associated medically relevant biosystem issues (Sampson et al., 2022; Borah and Sun, 2022). Specific configurations in various microscopic techniques, such as widefield microscopy, light-field microscopy, confocal microscopy, light-sheet microscopy, super-resolution microscopy, etc., have been found highly applicable in imaging structures in deep cellular compartments and addressing life's crucial processes.

12.1.1 Principles and Practices Used in Microscopy

Fundamental principles of light, compositions of instruments, and fixation of samples to be imaged are set in microscopy (Chen et al., 2011; Elliott, 2020). In light microscopy, illuminating light is passed through the sample as uniformly as possible over the field of view. For thicker samples, where the objective lens does not have sufficient depth of focus, light from sample planes above and below the focal plane will also be detected. The out-of-focus light will add blur to the image, reducing the resolution. In fluorescence microscopy, any dye molecules in the field of view will be stimulated, including those in out-of-focus planes. Confocal microscopy provides a means of rejecting the out-of-focus light from the detector such that it does not contribute blur to the images being collected. This technique allows for high-resolution imaging in thick tissues.

Different types of microscopes have so far been developed, but the basic imaging concept and structures can be simply illustrated in Figure 12.1. The optical microscope system consists mainly of an objective lens and eyepieces. The objective lens magnifies an object so that it can be clearly observed by the user. During the observation, the specimen is placed near the focal plane of the objective lens in the object space, and a magnified real image of the specimen is first created on the intermediate plane. The intermediate plane is located on the focal plane of the eyepiece, thus the eyepiece is working as a magnifier to further magnify the image projected on the intermediate image plane. A magnified, virtual, inverted image is finally provided.

12.1.2 Current Status of Microscopy

I wish to provide a few example images to let us understand the modern status of imaging. As an example case, I consider confocal microscopy (see Figure 12.2), in which the illumination and detection optics are focused on the same diffraction-limited spot in the sample, which is the only spot imaged by the detector during a confocal scan. For generating a complete image, the spot must be moved over the sample, and data collected point by point. An advantage of the confocal microscope is the optical sectioning provided, which allows for the 3D reconstruction of a sample from

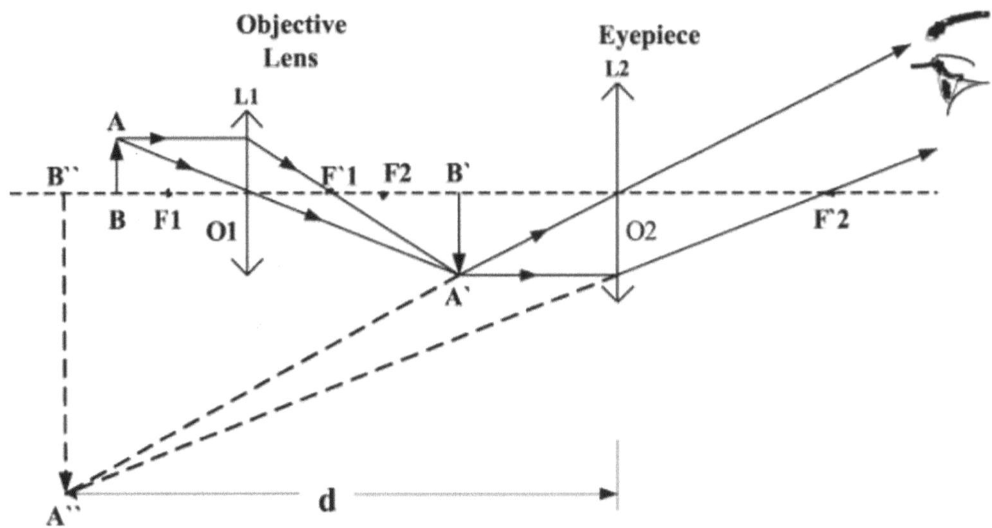

FIGURE 12.1 The optical principle of microscope imaging.

Source: Chen et al. (2011).

high-resolution stacks of images. Elliott and colleagues (2020) have provided a concise introduction to several types of confocal microscopy.

A confocal microscope primarily produces a point source of light and rejects out-of-focus light, providing the ability to image deep into tissues with high resolution, and optical sectioning for 3D reconstructions of imaged samples. The idea of rejecting out-of-focus light was first discovered by Marvin Minsky almost seven decades ago (Minsky, 1961). This was achieved by the use of illumination- and detection-side pinhole apertures in the same conjugate image plane, making them "confocal." In the configuration, one pinhole was placed in front of a zirconium arc light source for providing a point of light focused on the sample by an objective lens. The second objective lens was used to focus the illuminated sample point onto the second pinhole in front of the detector. The dual focusing process rejects the out-of-focus rays from the illuminated sample, thus avoiding reaching out to the detector, which was a low-noise photomultiplier. The stage could be moved on two dimensions (x, y plane) for scanning the sample through the illumination point to build the resulting image. As presented in Figure 12.2, we see a schematic of the core optics in a modern confocal microscope, many of which remain similar to the original Minsky design (Minsky, 1961).

For making a simple comparison, e.g. between two popular microscopy techniques, widefield fluorescence, and confocal microscopy, it may be mentioned that the lateral resolution of the latter (confocal microscope) is improved over a conventional widefield fluorescence microscope when the pinholes are closed to the minimum size providing a diffraction-limited imaging system. The resolution that can best be obtained is generally ~ 0.2 μm laterally and ~ 0.6 μm axially. Despite the pinholes, the axial resolution in both confocal and widefield microscopes is still worse than the lateral resolution. Confocal microscopy, however, is claimed to have achieved imaging resolution improvement over widefield imaging (see Figure 12.3).

Continued technological innovations since Hooke's first demonstration of cell imaging have accumulated huge advancements in the field of imaging. Hooke could detect micrometer (μm) resolution cells using his compound microscope (Hooke, 1665). Hooke's cork was an alveolar material composed of dead and empty closed cells corresponding to cork tissue including lenticels, having

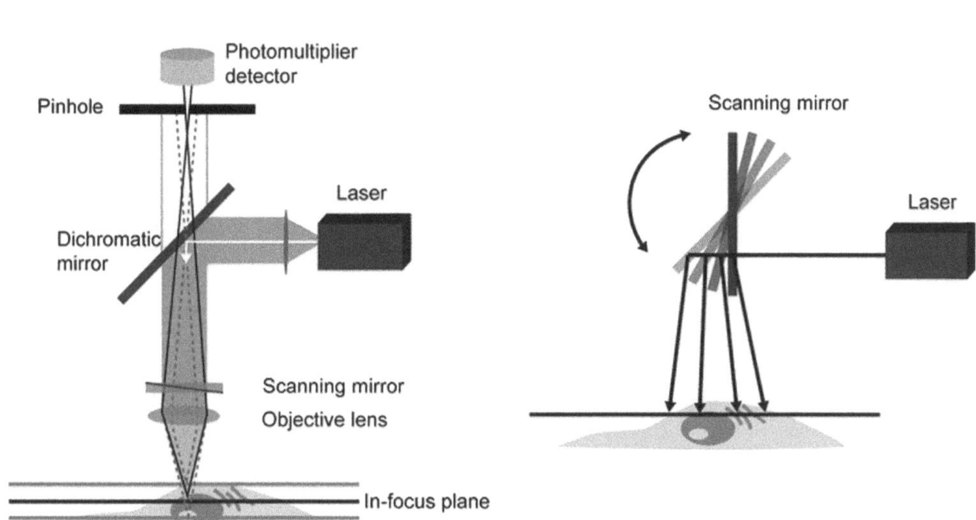

FIGURE 12.2 Components of a confocal microscope. (A) Light from a laser source is passed through collimating optics to a variable dichromatic mirror or AOBS and reflected to the objective lens which focuses the beam on a point in the sample. Scanning mirrors sweep the excitation beam over the sample point by point to build the image. Emitted fluorescence passes back through the objective lens, the dichromatic mirror or AOBS, and is detected by the PMT(s). A pinhole placed in the conjugate image plane to the focal point in the sample serves to reject out-of-focus light, which is not picked up by the detector. In this epifluorescence configuration, the illumination and emission light both pass through the same lens, thus requiring only the detector-side pinhole. Varying the size of the pinhole changes the amount of light collected and the optical section thickness. Spectral imaging can be achieved with an array of PMTs and a diffraction grating, or prism, placed in the emission light path. (B) A schematic of the scanning mirrors employed by confocal microscopes to sweep the excitation light across the sample. Copied from ref. Elliott (2020).

Source: **Elliott (2020).**

open macroporous channels. A finer description was reported by Gibson and colleagues (1981) using scanning electron microscopy (SEM). Hooke reported the 2D orientation of cork cells while Gibson highlighted their 3D structure. The structure of phellem cells (cork tissue is referred to as the phellem in Crouvisier-Urion et al., 2019, from which some of the information mentioned here is taken) varies according to the observation plane (see Figure 12.4f). In cork industries, cork stoppers are generally punched out from the tree bark perpendicularly to lenticels (see Fig. 12.4a, b, d). The phellem is sprinkled with lenticular phellogen leading to the formation of lenticels (Lendzian, 2006; Pereira, 2007; see Figure 12.4c).

Now we can even image nanometer (nm) resolution cellular constituents, such as plasma membrane proteins (Mateos-Gil et al., 2016). Figure 12.5 shows how imaging resolution has become enhanced over the last four centuries. We have achieved remarkable progress by gaining three orders of magnitude in terms of the resolution of light-utilized imaging. Although Figure 12.5 presents a historical progression in microscopy imaging and associated enhancement in obtaining the imaging resolution, it hasn't added the quantum mechanical imaging stories that have recently emerged (see Chapter 15 of this book, and Taylor, 2015; Schnell, 2019; PhysicsWorld, 2021).

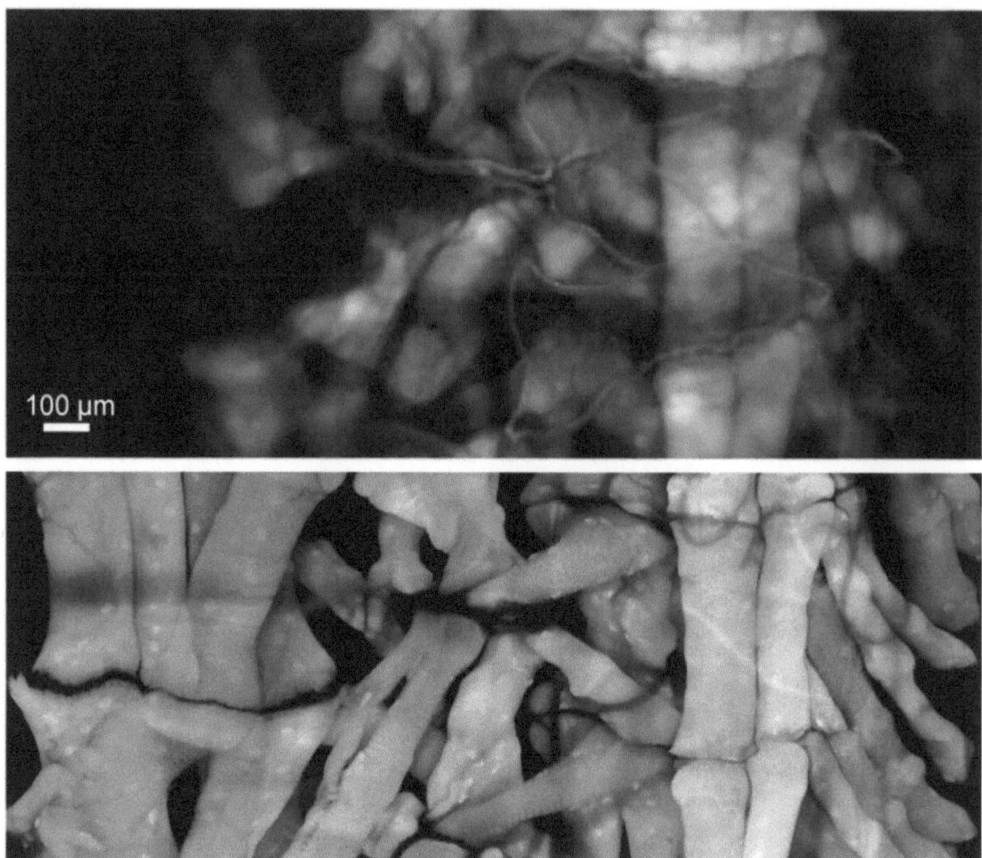

FIGURE 12.3 Widefield vs confocal microscopy. One hemisegment of a Drosophila larval fillet stained with AlexaFluor 647-conjugated phalloidin to label the musculature. In the widefield image (top), data were collected on a widefield epifluorescence microscope. The confocal image was taken with the pinhole set to 1 Airy Unit. Both images were collected with 20x objective lenses. The confocal image required ~2 hours to build in a point scanning system and the widefield image was collected with an integration time of 1 second.

Source: **Elliott (2020).**

12.2 DISCOVERY AND APPLICATIONS OF X-RAYS REVOLUTIONIZED BIOMEDICAL IMAGING, LEADING TO UNDERSTANDING OF BIOMOLECULES' ROLES IN LIFE

The first X-ray image showed us the fingers and ring of the wife of the discoverer Wilhelm Röntgen (see Figure 12.6). This great discovery brought him the first-ever Nobel Prize in physics in 1901. Wilhelm Röntgen discovered X-rays in 1895.

Following the great discovery of X-rays by Röntgen, scientists started attempting to image biomolecules using the power of X-rays. The first ever success appeared remarkably in the case of understanding the DNA molecule (see Figure 12.7), which led to the discovery of DNA's double helical structure (Watson and Crick, 1953). Within just five decades due to technological advancements we went from the bone dimension ~millimeter (mm) to a nm resolution DNA structure. Today we can even X-ray image the nucleotides of DNA and RNA (Zubavichus et al., 2008). Chemical

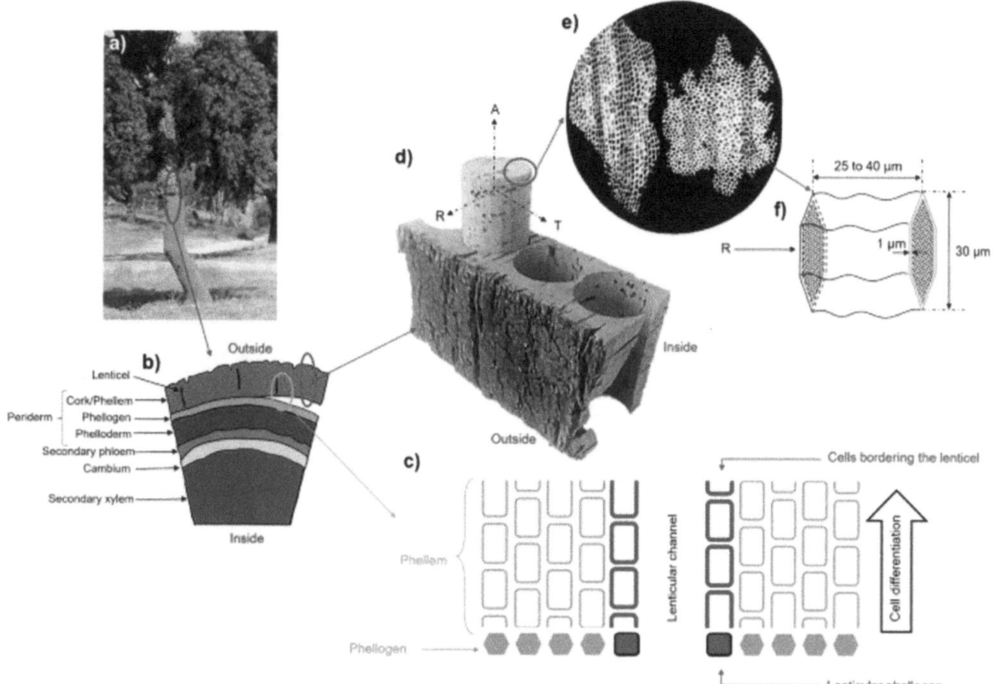

FIGURE 12.4 (a) *Quercus suber L.* tree after cork bark harvesting. (b) Representation of the transverse section of cork tree. (c) Zoom on the phellogen region with cellular differentiation. (d) Tubbing of cork stopper from the cork bark. Letters A, R, and T refer to the axial, radial, and tangential directions, respectively. (e) First observation of cork cells by Robert Hooke in 1665. (f) Characteristic shape and dimensions of a phellem cell. Copied from ref. Crouvisier-Urion et al. (2019.

***Source*: Crouvisier-Urion et al. (2019).**

effects related to the molecular structures of these heterocyclic compounds with extended π-electron systems are considered here.

Following the achievement of the X-ray crystallographic image of DNA (Figure 12.7), X-ray crystallography appeared as an important technique for the general structure determination of proteins and other biological macromolecules. The extension of this technique to systems such as viruses, immune complexes, and protein–nucleic acid complexes widens the appeal of crystallography. Structure-based drug design, site-directed mutagenesis, elucidation of enzyme mechanisms, and specificity of protein–ligand interactions are a few of the areas where X-ray crystallography has provided clarification (Smyth and Martin, 2000).

Knowledge of how to obtain phase information remains a critical challenge for the determination of nucleic acid structures. An X-ray synchrotron beamline designed to be tunable to long wavelengths at Diamond Light Source has been introduced, with the possibility of determining native de novo structures using intrinsic scattering elements. This helps overcome the limitations of introducing modifying nucleotides, often required to derive phasing information. Zhang and colleagues have recently generated new tools for nucleic acid structure determinations (for details on the methodology, see Zhang et al., 2020), and reported on the use of (i) native intrinsic potassium single-wavelength anomalous dispersion methods (K-SAD), (ii) use of anomalous scattering elements integral to the crystallization buffer (extrinsic cobalt and intrinsic potassium ions), (iii) extrinsic

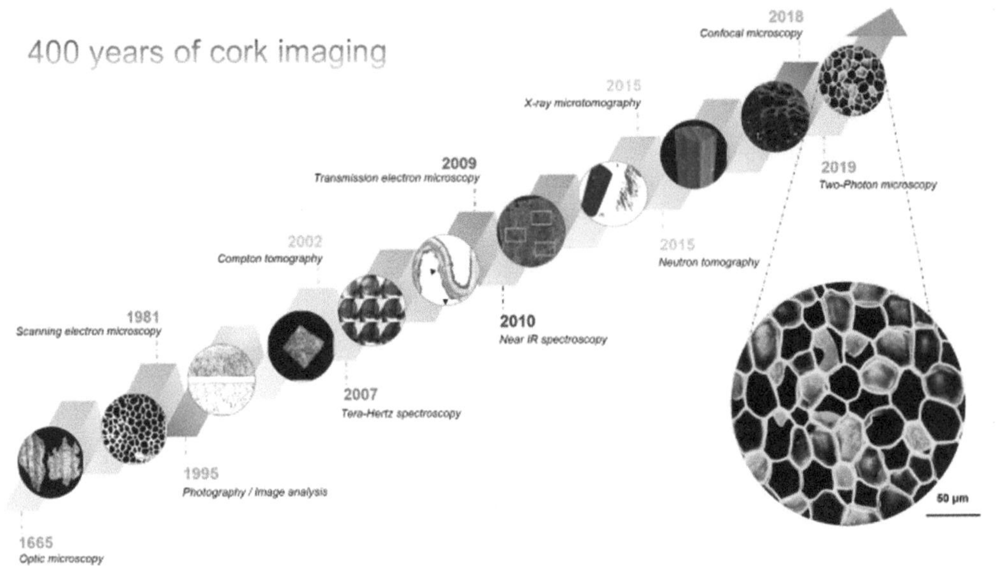

FIGURE 12.5 Overview of 400 years of cork imaging.

Source: **Crouvisier-Urion et al. (2019).**

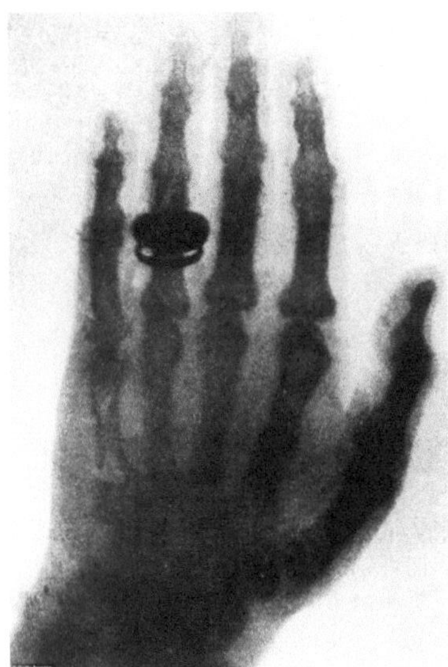

FIGURE 12.6 The X-ray image of Anna Bertha Röntgen's left hand and wedding ring.

Source: **Enoch (2023).**

bromine and intrinsic phosphorus SAD to solve complex nucleic acid structures. Using the reported methods in Zhang et al. (2020), the following structures were solved:

1. Pseudorabies virus (PRV) RNA G-quadruplex and ligand complex
2. PRV DNA G-quadruplex
3. an i-motif of human telomeric sequence

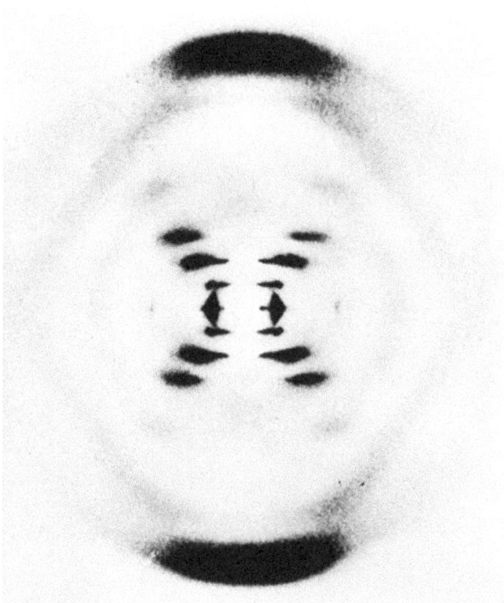

FIGURE 12.7 X-ray cryatallographic image of DNA, taken by Rosalind Franklin and Raymond Gosling in 1952.

Source: **www.pbs.org/wnet/americanmasters/american-masters-decoding-watson-james-watson-on-x-ray-crystallographer-rosalind-franklin/10930 (accessed November 22, 2022).**

The results highlight the utility of using intrinsic scattering as a pathway for solving and determining non-canonical nucleic acid motifs and revealing the variability of topology, influence of ligand binding, and glycosidic angle rearrangements seen between RNA and DNA G-quadruplexes of the same sequence.

The utility of using long-wavelength crystallography based on intrinsic scattering elements for experimental phasing and de novo structure determination of DNA and RNA G4s, along with its ability to unambiguously assign metal ions in the lattice, has been confirmed in Zhang et al. (2020). Access to long-wavelength X-ray diffraction is found to provide opportunities for tuning the wavelength for identifying lighter anomalous scatterers for the characterization of structures containing mixed ions.

Regarding the role(s) of DNA in biological life, the mechanism of DNA synthesis requires upfront understanding. Recently, Chim and colleagues (2021) reported having followed the reaction pathway of a replicative DNA polymerase with the use of time-resolved X-ray crystallography to elucidate the order and transition between intermediates. The structural changes observed in the time-lapsed images are found to reveal a catalytic cycle in which translocation precedes catalysis. The translocation step follows a push-pull mechanism where the O-O1 loop of the finger subdomain acts as a pawl to facilitate unidirectional movement along the template with conserved tyrosine residues 714 and 719 functioning as tandem gatekeepers of DNA synthesis. The details, presented in Figures 12.8 and 12.9, enhance our understanding of the nucleic acid synthesis level. The structures are found to capture the precise order of critical events that may be a general feature of enzymatic catalysis among replicative DNA polymerases.

12.3 TECHNIQUES AND TOOLS ASSOCIATED WITH IMAGING LARGE BIOLOGICAL SYSTEMS INCLUDING TISSUES

Versatile system biology imaging techniques and their utility are available now in applied biomedical research (Kherlopian et al., 2008). During over three decades systems biology has matured

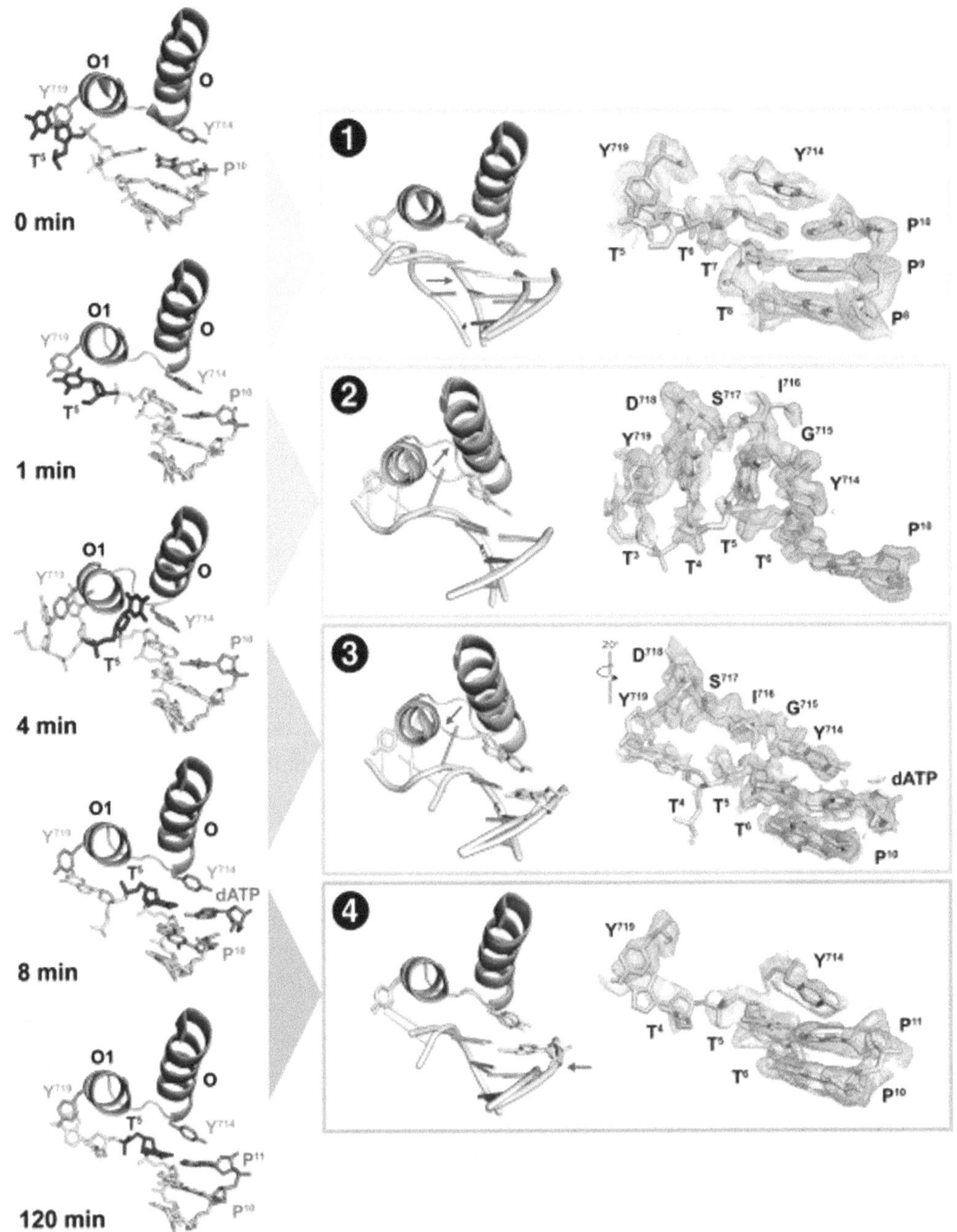

FIGURE 12.8 Time-ordered images capture the initiation step of DNA synthesis. X-ray crystal structures of polymerase intermediates observed between 0–120 min (left column). Cartoon overlays and polder maps contoured at 2–4 σ (blue boxes). Red arrow indicates conformational changes between structures. The transitions are labeled: (1) movement of the DNA duplex, (2) opening of the O-O1 loop to form a hydrophobic pocket, (3) closing of the O-O1 loop, and (4) chemical bond formation. Color scheme: 5' templating base (red), Y714 and Y719 (purple), 3' nucleotide of DNA primer (orange), dATP (blue), earlier reaction time (gray), and later reaction time (yellow). Abbreviations: O (O helix), O1 (O1 helix), T (template), P (primer), Y (tyrosine), D (aspartate), S (serine), I (isoleucine), and G (glycine). The color contrast in the figure here and all subsequent ones will be clear in the online version of the book.

Source: **Chim et al. (2021).**

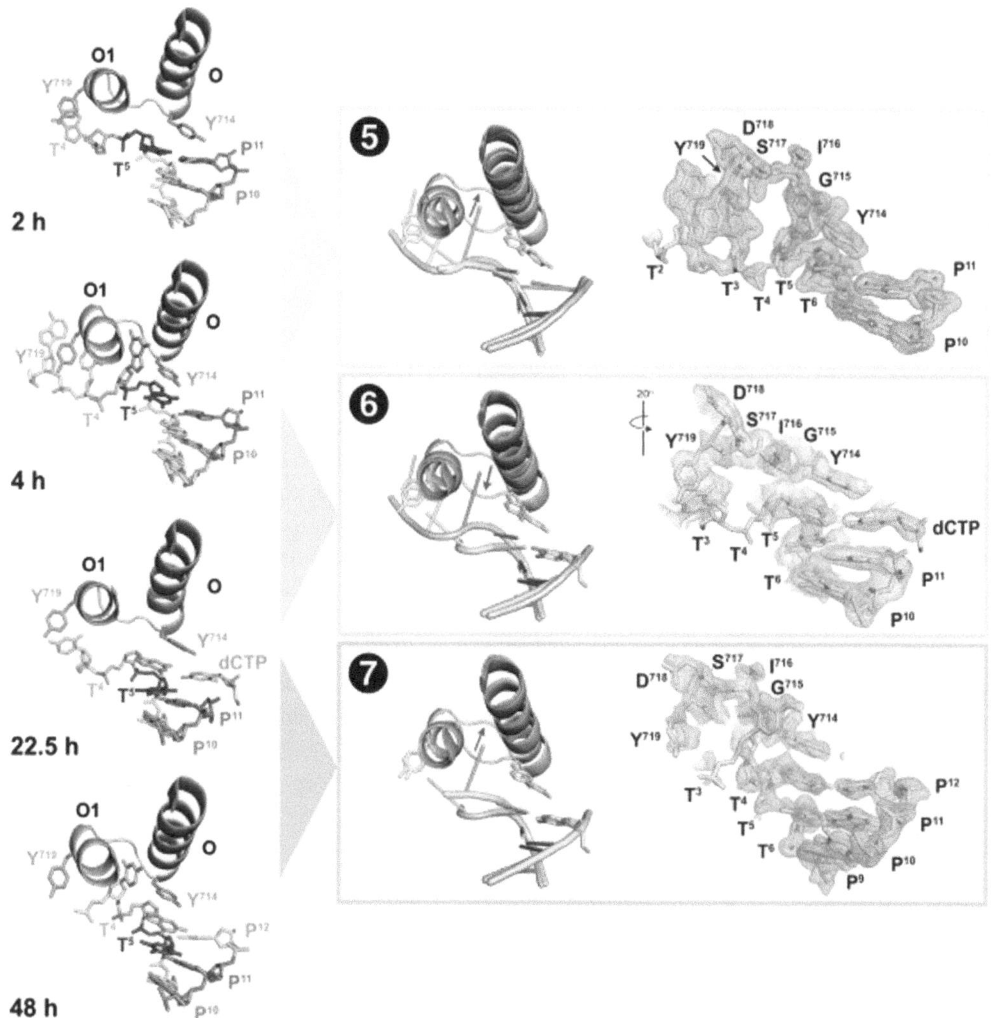

FIGURE 12.9 Time-lapsed images capture the elongation step of DNA synthesis. X-ray crystal structures of polymerase intermediates observed between 2–48 h (left column). Cartoon overlays and polder maps contoured at 2–4 σ (blue boxes). Red arrow indicates conformational changes between structures. The transitions are labeled: (5) opening of the O-O1 loop to form a hydrophobic pocket, (6) closing of the O-O1 loop, and (7) reopening of the O-O1 loop following chemical bond formation. Color scheme: 5' templating base (cyan), Y714 and Y719 (purple), 3' nucleotide of DNA primer (blue, dCTP (green), earlier reaction time (gray), and later reaction time (yellow). Abbreviations: O (O helix), O1 (O1 helix), T (template), P (primer), Y (tyrosine), D (aspartate), S (serine), I (isoleucine), and G (glycine).

Source: **Chim et al. (2021).**

into a distinct field and imaging has been increasingly used to enable the interplay of experimental and theoretical biology. The roles of microscopy, ultrasound, computed tomography (CT), magnetic resonance imaging (MRI), positron emission tomography (PET), and molecular probes such as quantum dots and nanoshells in systems biology have proven to be enormous (Megason and Fraser, 2007; Kherlopian et al., 2008). Imaging-based information has proved highly applicable in developing therapies (Chen et al., 2014; Hu et al., 2014; Chen et al., 2019).

The next generation of imaging tools includes innovative microscopy methods, ultrasound, CT, MRI, and PET. The improvements in temporal sampling and spatial resolution will certainly continue. With the advent of molecular probes, imaging can be conducted for visualizing both gross anatomical structures and substructures of cells and for monitoring molecule dynamics. The imaging modalities of microCT, microMRI, fMRI, MRS, and microPET are expected to play important roles. Table 12.1 presents a comparison of these imaging technologies. Each imaging technique is profiled with its respective underlying principle, a description of selected current applications, and a discussion of advantages and known limitations.

Systems biology approaches involve determining the structure of biological circuits using genomewide analyses. Imaging is known to offer the unique advantage of monitoring biological circuits function over time at single-cell resolution in the intact animal. The power of integrating imaging tools with more conventional genomic approaches has been detailed to analyze the biological circuits of microorganisms, plants, and animals (Megason and Fraser, 2007). Conventional genomic approaches excel at the first goal of systems biology: characterizing the structure of biological networks. The systems-level analysis works best when all the components and interactions associated with the biological circuitry are defined clearly. For this purpose, comprehensive approaches including sequencing, microarrays, and interactomic approaches are found ideal. As the network structure has been elucidated, how it functions as a circuit may be investigated. There are four important considerations about how biological circuits function that illustrate the advantages of imaging for systems biology (see Figure 12.10).

For understanding biological systems, including both molecular-level active and large systems such as cells and tissue-level active processes and principles, many imaging techniques have so far been developed. In 2004 Partain summarized Nobel Prize-winning discoveries on just one highly

TABLE 12.1
Comparison of Imaging Technology for Systems Biology

Imaging technique	Spatial resolution	Scan time	Contrast agents and molecular probes	Key use
Multi-photon microscopy	15 – 1000 nm	Secs	Fluorescent proteins, dyes, rhodamine amide, quantum dots	Visualization of cell structures
Atomic force microscopy	10 – 20 nm	Mins	Intermolecular forces	Mapping cell surface
Electron microscopy	~5 nm	Secs	Cyrofixation	Discerning protein structure
Ultrasound	50 µm	Secs	Microbubbles, nanoparticles	Vascular imaging
CT/microCT	12 – 50 µm	Mins	Iodine	Lung and bone tumor imaging
MRI/microMRI	4 – 100 µm	Mins – Hrs	Gadolinium, dysprosium, iron oxide particles	Anatomical imaging
fMRI	~1 mm	Secs – Mins	Oxygenated hemoglobin (HbO_2) deoxygenated hemoglobin (Hb)	Functional imaging of brain activity
MRS	~2 mm	Secs	N-acetylaspartate (NAA), creatine, choline, citrate	Detection of metabolites
PET/microPET	1 – 2 mm	Mins	Fluorodeoxyglucose (FDG), ^{18}F, ^{11}C, ^{15}O	Metabolic imaging

Source: Kherlopian et al. (2008).

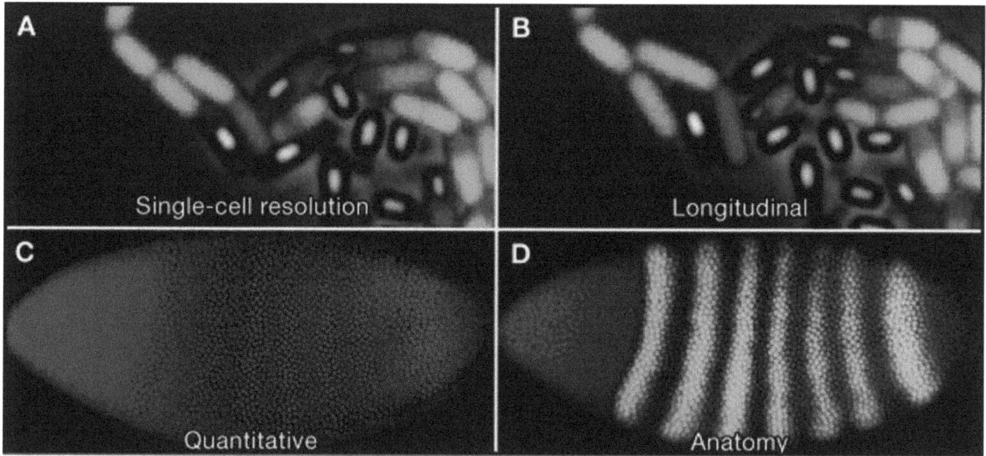

FIGURE 12.10 Model of how biological circuits function. (A) Biological circuits function at single-cell levels and thus require single-cell resolution for their analysis. Genetically identical cells grown under the same conditions, such as B. subtillis in this microcolony, can acquire different phenotypes, such as vegetative growth (green cells) or sporulation (small, light, refractile cells). (B) Biological circuits function on a time-dependent basis and thus require longitudinal analysis such as time-lapse imaging. The same microcolony as in (A) is shown 2 hr later in a time-lapse movie. The circuit for competence (as indicated with a red fluorescent reporter) has been activated in the central cell. (C) The concentration of components in biological circuits is important for their function and thus requires a technique that is quantitative (at protein levels). Shown is a fly embryo with opposing gradients of Bicoid (blue) and Caudal (red) protein. Bicoid and Caudal are morphogens known to activate circuits differentially as a function of their concentration. (D) The function of biological circuits may vary across the space of an organism, meaning that their analysis should preserve the anatomical context of the data. The imaging may provide data with high spatial resolution from anatomically intact systems. This is the same embryo as in (C) but stained to reveal expression of Even-skipped, whose seven precise stripes of expression result from the spatial distribution of morphogens.

***Source*: (A) and (B) reproduced with permission from Süel et al. (2006), (C) and (D) from Poustelnikova et al. (2004).**

utilized technique, magnetic resonance (MR), which has enormous applications in medical research, and reported that seven MR technique discoveries have already been recognized with the highest scientific prize (Partain, 2004). Similarly, other areas of imaging techniques have excelled in many physical, biological, and medical areas of discovery (Kherlopian et al., 2008; Volpe et al., 2018a). Volpe and colleagues (2018a) recently reviewed integration of medical and optical imaging for tracking multi-modal multi-scale *in vivo* cells. They rigorously describe which imaging modalities have been successfully used for *in vivo* cell tracking and how this challenging task has benefited from combining macroscopic with microscopic techniques. Figures 12.11 and 12.12 show combinations of a few demonstrations and comparisons thereof.

12.4 CONCLUDING REMARKS

The imaging field is growing every day through a combination of various unitary techniques. This is an ever-expanding physics tool and technique-based field of biological imaging. This area of innovative research is expected to uncover new aspects of life over time as technology experiences breakthrough advances. Besides, like many fields of research following the modern trend, consideration

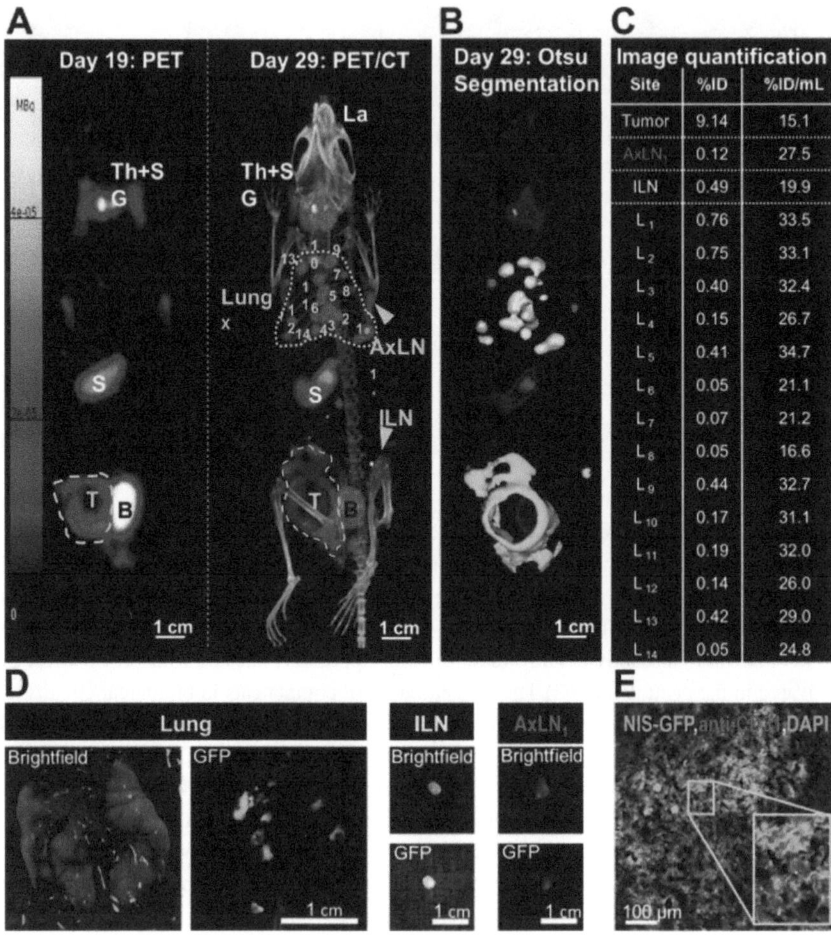

Site	%ID	%ID/mL
Tumor	9.14	15.1
AxLN1	0.12	27.5
ILN	0.49	19.9
L_1	0.76	33.5
L_2	0.75	33.1
L_3	0.40	32.4
L_4	0.15	26.7
L_5	0.41	34.7
L_6	0.05	21.1
L_7	0.07	21.2
L_8	0.05	16.6
L_9	0.44	32.7
L_{10}	0.17	31.1
L_{11}	0.19	32.0
L_{12}	0.14	26.0
L_{13}	0.42	29.0
L_{14}	0.05	24.8

FIGURE 12.11 Dual-mode radionuclide-fluorescence metastasis tracking is quantitative and provides data across multiple length scales. Representative results of metastasis tracking in a murine model of inflammatory breast cancer using the radionuclide-fluorescence fusion reporter NIS-GFP are shown. NIS served as an *in vivo* reporter and was imaged by PET/CT using the NIS tracer [^{18}F]BF_4^-. (A/left) On day 19 post tumor inoculation, the primary tumor (yellow dashed line) was clearly identified but no metastasis. It is noteworthy that endogenous NIS signals (white descriptors) were also recorded, i.e. the thyroid and salivary glands (Th + SG), the stomach (S), and, at very low levels, some parts of the mammary and lachrymal glands. Neither of these endogenous signals interfered with sites of expected metastasis in this tumor model. The bladder (B) signal stems from tracer excretion. (A/right) On day 29 post tumor inoculation, metastases were clearly identified in the lung (yellow dotted line; numbered individual metastases) and in some lymph nodes (inguinal (ILN), axillary (AxLN); yellow arrowheads). The primary tumor (yellow dashed line) had also invaded the peritoneal wall. Images presented are maximum intensity projections (MIP). (B) A 3D implementation of the Otsu thresholding technique enabled 3D surface rendering of cancerous tissues; these are superimposed onto a PET MIP. Lung metastases are shown in white, metastatic axillary lymph nodes in red, the metastatic inguinal lymph node in yellow, and the primary tumor that invaded the peritoneal wall in turquoise. (C) Radiotracer uptake into cancerous tissues was quantified from 3D images (% injected dose (ID)) and normalized by the corresponding volumes (% ID/mL). Individual lung metastases correspond to the numbers in (A). (D) NIS-GFP's fluorescence properties guided animal dissection. As exemplars brightfield and fluorescence images of the lung with several metastatic lesions and two positive lymph nodes are shown. (E) Immunofluorescence histology of the primary tumor. NIS-GFP expressing cancer cells were directly identified without the need for antibody staining. Blood vessels were stained with a rabbit antibody against mouse PECAM-1/CD31 and for nuclei (DAPI) before being imaged by confocal fluorescence microscopy. Data demonstrated vascularization heterogeneity of the primary tumor. The image also shows that the NIS-GFP

FIGURE 12.11 (*Continued*)

reporter predominantly resides in the plasma membranes of the tumor cells, demonstrating its correct localization to be functional *in vivo* and enabling tumor cell segmentation. (For interpretation of the references to color in this figure legend, the reader is referred to the web version of the source).

***Source*: reproduced with permission and minor rearrangements from Volpe et al. (2018b).**

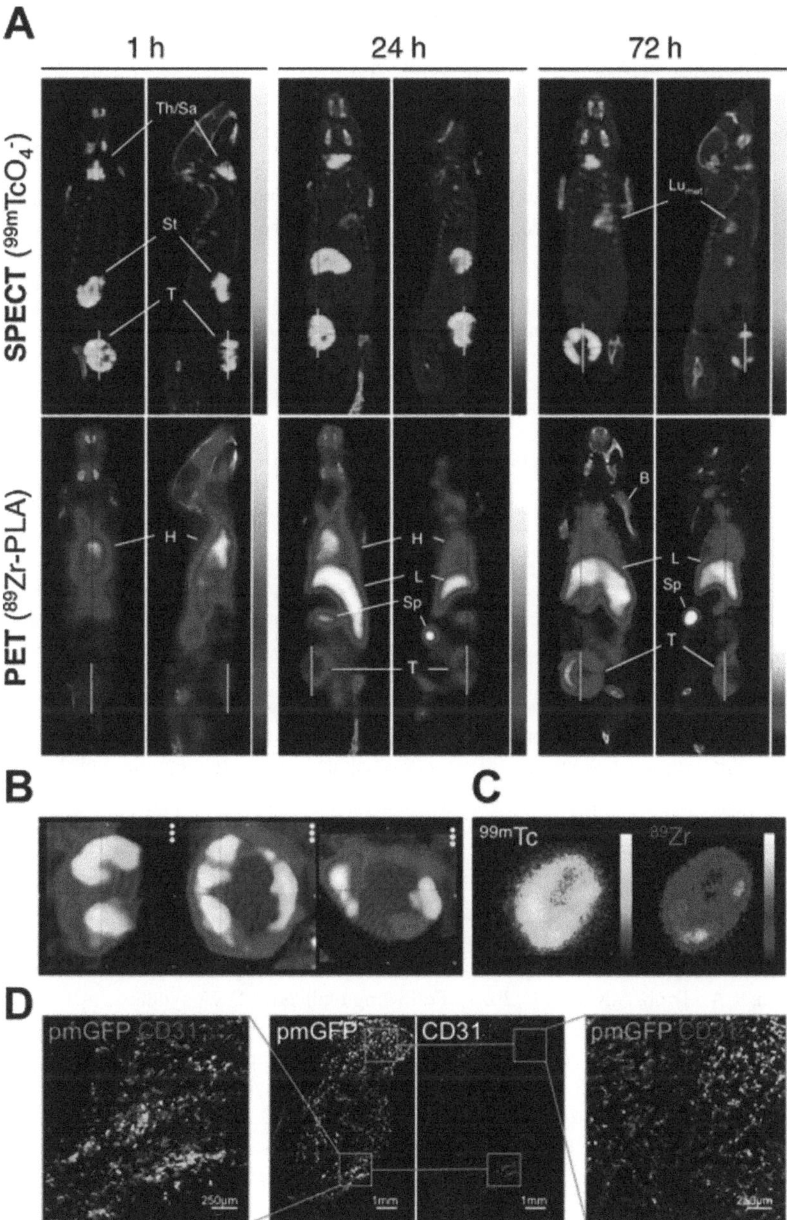

FIGURE 12.12 Tracking a nanomedicine to primary and secondary cancer lesions. Liposomal alendronate was radiolabeled with the PET isotope ^{89}Zr (^{89}Zr-PLA) and administered to animals bearing primary breast tumors that had already spontaneously metastasized (as determined by $^{99m}TcO4^-$-afforded NIS-SPECT/CT).

FIGURE 12.12 *(Continued)*

(A) Coronal and sagittal SPECT-CT (top; cancer cells) and PET-CT (bottom; nanomedicine) images centered at the tumors of the same animal are shown at indicated time points after intravenous administration of ^{89}Zr-PLA. SPECT-CT images show identical biodistribution over time with high uptake in endogenous NIS-expressing organs (stomach, thyroid) and NIS-FP-expressing cancer cells in the primary tumor (T) and metastases (LN_{met} and Lu_{met}). PET-CT images show the increasing uptake of ^{89}Zr-PLA over time in the primary tumor (T), spleen (Sp), liver (L), and bone (B) and decreasing amounts in the blood pool/heart (H). For corresponding time–activity curves, refer to Edmonds et al. (2016). (B) Co-registered SPECT/PET/CT images of the primary tumor (from left to right: sagittal, coronal, transverse) showing a high degree of colocalization but also intratumoral heterogeneity of ^{89}Zr-PLA (purple scale); $^{99m}TcO_4^-$NIS signals (green scale) show live cancer cells. (C) Autoradiography images (left, ^{99m}Tc; right, ^{89}Zr) of a coronal slice from the same tumor as in (B) showing a high degree of colocalization and heterogeneity. (D) Fluorescence microscopy of an adjacent slice of the same tumour as in (B/C) showing areas of high and low microvascular density (determined by anti-CD31 staining). (For interpretation of the references to color in this figure legend, the reader is referred to the web version of the source).

***Source*: reproduced with permission and minor rearrangements from Volpe et al. (2018b), which was reproduced with permission and minor rearrangements from Edmonds et al. (2016).**

of machine-learning algorithms is expected to help create additional advancements in biomolecular imaging (Yin et al., 2021).

REFERENCES

Ajoy, A., Liu, Y.X., Saha, K., Marseglia, L., Jaskula, J.C., Bissbort, U., & Cappellaro, P. (2017). Quantum interpolation for high-resolution sensing. *Proceedings of the National Academy of Sciences*, *114*(9), 2149–2153

Borah, B.J., & Sun, C.K. (2022). A rapid denoised contrast enhancement method digitally mimicking an adaptive illumination in submicron-resolution neuronal imaging. *Iscience*, *25*(2), 103773

Chen, L., Zhou, S.F., Su, L., & Song, J. (2019). Gas-mediated cancer bioimaging and therapy. *ACS Nano*, *13*(10), 10887–10917

Chen, X., Zheng, B., & Liu, H. (2011). Optical and digital microscopic imaging techniques and applications in pathology. *Anal. Cell Pathol.*, *34*(1–2), 5–18. doi: 10.3233/ACP-2011-0006; PMID: 21483100; PMCID: PMC3310926

Chen, Z.Y., Wang, Y.X., Lin, Y., Zhang, J.S., Yang, F., Zhou, Q.L., & Liao, Y.Y. (2014). Advance of molecular imaging technology and targeted imaging agent in imaging and therapy. *BioMed Research International*, 2014, 819324

Chim, N., Meza, R.A., Trinh, A.M., et al. (2021). Following replicative DNA synthesis by time-resolved X-ray crystallography. *Nat. Commun.*, *12*, 2641. https://doi.org/10.1038/s41467-021-22937-z

Corkum, P.B. (2005). Dynamic molecular imaging. In Ultrafast Phenomena XIV, ed. T. Kobayashi, T. Okada, T. Kobayashi, K.A. Nelson, & S. De Silvestri, pp. 139–143. Springer Series in Chemical Physics 79. Berlin and Heidelberg: Springer. https://doi.org/10.1007/3-540-27213-5_44

Crouvisier-Urion, K., Chanut, J., Lagorce, A., et al. (2019). Four hundred years of cork imaging: new advances in the characterization of the cork structure. *Sci Rep.*, *9*, 19682. https://doi.org/10.1038/s41598-019-55193-9

Edmonds, S., Volpe, A., Shmeeda, H., Parente-Pereira, A.C., Radia, R., Baguña-Torres, J., … & TM de Rosales, R. (2016). Exploiting the metal-chelating properties of the drug cargo for *in vivo* positron emission tomography imaging of liposomal nanomedicines. *ACS Nano*, *10*(11), 10294–10307

Elliott, A.D. (2020). Confocal microscopy: principles and modern practices. *Curr. Protoc. Cytom.*, *92*(1), e68. doi: 10.1002/cpcy.68; PMID: 31876974; PMCID: PMC6961134

Enoch, N. (2019). The doctor will see through you now. December 29, 2019, www.dailymail.co.uk/news/article-6491287/Roentgens-human-X-ray-wifes-hand-1895.html (accessed April 6, 2023)

Gibson, L.J., Easterling, K.E., & Ashby, M.F. (1981). The structure and mechanics of cork. *Proc. R. Soc. A*, *377*, 99–117

Hooke, R. (1665). *Micrographia: or some physiological descriptions of minute bodies made by magnifying glasses. With observations and inquiries thereupon.* London: Royal Society

Hu, F., Huang, Y., Zhang, G., Zhao, R., Yang, H., & Zhang, D. (2014). Targeted bioimaging and photodynamic therapy of cancer cells with an activatable red fluorescent bioprobe. *Analytical Chemistry*, *86*(15), 7987–7995

Kaye, G.W.C. (1934). Wilhelm Conrad Röntgen and the early history of the roentgen rays. Yale J. Biol. Med., *6*(4), 482. https://doi.org/10.1038/133511a0

Kherlopian, A.R., Song, T., Duan, Q., Neimark, M.A., Po, M.J., Gohagan, J.K., & Laine, A.F. (2008). A review of imaging techniques for systems biology. *BMC Systems Biology*, *2*(1), 1–18

Lendzian, K.J. (2006). Survival strategies of plants during secondary growth: barrier properties of phellemes and lenticels towards water, oxygen, and carbon dioxide. *J. Expe. Biol.*, *57*, 2535–2546

Mateos-Gil, P., Letschert, S., Doose, S., & Sauer, M. (2016). Super-resolution imaging of plasma membrane proteins with click chemistry. *Frontiers in Cell and Developmental Biology*, *4*, 98

Megason, S.G., & Fraser, S.E. (2007). Imaging in systems biology. *Cell*, *130*(5), 784–795

Minsky, M. (1961). U.S. Patent No. 30,130,467. Washington, DC: US Patent and Trademark Office

MIT News. (2017). High-res biomolecule imaging. February 14, 2017, https://news.mit.edu/2017/high-res-biomolecule-imaging-single-molecules-0214 (accessed on April 1, 2023)

National Geographic. (2023). History of the cell: discovering the cell. https://education.nationalgeographic.org/resource/history-cell-discovering-cell (accessed on April 5, 2023)

Paddock, S.W., & Eliceiri, K.W. (2014). Laser scanning confocal microscopy: history, applications, and related optical sectioning techniques. Methods Mol. Biol., *1075*, 9–47

Partain, C.L. (2004). The 2003 Nobel Prize for MRI: significance and impact. *Journal of Magnetic Resonance Imaging*, *19*(5), 515–526

Pereira, H. (ed.). (2007). *Cork: biology, production and uses*. Amsterdam: Elsevier

PhysicsWorld. (2021). Quantum microscope uses entanglement to reveal biological structures. June 15, 2021, https://physicsworld.com/a/quantum-microscope-uses-entanglement-to-reveal-biological-structures (accessed April 5, 2023)

Poustelnikova, E., Pisarev, A., Blagov, M., Samsonova, M., & Reinitz, J. (2004). A database for management of gene expression data in situ. *Bioinformatics*, *20*(14), 2212–2221

Sampson, D.M., Dubis, A.M., Chen, F.K., Zawadzki, R.J., & Sampson, D.D. (2022). Towards standardizing retinal optical coherence tomography angiography: a review. *Light: Science & Applications*, *11*(1), 1–22

Schnell, C. (2019). Quantum imaging in biological samples. *Nature Methods*, *16*, 214. https://doi.org/10.1038/s41592-019-0346-6

Shi, S., Kumar, P., & Lee, K.F. (2017). Generation of photonic entanglement in green fluorescent proteins. *Nature Communications*, *8*, 1934. https://doi.org/10.1038/s41467-017-02027-9

Smyth, M.S., & Martin, J.H. (2000). X ray crystallography. *Mol. Pathol.*, *53*(1), 8–14. doi: 10.1136/mp.53.1.8; PMID: 10884915; PMCID: PMC1186895

Süel, G.M., Garcia-Ojalvo, J., Liberman, L.M., & Elowitz, M.B. (2006). An excitable gene regulatory circuit induces transient cellular differentiation. *Nature*, *440*(7083), 545–550

Taylor, M. (2015). Quantum microscopy of biological systems. Cham: Springer

Volpe, A., Kurtys, E., & Fruhwirth, G.O. (2018a). Cousins at work: how combining medical with optical imaging enhances *in vivo* cell tracking. *International Journal of Biochemistry & Cell Biology*, *102*, 40–50

Volpe, A., Man, F., Lim, L., Khoshnevisan, A., Blower, J., Blower, P.J., & Fruhwirth, G.O. (2018b). Radionuclide-fluorescence reporter gene imaging to track tumor progression in rodent tumor models. *Journal of Visualized Experiments*, *133*, e57088

Wang, D., & Xia, J. (2019). Optics based biomedical imaging: principles and applications. *Journal of Applied Physics*, *125*(19), 191101

Watson, J.D., & Crick, F.H. (1953). Molecular structure of nucleic acids: a structure for deoxyribose nucleic acid. *Nature*, *171*(4356), 737–738

Yin, L., Cao, Z., Wang, K., Tian, J., Yang, X., & Zhang, J. (2021). A review of the application of machine learning in molecular imaging. *Annals of Translational Medicine*, *9*(9), 825

Zhang, Y., El Omari, K., Duman, R., Liu, S., Haider, S., Wagner, A., … & Wei, D. (2020). Native de novo structural determinations of non-canonical nucleic acid motifs by X-ray crystallography at long wavelengths. *Nucleic Acids Research*, *48*(17), 9886–9898

Zubavichus, Y., Shaporenko, A., Korolkov, V., Grunze, M., & Zharnikov, M. (2008). X-ray absorption spectroscopy of the nucleotide bases at the carbon, nitrogen, and oxygen K-edges. *Journal of Physical Chemistry B*, *112*(44), 13711–13716

13 Quantum Biophysics

Quantum biophysics is a multidisciplinary area that utilizes molecular biophysics principles and techniques to describe biological systems at quantum levels. The kingdoms of quantum effects and living organisms have long been thought to work broadly on different scales utilizing distinctive principles. The former is generally found to work on the scale spanning up to nanometer (nm) dimensions. although by cooling an aluminum drum, which has a diameter of just 20 micrometers (μms) and a thickness of 100 nm, to near absolute zero, researchers could reduce the noise from heat, allowing quantum effects to emerge (Ornes, 2019). In recent times, physicists have succeeded in taking entanglement (through which particles can have an intrinsic connection that endures no matter the distance between them) and other quantum effects to new extreme dimensions by observing them in large systems, including clouds of atoms, quantum drums, wires, and etched silicon chips. Thus the quantum world is being slowly pushed into macroscopic territory (Ornes, 2019). Quantum effects are known to work in a vacuum, ultra-low temperatures, and in a controlled environment, but the organisms that we call living systems require a macroscopic world which is a warm, messy, and uncontrolled environment (Ball, 2011). Quantum biology is emerging fast to deal with the functions of enzyme-catalyzed reactions, spin-dependent reactions, photosynthesis, deoxyribonucleic acid (DNA), fluorescent proteins, and ion channels, etc. which are key processes determining biological life (Kim et al., 2021). Life's building blocks are mostly found in cells, where quantum phenomena, e.g. coherence, in which the wave patterns of every part of a system stay in step, wouldn't last even a microsecond (μs). Quantum coherence in biology is conserved as long as the interactions of the coherent wave package with other degrees of freedom appear weak, as otherwise the population relaxation is known to cause a loss of this coherence (Khmelinskii and Makarov, 2020). Application of quantum mechanics in biological systems is thus subjective, depending on specific conditions and structural pockets within which the hazy biological systems contain certain processes and phenomena in isolated conditions (McFadden and Al-Khalili, 2018). Biological functions such as photosynthesis, enzyme catalysis, avian navigation, or olfaction may not only operate within the realm of "classical physics" but also utilize many non-trivial quantum mechanical features, such as coherence, tunneling, entanglement, etc. In this chapter, I am going to briefly explain most of the vital quantum mechanical effects that are of great interest for understanding natural processes of biology associated with defining vital parameters of life.

13.1 EMERGENCE OF QUANTUM BIOLOGY

Today's quantum biology (Mohseni et al., 2013; Fleming et al., 2011; Scholes et al., 2017) emerged through breakthrough discoveries over a century (McFadden and Al-Khalili, 2018; see Figure 13.1). This timeline is based on various sources (Kalckar, 2013; Jordan, 1932; Jordan, 1941; Watson and

DOI: 10.1201/9781003287780-13

Crick, 1953; Löwdin et al., 1966; Hopfield, 1974; Wiltschko and Wiltschko, 1972; Emlen et al., 1976; Schulten et al., 1976; Schulten and Weller, 1978; Cha et al., 1989; Ritz et al., 2000; Engel et al., 2007). This indicates that there has been roughly a century during which this quantum biology has experienced continuous enrichment, although it has advanced dramatically in the last three decades, during which experimental evidence for quantum coherence in photosynthesis and quantum tunneling in enzyme action has emerged.

The origins of quantum biology may be traced back to 1944 and the famous book of Erwin Schrödinger, *What is Life?* (1944). In contrast to the traditional consideration of living cells as biological units consisting just of life's biological and biochemical processes, the penetration of physics principles and physical techniques into exploring the state of life had already happened and had started producing preliminary understanding in the period from the seventeenth century to the nineteenth (see Chapter 1 of this book and, e.g., Welch, 1992). Biophysics of life involving especially its quantum mechanical aspects were seriously considered by the great physicist Schrödinger (1944). Several quantum physicists also thought about and worked on this area even before the publication of Schrödinger's book. Pascual Jordan, a German physicist, published a book entitled *Physics and the Secret of Organic Life* (Jordan, 1941), where we see him posing his question "Sind die Gesetze der Atomphysik und Quantenphysik für die Lebensvorgänge von wesentlicher Bedeutung?" ("Are the laws of atomic and quantum physics of essential importance for life?"). This book appeared as a result of Jordan's decade-long work, and since the late 1930s he had been using the term *Quantenbiologie*. The great physicist Bohr is known to have floated ideas of linking quantum physics with biology about a century ago in 1929 (Kalckar, 2013), which encouraged scientists to think further.

All of the early quantum physics ideas around biology gradually became enriched with developed concepts, but attention to this area experienced intense growth with the appearance of the concept of tunneling in biology during the last four decades of the twentieth century. During this time theoretical predictions on proton tunneling (Löwdin, 1963) and electron tunneling (DeVault and Chance, 1966) emerged. As tunneling is one of the fundamental quantum phenomena (Rubakov, 1984; Razavy, 2013), the idea of possibly finding quantum tunneling in biology truly engaged quantum physics with biology.

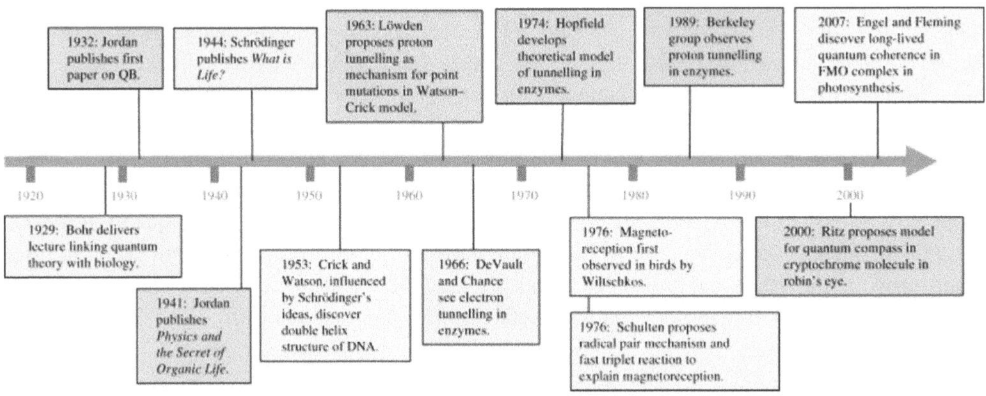

FIGURE 13.1 Timeline of landmarks in the development of quantum biology throughout the twentieth and early twenty-first centuries.

***Source*: McFadden and Al-Khalili (2018).**

13.2 QUANTUM BIOLOGY ENTERS INTO EXPERIMENTAL ERA AND MODELING

Concepts of quantum biology mainly evolved as theoretical propositions and predictive interpretations, as we find them in many of the early to middle periods of the timeline in Figure 13.1. Recent advancement in experimentation promises to enrich quantum biology further. I shall briefly address a few examples.

13.2.1 Proton Tunneling in Biological Systems

Proton tunneling in enzymes was observed in 1989 (Cha et al., 1989). Here primary and secondary protium-to-tritium (H/T) and deuterium-to-tritium (D/T) kinetic isotope effects for the catalytic oxidation of benzyl alcohol to benzaldehyde by yeast alcohol dehydrogenase at 25 degrees Celsius were determined. This reaction was known to be nearly or fully rate-limited by the hydrogen-transfer step. Semi-classical mass considerations that do not include tunneling effects would predict that $k_H/k_T = (k_D/k_T)$ (Bell, 1980; Huskey and Schowen, 1983), where k_H, k_D, and k_T are the rate constants for the reaction of protium, deuterium, and tritium derivatives, respectively. A considerable number of deviations from this relation were observed for both primary and especially secondary effects, such that experimental H/T ratios are reported to be much greater than those calculated. These deviations, holding over 0–40 degrees Celsius range, were predicted to result from a reaction coordinate that contains a significant contribution from hydrogen tunneling.

The role of proton tunneling in the generation of DNA base tautomers has recently been theoretically demonstrated (Godbeer et al., 2015). Density functional theory (DFT) was used for calculating the energies of the canonical (standard, amino-keto) and tautomeric (non-standard, imino-enol) charge-neutral forms of the adenine–thymine base pair (A–T and A*–T*). For computing the reaction pathway, a transition state search was used to provide the asymmetric double-well potential minima along with the barrier height and shape, which are combined to create the potential energy surface using a polynomial fit. The influence of quantum tunneling on proton transfer within a base pair H-bond, modeled as the DFT deduced double-well potential, was investigated by solving the time-dependent density matrix explaining the master equation. Quantum tunneling, due to transitions to higher energy eigenstates with significant amplitudes in the shallow (tautomeric) side of the potential, was found unlikely to be a significant mechanism for the creation of adenine–thymine tautomers within DNA, with the thermally assisted coupling of the environment only able to boost the tunneling probability up to 2×10^{-9}.

Various theoretical and computational treatments of hydrogen tunneling in enzymatic and biomimetic systems, and hence in general condensed phases, were summarized in a review article (Layfield and Hammes-Schiffer, 2014). Each method was reported to involve certain types of approximations and had been found valid only within the specified regime, such as electronically/vibrationally adiabatic or nonadiabatic, low-frequency or high-frequency proton donor-acceptor mode, small or large solvent reorganization energy, and so forth.

13.2.2 Quantum Entanglement and Tunneling Effects that Species May Grossly Perceive

Experimental evidence has been found to support the role of quantum entanglement in avian navigation and quantum tunneling in olfaction. A molecule within the eyes of migratory birds has been predicted to sense magnetic fields, helping it act as a quantum compass (Thompson and Howe, 2021). The mechanism of this compass may rely on the quantum spin dynamics of photoinduced radical pairs in cryptochrome flavoproteins in the eye's retina (Hore and Mouritsen, 2016; Ritz et al., 2000).

The hypothesis that migrating birds might utilize the geomagnetic field for orientation was proposed in 1859 (von Middendorff, 1859). The demonstration of this phenomenon was made just six decades ago, for European robins, in 1966 (Wiltschko and Merkel, 1966), to which 17 more species have been added (Wiltschko and Wiltschko, 1996).

Xu and colleagues recently showed that the photochemistry of cryptochrome 4 (CRY4) from the night-migratory European robin (*Erithacus rubecula*) is magnetically sensitive, and more so than CRY4 from two non-migratory bird species, chicken (*Gallus gallus*) and pigeon (*Columba livia*) (Xu et al., 2021). Four successive flavin–tryptophan radical pairs participate in generating magnetic field effects and in stabilizing potential signaling states, helping to sense and create signal functions to be independently optimized in night-migratory birds.

Figure 13.2 presents the magnetic field effects on the yields of photoinduced radicals in CRY4 proteins. Figure 13.2a presents the applied magnetic field effects on the absorbance of FAD and Trp radicals in wild-type *Er*CRY4, measured at 5 °C. Millitesla fields are reported to suppress radical yields by favoring a return to the ground state, as expected for radical pairs that are formed in a spin-correlated singlet state (Maeda et al., 2012). It is expected that if the magnetic field effect on CRY4 acts behind the magnetic compass sense in night-migratory robins, evolution should optimize this effect. Xu and colleagues compared the magnetic field effects on *Er*CRY4 with those on CRY4 proteins from two non-migratory birds (Qin et al., 2016), pigeon (*C. livia*) and chicken (*G. gallus*), in which there is less evolutionary pressure which would optimize any light-dependent magnetic compass sense. The change in the optical absorption spectra of the three proteins due to the effects of 30-mT magnetic field under conditions of continuous illumination at 450 nm is presented in Figure 13.2c. The 500 nm – 550 nm band suggests a considerably bigger magnetic field effect on the yield of neutral $Trp^{\bullet}$ radicals in *Er*CRY4 than in *Cl*CRY4 or *Gg*CRY4; the difference between the three proteins was supported by cavity ring-down spectroscopy (CRDS) (see Figure 13.2d) by measuring the effect of a 30-mT magnetic field on the radical absorption signal as a function of time after irradiation with blue light. The field produces a significant reduction in the yield of radicals in *Er*CRY4 that persists for more than 10 μs. Apart from one of the earliest data points for *Cl*CRY4, no such change could be observed above the noise level for either *Cl*CRY4 or *Gg*CRY4. Taking the results of the two experiments together, the magnetic sensitivity of CRY4 from the night-migratory robin is substantially larger than that of the CRY4 proteins from the non-migratory – primarily diurnal – pigeon and chicken. This *in vitro* experiment of Xu and colleagues hints at nature finding ways to independently optimize the magnetic sensing and signaling properties of CRY4 through control of its photochemistry.

The study by Xu and colleagues reports a reaction scheme and simulated magnetic field effects for ErCRY4 (see Figure 13.3; various features are detailed in the caption; for more detail, see Xu et al., 2021).

Does CRY4 act as a magnetoreceptor molecule *in vivo*? To find an answer to this question, direct manipulations of this protein in the eyes of night-migratory songbirds would be required. Then monitoring the large-scale, long-term movement of these birds would help conclude whether and how these fine quantum mechanical effects may work in biology. This study would need to monitor associated biochemical reactions influenced by the magnetic field effect. Although migratory birds' ability to detect the direction of the Earth's magnetic field is a half-century-old concept, the associated primary sensory mechanism is still mysterious. The most favored hypothesis centers on radical pairs: magnetically sensitive chemical intermediates formed by photoexcitation of cryptochrome proteins in the retina (Hore and Mouritsen, 2016). Future *in vivo* studies may reveal hidden clues to support this remarkable claim.

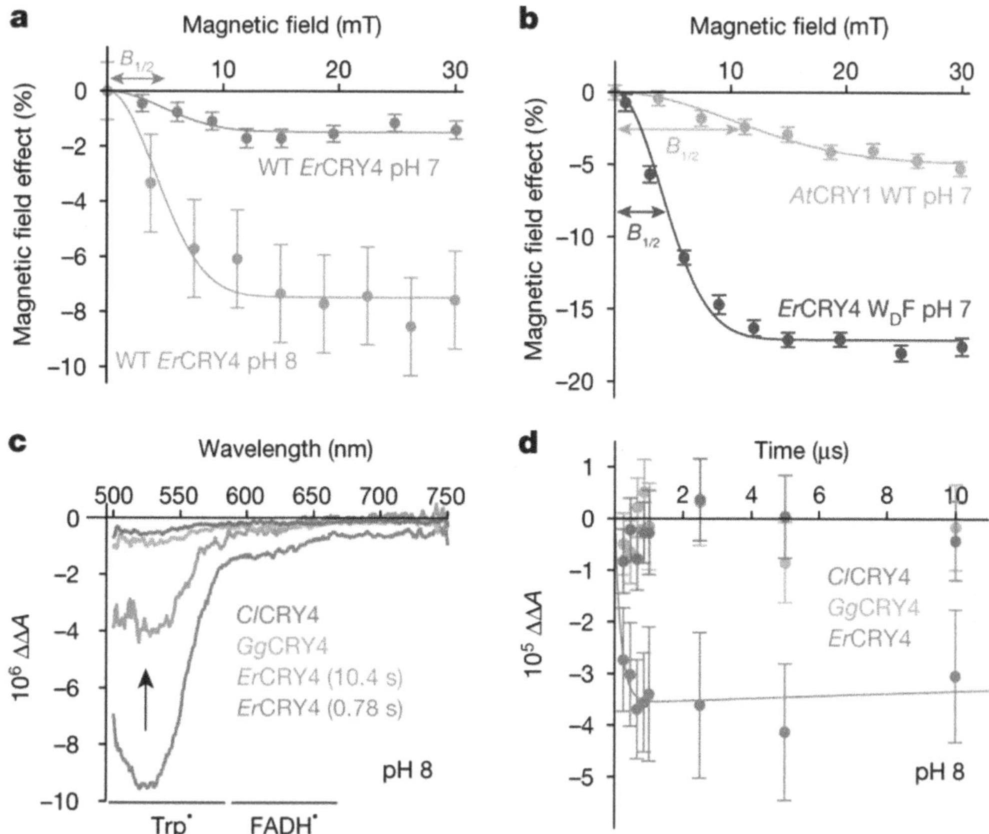

FIGURE 13.2 Magnetic field effects on the yields of photoinduced radicals in CRY4 proteins. (a, b) Magnetic field effect (percentage change in absorbance induced by the magnetic field) on the optical absorbance of photoinduced radicals 2 μs after a 450-nm laser pulse in wild-type *Er*CRY4 at pH 7 and pH 8 (a) and in W_DF and in wild-type *At*CRY1 at pH 7 (b). Data were measured by CRDS at 530 nm. The smaller value of $B_{1/2}$ (the magnetic field that produces 50 percent of the limiting change seen at high field) for wild-type *Er*CRY4 (4.9 ± 1.6 mT) compared to wild-type *At*CRY1 (14.3 ± 1.7 mT) suggests that, other factors being equal, the former would be more sensitive to weak magnetic fields. (c) Change in the optical absorbance of photoinduced radicals in three avian CRY4 samples induced by a 30-mT magnetic field. All four broadband cavity-enhanced absorption spectra measured using continuous illumination at 450 nm are dominated by the field-induced reduction in the yield of Trp• radicals, which absorb in the 500–600 nm range. *Er*CRY4 spectra are shown at two different times after the start of illumination. The equivalent spectra of *Cl*CRY4 and *Gg*CRY4 showed no time dependence and are averages over the first 9 s. The weak signals at wavelengths greater than 600 nm are thought to arise from magnetic field effects on the formation of FADH•. (d) Change in the photoinduced optical absorbance of three avian CRY4 samples upon application of a 30 mT magnetic field, measured by CRDS at 530 nm. Within error, the magnetic field effects on the CRY4 proteins from the non-migratory pigeon and chicken could not be distinguished from zero (apart from one very early data point for *Cl*CRY4).

***Source*: Xu et al. (2021).**

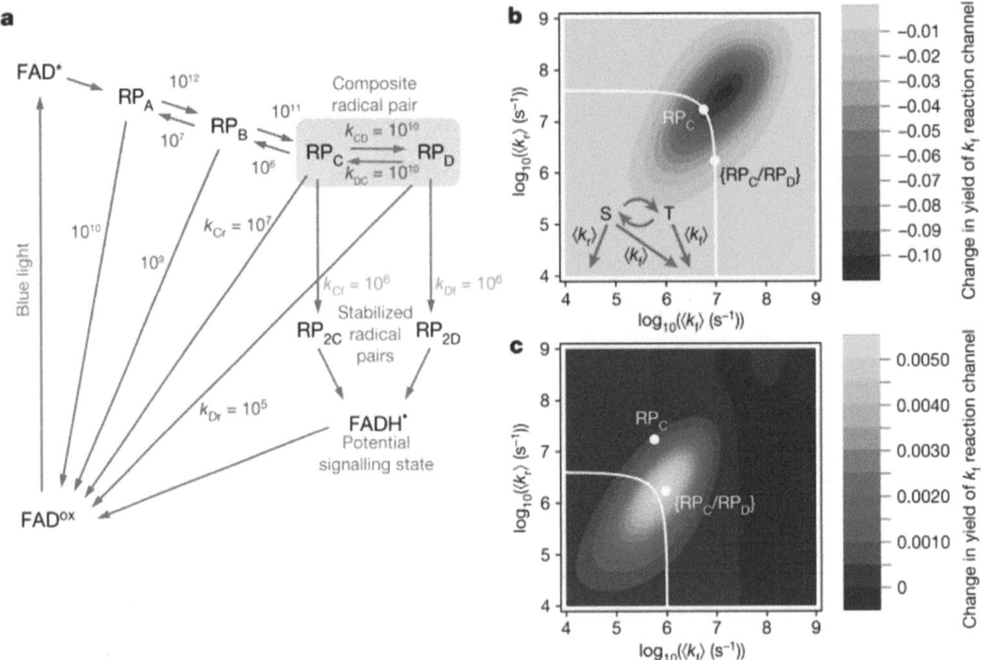

FIGURE 13.3 Reaction scheme and simulated magnetic field effects for ErCRY4. (a) Proposed reaction scheme for magnetic sensing and signaling by CRY4 in night-migratory songbirds. The arrows are labeled with approximate rate constants (in s^{-1}): red, calculations based on molecular dynamics simulations; purple, transient absorption measurements; green, estimated from magnetic field effects. (b) Spin dynamics simulation for the reaction scheme shown at the bottom left, in which the curved arrows represent the coherent interconversion of the singlet (S) and triplet (T) states of a composite radical pair and the straight arrows indicate the competing reaction pathways. $\langle k_f \rangle$ and $\langle k_r \rangle$ are weighted averages of the corresponding rate constants for RP_C and RP_D. The contour plot shows the change in the yield of the $\langle k_f \rangle$ pathway caused by a 30-mT magnetic field as a function of $\langle k_f \rangle$ and $\langle k_r \rangle$. On the white line the lifetime of the composite radical pair is approximately 100 ns. The positions of the white spots correspond to the rate constants estimated for RP_C and the composite pair with $f_C = 0.1$. (c) As (b) except that the magnetic field is 50 μT and the white line corresponds to a lifetime of the composite radical pair of approximately 1 μs. The white spots have been moved −1.0 log units along the $\langle k_f \rangle$ axis. Color contrast here and in all subsequent figures will be visible in the digital version of the book.

Source: **Xu et al. (2021).**

13.2.3 Electron Tunneling in Cells

An important quantum biology problem involves electron tunneling (Xin et al., 2019). Quantum biological electron transfer (ET) involves many important biological processes such as cellular respiration, DNA repair, cellular homeostasis, cell death, photosynthesis, etc. Xin and colleagues (2019) reported a quantum biological electron tunneling (QBET) junction and its application in real-time optical detection of QBET and the dynamics of ET in mitochondrial cytochrome c during the cell life and death process. QBET spectroscopy helped here to optically capture real-time ET in cytochrome c redox dynamics during cellular apoptosis and necrosis in living cells.

Optical antennas, which enable real-time spectroscopic molecular imaging in living cells (Xin et al., 2018), were found applicable to perform molecular ET imaging of enzymes and the transmission of QBET dynamics in live cells (see Figure 13.4a, and Xin et al., 2019). Cytochrome c (Cyt c) was taken here as an example of enzymes for demonstrating the capture of ET. Within the electron

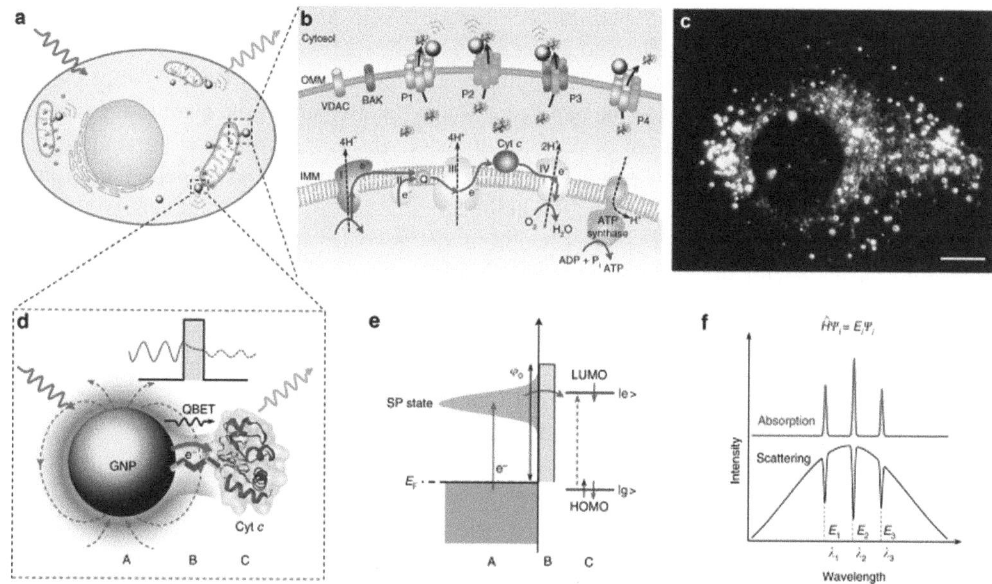

FIGURE 13.4 QBET imaging in a tunnel junction. (a) Schematic of gold nanoparticles (GNPs) used as optical antennas for intracellular QBET imaging in living cells. (b) Schematic illustration of GNPs for the detection of Cyt *c* released from mitochondria to cytosol. Cyt *c* is originally involved in the electron transport chain on the inner mitochondrial membrane (IMM), and transfers electrons between complexes III (Cyt *c* reductase) and IV (Cyt *c* oxidase, COX). After the formation of different pores or channels on outer mitochondrial membrane (OMM), Cyt *c* is released through these pores to cytosol. P1 to P4 represent pores based on oligomeric voltage-dependent anion channels (VDAC), BAX oligomer, BAX-BAK oligomer, and VDAC-BAX oligomer, respectively. BAX and BAK are two B-cell lymphoma protein-2 (BCL2) effector proteins. (c) A representative dark-field image of a living cell with GNP uptake. Scale bar: 10 μm. (d) Schematic of QBET in A/B/C tunnel junction. Single GNP acts as an Au plasmonic optical antenna (A) with light irradiation. Surface electrons collectively oscillate at the resonant light frequency. Electrons tunnel through the potential barrier via linker molecule (barrier molecular, B) to Cyt *c* (C). (e) Schematic illustration of electron tunneling in the A/B/C tunnel junction. Electron is excited from Fermi level (E_F) to a surface plasmon (SP) state, and tunnels through the potential barrier via the barrier molecule to Cyt *c*. The molecule is excited to the excited state (|e >) after the electron tunneling and transfer. |e >: excited state, |g >: ground state, LUMO: lowest unoccupied molecular orbital, HOMO: highest occupied molecular orbital. (f) Scattering spectra with quantised dips for QBET imaging. QBET results in quantised dips in the scattering spectra of GNP, which matches the frequencies of electronic transitions of Cyt *c* (absorption peaks). These dips correspond to the quantised eigenvalues E_i at the electron state ψ_i in the tunneling process. The cartoons for Cyt *c* molecules in (a), (b), and (d) were created from the RSCB protein data bank (Rose et al., 2016).

***Source*: Xin et al. (2019).**

transport chain in mitochondria (Wisnovsky et al., 2016) Cyt *c* is functionally involved in ET (see Figure 13.4b). In Figure 13.4c a dark-field image of an intact living cell with gold nanoparticles (GNPs) (50 nm, green) uptake has been presented, helping to detail QBET imaging using GNP-based optical antennas.

For visualizing QBET in living cells, electron tunneling across an A/B/C tunnel junction was captured (see Figure 13.4d). Cyt *c* I is conjugated with a GNP (Au, A) via linker molecules (barrier molecule, B), forming a quantum tunnel junction with a potential barrier. GNPs serve as an optical antenna with wavelength-specific surface plasmon (SP) resonance with strong light scattering when excited with white light irradiation. The free conduction electrons are elevated from Fermi level (E_F) to a higher energy level, SP state, and collectively oscillate at the GNP surface at the resonant light frequency. This system allows the plasmon resonance energy transfer (PRET) from the GNP to the conjugated Cyt *c* molecules. As the gap in the tunnel junction enters the quantum regime with a subnanometer distance, the spill-out of plasmon-induced surface electrons can tunnel across the gap at optical resonant frequencies (Zhu et al., 2016), and quantum electron tunneling and transfer will coexist with PRET, playing a key function in the modulation of the spectrum. Due to the linker molecules in the junction, the tunnel barrier height in the gap between the GNP and Cyt *c* is lower than that in vacuum without linker molecules, which facilitates electron tunneling (Tan, 2018), which allows ET from GNP to Cyt *c* (see Figure 13.4e). The height of the potential barrier (φ_0) is the offset between the GNP Fermi level and the lowest unoccupied molecular orbital (LUMO) of the linker molecule (Tan, 2018). The tunneling barrier width is the length of the linker molecule. The molecule is excited to the excited state ($|e>$) after the electron tunneling and transfer. As a result of the electron tunneling, we observe quantized dips located at Cyt *c* absorption peaks in the scattering spectrum of the GNP (see Figure 13.4f), which enable the optical capture of QBET in the tunnel junction.

Xin and colleagues used the tunnel junction to capture the real-time ET in Cyt *c* during reduction via QBET spectroscopy (Xin et al., 2019). Cyt *c*, in intact cells, resides in the inner mitochondrial membrane, and cytosolic Cyt *c* level is very low, thus Xin et al. were unable to detect any QBET-induced dips at the scattering spectra of the GNP (see Figure 13.5a). By stimulating apoptosis via ethanol treatment, the mitochondrial outer membrane permeabilization was increased via the outer mitochondrial membrane pore or channel formation, which promoted the Cyt *c* release from mitochondria. The released Cyt *c* was quickly reduced by cytosolic reductants and enzymes. Accordingly, the QBET in A/B/C (red.) tunnel junction with dips at 520 and 550 nm was captured. Observation of distinct morphologic changes was reported during the apoptosis process, including cell shrinkage, cell membrane blebbing to form apoptotic bodies and nucleus rupture (see Figure 13.5b). Significant QBET signal changes were captured in the cytosol after the ethanol exposure (see Figure 13.5c). Dip changes in the whole apoptosis process were observed (see Figure 13.5d).

The molecular dynamics of Cyt *c* during both apoptosis and necrosis are presumably different due to the difference in the inherent biological mechanisms. Xin et al. (2019) used Triton X-100 (0.17 mM) to stimulate cells and induce necrosis and utilized QBET junction spectroscopy to image Cyt *c* redox dynamics in necrosis (see Figure 13.5e). The cell membrane was gradually damaged and cellular organelles were released outside the cell during necrosis (see Figure 13.5f). Considerable changes in the real-time QBET signals at different locations inside the cell were captured with Cyt *c* redox changes (Figure 13.5g). Heterogeneity due to different cellular environments, Cyt *c* concentrations, and the total amount of conjugated Cyt *c* contribute to different dip depths detected in QBET signals. Figure 13.5h shows dip changes in the necrosis process in a cyclic behavior. Due to the different dynamics of enzymatic oxidation and reduction inside the cell, this cyclic behavior of Cyt *c* redox is different from the situation in apoptosis. Using QBET spectroscopy, the real-time ET in cytochrome c redox dynamics during cellular apoptosis and necrosis in living cells was optically captured.

13.3 QUANTUM COHERENCE REVEALING EXCITED-STATE DYNAMICS OF PHYSICOCHEMICAL SYSTEMS

Quantum coherence in photosynthetic systems as electronic, vibronic, or vibrational was recently reviewed by describing the latest experimental efforts toward unraveling the nature of the coherences, in particular ultrafast, two-dimensional electronic spectroscopy (Wang et al., 2019). Associated theoretical and computational results have been presented here. The measurement of coherences can inform us about the excited-state dynamics of biophysical and chemical systems relevant to natural

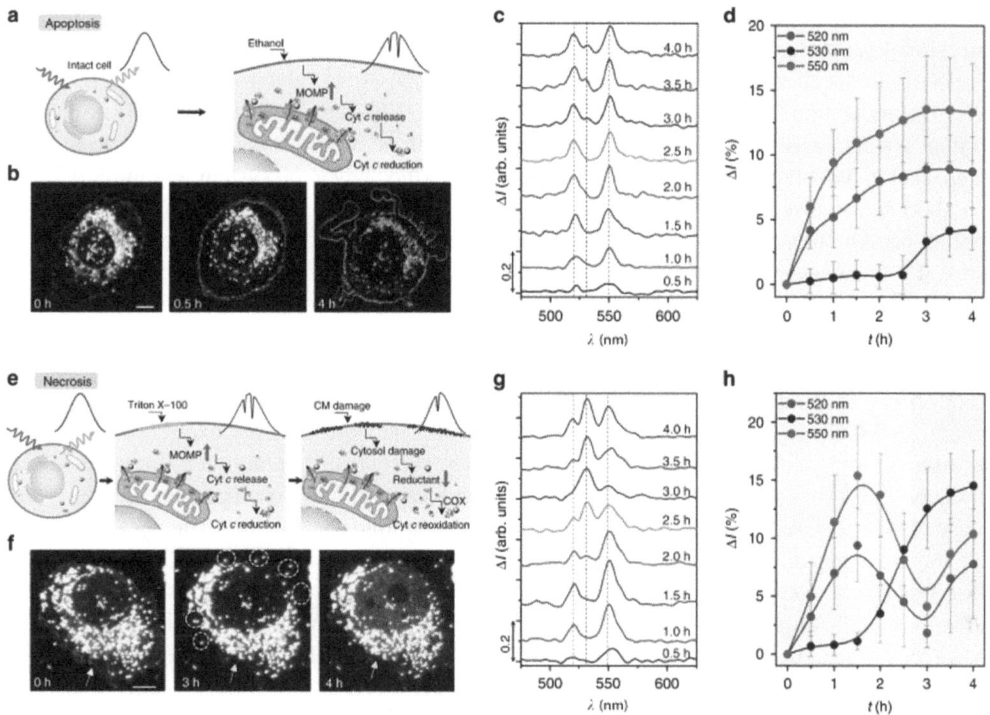

FIGURE 13.5 Real-time intracellular QBET imaging of ET during cellular apoptosis and necrosis.(a–d) Non-invasive QBET imaging for cell apoptosis. (a) Schematic of cellular stimulus with ethanol for apoptosis. Major QBET signals for Cyt *c* (red.) are captured during apoptosis. MOMP: mitochondrial outer membrane permeabilisation. (b) Dark-field images of the cell at different times with the stimulus of ethanol, scale bar: 10 μm. $t = 0$ h, the beginning of apoptosis measurement. $t = 0.5$ h, cell shrinkage was observed. $t = 4$ h, nucleus was ruptured and apoptosis bodies were formed. (c) Scattering spectra difference showing QBET in A/B/C tunnel junction in the process of cell apoptosis. The QBET spectra differences were obtained by subtracting the spectra at different measuring times from the spectra at the beginning of the measurement ($t = 0$ h). (d) Quantized dips at 520, 530, and 550 nm in the process of cell apoptosis. (e–h) QBET imaging with triton X-100 stimulus for cell necrosis. (e) Schematic of cellular stimulus with triton X-100 for necrosis. QBET is captured from Cyt *c* (red.) to Cyt *c* (Ox.) during necrosis. CM: cell membrane, COX: Cyt *c* oxidase. (f) Dark-field images of the cell with the stimulus of triton X-100. $t = 0$ h, white arrow indicates the intact cell membrane; $t = 3$ h, yellow arrow indicates partial lysis of cell membrane, yellow circles indicate organelles are released out to extracellular environment; $t = 4$ h, yellow arrow indicates the damage and lysis of the whole membrane. Scale bar: 10 μm. (g) Scattering spectra difference showing QBET in the process of cell necrosis. The dip changes show the change from Cyt *c* (Red.) to Cyt *c* (Ox.), and finally both Cyt *c* (Red.) and Cyt *c* (Ox.). (h) Quantized dips at 520, 530, and 550 nm in the process of cell necrosis with cyclic behaviour. The curves in (d), (h) are cubic B-spline connection of the experimental data. The cartoons for Cyt *c* molecules in (a), (e) were created from the RSCB protein data bank (Rose et al., 2016).

***Source*: Xin et al. (2019).**

light harvesting. These measurements reveal electronic structure beyond that captured by simplistic models.

Quantum coherent oscillations or quantum beats in ultrafast spectroscopic measurements of photosynthetic pigment–protein complexes consist of coupled chromophores embedded in the protein scaffold (Engel et al., 2007; Savikhin et al., 1997; Collini et al., 2010; Panitchayangkoon et al.,

2010). The quantum beating was observed in a photosynthetic pigment–protein complex with two-dimensional (2D) electronic spectroscopy (Engel et al., 2007). Here Engel and colleagues used 2D electronic spectroscopy for observing oscillations caused by electronic coherence evolving during the population time in the Fenna–Matthews–Olson (FMO) bacteriochlorophyll complex. The quantum coherence, which is a coherent superposition of electronic states analogous to a nuclear wavepacket in the vibrational regime, was found to be formed when the system was initially excited by a short light pulse with a spectrum that spanned multiple exciton transitions. The electronic quantum beats arising from quantum coherence in photosynthetic complexes were predicted earlier (Knox, 1996; Leegwater, 1996), and observed indirectly in 1997 (Savikhin et al., 1997). Engel and colleagues (2007) extended previous 2D electronic spectroscopy investigations of the FMO bacteriochlorophyll complex, and obtained direct evidence for remarkably long-lived electronic quantum coherence playing an important part in energy transfer processes within this system.

For observing the quantum beats, 2D spectra were taken at 33 population times T, ranging from 0 to 660 fs. Figure 13.6 presents representative spectra. The lowest-energy exciton is seen here to give rise to a diagonal peak near 825 nm that clearly oscillates: its amplitude grows, fades, and subsequently grows again. The shape of the peak evolves with these oscillations, becoming more elongated when weaker and rounder when the signal amplitude intensifies. The associated cross-peak amplitude appears to oscillate. The quantum beating was reported to last for 660 fs. This observation was found to contrast with the general assumption that the coherences responsible for such oscillations are destroyed very rapidly and that population relaxation proceeds with the destruction of coherence (van Amerongen et al., 2000) so that the transfer of electronic coherence between excitons during relaxation is usually ignored (van Amerongen et al., 2000; Abramavicius et al., 2004; Renger et al., 2001).

There have been theoretical models that include coherence among both donors and acceptors (Jang et al., 2004). The coherence needs to be treated between all chromophores to ensure that the theoretical models accurately reproduce the dynamics of the system (Novoderezhkin et al., 2003).

Theoretical models in FMO indicate that electronic coherence should dephase on the timescale of the initial population transfer, which was established experimentally to be less than 250 fs for all but excitons 1 and 3 (Vulto et al., 1999; van Amerongen et al., 2000). The observed strong quantum beating lasting for at least 660 fs exceeds the model predictions (Engel et al., 2007). Engel et al. believed that to account for this long-lived coherence and provide an accurate description of the system, the protein must have a more active role in a realistic bath model, so that it must be allowed to interact with both donors and acceptors, to enable coherence transfer and possibly the generation of new coherences.

In Figure 13.7, Engel et al. (2007) showed the amplitude oscillations with time along the main diagonal of the spectrum. The data were mapped onto a Fourier subspace using a non-uniform fast Fourier transform algorithm (Kunis and Potts, 2007a, 2007b). The sampled amplitude variations of the lowest-energy exciton are shown with a Fourier interpolation from the subspace that was selected to be maximally consistent with the excitonic model. The theoretical excitonic coherence line spectrum was calculated using the exciton energies from the Hamiltonian (Brixner et al., 2005), and the relative amplitudes were calculated using orientationally averaged response magnitudes for the associated cross-peak pathways (Cho and Fleming, 2005; Dreyer et al., 2003; Hochstrasser, 2001).

The quantum beating observed here (see Figure 13.7) is consistent with electronic coherence. Analyses of beating in more peaks and comparisons thereof are presented using an independent theoretical lineshape prediction to confirm this conclusion. As the predicted orientational factor is identical for a diagonal exciton beating and for a cross-peak, that the amplitude of the beat is of the order of the amplitude of the cross-peak even at long times (>500 fs) indicates that the electronic coherence can play a significant role in determining the overall relaxation dynamics within the protein complex, suggesting that coherence relaxation pathways, including coherence transfer,

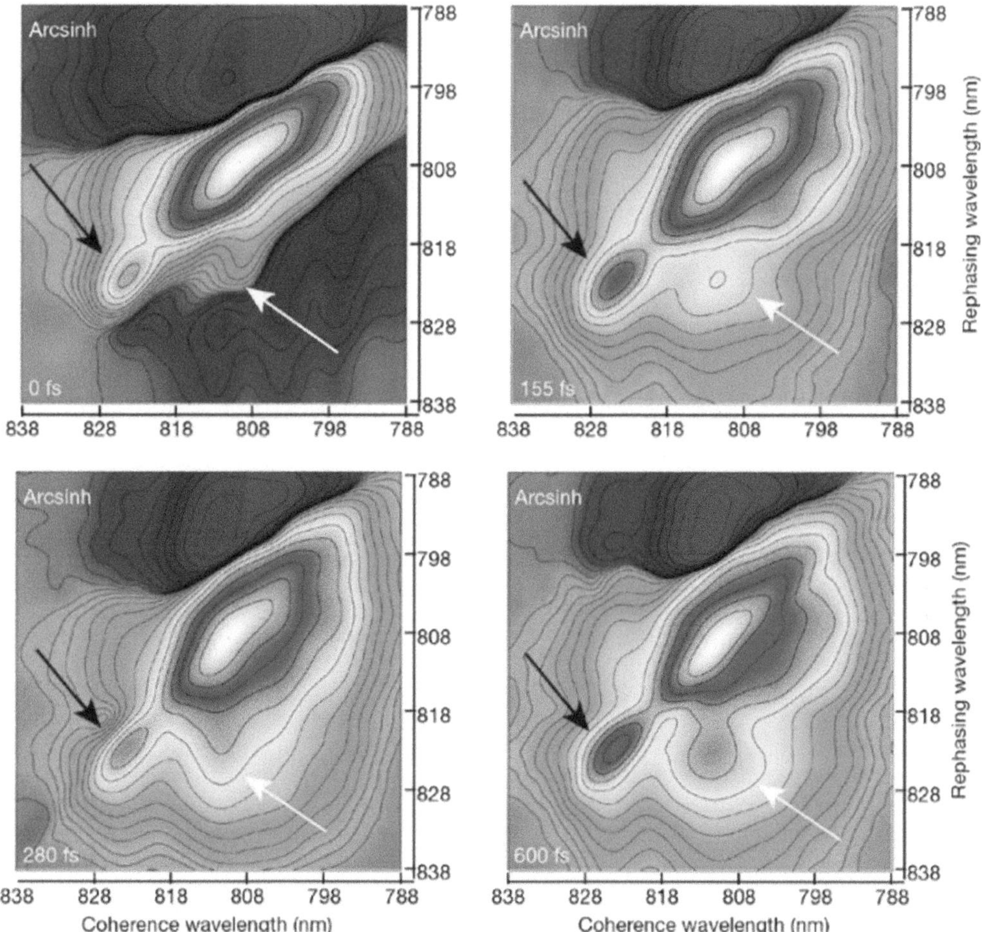

FIGURE 13.6 Two-dimensional electronic spectra of FMO. Selected two-dimensional electronic spectra of FMO are shown at population times from T = 0 to 600 fs demonstrating the emergence of the exciton 1–3 cross-peak (white arrows), amplitude oscillation of the exciton 1 diagonal peak (black arrows), the change in lowest-energy exciton peak shape and the oscillation of the 1–3 cross-peak amplitude. The data are shown with an arcsinh coloration to highlight smaller features: amplitude increases from blue to white.

***Source*: Engel et al. (2007).**

should no longer be disregarded in theoretical models of photosynthetic protein complexes (Engel et al., 2007).

Mohseni and colleagues (2014) showed that the FMO energy transfer efficiency was optimum and stable concerning important parameters of environmental interactions, including reorganization energy λ, bath frequency cutoff γ, temperature T, and bath spatial correlations. This group identified the ratio of kBλT/ħγg as a single key parameter governing quantum transport efficiency, where g is the average excitonic energy gap. An efficient technique was applied for estimating the energy transfer efficiency of the complex excitonic systems.

Wang and colleagues (2019) recently addressed implications of coherences for biology. We know nature evolved photosynthetic light-harvesting complexes for optimizing survivability and

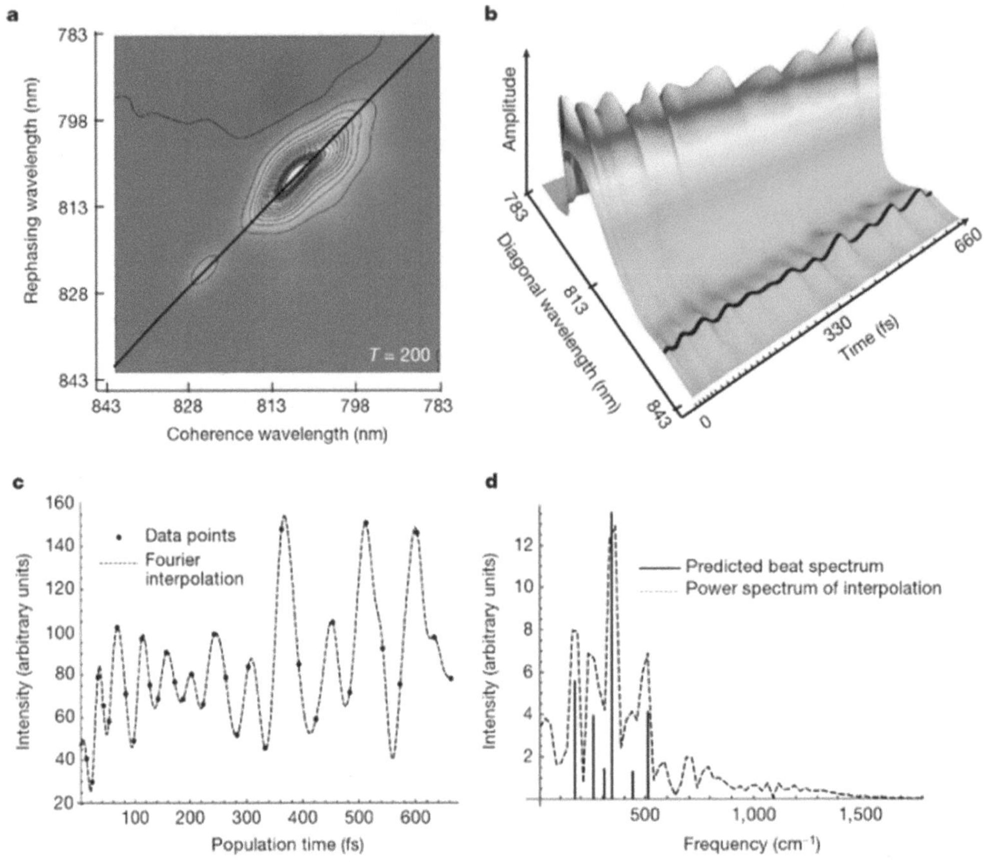

FIGURE 13.7 Electronic coherence beating. (a) A representative two-dimensional electronic spectrum with a line across the main diagonal peak. The amplitude along this diagonal line is plotted against population time in (b) with a black line covering the exciton 1 peak amplitude; the data are scaled by a smooth function effectively normalizing the data without affecting oscillations. A spline interpolation is used to connect the spectra; the times at which spectra were taken are denoted by tick marks along the time axis. (c) The amplitude of the peak corresponding to exciton 1 shown with a dotted Fourier interpolation. (d) The power spectrum of the Fourier interpolation in (c) is plotted with the theoretical spectrum showing beats between exciton 1 and excitons 2–7.

Source: **Engel et al. (2007).**

maintaining growth rates, which demand rapid and efficient transport of captured energy from the solar source. Photosynthetic organisms can leverage exquisite control over the structure of a pigment–protein complex to achieve high quantum efficiency in light-starved environments and exquisite control over long-range energy transfer. Pigment–protein complexes exhibit convergence of energy and timescales to build robust quantum mechanical systems with highly efficient energy transfer, known as the quantum Goldilocks principle (Mohseni et al., 2014). This convergence means that all the different parameters that influence energy transfer – interchromophore coupling, system–bath coupling (reorganization energy), decoherence timescale, and bath equilibration time – lie in the intermediate coupling regime that appears to be responsible for optimized transfer efficiency. An example of these energy-transfer dynamics can be seen in Figure 13.8 (see also Wang et al., 2019).

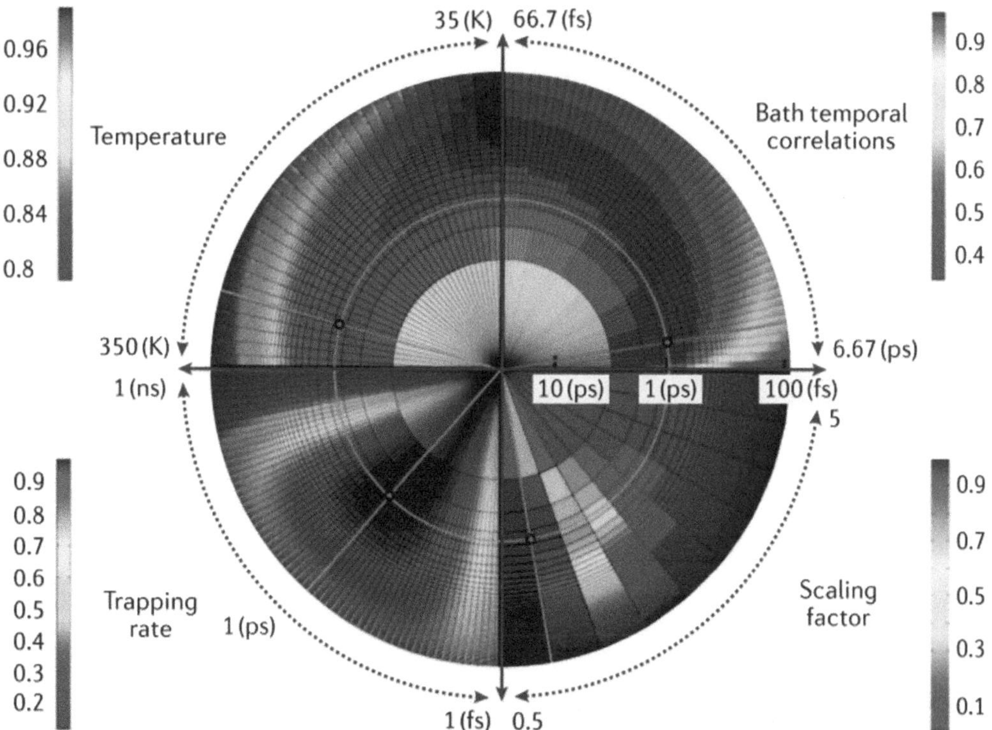

FIGURE 13.8 The radial axes plot the reorganization energy on a logarithmic scale, which essentially quantifies system–bath interactions. Angular coordinates in each quarter indicate different physical properties: temperature, bath correlation timescale, trapping rate, and the scaling factor corresponding to the relative distances between chromophores. All of these factors are significant contributors to the environment in which the chromophores sit. The color bars indicate energy-transfer efficiencies, with red colors indicating the most efficient processes. The estimated values of the FMO complex reside in the intermediate regime of both angular and radial coordinates (indicated by the white lines). Expressed in units of energy, these variables have similar scales ($kT = 207\ \text{cm}^{-1}$; reorganization, $\lambda = 35\ \text{cm}^{-1}$; coupling, $\gamma = 50\text{–}166\ \text{cm}^{-1}$; trapping, $r/hc = 34\ \text{cm}^{-1}$).

***Source*: Mohseni et al. (2014).**

13.4 QUANTUM IMAGING IN BIOLOGY

Quantum microscopy of biological systems has recently evolved to address crucial biological phenomena (PhysicsWorld, 2021). Michael Taylor (2015) reported on the development of the first quantum-enhanced microscope and on its applications in biological microscopy. This microscope was used to perform the first quantum-enhanced biological measurements, which was an important and long-standing goal in quantum measurements. Sub-diffraction-limited quantum imaging was achieved, claimed to be for the first time, with a scanning probe imaging configuration allowing 10-nanometer resolution.

Image scanning microscopy (ISM) and quantum imaging together may improve the resolution up to four-fold compared with the classical diffraction barrier (Schnell, 2019). Compared to confocal microscopy, in ISM the confocal pinhole is replaced by a detector array with individual pinholes for each detector, which results in a two-fold enhancement in resolution (Sheppard, 1988; Müller and Enderlein, 2010) compared with the diffraction barrier. In contrast to microscopy techniques treating light as a wave, quantum imaging treats light as quantum particles, "photons." Recent improvements in detector hardware, such as single-photon avalanche detectors (SPADs), have enabled researchers to demonstrate how the combination of quantum imaging and ISM (Q-ISM) can improve the resolution of biological samples (Tenne et al., 2019). Q-ISM set-up is demonstrated in Figure 13.9a. Tenne and colleagues used a standard confocal excitation scheme in which a pulsed blue laser beam (473 nm) was focused through a high-numerical-aperture (NA) objective lens while the sample was scanned with a piezo stage. The sample fluorescence was collected through the same

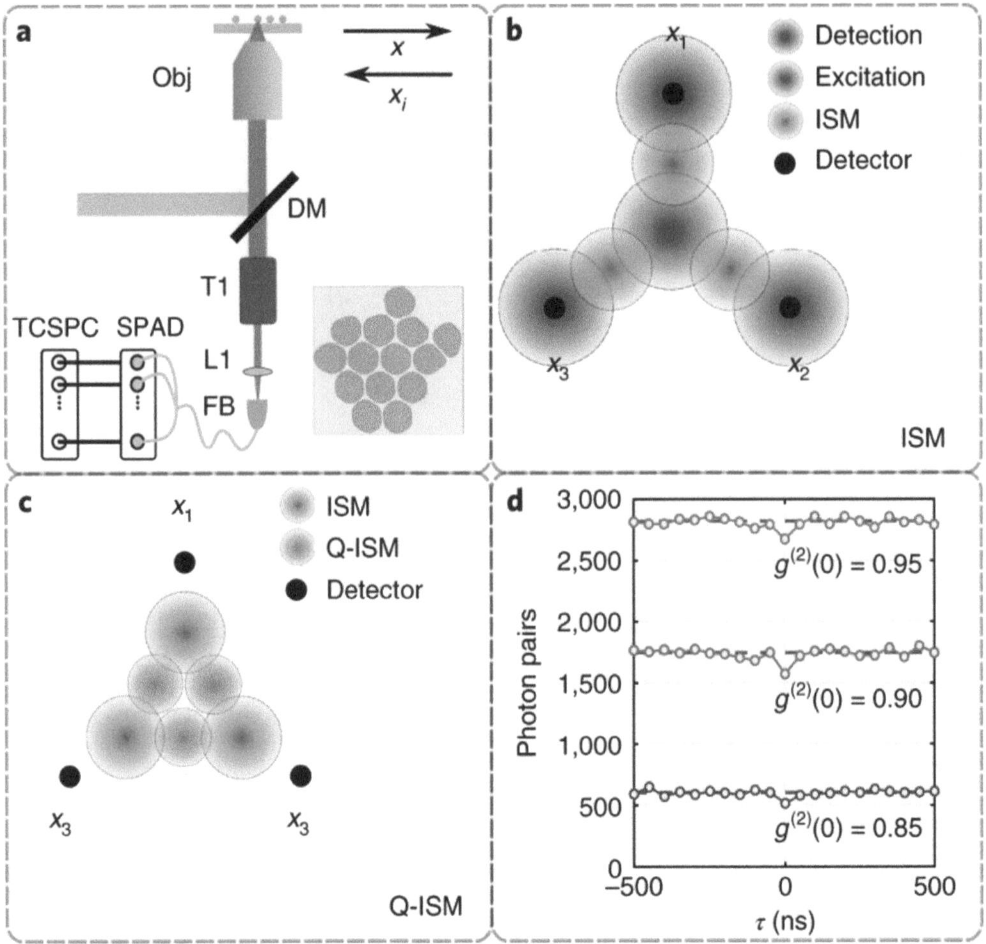

FIGURE 13.9 Q-ISM principle of operation. (a) The optical set-up used in this work. The pinhole in a standard confocal set-up is replaced with a fiber bundle (FB, shown in the inset) routing light into 14 individual SPADs. Obj, objective lens; DM, dichroic mirror; T1, variable telescope; L1, lens. (), Schematic of the ISM method. The ISM PSF for each detector (green circles) is a product of the excitation laser beam profile (blue circle) and the detection probability distribution (red circles) centered on the detector's position (black circles). For simplicity, the PSFs are drawn as cylindrically symmetric, although the principle of operation of ISM and Q-ISM does not require this symmetry. (c) Schematic of Q-ISM. The effective PSF for each detector pair (orange circles) is a product of the two ISM PSFs of the two detectors (green circles). (d) The unnormalized second-order correlation function, $G^{(2)}(\tau)$, generated for three positions in the scan shown in Figure 13.7. Black dashed lines indicate the average value of incident photon pairs with a time delay longer than one pulse. The calculated normalized second-order correlation function, $g^{(2)}(\tau = 0)$, is shown next to each of the curves.

Source: **Tenne et al. (2019).**

objective, further magnified, and then imaged onto a honeycomb lattice fibre bundle (see inset of Figure 13.9a). Each of all 14 fibers routed the light impinging on it to an individual singe-photon avalanche detector (SPAD) (SPCM-AQ4C, Perkin-Elmer) feeding a time-correlated single-photon counting (TCSPC) card logging the arrival times of the detected photons. For details on methods, see Tenne and colleagues (2019).

Quantum microscopy of a fixed cell sample is demonstrated in Figure 13.10 (Tenne et al., 2019). Q-ISM imaging was performed for a biological sample of microtubules labeled with quantum dots in fixed 3T3 cells. Figure 13.10 presents a super-resolved FR Q-ISM image of a 3 μm × 3 μm area as well as CLSM, ISM, FR ISM, and Q-ISM images analysed from the same data set. Although the SNR is lower in the Q-ISM and FR Q-ISM images (Figure 13.10d, e), all the visible features in the CLSM (Figure 13.10a) and ISM (Figure 13.10b) images are present with a finer resolution.

As expected from theoretical considerations, the FR ISM image (see Figure 13.10c) provides a similar transverse resolution to that of the unfiltered Q-ISM image (see Figure 13.10d). However, in the rather dense scene, the addition of a pronounced ringing artefact by Fourier reweighting can obscure faint features adjacent to bright ones. Another advantage of Q-ISM over CLSM, ISM, and FR ISM is an improved axial resolution.

Quantum entanglement, which was first experimentally demonstrated half a century ago (Freedman and Clauser, 1972), is found to overcome a key limitation on the speed, sensitivity, and resolution of a bioimaging technique called stimulated Raman scattering (SRS) gain microscopy (see (PhysicsWorld, 2021). Warwick Bowen and colleagues could show how correlations between the detection times of photons from a bright laser could greatly improve the signal-to-noise ratio of SRS, which allowed the detection of molecular samples with 14 percent lower concentrations than were previously possible.

The work of Casacio and colleagues (2021) demonstrated how to enable order-of-magnitude improvements in the signal-to-noise ratio and the imaging speed by using quantum photon correlations (Taylor and Bowen, 2016; Slusher, 1990). The microscope here was built to function as a coherent Raman microscope offering subwavelength resolution and incorporating bright quantum correlated illumination. The correlations allowed imaging of molecular bonds within a cell with a 35 percent improved signal-to-noise ratio compared with conventional microscopy, corresponding to a 14 percent improvement in concentration sensitivity.

13.5 QUANTUM BIOLOGY IN GENERAL MEDICAL TREATMENTS AND SPECIFIC THERAPIES

Quantum mechanics techniques that we have discussed here, and others in various stages of development, may emerge with promises and opportunities to help us deal with diseases and discover agents capable of fighting disease-associated disorders in physiology. Quantum mechanics may also play vital roles in therapies, e.g. in photodynamic therapy (Rossi, 2023), dental treatments (Fujii, 2021), etc.

Quantum techniques could allow researchers to understand the minute workings of our body at the subatomic level. We may be able to input a patient's medical and anatomical data into a quantum computer to create a "digital twin," a virtual replica of the person (Emani et al., 2021). The potential for quantum computing (QC) to aid in the merging of insights across different areas of biological sciences is considered to be enormous. Figure 13.11 presents regions of our body, from organs all way down to molecular levels, where various quantum mechanical approaches and associated algorithms are applicable. In this figure we see the chain of factors that leads from genetic variation to higher-level behaviors such as cognitive traits, including complex intermediate links, e.g. the molecular regulatory framework within cells, cell-to-cell interactions,

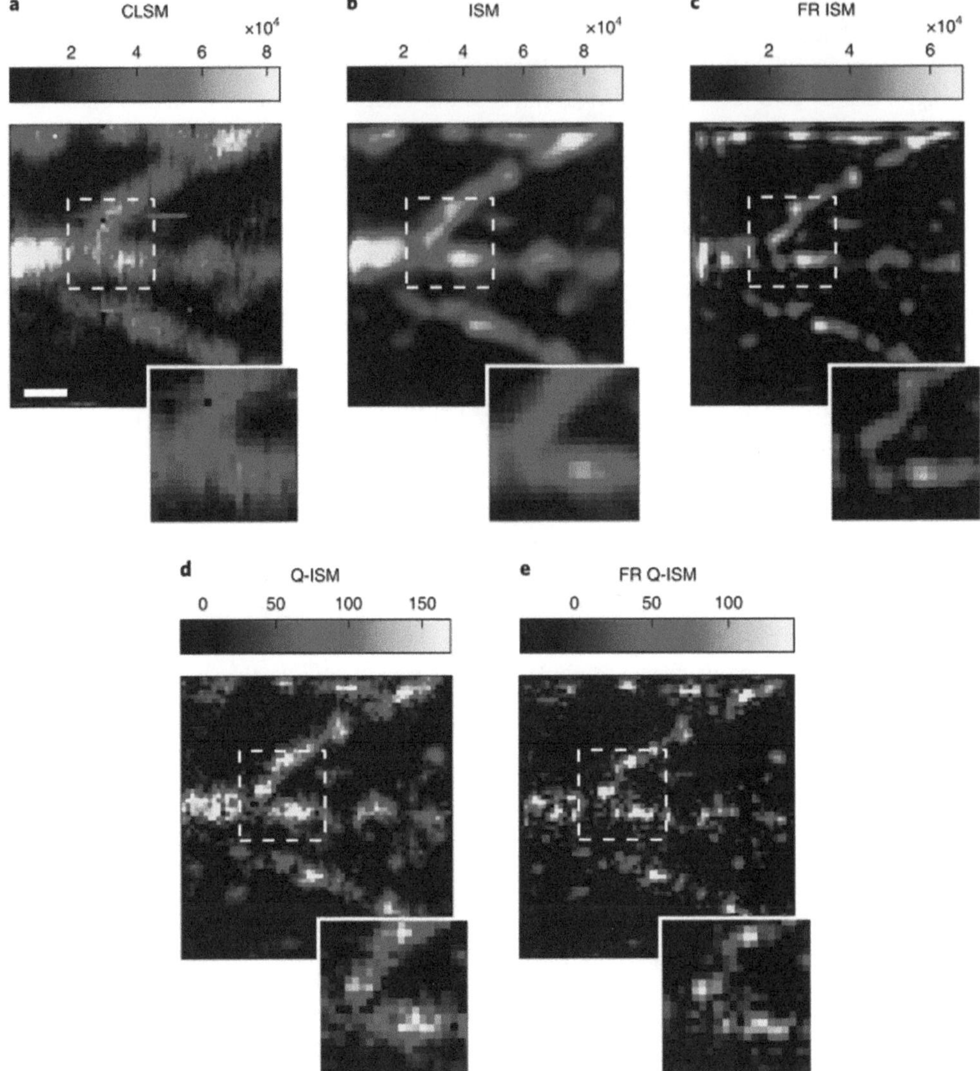

FIGURE 13.10 Q-ISM of labelled microtubule cell samples. (a–e) Images analysed from a confocal scan (50 nm steps, 100 ms pixel dwell time) of a 3 µm × 3 µm section of microtubules in a fixed 3T3 cell labeled with fluorescent quantum dots (QDot 625, Thermo Fisher): CLSM image (a), ISM image (b), FR ISM image (c), Q-ISM image (d), and FR Q-ISM image (e). The color bars for (a)–(c) represent the number of detected photon counts. The color bars for (d) and (e) represent the number of missing detected photon pairs. For clarity, the maximum intensity of the Fourier reweighted images shown in (c) and (e) is scaled to the maximum intensity of their source images in (b) and (d), respectively. Enlarged images of a section in each image, framed by the white dashed line, are shown below each image. Scale bar, 0.5 µm.

***Source*: Tenne et al. (2019).**

heterogeneity in cellular composition and behavior in tissues, and inter-regional connectivity patterns in the brain, etc.

This digital twin could be used as a test subject and doctors could run simulations with different drugs to see which one works best for individual patients. The digital twin, with the use of quantum

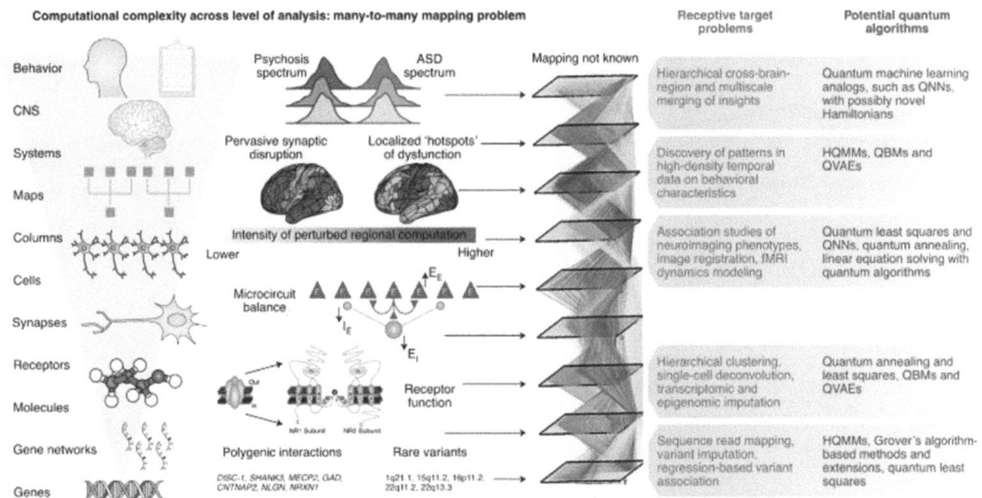

FIGURE 13.11 Complexity of linking levels of analyses from genetics to human behavior. The challenge consists, in part, of the need to interrogate the enormous search space for determining the mapping across levels, which constitutes a many-to-many probabilistic problem. Computational innovation will be a key effort to help close these gaps. Also shown are some of the ways in which QC can aid in the interrogation of these levels.

***Source*: portion adapted with permission from Foss-Feig et al. (2017); portion reproduced from Emani et al. (2021).**

technology, could do all the experimenting for our bodies and help us obtain personalized drugs (Pflitsch, 2022).

Quantum biology has been found to provide a promising biophysical theoretic system, on which to base pathophysiology and hopefully therapeutic strategies (Calvillo et al., 2022). Various laboratory studies related to electromagnetic fields, proton pumping in the mitochondrial respiratory chain, quantum theory of T-cell receptor (TCR)-degeneracy, theories on biophotons, pyrophosphates, or tubulin as possible carriers for neural information, and quantum properties of ions and protons, might be useful for understanding mechanisms of some serious immune, cardiovascular, and neural pathologies for which classic biomedical research, based on a biochemical approach, is struggling to find new therapeutic strategies. Figure 13.12 presents a flowchart of quantum biology studies and associated clinical issues (Calvillo et al., 2022).

Various studies have addressed the possibility of applications of quantum mechanical approaches in therapies. These studies demonstrate different biophysical aspects in biological systems that are associated with understanding diseases, imaging disease sites, and addressing drug binding to the concerned regions (Sommer et al., 2020; AbdullGaffar, 2022; Devi et al., 2022; Bordonaro, 2019; Heifetz, 2020; Blunt et al., 2022). For example, quantum dots (Bera et al., 2009; Bera et al., 2010) have been found useful in early diagnosis of cancer (Dong et al., 2017; see Figure 13.13).

Quantum dots have amazing applications in biological imaging, both *in vitro* and *in vivo* (Jin et al., 2011), as they are particularly significant for optical applications owing to bright, pure colors and ability to emit a rainbow of colors coupled with their high efficiencies and longer lifetimes along with high extinction coefficient. When semiconductor particles are made small enough, they exhibit quantum effects restricting the energies at which electrons and holes can exist within the particle. The atom-like energy states of quantum dots contribute to special optical properties, such as

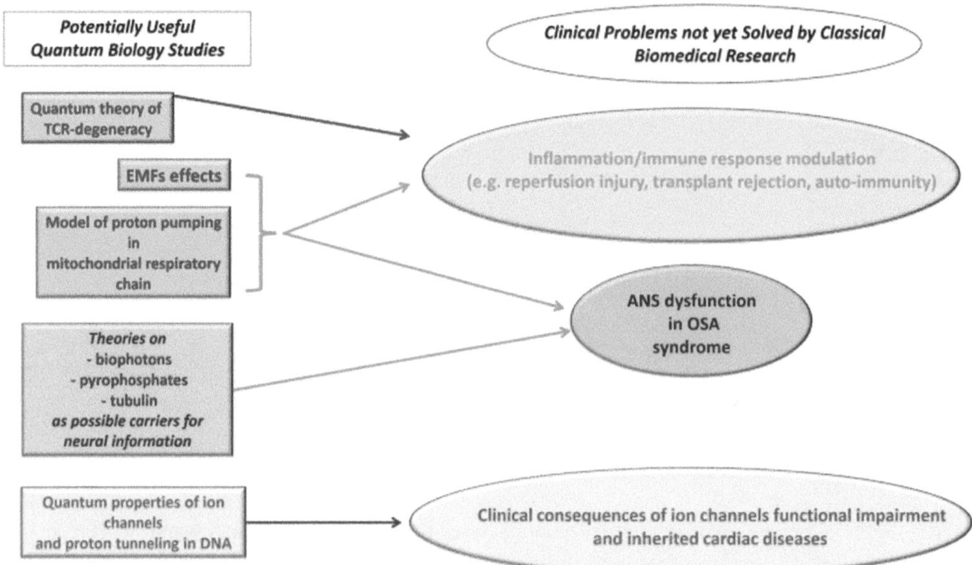

FIGURE 13.12 A flowchart demonstrating quantum biology studies and associated clinical issues.

Source: **Calvillo et al. (2022).**

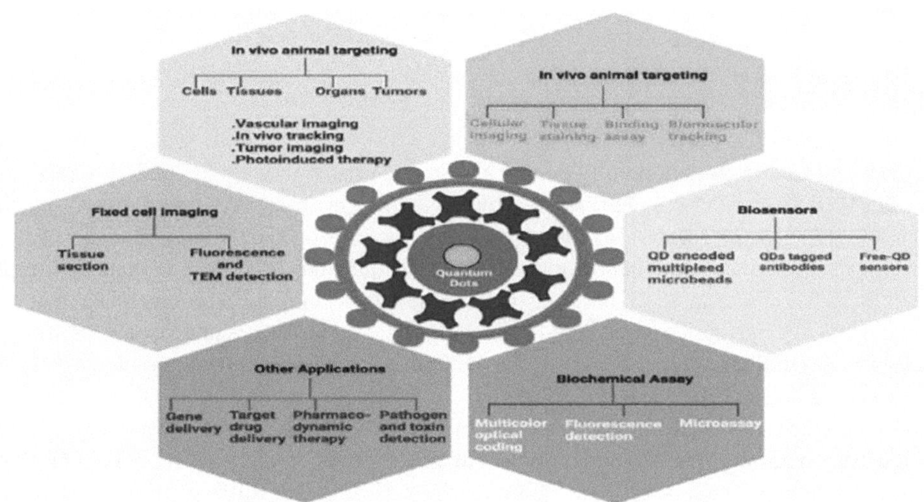

FIGURE 13.13 Various applications of quantum dots in bioimaging and diagnosis.

Source: **Devi et al. (2022).**

a particle-size-dependent wavelength of fluorescence, which is an effect used in fabricating optical probes for biological and medical imaging. A demonstration of biological imaging using quantum dots is presented in Figure 13.14, which may be utilized to demarcate between cancer cells and normal cells. Thus quantum dots may help diagnose cancer tissues.

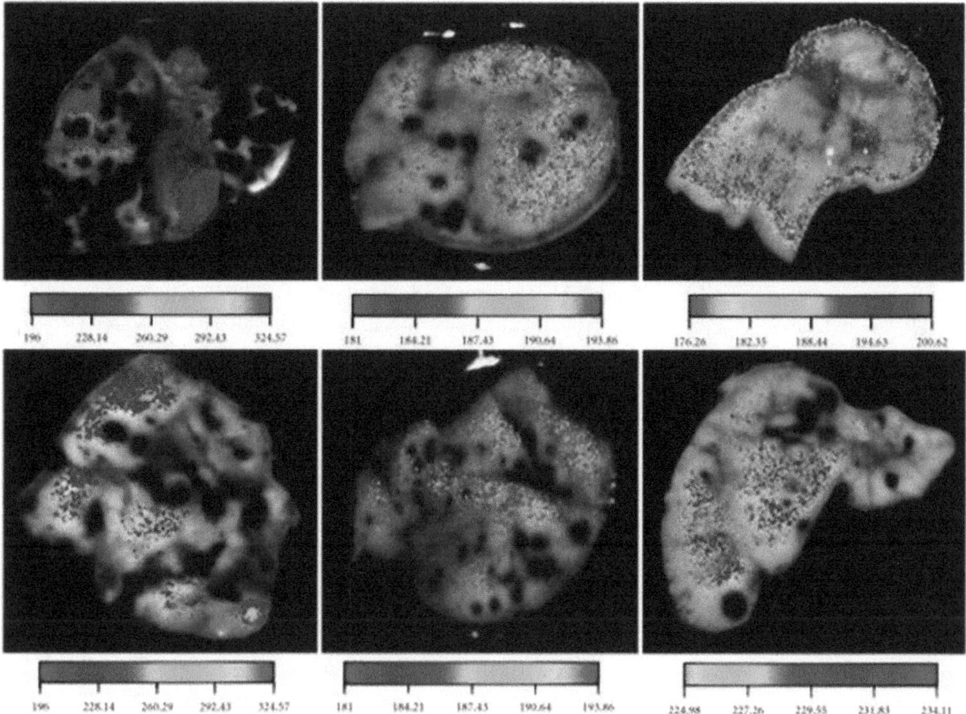

FIGURE 13.14 Quantum dot (QD)-loaded PEG-PE micelles (QD-Mic) was used with the monoclonal antibody (mAb) 2C5. Targeting of lung melanoma tumor by injecting 2C5 QD-Mic in two mice. Composite *ex vivo* images (white field image superimposed with the fluorescence intensity) of lungs from mice bearing metastatic B16F-10 lung melanoma tumor.

***Source*: Papagiannaros et al. (2010).**

REFERENCES

AbdullGaffar, B. (2022). Quantum mechanics and surgical pathology: a brief introduction. *Adv. Anat. Pathol.*, *29*(2), 108–116. doi: 10.1097/PAP.0000000000000328; PMID: 34799487

Abramavicius, D., Valkunas, L., & van Grondelle, R. (2004). Exciton dynamics in ring-like photosynthetic light-harvesting complexes: a hopping model. *Phys. Chem. Chem. Phys.*, *6*, 3097–3105

Ball, P. (2011). Physics of life: the dawn of quantum biology. *Nature*, *474*, 272–274. https://doi.org/10.1038/474272a

Bell, R.P. (1980). The tunnel eject in chemistry. New York: Chapman & Hall

Bera, D., Qian, L., & Holloway, P.H. (2009). Semiconducting quantum dots for bioimaging. In *Drug delivery nanoparticles formulation and characterization*, ed. Y. Pathak & D. Thassu. London: Informa UK

Bera, D., Qian, L., Tseng, T.K., & Holloway, P.H. (2010). Quantum dots and their multimodal applications: a review. *Materials*, *3*(4), 2260–2345. doi: 10.3390/ma3042260; PMCID: PMC5445848

Blunt, N.S., Camps, J., Crawford, O., Izsák, R., Leontica, S., Mirani, A., … & Holzmann, N. (2022). Perspective on the current state-of-the-art of quantum computing for drug discovery applications. *Journal of Chemical Theory and Computation*, *18*(12), 7001–7023

Bordonaro, M. (2019). Quantum biology and human carcinogenesis. *Biosystems*, *178*, 16–24. doi: 10.1016/j.biosystems.2019.01.010; PMID: 30695703

Brixner, T., Stenger, J., Vaswani, H.M., Cho, M., Blankenship, R.E., & Fleming, G.R. (2005). Two-dimensional spectroscopy of electronic couplings in photosynthesis. *Nature*, *434*(7033), 625–628

Calvillo, L., Redaelli, V., Ludwig, N., Qaswal, A.B., Ghidoni, A., Faini, A., … & Parati, G. (2022). Quantum biology research meets pathophysiology and therapeutic mechanisms: a biomedical perspective. *Quantum Reports*, *4*(2), 148–172

Casacio, C.A., Madsen, L.S., Terrasson, A., et al. (2021). Quantum-enhanced nonlinear microscopy. *Nature*, 594, 201–206. https://doi.org/10.1038/s41586-021-03528-w

Cha, Y., Murray, C.J., & Klinman, J.P. (1989). Hydrogen tunnelling in enzyme-reactions. *Science*, *243*, 1325–1330. doi: 10.1126/science.2646716

Cho, M., & Fleming, G.R. (2005). The integrated photon echo and solvation dynamics. II. Peak shifts and two-dimensional photon echo of a coupled chromophore system. *Journal of Chemical Physics*, *123*(11), 114506

Collini, E., Wong, C.Y., Wilk, K.E., Curmi, P.M., Brumer, P., & Scholes, G.D. (2010). Coherently wired light-harvesting in photosynthetic marine algae at ambient temperature. *Nature*, *463*(7281), 644–647

DeVault, D., & Chance, B. (1966). Studies of photosynthesis using a pulsed laser: temperature dependence of cytochrome oxidation rate in chromatium: evidence for tunneling. *Biophys. J.*, *6*, 825–847. doi: 10.1016/S0006-3495(66)86698-5

Devi, S., Kumar, M., Tiwari, A., Tiwari, V., Kaushik, D., Verma, R., … & Batiha, G.E.S. (2022). Quantum dots: an emerging approach for cancer therapy. *Frontiers in Materials*, *8*, 585

Dong, X., Moyer, M.M., Yang, F., Sun, Y.-P., and Yang, L. (2017). Carbon dots' antiviral functions against noroviruses. *Sci. Rep.*, *7*, 519. doi: 10.1038/s41598-017-00675-x

Dreyer, J., Moran, A.M., & Mukamel, S. (2003). Tensor components in three pulse vibrational echoes of a rigid dipeptide. *Bulletin of the Korean Chemical Society*, *24*(8), 1091–1096

Emani, P.S., Warrell, J., Anticevic, A., et al. (2021). Quantum computing at the frontiers of biological sciences. *Nat. Methods*, *18*, 701–709. https://doi.org/10.1038/s41592-020-01004-3

Emlen, S.T., Wiltschko, W., Demong, N.J., Wiltschko, R., & Bergman, B. (1976). Magnetic direction finding: evidence for its use in migratory indigo buntings. *Science*, *193*, 505–508. doi: 10.1126/science.193.4252.505

Engel, G.S., Calhoun, T.R., Read, E.L., Ahn, T.K., Mancal, T., Cheng, Y.C., Blankenship, R.E., & Fleming, G.R. (2007). Evidence for wavelike energy transfer through quantum coherence in photosynthetic systems. *Nature*, *446*, 782–786. doi: 10.1038/nature05678

Fleming, G.R., Scholes, G.D., & De Wit, A. (eds.). 2011. Proceedings of the 22nd Solvay conference in chemistry on quantum effects in chemistry and biology. *Proc. Chem.*, *3*, 1–355. doi: 10.1016/j.proche.2011.08.039

Foss-Feig, J.H., Adkinson, B.D., Ji, J.L., Yang, G., Srihari, V.H., McPartland, J.C., … & Anticevic, A. (2017). Searching for cross-diagnostic convergence: neural mechanisms governing excitation and inhibition balance in schizophrenia and autism spectrum disorders. *Biological Psychiatry*, *81*(10), 848–861

Freedman, S.J., & Clauser, J.F. (1972). Experimental test of local hidden-variable theories. *Physical Review Letters*, *28*(14), 938

Fujii, Y. (2021). Dental treatment and quantum mechanics. *Case Reports in Clinical Medicine*, *10*(7), 177–184

Godbeer, A.D., Al-Khalili, J.S., & Stevenson, P.D. (2015). Modelling proton tunneling in the adenine-thymine base pair. *Phys. Chem. Chem. Phys.*, *17*, 13034. doi: 10.1039/c5cp00472a

Heifetz, A. (ed.). (2020). *Quantum mechanics in drug discovery*. New York: Humana Press. https://doi.org/10.1007/978-1-0716-0282-9, ISBN978-1-0716-0281-2

Hochstrasser, R.M. (2001). Two-dimensional IR-spectroscopy: polarization anisotropy effects. *Chemical Physics*, *266*(2–3), 273–284

Hopfield, J.J. (1974). Electron transfer between biological molecules by thermally activated tunneling. *Proc. Natl Acad. Sci. USA*, *71*, 3640–3644. doi: 10.1073/pnas.71.9.3640

Hore, P.J., & Mouritsen, H. (2016). The radical-pair mechanism of magnetoreception. *Annu. Rev. Biophys.*, *45*, 299–344

Huskey, W.P., & Schowen, R.L. (1983). Reaction-coordinate tunneling in hydride-transfer reactions. *Journal of the American Chemical Society*, *105*(17), 5704–5706

Jang, S.J., Newton, M.D., & Silbey, R.J. (2004). Multichromophoric Forster resonance energy transfer. *Phys. Rev. Lett.*, *92*, 9312–9323

Jin, S., Hu, Y., Gu, Z., Liu, L., & Wu, H.C. (2011). Application of quantum dots in biological imaging. *Journal of Nanomaterials*, *2011*, 1–13

Jordan, P. (1941). *Die Physik und das Geheimnis des organischen Lebens*. Braunschweig: Friedrich Vieweg & Sohn

Jordan, P. (1932). Die Quantenmechanik und die Grundprobleme der Biologie und Psychologie. *Naturwissenschaften*, *20*, 815–821

Kalckar, J. (2013). *Foundations of quantum physics II (1933–1958)*. Amsterdam: Elsevier

Khmelinskii, I., & Makarov, V.I. (2020). Analysis of quantum coherence in biology. *Chemical Physics*, *532*, 110671

Kim, Y., Bertagna, F., D'Souza, E.M., Heyes, D.J., Johannissen, L.O., Nery, E.T., ... & McFadden, J. (2021). Quantum biology: an update and perspective. *Quantum Reports*, *3*(1), 80–126

Knox, R.S. (1996). Electronic excitation transfer in the photosynthetic unit: reflections on work of William Arnold. *Photosynth. Res.*, *48*, 35–39

Kunis, S., & Potts, D. (2007a). Stability results for scattered data interpolation by trigonometric polynomials. *SIAM Journal on Scientific Computing*, *29*(4), 1403–1419

Kunis, S., & Potts, D. (2007b). Stability results for scattered data interpolation by trigonometric polynomials. ArXiv.org, 0702019, https://arxiv.org/pdf/math/0702019.pdf

Layfield, J.P., & Hammes-Schiffer, S. (2014). Hydrogen tunneling in enzymes and biomimetic models. *Chem Rev.*, *114*(7), 3466–3494. doi: 10.1021/cr400400p; PMID: 24359189; PMCID: PMC3981923

Leegwater, J.A. (1996). Coherent versus incoherent energy transfer and trapping in photosynthetic antenna complexes. *J. Phys. Chem.*, *100*, 14403–14409

Löwdin, P.O. (1963). Proton tunneling in DNA and its biological implications. *Rev. Mod. Phys.*, *35*, 724–732. doi: 10.1103/RevModPhys.35.724

Maeda, K., Robinson, A.J., Henbest, KB., Hogben, H.J., Biskup, T., Ahmad, M., ... & Hore, P.J. (2012). Magnetically sensitive light-induced reactions in cryptochrome are consistent with its proposed role as a magnetoreceptor. *Proceedings of the National Academy of Sciences*, *109*(13), 4774–4779

McFadden, J., & Al-Khalili, J. (2018). The origins of quantum biology. *Proceedings of the Royal Society A*, *474*(2220), 20180674

Mohseni, M., Omar, Y., Engel, G.S., & Plenio, M.B. (eds). 2013. *Quantum effects in biology*. Cambridge: Cambridge University Press

Mohseni, M., Shabani, A., Lloyd, S., & Rabitz, H. (2014). Energy-scales convergence for optimal and robust quantum transport in photosynthetic complexes. *J. Chem. Phys.*, *140*, 035102

Müller, C.B., & Enderlein, J. (2010). Image scanning microscopy. *Phys. Rev. Lett.*, *104*, 198101

Novoderezhkin, V., Wendling, M., & van Grondelle, R. (2003). Intra- and interband transfers in the b800-b850 antenna of *Rhodospirillum molischianum*: Redfield theory modeling of polarized pump-probe kinetics. *J. Phys. Chem. B*, *107*, 11534–11548

Ornes, S. (2019). Quantum effects enter the macroworld. *Proceedings of the National Academy of Sciences*, *116*(45), 22413–22417

Panitchayangkoon, G., Hayes, D., Fransted, K.A., Caram, J.R., Harel, E., Wen, J., ... & Engel, G.S. (2010). Long-lived quantum coherence in photosynthetic complexes at physiological temperature. *Proceedings of the National Academy of Sciences*, *107*(29), 12766–12770

Papagiannaros, A., Upponi, J., Hartner, W., et al. (2010). Quantum dot loaded immunomicelles for tumor imaging. *BMC Med. Imaging*, *10*, 22. https://doi.org/10.1186/1471-2342-10-22

Pflitsch, M. (2022). Quantum biology: how quantum computing can unlock a new dimension of treating diseases. Forbes, December 12, 2022, www.forbes.com/sites/forbestechcouncil/2022/12/12/quantum-biology-how-quantum-computing-can-unlock-a-new-dimension-of-treating-diseases/?sh = 50cbe8012 33a (accessed January 15, 2023).

PhysicsWorld. (2021). Quantum microscope uses entanglement to reveal biological structures. June 15, 2021, https://physicsworld.com/a/quantum-microscope-uses-entanglement-to-reveal-biological-structures (accessed April 5, 2023)

Qin, S., Yin, H., Yang, C., Dou, Y., Liu, Z., Zhang, P., ... & Xie, C. (2016). A magnetic protein biocompass. *Nature Materials*, *15*(2), 217–226

Razavy, M. (2013). *Quantum theory of tunneling*. River Edge, NJ: World Scientific. https://doi.org/10.1142/4984

Renger, T., May, V. & Kuhn, O. (2001). Ultrafast excitation energy transfer dynamics in photosynthetic pigment-protein complexes. *Phys. Rep. Rev. Phys. Lett.*, *343*, 138–254

Ritz, T., Adem, S., & Schulten, K. (2000). A model for photoreceptor-based magnetoreception in birds. *Biophys. J.*, *78*, 707–718. doi: 10.1016/S0006-3495(00)76629-X

Rose, A.S., Bradley, A.R., Valasatava, Y., Duarte, J.M., Prlić, A., & Rose, P.W. (2016). Web-based molecular graphics for large complexes. In *Proceedings of the 21st international conference on Web3D technology*, pp. 185–186. New York: ACM.

Rossi, V.M. (2023). A quantum mechanical description of photosensitization in photodynamic therapy using a two-electron molecule approximation. arXiv preprint arXiv:2301.03653

Rubakov, V.A. (1984). Quantum mechanics in the tunneling universe. *Physics Letters B*, *148*(4–5), 280–286

Savikhin, S., Buck, D.R., & Struve, W.S. (1997). Oscillating anisotropies in a bacteriochlorophyll protein: evidence for quantum beating between exciton levels. *Chem. Phys.*, *223*, 303–312

Schnell, C. (2019). Quantum imaging in biological samples. *Nat. Methods*, *16*, 214. https://doi.org/10.1038/s41592-019-0346-6

Scholes, G.D., Fleming, G.R., Chen, L.X., Aspuru-Guzik, A., Buchleitner, A., Coker, D.F., ... & Zhu, X. (2017). Using coherence to enhance function in chemical and biophysical systems. *Nature*, *543*(7647), 647–656

Schrödinger, E. (1944). *What is life?* Cambridge: Cambridge University Press

Schulten, K., Staerk, H., Weller, A., Werner, H.-J., & Nickel, B. (1976). Magnetic field dependence of the geminate recombination of radical ion pairs in polar solvents. *Z. Phys. Chem*, *NF101*, 371–390. doi: 10.1524/zpch.1976.101.1-6.371

Schulten, K., & Weller, A. (1978). Exploring fast electron transfer processes by magnetic fields. *Biophys. J.*, *24*, 295–305. doi: 10.1016/S0006-3495(78)85378-8

Sheppard, C.J.R. (1988). Superresolution in confocal imaging. *Optik*, *80*, 53–54

Slusher, R.E. (1990). Quantum optics in the '80s. *Opt. Photon. News*, *1*, 27–30

Sommer, A.P., Schemmer, P., Pavláth, A.E., Försterling, H.D., Mester, Á.R., & Trelles, M.A. (2020). Quantum biology in low level light therapy: death of a dogma. *Annals of Translational Medicine*, *8*(7), 440

Tan, S.F. (2018). *Molecular electronic control over tunneling charge transfer plasmons modes*. Singapore: Springer.

Taylor, M. (2015). *Quantum microscopy of biological systems*. Cham: Springer

Taylor, M.A., & Bowen, W.P. (2016). Quantum metrology and its application in biology. *Phys. Rep.*, *615*, 1–59

Tenne, R., Rossman, U., Rephael, B., Israel, Y., Krupinski-Ptaszek, A., Lapkiewicz, R., ... & Oron, D. (2019). Super-resolution enhancement by quantum image scanning microscopy. *Nature Photonics*, *13*(2), 116–122

Thompson, B., & Howe, N.P. (2021). Quantum compass might help birds 'see' magnetic fields. *Nature Podcast*, June 23, 2021, www.nature.com/articles/d41586-021-01715-3 (accessed January 20, 2023)

van Amerongen, H., Valkunas, L., & van Grondelle, R. (2000). *Photosynthetic excitons*. Singapore: World Scientific

von Middendorff, A. (1859). Die Isepiptesen Russlands. *Mem. Acad. Sci. St. Petersbourg VI*, *8*, 1–143

Vulto, S.I., de Baat, M.A., Neerken, S., Nowak, F.R., van Amerongen, H., Amesz, J., & Aartsma, T.J. (1999). Excited state dynamics in FMO antenna complexes from photosynthetic green sulfur bacteria: a kinetic model. *Journal of Physical Chemistry B*, *103*(38), 8153–8161

Wang, L., Allodi, M.A., & Engel, G.S. (2019). Quantum coherences reveal excited-state dynamics in biophysical systems. *Nature Reviews Chemistry*, *3*(8), 477–490. https://doi.org/10.1038/s41570-019-0109-z

Watson, J.D., & Crick, F. H. (1953). Molecular structure of nucleic acids: a structure for deoxyribose nucleic acid. *Nature*, *171*, 964–967. doi: 10.1038/171964b0

Welch, G.R. (1992). An analogical "field" construct in cellular biophysics: history and present status. *Progress in Biophysics and Molecular Biology*, *57*(2), 71–128

Wiltschko, W., & Merkel, F.W. (1966). Orientierung zugunruhiger Rotkehlchen im statischen Magnetfeld. *Verh. Dtsch. Zool. Ges.*, *59*, 362–367

Wiltschko, W., & Wiltschko, R. (1996). Magnetic orientation in birds. *J. Exp. Biol.*, *199*, 29–38

Wiltschko, W, & Wiltschko, R. (1972). The magnetic compass of European robins. *Science*, *176*, 62–64. doi: 10.1126/science.176.4030.62

Wisnovsky, S., Lei, E.K., Jean, S.R., & Kelley, S.O. (2016). Mitochondrial chemical biology: new probes elucidate the secrets of the powerhouse of the cell. *Cell Chem. Biol.*, *23*, 917–927

Xin, H., Namgung, B., & Lee, L.P. (2018). Nanoplasmonic optical antennas for life sciences and medicine. *Nat. Rev. Mater.*, *3*, 228–243

Xin, H., Sim, W.J., Namgung, B., et al. (2019). Quantum biological tunnel junction for electron transfer imaging in live cells. *Nat Commun.*, *10*, 3245. https://doi.org/10.1038/s41467-019-11212-x

Xu, J., Jarocha, L.E., Zollitsch, T., et al. (2021). Magnetic sensitivity of cryptochrome 4 from a migratory songbird. *Nature*, *594*, 535–540. https://doi.org/10.1038/s41586-021-03618-9

Zhu, W., Esteban, R., Borisov, A.G., Baumberg, J.J., Nordlander, P., Lezec, H.J., ... & Crozier, K.B. (2016). Quantum mechanical effects in plasmonic structures with subnanometre gaps. *Nature Communications*, *7*(1), 1–14

14 Computational Biomodeling

Biomodeling helps to address biological systems simplistically, yet realistically enough to represent the real biological complexity. Biomodeling is a computational field within mainly the biophysics branch of science. It's just half a century old. But the progress made in modeling is enormous. It addresses both structural and functional aspects of independent biomolecules and biomolecular complexes where physics and mathematics formulas are utilized as the basis. The recent trend is to also include advanced algorithms, bioinformatics, and machine-learning techniques to process huge amounts of data and predict the geometry of biological systems, address their physical structural evolution, and understand their functions in biology.

14.1 BACKGROUND OF BIOMOLECULAR MODELING

Quantum computing (Ollitrault et al., 2021; Cheng et al., 2020), ab initio predictions (Rimola et al., 2012; Klepeis et al., 2005), neuromorphic computing (Christensen et al., 2022; Petruţ et al., 2021) and various other engineering and architectural modeling strategies and programs entered the biomolecular modeling arena (Schlick and Portillo-Ledesma, 2021) at a faster pace than other scientific fields. Recently, breathtaking results from AlphaFold on protein structure predictions (Jumper et al., 2021; Callaway, 2020) have encouraged scientists to consider the algorithm-based computational modeling of biomolecules more seriously besides (if not totally as an alternative to) applying traditional physics-principle-based technologies (Outeiral et al., 2022; Ołdziej et al., 2005). Although the structural modeling driven by deep learning (DL) methods has helped us achieve unprecedented success at predicting a protein's crystal structure, it is not clear whether these models can help us learn the physics of how proteins dynamically fold into their equilibrium structures (Outeiral et al., 2022). In modeling biomolecules' various energy states, transitions between energy states in physiological conditions may also be theoretically and computationally addressed using fundamental physics laws that can connect the predicted scenario with the experimentally observed results (Ashrafuzzaman et al., 2014; Ashrafuzzaman and Tuszynski, 2012a, 2012b). Biomolecular modeling deals with versatile techniques, most of which complement each other (van der Kamp et al., 2008). Modeling biomolecules started with simplified approaches for addressing simple aspects involving mainly physics principles (Warshel and Karplus, 1972; Warshel and Levitt, 1976), but the field later grew to address diversified structural and functional phenomena associated with the complex biological environment, including disease conditions (Schlick, 2010; Soto-Ospina et al., 2021; Hinbest et al., 2020). I shall address them in this chapter.

DOI: 10.1201/9781003287780-14

14.2 MODELING AND SIMULATING SMALL AND LARGE BIOMOLECULES

Modeling biomolecules has been enriched with versatile physical techniques and computational algorithms. Although simple principles of physics were applied to initiate this field, it has evolved such that now we can model not only the structure of both small and large biomolecules, but we can also understand various properties of the biomolecules, including their roles in creating biological systems. I shall address a few example cases here.

14.2.1 Modeling Biomolecular Structures

2013 Chemistry Nobel laureates Warshel, Levitt, and Karplus began their computer program-based molecular modeling in the 1970s, which later proved to have been the prelude to subsequent ground-breaking works helping scientists to understand crucial unsolved clues about biomolecular structures and functions. Modern molecular modeling saw its foundation created by this trio.

Merging a quantum mechanics-based program and classical calculations, Warshel and Karplus succeeded in calculating the energy changes of electrons within the p-bonds of simple planar molecules such as 1,6-diphenyl-1,3,5-hexatriene, five decades ago. Their program used quantum mechanics to model the effects of p-electrons, and classical mechanics on the atomic nuclei and s-electrons (Warshel and Karplus, 1972). Thus, they appeared to be the first to successfully demonstrate that it is possible to combine quantum and classical mechanics in a single model. Warshel and Levitt would later model a more complex system than ever before: an enzyme reaction involving making and breaking bonds, but the reactive site was part of a large protein containing thousands of atoms. They published the results of their simulation of the lysozyme enzyme cleaving a glycoside chain in 1976 (Warshel and Levitt, 1976). This model treated atoms within the enzyme's active site with quantum mechanics but dealt with the rest of the system more efficiently using molecular mechanics. This was more or less the beginning of computer-initiated modeling of molecular structures and interactions in biological systems, which later experienced so many developments involving versatile novel techniques and renewed algorithms during the last five decades (Phillips et al., 2020; Matlock et al., 2018; Vaidehi et al., 2009; Brooks et al., 2009; Meiler and Baker, 2006; Hamelberg et al., 2004; Martí-Renom et al., 2000). After the initiation of biomolecular modeling, trust in *in silico* simulation results and predictions on biomolecular structures and other related parameters was an issue. That's perhaps the reason why it took four decades for the trio to get the attention of the Nobel Committee.

Biomodeling started five decades ago. This field waas then gradually enriched by the addition of versatile modeling algorithms addressing complex aspects of biomolecules, utilizing various physical, mathematical, engineering, bioinformatics, genomics, machine-learning methods, etc. (Haghighatlari and Hachmann, 2019; Schlick, 2010).

14.2.2 Physical Modeling and Understanding How Biomolecules Work and Participate in Creating Biology

Biomolecular simulations emerged to exert huge impacts on our understanding of biology, as such studies help us understand the dynamics of biological macromolecules (Karplus and Kuriyan, 2005). Understanding the way biological macromolecules work requires knowledge of both structure and dynamics of the molecules in a biological environment. Molecular dynamics (MD) simulations provide powerful tools for the exploration of the conformational energy landscape accessible to these molecules. The father of modeling, Karplus, and his colleague Kuriyan surveyed two areas, protein folding and enzymatic catalysis, in which simulations were found to contribute to a general understanding of the mechanism (Karplus and Kuriyan, 2005).

In 1977, the first MD simulation of a protein was reported by Karplus himself. It consisted of a 9.2-ps trajectory for a small protein in a vacuum (Karplus, 1978). Within another decade a 210-ps

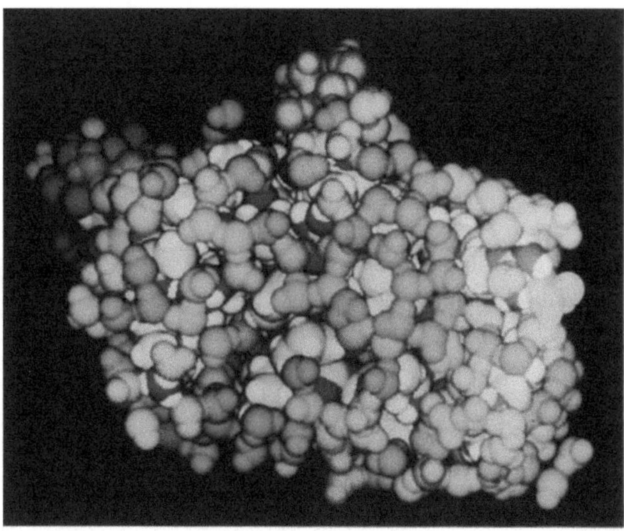

FIGURE 14.1 2,607 water molecules within 4 Å of the protein surface at 61.5 ps. The water molecules are distributed as follows: magenta, close to polar atom of the protein; green, close to nonpolar atom of the protein; cyan, not belonging to either class as mentioned here; yellow, those fitting inside the box. The color contrast in the figure here and all subsequent ones will be clear in the online version of the book.

***Source*: Levitt and Sharon (1988).**

simulation of the same small-size protein, bovine pancreatic trypsin inhibitor, in water was reported (Levitt and Sharon, 1988); see a snapshot of the simulated data in Figure 14.1.

By 2005 when an article by Karplus and Kuriyan focused on the chronological advancements in simulations (Karplus and Kuriyan, 2005), a phenomenal increase in computing power had already happened, and it had become routine to run simulations of much larger proteins, 1,000–10,000 times (≈10–100 ns) longer than the original simulation in 1977, in which the protein was surrounded by water and salt (see Figure 14.2).

Today we can deal with a much larger timescale (~microsecond, μs) and larger atoms in biomolecules; as an example, see Perilla and Schulten (2017). Simulations of over 64 million atoms for over 1 μs allowed Perilla and colleagues to conduct a comprehensive study of the chemical–physical properties of an empty HIV-1 capsid, including its electrostatics, vibrational and acoustic properties, and the effects of solvent (ions and water) on the capsid. The long-time simulations revealed critical details about the capsid with implications for biological function. Here a 1.2 μs MD trajectory of 64,423,983 atoms probed high spatial and temporal resolution characteristics of an empty HIV-1 capsid embedded in its native environment. This level of simulation-based detail is currently inaccessible to independent experimental methods (Perilla et al., 2015; Reddy and Sansom, 2016). The stability of the individual constituents of the capsid, namely CA pentamers and hexamers, was evaluated using root mean squared deviation (r.m.s.d.) from the starting structure throughout the 1.2 μs simulation (see Figure 14.3a). Changes in the cross-sectional area and height of the capsid are related to its global stability (see Figure 14.3b–d). Both the height and cross-sectional area were therefore calculated along the three principal axes of inertia of the capsid (see Figure 14.3b). Interestingly, during the first 400 ns of simulation, a shrinking of the capsid was observed, evidenced by a reduction in both the height and cross-sectional area at a rate of 0.025 nm ns−1 and 0.113 nm2 ns−1, respectively (Figure 14.3b, c). After 400 ns, the capsid height and cross-section reach a plateau for the remaining 800 ns of the simulation.

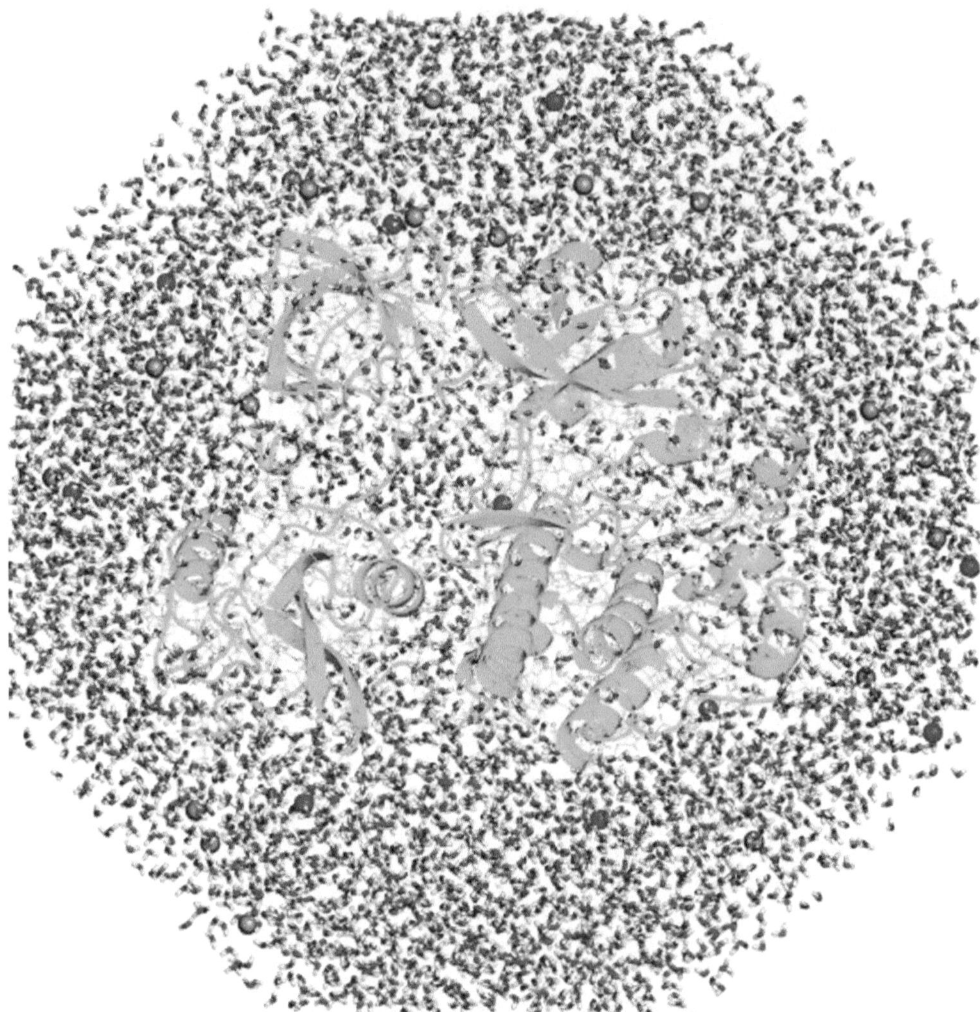

FIGURE 14.2 Simulation of a solvated protein. This slice through a simulation system shows a Src kinase protein (green) surrounded by ≈15,000 water molecules (oxygen atoms are red and hydrogen atoms are white). The simulation system consists of ≈50,000 atoms, including potassium and chloride ions (purple and orange spheres, respectively). A 1-ns MD trajectory for this system can be generated in four days by using a cluster of four inexpensive Linux-based computers.

***Source*: Karplus and Kuriyan (2005), courtesy of Olga Kuchment.**

The main objective of performing simulations is to help us address underlying physical properties associated with biomolecular structures and functions in biological systems. For example, see electrostatic (Figure 14.4) and acoustic properties (Figure 14.5) averaged over the final 400 ns of the HIV-1 capsid simulation. Long-time simulation may produce realistic values of physical parameters associated with biomolecular structures and functions.

Simulation of biomolecular structures for addressing associated physical conformations and biological functions requires the necessary computing power of the simulating facilities. One of the key elements in this success is the relentless pursuit and exploitation of state-of-the-art technology by the biomolecular simulation community, often in collaboration among experts in various fields.

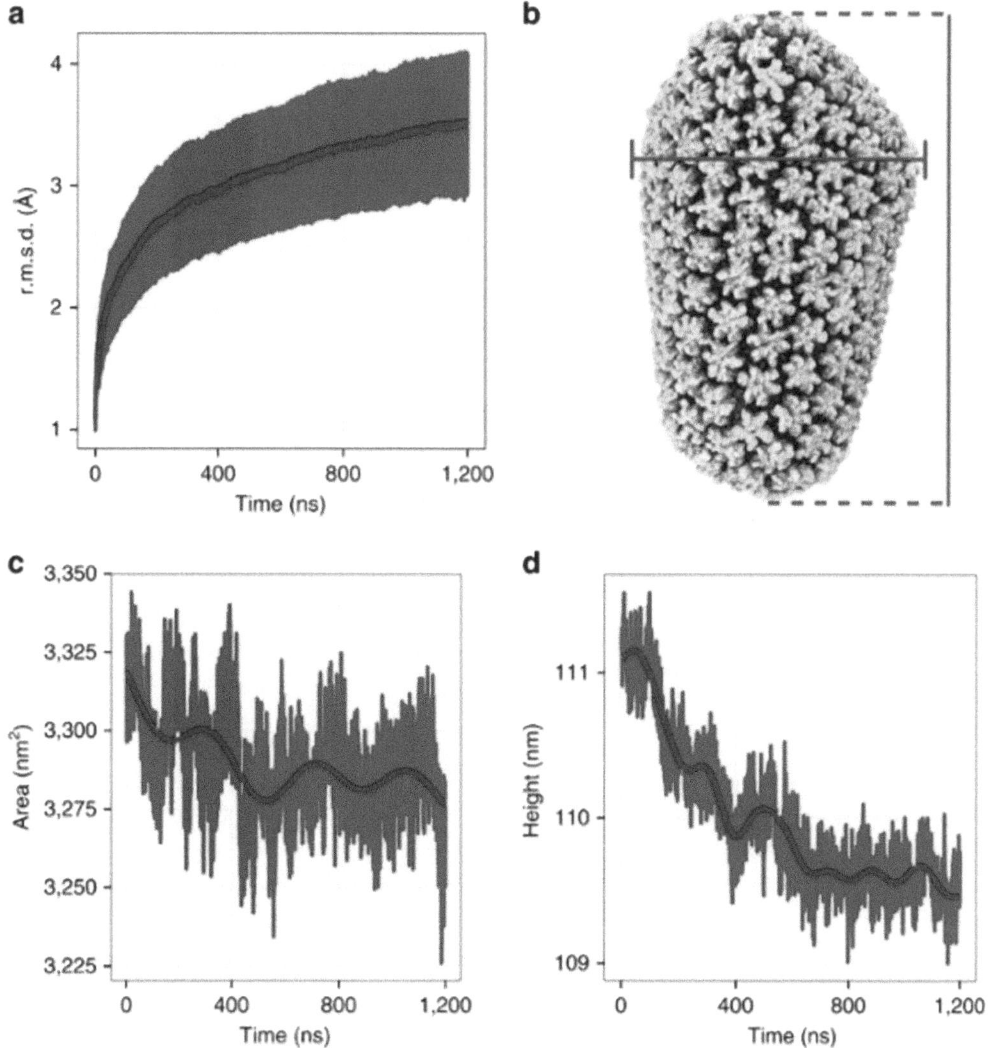

FIGURE 14.3 Stability of the HIV-1 capsid. (a) Time evolution of the root mean squared deviation (r.m.s.d.) for hexamers and pentamers, and the moving average with a window size of 10 ns is shown in blue. (b) For all area and height calculations, the three principal moments of inertia of the entire capsid define the x, y, and z axes. The cross-sectional area is estimated as the area of an ellipse where the major and minor axes are the maximal distance between parts of the capsid along the axes. The height of the capsid is defined as the longest distance from the tip (bottom) to the base (top) along the axis. (c) Time evolution of the capsid's cross-section, and the moving average with a 10 ns window size is shown in blue. (d) Time evolution of the height of the HIV-1 capsid; the moving average with a 10 ns window size is shown in blue.

***Source*: Perilla and Schulten (2017).**

The excellent utilization of supercomputers and technology by modelers led to comparable performance for landmark simulations with the world's fastest computers (Schlick and Portillo-Ledesma, 2021; see Figure 14.6). For the system size and the simulation time of each landmark simulation, see Figure 14.7 (for details see Schlick and Portillo-Ledesma, 2021). The time of simulation of biomolecular complexes scales up by about three orders of magnitude every decade (Vendruscolo

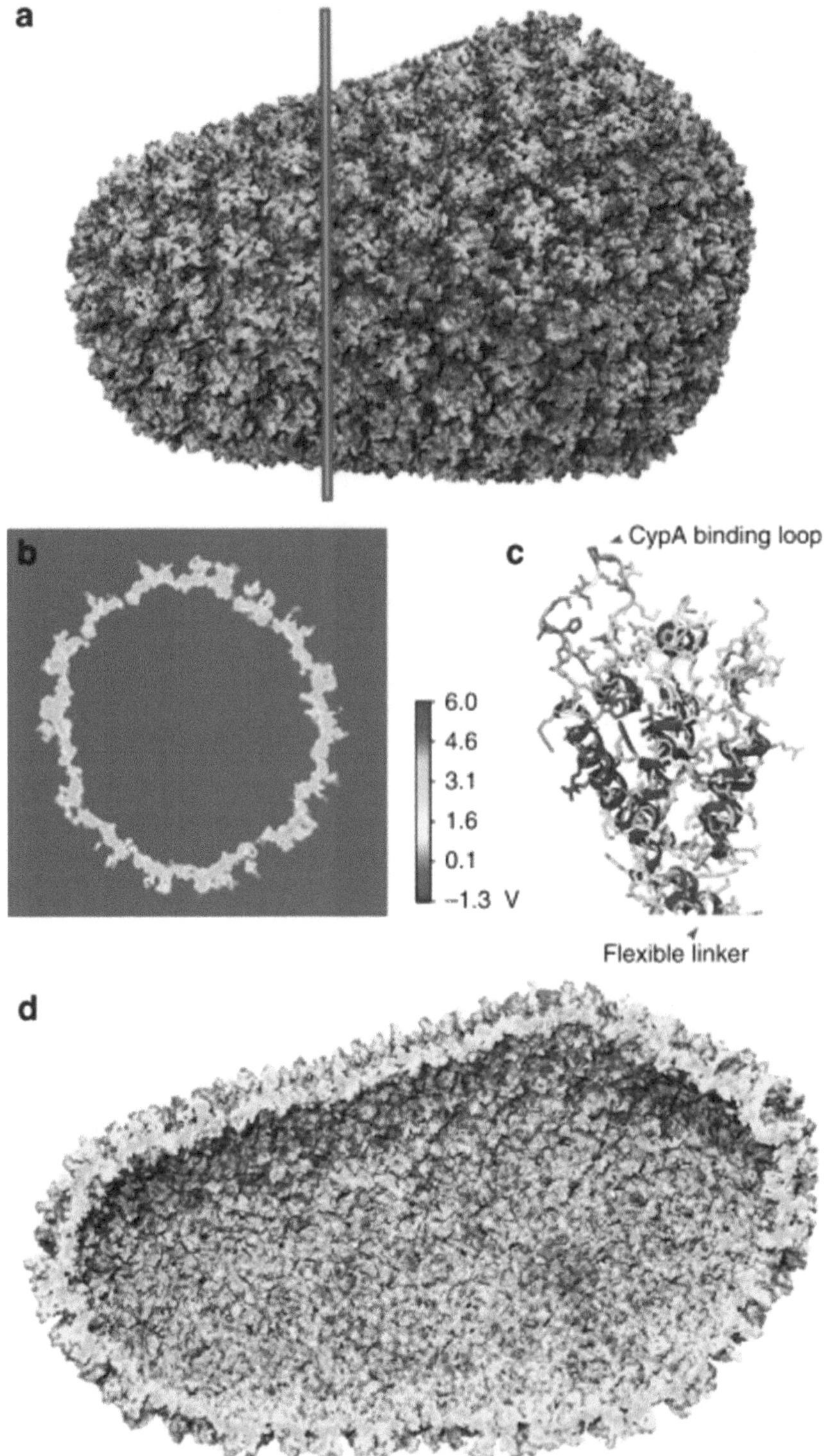

FIGURE 14.4 Electrostatic potential averaged over the final 400 ns of the HIV-1 capsid simulation. The electrostatic calculation includes all capsid atoms and all solvent molecules for a total of 64,423,983 atoms. The bar scale indicates the magnitude of the electrostatic potential in Volts, ranging from −1.3 V (red) to 6.0

FIGURE 14.4 *(continued)*

V (blue). (a) Exterior view of the HIV-1 electrostatic potential. The red line indicates the location of the cross-section shown in (b). (b) Cross-section of the electrostatic potential of the HIV-1 capsid. The bulk in the interior and exterior of the capsid assume the same electrostatic potential values, namely −1.3 V. (c) Electrostatics of the N-terminal domain of CA. The cypA binding loop and α-helix 4 show a significant potential difference to the inner core of the capsid in (d).

***Source*: Perilla and Schulten (2017).**

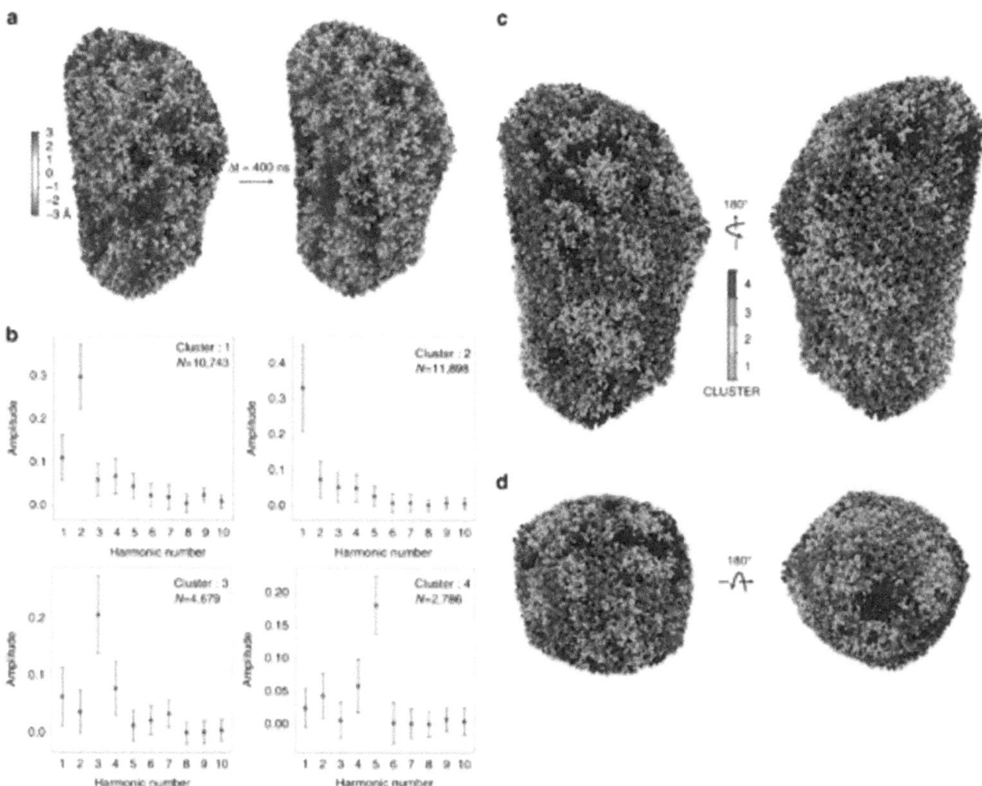

FIGURE 14.5 Acoustic properties of the HIV-1 capsid during the last 400 ns of simulations. (a) Structural fluctuations of the capsid observed between two states separated 400 ns apart, ranging from −3 to +3 Å (red to blue). (b) Periodograms of the time series of the capsid motions. Four classes of periodograms were found: the two largest classes ($N = 10{,}743$ and $N = 11{,}898$) are dominated by the two lower fundamental frequencies (2.38 and 4.76 MHz); conversely, the two smallest classes ($N = 4{,}679$ and $N = 2{,}786$) are dominated by two higher fundamental frequencies (7.14 and 11.9 MHz). The medoids of each class are represented in red dots in (b). The s.d. within each class is represented in blue error bars. (c) Projection of the periodogram clusters onto the structure of the capsid indicating the location of each cluster. (d) Top (base) and bottom (tip) view of the capsid colored by the periodogram clusters. The four clusters are coloured orange, green, cyan, and dark blue, respectively.

***Source*: Perilla and Schulten (2017).**

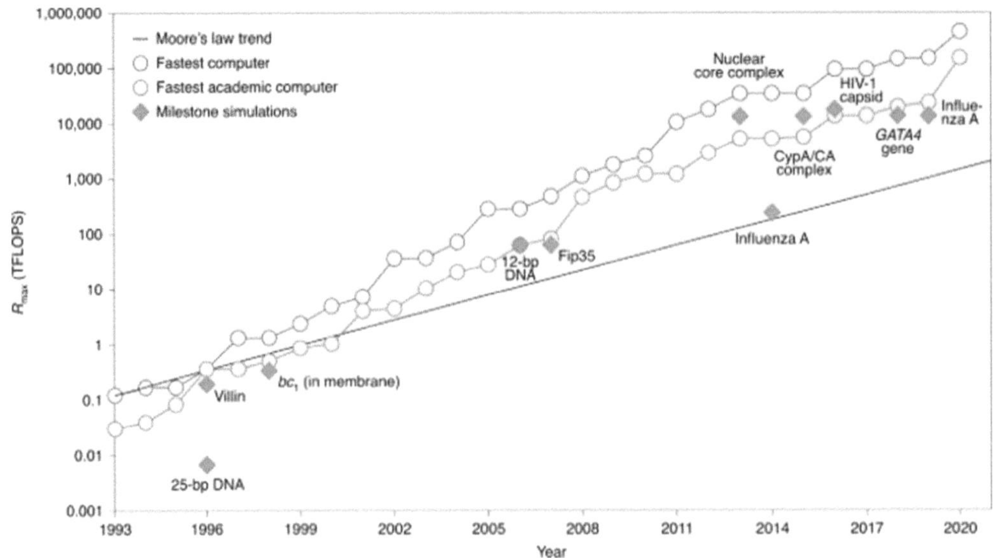

FIGURE 14.6 Performance of landmark simulations compared with the world's fastest supercomputers and Moore's law trend. Plot of the computational system ranked first (blue) and the highest ranked academic computer (orange) as reported in R_{max} according to the LINPACK benchmark as assembled in the Top500 supercomputer lists (www.top500.org). R_{max} is the unit used to define computer performance in TFLOPS (trillion floating point operations per second). Landmark simulations (green diamonds) are dated assuming calculations were performed about a year before publication, except for the publications in 1998, which we assumed were performed in 1996. These include, from 1996 to the present, 25-bp DNA using National Center for Supercomputing Applications (NCSA) Silicon Graphics Inc. (SGI) machines; villin protein using the Cray T3E900; bc_1 membrane complex using the Cray T3E900; 12-bp DNA using MareNostrum/Barcelona; Fip35 protein using NCSA Abe clusters; nuclear core complex using Blue Waters; influenza A virus using the Jade Supercomputer; CypA/CA complex using Blue Waters; HIV-1 capsid using Titan Cray XK7; *GATA4* gene using Trinity Phase 2; and influenza A virus H1N1 using Blue Waters. As Blue Waters has opted out of the Top500, we use estimates of sustained system performance/sustained petascale performance (SSP/SPP) from 2012 and 2020. For system size and simulation time of each landmark simulation, see Fig. 14.7.

***Source*: Schlick and Portillo-Ledesma (2021).**

and Dobson, 2011), and this progress is faster than Moore's law (Moore, 1965), projecting a doubling of the time scale every two years. Therefore, within just five decades of the first-ever modeling of biomolecular structures, we have already started seeing a gray future due to the computing power issue. Technological developments have to progress in parallel with sectors of scientific thought and developing algorithms.

14.3 MODELING-BASED ADDRESS OF BIOMOLECULAR INTERACTION ENERGETICS

Computer modeling can provide molecular details of processes such as conformational change, binding, and transportation of small molecules and proteins, that are hard to capture in experimental

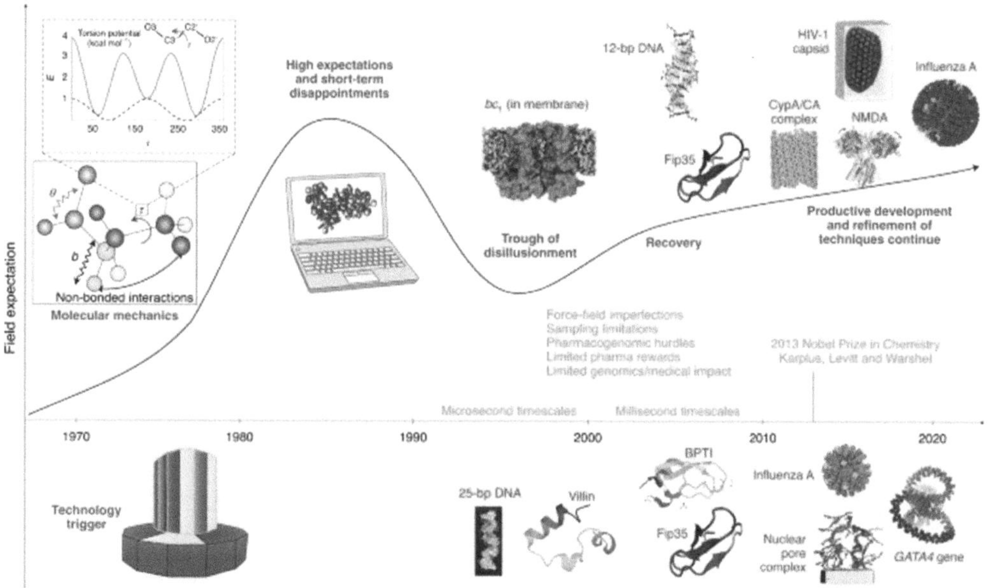

FIGURE 14.7 Expectation curve for the field of biomolecular modeling and simulation. The field started with comprehensive molecular mechanics efforts, and it took off with the increasing availability of fast workstations and later supercomputers. In the molecular mechanics illustration (top left panel), symbols b, θ, and τ represent bond, angle, and dihedral angle motions, respectively, and non-bonded interactions are also indicated. The torsion potential (E) contains two-fold (dashed black curve) and three-fold (solid violet curve) terms. Following unrealistically high short-term expectations and disappointments concerning the limited medical impact of modeling and genomic research on human disease treatment, better collaborations between theory and experiment have ushered the field to its productive stage. Challenges faced in the decade 2000–2010 include forcefield imperfections, conformational sampling limitations, some pharmacogenomics hurdles, and limited medical impact of genomics-based therapeutics for human diseases. Technological innovations that have helped drive the field include distributed computations and the advent of the use of GPUs for biomolecular computations. The molecular-dynamics-specialized supercomputer Anton made it possible in 2009 to reach the millisecond timescale for explicit-solvent all-atom simulations. The 2013 Nobel Prize in Chemistry awarded to Levitt, Karplus, and Warshel helped validate a field that lagged behind experiment and propel its trajectory. Along the timeline, we depict landmark simulations: 25-bp DNA (5 ns and ~21,000 atoms), villin protein (1 μs and 12,000 atoms); bc_1 membrane complex (1 ns and ~91,000 atoms); 12-bp DNA (1.2 μs and ~16,000 atoms); Fip35 protein (10 μs and ~30,000 atoms); Fip35 and bovine pancreatic trypsin inhibitory (BPTI) proteins (100 μs for Flip35 and 1 ms for BPTI, and ~13,000 atoms); nuclear pore complex (1 μs and 15.5 million atoms); influenza A virus (1 μs and >1 million atoms); *N*-methyl-D-aspartate (NMDA) receptor in membrane (60 μs and ~507,000 atoms); tubular cyclophilin A/capsid protein (CypA/CA) complexes (100 ns and 25.6 million atoms); HIV-1 fully solvated empty capsid (1 μs and 64 million atoms); *GATA4* gene (1 ns and 1B atoms); and influenza A virus H1N1 (121 ns and ~160 million atoms).

***Source*: Schlick and Portillo-Ledesma (2021). Adapted with permission from Cambridge University Press (timeline); Angela Barragan Diaz, University of Chicago, NIC Center for Macromolecular Modeling and Bioinformatics (*bc1* (in membrane)), American Chemical Society (12-bp DNA); under a Creative Commons license (CypA/CA complex), (HIV-1 capsid), Springer Nature Ltd (NMDA); courtesy of Lorenzo Casalino, UC San Diego (influenza A, top), Elsevier (25-bp DNA), Wiley (Fip35), Cell Press (influenza A, bottom), PLoS (Nuclear pore complex), Wiley (*GATA4* gene).**

studies. Yang et al. (2016) review simulation studies of both proteins and small molecules, considering three aspects:

- conformation sampling
- transportations of small molecules in enzymes
- enzymatic reactions involving small molecules

Detailed methodology developments and examples of simulation case studies in the field have been presented in various publications (e.g. Wang et al., 2019; Bruce et al., 2018; Yang et al., 2016).

Small molecule interaction energetics may be understood in predicted simulation modes (Bennett et al., 2020). Bennett and colleagues utilized both MD simulations and machine-learning (ML) techniques for accomplishing and comparing some important information on specific biomolecular energetics, besides accomplishing ML-based general modeling of biomolecules for understanding molecular structures, associated reaction coordinates, and other necessary biophysical properties active in biological systems (Brandt et al., 2018; Giannakoulias et al., 2021; Rickert and Lieleg, 2022). In understanding drug effects, small molecular interactions with target biomolecules or sections within a large biomolecular structure require us to address pinpointed interaction energetics. The chemical environment inside cells is mostly heterogeneous, so drugs must cross the hydrophobic cellular membrane to reach their intracellular targets, and hydrophobicity is an important driving force for drug–protein binding. Atomistic MD simulations are routinely used to calculate free energies of small molecules binding to proteins, crossing lipid membranes, and solvation, but are computationally expensive. ML and empirical methods are also used throughout drug discovery but rely on experimental data, limiting the domain of applicability. Bennett and colleagues (2020) performed atomistic MD simulations calculating 15,000 small molecule free energies of transfer from water to cyclohexane. The large dataset was then used to train ML models that predict the free energies of transfer.

To the end of integrating MD and ML calculations, ML models were constructed for predicting atomistic MD free energies. As a simple test system, the free energy was chosen for small molecule partitioning between bulk solvents (water and cyclohexane). MD simulations were used to calculate 15,000 small molecule free energies of transfer for training the model. Figure 14.8 depicts a schematic representation of the calculations and the ML model.

From the MD simulations, 3D atomic features were extracted that were used for training an ML model to predict the free energy of transfer from water to an interface ($\Delta G_{\text{Water-Interface}}$ or $\Delta G_{\text{W–I}}$) and to bulk cyclohexane ($\Delta G_{\text{Water-Cyclohexane}}$ or $\Delta G_{\text{W–C}}$). Figure 14.9 shows the process used for calculating free energies of transfer for 15,000 molecules with atomistic MD simulations. For a base molecule the umbrella sampling was used across the water–cyclohexane system (see Figure 14.9A). Relative thermodynamic integration (TI) free energy calculations were then used for perturbing the base molecule into 10 different molecules (differing by one functional group) in the three chemical environments: water, interface, and cyclohexane (see Figure 14.9B). The free energy profiles show that almost all molecules have a free energy minimum at the interface (see Figure 14.9C). Because these are drug-like small molecule fragments, most are expected to be amphipathic. In contrast to the interface, about one-half of the molecules prefer water and the other half prefer cyclohexane, with a maximum +40 and a minimum −40 kJ/mol (see Figure 14.9C). Figure 14.9D shows representative molecules with different transfer free energies.

From the 15,000 MD free energy calculations, Bennett and colleagues tested numerous ML methods for predicting the transfer free energies. A 2D Morgan molecular fingerprint with shallow ridge regression was tested. More complex 3D-CNN and SG-CNN were also built from the MD data (see Figure 14.10). The initial step for ML model building is splitting the data between training data and testing data. For small molecule cheminformatics, data splitting is crucial because chemical space is extremely large, whereas training data (usually from experiments) can be sparse, relatively

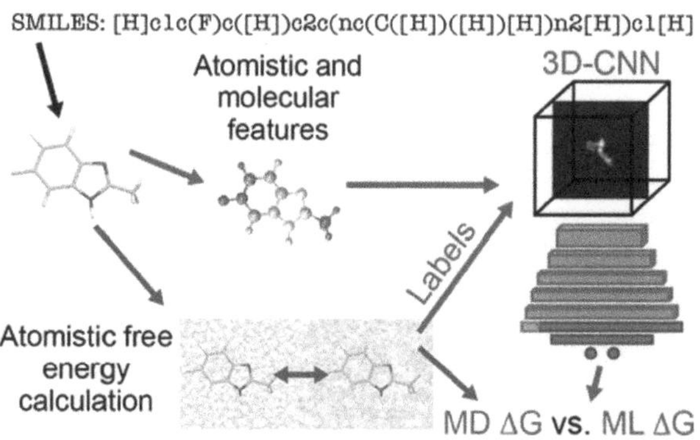

FIGURE 14.8 Scheme of the simulation to deep convolutional neural network (CNN) workflow. The relative free energy for moving 15,000 small molecules between water and cyclohexane was computed with atomistic MD simulations. From the simulations, features were extracted, such as each atom's partial charge, average number of water contacts, and molecular features, including number of hydrogen bonds and size/shape. A 3D-CNN and spatial graph CNN (SG-CNN) were then constructed using the atomic and molecular features to predict the free energies of transfer.

Source: **Bennett et al. (2020).**

small, and/or correlated. Data splitting was performed here using a scaffold split (30) and compared against a random split. Scaffold splitting divides the data into training, validation, and test partitions by binning molecules based on scaffold and assigning entire bins to one of the three partitions. As a result, the distribution between test and training partitions is more diverse based on the Tanimoto distance, better mimicking use cases and highlighting any failures of the model to generalize. For details on ML modeling, readers may consult Bennett et al. (2020).

The best model is expected to achieve an accuracy similar to that expected from the MD free energy calculations. Comparing 3D atomistic molecular features to the traditional molecular fingerprints for the ML model shows improvement using the atomic data. The model transferability of the ML model (Figure 14.10) is shown by testing the model on a unique set of molecules. The thermodynamic properties of small molecules are explained using MD simulation data associated with ML modeling (Bennett et al., 2020). A spatial graph neural network model was found to achieve the highest accuracy, followed closely by a 3D-convolutional neural network, and shallow learning based on the chemical fingerprint is significantly less accurate. A mean absolute error of ~4 kJ/mol compared to the MD calculations was achieved for the ML model.

Besides associating MD simulation and ML modeling data as seen in the study of Bennett and colleagues (2020), an interesting recent ML-based study associated predictions with highly abstract experimental parameters in model biological assays (Giannakoulias et al., 2021). Here predictions on the effects of acridon-2-yl-alanine (Acd) incorporation on protein yield and solubility were made using a combination of modeling assays fed by ML algorithms. Acd-containing mutants were simulated in PyRosetta, an interactive Python-based interface to the powerful Rosetta molecular modeling suite (Hostetler et al., 2018), and ML was performed using either the decomposed values of the Rosetta energy function or changes in residue contacts and bioinformatics. Using these feature sets, which represent Rosetta score function specific (RSFS) and bioinformatics-derived terms, ML models were trained to predict highly abstract experimental parameters such as mutant protein yield and solubility and displayed robust performance on well-balanced holdouts.

Hostetler and colleagues (2018) focused on establishing an accurate method for predicting Acd mutant protein soluble fraction, which would help to report on whether mutation of a residue to Acd should be tolerated and represents a class of experiments. The same group demonstrated earlier

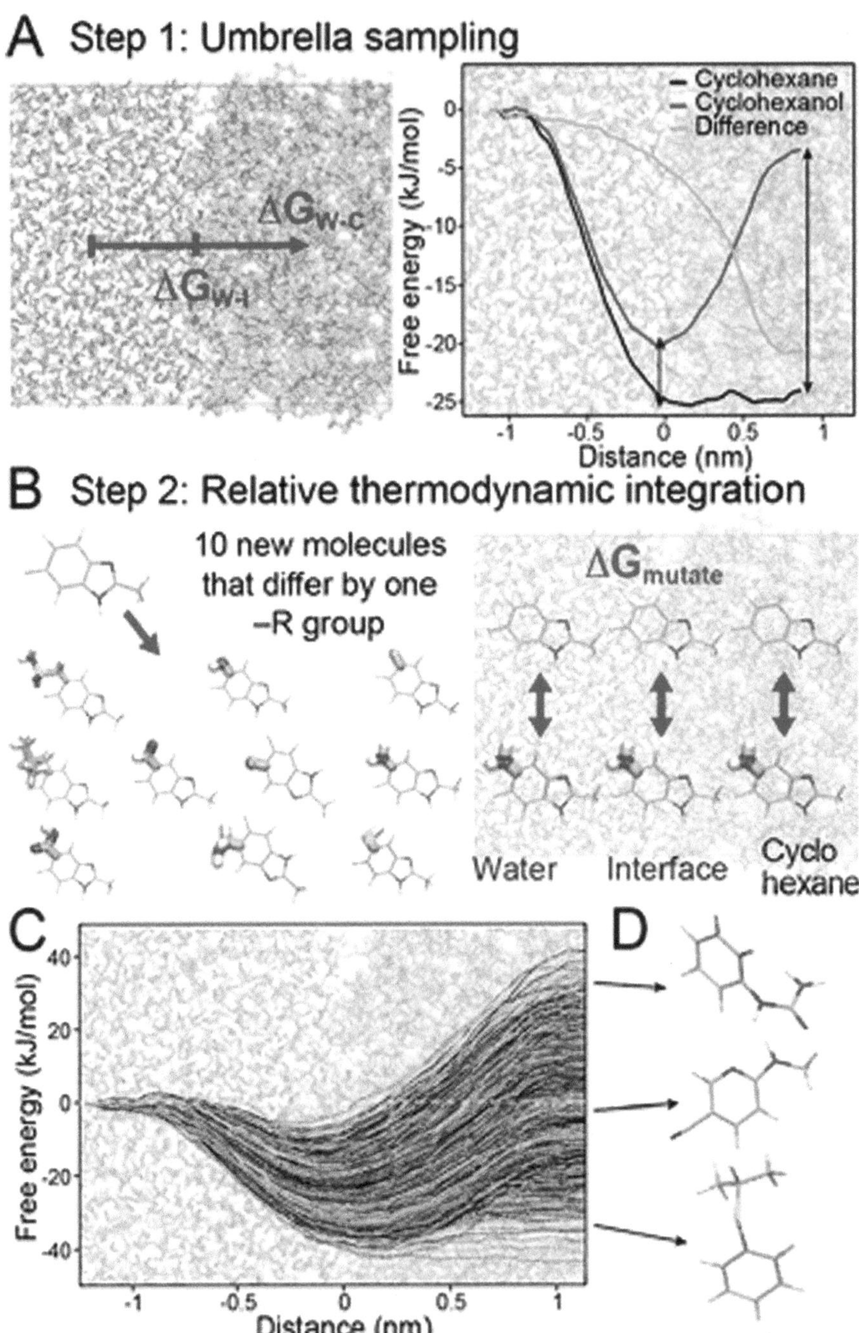

FIGURE 14.9 Atomistic molecular dynamics free energy calculations for water, interface, and cyclohexane (W–I–C) transfer. (A) Umbrella sampling calculations for base molecules moving from water to cyclohexane, across the interface. (B) The base molecule is then perturbed into 10 different molecules, and the free energy for the perturbation in W–I–C is computed. (C) Free energy profiles from 1,500 umbrella sampling calculations. Behind the curves, the molecular system is shown semitransparently, with water oxygens colored blue and cyclohexane carbons gray. (D) Molecular structure for three molecules.

Source: **Bennett et al. (2020).**

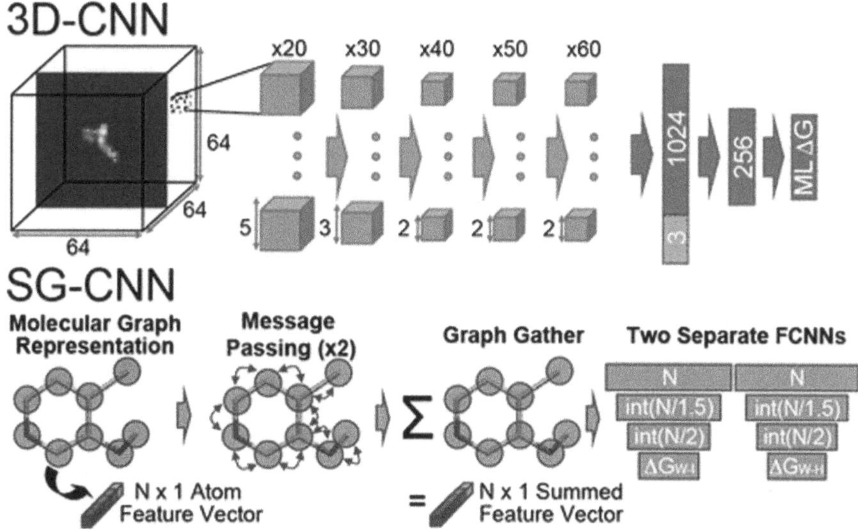

FIGURE 14.10 Architecture of the 3D-CNN and the SG-CNN used to predict the free energy. Predictions for $\Delta G_{W\text{–}C}$ and $\Delta G_{W\text{–}I}$ for the different ML models are shown in Table 14.1. Overall, the predictions for $\Delta G_{W\text{–}C}$ have a higher correlation than $\Delta G_{W\text{–}I}$ predictions but also higher MAE. This is likely because $\Delta G_{W\text{–}C}$ has much more variability than $\Delta G_{W\text{–}I}$, leading to larger error. The better correlation is likely due to the more complicated chemical environment of the interface. Overall, the best performing model is the SG-CNN followed by the 3D-CNN, with the shallow ECFP model performing the worst. Comparing Tables 14.1 and 14.2 including features from the MD simulations improves the predictions for all the models and metrics, although by a small amount.

Source: **Bennett et al. (2020).**

TABLE 14.1
Summary of ML Results for Predicting the $\Delta G_{W\text{–}C}$ and $\Delta G_{W\text{–}I}$ with Features from the MD Simulations

		SG-CNN	joined 3D-CNN	ECFP
interface	R^2	**0.70**	0.64	0.51
	MAE	**3.60**	3.91	4.70
	Pearson	**0.84**	0.81	0.72
cyclohexane	R^2	**0.84**	0.82	0.79
	MAE	**5.42**	5.67	6.15
	Pearson	**0.92**	0.91	0.89

that the predictivity of Rosetta methods could be dramatically improved through the use of RCSFs (Giannakoulias et al., 2020). RCSFs, or Rosetta Custom Score Functions, are known to rely on the generation of structural models in PyRosetta, which were subsequently scored with the Rosetta full atom score function (beta_nov_16) (Park et al., 2021), a linear combination of energetic score terms such as Lennard–Jones potential, electrostatics, implicit solvation, etc., that serves an analogous role to forcefields in MD simulations. Isolated score terms were then subsequently recombined through ML for generating an RCSF (see Figure 14.11A). Given the adaptability of RCSFs, their utility in this problem was then investigated. Besides focusing on determining whether the constitutive energies of the Rosetta score function were more correlative than the structure-independent

TABLE 14.2
Summary of ML Results for Predicting the ΔG_{W-C} and ΔG_{W-I} without Features from the MD Simulations

		SG-CNN	joined 3D-CNN	ECFP
interface	R^2	**0.68**	0.65	0.40
	MAE	**3.72**	3.93	5.34
	Pearson	**0.83**	0.81	0.34
cyclohexane	R^2	**0.82**	0.78	0.60
	MAE	**5.71**	6.38	8.41
	Pearson	**0.91**	0.89	0.77

bioinformatics terms, the descriptive capacity of combining these terms was investigated through multiple linear regression (MLR). Then a set of Empirical Score Terms (ESTs), which are based on contacts and structure independent bioinformatics, was investigated. Following the identification of both Rosetta and EST features that demonstrated significantly improved correlation, ML was used to train RCSFs and Empirical Score Functions (ESFs) and compare their ability to predict Acd mutant protein solubility and yield. Feature importance analysis of the most predictive models from both RCSF and ESF methods was then performed in order to see which features inspired the predictivity for a better understanding of the system. The whole effort demonstrated that ML approaches could help predict complex phenomena related to the unnatural amino acid (Uaa) incorporation.

For simulating the Acd mutant LexA and RecA proteins, LexA and RecA protein structures – from PDB IDs 1JHH (Luo et al., 2001) and 3CMW (Mechanism of homologous recombination from the RecA-ssDNA/dsDNA structures) – were preprocessed and energy minimized (see details Chaudhury et al., 2010). The parent structures of LexA and RecA (energy minimized version) were then mutated to incorporate Acd at experimentally-tested positions using PyRosetta. The Rosetta amino acid params and side chain rotamer library files used to make Acd mutant proteins are found in Hostetler et al. (2018). Following mutation to Acd, the structures were subjected to five independent cartesian FastRelax simulations, where only residues with an alpha carbon to alpha carbon (C_α-C_α) distance within 8 Å of Acd were allowed to be refined (Shringari et al., 2020).

For ML modeling the experimental dataset was prepared by first assigning a response class to each sample based on the distribution of the dependent variable. Figure 14.11B shows the spatial distribution of the Acd mutants and the effect they have on LexA and RecA solubility. For soluble yield, total yield, and soluble fraction, the response classes were balanced at 520 nM, 1600 nM, and 39 percent, respectively. The complete dataset contains 51 data points, of which 32 points are mutations in LexA and the remaining 19 are from RecA10. The ML unseen holdout dataset during hyperparameter tuning represented 20 percent of the total dataset. The holdout datasets may be found on GitHub (https://github.com/ejp-lab/EJPLab_Computational_Projects/tree/master/RML_ACD/Dataset).

We avoid presenting the detailed analysis of the huge volume results related to ML model-based predictions, but this will be found in Giannakoulias et al. (2021). The study reveals that the construction of custom scoring functions in order to predict a specific phenomenon is a superior strategy compared to the development of a singular generalized scoring function-forcefield for a Uaa such as Acd. It reveals that there may be a method for predicting current datasets, suggesting that the construction of a dataset that includes both different Uaas and multiple proteins may yield a generally predictive system of interest to the field. The ML model feature importance analyses demonstrated that various terms corresponding to hydrophobic interactions, desolvation, and amino acid angle preferences played a pivotal role in predicting tolerance of mutation to Acd. It is thus evident that the application of ML to features extracted from the simulated structural models might allow for the

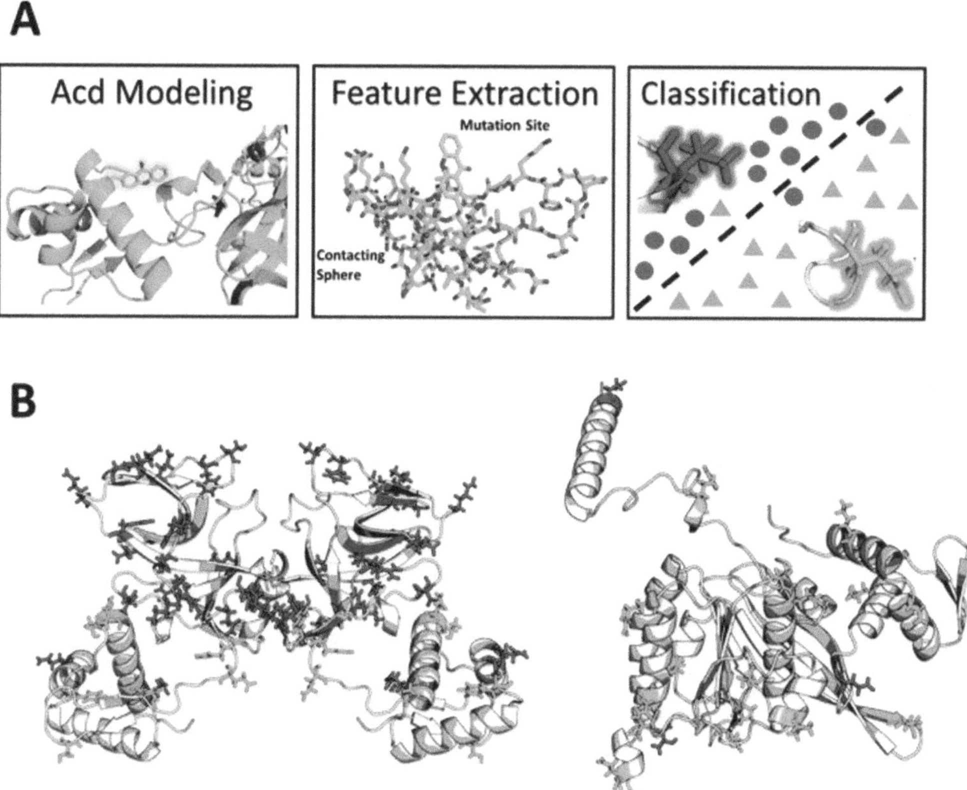

FIGURE 14.11 Schematic of the computational workflow for developing a Rosetta Custom Score Function or Empirical Score Function (A), spatial distribution and effect on soluble fraction of Acd mutants (B). LexA homo-dimer (left), RecA monomer (right). Note: red corresponds to soluble fraction percentage equal to or below 39 percent, and green above 39 percent.

Source: **Giannakoulias et al. (2021).**

accurate prediction of diverse and abstract biological phenomena, beyond the predictivity of traditional modeling and simulation approaches.

14.4 MODELING THE CELL

The whole-cell model accounted for all annotated gene functions and was validated against a broad range of data (Karr et al., 2012). This computational model of the life cycle of the human pathogen *Mycoplasma genitalium* includes all of its molecular components and their interactions. The model provides insights into various cellular behaviors, including *in vivo* rates of protein-DNA association and an inverse relationship between the durations of DNA replication initiation and replication. Figure 14.12 presents a whole-cell model integrating 28 submodels of diverse cellular processes (Karr et al., 2012).

While developing the cell model, specific approaches were taken by Karr and colleagues (2012) to divide the total functionality of the cell into modules, model each independently of the others, and integrate these submodels together. Here 28 modules were defined (see Figure 14.12) and they independently built, parameterized, and tested a submodel of each. Each module was modeled using the most appropriate mathematical representation, e.g. the metabolism aspect was modeled using

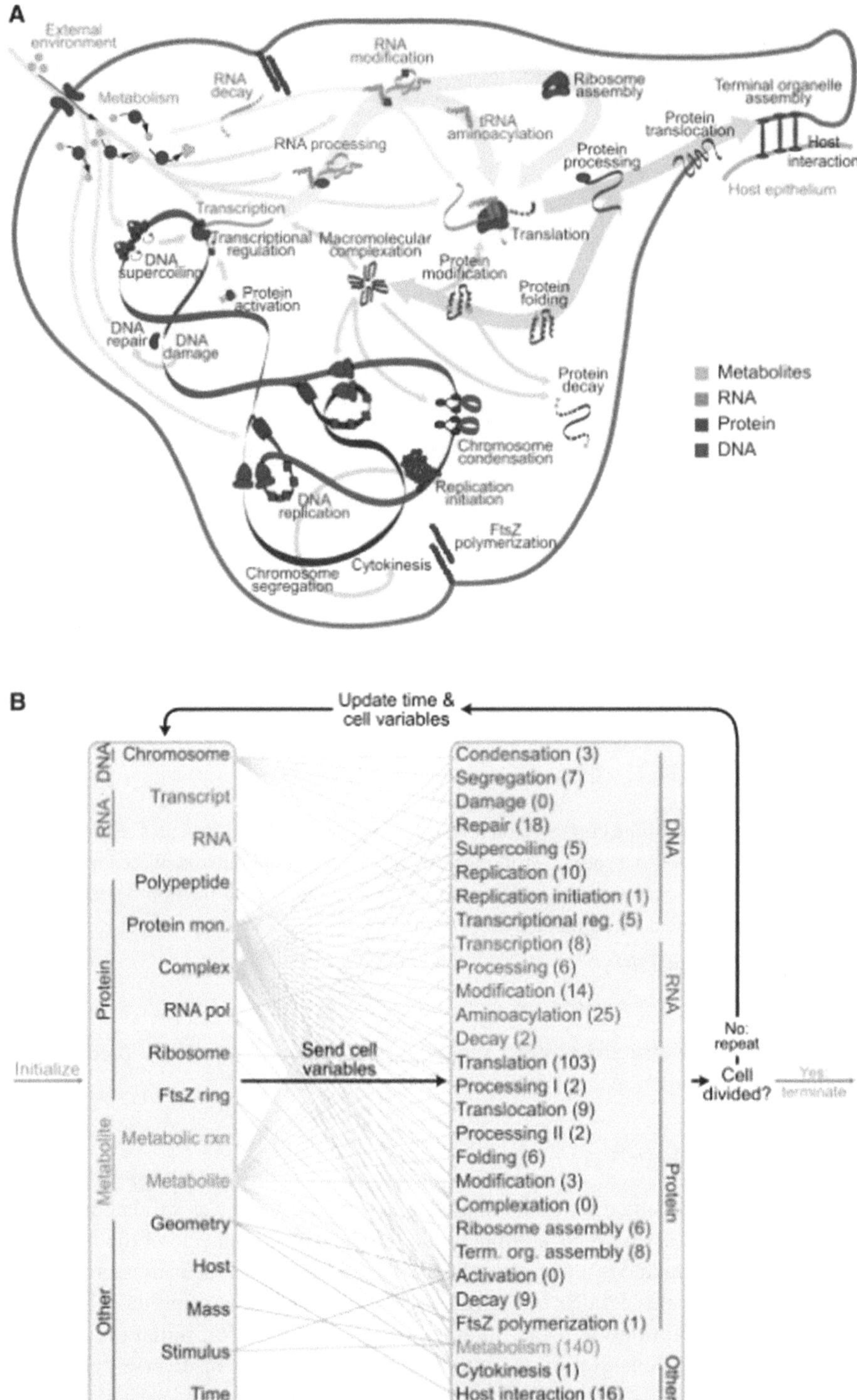

FIGURE 14.12 *M. genitalium* whole-cell model integrates 28 submodels of diverse cellular processes. (A) Diagram schematically depicts the 28 submodels as colored words – grouped by category as metabolic (orange), RNA (green), protein (blue), and DNA (red) – in the context of a single *M. genitalium* cell with

FIGURE 14.12 (*continued*)

its characteristic flask-like shape. Submodels are connected through common metabolites, RNA, protein, and the chromosome, which are depicted as orange, green, blue, and red arrows, respectively. (B) The model integrates cellular function submodels through 16 cell variables. First, simulations are randomly initialized to the beginning of the cell cycle (left gray arrow). Next, for each 1 s time step (dark black arrows), the submodels retrieve the current values of the cellular variables, calculate their contributions to the temporal evolution of the cell variables, and update the values of the cellular variables. This is repeated thousands of times during the course of each simulation. For clarity, cell functions and variables are grouped into five physiologic categories: DNA (red), RNA (green), protein (blue), metabolite (orange), and other (black). Colored lines between the variables and submodels indicate the cell variables predicted by each submodel. The number of genes associated with each submodel is indicated in parentheses. Finally, simulations are terminated upon cell division when the septum diameter equals zero (right gray arrow).

***Source*: reproduced with permission from Karr et al. (2012).**

the flux-balance analysis (Suthers et al., 2009). The RNA and protein degradation were modeled as Poisson processes.

In order to integrate the 28 submodels into a unified model, Karr and colleagues began with the assumption that the submodels are approximately independent on short timescales (<1 s), although previously several methods were developed for integrating ordinary differential equations (ODEs) (Atlas et al., 2008; Browning et al., 2004; Castellanos et al., 2007; Domach et al., 1984) with Boolean, probabilistic, and constraint-based submodels (Chandrasekaran and Price, 2010; Covert et al., 2008). *In silico* studies are performed by performing simulations by running through a loop in which the submodels are run independently at each time step but depend on the values of variables determined by the other submodels at the previous time step. Figure 14.11B presents the simulation algorithm and the relationships between the submodels and the cell variables.

Karr and colleagues (2012) have summarized the whole-cell model as follows:

- An entire organism may be modeled in terms of its molecular components.
- Complex phenotypes may be modeled by integrating cell processes into a single model.
- Unobserved cellular behaviors and phenomena may be predicted by the model, as found for *M. genitalium*.
- Novel biological processes and parameters may be predicted using this modeling, as done for *M. genitalium*.

The features of the model are provided in a sketch (Figure 14.13).

Whole-cell modeling addresses various cellular phenomena and specific mechanisms, helping us to address disease states, drug discovery, etc. As an example, we may mention that the model may help predict the DNA-binding protein interactions. Models are often used to predict molecular interactions that may be difficult or prohibitive to investigate experimentally (especially in animal model cases). The model therefore is popularly used to create opportunities for making such predictions, often computationally, in the context of the entire cell. The genomic distribution of DNA-binding proteins (Vora et al., 2009) or the detailed diffusion dynamics of specific DNA-binding proteins (Bratton et al., 2011) have previously been addressed, but the whole-cell model is found to predict both the instantaneous protein chromosomal occupancy as well as the temporal dynamics and interactions of every DNA-binding protein at the genomic scale at single-cell resolution (see Figure 14.14). Figure 14.14A illustrates the average predicted chromosomal protein occupancy and the predicted chromosomal occupancies for DNA and RNA polymerase and the replication initiator DnaA, which are 3 of the 30 DNA-binding proteins represented in the model.

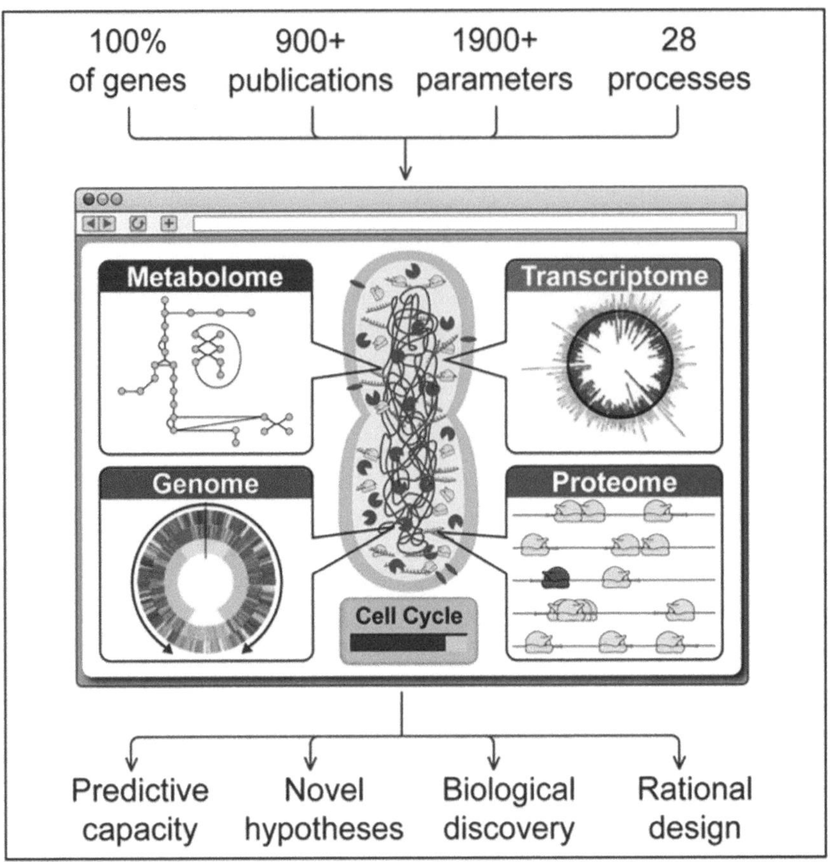

FIGURE 14.13 Various mechanistic contributions to whole-cell modelling and associated information, including research accomplishments, are depicted here.

***Source*: reproduced with permission from Karr et al. (2012).**

Consistent with another experimental study (Vora et al., 2009), the predicted high-occupancy RNA polymerase regions are found to correspond to highly transcribed ribosomal RNAs (rRNAs) and transfer RNAs (tRNAs). In contrast, the predicted DNA polymerase chromosomal occupancy was reported to be significantly lower and biased toward the terC.

The progress made by Karr and colleagues (2012) in their computational whole-cell model was tremendous over the previously initiated one by Tomita (2001). However, Karr and colleagues later acknowledged that their model was indeed incomplete and they proposed future work to be done to merge functional and structural modeling and expand their scopes (Karr et al., 2015). They outlined 11 fundamental and practical principles of whole-cell modeling to illuminate a path toward complete models (see Figure 14.15). I avoid explaining all these 11 terminologies here, but they are briefly explained in Karr et al. (2015). Karr and colleagues also outlined a strategy here to construct future whole-cell models (see Figure 14.16), highlighting important areas for further research.

Recently, the whole-cell model was further improved and subjected to computer-aided design, which included additional in-depth functionalities and perspectives (Marucci et al., 2020). A perspective on how whole-cell, multiscale models could transform design-build-test-learn cycles in synthetic biology was presented here (see Figure 14.17). These models were expected to significantly aid in the design and learning phases while reducing experimental testing by presenting case studies spanning from genome minimization to various cell-free systems. Unavoidable challenges

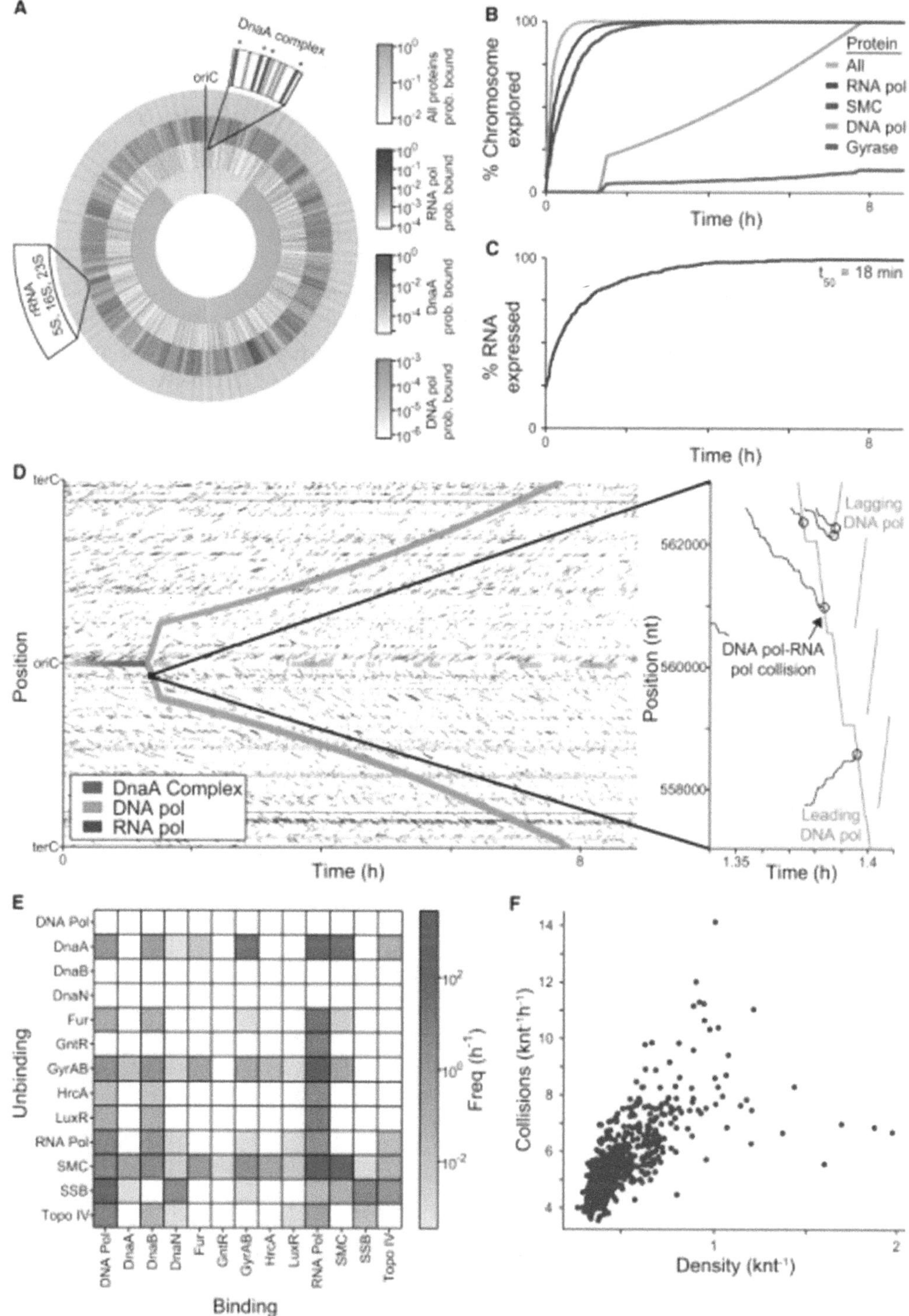

FIGURE 14.14 The model highlights the central physiological role of DNA-protein interactions. (A) Average density of all DNA-bound proteins and of the replication initiation protein DnaA and DNA and RNA polymerase of a population of 128 *in silico* cells. Top magnification indicates the average density of DnaA at several sites near the oriC; DnaA forms a large multimeric complex at the sites indicated with asterisks, recruiting DNA

FIGURE 14.14 (*Continued*)

polymerase to the oriC to initiate replication. Bottom left indicates the location of the highly expressed rRNA genes. (B and C) Percentage of the chromosome that is predicted to have been bound (B) and the number of genes that are predicted to have been expressed (C) as functions of time. SMC is an abbreviation for the name of the chromosome partition protein (MG298). (D) DNA-binding and dissociation dynamics of the oriC DnaA complex (red) and of RNA (blue) and DNA (green) polymerases for one *in silico* cell. The oriC DnaA complex recruits DNA polymerase to the oriC to initiate replication, which in turn dissolves the oriC DnaA complex. RNA polymerase traces (blue line segments) indicate individual transcription events. The height, length, and slope of each trace represent the transcript length, transcription duration, and transcript elongation rate, respectively. The inset highlights several predicted collisions between DNA and RNA polymerases that lead to the displacement of RNA polymerases and incomplete transcripts. (E) Predicted collision and displacement frequencies for pairs of DNA-binding proteins. (F) Correlation between DNA-binding protein density and frequency of collisions across the chromosome. Both (E) and (F) are based on 128 cell-cycle simulations.

***Source*: reproduced with permission from Karr et al. (2012).**

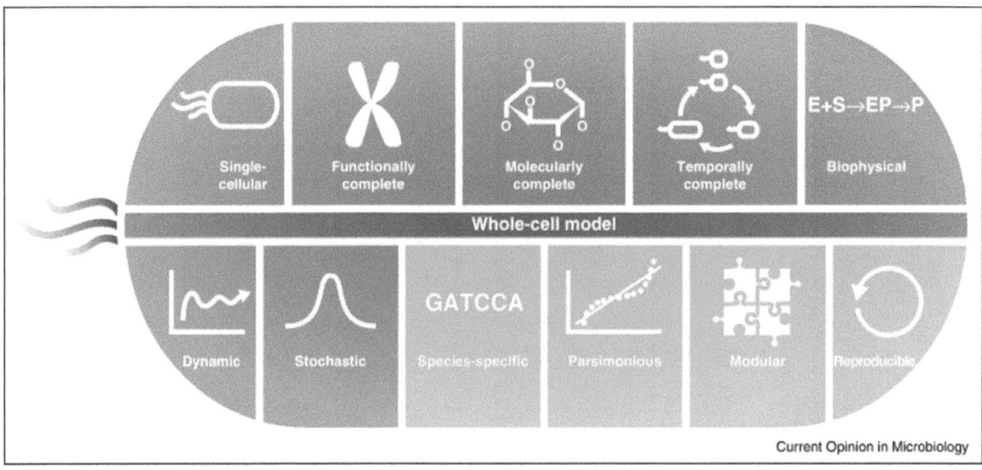

FIGURE 14.15 Fundamental (blue) and practical (green) principles of whole-cell modeling. No existing model satisfies every principle. The most advanced functional models are incomplete and do not fully represent molecular biophysics. The most advanced structural models do not represent cellular-scale behavior. Further work is needed to merge functional and structural modeling and expand their scope. Copied from ref. Karr et al. (2015).

***Source*: Karr et al. (2015).**

for the realization of this vision were also discussed in this article. The possibility of describing and building whole cells *in silico* was also expected to offer an opportunity for developing increasingly automatized, precise, and accessible computer-aided designing tools and strategies. Application of advanced bioinformatics and machine-learning tools and algorithms in future may also aid heavily in achieving a complete whole-cell model (Al-Kofahi et al., 2018; Greenwald et al., 2022; Eslami et al., 2022). This modeling issue is, however, a subject that needs renormalizing at least every decade due to tremendous progress in our understanding of the cell, especially in recent years in cell-based proteomics (Perkel, 2021), genomics (Adey, 2021; Hong and Park, 2020), metabolomics (Heinemann and Zenobi, 2011; Ali et al., 2019; Liu and Yang, 2021), etc.

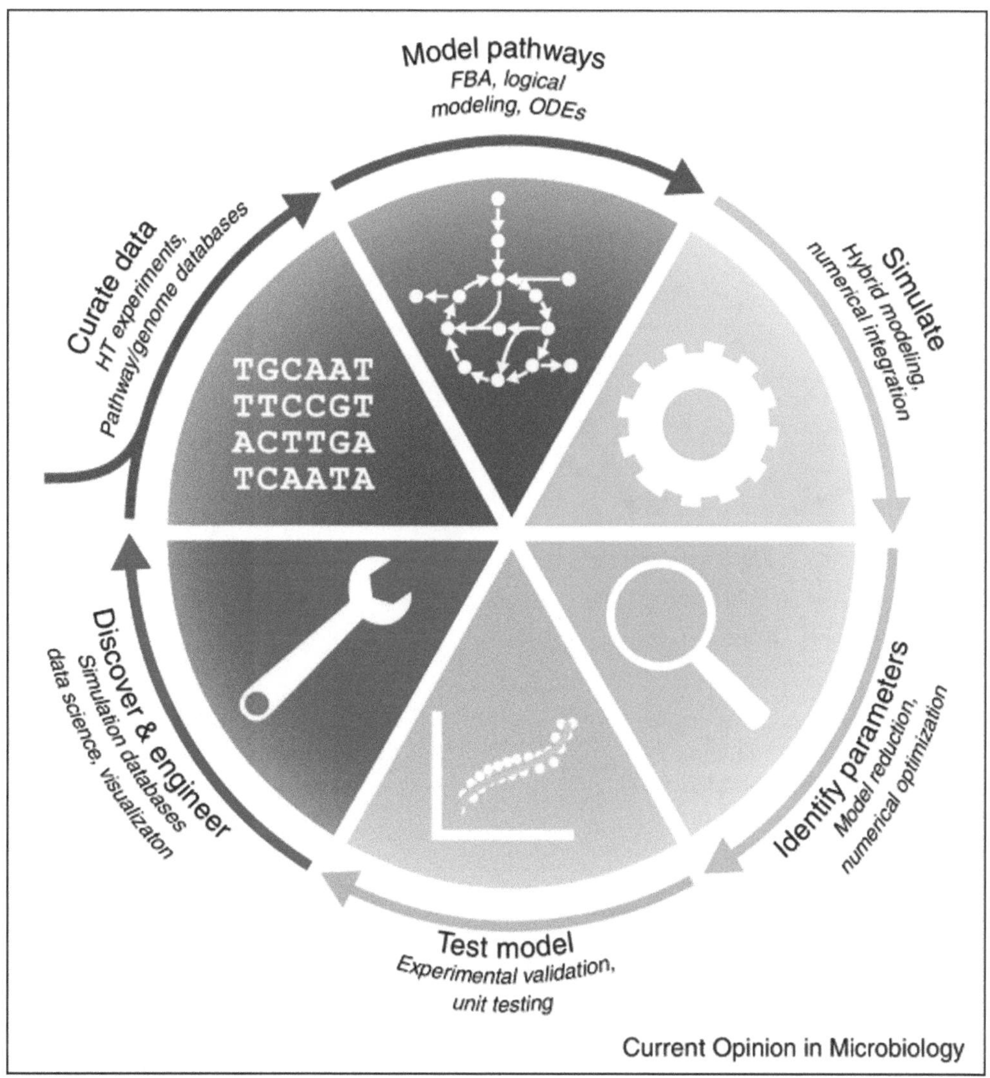

FIGURE 14.16 Whole-cell modeling process. Experimental data is organized into a database, pathway submodels are constructed, submodels are combined, parameters are identified, the model is simulated and tested, and the model is used to guide discovery and bioengineering. The process is iterated using additional data to refine the model until an accurate model is achieved.

Source: **Karr et al. (2015).**

14.5 REPOSITORY SERVERS FOR STORING BIOMODELS' INFORMATION

Understanding complex biological systems using modeling is a fascinating alternative helping us see the biological systems simplistically and progress fast. Versatile modeling approaches are used in the applied sciences, including physical, animal, conceptual, and mathematical models (Sacco et al., 2019). A huge number of *in silico*, physical, and mathematical approaches have been emerging popularly to address biological systems and processes. Various repositories are created to permanently save the information and data emerging from modeling studies. For example, BioModels

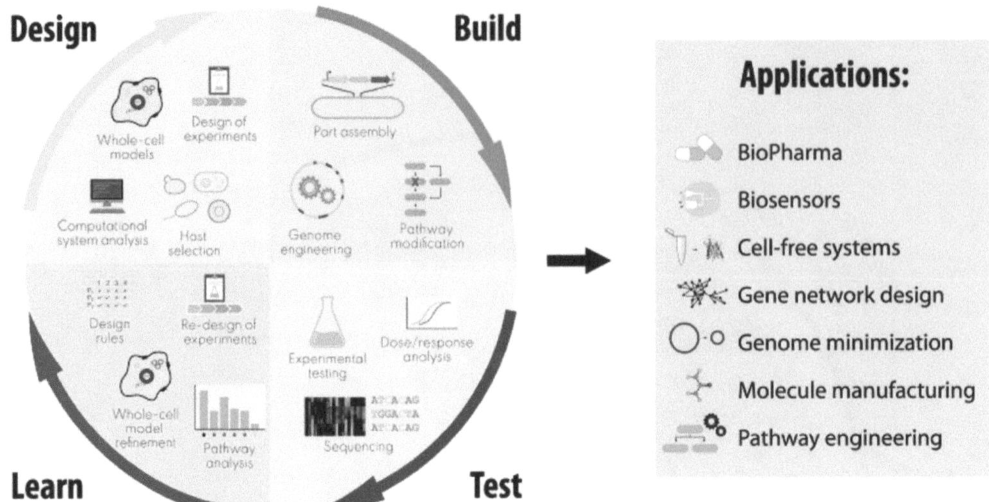

FIGURE 14.17 Integrated design-build-test-learn cycles in synthetic biology encompassing whole-cell model-guided approaches, and relative applications.

Source: **Marucci et al. (2020).**

(www.ebi.ac.uk/biomodels), created about two decades ago, is a reference repository of mathematical models associated with biology, where the models are stored as SBML files on a file system and metadata is provided in a relational database (Wimalaratne et al., 2014). Models can be retrieved through a web interface and programmatically via web services. Linked Data using Semantic Web technologies (such as the Resource Description Framework, RDF) is also becoming an increasingly popular means to describe and expose relevant biological data. BioModels provides several methods that can be used to retrieve all models annotated with commonly used resources such as GO, UniProt, Taxonomy, ChEBI, and Reactome (Li et al., 2010). Many textbooks provide information on mathematical and physical models for biosystems, e.g., *Physical Models of Living Systems* is a textbook presenting biological experiments, physical models, and computational approaches (Nelson et al., 2015). There have also been a large number of publications on biomodels using various physical and mathematical techniques (Montévil, 2019; Gross and Green, 2017). All of these databases are preserving models and associated information on biological systems that are helping biological scientists and modeling-performing biophysicists as reference resources.

REFERENCES

Adey, A.C. (2021). Tagmentation-based single-cell genomics. *Genome Research*, *31*(10), 1693–1705

Ali, A., Abouleila, Y., Shimizu, Y., Hiyama, E., Emara, S., Mashaghi, A., & Hankemeier, T. (2019). Single-cell metabolomics by mass spectrometry: advances, challenges, and future applications. *TrAC*, *120*, 115436

Al-Kofahi, Y., Zaltsman, A., Graves, R., Marshall, W., & Rusu, M. (2018). A deep learning-based algorithm for 2-D cell segmentation in microscopy images. *BMC Bioinformatics*, *19*(1), 1–11

Ashrafuzzaman, M., Tseng, C.Y., & Tuszynski, J.A. (2014). Regulation of channel function due to physical energetic coupling with a lipid bilayer. *Biochemical and Biophysical Research Communications*, *445*(2), 463–468

Ashrafuzzaman, M., & Tuszynski, J.A. (2012a). *Membrane biophysics*. Berlin and Heidelberg: Springer. https://doi.org/10.1007/978-3-642-16105-6

Ashrafuzzaman, M., & Tuszynski, J. (2012b). Regulation of channel function due to coupling with a lipid bilayer. *Journal of Computational and Theoretical Nanoscience*, *9*(4), 564–570

Atlas, J.C., Nikolaev, E.V., Browning, S.T., & Shuler, M.L. (2008). Incorporating genome-wide DNA sequence information into a dynamic whole-cell model of *Escherichia coli*: application to DNA replication. *IET Syst. Biol.*, *2*, 369–382

Bennett, W.D., He, S., Bilodeau, C.L., Jones, D., Sun, D., Kim, H., ... & Ingólfsson, H.I. (2020). Predicting small molecule transfer free energies by combining molecular dynamics simulations and deep learning. *Journal of Chemical Information and Modeling*, *60*(11), 5375–5381

Brandt, S., Sittel, F., Ernst, M., & Stock, G. (2018). Machine learning of biomolecular reaction coordinates. *Journal of Physical Chemistry Letters*, *9*(9), 2144–2150

Bratton, B.P., Mooney, R.A., & Weisshaar, J.C. (2011). Spatial distribution and diffusive motion of RNA polymerase in live *Escherichia coli*. *J. Bacteriol.*, *193*, 5138–5146

Brooks, B.R., Brooks III, C.L., Mackerell Jr, A.D., Nilsson, L., Petrella, R.J., Roux, B., ... & Karplus, M. (2009). CHARMM: the biomolecular simulation program. *Journal of Computational Chemistry*, *30*(10), 1545–1614

Browning, S.T., Castellanos, M., & Shuler, M.L. (2004). Robust control of initiation of prokaryotic chromosome replication: essential considerations for a minimal cell. *Biotechnol. Bioeng.*, *88*, 575–584

Bruce, N.J., Ganotra, G.K., Kokh, D.B., Sadiq, S.K., & Wade, R.C. (2018). New approaches for computing ligand–receptor binding kinetics. *Current Opinion in Structural Biology*, *49*, 1–10. doi: 10.1016/j.sbi.2017.10.001; PMID: 29132080

Callaway, E. (2020). "It will change everything": DeepMind's artificial intelligence (AI) makes gigantic leap in solving protein structures. *Nature*, *588*, 203–204

Castellanos, M., Kushiro, K., Lai, S.K., & Shuler, M.L. (2007). A genomically/chemically complete module for synthesis of lipid membrane in a minimal cell. *Biotechnol. Bioeng.*, *97*, 397–409

Chandrasekaran, S., & Price, N.D. (2010). Probabilistic integrative modeling of genome-scale metabolic and regulatory networks in *Escherichia coli* and *Mycobacterium tuberculosis*. *Proc. Natl. Acad. Sci. USA*, *107*, 17845–17850

Chaudhury, S., Lyskov, S., & Gray, J.J. (2010). PyRosetta: a script-based interface for implementing molecular modeling algorithms using Rosetta. *Bioinformatics*, *26*(5), 689–691

Cheng, H.P., Deumens, E., Freericks, J.K., Li, C., & Sanders, B.A. (2020). Application of quantum computing to biochemical systems: a look to the future. *Frontiers in Chemistry*, *8*, 587143

Christensen, D.V., Dittmann, R., Linares-Barranco, B., Sebastian, A., Le Gallo, M., Redaelli, A., ... & Pryds, N. (2022). 2022 roadmap on neuromorphic computing and engineering. *Neuromorphic Computing and Engineering*, *2*(2), 022501

Covert, M.W., Xiao, N., Chen, T.J., & Karr, J.R. (2008). Integrating metabolic, transcriptional regulatory and signal transduction models in *Escherichia coli*. *Bioinformatics*, *24*, 2044–2050

Domach, M.M., Leung, S.K., Cahn, R.E., Cocks, G.G., & Shuler, M.L. (1984). Computer model for glucose-limited growth of a single cell of *Escherichia coli* B/r-A. *Biotechnol. Bioeng.*, *26*, 1140

Eslami, M., Borujeni, A.E., Eramian, H., Weston, M., Zheng, G., Urrutia, J., ... & Yeung, E. (2022). Prediction of whole-cell transcriptional response with machine learning. *Bioinformatics*, *38*(2), 404–409

Giannakoulias, S., Shringari, S.R., Ferrie, J.J., et al. (2021). Biomolecular simulation based machine learning models accurately predict sites of tolerability to the unnatural amino acid acridonylalanine. *Sci. Rep.*, *11*, 18406. https://doi.org/10.1038/s41598-021-97965-2

Giannakoulias, S., Shringari, S.R., Liu, C., Phan, H.A.T., Barrett, T.M., Ferrie, J.J., & Petersson, E.J. (2020). Rosetta machine learning models accurately classify positional effects of thioamides on proteolysis. *Journal of Physical Chemistry B*, *124*(37), 8032–8041

Greenwald, N.F., Miller, G., Moen, E., Kong, A., Kagel, A., Dougherty, T., ... & Van Valen, D. (2022). Whole-cell segmentation of tissue images with human-level performance using large-scale data annotation and deep learning. *Nature Biotechnology*, *40*(4), 555–565

Gross, F., & Green, S. (2017). The sum of the parts: large-scale modeling in systems biology. *Philosophy, Theory, and Practice in Biology*, *9*(10).

Haghighatlari, M., & Hachmann, J. (2019). Advances of machine learning in molecular modeling and simulation. *Current Opinion in Chemical Engineering*, *23*, 51–57

Hamelberg, D., Mongan, J., & McCammon, J.A. (2004). Accelerated molecular dynamics: a promising and efficient simulation method for biomolecules. *Journal of Chemical Physics*, *120*(24), 11919–11929

Heinemann, M., & Zenobi, R. (2011). Single cell metabolomics. *Current Opinion in Biotechnology*, *22*(1), 26–31

Hinbest, A.J., Eldirany, S.A., Ho, M., & Bunick, C.G. (2020). Molecular modeling of pathogenic mutations in the keratin 1B domain. *Int. J. Mol. Sci.*, *21*(18), 6641. doi: 10.3390/ijms21186641; PMID: 32927888; PMCID: PMC7555247

Hong, T.H., & Park, W.Y. (2020). Single-cell genomics technology: perspectives. *Experimental & Molecular Medicine*, *52*(9), 1407–1408

Hostetler, Z.M., Ferrie, J.J., Bornstein, M.R., Sungwienwong, I., Petersson, E.J., & Kohli, R.M. (2018). Systematic evaluation of soluble protein expression using a fluorescent unnatural amino acid reveals no reliable predictors of tolerability. *ACS Chemical Biology*, *13*(10), 2855–2861

Jumper, J., Evans, R., Pritzel, A., et al. (2021). Highly accurate protein structure prediction with AlphaFold. *Nature*, *596*, 583–589. https://doi.org/10.1038/s41586-021-03819-2

Karplus, M. (1978). Dynamics of folded proteins. *Abstracts of Papers of the American Chemical Society*, *175*, 70

Karplus, M., & Kuriyan, J. (2005). Molecular dynamics and protein function. *Proceedings of the National Academy of Sciences*, *102*(19), 6679–6685

Karr, J.R., Sanghvi, J.C., Macklin, D.N., Gutschow, M.V., Jacobs, J.M., Bolival Jr, B., … & Covert, M.W. (2012). A whole-cell computational model predicts phenotype from genotype. *Cell*, *150*(2), 389–401

Karr, J.R., Takahashi, K., & Funahashi, A. (2015). The principles of whole-cell modeling. *Current Opinion in Microbiology*, *27*, 18–24

Klepeis, J.L., Wei, Y., Hecht, M.H., & Floudas, C.A. (2005). Ab initio prediction of the three-dimensional structure of a de novo designed protein: a double-blind case study. *Proteins*, *58*(3), 560–570. doi: 10.1002/prot.20338; PMID: 15609306

Levitt, M., & Sharon, R. (1988). Accurate simulation of protein dynamics in solution. *Proceedings of the National Academy of Sciences*, *85*(20), 7557–7561

Li, C., Courtot, M., Le Novère, N., & Laibe, C. (2010). BioModels.net Web Services, a free and integrated toolkit for computational modelling software. *Brief Bioinform.*, *11*, 270–277. doi: 10.1093/bib/bbp056

Liu, R., & Yang, Z. (2021). Single cell metabolomics using mass spectrometry: techniques and data analysis. *Analytica Chimica Acta*, *1143*, 124–134

Luo, Y., Pfuetzner, R.A., Mosimann, S., Paetzel, M., Frey, E.A., Cherney, M., … & Strynadka, N.C. (2001). Crystal structure of LexA: a conformational switch for regulation of self-cleavage. *Cell*, *106*(5), 585–594

Martí-Renom, M.A., Stuart, A.C., Fiser, A., Sánchez, R., Melo, F., & Šali, A. (2000). Comparative protein structure modeling of genes and genomes. *Annual Review of Biophysics and Biomolecular Structure*, *29*(1), 291–325

Marucci, L., Barberis, M., Karr, J., Ray, O., Race, P.R., de Souza Andrade, M., … & Woods, C. (2020). Computer-aided whole-cell design: taking a holistic approach by integrating synthetic with systems biology. *Frontiers in Bioengineering and Biotechnology*, *8*, 942

Matlock, M.K., Hughes, T.B., Dahlin, J.L., & Swamidass, S.J. (2018). Modeling small-molecule reactivity identifies promiscuous bioactive compounds. *Journal of Chemical Information and Modeling*, *58*(8), 1483–1500

Meiler, J., & Baker, D. (2006). ROSETTALIGAND: protein–small molecule docking with full side-chain flexibility. *Proteins: Structure, Function, and Bioinformatics*, *65*(3), 538–548

Montévil, M. (2019). Which first principles for mathematical modelling in biology? *Rendiconti di Matematica e delle sue Applicazioni*, *40*, 177–189

Moore, G.E. (1965). Cramming more components onto integrated circuits. *Electronics*, *38*, 114–117

Nelson, P.C., Bromberg, S., Hermundstad, A., & Prentice, J. (2015). *Physical models of living systems.* New York: W.H. Freeman

Ołdziej, S., Czaplewski, C., Liwo, A., Chinchio, M., Nanias, M., Vila, J.A., … & Scheraga, H.A. (2005). Physics-based protein-structure prediction using a hierarchical protocol based on the UNRES force field: assessment in two blind tests. *Proceedings of the National Academy of Sciences*, *102*(21), 7547–7552

Ollitrault, P.J., Miessen, A., & Tavernelli, I. (2021). Molecular quantum dynamics: a quantum computing perspective. *Accounts of Chemical Research*, *54*(23), 4229–4238

Outeiral, C., Nissley, D.A., & Deane, C.M. (2022). Current structure predictors are not learning the physics of protein folding. *Bioinformatics*, *38*(7), 1881–1887. https://doi.org/10.1093/bioinformatics/btab881

Park, H., Zhou, G., Baek, M., Baker, D., & DiMaio, F. (2021). Force field optimization guided by small molecule crystal lattice data enables consistent sub-angstrom protein–ligand docking. *Journal of Chemical Theory and Computation*, *17*(3), 2000–2010

Perilla, J.R., Goh, B.C., Cassidy, C.K., Liu, B., Bernardi, R.C., Rudack, T., … & Schulten, K. (2015). Molecular dynamics simulations of large macromolecular complexes. *Current Opinion in Structural Biology*, *31*, 64–74

Perilla, J., & Schulten, K. (2017). Physical properties of the HIV-1 capsid from all-atom molecular dynamics simulations. *Nat Commun.*, *8*, 15959. https://doi.org/10.1038/ncomms15959

Perkel, J.M. (2021). Single-cell proteomics takes centre stage. *Nature*, *597*(7877), 580–582

Petruţ, B., Beatrice, M., Casellato, C., Casali, S., Andrew, R., Michael, H., … & Oliver, R. (2021). Towards a bio-inspired real-time neuromorphic cerebellum. *Frontiers in Cellular Neuroscience*, *15*, 622870

Phillips, J.C., Hardy, D.J., Maia, J.D., Stone, J.E., Ribeiro, J.V., Bernardi, R.C., … & Tajkhorshid, E. (2020). Scalable molecular dynamics on CPU and GPU architectures with NAMD. *Journal of Chemical Physics*, *153*(4), 044130

Reddy, T., & Sansom, M.S. (2016). Computational virology: from the inside out. *Biochimica et Biophysica Acta (BBA) – Biomembranes*, *1858*(7), 1610–1618

Rickert, C.A., & Lieleg, O. (2022). Machine learning approaches for biomolecular, biophysical, and biomaterials research. *Biophysics Reviews*, *3*(2), 021306

Rimola, A., Corno, M., Garza, J., & Ugliengo, P. (2012). Ab initio modeling of protein–biomaterial interactions: influence of amino acid polar side chains on adsorption at hydroxyapatite surfaces. *Philosophical Transactions of the Royal Society A: Mathematical, Physical and Engineering Sciences*, *370*(1963), 1478–1498

Sacco, R., Guidoboni, G., & Mauri, A.G. (2019). Mathematical and physical modeling principles of complex biological systems. In *Ocular Fluid Dynamics. Modeling and Simulation in Science, Engineering and Technology*, ed. G. Guidoboni, A. Harris, & R. Sacco, pp. 3–20. Cham: Birkhäuser. https://doi.org/10.1007/978-3-030-25886-3_1

Schlick, T. (2010). *Molecular modeling and simulation: an interdisciplinary guide* (Vol. 2). New York: Springer

Schlick, T., & Portillo-Ledesma, S. (2021). Biomolecular modeling thrives in the age of technology. *Nat. Comput. Sci.*, *1*, 321–331. https://doi.org/10.1038/s43588-021-00060-9

Shringari, S.R., Giannakoulias, S., Ferrie, J.J., & Petersson, E.J. (2020). Rosetta custom score functions accurately predict ΔΔ G of mutations at protein–protein interfaces using machine learning. *Chemical Communications*, *56*(50), 6774–6777

Soto-Ospina, A., Araque Marin, P., Bedoya, G., Sepulveda-Falla, D., & Villegas Lanau, A. (2021). Protein predictive modeling and simulation of mutations of presenilin-1 familial Alzheimer's disease on the orthosteric site. *Frontiers in Molecular Biosciences*, *8*, 387

Suthers, P.F., Dasika, M.S., Kumar, V.S., Denisov, G., Glass, J.I., & Maranas, C.D. (2009). A genome-scale metabolic reconstruction of *Mycoplasma genitalium*, iPS189. *PLoS Comput. Biol.*, *5*, e1000285

Tomita, M. (2001). Whole-cell simulation: a grand challenge of the 21st century. *Trends Biotechnol.*, *19*(6), 205–210. doi: 10.1016/s0167-7799(01)01636-5; PMID: 11356281

Vaidehi, N., Pease, J.E., & Horuk, R. (2009). Modeling small molecule-compound binding to G-protein-coupled receptors. *Methods in Enzymology*, *460*, 263–288

van der Kamp, M.W., Shaw, K.E., Woods, C.J., & Mulholland, A.J. (2008). Biomolecular simulation and modelling: status, progress and prospects. *J. R. Soc. Interface*, *5*(Suppl 3), S173–190. doi: 10.1098/rsif.2008.0105.focus; PMID: 18611844; PMCID: PMC2706107

Vendruscolo, M., & Dobson, C.M. (2011). Protein dynamics: Moore's law in molecular biology. *Current Biology*, *21*(2), R68–R70

Vora, T., Hottes, A.K., & Tavazoie, S. (2009). Protein occupancy landscape of a bacterial genome. *Mol. Cell.*, *35*, 247–253

Wang, L., Chambers, J., & Abel, R. (2019). Protein–ligand binding free energy calculations with FEP+. *Biomolecular Simulations: Methods and Protocols*, *2022*, 201–232. doi: 10.1007/978-1-4939-9608-7_9; PMID: 31396905

Warshel, A., & Karplus, M. (1972). Calculation of ground and excited state potential surfaces of conjugated molecules: I. Formulation and parametrization. *Journal of the American Chemical Society*, *94*(16), 5612–5625

Warshel, A., & Levitt, M. (1976). Theoretical studies of enzymic reactions: dielectric, electrostatic and steric stabilization of the carbonium ion in the reaction of lysozyme. *Journal of Molecular Biology*, *103*(2), 227–249

Wimalaratne, S.M., Grenon, P., Hermjakob, H., et al. (2014). BioModels linked dataset. *BMC Syst. Biol.*, *8*, 91. https://doi.org/10.1186/s12918-014-0091-5

Yang, L., Zhang, J., Che, X., & Gao, Y.Q. (2016). Simulation studies of protein and small molecule interactions and reaction. *Methods in Enzymology*, *578*, 169–212. doi: 10.1016/bs.mie.2016.05.031; PMID: 27497167

15 Biophysics Tools in Bioinformatics and Genomics

Bioinformatics and genomics have emerged, in recent decades, as fast-growing fields in biology. Both fields utilize various techniques, many of which are associated with utilizing many principles of biophysics. Understanding the underlying laws of physics, mathematics, biology, and chemistry in shaping biological structures is crucial and partly doable using various algorithms. This new type of science started just a little more than half a century ago, but amazing progress has already been achieved. In this chapter, I shall provide a brief report on the stepwise progressions of the field.

15.1 BACKGROUND OF EMERGING BIOINFORMATICS AND GENOMICS FIELDS

Bioinformatics emerged as an interdisciplinary research field more than six decades ago when Margaret Dayhoff and Robert Ledley released the first de novo sequence assembler (Moody, 2004; Gauthier et al., 2019; Dayhoff and Ledley, 1962b). They were the first scholars to recognize, in 1962, the potential of computational methods in solving biomedical problems, which got the attention of both academic and industry researchers equally and initiated huge investments in translational research. As of today, we are in the process of entering a new era in data-driven health care, where translational bioinformatics methods continue to make practical differences to patients' lives (Tenenbaum, 2016). Figure 15.1 depicts the way translational bioinformatics fits within the bigger biomedical informatics picture as data is transformed into a knowledge process. The optimal goal of huge investment-based medical research is to carry the research accomplishments all the way to benefiting humans who seek advanced solutions to critical medical issues. Here bioinformatics in general and especially translational bioinformatics are capable of contributing greatly.

The background of today's bioinformatics glories was established during the 1950–1970 era. During this time the first-ever solutions to DNA and protein structures appeared (Watson and Crick, 1953; Muirhead and Perutz, 1963; Perutz et al., 1960; Kendrew et al., 1958), and various other methods helped us probe further; an important one was the famous Ramachandran plots (Ramachandran et al., 1963). Following these structural successes, scientists became ambitious to explore easy ways that might help us address biomolecular structures even if that resulted in predictions, instead of any complete picture like usually achieved using the time-consuming X-ray crystallography technique.

In the 1960s a computer program COMPROTEIN was developed by Dayhoff and Ledley for the IBM 7090 to determine specifically the protein primary structure (Dayhoff and Ledley, 1962a) using Edman peptide sequencing data (Edman, 1949). This FORTRAN-coded software was the first success that initiated the large-scale bioinformatics achievements of today. The de novo sequence assembler is shown in Figure 15.2. The three-letter abbreviated input and output amino

DOI: 10.1201/9781003287780-15

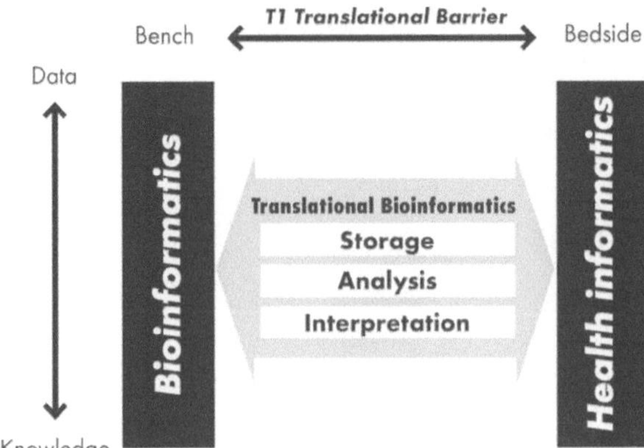

FIGURE 15.1 The Y axis depicts the "central dogma" of informatics, converting data to information and information to knowledge. Along the X axis is the translational spectrum from bench to bedside. Translational bioinformatics spans the data to knowledge spectrum, and bridges the gap between bench research and application to human health.

***Source*: reproduced with permission from Tenenbaum et al. (2014) and Tenenbaum (2016).**

acid sequences (e.g. Lys for lysine, Ser for serine) in the COMPROTEIN software were later simplified by Dayhoff by replacing them with the one-letter amino acid code that is still in use today in protein sequencing (IUPAC-IUB, 1968). This one-letter code was first used in the first-ever biological sequence database created by Dayhoff and Eck in 1965 (see *Atlas of Protein Sequence and Structure*, Dayhoff, 1965).

DNA bioinformatics analysis lagged two decades behind the analysis of proteins. It took more than a decade to obtain the genetic code (Nirenberg and Leder, 1964), following the discovery of the double-helix DNA structure (Watson and Crick, 1953). The first DNA sequencing methods of course took more time and were discovered in 1977 (Sanger et al., 1977; Maxam and Gilbert, 1977). Here a method for determining nucleotide sequences in DNA was developed which is similar to the "plus and minus" method (Sanger and Coulson, 1975), but makes use of the 2',3'-dideoxy, and arabinonucleoside analogues of the normal deoxynucleoside triphosphates, thus acting as specific chain-terminating inhibitors of DNA polymerase. Sanger and colleagues (1977) applied their technique to the DNA of bacteriophage varphiX174, and the technique appeared to be more rapid and more accurate than either the plus or the minus method of Sanger and Coulson (1975).

Fast forward to current days, when we also deal with genomics using bioinformatics approaches (Wee et al., 2019; Horner et al., 2010). Genomics is the study of all of a person's genes (the genome). This field also includes studies on interactions of genes with each other and with the person's surrounding environment. The first publicly funded genome project was established in 2001 (IHGSC, 2001). The field has continually improved over the past two decades (Nurk et al., 2022; Schneider et al., 2017; Venter et al., 2001; Myers et al., 2000; Eichler et al., 2004).

15.2 CURRENT STATUS OF BIOINFORMATICS AND ITS TECHNIQUES ASSOCIATED WITH BIOPHYSICS

Progress in bioinformatics research relies on two major aspects, the innovation of tuned technologies (Brusic and Ranganathan, 2008; Cheng et al., 2018) and social connectivity among exploring parties and organizations (Morrison-Smith et al., 2022). Bioinformatics techniques are always in progress through continued renormalization over existing approaches, and adoption of strategic planning for future developments (Bayat, 2002; Cheng et al., 2018; Abdurakhmonov, 2016; Meisel

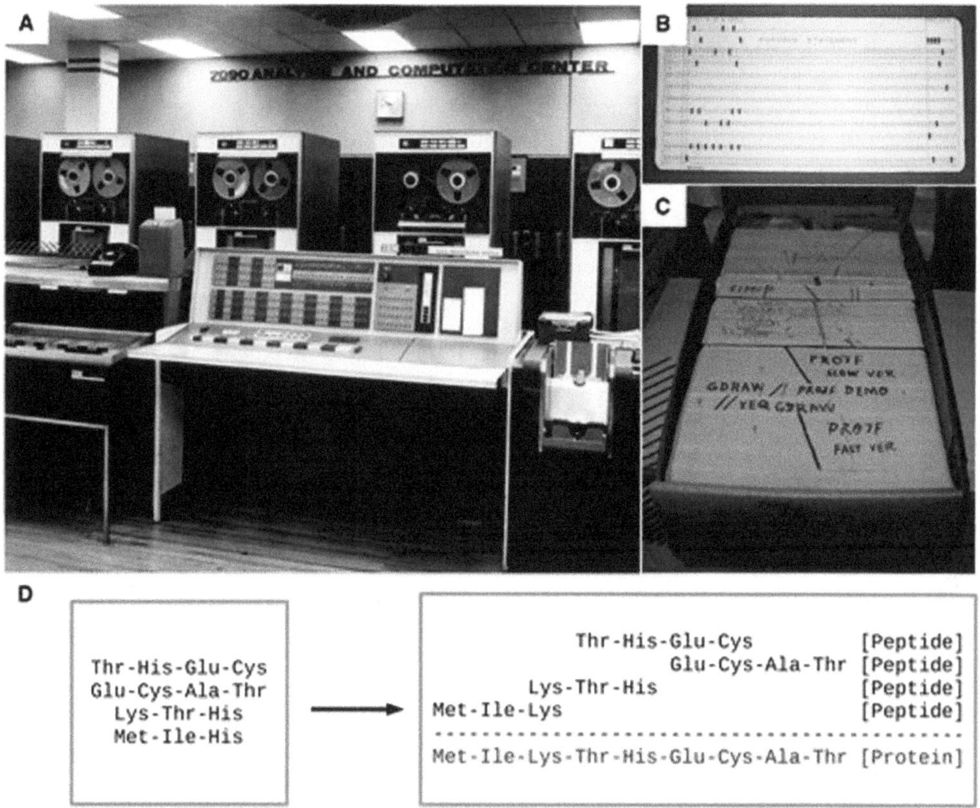

FIGURE 15.2 COMPROTEIN, the first bioinformatics software. (A) An IBM 7090 mainframe, on which COMPROTEIN was made to run. This famous computer was one of the first solid-state machines and was used widely in business and defense settings, as well as scientific applications. (B) A punch card containing one line of FORTRAN code (the language COMPROTEIN was written with). (C) An entire program's source code in punch cards. (D) A simplified overview of COMPROTEIN's input (i.e. Edman peptide sequences) and output (a consensus protein sequence).

***Source*: reproduced with permission from Gauthier et al. (2019).**

et al., 2018). Unlike traditional research techniques, such as experimental methodologies, which often require the creation of large-scale laboratory facilities, bioinformatics utilizes the power of computers, specialized programming algorithms, and digital connectivity to achieve computational research goals. Teaming up among scientists working anywhere and having various types of backgrounds is pretty achievable in bioinformatics research. *In silico* research also often requires just modest funding support.

There has been tremendous progress made over recent decades in developing bioinformatics techniques. My objective is certainly not to summarize all of them. There are many review articles and textbooks available (e.g. Sotirov et al., 2022; Ewens, 2005; Matthiesen, 2010). I shall rather pinpoint a few example cases in which breakthrough developments have been demonstrated and thus focus on how bioinformatics techniques may help generate easy, fast, and futuristic achievements in biomedical research especially.

Large-scale bioinformatics research projects face considerable challenges (Morrison-Smith et al., 2022). Life science research, these days, is usually conducted in large and multi-institutional collaborations. Although these large groups rely on "mutual respect" and collaboration (for participating agents/parties, see Figure 15.3), it is reported that the interdisciplinary nature of these projects causes technical language barriers and differences in methodology and lack of trust (Morrison-Smith et al., 2022). Morrison-Smith and colleagues have provided a few recommendations for technology to support life science, as well as life science research training programs, and noted the necessity for incorporating training in project management, multiple languages, and discipline culture. This study is suggestive of the necessity of multicultural collaborative initiatives having open channel connectivity. Figure 15.3 presents a scenario demonstrating how a collaborative initiative may go beyond the academic level and ensure the participation of parties from governments and industries. This way we may ensure that problem-solving academic minds cross-talk with policymakers and investors who help research findings reach the medical-benefit-seeking general public.

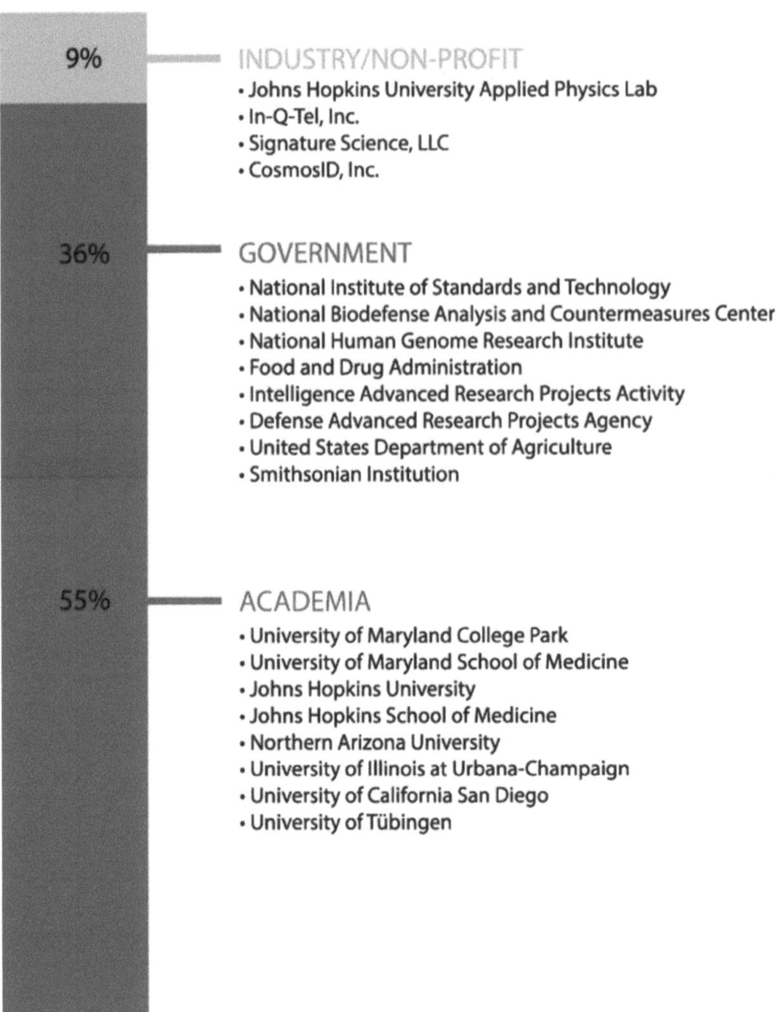

FIGURE 15.3 Different sectors and institutions represented at the January 2018 M3 Meet-up.

Source: **reproduced with permission from Meisel et al. (2018).**

The main component behind ensuring bioinformatics success is to continuously update bioinformatics techniques. As both experimental data and associated theoretical understanding often grow exponentially (Stratton et al., 2009; Shendure and Ji, 2008; Metzker, 2010; Nielsen et al., 2011; Zhang et al., 2011; Lynch, 2008), we need always to revise existing methods and/or invent new technologies to deal with such big data scenarios (Luo et al., 2016). Bioinformatics is the primary field in which big data analytics are currently being applied, largely due to the massive volume and complexity of bioinformatics data (Luo et al., 2016). This review also demonstrated the following areas:

1. Integrating different sources of information enables clinicians to depict a new view of patient care processes that consider a patient's holistic health status, from genome to behavior.
2. The availability of novel mobile health technologies facilitates real-time data gathering with more accuracy.
3. The implementation of distributed platforms enables data archiving and analysis, which will further be developed for decision support.
4. The inclusion of geographical and environmental information may further increase the ability to interpret gathered data and extract new knowledge.

In biomedical discoveries related to disease understanding and drug discovery, both biophysics and bioinformatics techniques provide complementary support. Bioinformatics tools have been found to provide a means for distinguishing heterogeneity in the biophysical features of tumor cells. Recently, a bioinformatics approach was demonstrated to analyse the phenotypic variables extracted from the spontaneous migration of tumor cells treated with drugs in various confined geometric conditions (Zhang and Mak, 2021). This study demonstrated applications to analyse complex biophysical data and provide the biophysical signatures for distinct subpopulations under different external stimuli at single-cell resolution.

A computational approach that uses latent variable models to account for hidden factors, such as the cell cycle, in the heterogeneity of gene expression, and to help identify subpopulations, was recently developed (Buettner et al., 2015), being named the single-cell latent variable model (scLVM). In this study, while dealing with the assay of the transcriptomes of hundreds of cells, Buettner and colleagues found that the bioinformatics technique scLVM allows the identification of undetectable subpopulations of cells that correspond to different stages during the differentiation of naive T cells into T helper 2 cells. This bioinformatics technique is therefore capable of not only identifying cellular subpopulations but also teasing apart different sources of gene expression heterogeneity in single-cell transcriptomes.

The two-stage scLVM procedure is explained in Figure 15.4, which was developed for accounting for the effects of the cell cycle. Specifically, scLVM was used for addressing the confounding effects of the cell cycle. In this bioinformatics approach, one first reconstructs the cell-cycle state (or other unobserved factors), then uses this information to infer "corrected" gene expression levels. This two-step approach helps understand the effect of unobserved factors on gene expression heterogeneity being accounted for in downstream analyses, thereby allowing us to study variation in gene expression levels which is independent of the cell cycle. For each gene whose expression is analysed, this method also allows the relative contribution of any reconstructed factors that affect cell-to-cell variation in expression to be determined. Figure 15.4b presents a schematic overview of the approach.

Patel et al. (2014) is another computational study, which applied bioinformatics technique to scRNA-seq data from primary glioblastoma cells and found underlying relationships between the expression of transcriptional programs and tumor phenotypes, such as oncogenic signaling, tumor proliferation, and immune response. The single-cell RNA-seq was used here for profiling 430 cells from 5 primary glioblastomas, found to be inherently variable in their expression of diverse transcriptional programs related to oncogenic signaling, proliferation, complement/immune response, and hypoxia.

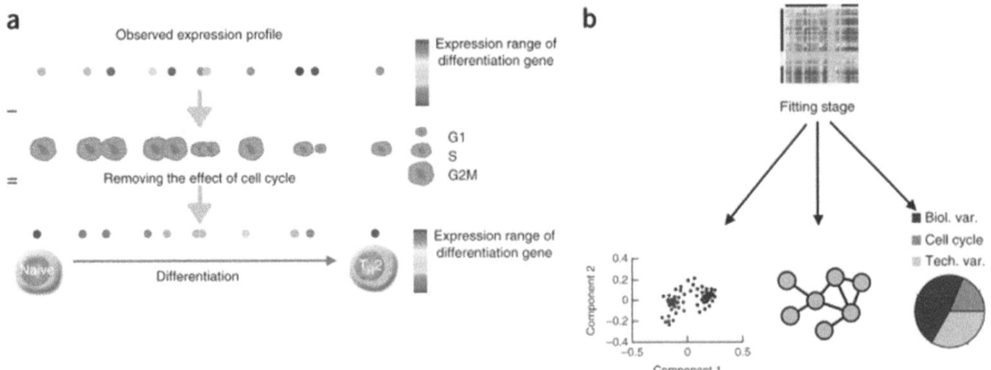

FIGURE 15.4 (a) The observed expression profile of differentiation marker genes (upper panel) is the result of the differentiation process of interest together with the effects of the cell cycle and other confounding sources of variation. After accounting for cell-cycle effects (middle panel), one can uncover gene expression signatures that contribute to the continuous differentiation process more clearly (lower panel). (b) scLVM two-stage procedure. First, in the fitting stage, the cell-to-cell covariance matrix that corresponds to the cell cycle is inferred from the gene expression profiles of genes with cell-cycle annotation (upper panel). The learnt covariance is then used in downstream analyses, including the detection of substructure, the detection of gene-to-gene correlations and the analysis of variance (lower panel). Biol. var., biological variance; Tech. var., technical variance.

***Source*: reproduced with permission from Buettner et al. (2015).**

Single-cell transcriptome analysis by the data analysing tool RNA-seq (Ramsköld et al., 2012; Shalek et al., 2013) should enable functional characterization from landmark genes and annotated gene sets, relate *in vivo* states to *in vitro* models, inform transcriptional classifications based on bulk tumors, and even capture genetic information for expressed transcripts. For interrogating intratumoral heterogeneity systematically, Patel and colleagues (2014) isolated individual cells from five freshly resected and dissociated human glioblastomas and used bioinformatics techniques to generate single-cell full-length transcriptomes utilizing the power of SMART-seq in the following physiological condition: 96–192 cells/tumor, total 672 cells (see Figure 15.5A). SMART-seq provides an integrated analysis of next-generation sequencing (NGS) data and develops bioinformatics tools for the analysis, visualization, and exploration of data. Prior to sorting, the suspension was depleted for $CD45^+$ cells to remove inflammatory infiltrate. As a control, population (bulk) RNA-seq profiles were generated from the CD45-depleted tumor samples. All tumors were IDH1/2 wild-type primary glioblastomas and three were EGFR amplified as determined by routine clinical tests. Genes and cells with low coverage were excluded, retaining ~6,000 genes quantified in 430 cells from 5 patient tumors and population controls. The population-level controls correlated with the average of the single cells in that tumor, supporting the accuracy of the single-cell data. Individual cells from the same tumor were more correlated to each other than cells from different tumors. Nevertheless, correlations between individual cells from the same tumor showed a broad spread (R~0.2–0.7), consistent with intratumoral heterogeneity.

The hierarchical clustering of all single cells and normal brain samples identified seven groups with concordant copy number variation (CNV) profiles (Figure 15.5B). The normal brain sample clustered with 10 single cells that have a "normal" copy number. The unsupervised transcriptional analysis identified 9 outlier cells with increased expression of mature oligodendrocyte genes and down-regulation of glioblastoma genes. All 9 of these expression outliers clustered with the normal brain in the CNV analysis (Figure 15.5B). The one additional "normal" cell inferred from this CNV cluster correlated with a

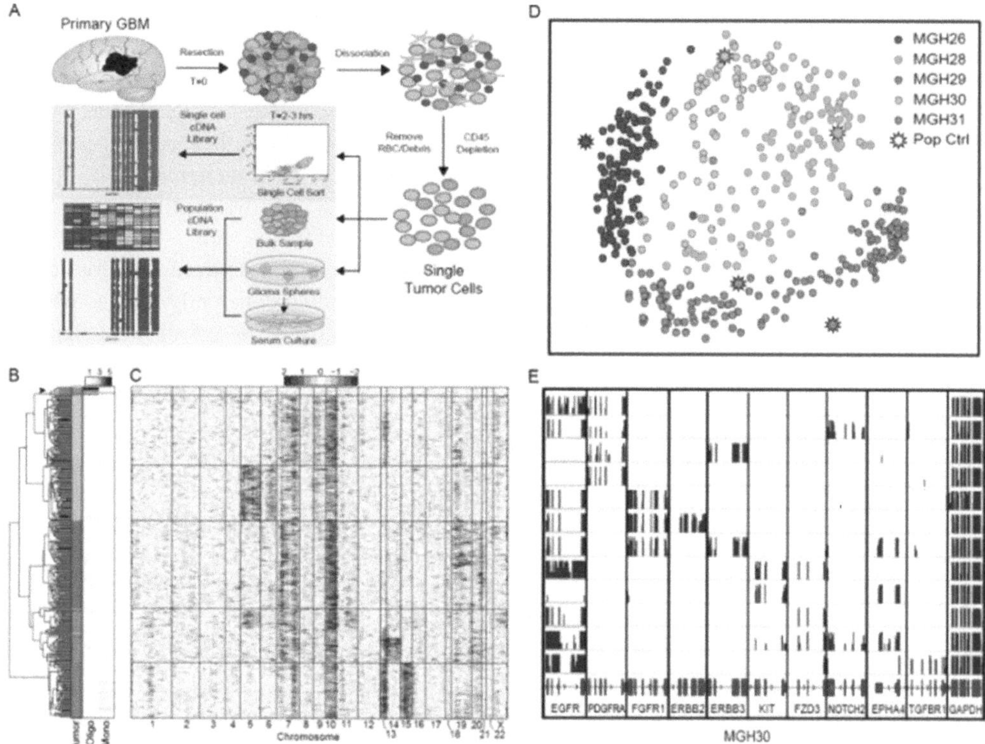

FIGURE 15.5 Intratumoral glioblastoma heterogeneity quantified by single-cell RNA-seq. (A) Workflow depicts rapid dissociation and isolation of glioblastoma cells from primary tumors for generating single-cell and bulk RNA-seq profiles and deriving glioblastoma culture models. (B) Clustering of CNV profiles inferred from RNA-seq data for all single cells and a normal brain sample. Clusters (dendrogram) primarily reflect tumor-specific CNV (colored bar coded as in panel D). Topmost cluster (red, arrow) contains the normal brain sample and 10 single cells, 9 of which correlate with normal oligodendrocyte expression profiles and one with normal monocytes ("Oligo" and "Mono," black and white heatmap). (C) Heatmap of CNV signal normalized against the "normal" cluster defined in (B) shows CNV changes by chromosome (columns) for individual cells (rows). All cells outside the normal cluster exhibit chromosome 7 gain (red) and chromosome 10 loss (blue), which are characteristic of glioblastoma. (D) Multidimensional scaling illustrates the relative similarity between all 430 single cells and population controls. The distance between any two cells reflects the similarity of their expression profiles. Cells group by tumor (color code), but each tumor also contains outliers that are more similar to cells in other tumors. (E) RNA-seq read densities (vertical scale of 10) over surface receptor genes are depicted for individual cells (rows) from MGH30. Cell-to-cell variability suggests a mosaic pattern of receptor expression, in contrast to constitutively expressed GAPDH. The color contrast in the figure here and all subsequent ones will be clear in the online version of the book.

***Source*: reproduced with permission from Patel et al. (2014).**

monocytic expression signature (Figure 15.5B). None of the remaining 420 cells show similarity to the transcriptional programs of a non-malignant brain or immune cell types. While non-malignant cells are critical components of the tumor microenvironment, the combination of dissociation methods, CD45^{+} depletion, flow cytometry gating, and computational filtering used in this study largely excluded non-tumor cells. The CNV profile normalization using the signal from the "normal" cluster revealed coherent chromosomal aberrations in each tumor (see Figure 15.5C). The gain of chromosome 7 and loss of chromosome 10, the two most common genetic alterations in glioblastoma (Brennan et al.,

2013), were consistently inferred in every tumor cell. Chromosomal aberrations were relatively consistent within tumors, with the exception that MGH31 appears to contain two genetic clones with discordant copy number changes on chromosomes 5, 13, and 14. While this data suggests large-scale intratumoral genetic homogeneity, we recognize that heterogeneity generated by focal alterations and point mutations will be grossly underappreciated using this method. Nevertheless, such panoramic analysis of the chromosomal landscape effectively separated normal from malignant cells. To interrogate global transcriptional interrelationships, multidimensional scaling was used to represent the degree of similarity between the cells in the dataset (see Figure 15.5D).

The hierarchical clustering and the principal component analysis (PCA) were used for defining four meta-signatures, each comprised of multiple related clusters that coherently vary across individual cells from a given tumor or the full dataset (see Figure 15.6A). These four meta-signatures were enriched for genes related to the cell cycle (see Figure 15.6B) and hypoxia (see Figure 15.6C). The direct examination of additional cells scoring for the complement/immune signature confirmed chromosomal aberrations characteristic of glioblastoma (Figure 15.6C).

The study of Patel et al. considered the classification scheme established by the Cancer Genome Atlas (TCGA) (Verhaak et al., 2010) for distinguishing four glioblastoma subtypes: proneural, neural, classical, and mesenchymal. The aim was to explore whether individual cells in a tumor vary in their classification. Based on population-level (bulk) expression data, the tumors in this study scored as proneural (MGH26), classical (MGH30), or mesenchymal (MGH28, MGH29) subtypes. To examine the distribution of subtype signatures across individual cells, subtype scores were calculated for each cell using the classifier gene sets. All five tumors consisted of heterogeneous mixtures with individual cells corresponding to different glioblastoma subtypes (see Figure 15.7A, B). All tumors had some cells conforming to a proneural subtype regardless of the dominant subtype of the tumor, whereas each of the other subtypes was below detection in at least one tumor.

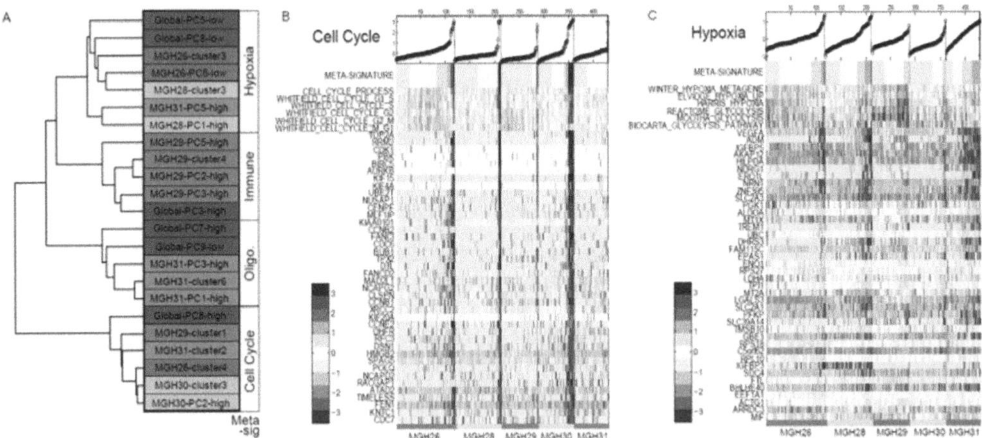

FIGURE 15.6 Unbiased analysis of intratumoral heterogeneity reveals coherent transcriptional modules. (A) Gene sets that vary coherently between cells in specific tumors or across the global dataset (colored boxes) were identified by principal component analysis or clustering (24). Hierarchical clustering of these gene sets across all cells (tree) reveals four meta-signatures related to hypoxia, complement/immune response, oligodendrocytes, and cell cycle. (B) Heatmap shows expression of the cell cycle meta-signature, selected cell cycle gene sets, and representative genes from the signature (rows) in individual glioblastoma cells (columns). Cells were grouped by tumor and ordered by meta-signature score. (C) Heatmap depicts hypoxia meta-signature as in (B).

***Source*: reproduced with permission from Patel et al. (2014).**

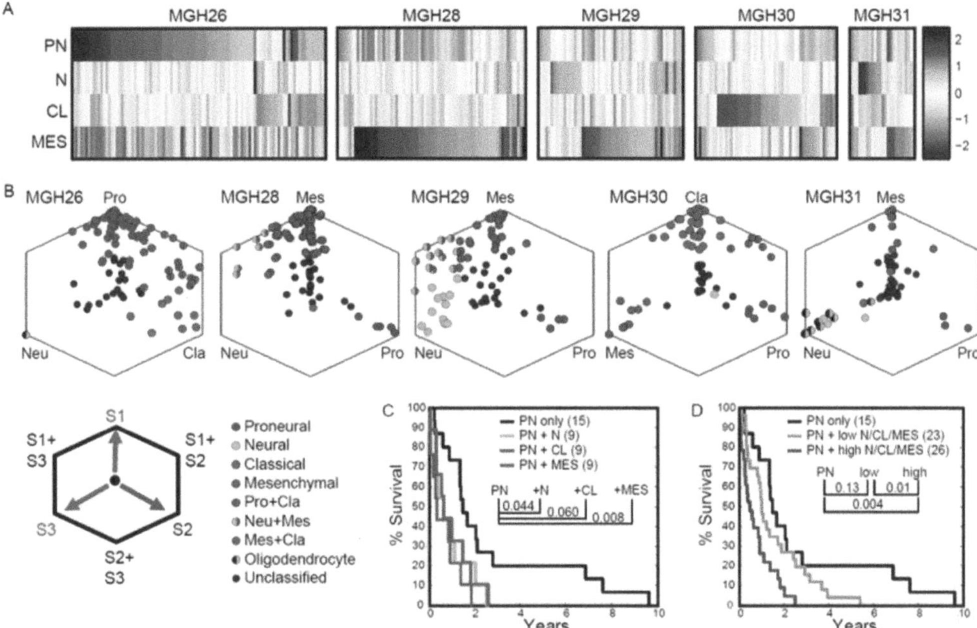

FIGURE 15.7 Individual tumors contain a spectrum of glioblastoma subtypes and hybrid cellular states. (A) Heatmap depicts average expression of classifier genes for each subtype (rows) across all classifiable cells grouped by tumor (columns). PN: proneural, CL: classical, MES: mesenchymal, N: neural. Each tumor contains a dominant subtype, but also has cells that conform to alternative subtypes. (B) Hexagonal plots depict bootstrapped classifier scores for all cells in each tumor. Each data point corresponds to a single cell and is positioned along three axes according to its relative scores for the indicated subtypes. Cells corresponding to each subtype are indicated by solid color, while hybrid cells are depicted by two colors. (C) Kaplan-Meier survival curves are shown for proneural tumors from the Cancer Genome Atlas (21). Intratumoral heterogeneity was estimated based on detected signal for alternative subtypes, and used to partition the tumors into a pure proneural group and three groups with the indicated additional subtype. Tumors with mesenchymal signal had significantly worse outcome than pure proneural ($p<0.05$). (D) Kaplan-Meier survival curves shown for proneural tumors partitioned based on the relative strength of alternative subtype signatures in aggregate (24). Tumors with high signal for alternative subtypes had significantly worse outcome than pure proneural ($p<0.05$).

***Source*: reproduced with permission from Patel et al. (2014).**

Single cell qPCR of 30 classifier genes in 167 additional cells from MGH26 and MGH30 confirmed the presence of multiple subtypes within these tumors in proportions similar to those identified by single-cell RNA-seq. While population-level data detects the dominant transcriptional program, it doesn't actually capture the true diversity of transcriptional subtypes within the tumor. Cells of the neural subtype do not correspond to the *in vitro* model (Figure 15.7C), but are more similar to normal oligodendrocytes (Figure 15.7B). Increased heterogeneity was found to be associated with decreased survival (see Figure 15.7C, D).

Application of the bioinformatics techniques associated with biological assays leads to promising conclusions. The analysis reveals that tumors contain multiple cell states with distinct transcriptional programs and provides inferential evidence for dynamic transitions. A better understanding of the spectrum and dynamics of cellular states in glioblastoma is thus critical for establishing faithful models and advancing therapeutic strategies that address the complexity of this disease. For detailed understanding of both techniques and scientific conclusions, readers may consult the original article (Patel et al., 2014). We have just presented here the key information to demonstrate how bioinformatics techniques may help reveal vital information on biological processes.

15.3 CURRENT STATUS OF GENOMICS AND ITS TECHNIQUES ASSOCIATED WITH BIOPHYSICS

As with bioinformatics, genomics also relies on both innovative technologies and social connectivity. If we carefully inspect recent breakthrough discoveries in genomics, it will become clear that both these technological and social aspects played vital roles in achievements such as the full human genome (Nurk et al., 2022), which involved scientists with various backgrounds from many countries.

Considerable progress has been made over recent decades in revising existing and developing novel genomics techniques. I do not wish to summarize all of them: there are many review articles and textbooks available (e.g. Kennedy, 2018; Saraswathy and Ramalingam, 2011; Brown, 2002). I shall rather pinpoint a few example cases in which breakthrough developments have been demonstrated and thus focus on how genomics techniques, including those of bioinformatics, may be applied to help us create breakthroughs in biomedical research.

Computational techniques are utilized to address many large-scale biological issues. High-throughput RNA sequencing (RNA-seq) promises a comprehensive picture of the transcriptome, which allows for the complete annotation and quantification of all genes and their isoforms across samples. Realizing this promise requires increasingly complex computational methods (Garber et al., 2011). Garber and colleagues have found these computational challenges fall into three main categories:

1. read mapping
2. transcriptome reconstruction
3. expression quantification

Garber et al. (2011) explained the major conceptual and practical challenges, and the general classes of solutions for each category. Finally, they highlighted the interdependence between these categories and discussed the benefits of different biological applications.

A recent article addressed bioinformatics tools applied to genome assembly and analysis based on third-generation sequencing (TGS) (Wee et al., 2019). The application of TGS technology in genetics and genomics provides opportunities to categorize and explore the individual genomic landscapes and mutations relevant for diagnosis and therapy using whole genome sequencing and de novo genome assembly. TGS technology can produce high-quality long reads for the determination of overlapping reads and transcript isoforms. But the technology still faces challenges, such as the accuracy of the identification of nucleotide bases and high error rates. Wee and colleagues surveyed 39 TGS-related tools for de novo assembly and genome analysis for identifying the differences among their characteristics, such as the required input, the interaction with the user, sequencing platforms, type of reads, error models, the possibility of introducing coverage bias, the simulation of genomic variants, and outputs provided. Figure 15.8 highlights the major development of the TGS tools. Most of the tools are involved in de novo and genome-based sequencing analysis such as RefAligner, Canu, and Nanopore Synthetic-Long (NaS). The decision trees are presented here to help researchers find the most suitable tools to analyse the TGS data (See Figures 15.9 and 15.10). I shall not provide details on these methods but readers may consult the original article (Wee et al., 2019) for in-depth understanding.

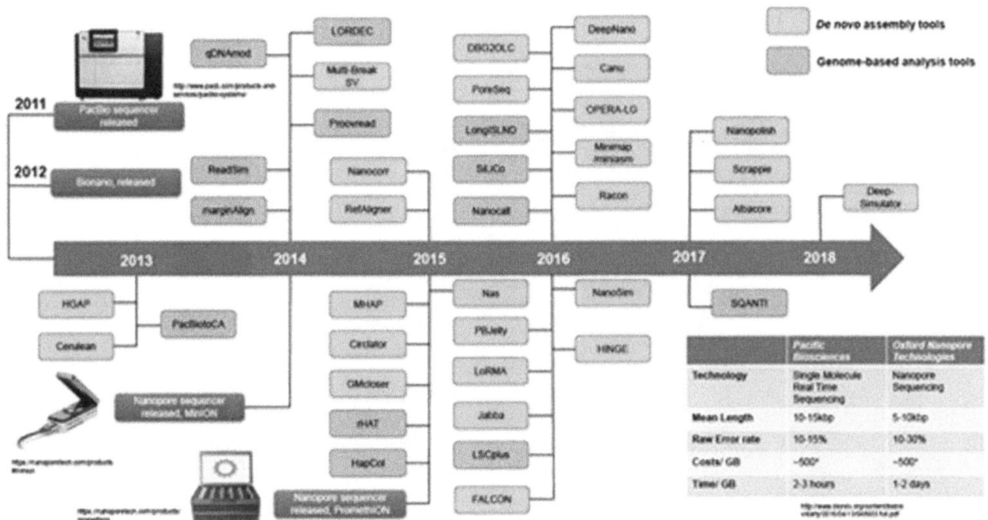

FIGURE 15.8 Milestones in TGS analysis software development. The green box refers to the de novo assembly tool while the orange box refers to the genome-based analysis tool.

Source: **reproduced with permission from Wee et al. (2019).**

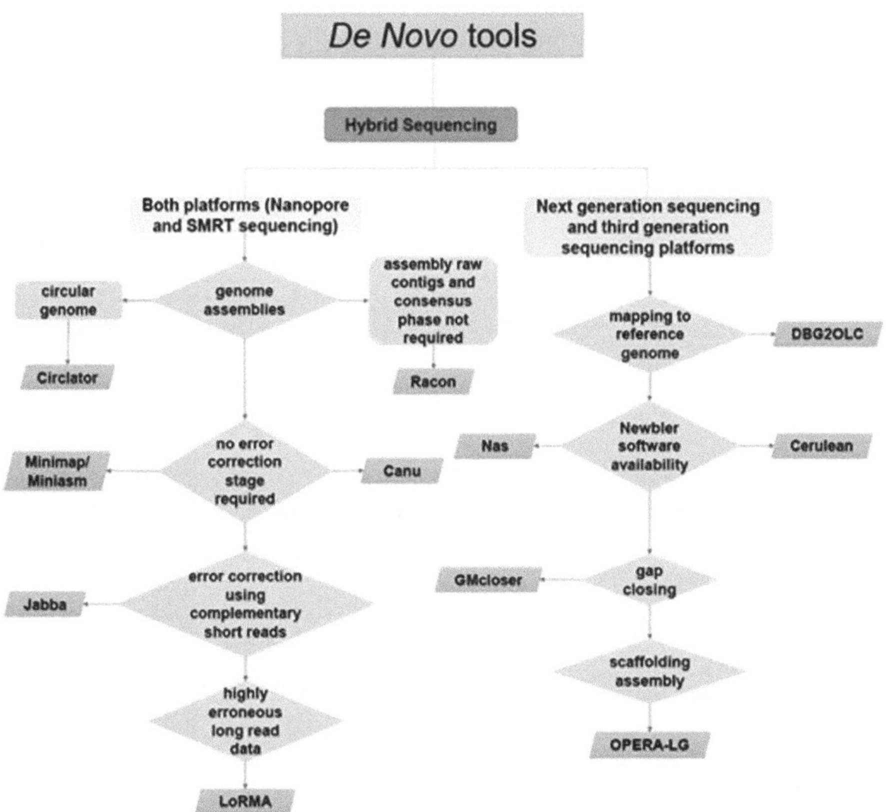

FIGURE 15.9 Decision tree for the selection of suitable TGS de novo sequencing analysis tools in hybrid sequencing platform. If the read used both platforms (nanopore and SMRT sequencing), one must decide whether the reads are from a circular genome and whether no error correction is required. If the reads are

FIGURE 15.9 (*Continued*)

applied from both NGS and the TGS platform, one must identify whether the reads are from AB eukaryotic-sized or bacterial-sized genome and determine the availability of the Newbler software. In addition, one must determine whether the reads from both the NGS and TGS platform need gap closing or scaffolding assembly.

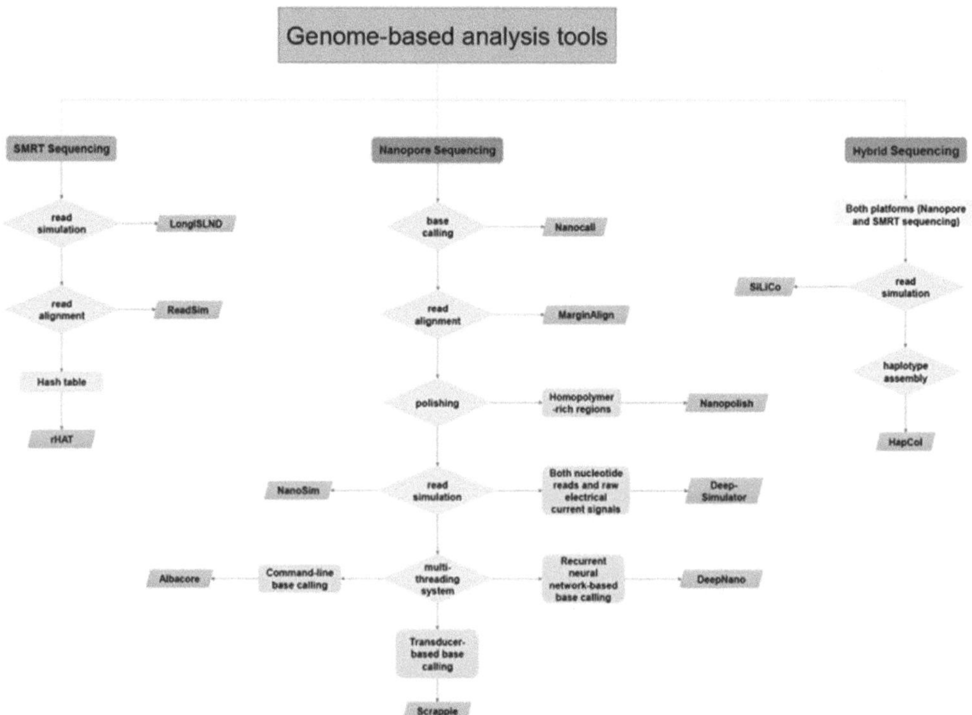

FIGURE 15.10 Decision tree for the selection of a suitable TGS genome-based sequencing analysis tool in SMRT, nanopore, and hybrid sequencing platform. A set of sequential decisions has to be made when performing genome-based sequencing analyses. First, one must determine whether the reads are generated from the two TGS platforms nanopore or ONT and SMRT sequencing or PacBio technologies and whether it is a hybrid or non-hybrid read. If the reads from the SMRT platform are used for read alignment, one must identify the length of the reads and whether a hash table is required. For the reads from the nanopore platform, one must decide whether the reads should be performed for base calling or read alignment. For the hybrid reads that utilize both platforms, one must decide whether the analyses should be carried out for read simulation or haplotype assembly.

15.4 BIOINFORMATICS AND GENOMICS IN BIOMEDICAL BREAKTHROUGHS

Advanced knowledge of mutated biomolecular states, especially physical properties of the states, and related statistics help us address vital medical problems requiring the discovery of means of recovery. Both bioinformatics (Cheng et al., 2018; Saeb, 2018; Mahmud et al., 2021; Meisel et al., 2018) and genomics (Bloss et al., 2011; Cain and Lees, 2015) techniques have recently been rigorously applied to combat diseases.

As said earlier, the Mid-Atlantic Microbiome Meet-up (M3) organization recently brought together interested stakeholders, including academic, government, and industry groups, to share ideas and develop best practices for microbiome research. They focused discussion on recent advances in biodefense, specifically those relating to infectious disease, and the use of metagenomic methods for pathogen detection (Meisel et al., 2018). These kinds of large-scale cross-talks lead to interactions targeting discoveries that will benefit large populations or even the entire world, e.g. the whole Human Genome Project, which ensures benefits for all of humanity (Nurk et al., 2022). 200 million DNA base pairs and 115 protein-coding genes have been added, but the Y chromosome is yet to be entirely sequenced (Reardon, 2021).

Since 2000, the human reference genome has covered only the euchromatic fraction of the genome, leaving important heterochromatic regions unfinished. Addressing the remaining 8 percent of the genome, the Telomere-to-Telomere (T2T) Consortium presented a complete 3.055 billion base pair sequence of a human genome, T2T-CHM13, that included gapless assemblies for all chromosomes except Y, corrected errors in the prior references, and introduced nearly 200 million base pairs of sequence containing 1,956 gene predictions, 99 of which were predicted to be protein-coding. Figure 15.11 presents the historical progression of the genome project since its inception. Figure 15.12 presents the summary of the complete T2T-CHM13 human genome assembly.

During the two decades since the initiation of the Human Genome Project (Venter et al., 2001; IHGSC, 2001), huge opportunities have been created that have empowered research to penetrate the genetic roots of human disease, change drug discovery, and help to revise the idea of the gene itself. Gates et al. (2021) merged several datasets to quantify the different types of genetic elements that

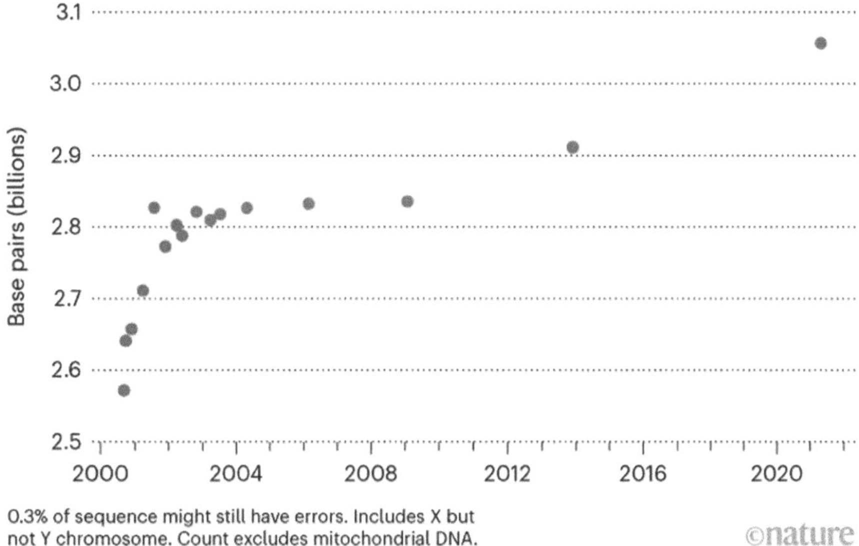

FIGURE 15.11 Time-dependent progression in the genome project since its start at the beginning of the current century.

***Source*: reproduced with permission from Reardon (2021).**

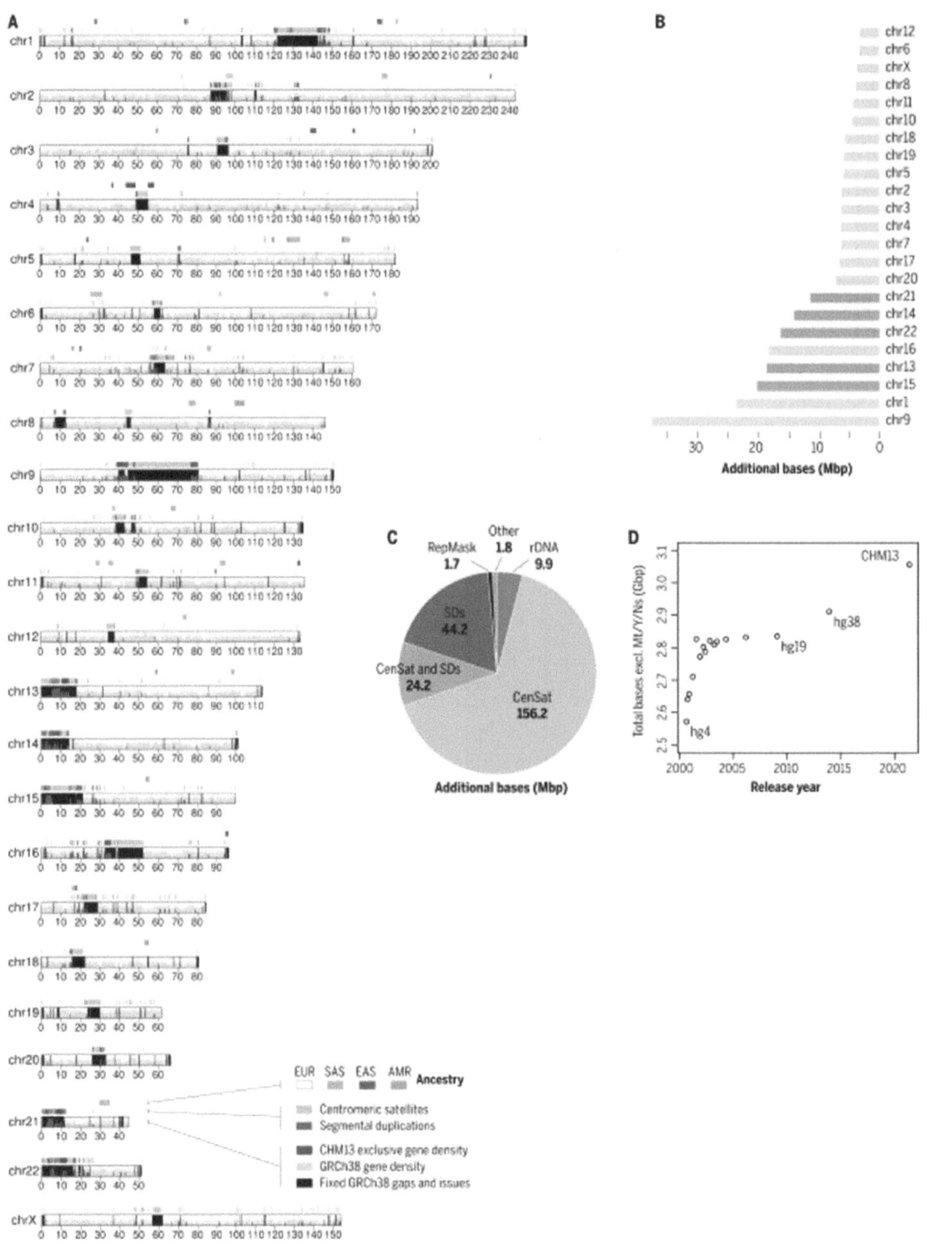

FIGURE 15.12 Summary of the complete T2T-CHM13 human genome assembly. (A) Ideogram of T2T-CHM13v1.1 assembly features. For each chromosome (chr), the following information is provided from bottom to top: gaps and issues in GRCh38 fixed by CHM13 overlaid with the density of genes exclusive to CHM13 in red; segmental duplications (SDs) (Vollger et al., 2022) and centromeric satellites (CenSat) (Altemose et al., 2022); and CHM13 ancestry predictions (EUR, European; SAS, South Asian; EAS, East Asian; AMR, admixed American). Bottom scale is measured in Mbp. (B and C) Additional (nonsyntenic) bases in the CHM13 assembly relative to GRCh38 per chromosome, with the acrocentrics highlighted in black (B) and by sequence type (C). (Note that the CenSat and SD annotations overlap.) RepMask, RepeatMasker. (D) Total nongap bases in UCSC reference genome releases dating back to September 2000 (hg4) and ending with T2T-CHM13 in 2021. Mt/Y/Ns, mitochondria, chrY, and gaps.

Source: **reproduced with permission from Nurk et al. (2022).**

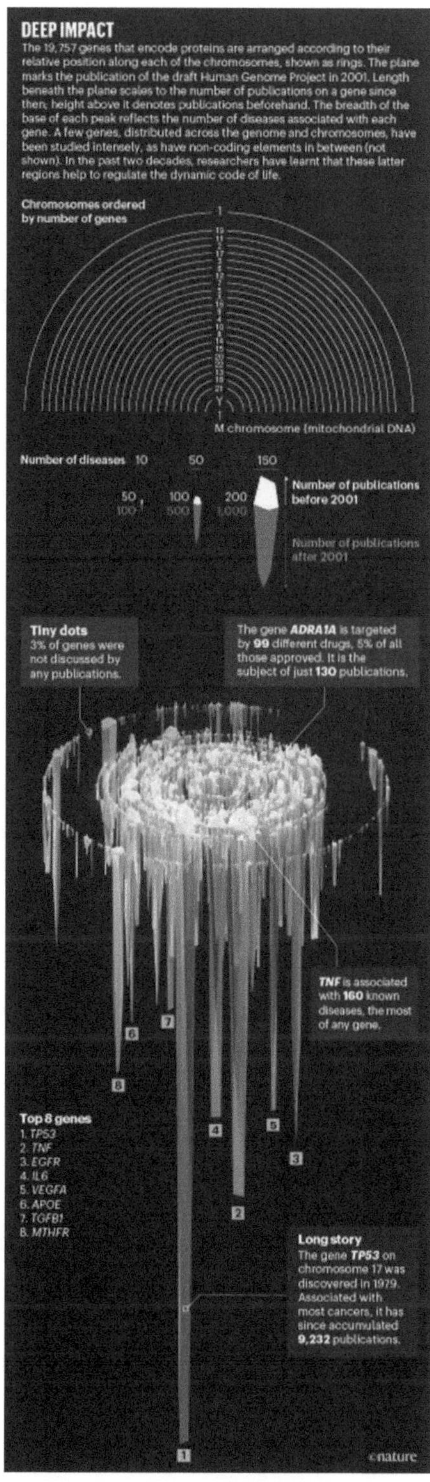

FIGURE 15.13 The medical impacts of the genome project.

***Source*: reproduced with permission from Gates et al. (2021).**

TWENTY YEARS OF JUNK, STARS AND DRUGS

What genomics researchers have studied, when and why — as traced by bibliometric analysis.

Non-coding elements

Most protein-coding genes were discovered before the first draft of the Human Genome Project (HGP) in 2001. Many other genomic elements, previously called junk DNA, came in for scrutiny after that.

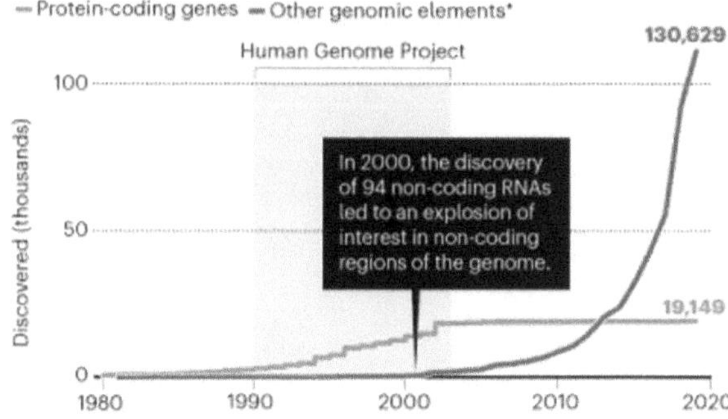

Star genes

The HGP gave rise to an explosion in research concentrated on just a few genes.

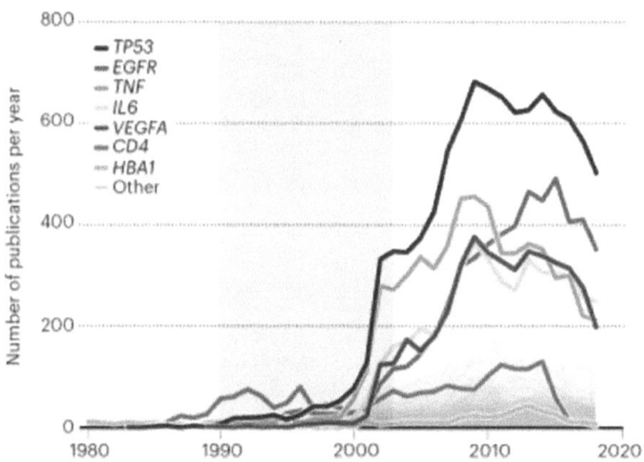

Drug targets

Since 2001, nearly 100% of US drugs licensed in any given year have had all their potential protein targets identified.

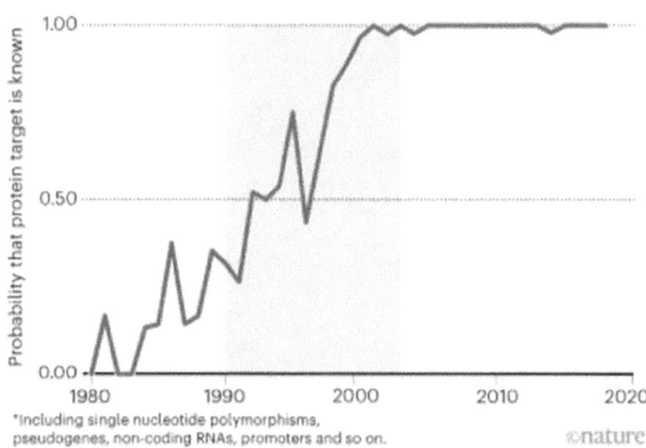

FIGURE 15.14 Statistics on drug discovery research and associated data.

***Source*: reproduced with permission from Gates et al. (2021).**

have been discovered and that have generated publications, and how the pattern of discovery and publishing has changed over the years. This analysis linked together data including 38,546 RNA transcripts; around 1 million single nucleotide polymorphisms (SNPs); 1,660 human diseases with documented genetic roots; 7,712 approved and experimental pharmaceuticals; and 704,515 scientific publications between 1900 and 2017. The medical impacts of the genome project are somewhat summarized in Figure 15.13.

Before the 1980s, drugs in drug discovery research were often found randomly, yet appeared occasionally as fortunate discoveries. Their molecular and protein targets were largely unknown. Until 2001 when the Human Genome Project started, the probability of knowing all of a drug's protein targets was less than 50 percent in any given year. The Human Genome Project changed the scenario. The targets are now known for almost all drugs licensed in the United States each year (see Figure 15.14). Of the roughly 20,000 proteins revealed by the Human Genome Project as potential drug targets, only about 10 percent (2,149) have so far been targeted by approved drugs, leaving 90 percent of the proteome untouched by pharmacology (Wishart et al., 2018). Experimental drugs in the dataset of Gates and colleagues (2021) increase this number to 3,119.

The Human Genome Project has exemplified the power, necessity, and success of large, integrated, cross-disciplinary efforts – so-called "big science" – directed toward complex major objectives (Hood and Rowen, 2013). The genome project led to the development of novel technologies and analytical tools, and brought the expertise of engineers, computer scientists, biophysicists, and mathematicians together with biologists. The project established especially an open approach to data sharing and open-source software, thereby making the data resulting from the project accessible to all interested parties and stakeholders. The deeper knowledge of human sequence variation has started to alter approaches to medicine, especially in the development of drugs for targets identified by the project. The project also inspired subsequent large-scale data acquisition initiatives, as follows: the International HapMap Project, 1000 Genomes, The Cancer Genome Atlas, Human Brain Project, Human Proteome Project, etc. Detailed analysis may be found in Hood and Rowen (2013).

REFERENCES

Abdurakhmonov, I.Y. (2016). Bioinformatics: basics, development, and future. In *Bioinformatics – updated features and applications*, ed. I.Y. Abdurakhmonov. Rijeka, Croatia: IntechOpen. https://doi.org/10.5772/63817

Altemose, N., Logsdon, G.A., Bzikadze, A.V., Sidhwani, P., Langley, S.A., Caldas, G.V., ... & Miga, K.H. (2022). Complete genosmic and epigenetic maps of human centromeres. *Science*, *376*(6588), eabl4178

Bayat, A. (2002). Science, medicine, and the future: bioinformatics. *BMJ*, *324*(7344), 1018–1022. doi: 10.1136/bmj.324.7344.1018; PMID: 11976246; PMCID: PMC1122955

Bloss, C.S., Jeste, D.V., & Schork, N.J. (2011). Genomics for disease treatment and prevention. *Psychiatric Clinics*, *34*(1), 147–166. doi: 10.1016/j.psc.2010.11.005; PMID: 21333845; PMCID: PMC3073546

Brennan, C.W., Verhaak, R.G., McKenna, A., Campos, B., Noushmehr, H., Salama, S.R., ... & Davidsen, T. (2013). The somatic genomic landscape of glioblastoma. *Cell*, *155*(2), 462–477

Brown, T.A. (2002). *Genomes* (2nd edn). Oxford: Wiley-Liss.

Brusic, V., & Ranganathan, S. (2008). Critical technologies for bioinformatics. *Briefings in Bioinformatics*, *9*(4), 261–262. https://doi.org/10.1093/bib/bbn025

Buettner, F., Natarajan, K., Casale, F., et al. (2015). Computational analysis of cell-to-cell heterogeneity in single-cell RNA-sequencing data reveals hidden subpopulations of cells. *Nat. Biotechnol.*, *33*, 155–160. https://doi.org/10.1038/nbt.3102

Cain, A.K., and Lees, J.A. (2015). Using genomics to combat infectious diseases on a global scale. *Genome Biol.*, *16*, 250. doi: 10.1186/s13059-015-0822-y; PMID: 26581564; PMCID: PMC4652341

Cheng, L., Hu, Y., Sun, J., Zhou, M., & Jiang, Q. (2018). DincRNA: a comprehensive web-based bioinformatics toolkit for exploring disease associations and ncRNA function. *Bioinformatics*, *34*(11), 1953–1956. doi: 10.1093/bioinformatics/bty002; PMID: 29365045

Dayhoff, M.O. (1965). Atlas of Protein Sequence and Structure, Vol. 1. Silver Spring, MD: National Biomedical Research Foundation

Dayhoff, M.O., and Ledley, R.S. (1962a). Comprotein: a computer program to aid primary protein structure determination. In *Proceedings of the December 4–6, 1962, Fall Joint Computer Conference*, pp. 262–274. New York: ACM

Dayhoff, M.O., and Ledley, R.S. (1962b). Managing requirements knowledge: International workshop on COMPROTEIN, a computer program to aid primary protein structure determination. www.computer.org/csdl/proceedings-article/afips/1962/50610262/12OmNyOq4Uy

Edman, P. (1949). A method for the determination of amino acid sequence in peptides. *Arch. Biochem.*, *22*, 475

Eichler, E.E., Clark, R.A., & She, X. (2004). An assessment of the sequence gaps: unfinished business in a finished human genome. *Nat. Rev. Genet.*, *5*, 345–354

Ewens, W.J. (2005). *Statistical methods in bioinformatics*. New York: Springer

Garber, M., Grabherr, M.G., Guttman, M., & Trapnell, C. (2011). Computational methods for transcriptome annotation and quantification using RNA-seq. *Nature Methods*, *8*(6), 469–477. https://doi.org/10.1038/nmeth.1613

Gates, A.J., Gysi, D.M., Kellis, M., & Barabási, A.L. (2021). A wealth of discovery built on the Human Genome Project – by the numbers. *Nature*, *590*(7845), 212–215. doi: https://doi.org/10.1038/d41586-021-00314-6

Gauthier, J., Vincent, A.T., Charette, S.J., & Derome, N. (2019). A brief history of bioinformatics. *Briefings in Bioinformatics*, *20*, 1981–1996. https://doi.org/10.1093/bib/bby063

Hood, L., & Rowen, L. (2013). The Human Genome Project: big science transforms biology and medicine. *Genome Medicine*, *5*, 1–8. https://doi.org/10.1186/gm483

Horner, D.S., Pavesi, G., Castrignano, T., De Meo, P.D.O., Liuni, S., Sammeth, M., ... & Pesole, G. (2010). Bioinformatics approaches for genomics and post genomics applications of next-generation sequencing. *Briefings in Bioinformatics*, *11*(2), 181–197

IHGSC. (2001). Initial sequencing and analysis of the human genome. *Nature*, *409*, 860–921

IUPAC-IUB. (1968). A one-letter notation for amino acid sequences*. *Eur. J. Biochem.*, *5*, 151–153

Kendrew, J.C., Bodo, G., Dintzis, H., et al. (1958). A three-dimensional model of the myoglobin molecule obtained by X-ray analysis. *Nature*, *181*, 662–666. https://doi.org/10.1038/181662a0

Kennedy, V. (2018). *Fundamentals of genomics*. New York: Larsen and Keller Education

Luo, J., Wu, M., Gopukumar, D., & Zhao, Y. (2016). Big data application in biomedical research and health care: a literature review. *Biomed. Inform. Insights*, *8*, 1–10. doi: 10.4137/BII.S31559; PMID: 26843812; PMCID: PMC4720168

Lynch, C. (2008). Big data: how do your data grow? *Nature*, *455*(7209), 28–29

Mahmud, S.H., Al-Mustanjid, M., Akter, F., Rahman, M.S., Ahmed, K., Rahman, M.H., ... & Moni, M.A. (2021). Bioinformatics and system biology approach to identify the influences of SARS-CoV-2 infections to idiopathic pulmonary fibrosis and chronic obstructive pulmonary disease patients. *Briefings in Bioinformatics*, *22*(5), bbab115. https://doi.org/10.1093/bib/bbab115

Matthiesen, R. (2010). *Bioinformatics methods in clinical research*. New York: Humana Press

Maxam, A.M., & Gilbert, W. (1977). A new method for sequencing DNA. *Proceedings of the National Academy of Sciences*, *74*(2), 560–564

Meisel, J.S., Nasko, D.J., Brubach, B., et al. (2018). Current progress and future opportunities in applications of bioinformatics for biodefense and pathogen detection: report from the Winter Mid-Atlantic Microbiome Meet-up, College Park, MD. *Microbiome*, *6*, 197. https://doi.org/10.1186/s40168-018-0582-5

Metzker, M.L. (2010). Sequencing technologies – the next generation. *Nat. Rev. Genet.*, *11*(1), 31–46

Moody, G. (2004). *Digital code of life: how bioinformatics is revolutionizing science, medicine, and business*. Hoboken, NJ: John Wiley & Sons

Morrison-Smith, S., Boucher, C., Sarcevic, A., et al. (2022). Challenges in large-scale bioinformatics projects. *Humanit. Soc. Sci. Commun.*, *9*, 125. https://doi.org/10.1057/s41599-022-01141-4

Muirhead, H., & Perutz, M.F. (1963). Structure of hæmoglobin: a three-dimensional Fourier synthesis of reduced human haemoglobin at 5.5 Å resolution. *Nature*, *199*, 633–638

Myers, E.W., Sutton, G.G., Delcher, A.L., et al. (2000). A whole-genome assembly of Drosophila. *Science*, *287*, 2196–2204

Nielsen, R., Paul, J.S., Albrechtsen, A., et al. (2011). Genotype and SNP calling from next-generation sequencing data. *Nat. Rev. Genet.*, *12*(6), 443–451

Nirenberg, M., & Leder, P. (1964). RNA codewords and protein synthesis: the effect of trinucleotides upon the binding of sRNA to ribosomes. *Science*, *145*(3639), 1399–1407

Nurk, S., Koren, S., Rhie, A., Rautiainen, M., Bzikadze, A.V., Mikheenko, A., ... & Phillippy, A.M. (2022). The complete sequence of a human genome. *Science*, *376*(6588), 44–53

Patel, A.P., Tirosh, I., Trombetta, J.J., Shalek, A.K., Gillespie, S.M., Wakimoto, H., ... & Bernstein, B.E. (2014). Single-cell RNA-seq highlights intratumoral heterogeneity in primary glioblastoma. *Science*, *344*(6190), 1396–1401

Perutz, M.F., Rossmann, M.G., Cullis, A.F., Muirhead, H., Will, G., & North, A.C.T. (1960). Structure of hæmoglobin: a three-dimensional Fourier synthesis at 5.5-Å. resolution, obtained by X-ray analysis. *Nature*, *185*, 416–422

Ramachandran, G.N., Ramakrishnan, C., & Sasisekharan, V. (1963). Stereochemistry of polypeptide chain configurations. *J. Mol. Biol.*, *7*, 95–99. doi: 10.1016/s0022-2836(63)80023-6; PMID: 13990617

Ramsköld, D., Luo, S., Wang, Y.C., Li, R., Deng, Q., Faridani, O.R., ... & Sandberg, R. (2012). Full-length mRNA-seq from single-cell levels of RNA and individual circulating tumor cells. *Nature Biotechnology*, *30*(8), 777–782

Reardon, S. (2021). A complete human genome sequence is close: how scientists filled in the gaps. *Nature*, *594*(7862), 158–159

Saeb, A.T. (2018). Current bioinformatics resources in combating infectious diseases. *Bioinformation*, *14*(1), 31–35. doi: 10.6026/97320630014031; PMID: 29497257; PMCID: PMC5818640

Sanger, F., & Coulson, A.R. (1975). A rapid method for determining sequences in DNA by primed synthesis with DNA polymerase. *J. Mol. Biol.*, *94*(3), 441–448. doi: 10.1016/0022-2836(75)90213-2; PMID: 1100841

Sanger, F., Nicklen, S., & Coulson, A.R. (1977). DNA sequencing with chain-terminating inhibitors. *Proceedings of the National Academy of Sciences*, *74*(12), 5463–5467

Saraswathy, N., & Ramalingam, P. (2011). *Concepts and techniques in genomics and proteomics*. Oxford: Woodhead

Schneider, V.A., Graves-Lindsay, T., Howe, K., Bouk, N., Chen, H.C., Kitts, P.A., ... & Church, D.M. (2017). Evaluation of GRCh38 and de novo haploid genome assemblies demonstrates the enduring quality of the reference assembly. *Genome Research*, *27*(5), 849–864

Shalek, A.K., Satija, R., Adiconis, X., Gertner, R.S., Gaublomme, J.T., Raychowdhury, R., ... & Regev, A. (2013). Single-cell transcriptomics reveals bimodality in expression and splicing in immune cells. *Nature*, *498*(7453), 236–240

Shendure, J., & Ji, H. (2008). Next-generation DNA sequencing. *Nat. Biotechnol.*, *26*(10), 1135–1145

Sotirov, S.S., Pencheva, T., Kacprzyk, J., Atanasov, K., Sotirova, E., & Staneva, G. (eds.). (2022). *Contemporary methods in bioinformatics and biomedicine and their applications*. Cham: Springer

Stratton, M.R., Campbell, P.J., & Futreal, P.A. (2009). The cancer genome. *Nature*, *458*(7239), 719–724

Tenenbaum, J.D. (2016). Translational bioinformatics: past, present, and future. *Genomics Proteomics Bioinformatics*, *14*(1), 31–41. doi: 10.1016/j.gpb.2016.01.003; PMID: 26876718; PMCID: PMC4792852

Tenenbaum, J.D., Shah, N.H., & Altman, R.B. (2014). Translational bioinformatics. In *Biomedical informatics*, ed. E.H. Shortliffe, & J.J. Cimino, pp. 721–754. London: Springer

Venter, J.C., Adams, M.D., Myers, E.W., et al. (2001). The sequence of the human genome. *Science*, *291*, 1304–1351

Verhaak, R.G., Hoadley, K.A., Purdom, E., Wang, V., Qi, Y., Wilkerson, M.D., ... & Cancer Genome Atlas Research Network. (2010). Integrated genomic analysis identifies clinically relevant subtypes of glioblastoma characterized by abnormalities in PDGFRA, IDH1, EGFR, and NF1. *Cancer Cell*, *17*(1), 98–110

Vollger, M.R., Guitart, X., Dishuck, P.C., Mercuri, L., Harvey, W.T., Gershman, A., ... & Eichler, E.E. (2022). Segmental duplications and their variation in a complete human genome. *Science*, *376*(6588), eabj6965

Watson, J.D., & Crick, F.H. (1953). Molecular structure of nucleic acids: a structure for deoxyribose nucleic acid. *Nature*, *171*(4356), 737–738

Wee, Y., Bhyan, S.B., Liu, Y., Lu, J., Li, X., & Zhao, M. (2019). The bioinformatics tools for the genome assembly and analysis based on third-generation sequencing. *Briefings in Functional Genomics*, *18*(1), 1–12

Wishart, D.S., Feunang, Y.D., Guo, A.C., Lo, E.J., Marcu, A., Grant, J.R., … & Wilson, M. (2018). DrugBank 5.0: a major update to the DrugBank database for 2018. *Nucleic Acids Research*, *46*(D1), D1074–D1082

Zhang, J., Chiodini, R., Badr, A., et al. (2011). The impact of next-generation sequencing on genomics. *J Genet. Genomics*, *38*(3), 95–109

Zhang, X., & Mak, M. (2021). Biophysical informatics approach for quantifying phenotypic heterogeneity in cancer cell migration in confined microenvironments. *Bioinformatics*, *37*(14), 2042–2052

16 Biophysical Economics

Economics belongs to social sciences and biophysics to natural sciences. We deal with these distinguishable subjects in individual areas. Economics engages academicians and policymakers to construct choices and preferences for managing resources. With the help of adopted choices or rules that are set in the system, there is a social balance maintained between the available goods, and services, and their consumption by the people in an environment of finite resources. Failed policy may cause social crises, often leading to chaos. Biophysical and social systems are linked to form social-ecological systems whose sustainability depends on their capacity to absorb uncertainty and cope with disturbances (García-Llorente et al., 2015). For maintaining this system, ecological economics is required to provide institutional and methodical insights, based on biophysical foundations, to fully disclose how human society extracts material and energy from the dynamical ecosystem, transforms them into physical wealth, and distributes the physical wealth among societies with diversified ethnicities and religious and cultural differences (Ji and Luo, 2020).

Biophysics, as a natural science subject, is mostly involved in addressing energetics and informatics within complex, dynamic, and fluctuating biosystems. The biophysical version of any economic theory uses mass and energy flows, as are found to happen in any biophysical system having environmental constraints, to describe the delivery of goods and services in a society (Yan et al., 2019). Biophysical economics can illustrate the laws that correlate energy and economic developments. Energy is associated with producing the economic driving force. On an applied side, biophysical economics focuses on the transformation of material and energy to maintain resources required for human survival and promote timely social development. Students, politicians, academics, and other parties who are interested to learn about how to create policies and develop methodologies for integrating energy and economics (Hall and Klitgaard, 2018; Peniche Camps et al., 2020) have to be open to welcoming combined natural and social sciences approaches. This chapter will provide brief guidelines by addressing theoretical and practical cases.

16.1 INTRODUCTION

Sociocultural and socioeconomic progress in developing and developed societies continues through the adoption of policies that always undergo renormalization with time. Underdeveloped societies may often fail to cope with the pace of changes that are recommended as per international standards. The economy is the main issue. Due to the lack of appropriate resources, the policymakers in poverty-affected societies often rely on aid (mostly financial and logistic support) ensured by mostly the rich nations. But there is a gap between the approaches of these two quite different classes of economic societies (Stapelbroek and Marjanen, 2012).

DOI: 10.1201/9781003287780-16

The adoption of strategies on how to implement the required changes in underdeveloped societies requires both participating parties (receiving and supporting parties) to consider the effects the new policies may have on social values, religious beliefs, and common practices involving attachment to the local environment. Ambitious changes require modern methods to be adopted and implemented. These methods are primarily developed to serve the rich world as most of these innovations are modeled considering their needs. However, the economy of the whole world is going through phenomenal changes (World Bank, 2010; Desai et al., 2011). Connections regarding these changes between societies with a wide distribution of wealth and facilities can only be made by adopting some common principles and methodologies. Instead of using traditional economics theories, policymakers look for new ideas and methodologies merging social science and natural science techniques, such as economic policies involving biophysical principles to tackle environmental issues and in parallel achieve sustainable growth in certain economic indicators involved in developing the lives of the people (Yan et al., 2019). Specifically, biophysical economics has recently been found adaptable to various sectors that are considered vital in modern norms of development among underdeveloped, developing, and developed societies. See the framework of biophysical economics proposed in Figure 16.1.

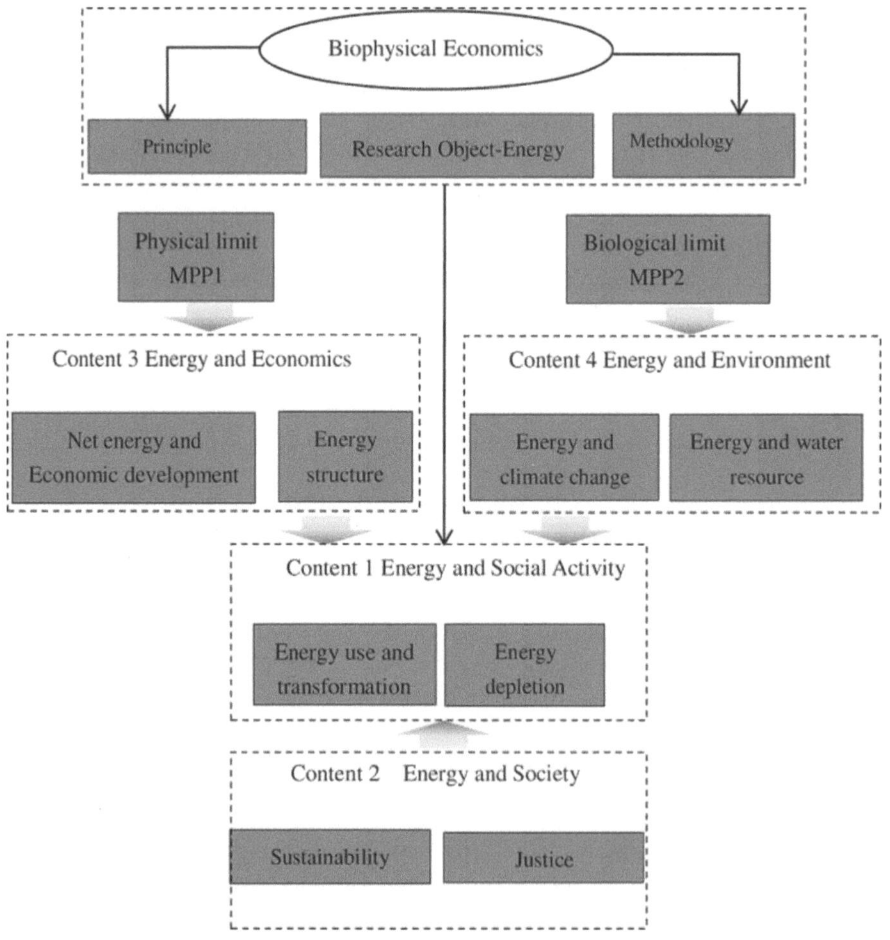

FIGURE 16.1 Biophysical economics framework.

Source: Yan et al. (2019).

The progress of historical economic arguments, exploration of hypotheses, and developments of proof of principles are often compared to the currently developing biophysical economics framework, which, instead of focusing on just investment, debt, and growth, focuses on sustainable energy and mass flows to deliver goods and services to civilization. Physical parameter entropy is considered irrelevant to the economics of the use of resources in the neoclassical view but relevant in the ecological view (McMahon and Mrozek, 1997). Entropy is a measure of disorders of a system, such as the environment as a whole or components thereof. If the system is open without a classified closed boundary surrounding it, the entropy increases (for a general understanding of the concept, see Sewell, 2013). For any reversible process, there won't be any net change in entropy, but in any irreversible process, entropy always increases. The environment is as a whole an irreversible process, so its total entropy is supposed to increase over time. The spectacular increase in order occurring on Earth is argued to be consistent with the second law of thermodynamics because the Earth itself is not an isolated system. In any non-isolated system, things happen randomly and contribute to the entropic increase as long as the entropy increases outside the system compensate for the entropy decreases inside the system (Sewell, 2013). A hypothesis or a probable theory is presented here suggesting that "if an increase in order is extremely improbable when a system is isolated, it is still extremely improbable when the system is open unless something is entering (or leaving) which makes it not extremely improbable." As an example, this report has argued in favor of using the influx of energy from the ultimate energy reservoir "solar energy" to downregulate the increase of environmental disorders due to the actions humans take for day-to-day cosmetic developments.

The entropic increase outside the system and decrease inside are not mutually compensating mechanisms, so together they lead to the rise of inequilibration in environmental health over time, hence requiring economic policies to tackle the components responsible for regulating the environmental alterations. Therefore, sustainability is always an issue while dealing with the environment. However, the amount of induction of disorder into the environment partly depends on how we handle the environment. Political processes, socioeconomic conditions, sociocultural practices, religious beliefs, availability of resources, the health of the economy, etc. contribute to the speed of the induction of disorders in the environment where we live. Economics dealing with environmental sustainability needs to tackle the physical entropy or intensity of disorders with appropriate strategies and engineering technologies. Economic growth or scarcity is associated with thermodynamic entropy and energy of the environmental systems through various hypotheses and models (Georgescu-Roegen, 1971).

Besides thermodynamic entropy, systems' organization and sustainability may also rely on the processing of information. For example, information is gained on assembling and maintaining a living state (Davies et al., 2013). This is a time-dependent issue, mostly applicable to the dynamic system. The environment may also be considered overall a dynamic system, though there are a lot of static components inherited permanently from the past. So information processing, like that of biological systems, especially the cellular processes, may also be considered while addressing the sustainability of the rather dynamic environmental economics (Moser et al., 2014). I shall briefly discuss the biophysical aspects of sustainable economics in this chapter.

16.2 ECONOMIC CONDITION VERSUS LIFE EXPECTANCY

Dayanikli et al. (2016) have constructed an important model to derive life expectancy depending on various socioeconomic parameters, as follows:

$$\text{Predicted Life Expectancy} = 26.34697 + 3.843959 \times \log(\text{GDP per capita}) + 0.2454119 \times \text{public health expenditure} + 0.7925244 \times \text{average years of schooling} \quad (16.1)$$

Prediction is made using the above equation that GDP per capita affects life expectancy up until a certain threshold. Beyond a certain amount of GDP, the correlation between the variables weakens. The below-median GDP to life expectancy regression is stronger than the above-median GDP to life expectancy regression. The plot of life expectancy (Figure 16.2) versus GDP per capita suggests that there is a clear curve shape at the lower regime (Freeman et al., 2020). As the GDP per capita gets higher, the slope decreases, suggesting that it affects life expectancy less. This suggests that GDP per capita (economic parameter) is more significant up until a certain maximum GDP point, then other factors (social conditions) such as public health expenditure, years of schooling, etc. become important. Biophisical factors are heavily involved in these social conditions, e.g. biomedical research standards, preference, and funding, biomedical research excellence, medical products availability, access to medical services, etc. The reflection of these parameters is found in Figure 16.3. In Chetty et al. (2016), the association between income and life expectancy in the USA has been demonstrated. A huge household income-dependent distribution has been found. Between 2001 and 2014, higher income was found to be associated with greater longevity throughout the income distribution, and differences in life expectancy across income groups, reflecting ethnic and racial group contributions, increased during this period. E.g., in 2020, the average white American household income was $70,000 while black American household income was $41,000. The ratio between the incomes of these two income groups is approximately 1.71. Consequently, the life expectancies differ, though the gap has been narrowing during the last three decades. See Figure 16.4, where besides these two ethnic groups, Europe's average (over the life expectancies of all people of England (ENG), France (FR), Germany (GER), Netherlands (NL), Norway (NO), Spain (SP)) is also presented for comparison (Schwandt et al., 2021).

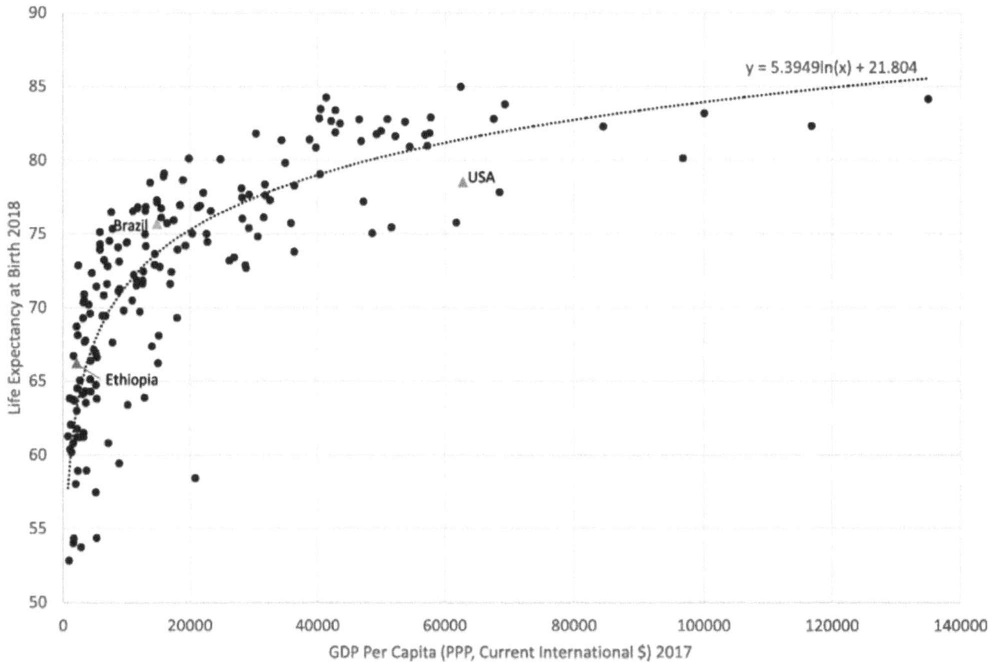

FIGURE 16.2 The relationship (the Preston Curve) between life expectancy in years and gross domestic product per capita in 2017 USD.

Source: **Freeman et al. (2020).**

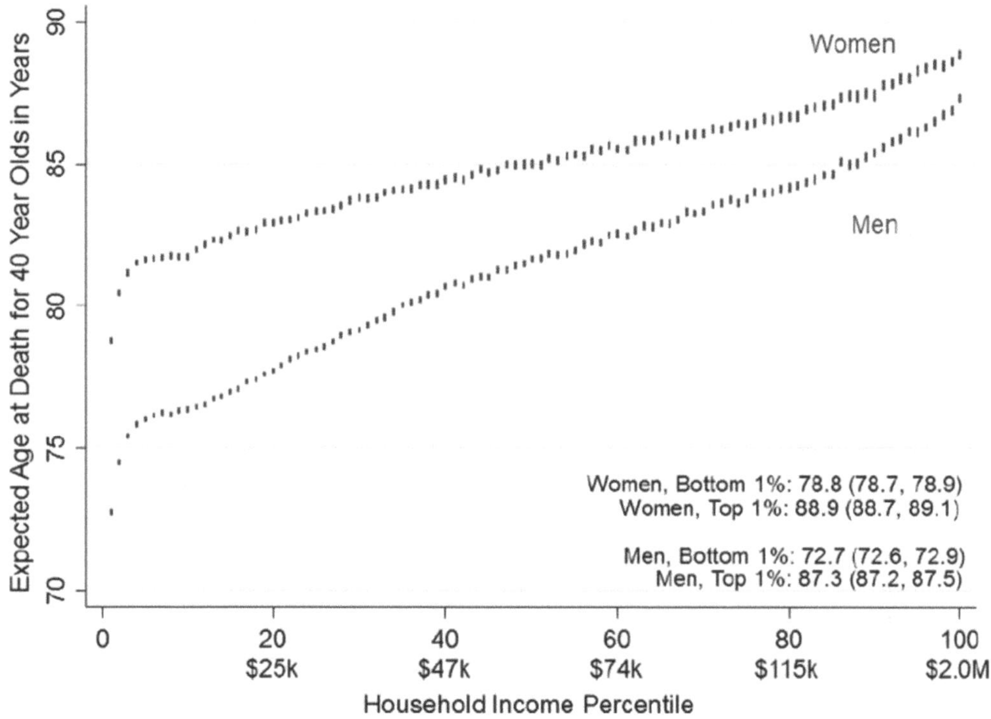

FIGURE 16.3 Race- and ethnicity-adjusted life expectancy for 40-year-olds by household income percentile, 2001–2014. The vertical height of each bar depicts the 95 percent confidence interval. The difference between expected age at death in the top and bottom income percentiles is 10.1 years (95 percent CI, 9.9–10.3 years) for women and 14.6 years (95 percent CI, 14.4–14.8 years) for men. To control the differences in life expectancies across racial and ethnic groups, race and ethnicity adjustments were calculated using data from the National Longitudinal Mortality Survey and estimates were reweighted so that each income percentile bin has the same fraction of black, Hispanic, and Asian adults.

***Source*: Chetty et al. (2016).**

16.3 BIOPHYSICAL INDICATORS AND ECONOMIC DEVELOPMENTS

Economic developments may partially be associated with specific biophysical indicators. For evaluating the impact of biophysical parameters on economic developments, the role of a few biophysical indicators for the economy of Catalan Spain was recently analysed (Manera et al., 2021). GDP and GDP per capita were inspected and certain biophysical variables were incorporated to check their cross-dependences. For the period 2000–2016, the figures for energy consumption, CO_2 emissions, energy intensity of the economy, and water consumption were collected. Energy consumption, including mode of energy intensity (the ratio of energy use or supply to GDP) (Remme et al., 2018; Niven, 2016), and CO_2 emission (Smith et al., 2016) are all biosystem-engaged biophysical parameters that are involved with environmental sustainability linked to the economic status of a society.

Among the ambitious objectives of social scientists, the understanding of the health of a society's economy in connection with its environment is important. To this end, theories of economics are often connected or merged with those of the natural sciences. That means a link is established between biology and thermodynamics and economics to model the strength of economic growth

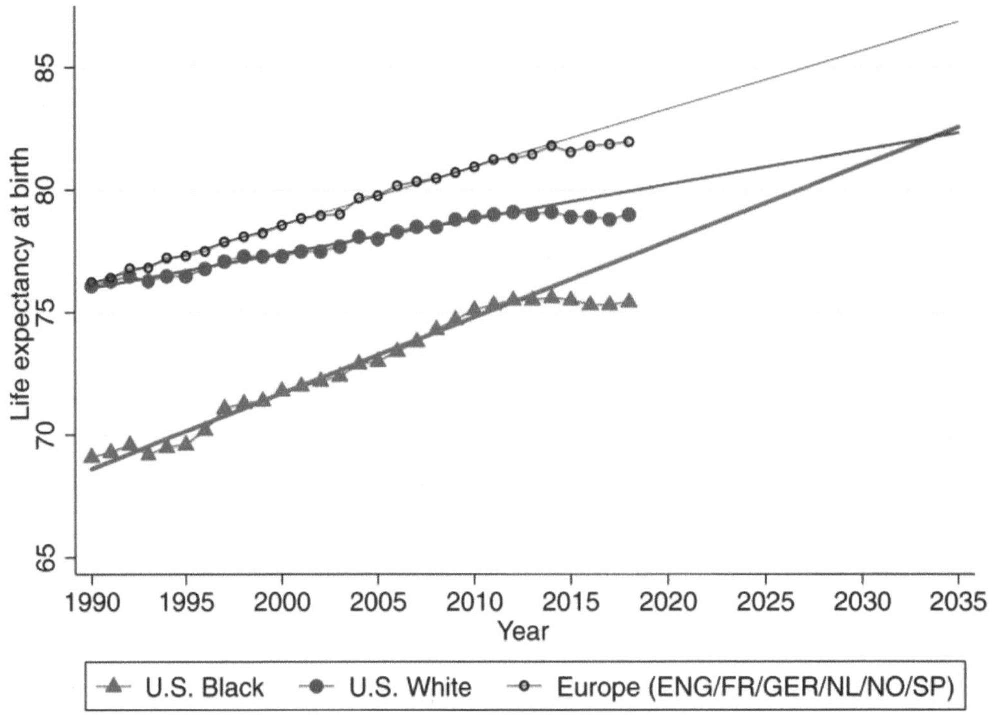

FIGURE 16.4 Life expectancy for black Americans, white Americans, and six European countries, extrapolated to 2035 fitting a linear trend through 1990 to 2012. Black circles show the population-weighted average life expectancy across England, France, Germany, the Netherlands, Norway, and Spain. For details, see Schwandt et al. (2021).

Source: **Schwandt et al. (2021).**

and create future predictions (Manera et al., 2021). In this combined attempt, two major aspects are covered, which are briefed here:

1. *Firstly, the issues that affect the environmental economy* (a sub-field of economics concerned with various environmental issues). In this case, the instruments applied are the neoclassically oriented aspects, such as the willingness to pay to maintain a certain natural or landscape resource and approaches to its economic valuation.
2. *Secondly, the issues that immerse themselves in the green economy.* The green economy adopts sustainable technologies to help generate economic development and improve people's lives in ways consistent with also advancing environmental and social well-being (Söderholm, 2020). In this case, non-chrematistic data are used, of a biophysical nature – territorial consumption, pollution, waste production, etc. – without creating any direct translation into prices.

The distinction between price and value (in the above two aspects) is significant. Therefore, in social science, various other indicators, such as ecological footprint, life expectancy, educational level, mortality, etc. are considered (to renormalize the state of overall benefits) that automatically include the status of GDP (Bericat and Jimenez-Rodrigo, 2019).

Many social scientists are scholarly enough to deal with the laws of thermodynamics. They can create a strong link between economics and the natural sciences by assimilating the fundamental thermodynamic principles into economics (Georgescu-Roegen, 1975). These thermodynamic rules function equally at the core of natural sciences fields, namely biology, physics, and chemistry, and now find an extension in the analysis of economic processes. Georgescu-Roegen proposed eight points to be considered in a thoughtful bioeconomy program to tackle environmental, biological, ecological issues, etc.

One of the major objectives of today's bioeconomy is to deal with the entropic nature of the economic process (Georgescu-Roegen, 1986). Entropy is a pure physics parameter, which represents the measure of disorder (Michaelides, 2008). In economics, entropy is associated with the process of the transformation of a useful energy into a low-grade energy. It's a measure of the dissipation of useful or high-grade energy. Although exotic parametric gains can be achieved due to this conversion, the environment is one factor that may be seriously affected due to the rise of disorder. Nicholas Georgescu-Roegen, a Romanian mathematician, statistician, and economist, who is best known for his 1971 magnum opus, argued by saying that all natural resources are irreversibly degraded when put to use in economic activity. His conceptual argument is now considered an important entropic law concerning economic processes. Recently, Antoine Missemer has argued differently on Georgescu-Roegen's ecological claim which has often been considered as a promotion of degrowth (Missemer, 2017). In this paper, he challenged this usual interpretation and concluded that Georgescu-Roegen might be a source of inspiration for degrowth defenders only in a very narrow sense. A cautious reading of his bioeconomic paradigm shows that Georgescu-Roegen's stance was different from the growth/degrowth debate, and might be more accurately linked with an "agrowth" option. Degrowth is known to emphasize reducing global consumption and production to achieve controlled social metabolism, and to advocate a socially just and ecologically sustainable society irrespective of any emphasis on GDP as an important prosperity indicator. In this process, economic entropy is considered to be kept under control.

We observe that economic growth and disorder in the environment are correlated. Therefore, it is now important to address the land structure, including the qualities of the soil, the possible impact of climate on economic structure, and the economic and ecological transformation of the landscape, etc. The adoption of a few biophysical indicators that are directly related to the environment may help determine the consequences that economic growth has on the environment (Manera et al., 2021). As mentioned earlier, we may consider the Catalan economy as an example case. Figure 16.5 represents how real GDP is associated with energy and water consumption and CO_2 emissions.

In Figure 16.6 (plotting the intensity of the use of resources), we see that a tendency toward an improvement in environmental indicators with time was achieved during the period 2000–2016. The indicators (energy consumption, CO_2 emission, and water consumption per GDP per capita) showed a downward trend. The intensity of the use of resources has been clearly demonstrated. The energy intensity decreased considerably from 2,000 to a base of 100 to 58 in 2016, which is a significant reduction and was constant throughout the entire period (except during 2003–2004). This contrasts with the evolution of the curves in Figure 16.7 (plotting the resource efficiency), where the so-called "productivity" of the resources was inserted in the production processes. These results altogether supported the idea of a better and more efficient use of various biophysical inputs for the achievement of Catalan GDP, which is also linked to its demographic variable, GDP per capita.

16.4 BIOPHYSICAL INDICATORS AND ECONOMIC SCARCITY

Thermodynamic entropy and economic scarcity are directly associated only with a thermodynamically isolated economy (Kovalev, 2016). If the economy (as an isolated system) is ergodic, at a macro level it is in stochastic equilibrium without time playing any crucial role. In this case, economic policies disturb, rather than contribute, to its natural growth (Anastasopoulos, 2020). This article has provided arguments and some evidence against the economy being a fully ergodic system. Instead, the economy may be considered a system experiencing ergodicity breaks. The economy may behave like an ergodic system occasionally or for a limited time, but frequently loses its independence from

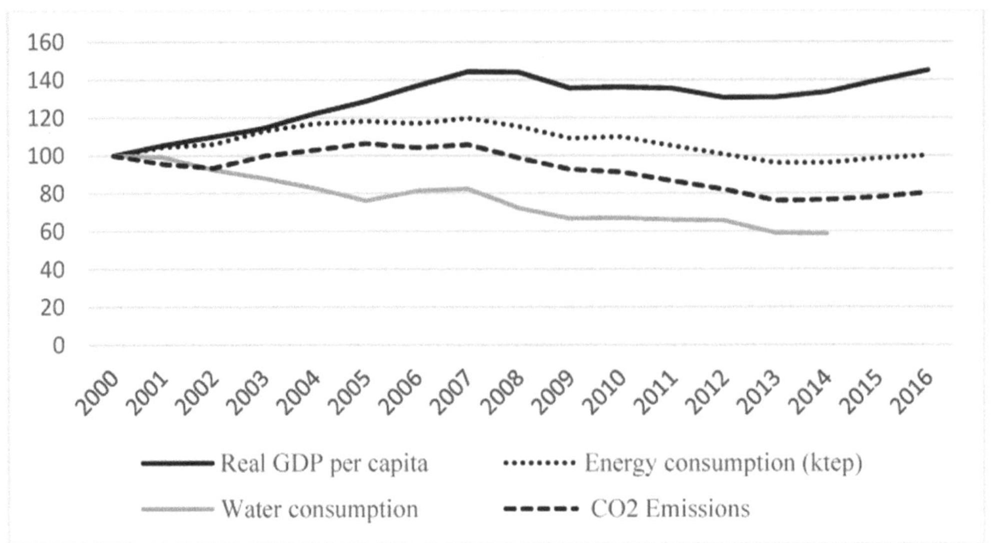

FIGURE 16.5 Descriptive evolution (2000 = 100). For details, see Manera et al. (2021).

Source: **Manera et al. (2021).**

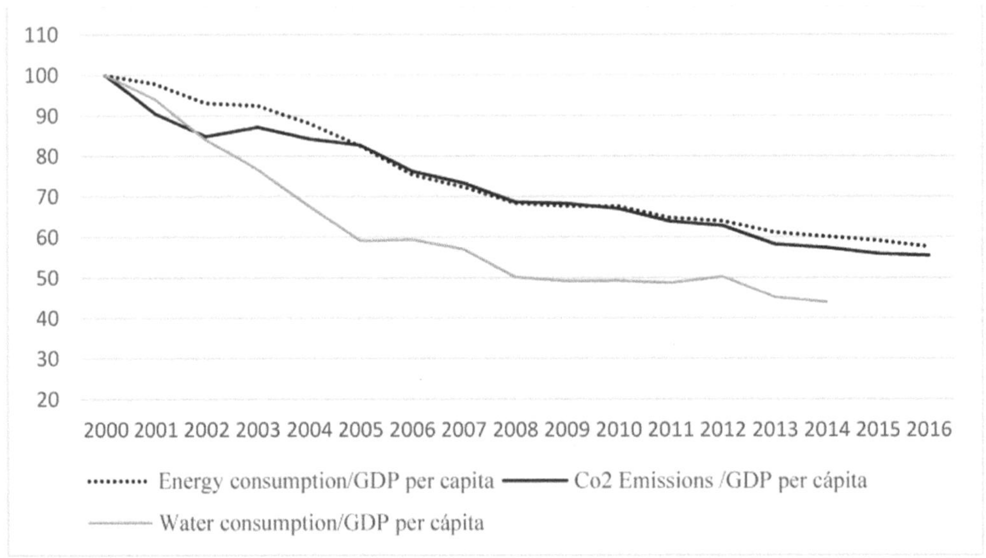

FIGURE 16.6 Use intensity of resources (2,000 = 100).

Source: **Manera et al. (2021).**

its historical time. External and internal disturbances contribute to this uncertainty in connectivity. The inevitable disturbances may alter the institutional framework, and be found to participate in modifying the behavior of economic agents while facing uncertainty. Therefore, the role of time cannot be denied in creating economic policies.

Disruption in the economy by natural resource depletion, due to our inability to recycle industrial waste and close the technological cycle, may happen while the ecosystem's total thermodynamic

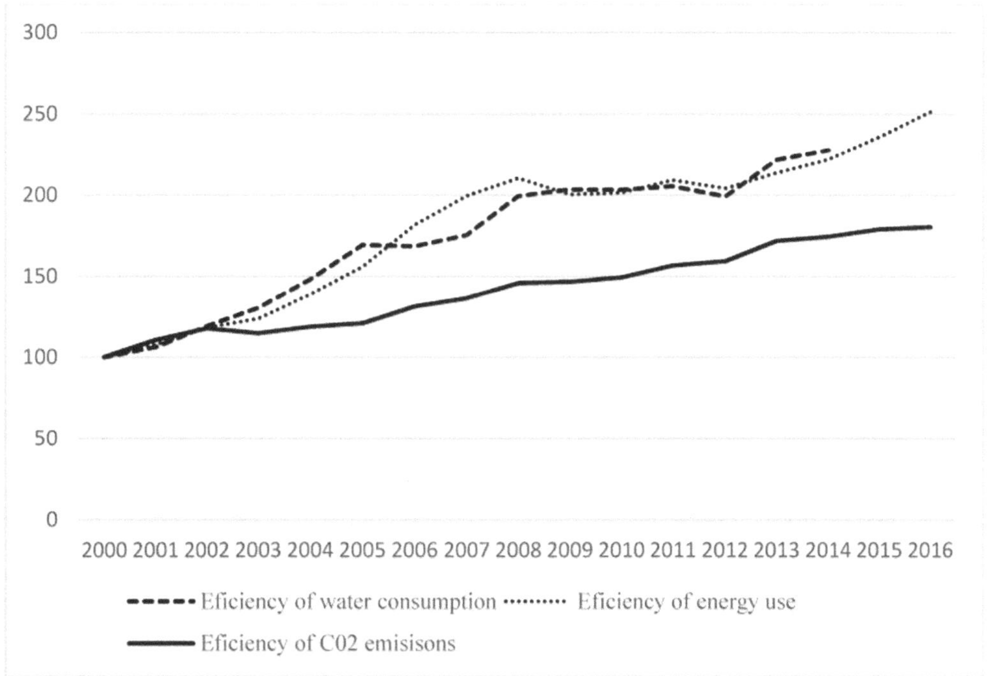

FIGURE 16.7 Resource efficiency (the inverse of the intensity).

Source: **Manera et al. (2021).**

entropy remains unchanged. This is because the transfer of chemically refined products may not contribute significantly to the increase of entropy, but it may decrease recyclability. The material entropy, which appears as a measure of the complexity and the economic dispersal of resources, may be considered a recyclability metric, but not a thermodynamic parameter, so its growth is not equivalent to the growth of thermodynamic entropy (Kovalev, 2016).

Among the environmental parameters, water is an important one. The economy is directly related to the use of water resources. Multi-scale economic teleconnections are found to mitigate or exacerbate water shortages. Major hydrologic basins can experience strongly positive or strongly negative economic impacts due to global trade dynamics and market adaptations to regional scarcity (Dolan et al., 2021). Water and wastewater processes are found to have a direct impact on the local economy. Oil-field-dependent economies (e.g. Alberta, Canada) directly rely on processing strategies. Here the management of the entropy of water and wastewater treatment plants has impacts on not only the local economy but also the environment. Appropriate entropy analysis of water and wastewater treatment processes is, therefore, a crucial factor (Tai and Goda, 1985). In a closed system, we know that entropy tends to increase. For example, if a compound (in this case mostly chemicals) is dumped in pure water, it is likely that the compound will be dissolved and diffused into the whole water body. Two impacts may be considered in this case:

1. increase in entropy of the solution (biophysical impact)
2. increase in the extent of pollution (environmental impact)

Input energy is required to remove the pollutants and purify the water and decrease the entropy of polluted water. These processes altogether impose a heavy economic penalty. The entropy production is associated with the efficiency of the water treatment plant and vice versa. This efficiency may

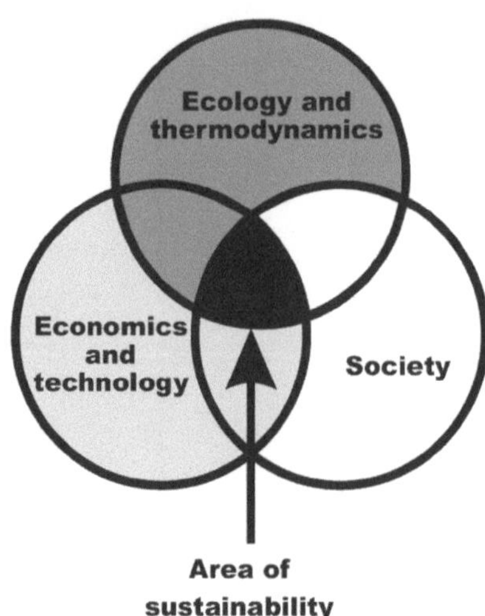

FIGURE 16.8 Venn diagram representation of "The Three Pillars" of socioeconomic sustainability.

be considered as the thermodynamic efficiency of the water treatment plant. For a thorough analysis of this environmental aspect of biophysical economics, the original article may be consulted (Tai and Goda, 1985). It is clear from this analysis that physical parameters, such as energy, entropy, and other thermodynamic indicators (Burness et al., 1980) may be associated with economic scarcity due especially to environmental issues.

Some would argue that energy is better understood as thermodynamic property than entropy to represent complex systems' irreversibilities, and that the behaviors of energy and matter are not equally mirrored by thermodynamic laws (Hammond and Winnett, 2009). The 'Entropy Law' of Nicholas Georgescu-Roegen was widely accepted as constructing a determinant of economic scarcity. However, the thermodynamic insights typically employed in ecological economics should be empirically tested against the real world, where achieving a sustainable economy is a natural desire among all nations. Sustainable development ensures that the global system satisfies the needs of the present without any sort of compromise on the ability of future generations to meet their own needs (WCED, 1987). This means that while developing we can never cause issues that might lead to damage to the economy and quality of lives of our future generations. The environmental economy is a prime example that needs to be sensibly built to ensure a sustainable future.

Engineering constraints and economic and social domains are interconnected in a sustainable society. The sustainability Venn diagram in Figure 16.8 shows the connectivity (Parkin, 2000; Hammond, 2004; Clift, 1995). Here thermodynamic limits are represented as underpinning the environmental sphere. But the notion of sustainable development is not without its critics as development per se cannot be truly sustainable. We need to strive for a creative (knowledge-based) and (sociopolitically) stable world with the aid of "equilibrium engineering" (Thring, 1990).

16.5 ECONOMIC SUSTAINABILITY RELIES ON ADAPTATION OF BIOPHYSICAL ECONOMICS PRINCIPLES

Biophysical economics is built upon the principle that nature poses limits to growth. The internal dynamics of the economy itself also has parameters that limit growth unless the boundaries are broken using applied technologies. A non-stop supply of high-quality fuels is one of the key factors behind achieving functional and sustainable economic growth. Any sort of interruption in this fuel supply chain, due to natural or artificial reasons, causes capital disbursement to experience distortions in the investment trajectories. Figure 16.9 demonstrates the embedded economy mode, which presents all the elements of energy flow and the embodied economy and adds the financial sector, government, and the commons (Raworth, 2017). Most importantly, this model demonstrates how the supply of energy participates in constructing the appropriate input phase, vitally contributing to regulating the market economy. This model also associates the managements of waste matter and waste heat production at the outlet phase of the economy. A conservative balance between these outlet and inlet phases of product management may be created by applying biophysics techniques (as explained earlier in this chapter) that would help maintain a sustained economy. There are various embedded economy models, e.g. that proposed by the anthropologist Karl Polanyi seven decades ago (Polanyi, 1957). Conventional neoclassical economics was criticized by Nicholas Georgescu-Roegen for the lack of attention to the vital physical parameters of energy and entropy. He built models that addressed these concerns (Georgescu-Roegen, 1971, 1975). Herman Daly (1996) successfully placed the economy within the ecosystem and drew a visual model of the embedded economy just two decades ago. Another successful model which included the economy within the society is found in Goodwin et al. (2019).

In the above-mentioned embedded economy mode, we see that the sustained energy influx into the economic scenario plays a vital role. However, the ongoing strategy of relying on unitary sources of fuels, such as fossil fuels, may pose a critical threat to the energy supply chain phase. Long before we physically run out of fossil fuels the world economies will feel the economic effects, and the potential disruption of economic growth is the prime factor advanced in opposition to efforts at sustainability. We, therefore, need to urgently start searching for alternative, yet sustainable sources of fuel. Solar and wind power sources may provide unlimited energy. Technological breakthroughs in these sectors may help achieve a sustained supply of energy to run the economic machineries. Another potential sector is moving toward adopting enteric engines instead of gas-powered engines. Biophysical economics focuses on this unique interaction of natural resources and economic growth limits in order to raise strategies helping to ensure the sustained economic growth required to fulfill

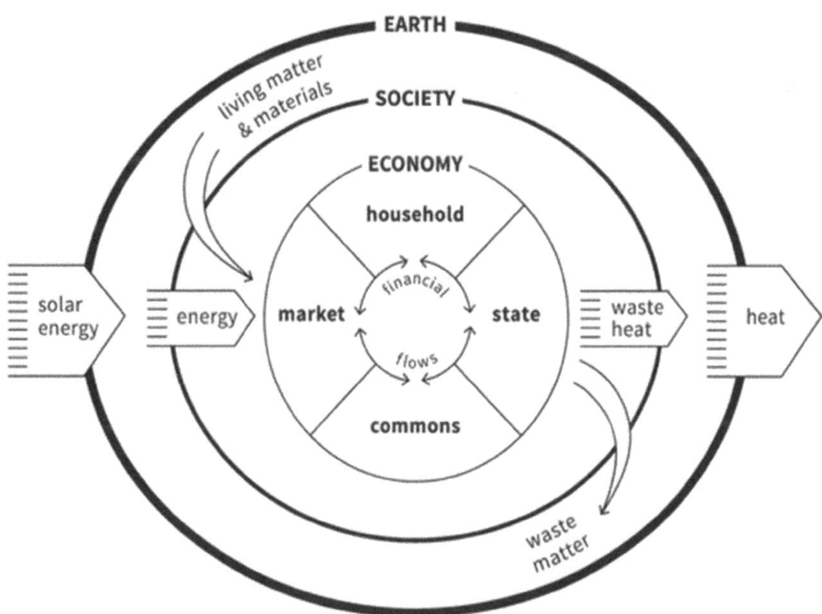

FIGURE 16.9 The embedded economy mode.

***Source*: Raworth (2017); reproduced from Klitgaard (2020).**

the increased needs of societies. Understanding the complex interactions among energy supply chains, versatile sectorial economic growths, and technology-engaged environmental management strategies, as done in biophysical economics, will allow us to craft economic institutions and policies that will allow us to live within nature's limits and cause no damage to nature.

16.6 BIOPHYSICAL INNOVATIONS MAY LEAD TO REGULATING ECONOMIC TURMOIL

Economics as we conventionally know it from the past is now a drastically changed subject. The recent trend is to incorporate a scientific understanding of the parameters that may be adopted to create policies so that societies can experience a sustained supply of their day-to-day living requirements through national framework-managed balancing between supply and demand. In this chapter, we have discussed in detail how various biophysical concepts and technologies may be adopted to create economic policies that would ensure humans live within the limits of nature, utilizing the available natural resources. We see that by applying physical technologies, we may either enhance natural resources' recycling process, or may employ conservatively expensive (mainly regarding fuel use) technologies to limit the use of natural resources and increase the production of biological products necessary for our everyday living. In doing so we may utilize the concept of energy and entropy to maintain checks and balances in available resources, products, and consumptions of the products. Pharmaceutical industries are also adopting more and more biophysical techniques to address issues that are capable of creating economic fluctuations or even catastrophes. As an example, we may consider the case of ongoing coronavirus-caused worldwide economic turmoil which has cost the world almost $20 trillion directly. The indirect costs are much greater and immeasurable. The total US cost is estimated at more than $16 trillion, or roughly 90 percent of the annual GDP of the United States (Cutler and Summers, 2020). In 2020, the global gross domestic product declined by 6.7 percent as a result of the COVID-19pandemic. One thing here is very important: high-income countries' economies shrank and recovered much faster than their low-income counterparts (see Figure 16.10). Among the major reasons, an important one is that the developed countries usually rely on high-tech industries utilizing knowledge and pieces of machinery that could be restarted soon after the discovery of vaccines. Here we need to consider that biophysical approaches played enormous roles in the vaccine discovery (Barrantes, 2021).

The discovered crystal structures of several SARS-CoV-2 proteins alone or in complex with their receptors or other ligands opened the doors to discovering treatments and vaccines (Mariano et al., 2020; Shang et al., 2020). The biophysical approaches to understanding coronavirus physiology and subsequent development of therapeutics and vaccines helped put a brake on the ongoing pandemic, and as a result we see the economy started recovering soon, and by the end of 2021 the world and the high-income countries' economies had crossed the pre-pandemic level (Figure 16.10). However, the low-income countries' economies were still way behind the full recovery line. Here if we carefully cross-examine, the biophysical understanding of the virus and subsequent vaccine and treatment developments immediately started helping the world economically. So biophysical innovations are found to substantially influence the economy (the latter is excelling on the excellence of the former) even though they work independently, unlike how their association has been modeled in the earlier-presented "embedded economy mode," where physical parameters are placed to associate with both inlet and outlet channels of the economy. If we had not achieved the scientific (especially biophysical) breakthrough in addressing the biological substance "virus" in a timely manner, the world might still (beyond 2021) have continued to deal with an annual loss of more than $20 trillion for an uncertain amount of time.

Another sector where biophysical innovations may lead to creating better economic sustainability and growth is the environment. Biofuel is a prime area where biophysical approaches may create benefits that may help strengthen the economy. Biofuel is produced through contemporary processes

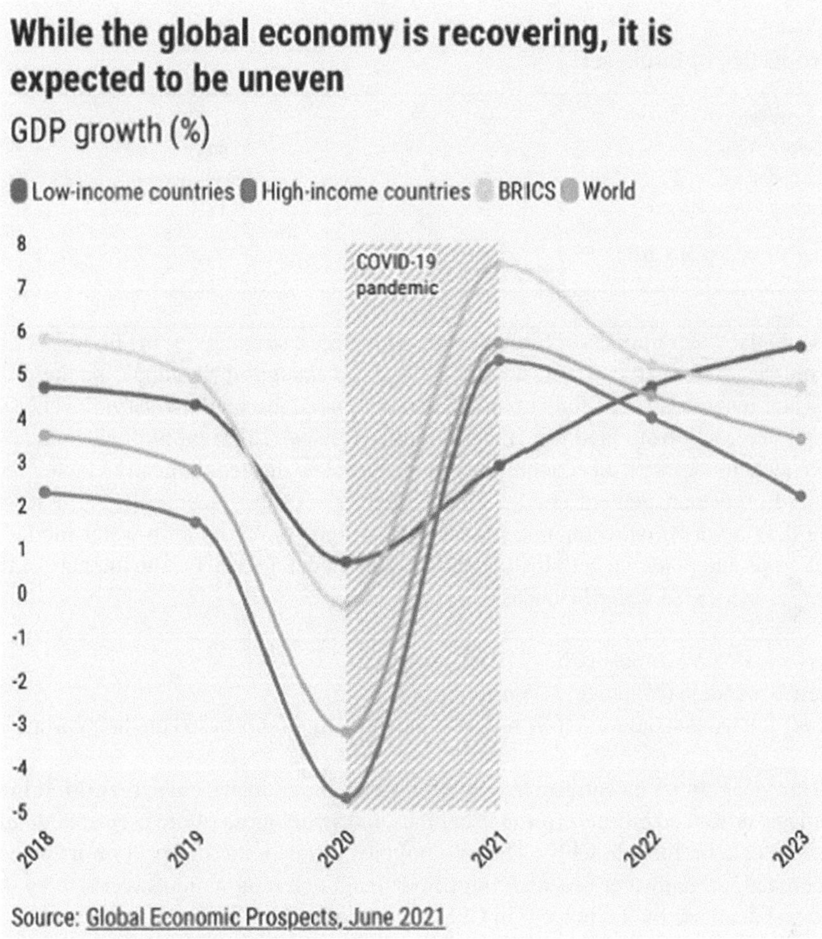

FIGURE 16.10 Global economic condition.

Source: **Gopalakrishnan et al. (2021).**

from biomass, rather than by the very slow geological processes involved in the formation of fossil fuels, such as oil. For protecting our environment, it is a well-established fact that we need to create strategies to avoid using fossil fuels. The economic benefits of using easy fossil fuel sources have been enormous, but only at the cost of the natural balance that has historically obtained between the geological integrity of the earth, the environment, and ecosystems. Alternatives are therefore at the forefront of searches. Biomass architecture and breakdown have offered significant new strategies to improve biomass conversion. Management of biodiesel, a harbinger of clean and sustainable energy using biophysical approaches, especially in characterization of their physical properties (a few are presented in Table 16.1), may contribute to creating an association between biophysical economics and renewable sources of energy as the primary sources in any industry for a promising future.

Water use in crop production is another cause of resource scarcity. Inappropriate use of excessive clean water for irrigation purposes may produce disastrous consequences in the near future. Water impacts of biofuels in the USA were assessed recently using combined economic and biophysical models (Teter et al., 2018). This study illustrated the importance of accounting for the overall land use changes and shifts in agricultural production and management practices in response to policies

TABLE 16.1
Properties of Biodiesel

Density (kg/m^3)	937
Viscosity (cSt)	35.44
Flashpoint (°C)	> 270
Heating value (MJ/kg)	39.19

Source: Patel et al. (2020).

when assessing the water impacts of biofuels. I provide here a summary of the findings of this study. Economic policies differ in the extent and type of land use changes they induce and therefore in their impact on water resources. This study qualified and compared the spatially varying water impacts of biofuel crops stemming from land use changes induced by two different biofuels policies by coupling a biophysical model with an economic model to simulate the economically viable mix of crops, land uses, and crop management choices under alternative policy scenarios. An assessment was made using the outputs of an economic model with a high-resolution crop-water model for major agricultural crops and potential cellulosic feedstocks in the US to analyse the impacts of three alternative policy scenarios on water balances:

- a counterfactual "no-biofuels policy" (BAU) scenario
- a volumetric mandate (Mandate) scenario
- a clean fuel-intensity standard (CFS) scenario incentivizing fuels based on their carbon intensities

While both biofuel policies incentivize more biofuels than the counterfactual, they differ in the mix of corn ethanol and advanced biofuels from miscanthus and switchgrass (more corn ethanol in Mandate and more cellulosic biofuels in CFS). The two policies differ in their impact on irrigated acreage, irrigation demand, groundwater use, and runoff. Net irrigation requirements increase by 0.7 percent in Mandate and decrease by 3.8 percent in CFS, but in both scenarios, increases are concentrated in regions of Kansas and Nebraska that rely upon the Ogallala aquifer for irrigation water. This study thus demonstrates that instead of using general social scientific or specific economic scientific policies that are often adopted for politically biased policies, if innovative biophysical methods and models are associated with applicable economic models to create applied biophysical economic models to deal with biofuel economy, the outcomes may help reduce the impacts in terms of adding scarcity into the environment. Continued research may unite both branches to help us achieve the benefits of utilizing biophysical excellence in achieving economic sustainability with low impacts on environments and the associated economy.

16.6 BIOPHYSICAL ECONOMICS MEETS POLITICAL ECONOMY

The economy as seen from a political perspective may have components that can be seen in the mirror of physics. Diemer and Guillemin (2011) show that Adam Smith, the father of modern economics, used the great physicist Newton's laws to resolve economic questions and that there were close connections between the material world, known as the world of bodies, as viewed by Newton, and the world of men as viewed by Smith.

Explicit references to Newton's natural philosophy are rare in Smith's valuable works of political economy and moral philosophy, and Newton appeared perhaps only once in *The Theory of Moral Sentiments* (Smith, 1822). However, historians of science have drawn parallels between Newtonian physics and the works of Smith in the field of political economy (Moscovici, 1956; Buchdahl,

1961; Greene, 1961; Thomson, 1965). Attempts to find similarities between political economics and physics may illustrate the following phenomena (Diemer and Guillemin, 2011): irrespective of the fields, scholars are secretly ambitious to imitate Newton, and Smith's efforts to discover general laws of economy, inspired by the success of Newton and his discovery of the natural laws of motion (Hetherington, 1983).

The biophysical economy may be associated with political economy in pinpointed areas. If biophysical economics helps determine the production of wealth, what potential mode of wealth distribution could then follow within political economy? Kennedy (2022) addressed this question by building upon his model of the British Industrial Revolution to approach the overall objective of understanding the intersection of biophysical economics and political economy. Biophysical economics is first assumed here to be able in its models to determine the labor, capital, and energy use in a society. The monetary transactions (prices, profits, and wages) were then added into the model. The transactions thus capture the trade-offs between the returns to labor and the returns to capital, which places the model in the political economy realm.

According to the conception of duality of flows (for details, see Kennedy, 2022), physical resources and financial transactions work in opposite directions, which is important for understanding the association, intersection, and interdependence between biophysical economics and political economy. It may be theoretically possible to describe the economy using just one of the two types of flow, as found in the input-output (IO) modeling, for example, where monetary input-output tables track financial flows – whereas physical IO tables alternatively track resources. In practice, there tends to be a cross-over between the two types of flows, e.g. such as environmentally extended IO models (Victor, 2017).

Ecological economics is found to be concerned with both the production and the distribution of wealth. As economies are subject to physics laws (Georgescu-Roegen, 1986), ecological economics is found to encompass both biophysical economics and political economy. Ecological economics is found to place the macro-economy within the context of the wider biophysical or ecological systems upon which it depends (Kennedy, 2022).

Kennedy (2022) explored the theoretical range of trade-offs between profits and wages for each sector represented in his model, instead of assuming that any specific theory or format of capitalism may apply (Buch-Hansen, 2014; Acemoglu and Robinson, 2015; Svartzman et al., 2019; Cahen-Fourot, 2020). Drawing upon data from economic history, Kennedy showed how Britain's economy fared in this distributional trade-off for different years of the Industrial Revolution during its first and second periods. A four-sector model of Great Britain's economy from 1760 to 1913 was used for analysis, which included agriculture, coal-mining, construction of capital, and production of goods and services (Kennedy, 2022). Energy price was considered a key variable in the model that would influence the distribution of income between different sectors. Taking the price of coal at historically observed values, the distribution of total factor income per worker was plotted as a trade-off between annual wages and profits per worker for example years of 1761, 1817, and 1871. The ways in which political economy parameters may influence the biophysical economy were discussed here. The conclusions were based on the contemporary challenges of ecological economics of developing environmentally sustainable economies with a just and equitable sharing of resources and services (see summarizing model in Figure 16.11).

Britain's Industrial Revolution had several underlying biophysical characteristics (Kennedy, 2021), such as:

1. improvements in agricultural productivity
2. large increases in the use of coal-energy supply
3. the physical construction of infrastructure for industrialization and urbanization

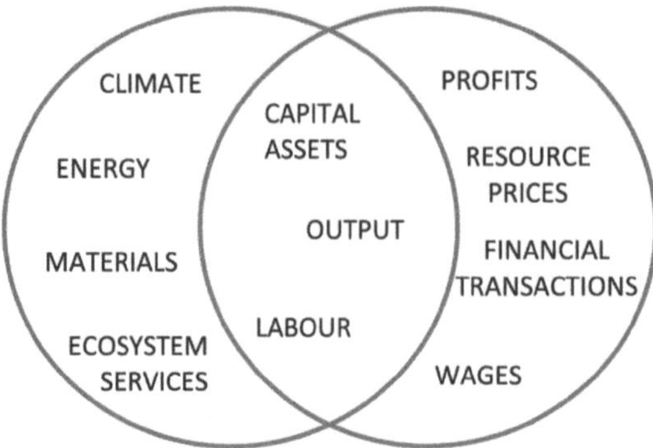

FIGURE 16.11 The circular flow diagram showing independent and sharing components, many of which are associated with biophysical parameters that influence political economy.

Source: **Kennedy (2022).**

These characteristics have been represented in a four-sector model of Britain's economy during the 1760–1913 era, including mining, agriculture, production of goods and services, and construction of capital. Kennedy's model has a novel mathematical representation of a dynamic general equilibrium between labor, capital, and energy in an economy. Using historical data, the model was calibrated for the growth of Britain's capital stock, the use of coal, and employment during the first and second Industrial Revolution periods. The impacts of two biophysical constraints were explored in model simulations, such as stagnation in agricultural productivity and reduced efficiency in the coal-mining industry without the presence of steam engines. Both scenarios were found to exhibit substantial reductions in the growth of capital stock and significant changes to the distribution of labor.

Kennedy's biophysical model of the industrial revolution was created as an endogenous growth model determining the distribution of labor, growth of capital stock, and use of energy in the economy. The model is found to resolve both balance equations for each of the core variables and a standard equation for the accumulation of capital stock. For details, see Kennedy (2021, 2022). As outlined here, Britain's economy during 1760–1913 was used for demonstrating the potential variation in the distribution of wealth considering biophysical economic constraints.

An obvious question – "how does political economy change the biophysical economy?" – was addressed. The analysis predicted a dominant role for the biophysical economy, but could there be definite ways in which the political economy changes the biophysical economy? Kennedy argues that in the biophysical view of the economy, energy and labor come together to produce materials. These materials are used for constructing physical capital. The political economy certainly plays a decisive role in the use and distribution of the physical wealth obtained as a result of the application of appropriate biophysical economics parameters. Continued timely policy innovations are required. As an example, we may mention the effect of oil and natural gas rents (which is a benefit-sharing compassionate political policy) on economic growth in Gulf Cooperation Council (GCC) member countries (Inuwa et al., 2022). As our resources on planet Earth are limited, we need to make the right political decisions both in applying sophisticated biophysics economics formulas to the exploitation of natural resources, especially biomass and other related ones, and in producing wealth and utilizing it to fulfill the minimum necessary demands. But at the same time, we need to work toward connecting all these productions and use processes to a cyclical process so that we keep recycling our resources repeatedly and avert any artificially induced natural disaster due to inappropriate human interventions.

REFERENCES

Acemoglu, D., & Robinson, J.A. (2015). The rise and decline of general laws of capitalism. *Journal of Economic Perspectives*, *29*(1), 3–28

Anastasopoulos, A. (2020). Macro-thermodynamic economics: the human factor. *SSRN*, February 20, 2020) http://dx.doi.org/10.2139/ssrn.3542013

Barrantes, F.J. (2021). The contribution of biophysics and structural biology to current advances in COVID-19. *Annual Review of Biophysics*, *50*, 493–523

Bericat, E., & Jimenez-Rodrigo, M.L. (2019). The quality of European Societies: a compilation of composite indicators. Berlin and Heidelberg: Springer

Buchdahl, G. (1961). *The image of Newton and Locke in the Age of Reason.* London: Sheed and Ward

Buch-Hansen, H. (2014). Capitalist diversity and de-growth trajectories to steady-state economies. *Ecological Economics*, *106*, 167–173

Burness, S., Cummings, R., Morris, G., & Paik, I. (1980). Thermodynamic and economic concepts as related to resource-use policies. *Land Economics*, *56*(1), 1–9. https://doi.org/10.2307/3145824

Cahen-Fourot, L. (2020). Contemporary capitalisms and their social relation to the environment. *Ecological Economics*, *172*, 106634

Chetty, R., Stepner, M., Abraham, S., Lin, S., Scuderi, B., Turner, N., ... & Cutler, D. (2016). The association between income and life expectancy in the United States, 2001–2014. *JAMA*, *315*(16), 1750–1766

Clift, R. (1995). The challenge for manufacturing. In *Engineering for sustainable development*, ed. J. McQuaid, pp. 82–87. London: Royal Academy of Engineering

Cutler, D.M., & Summers, L.H. (2020). The COVID-19 pandemic and the $16 trillion virus. *JAMA*, *324*(15), 1495–1496

Daly, H. (1996). *Beyond growth.* Boston, MA: Beacon Press

Davies, P.C., Rieper, E., & Tuszynski, J.A. (2013). Self-organization and entropy reduction in a living cell. *Biosystems*, *111*(1), 1–10

Dayanikli, G., Gokare, V., & Kincaid, B. (2016). Effect of GDP per capita on national life expectancy. https://smartech.gatech.edu/bitstream/handle/1853/56031/effect_of_gdp_per_capita_on_national_life_expectancy.pdf (accessed March 31, 2023)

Desai, D., Lange, G.M., Hamilton, K., Ruta, G., Chakraborti, L., Edens, B., ... & Li, H. (2011). The changing wealth of nations: measuring sustainable development in the new millennium. World Bank

Diemer, A., & Guillemin, H. (2011). Political economy in the mirror of physics: Adam Smith and Isaac Newton. *Journal of the History of Science*, *64*(1), 5–26

Dolan, F., Lamontagne, J., Link, R., Hejazi, M., Reed, P., & Edmonds, J. (2021). Evaluating the economic impact of water scarcity in a changing world. *Nature Communications*, *12*(1), 1915

Freeman, T., Gesesew, H.A., Bambra, C., et al. (2020). Why do some countries do better or worse in life expectancy relative to income? An analysis of Brazil, Ethiopia, and the United States of America. *Int. J. Equity Health*, *19*, 202. https://doi.org/10.1186/s12939-020-01315-z

García-Llorente, M., Iniesta-Arandia, I., Willaarts, B.A., Harrison, P.A., Berry, P., del Mar Bayo, M., ... & Martín-López, B. (2015). Biophysical and sociocultural factors underlying spatial trade-offs of ecosystem services in semiarid watersheds. *Ecology and Society*, *20*(3), 39

Georgescu-Roegen, N. (1986). *The entropy law and the economic process* in retrospect. *Eastern Economic Journal*, *12*(1), 3–25

Georgescu-Roegen, N. (1975). Energy and economic myths. *South. Econ. J.*, *41*, 347–381

Georgescu-Roegen, N. (1971). *The entropy law and the economic process.* Cambridge, MA: Harvard University Press. https://doi.org/10.1016/j.energy.2016.01.071

Goodwin, N., Harris, J.M., Nelson, J.A., Roach, B., & Torras, M. (2019). *Microeconomics in context* (4th edn). New York: Routledge

Gopalakrishnan, V., Wadhwa, D., Haddad, S., & Blake, P. (2021). 2021 year in review in 11 charts: the inequality pandemic. World Bank, December 21, 2021, www.worldbank.org/en/news/feature/2021/12/20/year-2021-in-review-the-inequality-pandemic (accessed April 1, 2023)

Greene, J. (1961). *Darwin and the modern world view* (Baton Rouge: Louisiana State University

Hall, C.A., & Klitgaard, K. (2018). *Energy and the wealth of nations: an introduction to biophysical economics.* Berlin and Heidelberg: Springer. https://doi.org/10.1007/978-3-319-66219-0

Hammond, G.P. (2004). Engineering sustainability: thermodynamics, energy systems, and the environment. *Int. J. Energ. Res.*, *28*, 613–639

Hammond, G.P., & Winnett, A.B. (2009). The influence of thermodynamic ideas on ecological economics: an interdisciplinary critique. *Sustainability*, *1*(4), 1195–1225. doi: 10.3390/su1041195

Hetherington, N. (1983). Isaac Newton's influence on Adam Smith's natural laws in economics. *Journal of the History of Ideas*, *44*(3), 497–505

Inuwa, N., Adamu, S., Sani, M.B., et al. (2022). Resource curse hypothesis in GCC member countries: evidence from seemingly unrelated regression. *Biophys. Econ. Sust.*, *7*, 13. https://doi.org/10.1007/s41247-022-00108-y

Ji, X., & Luo, Z. (2020). Opening the black box of economic processes: ecological economics from its biophysical foundation to a sustainable economic institution. *Anthropocene Review*, *7*(3), 231–247

Kennedy, C. (2022). The intersection of biophysical economics and political economy. *Ecological Economics*, *192*, 107272

Kennedy, C. (2021). A biophysical model of the Industrial Revolution. *Journal of Industrial Ecology*, *25*(3), 663–676

Klitgaard, K. (2020). Sustainability as an economic issue: a biophysical economic perspective. *Sustainability*, *12*(1), 364. https://doi.org/10.3390/su12010364

Kovalev, A.V. (2016). Misuse of thermodynamic entropy in economics. *Energy*, *100*, 129–136. https://doi.org/10.1016/j.energy.2016.01.071

Manera, C., Serrano, E., Pérez-Montiel, J., & Buil-Fabregà, M. (2021). Construction of biophysical indicators for the Catalan economy: building a new conceptual framework. *Sustainability*, *13*(13), 7462

Mariano, G., Farthing, R.J., Lale-Farjat, S.L., & Bergeron, J.R. (2020). Structural characterization of SARS-CoV-2: where we are, and where we need to be. *Frontiers in Molecular Biosciences*, *7*, 605236. https://doi.org/10.3389/fmolb.2020.605236

McMahon, G.F., & Mrozek, J.R. (1997). Economics, entropy and sustainability. *Hydrological Sciences Journal*, *42*(4), 501–512. doi: 10.1080/02626669709492050

Michaelides, E.E. (2008). Entropy, order and disorder. *Open Thermodynamics Journal*, *2*(1), 7–11

Missemer, A. (2017). Nicholas Georgescu-Roegen and degrowth. *European Journal of the History of Economic Thought*, *24*(3), 493–506. http://dx.doi.org/10.1080/09672567.2016.1189945

Moscovici, S. (1956). À propos de quelques travaux d'Adam Smith sur l'histoire et la philosophie des sciences. *Revue d'Histoire des Sciences et de leurs Applications*, *9*(1), 1–20

Moser, E., Semmler, W., Tragler, G., & Veliov, V.M. (eds.). (2014). *Dynamic optimization in environmental economics*. Berlin and Heidelberg: Springer. https://doi.org/10.1007/978-3-642-54086-8

Niven, J.E. (2016). Neuronal energy consumption: biophysics, efficiency and evolution. *Current Opinion in Neurobiology*, *41*, 129–135

Parkin, S. 2000. Sustainable development: the concept and the practical challenge. *Proc. Inst. Civil Eng. – Civil Eng.*, *138*, 3–8

Patel, P., Patel, B., Vekaria, E., & Shah, M. (2020). Biophysical economics and management of biodiesel, a harbinger of clean and sustainable energy. *International Journal of Energy and Water Resources*, *4*, 411–423

Peniche Camps, S., Hall, C.A., & Klitgaard, K. (2020). Biophysical economics for policy and teaching: Mexico as an example. *Sustainability*, *12*(7), 2580

Polanyi, K. (1957). The economy as instituted process. In *Trade and market in early empire*, ed. P. Karl, M.A. Conrad, & W.P. Harry. New York: Free Press

Raworth, K. (2017). *Doughnut economics: seven ways to think like a 21st-century economist*. White River Junction, VT: Chelsea Green

Remme, M.W., Rinzel, J., & Schreiber, S. (2018). Function and energy consumption constrain neuronal biophysics in a canonical computation: coincidence detection. *PLoS Computational Biology*, *14*(12), e1006612

Schwandt, H., Currie, J., Bär, M., Banks, J., Bertoli, P., Bütikofer, A., … & Wuppermann, A. (2021). Inequality in mortality between Black and White Americans by age, place, and cause and in comparison to Europe, 1990 to 2018. *Proceedings of the National Academy of Sciences*, *118*(40), e2104684118. https://doi.org/10.1073/pnas.2104684118

Sewell, G. (2013). Entropy, evolution and open systems. In *Biological information: new perspectives*, ed. R.J. Marks II, et al. (pp. 168–178). Singapore: World Scientific Publishing. https://doi.org/10.1142/9789814508728_0007

Shang, J., Ye, G., Shi, K., Wan, Y., Luo, C., Aihara, H., ... & Li, F. (2020). Structural basis of receptor recognition by SARS-CoV-2. *Nature*, *581*(7807), 221–224.

Smith, A. (1822). *The theory of moral sentiments*. London: J. Richardson

Smith, P., Davis, S.J., Creutzig, F., Fuss, S., Minx, J., Gabrielle, B., ... & Yongsung, C. (2016). Biophysical and economic limits to negative CO2 emissions. *Nature Climate Change*, *6*(1), 42–50

Söderholm, P. (2020). The green economy transition: the challenges of technological change for sustainability. *Sustainable Earth*, *3*(1), 1–11

Stapelbroek, K., & Marjanen, J. (eds.). (2012). *The rise of economic societies in the eighteenth century: patriotic reform in Europe and North America*. Basingstoke: Palgrave Macmillan. https://doi.org/10.1057/9781137265258

Svartzman, R., Dron, D., & Espagne, E. (2019). From ecological macroeconomics to a theory of endogenous money for a finite planet. *Ecological Economics*, *162*, 108–120

Tai, S., & Goda, T. (1985). Entropy analysis of water and wastewater treatment processes. *International Journal of Environmental Studies*, *25*(1–2), 13–21. https://doi.org/10.1080/00207238508710208

Teter, J., Yeh, S., Khanna, M., & Berndes, G. (2018). Water impacts of US biofuels: insights from an assessment combining economic and biophysical models. *PLoS One*, *13*(9), e0204298. https://doi.org/10.1371/journal.pone.0204298

Thomson, H. (1965). Adam Smith's philosophy of science. *Quarterly Journal of Economics*, *79*, 212–33

Thring, M.W. (1990). Engineering in a stable world. *Science, Technology and Development*, *8*, 107–121

Victor, P.A. (2017). *Pollution: economy and environment*. London: Routledge. https://doi.org/10.4324/9781315108483

WCED. (1987). *Our common future*. Oxford: Oxford University Press

World Bank. (2010). *The changing wealth of nations: measuring sustainable development in the new millennium*. https://openknowledge.worldbank.org/handle/10986/2252

Yan, J., Feng, L., Steblyanskaya, A., Kleiner, G., & Rybachuk, M. (2019). Biophysical economics as a new economic paradigm. *International Journal of Public Administration*, *42*(15–16), 1395–1407

Epilogue

Biophysics is an established subject that covers many disciplines in biology, medicine, and agriculture. Technological breakthroughs that took place primarily in physics, chemistry, and engineering fields have been found applicable in biophysics explorations of biomedical sciences. As a result of enhanced technological applications in biophysics explorations, the understanding of biological structures and physiological processes has been taken to advanced levels, helping us address crucial medical science problems, including understanding diseases and finding means to fight them. Understanding mutations in genes, proteins, and other biomolecules by applying biophysics laws and techniques has helped us model and design artificial means and molecules to regulate, repair, or delete vital mutated structures. We have already entered a visionary medical discovery phase and have been achieving breakthrough medical discoveries. Biophysics has participated considerably in this triumph. A huge amount of materials in the form of research publications, scientific news, patents, and books has accumulated. But a summary of most of them on a single platform is absent to date. Exactly for this purpose, I took the initiative with CRC Press of the Taylor & Francis Group and decided to write a textbook with the title *Introduction to Modern Biophysics*.

This book has provided both basic and advanced understandings of biological systems and processes. A special focus has been given to applied technological breakthroughs in the advance of biomedical science. A biophysical understanding of versatile biomolecular mutations and associated diseases has been detailed. Biophysics-technique-based development of natural and artificial means and drug molecules helps aid in the successful discovery of drugs. Nanotechnology appears as a sister subject to biophysics in the nanoscience of biology and both together have made considerable progress, covered in the book.

Going beyond presenting the traditional biology research, I have added a summary of the past developments and ongoing progress in applying biophysics laws and principles in economics, social science, and the environment on Earth, in space, and on exoplanets. In energy production, wealth management, GDP growth, etc., biophysical economics principles, models, predictions, and guidelines are now adopted, especially in research and creating technological innovations. We have experienced considerable progress in this area over the last century, summarized in the book.

Introduction to Modern Biophysics will help readers by summarizing a huge body of information on biophysics research, since its inception about four centuries ago with the discovery of cells by applying optophysics principles. I hope that this textbook will be a catalyst for future studies of biology, medicine, agriculture, and associated fields using physics laws, principles, and modelings. It is my great hope that this book will become a source of empirical information and conceptual inspiration for students of biophysics, biology, biomedical sciences, engineering disciplines, and expert researchers, including those involved in pharmaceutical sciences and industries. *Introduction to Modern Biophysics* is expected to be adopted as a premier reference textbook in both academia and industry.

DOI: 10.1201/9781003287780-17

Index

A

B

C

D

E

F

G

H

N

O

P

Q

R

S

T

U

V

W

X

Z